Primates in Flooded Habitats

Ecology and Conservation

Nearly half the world's primate species use flooded habitats at one time or another, from swamp-going Congo gorillas and mangrove-eating proboscis monkeys, to uacaris in Amazonian riverside forests. This first-ever volume on the subject brings together experts from around the world in a ground-breaking volume spanning fossil history, current biology, and future research and conservation priorities. Flooded habitats are a vital part of tropical biology, both for the diversity of the species they house, and the complexity of their ecological interactions, but are often completely overlooked.

This book will set the stage for a new wave of research on primates in these extraordinarily productive and highly threatened areas, and is ideal for researchers and graduate students in primatology, zoology, ecology, and conservation.

Katarzyna Nowak is a Research Associate at the University of the Free State, Qwaqwa campus, South Africa and a Fellow at The Safina Center in New York. Her conservation research focuses on the behavioural flexibility of threatened species.

Adrian A. Barnett is Lecturer and Researcher at the National Institute for Amazon Research, Brazil, and the Federal University of Amazonas, Brazil, and a Research Fellow at Roehampton University, UK. He has worked with flooded habitats and their primates for more than 25 years.

Ikki Matsuda is an Associate Professor at the Chubu University Academy of Emerging Sciences and a member of both the Japan Monkey Centre (Advisor) and the Wildlife Research Center of Kyoto University (specially appointed Associate Professor) in Japan. He focuses on a synthetic understanding of the evolution and ecology of colobine species, in particular of the proboscis monkey.

Primates in Flooded Habitats

Ecology and Conservation

Edited by

Katarzyna Nowak
University of the Free State, South Africa

Adrian A. Barnett
National Institute for Amazon Research, Brazil

Ikki Matsuda
Chubu University/Kyoto University/Japan Monkey Centre, Japan

CAMBRIDGE
UNIVERSITY PRESS

CAMBRIDGE
UNIVERSITY PRESS

University Printing House, Cambridge CB2 8BS, United Kingdom

One Liberty Plaza, 20th Floor, New York, NY 10006, USA

477 Williamstown Road, Port Melbourne, VIC 3207, Australia

314–321, 3rd Floor, Plot 3, Splendor Forum, Jasola District Centre, New Delhi – 110025, India

79 Anson Road, #06-04/06, Singapore 079906

Cambridge University Press is part of the University of Cambridge.

It furthers the University's mission by disseminating knowledge in the pursuit of education, learning, and research at the highest international levels of excellence.

www.cambridge.org
Information on this title: www.cambridge.org/9781107134317
DOI: 10.1017/9781316466780

First published 2019

Printed in the United Kingdom by TJ International Ltd. Padstow Cornwall

A catalogue record for this publication is available from the British Library.

ISBN 978-1-107-13431-7 Hardback

Additional resources for this publication at www.cambridge.org/9781107134317

Contents

Part III – Beach Primates

Part IV – Swamp Primates

Part V – Primates in Freshwater Flooded Forests

Part VI – Conservation Case Studies

Colour plates can be found between pages 174 and 175.

Contributors

Abdul Hamid Ahmad
Unit for Primate Studies-Borneo, Institute for Tropical Biology and Conservation, Universiti Malaysia Sabah, Kota Kinabalu, Sabah, Malaysia

Cleber J.R. Alho
University Uniderp – Postgraduate Programme in Environment and Regional Development, Campo Grande, MS, Brazil

Laurentius N. Ambu
Sabah Wildlife Department, Wisma MUIS, Sabah, Malaysia

Marc Ancrenaz
HUTAN, Sabah, Malaysia

Chieko Ando
Ecologic, Miyacho, Shizuoka, Japan

Herizo Andrianandrasana
Department of Zoology, University of Oxford, Oxford, UK

Andie Hui Fang Ang
Raffles Banded Langur Working Group, Singapore

Lynne R. Baker
Department of Natural and Environmental Sciences, American University of Nigeria, Yola, Nigeria

Adrian A. Barnett
Centre for Research in Evolutionary Anthropology, University of Roehampton, London, UK and Amazon Mammal Research Group, Biodiversity Sector, National Institute of Amazon Research, Manaus, Brazil

A.K. Basabose
Centre de Recherche en Sciences Naturelles, Bukavu, Republic of Congo

Monique Bastos
Departmento de Zoologia, Centro de Biociências, Universidade Federal de Pernambuco, Brazil

Raone Beltrão-Mendes
Centre for Biological Sciences and Health, Department of Ecology, Sergipe Federal University, SE, Brazil

Amy C. Bennett
UNEP World Conservation Monitoring Centre, Cambridge, UK

Henry Bernard
Unit for Primate Studies-Borneo, Institute for Tropical Biology and Conservation, Universiti Malaysia Sabah, Kota Kinabalu, Sabah, Malaysia

Bruna M. Bezerra
Departmento de Zoologia, Centro de Biociências, Universidade Federal de Pernambuco, Brazil

Hani R. El Bizri
Mamirauá Sustainable Development Institute. Estrada do Bexiga, Tefé, Brazil; School of Science and the Environment, Manchester Metropolitan University, Manchester, UK

Sarah A. Boyle
Department of Biology, Rhodes College, Memphis, USA

LeAndra Luecke Bridgeman
Department of Anthropology, Washington University in St Louis, St Louis, USA

Michael W. Bruford
Cardiff University School of Biosciences, Cardiff, UK

Neil D. Burgess
UN Environment World Conservation Centre (UNEP-WCMC), Cambridge, UK

Thomas M. Butynski
Eastern Africa Primate Diversity and Conservation Program & Lolldaiga Hills Research Programme, Nanyuki, Kenya

Nayara de Alcântara Cardoso
Department of Systematics and Ecology, Centre for Natural and Exact Sciences – Campus I, Paraíba Federal University, Cidade Universitária, João Pessoa, Brazil and Terrestrial Vertebrate Ecology Group, Mamirauá Sustainable Development Institute, Tefé, Brazil

Xyomara Carretero-Pinzon
Zocay Project, Bogotá, Colombia; School of Geography, Planning and Environmental Management, University of Queensland, Brisbane, Queensland, Australia; ARC Centre of Excellence for Environmental Decisions, University of Queensland, Brisbane, Queensland, Australia

Colin A. Chapman
Department of Anthropology, McGill University, Montreal, Canada

Susan M. Cheyne
Borneo Nature Foundation, Palangka Raya, Indonesia; Oxford Brookes University, UK

Lounès Chikhi
Laboratoire Évolution & Diversité Biologique (EDB UMR 5174), Université de Toulouse Midi-Pyrénées, CNRS, IRD, UPS, Toulouse, France; Instituto Gulbenkian de Ciência, Oeiras, Portugal

Janice Chism
Department of Biology, Winthrop University, Rock Hill, South Carolina, USA

Rebecca Coles
RCC Services, Toulouse, France

Herbert H. Covert
Department of Anthropology, University of Colorado Boulder, Boulder, USA

Thomas R. Defler
Universidad Nacional de Colombia, Bogotá, Colombia

Giuseppe Donati
Department of Social Sciences, Oxford Brookes University, Oxford, UK

Timothy M. Eppley
Institute for Conservation Research, San Diego Zoo Global, Escondido, USA

Shun Deng Fam
School of Archeology and Anthropology, Australian National University, Canberra, Australia

Phillip Fearnside
National Institute for Amazonian Research (INPA), Manaus, Brazil

Vanina A. Fernández
Museo Argentino de Cs. Naturales 'B. Rivadavia', Buenos Aires, Argentina

Eduardo Fernandez-Duque
Department of Anthropology, Yale University, New Haven, USA

Stephen F. Ferrari
Centre for Biological Sciences and Health, Department of Ecology, Sergipe Federal University, SE, Brazil

Leandro V. Ferreira
Museu Paraense Emílio Goeldi, Belém, Brazil

Renata Gonçalves Ferreira
Co-Lab: Studies on Coexistence. Psychobiology postgraduate program, Universidade Federal do Rio Grande do Norte, Natal, Brazil

Anthony Di Fiore
Department of Anthropology, University of Texas at Austin, Austin, USA

Dorothy M. Fragaszy
Department of Psychology, University of Georgia, Athens, USA

Takeshi Furuichi
Primate Research Institute, Kyoto University, Inuyama, Japan

Anh Galat-Luong
UICN, Commission de Survie des Espèces, Argelliers, France

Gerard Galat
UICN, Commission de Survie des Espèces, Argelliers, France

Jörg U. Ganzhorn
Biozentrum Grindel, Department of Animal Ecology and Conservation, University of Hamburg, Hamburg, Germany

Benoît Goossens
Danau Girang Field Centre, c/o Sabah Wildlife Department, Sabah, Malaysia; Cardiff University School of Biosciences, Cardiff, UK; Sabah Wildlife Department, Wisma MUIS, Sabah, Malaysia

Michael D. Gumert
Psychology Division, School of Humanities and Social Sciences, Singapore

Goro Hanya
Primate Research Institute, Kyoto University, Inuyama, Japan

Amy Harrison-Levine
Denver Zoo Department of Conservation and Research, Denver, USA

Torbjørn Haugaasen
Norwegian University of Life Sciences, Faculty of Environmental Sciences and Natural Resource Management, Ås, Norway

Joseph E. Hawes
Applied Ecology Research Group, School of Life Sciences, Anglia Ruskin University, Cambridge, UK

Josephine Head
Earthwatch Europe, Oxford, UK

Aoife Healy
Oxford Brookes University, Oxford, UK

Rose Marie Hoare
Department of Anthropology, Southern Illinois University, Carbondale, USA

Keith Hodges
Reproductive Biology Unit, German Primate Center, Göttingen, Germany

Michaela E. Howells
Department of Anthropology, University of North Carolina Wilmington, Wilmington, USA

Yuji Iwata
Kyoto Architectural College, Kamigyo Kyoto, Japan

Mohammaed Fairus B. Jalil
Unit for Primate Studies-Borneo, Institute for Tropical Biology and Conservation, Universiti Malaysia Sabah, Sabah, Malaysia

Caspian Johnson
Department of Biosciences, University of Swansea, Wallace Building, Singleton Park, Swansea, UK

Gareth Jones
School of Life Sciences, University of Bristol, Bristol, UK

Yvonne A. de Jong
Eastern Africa Primate Diversity and Conservation Program & Lolldaiga Hills Research Programme, Nanyuki, Kenya

Tomoko Kanamori
Wildlife Research Center of Kyoto University, Kyoto, Japan

Blair Kaufer
Department of Anthropology, Western Washington University, Bellingham, USA

Nobuko Kazahari
Field Science Center for Northern Biosphere, Hokkaido University, Kita-ku, Sapporo

Shiro Kohshima
Wildlife Research Center of Kyoto University, Kyoto, Japan

Martin M. Kowalewski
Estacion Biologica Corrientes (MACN-CONICET), Corrientes, Argentina

Takuya Kubo
Graduate School of Environmental Earth Science, Hokkaido University, Sapporo, Japan

Noko Kuze
Department of Anthropology, The National Museum of Nature and Science, Japan

Viviane M. Guedes Layme
Department of Botany and Ecology, Mato Grosso Federal University (UFMT), Cuiabá, Brazil

Teddy Marcelo Siles Lazzo
Wildlife Conservation Society, La Paz, Bolivia

Shawn Lehman
Department of Anthropology, University of Toronto, Toronto, Canada

Anna Lenoël
World Wildlife Fund, Washington, USA

Matthew C. Lewis
Institute for Communities and Wildlife in Africa, University of Cape Town, South Africa

Stanislav Lhota
Department of Animal Science and Food Processing, Faculty of Tropical AgriSciences, Czech University of Life Sciences Prague; Usti nad Labem Zoo, Czech Republic

James E. Loudon
Department of Anthropology, East Carolina University, Greenville, USA

Fiona Maisels
Global Conservation Program, Wildlife Conservation Society, Bronx, USA and School of Natural Sciences, University of Stirling, Scotland, UK

Suchinda Malaivijitnond
Department of Biology, Faculty of Science, Chulalongkorn University, Bangkok, Thailand

Jayanta Kumar Mallick
Wildlife Wing, Forest Department, Government of West Bengal, India

Jesus Martinez
Wildlife Conservation Society, La Paz, Bolivia

Ikki Matsuda
Chubu University Academy of Emerging Sciences, Kasugai-shi, Japan; Wildlife Research Center of Kyoto University, Kyoto, Japan; Japan Monkey Centre, Inuyama, Japan; Institute for Tropical Biology and Conservation, Universiti Malaysia Sabah, Sabah, Malaysia

Keriann McCoogan
Department of Anthropology, University of Toronto, Toronto, Canada

Karolina Medeiros
Departmento de Zoologia, Centro de Biociências, Universidade Federal de Pernambuco, Brazil

Mbangi Norbert Mulavwa
Research Center for Ecology and Forestry, Ministry of Scientific Research, Mbandaka, DR Congo

Janet Nackoney
Department of Geographical Sciences, University of Maryland, Maryland, USA

Miyabi Nakabayashi
Faculty of Science, University of the Ryukyus, Nishihara, Okinawa, Japan

Anna Nekaris
Oxford Brookes University, Oxford, UK

Caroline M. Nievergelt
School of Medicine, Department of Psychiatry, University of California, San Diego, USA

Vincent Nijman
Oxford Brookes University, Oxford, UK

Jan Nizinski
Institut de Recherche pour le Développement (IRD), Bondy, France

Marilyn A. Norconk
Department of Anthropology, Kent State University, Kent, USA

Katarzyna Nowak
Department of Zoology and Entomology, University of the Free State, Qwaqwa campus, South Africa; Safina Center, New York, USA

John F. Oates
Department of Anthropology, Hunter College–CUNY, New York, USA

Felicity Oram
Unit for Primate Studies-Borneo, Institute for Tropical Biology and Conservation, Universiti Malaysia Sabah, Kota Kinabalu, Sabah, Malaysia

Yosuke Otani
Cross-Boundary Innovation Program (CBI), Osaka University, Yamadaoka Suita, Japan

Fernanda P. Paim
Mamirauá Sustainable Development Institute, Tefé, Brazil

Erwin Palacios
Conservation International Colombia, Bogotá, DC, Colombia

Pia Parolin
Institut National de la Recherche Agronomique (INRA), Univ. Nice Sophia Antipolis, CNRS, France

Fernando C. Passos
Zoology Department, Universidade Federal do Paraná, Curitiba, PR, Brazil

Romina Pavé
Laboratorio de Biodiversidad y Conservación de Tetrápodos, Instituto Nacional de Limnología, CONICET-UNL, Ciudad Universitaria, Pje El Pozo s/n, Santa Fe, Argentina

Nelissa Peralta
Faculty of Social Sciences, Federal University of Pará, Pará, Brazil

Mui-How Phua
Faculty of Science and Natural Resources, Universiti Malaysia Sabah, Kota Kinabalu, Sabah, Malaysia

Alex K. Piel
School of Natural Sciences and Psychology, Liverpool John Moores University, Liverpool, UK

Antonio R. Mendes Pontes
Instituto Nacional de Pesquisas da Amazônia – INPA, Núcleo de Pesquisas de Roraima – NPRR, Boa Vista, Roraima, Brazil

Marcel Quinten
Reproductive Biology Unit, German Primate Center, Göttingen, Germany

Hugo Rainey
Global Conservation Program, Wildlife Conservation Society, Bronx, USA

Fidimalala B. Ralainasolo
Durrell Wildlife Conservation Trust, Jersey, Channel Islands, UK

José Ralison
Groupe d'Etude et de Recherche sur les Primates de Madagascar (GERP), Madagascar

Mariana Raño
Estacion Biologica Corrientes Museo Argentino de Cs. Naturales 'B. Rivadavia'-CONICET, Corrientes, Argentina

Jonah H. Ratsimbazafy
Groupe d'Etude et de Recherche sur les Primates de Madagascar (GERP), University of Antananarivo, and Durrell Wildlife Conservation Trust-Madagascar, Antananarivo, Madagascar

M. Justin O'Riain
Department of Biological Sciences, University of Cape Town, Cape Town, South Africa

Tetsuya Sakamaki
Primate Research Institute, Kyoto University, Inuyama, Japan

Milena Salgado-Lynn
Danau Girang Field Centre, c/o Sabah Wildlife Department, Kota Kinabalu, Malaysia; Sabah Wildlife Department's Wildlife Health, Genetic and Forensic Laboratory, Sabah, Malaysia; Cardiff University School of Biosciences, Cardiff, UK

Agus Salim
Center for International Forestry Research, Bogor (Barat), Indonesia

Roberta Salmi
Department of Anthropology / Center for Geospatial Research, Department of Geography, University of Georgia, Athens, USA

Ricardo Rodrigues dos Santos
Center for Agrarian and Environmental Sciences, Federal University of Maranhão, Chapadinha, Brazil

Katherine S. S. Scott
Oxford Brookes University, UK

John Chih Mun Sha
School of Sociology and Anthropology, Sun Yat-sen University, China

Christopher A. Shaffer
Department of Anthropology, Grand Valley State University, Allendale, USA

Myron Shekelle
Center for Biodiversity and Conservation Studies, Faculty of Mathematics and Natural Science,
University of Indonesia, Depok, Indonesia

Claudia Regina da Silva
Laboratório de Mamíferos, Instituto de Pesquisas Científicas e Tecnológicas do Estado do Amapá (IEPA), Macapá, Brazil

Rondang Siregar
Research Center for Climate Change, University of Indonesia, Depok, Indonesia

Arrilton Araújo de Sousa
Behavioral Biology Laboratory, Biosciences School, Universidade Federal do Rio Grande do Norte, Natal, Brazil

Antonio Souto
Departmento de Zoologia, Centro de Biociências, Universidade Federal de Pernambuco, Brazil

Denise Spaan
Universidad Veracruzana, Instituto de Neuroetología, Xalapa de Enríquez, Estado de Veracruz-Llave, Mexico

Matt Sponheimer
Department of Anthropology, University of Colorado, Boulder, USA

Joan Stevenson
Department of Anthropology, Western Washington University, Bellingham, USA

Fiona A. Stewart
Division of Biological Anthropology, University of Cambridge, Cambridge, UK

Steven Stilwell
Department of Anthropology, Western Washington University, Bellingham, USA

Jatna Supriatna
Research Center for Climate Change, University of Indonesia, Depok, Indonesia

Amanda Tan
Psychology Division, School of Humanities and Social Sciences, Singapore

Simon Tapper
Centre for Research in Evolutionary Anthropology, University of Roehampton, London, UK

Saeko Terada
National Institute for Environmental Studies, Ibaraki, Japan

Yamato Tsuji
Primate Research Institute, Kyoto University, Inuyama, Japan

Augustine Tuuga
Sabah Wildlife Department, Kota Kinabalu, Sabah, Malaysia

Bernardo Urbani
Centro de Antropología, Instituto Venezolano de Investigaciones Científicas, Caracas, Venezuela

João Valsecchi
Mamirauá Sustainable Development Institute, Tefé, Brazil

Alexander Vining
Department of Evolutionary Anthropology, Duke University, Durham, USA

Patrick O. Waeber
Forest Management and Development, Department of Environmental Sciences, Swiss Federal Institute of Technology Zurich, Zurich, Switzerland

Yap Sau Wai
Research and Development Division, Yayasan Sabah – Innoprise Corporation, Kota Kinabalu, Malaysia

Robert B. Wallace
Wildlife Conservation Society, San Miguel, La Paz, Bolivia

Serge A. Wich
School of Natural Sciences and Psychology, Liverpool John Moores University, Liverpool, UK

Nurul Winarni
Research Center for Climate Change, University of Indonesia, Depok, Indonesia

Anna Wong
Unit for Primate Studies-Borneo, Institute for Tropical Biology and Conservation, Universiti Malaysia Sabah, Kota Kinabalu, Malaysia

Barth Wright
Department of Anatomy, Kansas City University of Medicine and Biosciences, Kansas City, USA

Kristin Wright
University of Missouri Kansas City School of Medicine, Kansas City, USA

Bruno de Freitas Xavier
Programa de Pós Graduação em Ciências Biológicas – Zoologia; Universidade Federal da Paraíba, Cidade Universitária, João Pessoa – PB, Brazil

Juichi Yamagiwa
The President's Office, Kyoto University, Sakyo Kyoto, Japan

Jacky Youssouf
Faculté des Sciences, Université de Toliara, Madagascar

Takakazu Yumoto
Primate Research Institute, Kyoto University, Inuyama, Japan

Gabriel E. Zunino
Universidad Nacional de General Sarmiento, Instituto del Conurbano, Buenos Aires, Argentina

Foreword

Soggy But Separate, Why Flooded Habitats Need Special Conservation Attention

When people think of apes, monkeys or even lemurs the image that comes to mind is usually an animal in a tall tropical forest with a dense understory or maybe a baboon group in the savanna. However, as this book so superbly demonstrates, many primates, both in terms of numbers and species richness, occur in habitats that are flooded. For example, in Reserva de Desenvolvimento Sustentável Mamirauá, Brazil, the *várzeas* (flooded forests) can flood to a depth of 10 m for at least 5 months (Peralta *et al.* this volume) and this forest supports 11 primate species. For a human observer, it is difficult to follow primate groups in flooded forests (e.g. in Brazil's *várzeas* one has to canoe in the flooded season) and in some swamp habitats, it is literally impossible. For example, I have tried to follow chimpanzees and baboons through papyrus swamps, where they are eating either papyrus or fruits of trees in the swamp, and while they seem to be able to balance themselves well using four limbs, I was frequently plunging through the papyrus mat, sinking up to my waist in water and soon losing the groups I was trying to follow. Such difficulties have resulted in primate populations in flooded habitats being very poorly studied and, in some cases, these habitats not even being considered as primate habitats. From reading the compilation of chapters from flooded habitats from around the globe by Adrian Barnett, Ikki Matsuda, and Katarzyna Nowak in *Primates in Flooded Habitats: Ecology and Conservation*, we now know that flooded habitats can be very important for some primate species. This should ignite a burst of research and, by the end of the next decade, we will know much more about how important flooded habitats are for primates and their behavioural adaptations to them.

Unfortunately, just as we are discovering that these flooded habitats can be very important for primates, and as stated by Galat-Luong *et al.* and Nowak *et al.* (this volume), these habitats might be a temporary refuge from bushmeat hunting and habitat clearing, but they are being degraded and destroyed at an alarming rate. Flooded habitats can be very extensive in all continents; for example large wetlands in Africa cover 2 072 775 km², which is approximately 9% of the landmass (Mitchell 2013), thus they have the potential to support large numbers of primates. However, these areas are gravely threatened. In fact, many consider surface freshwaters, lakes, wetlands and rivers, to be among the most extensively altered ecosystems on earth (Carpenter *et al.* 2011). Let me provide you with some examples of different types of flooded habitats and how they are threatened. Since 1990, approximately 50% of the world's wetlands have been drained and converted to some other use (Michener *et al.* 1997). North America and Europe often lead the way in flooded habitat destruction. For example, approximately 90% of the floodplains in North America and Europe have been converted and are now cultivated lands; in the tropics, this conversion is occurring at an accelerating rate (Tockner and Stanford 2002). This rate is bound to increase as the human population expands, creating an increased demand for agricultural products.

In addition to direct anthropogenic disturbance, coastal flooded habitats are additionally threatened by global warming. In the next 50 years, average global temperature is projected to rise 2.5°C, which is predicted to lead to an 80–100 cm sea-level rise (Michener *et al.* 1997). This will have a significant effect on coastal flooded forests. For example, the Niger Delta is an important area for primates (Baker and Oates, this volume), but the estimated sea-level rise that is projected to occur in the next century would lead to 18 000 km² of the delta being inundated with salt water and lost (Brown and Thieme 2005). This sea-level rise will also negatively impact important mangrove forests, which a number of chapters in the book point out as an important primate habitat. Mangrove forests are already facing serious threats from shrimp fisheries and timber extraction and are declining by 1% each year (Polidoro *et al.* 2010).

In this volume, Cheyne and colleagues demonstrate that tropical peat swamp forest across Borneo and Sumatra supports 11 monkey species (possibly up to 14), 8 gibbon species (including the siamang), both orangutan species and 2–3 nocturnal species. Given this diversity, it is a tragedy that only 36% of the historical tropical peat swamp forest remains (Posa *et al.* 2011) and, as Cheyne and colleagues point out, annual deforestation rates in the region between 1990 and 2010 were 3% and 5% for Borneo and Sumatra, respectively.

As a final example, it should be recognized that global demand for water increased ten fold in the last century (Junk 2002). As more agricultural land is needed to feed the growing human population, there will be increased pressure to draw water away from wetlands for irrigation or to convert flooded habitats to agricultural land. Associated with increasing agricultural demand for water and a demand for energy there is a rapid increase in the number of dams built (see Harrison-Levine *et al.*, this volume, to discover primate responses to widespread flooding caused by dams), which often draws water away from wetlands and deprives them of nutrients. There are 1269 dams in Africa (Junk 2002). On the 26 August 2012, Aljazeera reported that globally 780 million people currently lack access to safe drinking water, according to the United Nations. By 2030, 47% of the world's population will be living in areas of high water stress, according to the Organization for Economic Co-operation and Development's 'Environmental Outlook to 2030' report. This creates strong incentives to draw water away from flooded habitats like wetlands, which often leads to the total loss of these important habitats.

The studies in this volume cover all types of flooded habitats where primates are found and do a remarkable job of covering all geographic areas. The book includes 45 chapters organized into seven parts. The first part includes six chapters and is an introductory part that gives the reader a greater understanding of why flooded habitats are of interest and some of the methodological challenges researchers have to deal with, such as how to survey flooded habitats and how do researchers assess fruit production when fruit floats away. The next series of chapters covers primates occupying or using different types of flooded habitats, including mangrove and coastal forests, beaches, swamps and freshwater flooded forests. The sixth part of this volume directly deals with conservation case studies involving primates that use flooded habitat. Topics in this part range from the responses of primates to dams, ecotourism, populations seeking refuge in flooded habitats to avoid hunting or because of habitat destruction in *terra firma* forest and the incorporation of genetic data into management plans of isolated populations. The final part deals directly with conservation issues and there are chapters discussing Africa, Southeast Asia and the Neotropics.

Ours is an era characterized by great destruction of primate habitat. Global forest loss between 2000 and 2012 was estimated at 2.3 million km^2, with an increase of 2101 km^2 every year in the tropics (Hansen *et al.* 2013). But this is also an era characterized by huge new conservation efforts and a wave of restoration through tree planting, corridors, re-wilding, as well as creative mitigation of human development impacts. For these efforts to succeed, knowledge becomes a key tool to wisely invest time, effort and money. Conservation and restoration at the scale needed will be no simple task: we need to identify key habitats, which species depend on and which are just being used as a refuge of last resort. We need the knowledge from different areas of the world and all types of habitats. This volume tackles these issues for flooded habitats and, given their geographic extent, will help managers in many areas construct informed conservation plans. Most importantly, for the first time, it significantly highlights the potential importance of flooded habitats and I predict this will lead to a great deal of future research of primates and other taxa in flooded habitats.

Colin A. Chapman
Professor, Canada Research Chair, Killam Fellow
McGill School of Environment and Department of Anthropology, McGill University, Montreal, Quebec, Canada and Wildlife Conservation Society, Bronx, NY, USA

Acknowledgements

We would like to take the opportunity to thank all those who at Cambridge University Press have helped shepherd this book through the various stages of production, but especially Ilaria Tassistro, Victoria Parrin, Jenny van der Meijden and also Dominic Lewis and Martin Griffiths, commissioning editors. We would also like to thank Janice Chism, Tørben Haughaasen and Marilyn Norconk for their help in organizing the initial primates in flooded habitats symposium at the 2012 International Primatological Congress, from which this book sprang. We would also like to extend our gratitude to all chapter authors for their patience, team spirit, dedication and kindness, and Stephen Nash for the original cover art. We acknowledge too the non-human primates that inspired and continue to inspire us.

Finally, we thank the following people for kindly peer reviewing chapters in this volume:

Kate Abernethy, Meredith Bastian, Bruna M. Bezerra, Sarah Ann Boyle, Tom Butynski, Susan Cheyne, Janice Chism, Marina Cords, Nabajit Das, Lorna Depew, Anh Galat-Luong, Gerard Galat, Mary Glenn, Colin Groves, Cyril C. Grüter, Michael Gumert, Olivier Hamerlynck, Torbjørn Haugaasen, Joseph Hawes, Keith Hodges, Hiroshi Ihobe, Eiji Inoue, Leandro Jerusalinsky, Shawn Lehman, Jean-Marc Lernould, Stanislav Lhota, Raone Beltrão-Mendes, Antonio Rossano Mendes-Pontes, Bill McGrew, Sangita Mitra, Vincent Nijman, John Oates, Mary Susan Pavelka, Helder L. Queiroz, Tacyana Ribeiro, Ricardo Rodrigues dos Santos, John C. M. Sha, Danica Schaffer-Smith, João Pedro Souza-Alves, Hideki Sugiura, Yamato Tsuji, Patrick Waeber, Serge A. Wich, Lucienne Wilmé.

Chapter 1

Editors' Introduction

Katarzyna Nowak, Adrian A. Barnett and Ikki Matsuda

Whether they are freshwater swamps, riverside forests, mangroves or any other of the great diversity of plant assemblages that have their roots in water for some of the year, flooded habitats are a major part of the tropics and subtropics. By land area alone, they cover over 1 million square kilometres of the land area occupied by primates (Sjögersten *et al.* 2014). As the chapters in this book reveal, well over 50% of the world's 695 primate taxa have some populations that use flooded habitats. Some species, like Africa's Allen's swamp monkey, Asia's proboscis monkey, Amazonia's white bald uacari and Madagascar's Lac Alaotra lemur use no other habitat type. In some cases, flooded habitat use appears to be very recent, with primates moving into swamps and mangroves as their former *terra firma* forest homes are destroyed. But whether they are refuges, permanent residences or occasional resources, flooded habitats are clearly important to primates. Yet until now there was no single book on the subject, nothing to bring together the range of species, topics and themes in a way that might shed light on what we know and what we do not, and how best to go about filling in the gaps as well as prioritizing conservation effort.

Tropical flooded habitats include some of the most productive ecosystems on the planet (Maltby 1986), and yet tropical wetlands also show the fastest loss rates of any habitat type in the tropics (Davidson *et al.* 2005). Despite this, flooded habitats represent one of nature's last great unknowns. Many aspects of their ecology have yet to be studied in detail. This book is about one such group: the primates of flooded habitats. The first-ever treatment of the subject, our volume brings together experts from around the world summarizing current knowledge of primates in habitats with great annual changes in water levels and productivity.

While research on primate behaviour and ecology in relation to environmental factors such as food availability has been thorough compared with research on many other mammalian taxa, there have been few attempts to study primates inhabiting flooded forests. This is due to the numerous practical problems and challenges that accompany studies in such swampy habitats. Nonetheless, given notable accumulation of natural history knowledge on primates in flooded habitats over the past 20 years, we believed that the time was right to collate these data into a single coherent volume for ease of reference.

The contributions represent all major flooded habitats including mangrove and peat swamp forests (Parts II and V), beaches (Part III), delta swamps, reed beds and seasonally flooded grasslands, as well as freshwater flooded forests, palm and raphia swamps (Part IV) (with habitat definitions in Chapter 2). We also include riparian forests (Chapters 24, 31, 33), which, given that they occur along the banks of rivers, can be seasonally flooded from several metres to several kilometres from the riverbank, and may include permanent marshes (Gautier-Hion & Brugiere 2005). Our coverage embraces Africa, Madagascar, Asia, Central and South America and involves authors from primate range states as well as those from Europe, North America and Japan. This volume shows the diversity of flooded habitats inhabited by primates, the breadth of their socioecologies and the nature of the threats they face from the perspectives of primatologists based around the world. Although the phrase terra firme is widely used to describe types of never-flooded lowland rainforest throughout the Neotropics, it is not used elsewhere. Therefore, for consistency, we use the term *terra firma* throughout the volume, in chapters referring to both the Paleo- and Neotropics. It has been used to refer to both unflooded forests and never-flooded land in general.

The scope of our volume ranges from fossil history (Chapter 3), to survey methods (Chapter 6), to comparative behaviour and biology (Chapters 7–8), to the role of flooded habitats as refuges (Chapter 39), to conservation priorities (Part VII), *Primates in Flooded Habitats* will, we anticipate, set the stage for a new wave of research on primates and other animals of these as yet still barely known areas of biodiversity. We hope that the book will provide an incentive, a jumping off point from which future ecological studies can spring, and be used by field biologists to plan projects and prioritize conservation initiatives for primates in the world's tropical and subtropical wetlands and flooded habitats. As the first step in such a coordinated effort to investigate and conserve, we offer both this book, and the web-based resource that accompanies it (www.cambridge.org/9781107134317).

Chapter 2

Flooded and Riparian Habitats in the Tropics

Community Definitions and Ecological Summaries

Amy C. Bennett, Pia Parolin, Leandro V. Ferreira, Ikki Matsuda and Neil D. Burgess

Introduction

Wetlands occupy the interface between fully terrestrial and aquatic ecosystems, making them inherently different from both systems yet dependent on both (Mitsch & Gosselink 2007). A number of distinct types of flooded and riparian wetland habitats occur in the tropics, including those with seasonal or permanent inundation (Kalliola *et al.* 1991), with or without the influence of salt. Riverine margins and river deltas can also contain flooded habitats. For example, seasonal floods can submerge riparian forests alongside rivers, sometimes extending several kilometres from the riverbank, with permanent and seasonal wetlands connected to the floodwater (Gautier-Hion & Brugière 2005).

Vascular plant species evolved and diversified in terrestrial ecosystems and, in general, die more readily in response to flooding than to desiccation (Larcher 1994). In tropical wetlands, the constraints imposed by flooding are even more pronounced as high ambient temperatures result in high rates of plant metabolism. Inundation causes drastic changes in the bioavailability of nutrients, oxygen levels, and concentrations of phytotoxins (Parolin *et al.* 2004), with oxygen rapidly consumed by respiring roots and microorganisms and soils becoming hypoxic or anoxic within a few hours of flooding (Visser *et al.* 2003). Oxygen depletion is accompanied by increased levels of carbon dioxide, anaerobic decomposition of organic matter, increased solubility of mineral substances and reduction of the soil redox potential (Vartapetian *et al.* 2003), followed by the accumulation of potentially toxic compounds (Ponnamperuma 1984). Only those plant species adapted to these conditions persist in these habitats (De Oliveira *et al.* 2014).

Tropical wetlands have characteristic vegetation zones, with the most tolerant species growing in areas subject to longer and deeper flooding. In the Amazon, flood waters can be 15 m deep and remain for up to 9 months, but trees survive and remain rooted in the soil (Junk *et al.* 1989). A regular periodicity of flooding is important to allow plants to evolve and adapt (the 'Flood Pulse Concept': Junk *et al.* 1989; Junk & Wantzen 2004), so in wetlands with unpredictable flooding, such adaptations are rare and plant diversity is lower.

Species richness and the number of endemic plant species in most tropical wetlands is low, with two exceptions: the Everglades and Amazonia (Junk *et al.* 2006; Slik *et al.* 2015; Wittmann *et al.* 2013).

Various kinds of tropical flooded and riparian wetland habitats are inhabited by members of the Primate order (the topic of this volume), but many primate studies have used inconsistent terms for flooded habitat types. Here, we summarize the ecology of flooded and riparian habitats in South and Central America, Africa and Asia and define habitats used by primates and employed throughout this volume. Habitats include lowland tropical floodplain forest, riverine and gallery forest, river delta, swamp forest(s), mangroves, oxbow lakes, reedbed/papyrus, seasonally flooded grassland and intertidal zones.

Definition of Habitats

In the following sections, we describe the major tropical wetland habitats of importance to primates.

Lowland Tropical Floodplain Forest

Definition

Floodplains are highly dynamic, where the aquatic and terrestrial transition zone (the whole floodplain area) alternates between wet and dry, with a moving littoral region (Junk *et al.* 1989). The natural disturbance caused (Leyer 2006), and materials deposited (Maddock 1976), by flooding produce diverse floodplain structures. Forests growing on floodplains often include three layers of vegetation: a canopy layer of mature trees, saplings and shrubs in the understory and herbaceous plants and tree seedlings covering the ground (Yin 1998).

Neotropics

Several extensive freshwater floodplains occur in tropical America: the Amazon (Toivonen *et al.* 2007), Orinoco (Rosales *et al.* 2002), Pantanal (Junk & Nunes da Cunha 2005) and Paraguay/Paraná (Galán de Mera & Perea 2008). The Amazon floodplain is mostly forested, whereas the Orinoco, Pantanal and Paraguay/Paraná are dominated by grasslands. The largest floodplain forests of the Amazon are characterized by immense cyclic water-level fluctuations; a flood pulse produces water depths up to 12–15 m and flood durations between 50 and 210 days per year (Junk *et al.* 1989). As a result, ecological processes, including plant reproductive cycles (e.g. flowering, fruiting) and animals' seasonal movements and migrations are synchronized with the flooding period (Schöngart *et al.* 2002).

In contrast, flood tolerance and responses to inundation cycles varies greatly between species (Parolin 2001), resulting in a mosaic of habitats within and between floodplain ecosystems (Wittmann *et al.* 2013).

Two types of floodplain forest exist in the Amazon, determined by the water quality of the flooding rivers (Prance 1979). Várzeas form along the banks of whitewater rivers, which are nutrient rich, originate in the Andes and have a high sediment load. *Igapós* form along the banks of blackwater and clearwater rivers, both with low sediment loads; their flooded soils have low nutrient contents (Furch 1997). Seven per cent of the Brazilian Amazon (400 000 km^2) can be categorized as floodplain forest, with approximately 300 000 km^2 covered by *várzea* and 100 000 km^2 covered by *igapó*. The Amazonian *várzea* is the most species-diverse floodplain forest in the world (Wittmann *et al.* 2006). Characteristic plant genera of both habitats include *Aldina* (Fabaceae: Papilionoidae), *Amanoa* (Euphorbiaceae), *Buchenavea* (Combretaceae), *Elaeoluma* (Sapotaceae), *Eschweilera* (Lecythidaceae), *Hevea* (Euphorbiaceae), *Hydrochorea* (Fabaceae: Mimoidoidae), *Mabea* (Euphorbiaceae), *Macrolobium* (Fabaceae: Caesalpinoidae) and *Pouteria* (Sapotaceae). Research on primates in these habitats includes studies of white uacaris (*Cacajao calvus calvus:* de Alecântara Cardoso *et al.* 2014), red uacaris (*Cacajao c. ucayalii*: Bowler & Bodmer 2011; Bowler *et al.* 2012), golden-backed uacaris (*Cacajao ouakary*: Barnett *et al.* 2005, 2012c; Bezerra *et al.* 2011; Defler 2001), squirrel monkeys (*Saimiri vanzolinii, S. sciureus*: Paim *et al.* 2013) and assemblage-level studies of *Alouatta, Ateles, Cebus, Pithecia, Saimiri, Saguinus* and *Sapajus* (e.g. Haugaasen & Peres 2005a, 2009).

Africa

In the Congo Basin, forested floodplains extend along the Congo River and its tributaries. A geographical depression, the *cuvette centrale congolaise*, located at the centre of the Congo Basin, supports swamps and wetlands intermixed with lowland rainforest. In the more permanently flooded areas, the canopy is 30–35 m and trees often have stilt roots. The upper floodplain is well drained in periods of low water, but is flooded once or twice a year. Its forest canopy reaches 20–25 m with a relatively low tree density, but many lianas. Species include *Didelotia unifoliata* (Fabaceae: Caesalpinioideae), *Diospyros* spp. (Ebenaceae), *Uapaca* spp. (Phyllanthaceae), *Guibouria demeusei* (Fabaceae: Caesalpinioideae), *Mitragyna stipulosa* (Rubiaceae) and *Oubanguia africana* (Lecythidaceae). Soil of the upper floodplain is sandy, while the lower floodplain has peat soils, often of considerable depth.

The Congo Basin supports one of the highest diversities of primates in the world, including some found nowhere else: the bonobo (*Pan paniscus*), found only south of the Congo River, and Allen's swamp monkey (*Allenopithecus nigrovirdis*; Maisels *et al.* 2006), Africa's only primate swamp specialist (Chapter 43).

Asia

In the absence of human impact, forest would form the natural vegetation of most wetland areas in Southeast Asia (Corlett 2009), although much has now been converted to oil palm (Abram *et al.* 2014; Runting *et al.* 2015). In south central Cambodia, the Mekong and Tonle Sap rivers join forming the lower Mekong floodplain (70 000 km^2) (Rainboth 1996). The Tonle Sap is highly seasonal and flooding in the wet season can inundate ten times the area of the dry season (Arias *et al.* 2013, 2014a). On the edge of the permanent Tonle Sap lake is a tall strip of forest which acts as a physical barrier between the open lake and the floodplain (Arias *et al.* 2014b; Kummu & Sarkkula 2008).

The forests of Borneo, are one of the most species-rich habitats on earth (Whitten *et al.* 2004), with 60% higher biomass than the Amazon (Slik *et al.* 2010). The Kinabatangan River, 560 km long, is the longest river in Sabah and is made up of a patchwork of different habitat types, i.e. riverine forest, seasonally flooded forest, swamp forest, dry dipterocarp forest and mangrove (Boonratana 2000a; Sha *et al.* 2011; Chapter 4). This area sustains one of the world's richest primate ecosystems inhabited by ten species; five of these endemic to Borneo: the Bornean orangutan (*Pongo pygmaeus*), Bornean gibbon (*Hylobates muelleri*), proboscis monkey (*Nasalis larvatus*), red leaf monkey (*Presbytis rubicunda*) and Hose's langur (*Presbytis hosei*) (see Chapter 29).

Riverine and Gallery Forest

Definition

The terms 'riverine' and 'gallery' forest are often used interchangeably as both form forested riparian corridors in otherwise grass-dominated or savanna-woodland ecosystems. However, these two habitat types can be distinguished (Kingdon *et al.* 2013). Riverine forests grow along the banks of rivers or streams where the soil is moister than the surrounding area. Immediately next to the river, the soils are fertile, the understory is green and moist with little grass, and hence these areas rarely burn. These factors are more important in determining the vegetation of the riverine edge than the surrounding climate (Emmerich 1990). Riverine forest strips are periodically flooded when the river bursts its banks; which can occur annually, more than once a year or every few years (Chapter 33).

Gallery forest is a type of forest outlier in a grassland region where the soils are moist enough and the conditions humid enough to support evergreen or semi-evergreen trees (Rosevear 1953a). Typically, gallery forest is found in narrow sheltered valleys and ravines on hillsides.

Neotropics

In the *cerrado* (tropical savanna of Brazil), gallery forest occurs where the water table is close to the surface. These forests can be dense and stratified, composed of evergreen trees and a sparse understory. These forests make up < 10% of the *cerrado*, but are found throughout the region (Redford & Fonseca 1986). There is a sharp and distinct boundary between gallery forest and xeromorphic *cerrado* vegetation (Mares *et al.* 1989). These forests can be important for local primates (Chapter 31).

Riverine forests of this region contain many species that are also found in Amazonian rainforest and along the Atlantic coast of Brazil (Oliveira-Filho & Ratter 1995). In the transition zone between gallery forest and *cerrado*, the waterlogging of the wet season excludes woody *cerrado* species, while desiccation of the soils in the dry season excludes gallery forest species (Pennington *et al.* 2000).

As an example, the Miranda riverine forest in the Pantanal is 50–200 m wide with an 8–13 m canopy (emergents up to 17 m). Inundation pulses occur from January to March and the water level in the gallery forest may reach 1.5 m. Many trees are deciduous, dropping their leaves from July to September, but evergreen species are also abundant including *Inga vera* (Fabaceae: Mimoisoideae), *Ocotea diospyrifolia* (Lauraceae) and *Vitex cymosa* (Lamiaceae). Fruiting has two peaks annually, one in the middle of the wet season and the other in the transition from wet to dry (Ragusa-Netto 2004).

Africa

Riverine forests are found along many of the continent's rivers. Within the rainforest zone, riverine forest can be distinguished from rainforest by a different species composition. In East Africa, riverine forests of the Tana floodplain are restricted to small stands that depend on the ground water from the river (Hughes 1985; Chapter 30). In savanna habitats of Africa, strips of riverine forest extend far into these drier regions.

Riverine forest strips provide movement corridors for forest animals, including many species of primates, and refuges for forest plants. They provide sources for forest expansion if the climate becomes wetter, or areas to retreat to if it becomes drier, which has happened cyclically in parts of tropical Africa over millions of years. For example, the rivers draining the Congo Basin often support riverine forest strips far outside the current rainforest zone. In wetter climatic periods, these forests could expand to form an expanded rainforest zone.

Asia

There are many seasonally flooded riverine forests in South Asia (e.g. Dudgeon 1992, 2000a, b; Chapter 44). Some of the most extensive are in Bangladesh and are situated in the natural levees of rivers subject to overflow during the monsoon.

The spatial extent of the riparian zone generally encompasses the terrestrial landscape between low- and high-water marks (Naiman *et al.* 1993, 2000; Naiman & Décamps 1997; Vellidis & Lowrance 2004). Riparian deforestation and pollution is common in many Asian rivers, especially in Southeast Asia, hence these forests are in decline (Dudgeon 2000a; Iwata *et al.* 2003).

River Delta

Definition

River deltas can support both forest and non-forest habitats. They are formed by the deposition of sediment carried by the river as it enters the sea, with deltas classified according to the main control of this deposition; river, wave or tide. The sediment deposition shapes the resulting delta and its vegetation with drier remnants of levees being more diverse than waterlogged areas. Species composition is also linked to the adjacent fringing vegetation, e.g. mangroves or upland forest.

Neotropics

The Amazon and Orinoco Delta swamp forests occur within a diverse matrix of coastal vegetation types and are characterized by permanent inundation and abundant waterways and canals (IUCN 1996), with the core a 'tropical ombrophilous swamp forest' (UNESCO 1981). These deltas support a number of endemic plant species (Dinerstein *et al.* 1995). Primates of the Orinoco Delta have recently been surveyed by Bernardo Urbani (unpublished data). The primates of the islands in the Amazon delta are also poorly known, with the few published surveys focusing on the larger islands (e.g. Marajó) (Fernandez *et al.* 1995; Peres 1989; Chapter 45).

Africa

The Niger Delta in southern Nigeria contains a diverse mix of swamps in a 'mosaic of edaphic grassland and aquatic vegetation' (White 1983) separated from the Atlantic Ocean by mangroves (Welcomme 1986; Chapter 40). Average tree height is relatively low (20–25 m) and the forests are dominated by *Hallea ledermannii* (Rubiaceae), *Ctenolophon englerianus* (Ctenolophonaceae), *Klaineanthus gaboniae* (Euphorbiaceae), *Hexalobus crispiflorus* (Euphorbiaceae), *Symphonia globulifera* (Clusiaceae) and *Pycnanthus marchalianus* (Myristicaceae) (Werre 2001b). Emergents vary with soil water content; in drier sections, red ironwood *Lophira alata* (Ochnaceae), *Sacoglottis gabonensis* (Humiriaceae) and *Irvingia gabonensis* (Irvingiaceae), are most common, while in the wetter sections it is *Alstonia boonei* (Apocynaceae). Oil palm (*Elaeis guineensis*) is common and the understory is often dominated by rattan palms (e.g. *Calamus deerratus, Raphia* spp.). Further detail on the Niger delta is provided in Chapters 30 and 40 of this volume.

The Tana Delta is the name loosely given to the flood plain ecosystem of the lower Tana River, a vast wetland complex on the Kenyan coast. The delta is roughly triangular in shape, with its apex at Lake Bilisa (north of Garsen) and its base a 50 km stretch of beach along Ungwana (or Formosa) Bay, stretching from Kipini in the northeast to Mto Kilifi in the southwest. Habitats include fresh and brackish lakes, freshwater and saline grasslands, and successional stages of forest and woodland on the riverbanks and the dune ridges parallel to the shore (see Chapter 30).

Asia

The Mekong River, one of the largest rivers in Southeast Asia, flows southward from the Tibetan Plateau to the South China Sea through the Indochina Peninsula and forms a wide delta plain at its mouth containing one of the largest deltas in the world. In the past, the pileated gibbon *Hylobates pileatus* may have occurred in Vietnamese territory to the west of the Mekong Delta, though east of the Mekong only the primate genus *Nomascus* occurs (Groves 2007a). The primates of the Sunderbans (the combined deltas of the Ganges, Brahmaputra and Meghna rivers) are covered by Mallick (Chapter 16).

Swamp Forest

A swamp forest is a forested waterlogged area where the physiognomy of the vegetation is clearly different from well-drained forest vegetation (Kalliola *et al.* 1991). Forested swamp types can often be distinguished by vegetation; palm swamp and forest swamp, while peat swamp and backswamp are usually defined by soil type.

Palm Swamp

Definition

Palm swamp is found mainly on alluvial soils subject to flooding and anaerobic conditions. Their defining feature is a high concentration or dominance of palms of various types.

Neotropics

South American palm swamp forests are dominated by *Mauritia flexuosa* (known by many names: South American moriche palm, *ite palm, ita, buruti, canangucho, aguaje*). Producing nearly monospecific stands (Penn 1999), *M. flexuosa* can reach up to 35 m at high densities. Many animal species – including primates (Bennett *et al.* 2001; Bowler & Bodmer 2011; Pontes 1999) – depend on the *M. flexuosa* fruit, which has high vitamin C content and fruits when many other trees don't (Chapters 27 and 31).

The permanently waterlogged palm swamp of the Pastaza–Maranon basin in the northern Peruvian Amazon (Roucoux 2013) is a closed-canopy forest dominated by *M. flexuosa* (most abundant species by basal area), *Mauritiella armata* and *Tabebuia insignis* (Bignoniaceae) (Atrium 2012).

Africa

Forest swamps cover huge areas of the central Congo Basin (Burgess *et al.* 2004; Thieme *et al.* 2005), where they are dominated by *Raphia* palm (Kingdon 1997; White 1993). These *Raphia* swamp forests support high densities of western lowland gorilla (*Gorilla gorilla gorilla*) (Rainey *et al.* 2010; Chapter 43). In Budongo forest, Uganda, chimpanzees compete for the pith of *Raphia farinifera*, which provides them with sodium (Reynolds *et al.* 2009).

Asia

Nipa palm (*Nypa fruticans*) is commonly found in the tidal wetlands of Southeast Asia, forming monodominant or mixed stands along the tidal section of rivers, as well as covering extensive low-lying areas in estuaries (Chapter 4). At the mouths of large rivers with a high freshwater discharge, mangroves are often replaced by nipa palm stands that perform the same ecological function, trapping sediments (Fong 1992). In Southeast Asia, Nibong palm (*Oncosperma tigillarium*) is also commonly found along the tidal section of rivers (near mangroves).

Forest Swamp

Definition

Freshwater swamp forests grow on fertile alluvial soils in areas of low relief. In contrast to palm swamp, they support a higher diversity of trees and not monospecific stands of palms.

Neotropics

Wallaba is the local name for *Eperupa falcata* (Fabaceae: Caesalpinioideae) which dominates Guyanese swamp forests (J. Chave, pers. comm., 2013). Wallaba swamp is both structurally and floristically distinct from lowland rainforest (Davis & Richards 1934). Three features produce its characteristic appearance: an extraordinary number of trees per unit area, a scarcity of very large trees and an almost complete absence of buttressed trees. Usually wallaba forests are adjacent to areas of white sand and there is a sharp boundary between the white sand and the brown wallaba soil. The use of these forests by primates in Guyana is described by Shaffer *et al.* (Chapter 28). In French Guiana, *E. falcata*-dominated swamp forests are called *wapa*. Though *E. falcata* is overwhelmingly dominant, there are sometimes tree species of Myristicaceae and Sapotaceae within these forest stands.

Africa

In southern Nigeria, forest swamps are the product of both fluviatile and marine sediment build up (Grubb 1990). Species include: *Lophira alta, Ricinodendron heudelotii, Sacoglottis gabonensis, Uapaca* spp., *Hallea ledermanii* (Rubiaceae) and *Ficus vogeliana* (Moraceae) (Werre 2001b).

Southeast Asia

Forests subject to flooding by freshwater are extremely floristically varied within the region and are lumped together into one category purely for convenience (Corlett 2005). In Sumatra, the dominant tree is *Adina polycephala* (Rubiaceae) but *Malaleuca leucadendron* (Myrtaceae) also covers extensive areas (FAO 1981). Swamp forests can range from species-poor stands dominated by the genus *Malotus* to floristically diverse forests similar to the surrounding lowland rainforest. The tall legumes *Koompassia, Callophyllum* and *Melanorrhoea* and *Metrixylon sagu* thrive in this habitat (MacKinnon 1987).

Peat Swamp

Peat swamp soils are made up primarily of organic matter (Driessen 1978), nutrient poor, often acidic (pH < 4) and hygromorphic, being almost permanently flooded or waterlogged. Peat swamp forests are species poor and support few endemic species (IUCN 1991). Despite extreme conditions, trees can grow above 70 m high, reliant upon nutrient inputs from rainfall, dust and marine aerosols (Yule 2010).

Africa

Recently, a huge bog (over 145 500 km^2) was discovered in the centre of the Congo Basin. The swamp found in Congo Brazzaville is up to 7 m thick and thought to contain billions of tonnes of peat (Dargie *et al.* 2017).

Asia

The edges of swamp forests in Malaysia are characterized by strangler figs, an important food for mammals (Payne & Andau 1994), and more than 30 palm species are found in the peat swamp forest, also providing food. On Sumatra, peat deposits are at least 50 cm thick (but up to 20 m); fires are common and prevent natural succession, promoting the development of extensive nearly monospecific stands of paperbark (*Melaleuca cajupiti*, Myrtaceae) (Whitten *et al.* 2000). The Tonle Sap–Mekong peat swamp forests of southern Vietnam and Cambodia are also dominated by *Melaleuca* (though it is *M. leucadendron*); they are low in plant diversity, but significant in maintaining ecosystem function, reducing water flow in the wet season and minimizing flooding, storing freshwater and reducing soil acidification.

Backswamp

Definition

Geologically, a backswamp is the section of a floodplain where deposits of fine silts and clays settle after a flood, usually behind the natural levee of the main river or one of the distributary rivers. Backswamp areas tend to have saturated soils throughout the dry season (Hamilton *et al.* 2007). These forests are less diverse than riverine forest; trees must grow in saturated soils that lack oxygen all year.

Neotropics

In the Peruvian part of the watershed of the Madre de Dios river, backswamp forest is particularly common on the left river bank. Forests are dominated by *aguaje* palms (*Mauritia flexuosa*, forming palm swamp forests known as *aguajales*) but other species, especially figs (*Ficus* spp.) – keystone species and important resources for primates – can also be abundant.

Africa

Backswamp areas in the Niger delta of Nigeria are relatively stable, consisting of constantly waterlogged heavy clay covered by peat (NEDECO 1961). The waterlogged backswamps are not as diverse in species as the drier areas of the swamp (Werre 2000). Forests of backswamps include members of the Euphorbiaceae (*Uapaca* spp., *Klaineanthus gaboniae, Macaranga* spp.), Clusiaceae (*Symphonia globulifera*) and Rubiaceae (*Hallea ledermannii, Rothmannia* spp.) families.

Asia

Many of the former backswamp areas in Asia were converted to agriculture centuries ago as they are fertile and amenable to irrigated agriculture – especially rice. Bowl-shaped depressions between the natural levees of rivers in Bangladesh are known as *haors* and, in eastern Bangladesh, they cover 2.5 million hectares (Poffenberger 2000). Forests in this area are dominated by tree species that can withstand inundation, including *Barringtonia acutangula* (Lecythidaceae), *Pongamia* (=*Millettia*) *pinnata* (Fabaceae: Papilionoideae) and *Crataeva nurvala* (Capparaceae).

Mangroves

Definition

Mangroves are largely restricted to the tropics and a few warm temperate regions (Chapter 7). Adapted to the intertidal zone, the physical environment of mangroves is defined by regular inundation of the soil and variable salinities, and although the soils range in type, most are anaerobic within a few millimetres of the soil surface. Mangrove trees are specifically adapted to this saline, low oxygen environment and are rarely found elsewhere. Characteristic physiological and morphological adaptations include active extrusion of salt from xylem (via deposits in bark, senescence in leaves, salt glands on leaves), and a number of secondary root adaptations: pneumatophores (upward extensions from subsurface roots into the air), knee roots (rounded knob-like extrusions) and buttress roots (important for structural support).

Mangroves can form extensive forests with canopy heights above 30 m, but in more arid or saline conditions they form dwarf or scrub communities. The most extensive mangrove forests occur in river deltas, but mangroves are also common along coastlines where wave energy is low. The coasts of Central America and Central Africa have hundreds of kilometres of almost continuous mangrove systems, which can extend into deltaic or estuarine systems, often far inland.

Mangroves have a vital role in coastal protection. They are also important for the breeding, passage and wintering of water birds. Sea turtles feed in mangrove ecosystems, which are important breeding and nursery grounds for fish and invertebrates while regulating water and nutrient flows to adjacent ecosystems.

Neotropics

Over 80% of mangroves are found in complex deltaic systems. In Brazil, the coastal region (Pará) is dominated by three mangrove species (Krause *et al.* 2001); *Rhizophora mangle* (Rhizophoraceae: the red mangrove), *Avicennia germinans* (Acanthaceae: the black mangrove, referring to the colour of the trunk and the heartwood) and *Laguncularia racemosa* (Combretaceae: the white mangrove, as bark appears white). Red mangroves grow closest to the sea, tolerating the deepest water, while white mangroves grow furthest inland. The primates of Neotropical mangroves are described by Santos and Bridgeman (Chapter 8).

Africa

Mangrove swamps in West Africa are becoming increasingly important refuges for large mammals, including primates, as human populations expand (Bi *et al.* 2009; Galat-Luong & Galat 2007). Nigeria has the largest extent of mangroves in Africa, mainly within the Niger delta, which extend up to 50 km inland (Chapter 40). In East Africa, the most extensive mangrove stands are in the Rufiji and Zambezi river deltas, and in Zanzibar, stands of *Rhizophora mucronata* are common, productive and aseasonal with continuous leafing, flowering and fruiting of trees providing food for endemic red colobus monkeys (*Procolobus kirkii*) and also Sykes monkeys (*Cercopithecus mitis albogularis*) (Nowak 2008). Primates of African mangroves are summarized by Head *et al.* (Chapter 12).

Asia

This region has the highest mangrove diversity in the world, with extensive mangrove stands in many countries (Spalding *et al.* 2007). Compared with other forest types in the region, mangroves have simpler structure and species composition (Corlett 2005). The Sundarbans in Bangladesh and India form the world's largest mangrove area, covering 20 400 km^2, deriving their name from the dominant mangrove species *Heritiera fomes* (Malvaceae), known locally as *sundri* or *sundari*. In coastal India and Pakistan, the Indus Delta mangrove habitat is adapted to extreme temperatures and salinity. There are also large stands of mangrove in the Mekong Delta, although they declined from 21 221 ha in 1965 to 12 797 ha in 2001 (Thu & Populus 2007), and have declined further since then.

The proboscis monkey is a known mangrove specialist, using the mangroves for sleeping (Bernard *et al.* 2011*a*), and also other activities (Bennett & Sebastian 1988; Boonratana 1993; Salter *et al.* 1985). In this volume, the ecology of primates in Asian mangroves are described by Bernard *et al.* (Chapter 13), Nijman (Chapter 14), Shekell *et al.* (Chapter 15) and Mallick (Chapter 16).

Oxbow Lakes

Definition

Oxbows are crescent-shaped lakes alongside meandering, braiding or reconnecting rivers. The erosion and deposition of sediments may alter the course of a river as hydraulic action, abrasion and corrosion can increase the curve of a meander, narrowing the neck until the river cuts the meander at the time of flooding (Hickin 2003). Many oxbow lakes can become reconnected to the main river channel in times of high water and hence are considered 'semi-open' ecosystems (Davenport 2008; Junk 1997) with fish, nutrients and floating macrophytes able to enter and exit the lake through a temporary connective channel (Osorio *et al.* 2011).

Neotropics

In the Neotropics, oxbow lakes are frequent in Amazonia, where the low relief and high sedimentation rate means that river systems are highly dynamic forming meanders and regularly changing their course (Toivonen *et al.* 2007). In the Manú Biosphere Reserve in Peru, the watershed of the Rio Manú, oxbow lakes are called *cochas* and vary notably in physical, chemical and biological properties (Davenport 2008). Some oxbow lakes quickly form low-lying *renacals* or swamp forests dominated by *Ficus trigona* (Moraceae) and others may temporarily become filled with grasses and tree seedlings, ultimately succeeding to other forest types (Gentry & Terborgh 1990). Some primate studies have been conducted along oxbow lakes, such as on the black-faced black spider monkey (*Ateles chamek*) at Lago Caiman, Noel Kempff Mercado National Park, Department Santa Cruz, Bolivia (Wallace 2006), and white uacari (*Cacajao calvus calvus*) at Lake Teiu, Brazil (Walker & Ayres 1996).

Africa

In Kenya, oxbow lakes have been created along the Tana River; these lakes influence forest dynamics (Hughes 1990) and are occupied by primates such as the Tana River crested mangabey (*Cercocebus galeritus galeritus*) and red colobus *Procolobus rufomitratus* (Chapter 30). In Madagascar, two gallery forests in the Berenty Reserve, the main Berenty forest (200 ha) and Bealoka (100 ha), are threatened forested islands formed by ancient oxbow lakes (Chapter 5). These forests are characterized by tamarind trees, *Tamarindus indica*. These unusual gallery forests, and the spiny forest that lies beyond the steep river banks, are used by lemurs, including *Lemur catta* (Jolly *et al.* 2006).

Asia

In Asia, oxbow lakes are found in various regions including in the Ganges river in India (local name: *baors*) (Hasan *et al.* 2001). In the dry season, most of these oxbow lakes become fully isolated from the main river leading to changes in water chemistry, aquatic plants and animals. Approximately 600 oxbow lakes are found in southwestern Bangladesh, covering around 5488 ha. There are also oxbow lakes in many floodplains of large rivers, e.g. the Kinabatangan, Sugut and Padas rivers, in Sabah, Malaysia (e.g. Azmi 2007; Jawan & Sumin 2012). Stretches of the Kinabatangan River contain *c.* 30 oxbow lakes with various stages of infilling (Boonratana & Sharma 1997; Davison 2002: noted as cited in Salgado-Lynn 2010).

Reedbed/Papyrus

Definition

Largely confined to Africa, papyrus swamp is named for its dominant species *Cyperus papyrus* (papyrus sedge, paper reed, Indian matting plant, Nile grass). An aquatic, herbaceous perennial sedge, it forms tall monoculture stands in shallow water (Thompson 1976). Papyrus swamps have a high growth rate (C4 photosynthesis; Jones 1988) and the ability to recycle nutrients. They provide structural refugia and habitats for several endangered wildlife species (Chapman *et al.* 1996; Maclean *et al.* 2006). *Cyperus papyrus* occupies two major habitats: the edges of freshwater lakes (a floating mat attached at its perimeter) and river valleys where flow is slow (spreads across the frequently flooded valley floor).

The natural distribution of *C. papyrus* is within a belt across equatorial Central Africa (17°N–29°S), with the most extensive swamps in Lake Victoria and the Nile basins (Thompson 1976). *Papyrus* sedge also dominates areas of the Okavango inland delta in Botswana.

One primate species on Madagascar, the Lac Alaotra bamboo lemur (*Hapalemur alaotrensis*), feeds almost exclusively on papyrus and reeds in marshland, the only primate adapted to marsh habitat (Chapter 34).

Seasonally Flooded Grassland

Definition

Flooding is an important ecological phenomenon of riverine systems as floodwaters regenerate the floodplain with nutrients and facilitate the development of a mosaic of aquatic plants, flooded grassland, riparian forest, savanna, woodland and dry forest (Pott *et al.* 2011).

Neotropics

The most extensive seasonally flooded grassland in South America is the Pantanal, a savanna wetland inundated by tributaries of the upper Paraguay River. Characterized by an indistinct and ever-changing boundary between water and land, seasonal hydrology patterns drive the productivity of this system (Junk 1993; Vinson & Hawkins 1998; Wantzen & Junk 2000). The flood discharge carries nutrients, sediment and inorganic and organic material which enriches the floodplain allowing the proliferation of microorganisms, invertebrates, fish, amphibians, reptiles, birds and mammals (Mamede & Alho 2006). The flooded Pampa grasslands in Argentina are inundated entirely by rainfall (Perelman *et al.* 2001).

Africa

A number of large, seasonally inundated wetlands exist in Africa, scattered throughout the seasonally wet and dry areas of the Sahel zone of western Africa, and also in eastern and southern Africa (Burgess *et al.* 2004). As an example, on the southern edge of the Sahara, the Lake Chad savanna is flooded when the lake expands in the wet season. Despite high evaporation the lake retains low salinity because inputs are riverine and saline waters sink, leaving the lake through subterranean conduits. Floating Nile lettuce (*Pistia stratiotes*, Araceae) is present alongside *Cyperus papyrus* and *Phragmites* spp. (Poaceae) with *yaere* grassland on the southern lakeshore, where flooding is prolonged and water depth reaches 1–2 m.

The inner Niger delta (Mali) and the Hadejia-Nguru wetlands (Nigeria) in the Sahelian belt are also large, seasonally flooded, wetlands that support extensive grassy wetland vegetation that expands and contracts in line with the flooding cycles (Thieme *et al.* 2005).

Further to the east in Africa, the nutrient-rich White Nile overflows in southern Sudan to form the extensive Sudd swamps (Seymour 2013). Within the Sudd swamps, there are a series of ecological zones from open water to submerged vegetation, including flooded grassland (Hickley & Bailey 1987), and it has been estimated that only 11% of the total flooded area is permanent water (Welcomme 1979).

The seasonally flooded grassland of the Okavango, Botswana is inundated to a variable extent every year when floodwaters from the Angolan highlands arrive between February and May (Gumbricht *et al.* 2004). The main species of grass are *Miscanthus junceus, Phragmites communis, Cyperus articulates* and *Schoenoplectus corymbosus* (all Cyperaceae). The Okavango supports some of the largest concentrations of wildlife in Africa (Thieme *et al.* 2005), including chacma baboons (*Papio ursinus*) behaviourally adapted to flooding in this region.

Asia

In Southeast Asia, natural, non-forest wetlands seem to be confined, in the lowlands, to areas with seasonal rainfall. In these climates, swamp grasses, sedges and ferns dominate the system (Corlett 2005).

Intertidal Zones

Definition

Tidal flats are a transitional zone between aquatic and terrestrial ecosystems. Marine and freshwater tidal zones differ; marine are often subject to high wave energy and may have large tides, while tidal effects in freshwater ecosystems tend to be smaller. Intertidal zones are often dispersal corridors for plants and animals that float as seeds, propagules or rafts of debris (Nilson & Svedmark 2002; Renofalt *et al.* 2005). These zones can provide rich foraging grounds, including for primates like the Burmese long-tailed macaque (*Macaca fascicularis aurea*) (Chapter 19), Japanese macaque (*M. fuscata*) (Chapter 18) and chacma baboon (*Papio ursinus*) (Chapter 20).

'Beach forest', not to be confused with mangroves, completely roots in freshwater and is almost never inundated by seawater. Beach forest tends to exist above the high-tide line, but is tricky to identify as similar species composition is found in the transition zone from mangroves to upland forest. Beach forests provide suitable habitat for primates especially in Southeast Asia where beach forests are called '*Barringtonia* formations'.

Neotropics

Neotropical beach vegetation consists of coastal forest on sandy soils. Called *restinga* in Brazil, they vary from being dominated by trees, shrubs or creeping vines. Salt marshes are also present (da Silva *et al.* 2010; Pimental *et al.* 2007). Important tree species include *Humiria balsamifera* (Humiriaceae), *Pouteria ramiflora* (Sapotaceae), *Anacardium occidentale* (Anacardiaceae), *Byrsonima crassifolia* (Malpighiaceae) and *Tapirira guianensis* (Anacardiaceae), while *Sesuvium portulacastrum, Ipomoea pes-caprae* (Convolvulaceae), *Sporobolus virginicus* (Poaceae) and *Blutaparon portulacoides* (Amaranthaceae) are important herbs and vines, serving to stabilize the sand (da Silva *et al.* 2010). Though several Neotropical primates regularly use mangroves (Chapter 8), there seem to be no published records of habitual beach use as in Africa (Chapter 20) or Asia (Chapters 18 and 19). In Costa Rica, white-headed capuchins (*Cebus capucinus*) visit beaches, but only to steal food from tourists (Edilton Rodrigues Santos and Stephen Ferrari, unpublished data; see Santos 2013). In South America, blond capuchins (*Sapajus flavius*) feed on the fruits of introduced oil palm (*Elaeis guineensis*) in forest fragments by Porto de Galinhas beach, Permanbuco state, northern Brazil. In addition, a group of common marmosets (*Callithrix jacchus*) regularly feed on fruit/gum/insects at the border between the Atlantic forest and the sandy beaches at Porto de Galinhas and São José da Coroa Grande (Antonio Mendes Pontes, unpublished data). In coastal Paraíba state, northern Brazil, *S. flavius* forage in vegetation and their footprints have been recorded on sand dunes (Monique Bastos and Bruna Bezerra, unpublished data). In Sao Paulo state, southern Brazil, the Superagui island population of the Critically Endangered black-faced lion tamarin (*Leontopithecus caissara*) uses arboreal restinga (coastal forest on sandy soils) as a primary habitat (Nascimento & Schmidlin 2011). Surprisingly, primates are not known predators of sea turtle eggs (Engeman *et al.* 2003; Stancyk *et al.* 1982), even though a number of other opportunistic mammalian foragers (e.g. raccoons, *Procyon* spp.) feed on them (Garmestani & Percival 2005), and capuchins may predate river turtle eggs (Barnett *et al.* 2005; Batistella & Vogt 2009).

Africa

The Jozani beach forest of Zanzibar is dominated by a species of *Eugenia* (Myrtaceae) and *Vitex doniana* (Robins 1976), but composition is not uniform with species distribution dependent upon water availability. Beach forest on Uzi Island, Zanzibar, has *Vitex trifolia, Hibiscus tiliaceus* (Malvaceae), *Terminalia boivinii* (Combretaceae), and *Rhus natalensis* (Anacardiaceae), all consumed by red colobus monkeys, which also use the adjacent habitats: mangroves and coral rag forest (Nowak 2007).

Asia

On accreting sandy beaches, a low community of creeping herbs, grasses and sedges occupies the zone between the sea and firm land (Corlett 2005). Here, beach forest is known as '*Barringtonia* formation' or *Barringtonia asiatica–Terminalia catappa* vegetation after the dominant tree species *Barringtonia asiatica* (though *B. asiatica* itself is not always present in the formation) (Whitten *et al.* 1997). *Barringtonia* formation is tolerant of salt spray, nutrient deficient soil and seasonal drought and merges with lowland rainforest inland. *Calophyllum inophyllum* (Sapotaceae) is also present and some of the plants found in this type of beach vegetation are typical of sandy shores throughout the tropics.

Conclusion

The range of flooded habitats across the tropics has undoubtedly influenced the evolution and persistence of non-human primates. The diversity of wetland habitats is mirrored by highly biodiverse inhabitants, including primates which may use flooded areas for refuge given these are not easily accessible to humans. When surrounding *terra firma* forest has been cleared, often much of the original floodplain forest remains standing because wet areas are not as ideal for agriculture and habitation. As a consequence, despite large-scale land-use change and habitat degradation, flooded habitats still host a large number of primates and other mammals that may otherwise be rare or even extinct.

Chapter 3

Fossil Primates from Flooded Habitats

The Antiquity of an Association

Matt Sponheimer, James E. Loudon and Michaela E. Howells

Introduction

Primates, big and small and across a wide array of taxonomic groups and geographic areas, live in varied associations with water today (e.g. Gautier-Hion & Brugière 2005; Peres 1997b; Poulsen & Clark 2004). There is considerable evidence that such associations were also prevalent in the past, although the fossil record often obscures their true nature. Terrestrial animals are often rent by predators or scavengers after death, so it is typically animals that are quickly buried (in a suitable chemical environment) that successfully transition from the biosphere to the geosphere (Carroll 1987). Timely interment and subsequent preservation is much more likely near water (see Feibel 2013), so it can be no surprise that much of the primate fossil record has been recovered from floodplain or lake margin environments (Fleagle 2013), and is often directly associated with aquatic elements that belie thoroughly terrestrial inclinations (see Godfrey & Jungers 2003 for early ideas about aquatic lemurs). Moreover, even where an association with water is undisputed, understanding of how individual species used reconstructed landscapes may remain elusive. For instance, what was the habitat of a primate for which sedimentological and faunal evidence suggests associations with both riverine forests and associated floodplain grasslands? Did the primate restrict its range to the riverine forest (e.g. *Cercocebus galeritus*) or did it use resources from both the open and closed part of its range (e.g. *Papio cynocephalus*: see Wahungu 1998). This is a recurrent problem for our interpretation of the hominin fossil record.

These caveats aside, the evidence for flooded environments in the primate fossil record is so prevalent that their general importance for understanding primate biogeography and evolution cannot be doubted, even if the specifics are sometimes murky. In this review, we provide a brief introduction to some of this evidence spanning about 60 million years, but with a particular focus on later time periods, and especially early hominins, given recent suggestions that they were variably dependent on underground storage organs (USOs) from aquatic or semi-aquatic plants (Conklin-Brittain *et al.* 2002; *van der* Merwe *et al.* 2008; Wrangham 2005; Wrangham *et al.* 2009). We start with a synopsis of modern primates in flooded environments to provide context for interpretation of the primate fossil record. It should be remembered that this chapter is far from exhaustive, as we will perforce ignore over 99% of known primate taxa (see Fleagle 2013). We hope, however, that the examples chosen are generally illustrative and interesting for those whose concerns are primary neontological.

Modern Primates

The free-ranging Japanese macaques (*Macaca fuscata*) are perhaps the poster children for primate–water interactions, as their bathing in hot springs during cold periods (Kawai 1965; Suzuki 1965) and their tendency to disport in water have been featured across the internet and in calendars and post cards. But they are far from alone. Multiple strepsirrhines live in aquatic environments. Populations of pottos and angwantibos (Perodicticinae) and bushbabies (Galaginae) live in swampy areas (Pimley 2009) and the slow loris (*Nycticebus coucang menagensis*) lives in peat swamps (Blackham 2005). The Alaotran gentle lemur (*Hapalemur griseus alaotrensis*) lives in reed and papyrus beds (Mutschler 2002) (see Chapters 34 and 37) and at least one population of ring-tailed lemurs (*Lemur catta*) lives on the shores of a large alkaline lake (LaFleur 2012).

Among the platyrrhines, there are large primate communities utilizing flooded habitats (see Chapters 8, 21, 22, 23, 27 and 41). Peres (1997b) compared the primate community structures of 12 unflooded forests (*terra firma*) and eight annually flooded forests (*várzea*) in the Amazon and found that the flooded forests exhibited a lower species richness but a higher population density (Peres 1997b; but see Gautier-Hion & Brugière 2005 for Central Africa). In a similar study, Haugaasen and Peres (2005a) examined the primate communities utilizing *terra firma* and seasonally flooded forests at Lago Uauaçú, Brazil and found that seasonal use of flooded habitats increased home ranges and reduced feeding competition for some species. Uacaris (genus *Cacajao*) and some species of *Saimiri* and *Cebus* regularly use flooded forest (e.g. Barnett *et al.* 2002, 2013b), while *Saimiri vanzolinii* appears restricted to seasonally flooded *várzea* forest of central Amazonia (Paim & Queiroz 2009). Beyond the Amazon basin, primates also inhabit the Pantanal (Chapter 22) and flooded habitats of northern Argentina (Bravo & Sallenave 2003; Chapter 32) and the seasonally flooded mixed grassland and shrub habitats (e.g. Bolivian Chaco, Chapter 23; Venezuelan llanos, Chapter 21).

Catarrhines are no less inclined to use aquatic environments (Chapters 30 and 38). Among the best studied in this regard are the proboscis monkeys (*Nasalis larvatus*) found throughout Borneo (Kawabe & Mano 1972; Yeager 1989) and Allen's swamp monkeys (*Allenopithecus nigroviridis*) which inhabit the Congo Basin (Maisels *et al.* 2006). Many catarrhines also use foods near waterways on a seasonal basis. For example, chacma baboons (*Papio ursinus*) in the Okavango Delta, Botswana consume a variety of aquatic roots, rhizomes, corms and tubers as fallback foods during the five months of flooding in their habitat (Wrangham *et al.* 2009). Catarrhines also use water as a tool and for thermoregulation as exemplified by Japanese macaques which wash dirt from grass roots (Nakamichi *et al.* 1998) and sweet potatoes (Hirata *et al.* 2001) with water, and chimpanzees which use water to cool themselves in extreme heat (Pruetz & Bertolani 2009; Suzuki 1965). In sum, primates across a broad spectrum of sizes, taxonomic groups, social systems, dietary preferences, and locomotor adaptations use water and, more broadly, environments that are at least seasonally inundated (Kempf 2009).

Non-Hominin Primates in the Fossil Record

Molecular evidence suggests that the Order Primates originated over 87 million years ago (Ma) (Perelman *et al.* 2011), but the primate fossil record begins in the Palaeocene and with much equivocation (Rasmussen 2007; Silcox *et al.* 2007). Some have argued that the first fossilized primate is *Altiaatlasius koulchii*, which was found in 60 million year old sediments in Morocco, but its relationship to the other primate groups remains uncertain (Rasmussen 2007; Silcox *et al.* 2007). It was recovered from a near shore marine setting, so it can be argued that primate associations with water were there from the beginning (Gingerich 1990). By the Eocene, we find unquestioned primates associated with flooded environments. For example, paleosol evidence has been used to argue that in the Bighorn and Green River Basins of North America omomyine primates (e.g. *Omomys* spp.) were primarily associated with lake margin and proximal floodplain habitats, while anaptomorphine primates (e.g. *Teilhardina* spp.) preferred basin margin and distal floodplain areas (Gunnell 1997). Thus, although both groups were broadly associated with flooded environments (as would be expected given the depositional context), there appears to be considerable evidence of habitat differentiation.

The famous late Eocene to early Oligocene age primates of the Fayum Depression, which is located on the eastern edge of the Sahara Desert, were also found in sediments deposited by large streams, although there is also evidence of marine incursions as evidenced by fossil sharks, rays, and sirenians (Bown *et al.* 1982). The Fayum primates include both stresirrhines (Seiffert *et al.* 2003) and haplorrhines (Rasmussen *et al.* 1998), and are believed to have lived in habitats ranging from near shore mangrove swamp to more inshore tropical forest (Bown *et al.* 1982; Gagnon 1997). Evidence for subtropical to tropical forest includes abundant fossilized wood lacking strong growth rings which indicates poor growth seasonality (as we find in the tropics), fossilized *Epipremnum* (Araceae) fruits, which are now common throughout the tropical rainforests of Southeast Asia (Bown *et al.* 1982), and a large mammal community structure approximating that found in lowland African rainforests (Gagnon 1997). The arboreal components of the environment are further supported by the primates themselves, such as *Aegyptopithecus* and *Apidium*, which functional morphology (Bown *et al.* 1982) and dental microwear (Teaford *et al.* 1996) suggest were arboreal frugivores.

Primate associations with flooded environments continue in the Miocene. *Sivapithecus* is a fossil hominoid found in the Siwalik sediments of Pakistan dating from about 13.0 Ma to 8.4 Ma (Begun 2007; Nelson 2007). The skull of *Sivapithecus* is remarkably similar to that of modern orangutans (Pilbeam 1982). However, other features, including its humeral shaft curvature, suggest that *Sivapithecus* lacked the suspensory specializations of orangutans (Richmond & Whalen 2001). Despite this, most researchers default to the view that there is a close relationship between *Sivapithecus* and modern *Pongo* (Fleagle 2013).

The Siwalik sediments were deposited by ancient rivers, comparable in size to the modern Indus or Ganges systems, and preserve a remarkable range of freshwater and terrestrial vertebrates (Barry *et al.* 2002). Fossils were preserved from a broad array of depositional contexts which appear to preserve meaningful ecological signals. For instance, channel fill assemblages are dominated by fish and crocdyloids, while the floodplain assemblages are dominated by dry substrate taxa such as suids and giraffids (Behrensmeyer *et al.* 2005). Although there was clearly environmental change, it is believed that a similar diversity of vegetation types existed through time, ranging from gallery forest to low relief swamps to grassy meadows and extensive upland forest (Barry *et al.* 2002). Exactly where *Sivapithecus* dwelt within this mélange remains unknown, although morphological, microwear and isotopic evidence suggests that it was an arboreal frugivore that probably went extinct (~8.4 mya) as the habitat opened and C_4 grass abundance increased (Nelson 2005, 2007; Teaford & Walker 1984). Other taxa associated with forest environments (e.g. glirid rodents, tree shrews) also disappeared from the Siwalik record at about this time, while others associated with more open or arid habitats (e.g. hypsodont bovids, gerbils) appeared (Flynn & Jacobs 1982). This implies that *Sivapithecus* required extensive wooded environments, and that gallery forest would not have been sufficient to sustain it long term.

As *Sivapithecus* waned in Asia, the enigmatic *Oreopithecus bambolii* appeared in Europe. It has been labelled a pig, a monkey, an ape and even an early hominin (Begun 2007), at least partly because its most important postcranial remains are crushed and difficult to interpret, but also because of its anatomical peculiarities. For example, it has small, peg-like, upper lateral incisors, lower molars with a unique central cusp from which crests radiate out to the four main cusps, and upper molars that appear to be bilophodont as is found in cercopithecoids (Begun 2007; Fleagle 2013). When coupled with traits typically associated with bipedalism, such as a well-developed anterior inferior iliac spine and short, broad iliac blades, its phylogenetic ambiguity is unsurprising (Wood & Harrison 2011).

Oreopithecus lived in the Tuscany area (northern Italy) at a time when the Italian peninsula was separated from mainland

Europe by the sea. Geological evidence suggests *Oreopithecus* was confined to island environments, which is supported by its associated faunas which demonstrate high levels of endemism, low taxonomic diversity, gigantism among rodents and an absence of carnivores (Harrison & Harrison 1989). *Oreopithecus* remains were initially found in lignite mines indicating wetland depositional conditions, but specimens have also been found in fluvial and lacustrine sediments (Casanovas-Vilar *et al.* 2011). A generally wet environment is further supported by its association with pollen from lilies, sedges, ferns, shrubs and trees often found in moist and swampy conditions. The broader environment was probably mesophytic forest roughly analogous to areas of east central China bordering the Yangtze River Valley, where warm temperatures and high precipitation predominate (Harrison & Harrison 1989).

Oreopithecus individuals had robust mandibles, well-developed chewing muscles, high shearing crests and molar microwear suggestive of folivorous diets (Begun 2007; Ungar 2006; Ungar & Kay 1996). Some have argued that *Oreopithecus* was bipedal and had hands capable of a human-like precision grip (Köhler & Moyà-Solà 1997; Moyà-Solà *et al.* 1999). This, coupled with evidence for wetland depositional environments, has been used to argue that *Oreopithecus* was a bipedal wader in swampy environments which regularly ingested aquatic plants. It was further suggested that *Oreopithecus* supplemented its diet with aquatic invertebrates which were caught using a pad-to-pad precision grip (Williams 2008). Weaknesses with this scenario include the overall pattern of its postcranium-long forelimbs, broad thorax, long curved phalanges and an abducted hallux, which has convinced most authorities that it had a suspensory adaptation (Begun 2007; Russo & Shapiro 2013), and evidence that it emphasized ape-like power grasping rather than a human-like precision grip (Susman 2004). Regardless, the association between *Oreopithecus* and wet and wooded environments remains unquestioned, and was potentially essential, as its demise is broadly coincident with regional changes in vegetation. Most believe, however, that an influx of competitors from mainland Europe at about 6.5 Ma was a decisive factor in its extinction (Agustí 2007; Matson *et al.* 2012).

Fossil Hominins

The earliest potential hominins (and we will avoid phylogenetic wrangling here) also have strong aquatic associations. *Sahelanthropus*, which is about 7 million years old, is associated with aquatic indicators including wading mammals, amphibians and fish, but also with savanna elements including hysodont bovids. This, together with geological evidence, led researchers to suggest that these hominins lived next to a lake that was fringed with trees (Vignaud *et al.* 2002). *Ardipithecus ramidus*, another contender for the mantle of earliest known hominin from 4.4 Ma, may have lived amidst a riverine forest within a broader grassy environment (Gani & Gani 2011; but see White *et al.* 2010).

The trend of strong water associations continues with the later *Australopithecus* spp. which existed from about 4.1 Ma until perhaps 2.0 Ma, and which ranged from what is now Chad to Ethiopia down to South Africa (Behrensmeyer & Reed 2013; Fleagle 2013). *Australopithecus* is generally considered to have been a eurytope and dietary generalist (Behrensmeyer & Reed 2013) ranging across riverine forests and edaphic grasslands (Alemseged 2003; Behrensmeyer & Reed 2013; Reed 1997). Its density may have been higher at wetter and more heavily wooded habitats (Su & Harrison 2008), although at Hadar it appears to have been more abundant in edaphic grassland than among riverine forest (Campisano 2007), but in virtually all cases it was closely associated with lakes, rivers, or streams (Verhaegen & Puech 2000). In a sense, all *Australopithecus* spp., and the earlier contenders for hominin status, were creatures of two worlds, for where postcranial material has been found, it seems to indicate adaptations for life in the trees (e.g. long arms, curved fingers) coupled with evidence for bipedal locomotion (e.g. anterior inferior illiac spine, adducted hallux), which is often associated with the need to traverse more open environments (Rodman & McHenry 1980; and see Ward 2013). Such a set of adaptations might make a great deal of sense for organisms forced to cope with increasingly open environments (see Cerling *et al.* 2011a), but which were at least seasonally dependent on fruits from riparian forests.

Arguably the most intriguing hominins with regard to flooded environments are the more recent (< 2.6 Ma) robust australopiths. *Paranthropus boisei*, in particular, has been found throughout eastern Africa, and is usually associated with wet environments including lake margins, swamps, and edaphic grasslands (Alemseged 2003; Feibel *et al.* 1991; Reed 1997; Shipman & Harris 1988). In addition, recent isotopic studies suggest that its diet was dominated by C_4 plants (predominantly grasses and/or sedges) (Cerling *et al.* 2011b; van der Merwe *et al.* 2008), which could reflect a strong dependence on aquatic or semi-aquatic resources. It is theoretically possible that the *P. boisei*'s C_4 signal was imparted secondarily through consumption of animals that ate C_4 plants, but this is unlikely given *P. boisei*'s dental morphology (Teaford & Ungar 2000), and because even carnivores rarely achieve the zebra-like carbon isotope compositions of *P. boisei* (Codron *et al.* 2007). If lions and hyenas rarely consume enough C_4 grazers to achieve this carbon isotope composition, it strains the bounds of credulity to believe that *P. boisei* was capable of doing so. It could be that *P. boisei*'s C_4 signal was derived from consumption of tropical grasses (see Jolly 1970), but this idea has received little attention in recent years, at least partially because the dental morphology of *Paranthropus* is suboptimal for such a diet (Teaford & Ungar 2000; but see below). And while rarely stated, most researchers regard grasses to be too nutrient poor to sustain such highly-encephalized mammals (see general arguments in Aiello & Wheeler 1995; Milton 1999).

So what C_4 plants (and plant parts) did *P. boisei* eat? Some feel that sedges fit the bill, especially their Underground Storage Organs (USOs) (Conklin-Brittain *et al.* 2002; Dominy 2012; Dominy *et al.* 2008; *van der* Merwe *et al.* 2008; Wrangham 2005; Wrangham *et al.* 2009). Such arguments tend to follow these broad lines. Early hominins had chimpanzee-like dietary requirements and a preference for ripe fruit, but were forced into savanna habitats where such resources were limited (see discussion in Wrangham 2005). Hominins would then have gravitated towards habitats, at least seasonally, where fruits (and preferred chimpanzee fallback foods) were more readily accessible, such as riverine forests or wetlands. Once focused on these habitats, it would have been a small step towards use of the abundant USOs of *Cyperus* (Cyperaceae), *Typha* (Typhaceae), *Nymphaea* (Nymphaeaceae) and other taxa therein (Conklin-Brittain *et al.* 2002; Wrangham 2005). The USOs might have been used initially as fallback foods, but later hominins with more derived craniodental morphologies (robust australopiths) would have relied on USOs more regularly (Laden & Wrangham 2005).

In accord with such scenarios, modern baboons emphasize subterranean foods when preferred foods are not available (Hill & Dunbar 2002), and in the Okavango Delta, USOs from aquatic plants are important fallback foods (Wrangham *et al.* 2009). Many sedges use C_4 photosynthesis, and some, like *Cyperus papyrus*, can be found in extensive stands (Muthuri *et al.* 1989) that might be suitable for large-scale exploitation by hominins. It has been argued that *Cyperus* USOs are high-value foods, so their consumption would make good energetic sense for tool-wielding hominins who would have little competition for such resources (Conklin-Brittain *et al.* 2002; van der Merwe *et al.* 2008).

But are there reasons to suspect that sedge USOs are not responsible for *P. boisei*'s zebra-like carbon isotope composition? There is no smoking gun one way or another, but there are things that dampen our enthusiasm for this idea, some which we enumerate below:

1. The dental microwear of *P. boisei* (Ungar *et al.* 2008) shows none of the pitting found in baboons that regularly consume USOs (Daegling & Grine 1991). While it is theoretically possible that the grit adherent to USOs was somehow removed (Dominy 2012), the energetic efficiency and efficacy of such behaviours is worth questioning.
2. Although the gross energy of *Cyperus papyrus* rhizomes is high (van der Merwe *et al.* 2008), these foods are rich in refractory materials (a quid is expectorated) greatly reducing their available energy (Schoeninger *et al.* 2001; but see Dominy *et al.* 2008 on corms).
3. Where the USOs of aquatic or semi-aquatic vegetation are used extensively as fallback foods, they predominantly use C_3 photosynthesis (e.g. *Nymphaea*), even where C_4 sedges otherwise predominate (Wrangham *et al.* 2009).
4. Consumption of enough USOs of aquatic and/or semi-aquatic plants to explain the observed carbon isotope composition of *P. boisei* is unknown among mammals, even if we assume all such USOs were from C_4 plants (which is a very unlikely assumption). The closest primate analogue might be the strepsirrhine *Hapalemur griseus alaotrensis* which eats sedges and grasses in swampy habitats, but which focuses on shoots, pith and leaf bases/tips rather than USOs (Mutschler 2002; Tan 2007; Chapter 34). The robust masticatory apparatus of *P. boisei* might be indicative of considerable repetitive loading (Daegling *et al.* 2011; Hylander 1988; Scott *et al.* 2014), as would be required for a diet high in pith from plants in swampy or flooded environments. It has also been suggested that the dental microwear of eastern African robust australopiths reflects a diet of 'aquatic plants and papyrus shoots' (Verhaegen & Puech 2000).

So what about C_4 grass consumption? Most authorities do not like this idea. However, every non-carnivorous mammal with a carbon isotope composition similar to that of *P. boisei* eats predominantly C_4 grass products. So if *P. boisei* did otherwise, its dietary ecology was unique among known mammals. The focus on grass among mammals makes sense given the relative abundance of grass on savanna landscapes relative to sedges. It is also worth mentioning that grass, including grass USOs, is consumed by many primates including baboons, and is generally more prominent in the diets of baboons than are sedges, even in the Okavango Delta (Alberts *et al.* 2005; Hamilton *et al.* 1978; Post 1982). In addition, *P. boisei* is strongly associated with the grass-consuming primate *Theropithecus* (Alemseged & Bobe 2009), which together with evidence that it inhabited edaphic grasslands (Alemseged 2003; Reed 1997), suggests that grasses were at least part of its fundamental niche. Given all of this, if not for the apparent discontinuity between its dental morphology and a grassy diet (excepting grass seeds for which many have argued the robust australopith dental morphology is well-suited), C_4 grass comprising the bulk of *P. boisei*'s diet would seem to be a certainty. Some might counter that *P. boisei* lacked the soft-tissue digestive apparatus required for such a diet, but it should be remembered that we know virtually nothing about its soft tissues, and that modern grass-eating geladas are not particularly efficient fermenters of fibre (Mau *et al.* 2011).

Of course, grass could have been consumed with sedges since they both offer above and below ground resources and have strong microhabitat overlaps. But even if grass was used exclusively, it does not necessarily mean any less of a focus on flooded environments. Dunbar (1993), for instance, argued that ancient *Theropithecus oswaldi* was restricted to eat grasses in the vicinity of permanent water, as less nutritious open grassland grasses could not have sustained them. This energetic argument could be readily applied to *P. boisei* given it and *Theropithecus*' close associations (Alemseged & Bobe 2009) and nearly identical carbon isotope compositions (Cerling *et al.* 2013; Codron *et al.* 2007).

The above should make it clear that we still have little idea which C_4 resources were consumed by *P. boisei* and other hominins. To solve this problem we will need a greater focus on the availability, abundance, potential harvesting rates and nutritional/mechanical properties of plant foods, as well as further analyses of the morphology, microwear and carbon isotopes of pith and USO eaters at the very least. What we can say is that there is no question that most, and possibly all, early hominins were associated with flooded habitats, although whether they were limited to such habitats or had far broader tolerances remains uncertain.

Conclusions

We hope to have demonstrated that fossil primates, like their modern counterparts, had important associations with flooded environments. Or perhaps that should be turned around, and we should emphasize that modern primates continue truly ancient, and frankly, unsurprising associations with water given the locomotor, feeding and thermoregulatory adaptations of primates (Fleagle 2013). There is still much to be done. Understanding the ways fossil primates used their worlds will always be problematic as directly observing ancient behaviour is impossible. Thus, much is forever lost to us. Nevertheless, we can take some comfort in the fact that techniques continue to emerge that make questions once relegated to the bins of science fiction legitimate targets for scientific inquiry. Such continued innovation should greatly advance our understanding of patterns and processes in primate evolution and, in so doing, better our ability to predict the multifarious effects of habitat loss and fragmentation on our living kin.

Chapter 4

Comparison of Plant Diversity and Phenology of Riverine and Mangrove Forests with Those of the Dryland Forest in Sabah, Borneo, Malaysia

Ikki Matsuda, Miyabi Nakabayashi, Yosuke Otani, Yap Sau Wai, Augustine Tuuga, Anna Wong, Henry Bernard, Serge A. Wich and Takuya Kubo

Introduction

Approximately 90% of all primate species found in tropical regions are dependent on rapidly disappearing forests (reviewed by Chapman & Peres 2001). Among the various forest types in the tropics, temporarily flooded (hereafter referred to as flooded) primate habitats such as riverine forests are considered important as relatively food-dense habitats for frugivorous primates (Ahumada *et al.* 1998; Gautier-Hion & Brugière 2005; McLennan & Plumptre 2012), and as habitats producing high-quality leaves for folivorous primates (Matsuda *et al.* 2013). In addition, peat swamp and mangrove forests appear to be crucial refuge habitats, particularly for threatened taxa because, at least currently, such forests are generally less impacted by human development than are riverine forests. This is due to their relative inaccessibility, impenetrable vegetation, low nutrient levels and the difficulties associated with converting them to agricultural land (reviewed by Nowak 2012). However, in Southeast Asia large areas of peat swamps and mangrove forests have been cleared/destroyed for commercial timber, agriculture (e.g. rice, pineapple, coconut and sago), shrimp/brackish fish ponds or charcoal (Corlett 2009). Despite this, there are fewer data on forest species composition and tree phenology from flooded forests than other forest types. Therefore, for the effective protection of primates, it is important to assess the structure of flooded forests such as plant assemblage composition and phenological patterns, with a comparison among different forest types.

Many studies have tried to generalize the factors affecting species richness and abundance of individual primate species. Seasonality in food availability (especially fruit productivity) has a negative effect on the biomass of frugivorous primates and their species richness, whereas total annual food availability has a positive impact on the same (Hanya & Aiba 2010; Hanya *et al.* 2011). However, primates living in flooded habitats, particularly in such floristically distinct habitats as mangrove forests, are rarely included in such studies, in part because of the difficulties associated with surveying them (Nowak 2012). The same is largely true for other flooded forests, and may be especially important when such habitats are used for only part of the year, when seasonal resource availability is especially favourable (see Chapter 27 for Neotropical examples). Thus, information on plant composition and fruit phenology in mangrove forests, and also other flooded forest types, may provide an insight on further understanding of primate abundance and richness and hence their effective conservation.

Plants in tropical forests display diverse patterns of fruiting seasonality. Lowland dipterocarp forests in Borneo, Sumatra and Peninsula Malaysia, are characterized by masting events (Appanah 1985; Corlett 2009), which occur at unpredictable intervals every 2–6 years and involve widespread general flowering/fruiting, with many plant species flowering simultaneously for a period of a few weeks to a few months and subsequently bearing large fruit crops. Many of the species involved rarely flower outside these periods (Brearley *et al.* 2007; Sakai 2002; Wich & Van Schaik 2000). This pattern of fruiting periodicity characterizes Southeast Asia forests and is believed to have strong impacts on the structure of animal communities. It may be responsible for the low species diversity observed in this region compared to other areas (Reed & Bidner 2004). In contrast, other Southeast Asian forest types, including peat swamp, only weakly conform to this otherwise widespread pattern of mass-flowering followed by mass-fruiting followed by a dearth. Instead, these forests maintain consistently high levels of fruit productivity (Cannon *et al.* 2007; van Schaik & Pfannes 2005; Wich *et al.* 2011), probably because of the relatively low density of trees of the Dipterocarpaceae and other masting tree families within them. For primates, this variation in phenological patterns (i.e. general fruiting versus non-general fruiting) between forest types can influence feeding strategy even within the same species. For example, the red langur (*Presbytis rubicunda*) is endemic, but widespread, in Borneo and occurs in both swamp forests and non-flooded lowland forests. A relatively constant availability of fruit provides *P. rubicunda* populations in swamp forests with regular access to nutritionally superior food and, therefore, there is no reliance on fallback foods (Ehlers-Smith *et al.* 2013b). In contrast, populations in dipterocarp-dominated lowland unflooded forest increase their fruit consumption in proportion to fruit availability, depending more on young leaves of a specific liana species (*Spatholobus macropterus*, Fabaceae: Papilionoideae) as a fallback food during fruit dearths (Hanya & Bernard 2012).

Thus, to understand the flexibility of primate feeding ecology (i.e. the mechanisms of coping with this strong seasonality in Southeast Asia), it is important to investigate the fruit phenology of flooded forests including riverine and mangrove forests, where masting is either less intense or absent, and compare this with the fruit phenology of masting dipterocarp (non-flooded) forests.

In this chapter, we describe the detailed plant composition of two different forest types (riverine and mangrove forests), describe the phenological patterns and provide a comparison of their plant diversity with those of the dryland forest in Sabah, Borneo, Malaysia, which are recognized to be among the richest tropical forests in terms of tree species number, and one of the most diverse ecosystems on earth (Whitmore 1984).

Study Sites and Vegetation Survey

Riverine Forest

Data were obtained from May 2005 to December 2013, as part of a project studying the primates of the riverine forests along the Menanggul River, a tributary of the Kinabatangan River in Sabah, Borneo, Malaysia (118°30′E, 5°30′N). The southern area of the Menanggul River is extensively covered by a secondary forest, whereas the northern area has been deforested for oil palm plantations, except for a protected zone along the river (Matsuda 2008). Regionally, the mean minimum and maximum daily temperatures were approximately 24°C (SD: 0.6) and 30°C (SD: 1.8), respectively (May 2005–2006), and the mean annual precipitation at the site was 2474 mm (SD: 606; May 2005–April 2006 and November 2009–October 2013). To assist observation and tracking of primates in the riverine forest, trails 200–500 m long and 1 m wide were established at 500 m intervals on both sides of the river by cutting forest floor undergrowth. Trees with a diameter breast height (DBH) of ≥ 10 cm and vines with a diameter of ≥ 5 cm located either on the trail or within 1 m from the trail edge were marked (a sample area width of 3 m). All labelled trees and vines were taxonomically identified with the support of the Forest Research Center, Sandakan, Sabah. A total of 16 trails were established. All were 500 m in length, with the exceptions of TR 2 (200 m long), TR 8 (250 m long), and TR 4, TR 6 and TR 10 (400 m long) (Matsuda *et al.* 2010a). The total survey area was 2.15 ha (i.e. 7150 m × 3 m). We divided each transect into 50 m long segments (i.e. 50 m × 3 m) to examine species accumulation curves. From May 2005 to December 2013, at the end of each monthly primate survey, the phenology of the 2142 labelled plants along the 16 trails was recorded by examining each plant for the presence or absence of fruits (both ripe and unripe). However, to assess the numbers of fruiting plants during the study period, we used the fruiting ratio (number of fruiting plants/total number of monitored plants). This statistic was chosen because the number of monitored plants decreased every month on account of natural mortality or illegal logging. In addition, it was impossible to record the status of all monitored plants in some months because of adverse weather conditions such as flooding or logistical problems.

Mangrove Forest

Data were also collected from July 2012 to January 2014, as part of a primate study of the mangrove forest along the Kinabatangan River (118°21′E, 5°45′N), situated approximately 30 km from the riverine forest study site. Along the Kinabatangan River (< 100 m from the forest–river margin), we established 63 quadrats of 20 m × 20 m (a total of 2.52 ha) in the mangrove forest in regions. Locations for quadrat placement were restricted to locations with relatively hard soil (termed *mixed mangrove forest* in this chapter), it being difficult to place such transects in riverine forest because mangrove forest terrain is frequently swampy and impassable on foot. In each quadrat, trees with a DBH of ≥ 10 cm were marked. Vines with a diameter of ≥ 5 cm were to be recorded, but none were found. All labelled trees were taxonomically identified with the support of the Forest Research Center, Sandakan, Sabah. In addition to these quadrats, we randomly marked five adult *Sonneratia caseolaris* (Sonneratiaceae) trees in each of 13 *S. caseolaris*-dominated forests (a total 65 trees) along the Kinabatangan River (termed *S. caseolaris mangrove forest* in this chapter). This was done because, although *S. caseolaris* is one of the dominant species in this region, few *S. caseolaris* trees were found in the quadrats, since this species is usually found in tidal mud flats. From July 2012 to January 2014, at the beginning of each monthly primate survey, the phenology of the 298 labelled plants within the 63 quadrats and the 65 labelled *S. caseolaris* trees, was recorded by examining each plant for the presence or absence of fruits (both ripe and unripe) as recorded in riverine forest.

Dryland Forest: Lowland Mixed Dipterocarp Forest

We used monthly tree phenology data, which has been collected at the Danum Valley Field Centre since July 2004 (see Yap 1998, 2013), using the same plot set as used by Norhayati (2001), with the same protocol as used during the census from August 1997 to December 2000 (Te Wong *et al.* 2005). Fruiting activities of trees with DBH of ≥ 10 cm were monitored every month in two different areas (504 and 533 individual trees). The plots were situated in the dryland forest. The two monitored areas consisted of five transects of 20 m × 100 m placed every 400 m along the 2 km trail, and two transects of 2 m × 2000 m along it.

Overall Vegetation in Flooded Forests

Riverine Forest

The vegetation survey of riverine forest recorded 2142 plants (1645 trees and 497 vines) of 180 species (125 genera, 52 families) along 16 trails covering a total area of 2.15 ha (Appendix A). The five most abundant families were Euphorbiaceae (19.6% of all plants), Fabaceae (11.1%), Rubiaceae (6.8%), Phyllanthaceae (6.5%) and Lophopyxidaceae (5.3%).[1] The five most abundant tree species were *Mallotus muticus* (Euphorbiaceae: 9.1% of all trees), *Excoecaria indica* (Euphorbiaceae: 6.7%), *Dillenia excelsa* (Dilleniaceae: 6.0%), *Croton oblongus* (Euphorbiaceae: 5.0%)

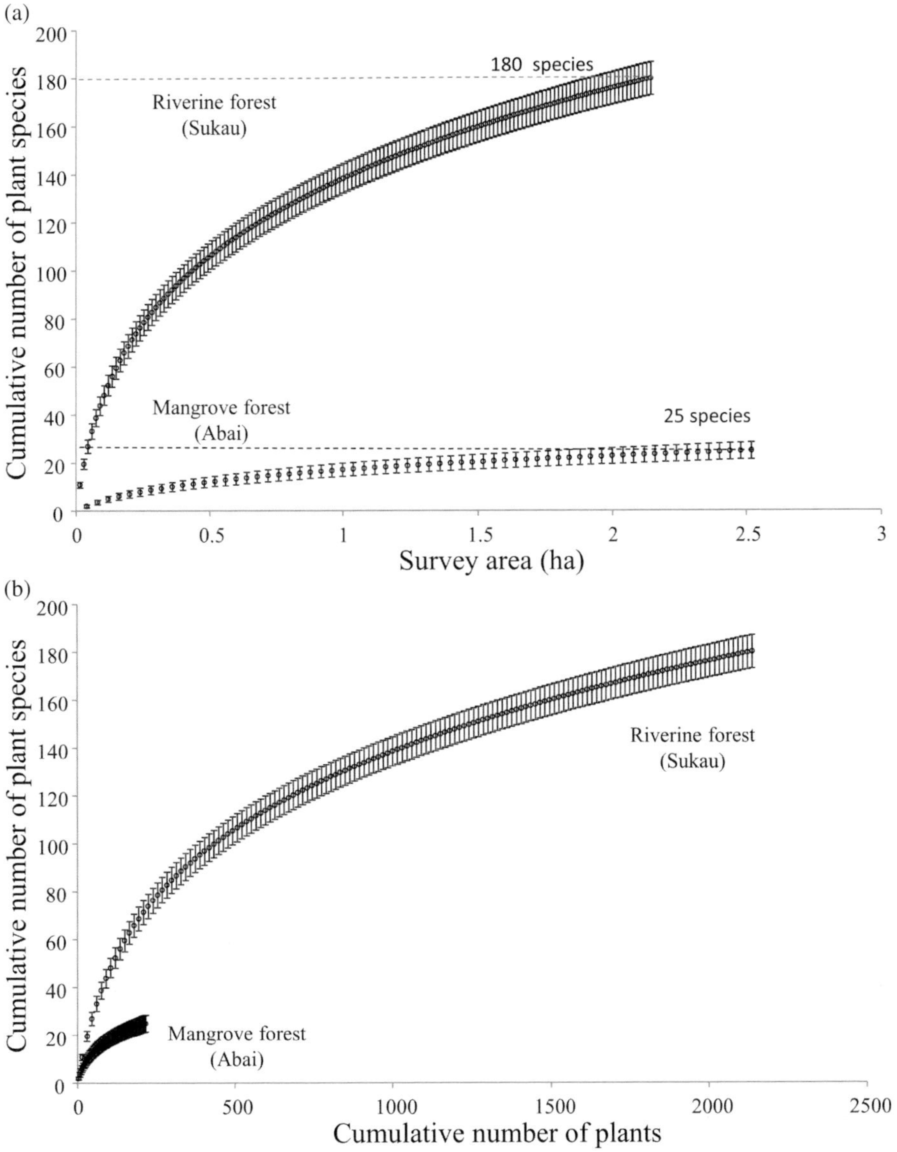

Figure 4.1 Species accumulation curves of the riverine and mangrove forests covering survey areas of 2.15 ha and 2.52 ha, respectively: (a) species versus area curves; (b) species versus number of plants curves. Error bars indicate standard deviation. We performed 100 randomizations using the data of the number of tree species with a DBH of ≥ 10 cm and vine species with a diameter of ≥ 5 cm at subplots of 3 × 50 m (riverine forest) and 20 × 20 m (mangrove forest) using EstimateS ver. 9 (Colwell 2013) to produce the curves.

and *Nauclea subdita* (Rubiaceae: 4.7%). The five most abundant vines were *Lophopyxis maingayi* (Lophopyxidaceae: 22.7% of all vines), *Croton caudatus* (Euphorbiaceae: 9.3%), *Dalbergia parviflora* (Fabaceae, Papilionoideae: 7.6%), *Entada rheedii* (Fabaceae, Mimosoideae: 6.4%) and *Bridelia stipularis* (Phyllanthaceae: 6.0%). The total basal area of each species is also shown in Appendix A.

Although the surveyed area was relatively large (2.15 ha), the cumulative number of plant species did not appear to reach an asymptote (Figure 4.1a). This is probably due to the extremely high species diversity in tropical riverine forests compared with that of mangrove forests (see the following section). Indeed, the total number of plant species used as a food source by proboscis monkeys (*Nasalis larvatus*) over 3506 h (13 months) at the same site was 188 (127 genera and 55 families: Matsuda *et al.* (2009a)), exceeding the total number of plant species described in the vegetation survey (180 species). Furthermore, the total number of plant species used as a food source by orangutans (*Pongo pygmaeus*) over a period of eight years at a forest near this study site was 179 (123 genera, 59 families: Russon *et al.* (2008)), which is almost equal to the total number of plant species found in this vegetation survey. Consistent with the results of high plant diversity in the riverine forest, various sympatric primate species, including proboscis monkeys, long-tailed macaques (*Macaca fascicularis*), pig-tailed macaques (*Macaca nemestrina*), silvered langurs (*Trachypithecus cristatus*), red langurs (*P. rubicunda*), Hose's langurs (*Presbytis hosei*), Bornean gibbons (*Hylobates muelleri*) and orangutans were found during the river census in the late afternoon (Matsuda *et al.* 2011, 2016 with personal observation by IM), which is one of the common survey methods to detect primate species in the Lower Kinabatangan region (Goossens *et al.* 2002; Matsuda *et al.* 2014; Sha 2006; Sha *et al.* 2008).

Figure 4.2 *Nypa fruticans*-dominated (a) and *Sonneratia caseolaris*-dominated (b) mangrove forests.

Mangrove Forest

The vegetation survey in mixed mangrove forest recorded 298 trees of 25 species (22 genera, 20 families) in 63 quadrats covering an area of 2.52 ha (Appendix B). Though highly abundant, *Nypa fruticans* (Arecaceae) was excluded from the survey as it occurred on very swampy tidal mud flats where it was impossible to count the exact number of individuals of this species. The five most abundant families were Arecaceae (33.0% of all plants), Moraceae (17.6%), Myrtaceae (12.0%), Euphorbiaceae (9.0%) and Meliaceae (6.4%). The five most abundant species were *Oncosperma tigillarium* (Arecaceae: 20.1%), *Ficus microcarpa* (Moraceae: 13.8%), *Eugenia tawahense* (Myrtacae: 6.7%), *Aglaia cucullata* (Meliaceae: 5.0%) and *Excoecaria agallocha* (Euphorbiaceae: 5.0%). Although *S. caseolaris* and *N. fruticans* were not among the five most abundant species, they dominated certain areas (Figure 4.2).

As depicted in Figure 4.1, the cumulative number of plant species almost reached an asymptote. This is not unexpected considering the lower plant species diversity of tropical mangrove forests when compared with that in riverine and non-flooded forests (see the following section). Concomitant with this low species diversity, a preliminary survey of proboscis monkeys feeding ecology at this site (approximately 120 observations for 5 months) showed that the monkeys fed only on seven plant species (7 genera, 7 families; Ikki Matsuda, unpublished data). In addition, proboscis monkeys were typically found in the mangrove forest dominated by *S. caseolaris*, and rarely in the mixed mangrove forest. Furthermore, during the late afternoon river censuses, we identified only four primate species (proboscis monkeys, long-tailed macaques, silvered langurs and orangutans), although silvered langurs and orangutans were rarely found (Matsuda *et al.* unpublished data). This lower primate species richness in the mangrove forest compared to the riverine forest may also relate to the lower plant richness in the former.

Dryland Forest: Lowland Mixed Dipterocarp Forest

In the dryland forest vegetation survey, 504 trees of 186 species (101 genera, 46 families) were recorded in an area covering 1.4 ha. The five most abundant families were Dipterocarpaceae (15.0% of all plants), Euphorbiaceae (10.6%), Meliaceae (6.7%), Malvaceae (5.6%) and Annonaceae (4.4%). The five most abundant species were *Shorea johorensis* (Dipterocarpaceae: 9.7%), *Macaranga hypoleuca* (Euphorbiaceae: 8.1%), *Shorea fallax* (Dipterocarpaceae: 7.5%), *Macaranga triloba* (Euphorbiaceae: 5.9%) and *Syzygium* sp. (Myrtaceae: 5.9%). More detailed information is available in Norhayati (2001) and Yap (1998).

Table 4.1 Comparison of plant density, species richness and species diversity. The values in parentheses for riverine forest indicate the results including vine species.

	Number of plants/ha	Numbers of plant species/ha	$\underline{H}'$ of species
Dryland forest (Danum Valley)	360	132	4.88
Riverine forest (Sukau)	765 (996)	65 (84)	3.98 (4.27)
Mixed mangrove forest (Abai)	118	10	2.5

Comparison Between Flooded and Dryland Forests

Plant Diversity

As shown in Figure 4.1, curves for both species versus area (a) and species versus number of plants (b) indicated lower plant density and diversity in the mangrove forest than in the riverine forest. However, because of limited data for the dryland forest, we could not compare the species versus area and species versus number of plants curves among the three forest types; nonetheless, the plant density in the dryland forest was 360 trees/ha, which was still lower than the value for the riverine forest (765 trees/ha excluding vine species; Table 4.1), and higher than that for the mangrove forest (118 trees/ha).

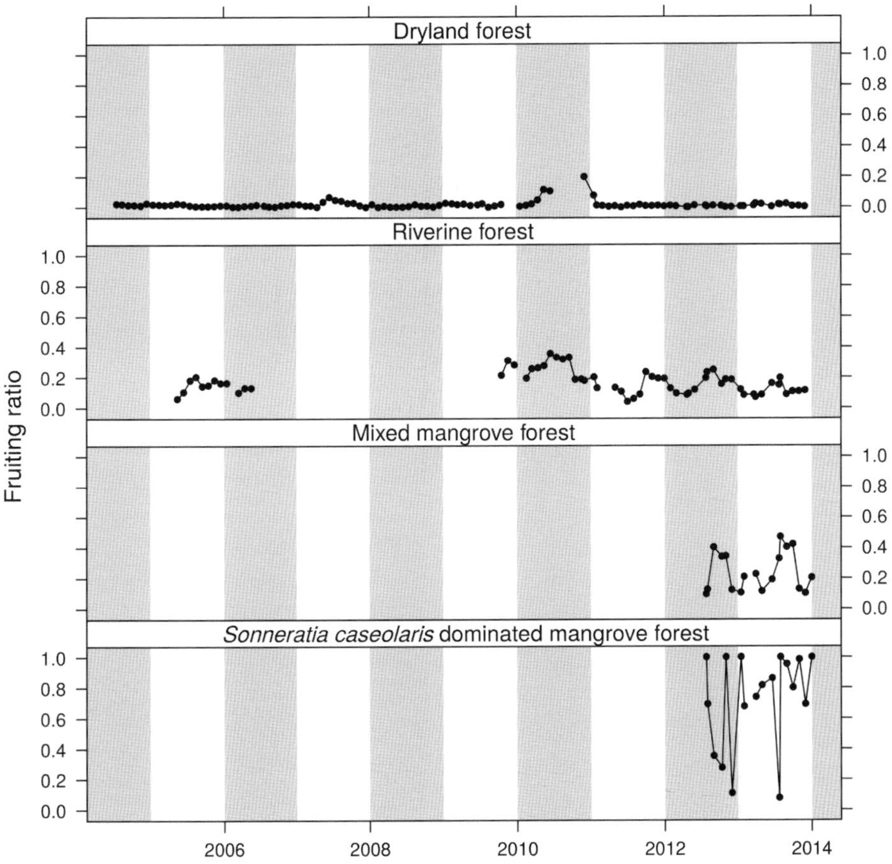

Figure 4.3 Seasonal fluctuations in the fruiting ratio in dryland, riverine, mixed mangrove and *S. caseolaris*-dominated mangrove forests.

On the other hand, species richness in the dryland forest (132 species/ha) was higher than in either riverine forests (65 species/ha) or mangrove (10 species/ha). Similar to the results obtained for species richness, species (trees only) diversity calculated using the Shannon–Wiener index (*H′*) (Pielou 1966) showed a higher value for dryland forest (4.88) than for the riverine (3.98) and mangrove forests (2.50).

Phenological Patterns of Fruits

Fruiting ratios in monthly phenological surveys fluctuated in all three forest types. The monthly fruiting ratio in the riverine forest ranged from 0.04 (July 2011) to 0.36 (June 2010, i.e. over one-third of individuals in fruit), with a general yearly peak from June to September (though in some years, the peak was from October to January) (Figure 4.3). In the mixed mangrove forest, the fruiting ratio ranged from 0.096 (July 2012) to 0.47 (August 2013), with a yearly peak from August to November (Figure 4.3). The fruiting ratio in the *S. caseolaris* mangrove forest was not as well defined as that in the mixed mangrove forest, with the ratio fluctuating substantially even within a year (range: 0.28–1.0). However, in general, *S. caseolaris* mangrove forest showed a higher fruiting ratio than the other three forest types. The fruiting ratio in the dryland forest showed less fluctuation throughout the year compared with the riverine and mangrove forests, except in 2010 when a mast fruiting event occurred, with the highest recorded value of fruiting ratio being 0.36. Although according to Hanya & Bernard (2012), a mast fruiting event occurred in 2007, the highest value of the fruiting ratio was 0.078, which is rather lower than the value in 2010. In general, the annual fruiting ratio in the dryland forest was much lower than that in the riverine and mangrove forests (Figure 4.4).

Discussion

This study demonstrates that the plant species diversity in mangrove forest is lower than that of the riverine forest. This corroborates results of a previous study conducted in 1990 in approximately the same areas (Boonratana 1993, 2000c), although that study recorded only those trees with a DBH of ≥ 30 cm. This lower plant diversity may be one of the reasons why only two sympatric diurnal primate species (proboscis monkey and long-tailed macaque), are commonly found in the mangrove forest, whereas the riverine forest is inhabited by eight sympatric diurnal primate species, similar to the dryland forest (10 primate species, probably including nocturnal species: Marsh & Greer (1992)). The lower plant diversity in the mangrove forest indicates that diet plants available for primates may also be limited, a circumstance which could influence primate distribution patterns. Researchers have suggested that, to prevent the accumulation of secondary compounds or to promote the efficiency of digestion, some primate species need to consume a variety of food items (Agetsuma 1995; Boonratana 2000b; Coelho *et al.* 1976; Marsh 1981; Matsuda *et al.* 2009b). Compared with other primate species, proboscis

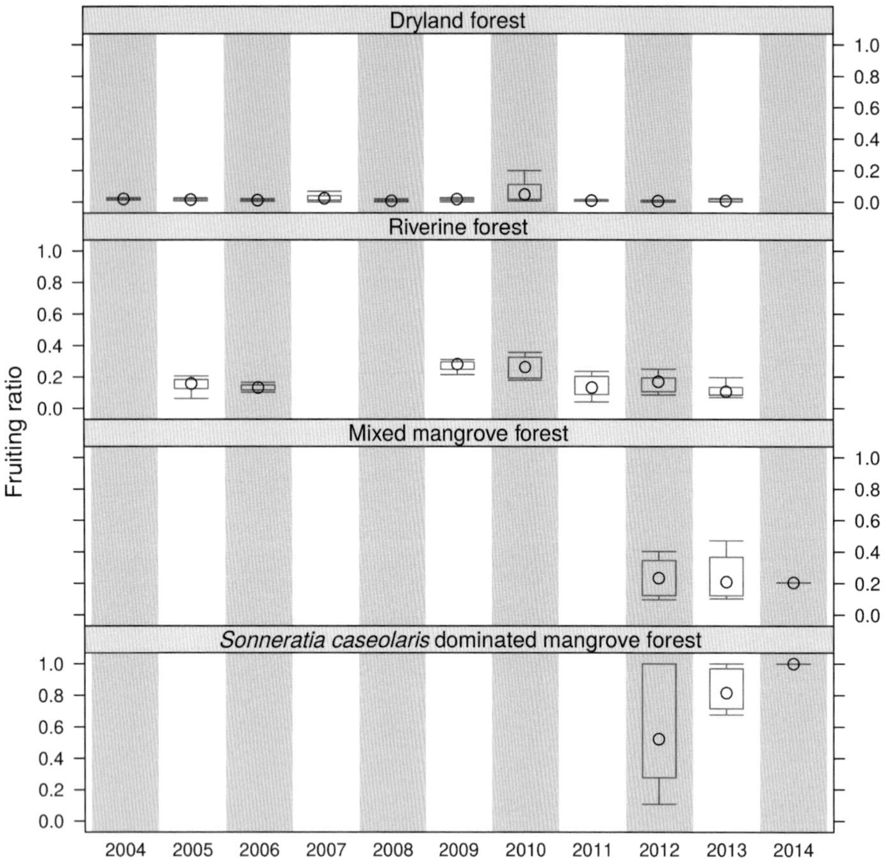

Figure 4.4 Comparison of annual fruiting ratio among the different forest types. Circles indicate the median. Boxes represent the range from 25% to 75% quartiles. Paired extensions show the approximate 95% intervals evaluated by the boxplot function in statistical software R. Note that full months' data were not available for some years (see Figure 4.3).

monkeys and long-tailed macaques may have higher tolerance for lower dietary diversity. Further research focusing on the comparisons of primate abundance, along with their detailed feeding ecology in the three different forest types, is needed to test this hypothesis and aid in better understanding of the factors affecting primate richness.

In addition to plant diversity, it is important to consider the nutritional quality of plants in different forest types, as this may be another prime determinant of primate abundance (Lambert 2011). Based on the finding that herbivorous primates prefer foods with more protein (e.g. Chapman & Chapman 2002; Davies *et al.* 1988; Hanya & Bernard 2012; Wasserman & Chapman 2003), it has been proposed that the protein–fibre ratio of leaves positively correlates with the abundance of colobine monkeys (Chapman *et al.* 2002, 2004). Such nutritional information is available for leaves from the riverine and dryland forests studied in this chapter (see Matsuda *et al.* 2013), but not for the mangrove forest. It has been shown that leaf quality (as measured by the protein–fibre ratio) is higher in the riverine forest than in the dryland forest (Figure 4.5; Matsuda *et al.* 2013). This has been linked to differences in soil quality between forest types, with the leaves of trees growing in nutrient-poor soils containing lower levels of protein and higher levels of fibre than those of trees growing in nutrient-rich soils (Janzen 1974; McKey *et al.* 1981). Alluvial areas exposed to seasonal flooding, such as riverine forests, frequently contain richer soils than areas away from the river and hence support increased forest productivity (van Schaik & Mirmanto 1985). Therefore, plants inhabiting such riverine forests may be expected to exhibit higher protein–fibre ratios (possibly with higher level of ash or minerals) than those in dryland forests. In addition to soil nutrient profiles, it is also important to consider plant life history. In contrast to dryland forest, riverine forest has potentially fewer shade-tolerant plants and more gap specialists, and these generally have leaves which contain more water and protein than shade-tolerant plant species (Coley 1987). Thus, although the plant diversity in riverine forest is lower than that in dryland forest, the number of primate species that can be supported by riverine forests approximates that of dryland forests, probably because of the better nutritional quality of young leaves, which are one of predominant primate food sources in riverine forests.

In addition to the better quality (higher protein–fibre ratio with higher levels of minerals) of leaves of common plants, the reduced supra-annual seasonality in fruit abundance is likely to be advantageous for primates living in the riverine forest. In this study, we have shown that the monthly fruiting ratio in the riverine forest was consistently higher than that in dryland forest; the fruiting ratio in the dryland forest during the time of mast fruiting was not as high as that in the riverine forest. Although primates mostly feed on fruits during a fruit masting in dryland forest, they revert to young leaves, figs and other food sources during non-masting periods (Marshall *et al.* 2009: and see Marshall & Wrangham 2007 for detailed review of fallback food). This has been reported for some primate taxa,

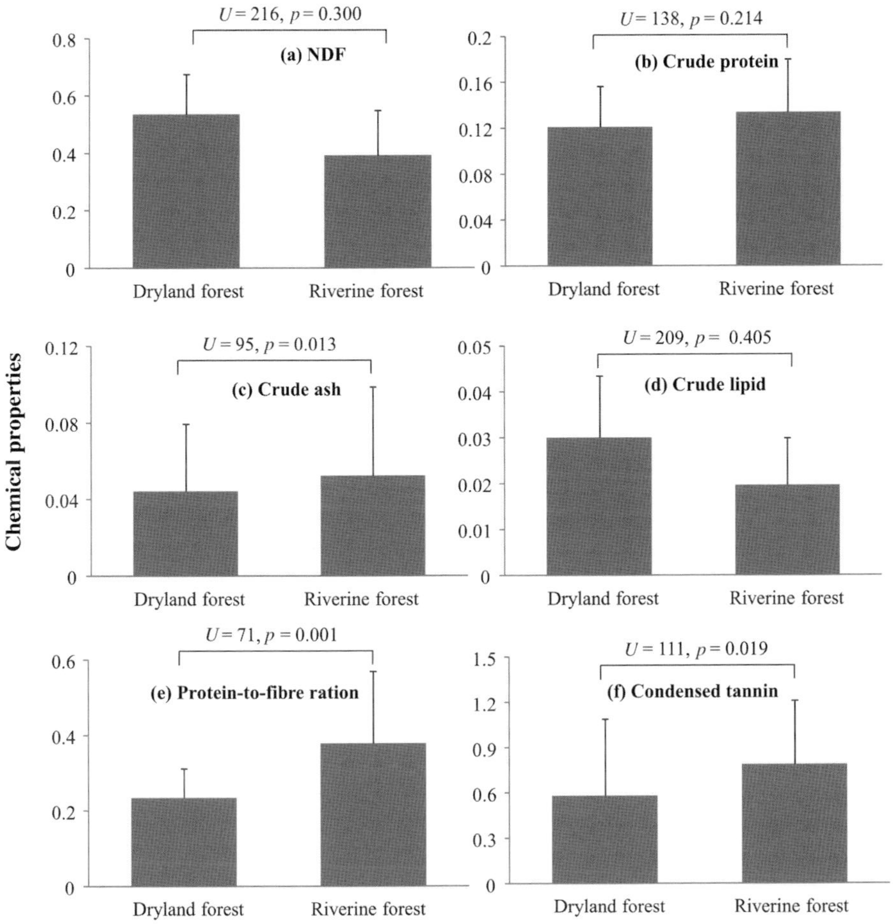

Figure 4.5 Comparison of the chemical properties (mean ± SD) of young leaves of a common plant species between unflooded and riverine forests. Values are presented as proportions of dry weight for (a–d). Presence/absence was used for the values for condensed tannin because it is difficult to detect subtle differences in leaf tannin content using the method (see Matsuda *et al.* 2013 in detail). Using the Mann–Whitney *U*-test, the mean values of chemical properties of the common plant species were compared between the forests.

including orangutans (Kanamori *et al.* 2010; Knott 1998; Wich *et al.* 2006) and red langurs (Hanya & Bernard 2012).

Similar to the riverine forest, relatively constant availability of fruits has been reported in the peat swamp forest in Borneo (reduced supra-annual seasonality in fruit abundance, e.g. Wich *et al.* 2011). This permits primates such as red langurs regular access to nutritionally superior food, so reducing reliance on fallback foods such as young leaves (Ehlers-Smith *et al.* 2013b). Indeed, compared with dryland forests, primate densities in such flooded habitats as riverine and peat swamp forests are often higher for orangutans (Husson *et al.* 2009), and almost similar for red langurs (Ehlers-Smith & Ehlers-Smith 2013). This information on seasonally flooded riverine and peat swamp forests indicates that flooded forests with relatively constant fruit availability and high-quality leaves are a prime and important habitat for primates. Therefore, protection of not only dryland forests but also flooded forests is essential for primate conservation.

Future research could consider the nutritional quality of the mangrove forest to provide a more complete general understanding of the value of flooded habitats for primates in Borneo. The mangrove forest had a much lower plant diversity than either riverine or dryland forests, although the fruiting ratios in monthly phenological surveys were consistently higher than those of dryland forest. In addition, although the phenological patterns and fruiting ratios in the mangrove forest and the riverine forest were similar, the fruiting ratio in *S. caseolaris* mangrove forest was much higher than that of riverine forest; fruits of *S. caseolaris* were available and abundant almost throughout the entire year. The fruits of *Sonneratia* spp. are one of the important food sources for proboscis monkeys, as briefly described in Boonratana (2003). Indeed, during some 120 h of focal data collected from July 2010 to February 2011 in the mangrove study area, adult male proboscis monkeys devoted 34% of their feeding time to fruit eating, with *S. caseolaris* being the sole fruit that the monkeys consumed during this period (Ikki Matsuda, unpublished data). Consistent with this observation, long-tailed macaques inhabiting the mangrove forest were also frequently observed to consume the fruits of *S. caseolaris* (Ikki Matsuda, unpublished data). Excluding the disadvantage of low plant diversity, the mangrove forest, especially *S. caseolaris* mangrove forest, may form a highly important habitat for some primate species, providing them with a highly stable supply of foods.

Conclusions and Recommendations

In conclusion, we report here on the detailed plant composition and phenological patterns of fruit availability in Bornean flooded forests, including riverine and mangrove forests, and compare these with the findings from dryland forest. A relatively constant availability of fruits, combined with greater plant diversity and higher leaf quality in the riverine forest, when compared with the dryland forest, indicates that the riverine habitat is as equally important for primate survival as dryland forest. Although the quality of the mangrove forest was not assessed as fully, at least for proboscis monkeys and long-tailed macaques, the mangrove forest appears to be an indispensable resource because of abundant availability of fruits throughout the year and reduced competition from other primate species. Future research should focus on how the spatial and temporal variation in fruit production is related to primate distribution and ranging. This remains a highly relevant topic that can aid, not only in a basic understanding of primate behavioural ecology, but also in the development of conservation strategies for primate species with ranges that include several habitat types. With the high and continuous pressure on seasonally flooded habitats, such research deserves high priority.

Acknowledgements

Some parts of this chapter were originally prepared for the symposium on 'Primates in flooded forests' which was held at the 24th Congress of the International Primatological Society in Cancun, Mexico, 2012. IM, MN and YO thank the Economic Planning Unit of the Malaysian Government, the Sabah Biodiversity Centre, the Sabah Wildlife Department and the Sabah Forestry Department. The Economy Planning Unit of Malaysia, the Sabah Biodiversity Centre and the Danum Valley Management Committee of Yayasan Sabah permitted our study. Advice and support has been generously supplied by G. Hanya, T. Kanamori, S.T. Wong and N. Imai. This study was partly financed by HOPE and Human Evolution Project of KUPRI, JSPS Grant-in-Aid for Challenging Exploratory Research (24657170) and for Young Scientist A (26711027) and the National Geographic Society (9254-13) to I. Matsuda and for Strategic Young Researcher Overseas Visits Program for Accelerating Brain Circulation, to KUPRI. Lastly, the authors would like to thank Enago (www.enago.jp) for the English language review.

Note

1 The values are slightly different from those described by Matsuda (2008) and Matsuda *et al.* (2009a) because some species have been categorized into new families.

Appendix A Floral composition along the 16 trails described in Matsuda *et al.* (2009a). The total distance of the trails was 7.15 km and the vegetation survey was conducted for trees with a DBH of > 10 cm and vines with a diameter of > 5 cm on the trail or within 1 m from the edge of the trail (i.e. survey area = 2.15 ha).

Family	Species	Life form	*N*	Density (/ha)	Total basal area (cm²)
Achariaceae	*Hydnocarpus polypetalus*	Tree	4	1.9	705
	Hydnocarpus sumatrana	Tree	36	16.8	11 386
	Hydnocarpus woodii	Tree	1	0.5	140
Anacardiaceae	*Androtium astylum*	Tree	1	0.5	105
	Buchanania arborescens	Tree	10	4.7	7361
	Dracontomelon dao	Tree	15	7.0	7532
	Koordersiodendron pinnatum	Tree	2	0.9	1541
	Mangifera parvifolia	Tree	3	1.4	2734
	Melanochyla auriculata	Tree	2	0.9	812
Annonaceae	*Artabotrys suaveolens*	Vine	26	12.1	1318
	Cananga odorata	Tree	4	1.9	2873
	Polyalthia sumatrana	Tree	1	0.5	191
	Polyalthia sp. 1	Tree	1	0.5	131
	Uvaria lobbiana	Vine	1	0.5	26
	Uvaria sp. 1	Vine	1	0.5	27
Apocynaceae	*Parameria polyneura*	Vine	1	0.5	24
	Tabernaemontana macrocarpa	Tree	1	0.5	575
	Urceola sp. 1	Vine	2	0.9	81
	Willughbeia angustifolia	Vine	1	0.5	27
Aquifoliaceae	*Ilex cymosa*	Tree	28	13.1	13 625
Burseraceae	*Canarium decumanum*	Tree	1	0.5	102
	Canarium denticulatum	Tree	2	0.9	214
Calophyllaceae	*Calophyllum blancoi*	Tree	1	0.5	251
	Calophyllum pisiferum	Tree	1	0.5	556
	Kayea oblongifolia	Tree	1	0.5	194
	Mesua elmeri	Tree	23	10.7	10 885
	Mesua macrantha	Tree	1	0.5	392
Carabidae	*Agelaea borneensis*	Vine	1	0.5	26
	Agelaea trinervis	Vine	1	0.5	46
Celastraceae	*Lophopetalum multinervium*	Tree	2	0.9	1589
Chrysobalanaceae	*Kostermanthus heteropetalus*	Tree	1	0.5	3317
	Maranthes corymbosa	Tree	3	1.4	596
	Parinari oblongifolia	Tree	3	1.4	6713
Clusiaceae	*Garcinia brevipes*	Tree	1	0.5	103
	Garcinia parvifolia	Tree	23	10.7	3317

(continued)

Appendix A *(cont.)*

Family	Species	Life form	*N*	Density (/ha)	Total basal area (cm²)
	Garcinia sp. 1	Tree	1	0.5	353
Combretaceae	*Combretum acuminatum*	Vine	5	2.3	129
	Terminalia citrina	Tree	2	0.9	390
Connaraceae	*Connarus grandis*	Vine	6	2.8	211
	Rourea mimosoides	Vine	1	0.5	58
	Rourea minor	Vine	11	5.1	1069
Convolvulaceae	*Erycibe grandifolia*	Vine	1	0.5	39
Dilleniaceae	*Dillenia excelsa*	Tree	98	45.7	16 778
	Tetracera scandens	Vine	3	1.4	93
Dipterocarpaceae	*Dipterocarpus validus*	Tree	14	6.5	12 665
	Parashorea malaanonan	Tree	2	0.9	898
	Vatica rassak	Tree	52	24.2	17 577
	Vatica umbonata	Tree	2	0.9	617
	Vatica venulosa	Tree	11	5.1	5092
Ebenaceae	*Diospyros curranii*	Tree	23	10.7	6079
	Diospyros elliptifolia	Tree	21	9.8	5057
	Diospyros euphlehia	Tree	3	1.4	405
	Diospyros macrophylla	Tree	1	0.5	407
	Diospyros wallichii	Tree	6	2.8	1459
	Diospyros sp. 1	Tree	6	2.8	1096
	Diospyros sp. 2	Tree	6	2.8	2921
	Diospyros sp. 3	Tree	5	2.3	207
Elaeocarpaceae	*Elaeocarpus macrocarpus*	Tree	1	0.5	539
	Elaeocarpus nitidus	Tree	15	7.0	4889
	Elaeocarpus stipularis	Tree	6	2.8	1969
Erythroxylaceae	*Erythroxylum cuneatum*	Tree	2	0.9	503
Euphorbiaceae	*Croton caudatus*	Vine	46	21.4	1572
	Croton oblongus	Tree	82	38.2	14 950
	Excoecaria indica	Tree	111	51.7	161 427
	Mallotus floribundus	Tree	1	0.5	47
	Mallotus muticus	Tree	149	69.5	95 716
	Mallotus penangensis	Tree	18	8.4	3275
	Macaranga conifera	Tree	5	2.3	1517
	Paracroton pendulus	Tree	8	3.7	1457
Fabaceae	*Acacia borneensis*	Vine	10	4.7	306
	Albizia corniculata	Vine	28	13.1	1768
	Bauhinia diptera	Vine	26	12.1	1231

Appendix A *(cont.)*

Family	Species	Life form	*N*	Density (/ha)	Total basal area (cm^2)
	Crudia ornata	Tree	1	0.5	144
	Crudia reticulata	Tree	27	12.6	7520
	Cassia indica	Tree	3	1.4	242
	Cynometra ramiflora	Tree	3	1.4	533
	Dalbergia parviflora	Vine	38	17.7	1163
	Derris elegans	Vine	2	0.9	56
	Derris sp. 1	Vine	5	2.3	159
	Derris sp. 2	Vine	1	0.5	58
	Dialium indum	Tree	1	0.5	4582
	Entada rheedii	Vine	32	14.9	2411
	Koompassia excelsa	Tree	1	0.5	611
	Millettia nieuwenhuisii	Vine	20	9.3	1152
	Millettia sp. 1	Vine	4	1.9	377
	Millettia sp. 2	Vine	2	0.9	31
	Ormosia sumatrana	Tree	2	0.9	596
	Pongamia pinnata	Tree	11	5.1	6312
	Sindora leiocarpa	Tree	2	0.9	488
	Spatholobus macropterus	Vine	18	8.4	1087
Gnetaceae	*Gnetum gnemonoides*	Vine	4	1.9	233
Hypericaceae	*Cratoxylum cochinchinense*	Tree	8	3.7	3472
	Cratoxylum formosum	Tree	3	1.4	1859
Lamiaceae	*Teijsmanniodendron bogoriense*	Tree	10	4.7	3924
	Vitex pinnata	Tree	55	25.6	36 293
Lauraceae	*Cryptocarya* sp. 1	Tree	5	2.3	636
	Cryptocarya sp. 2	Tree	1	0.5	123
	Dehaasia cuneata	Tree	9	4.2	3409
	Endiandra sp. 1	Tree	12	5.6	2978
	Endiandra sp. 2	Tree	8	3.7	2621
	Litsea sp. 1	Tree	1	0.5	542
	Litsea sp. 2	Tree	1	0.5	234
	Nothaphoebe sp. 1	Tree	2	0.9	1375
	Cinnamomum sp. 1	Tree	1	0.5	161
Lecythidaceae	*Barringtonia macrostachya*	Tree	32	14.9	5772
	Planchonia valida	Tree	1	0.5	3847
Loganiaceae	*Strychnos ignatii*	Vine	2	0.9	78
	Strychnos minor	Vine	3	1.4	136

(continued)

Appendix A (cont.)

Family	Species	Life form	*N*	Density (/ha)	Total basal area (cm²)
Lophopyxidaceae	*Lophopyxis maingayi*	Vine	113	52.7	5345
Lythraceae	*Lagerstroemia speciosa*	Tree	6	2.8	1364
Malvaceae	*Colona serratifolia*	Tree	9	4.2	4333
	Durio kutejensis	Tree	1	0.5	140
	Grewia acuminata	Vine	1	0.5	23
	Microcos crassifolia	Tree	27	12.6	6704
	Pterospermum macrocarpum	Tree	10	4.7	9036
Melastomataceae	*Memecylon paniculatum*	Tree	6	2.8	802
	Memecylon edule	Tree	1	0.5	117
	Pternandra galeata	Tree	65	30.3	89 638
Meliaceae	*Aphanamixis* sp. 1	Tree	1	0.5	107
Menispermaceae	*Haematocarpus validus*	Vine	1	0.5	26
Moraceae	*Artocarpus kemando*	Tree	2	0.9	477
	Ficus benjamina	Hemi-epiphyte (self-standing)	1	0.5	1256
	Ficus binnendijkii	Hemi-epiphyte (self-standing)	14	6.5	9315
	Ficus villosa	Vine	2	0.9	80
	Ficus crassiramea	Hemi-epiphyte (self-standing)	4	1.9	8518
	Ficus globosa	Hemi-epiphyte	1	0.5	98
	Ficus variegata	Tree	1	0.5	804
Myristicaceae	*Knema laurina*	Tree	4	1.9	463
	Myristica sp. 1	Tree	2	0.9	1487
	Myristica sp. 2	Tree	8	3.7	2048
Myrsinaceae	*Embelia philippinensis*	Vine	2	0.9	72
Myrtaceae	*Eugenia litseaefolia*	Tree	16	7.5	4081
	Eugenia sandakanensis	Tree	3	1.4	591
	Eugenia sp. 1	Tree	3	1.4	284
	Eugenia sp. 2	Tree	33	15.4	17 983
	Eugenia sp. 3	Tree	28	13.1	8901
	Eugenia sp. 4	Tree	1	0.5	754
	Eugenia sp. 5	Tree	1	0.5	100
	Syzygium fastigiatum	Tree	12	5.6	1605
Oleaceae	*Chionanthus pluriflorus*	Tree	2	0.9	1247
Phyllanthaceae	*Antidesma thwaitesianum*	Tree	33	15.4	13 980
	Aporosa acuminatissima	Tree	1	0.5	175
	Aporosa nigricans	Tree	4	1.9	794
	Baccaurea bracteata	Tree	3	1.4	722
	Baccaurea stipulata	Tree	18	8.4	2165

Appendix A *(cont.)*

Family	Species	Life form	*N*	Density (/ha)	Total basal area (cm²)
	Bridelia penangiana	Tree	3	1.4	502
	Bridelia stipularis	Vine	30	14.0	1598
	Cleistanthus myrianthus	Tree	19	8.9	5959
	Glochidion macrostigma	Tree	2	0.9	119
	Glochidion obscurum	Tree	20	9.3	3288
	Glochidion sp. 1	Tree	2	0.9	375
	Margaritaria indica	Tree	5	2.3	1730
Polygalaceae	*Xanthophyllum flavescens*	Tree	12	5.6	3700
Primulaceae	*Ardisia* sp. 1	Tree	1	0.5	134
Proteaceae	*Helicia robusta*	Tree	3	1.4	566
Putranjivaceae	*Drypetes castilloi*	Tree	15	7.0	2412
	Drypetes sp. 1	Tree	1	0.5	177
Rhamnaceae	*Ventilago dichotoma*	Vine	12	5.6	451
	Ziziphus borneensis	Vine	3	1.4	170
Rhizophoraceae	*Carallia brachiata*	Tree	13	6.1	6699
Rubiaceae	*Diplospora* sp. 1	Tree	1	0.5	159
	Gardenia tubifera	Tree	3	1.4	553
	Ixora elliptica	Tree	2	0.9	302
	Nauclea orientalis	Tree	3	1.4	1490
	Nauclea subdita	Tree	77	35.9	37 757
	Neolamarckia cadamba	Tree	7	3.3	2617
	Neonauclea excelsa	Tree	4	1.9	1110
	Pleiocarpidia sandahanica	Tree	18	8.4	2641
	Uncaria callophylla	Vine	10	4.7	334
Rubiaceae	*Uncaria ferrea*	Vine	20	9.3	609
	Clausena excavata	Tree	3	1.4	513
Salicaceae	*Xylosma sumatrana*	Tree	67	31.2	21 264
Sapindaceae	*Dimocarpus longan*	Tree	1	0.5	757
	Lepisanthes amoena	Tree	1	0.5	138
	Mischocarpus sundaicus	Tree	1	0.5	113
	Nephelium uncinatum	Tree	1	0.5	478
	Pometia sp. 1	Tree	1	0.5	142
Sapotaceae	*Madhuca dubardii*	Tree	9	4.2	1609
	Payena microphylla	Tree	1	0.5	1356
Symplocaceae	*Symplocos celastrifolia*	Tree	6	2.8	1491

Appendix B Floral composition in 88 quadrats (20 m × 20 m; i.e. 2.52 ha). The vegetation survey was conducted for trees with a DBH of > 10 cm and vines with a diameter of > 5 cm in the quadrats.

Family	Species	Life form	*N*	Density (/ha)	Total basal area (cm^2)
Meliaceae	*Aglaia cucullata*	Tree	15	6.0	28 824
Alangiaceae	*Alangium* sp.	Tree	1	0.4	138
Anacardiaceae	*Buchanania arborescens*	Tree	12	4.8	8423
Apocynaceae	*Cerbera odollam*	Tree	1	0.4	379
Tiliaceae	*Colona serrartifolia*	Tree	1	0.4	645
Hypericaceae	*Cratoxylum cochinchinense*	Tree	1	0.4	436
Leguminosae	*Cynometra ramiflora*	Tree	1	0.4	861
Myrtaceae	*Eugenia borneensis*	Tree	6	2.4	10 495
	Eugenia sp.	Tree	1	0.4	109
	Eugenia tawahense	Tree	20	7.9	8627
Euphorbiaceae	*Excoecaria agallocha*	Tree	15	6.0	28 187
	Excoecaria indica	Tree	6	2.4	15 835
Moraceae	*Ficus microcarpa*	Hemi-epiphyte (self-standing)	41	16.3	20 866
Sapindaceae	*Guioa bijuga*	Tree	1	0.4	147
Sterculiaceae	*Heritiera littoralis*	Tree	3	1.2	3976
Rubiaceae	*Nauclea subdita*	Tree	3	1.2	1287
Palmae	*Nypa fruticans*	Tree	17	6.7	2560
	Oncosperma tigillarium	Tree	60	23.8	11 848
Pandanaceae	*Pandanus* sp.	Tree	8	3.2	1982
Anonaceae	*Polyalthia* sp.	Tree	5	2.0	3370
Melastomataceae	*Pternadra* sp.	Tree	4	1.6	985
Myrsinaceae	*Rapanea* sp.	Tree	1	0.4	861
Sonneratiaceae	*Sonneratia caseolaris*	Tree	71	28.2	21 859
Symplocaceae	*Symplocos celastrifolia*	Tree	2	0.8	806
Myrtaceae	*Syzygium fastigiatum*	Tree	1	0.4	215
Verbenaceae	*Vitex pinnata*	Tree	1	0.4	140

Chapter 5

Lemurs in Mangroves and Other Flooded Habitats

Giuseppe Donati, Timothy M. Eppley, José Ralison, Jacky Youssouf and Jörg U. Ganzhorn

Introduction

Mangroves on Madagascar

Recent estimates indicate that mangroves in Madagascar occupy an area of approximately 2800 km^2, 98% of which lies along the western coast of the island, representing about 2% of global mangroves (Giri *et al.* 2011; Giri & Muhlhausen 2008). This asymmetry is due mainly to tidal range differences, wider in the west than in the east of the island, and to the presence of large estuaries (Roger & Andrianasolo 2003). Being marginal habitats with extreme temperatures, solar radiation levels, winds, salinity and flooding, mangrove forests are relatively poor in species diversity (Vannucci 2001). Compared to other continents, mangroves in Madagascar are even poorer in plant species diversity than other areas, with only 9 species compared to more than 45 species in Southeast Asia and approximately 17 in Africa (Ellison *et al.* 1999). This depauperate plant community matches a low diversity of animal groups, including lemurs (which have radiated so dramatically in other forest types in Madagascar). In fact, reports of lemur species in mangrove habitats are scarce (Roger & Andrianasolo 2003). This contrasts the Madagascan situation with that on other continents where mangroves are key resources for several primate species, although mangrove specialists are very rare within the order (Matsuda *et al.* 2011; Nowak 2012).

The unpalatability of mangrove leaves, with some mangrove tree species being rich in tannins, has been suggested as one possible reason for the lack of lemurs in mangroves (Birkinshaw & Colquhoun 2003). Other hypotheses propose that the open mangrove canopy may be associated with higher predation pressure on lemurs from birds of prey and/or higher thermoregulatory stress (Birkinshaw & Colquhoun 2003). However, at present, it is not clear whether the rare occurrence of lemurs in mangroves has ecological reasons or is simply the result of low survey effort and absence of systematic studies in this habitat.

In order to explore whether the occurrence of lemurs in mangroves is a rare event, we provide an updated review of observations of lemurs in this habitat, as well as in other inundated habitats. The information has been compiled using the IUCN Red List, a questionnaire sent to the IUCN Lemur Specialist Group members, and an additional list of researchers known to work or to have worked in areas where mangroves occur.

A case study is also used, focusing on the littoral forest of the Mandena Conservation Zone in southeastern Madagascar, to explore the hypothesis that mangroves represent low-quality food resources for lemurs as compared to upland forest and freshwater swamp plants. This is done by comparing phytochemical contents of leaves available in forests, swamps and mangroves.

Lemurs in Mangroves

Although few studies of mangrove use by lemurs have been published and reports often lack contextual information, lemurs have been observed in several mangrove localities along the western coast of Madagascar (Figure 5.1). Decary (1950) was the first to report the presence of a lemur, the aye-aye (*Daubentonia madagascariensis*), in an unspecified mangrove forest near Mahajanga. This was the only report of this species in this type of forest. Mouse lemurs (*Microcebus* spp.) have been observed several times in this habitat. *Microcebus* cf. *myoxinus* has been observed in Baie de Baly close to Antsakoamarovitiky (Hawkins *et al.* 1998), the Baie de Bombetoka, the largest area of mangroves in Madagascar (Schmid & Kappeler 1994), and in the mangroves of Besalampy, south of Cap Sainte André (J.U. Ganzhorn, unpublished data). Hawkins and co-authors (1998) speculate that it is possible that this species occurs regularly in mangroves. Another species of mouse lemur, *M. griseorufus* has been observed in the mangrove area of Saint Augustin, in the southwest of the island (J. Youssouf, unpublished data).

Sifaka species have been observed several times as frequent inhabitants of mangroves. The presence of two species of sifakas, *Propithecus deckeni* and *P. coquereli*, has been reported in the areas of Katsepy and Anjohibe (near Mahajanga), respectively (M. Markoff, pers. comm., 2013). More detailed reports come from the large mangrove patches of the Katsepy area, located between the estuaries of Betsiboka and the Mahavavy rivers, where *P. coronatus* has been observed regularly (Gauthier *et al.* 1999, 2000). Mangroves are used by this species as sleeping sites and for feeding. However, more systematic observations of two groups of *P. coronatus* revealed that they leave the mangrove sites for the upland, dry forest early in the morning and return late in the afternoon (Gauthier *et al.* 1999). *Propithecus coronatus* have also been reported in the area of Antrema (Roger & Andrianasolo 2003).

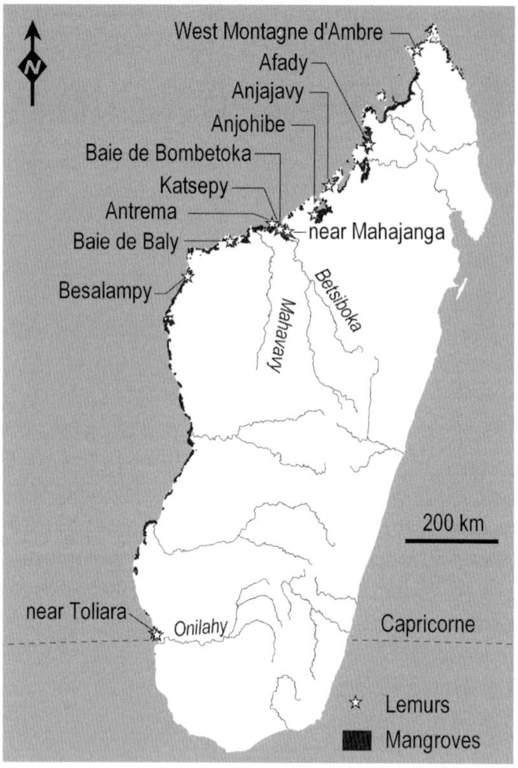

Figure 5.1 Mangrove distribution in Madagascar and locations where lemurs have been observed within this habitat. The observations labelled as 'near Toliara' include the following sites: Ambanilia, Andalo, Andoharano, Andriambe, Antanifotsy, Antsifotse, Lovokampy and Saint Augustin.

Aided by their well-known flexibility, several species of lemurids also appear to use the mangroves, at least as resting sites. In the Katsepy region *Eulemur rufus* and *E. mongoz* have been observed to rest in the areas used by the sifakas (Gauthier *et al.* 1999, 2000). The latter species has also been observed in Antrema (E. Roger, pers. comm., 2013). Further north, in the area of Afady, blue-eyed lemurs (*E. flavifrons*) have been observed in mangrove patches (A. Dumoulin, unpublished data). At the extreme north of the island, northwest of Montagne d'Ambre, crowned lemurs (*E. coronatus*) and Sanford's lemurs (*E. sanfordi*) have been observed at the edge of mangroves connecting patches of forest (B. Freed, pers. comm., 2013). The presence of identified groups of these species in areas only connected by mangroves implies their use of this habitat as a corridor (B. Freed, pers. comm., 2013). Finally, in the southwestern corner of the island, ring-tailed lemurs (*Lemur catta*) have been observed in several areas of mangroves north and south of Toliara, including the mangroves of Ambanilia, Andalo, Andoharano and Antsifotse, in the western portion of the new protected area of Tsinjoriake. Ring-tailed lemurs have also been recorded in the mangroves of Andriambe, Antanifotsy, and Lovokampy located on the south shore of the Onilahy River, near Saint Augustin (J. Youssouf, unpublished data).

Lemurs in Other Flooded Habitats

Despite the paucity of research conducted on lemurs in mangroves, a wealth of data has been collected on lemur species occupying other types of flooded habitats. The best known example is the Lake Alaotra gentle lemur (*Hapalemur alaotrensis*; see Chapters 34 and 37). The area surrounding Lac Alaotra is devoid of bamboo (subfamily Bambusoideae) (Mutschler 1999; Mutschler *et al.* 1998), a resource on which congeners rely heavily (Tan 1999). Instead, the strict folivore *H. alaotrensis* inhabits the surrounding marshes, resting and feeding on sedges and reeds (Cyperaceae and Poaceae) (Mutschler 1999; Mutschler *et al.* 1998). Why this lemur species chose to exploit such a habitat is still a major question, especially since the overall dietary quality in such an environment is low (Mutschler 1999; but see Chapter 34 for possible explanations and ecological adaptations).

Another seasonally flooded habitat is that of the Mandena Conservation Zone, located in southeast Madagascar. The area contains two fragments of littoral forest comprising 148 ha with 82 ha of interspersed swamps (Ganzhorn *et al.* 2007). Although some of this area remains wet year round, much of it becomes inundated only seasonally, sometimes with water up to 1.8 m (T.M. Eppley, unpublished data). Four lemur species (*Eulemur collaris, Hapalemur meridionalis, Avahi meridionalis* and *Microcebus ganzhorni*) have been observed within the swamp (Eppley *et al.* 2015), and it is possible that *Cheirogaleus major* and *C. medius* also use the area.

The collared brown lemur (*E. collaris*) is a cathemeral frugivore that in Mandena spends much of its feeding time in upland littoral forest where fruits are more available. However, they use the swamp extensively for travelling and resting (Donati *et al.* 2011). They have also been observed to feed there, although this represents a marginal 12.1% ($N = 7$ items) of their total diet. By comparison, approximately 40% ($N = 22$) of the diet of the southern bamboo lemur (*H. meridionalis*) comprises plants in the swamp habitat (Eppley *et al.* 2011; J. Ralison, unpublished data). At Mandena, *H. meridionalis* were observed to spend an average of 67.3% of their time feeding in swamp areas, with the remaining 32.7% of feeding being spent in the littoral forest (Eppley *et al.* 2011), with similar figures for their overall activity budget (Eppley *et al.* 2015). Although this species tends to graze on the sedges and grasses (Cyperaceae and Poaceae) available there during the dry austral winter (Eppley *et al.* 2011), they also access this habitat during full inundation to forage for arboreal fruits, flowers and leaves (J. Ralison, unpublished data). *Eulemur cinereiceps* also seems to spend a significant amount of time foraging in swamp areas in the eastern rainforest (Andriamaharoa *et al.* 2010).

Comparison of Nutritional Quality Between Upland Forest Plants, Swamps and Mangroves

To examine potential phytochemical differences between leaves available in different habitats, we contrasted leaves from randomly selected trees in the forest and in the swamp of Mandena. Overall, leaves analysed from both the upland forest and swamp showed no significant differences (Table 5.1). Leaves from the forest, however, showed a strong tendency to contain higher nitrogen content ($z = -1.90, p = .057$). Mangrove biochemical characteristics (from leaves of *Avicennia marina*,

Table 5.1 Phytochemical characteristics of mature leaves from upland littoral forest, inundated swamp, and mangroves in southeast Madagascar. N = nitrogen; SP = soluble protein; SC = soluble carbohydrates; NDF = neutral detergent fibre; ADF = acid detergent fibre. Values are medians and quartiles.

Habitat	N	SP	SC	NDF	ADF	Tannins	Phenol	Ash
Forest	**1.25**	**2.72**	**6.50**	**42.01**	**30.89**	**0.20**	**2.74**	**7.18**
(*n* = 11)	1.15–1.49	2.08–4.13	4.68–9.77	38.59–51.48	26.20–34.01	0.00–0.65	1.58–3.99	5.19–9.89
Swamp	**1.01**	**2.97**	**5.13**	**40.73**	**33.43**	**0.00**	**1.72**	**6.82**
(*n* = 10)	0.78–1.24	1.35–3.40	4.66–6.76	35.68–49.93	24.43–39.39	0.00–0.34	0.73–2.48	5.45–10.46
Mangroves	**1.36**	**3.87**	**12.01**	**30.79**	**19.20**	**1.45**	**4.63**	**9.24**
(*n* = 2)	1.23–1.50	3.08–4.65	11.08–12.22	28.92–32.66	18.75–19.64	0.72–2.1	4.49–4.76	8.64–9.85

Acanthaceae, and *Bruguiera gymnorrhiza*, Rhizophoraceae) are shown in the table but the data were not included in the statistical analysis due to small sample size. Mangrove leaves appear to be lower in fibre and higher in sugar, tannin and polyphenolic content.

Discussion

This brief review suggests that various lemur taxa use mangroves in one way or another. While no lemur species appear to specialize or display a preference for mangroves, subsets of the lemur community occurring in areas where mangroves are also present seem to utilize them. This first attempt to summarize available knowledge on lemur presence in mangroves indicates that the importance of this particular kind of forest as lemur habitat may have been underestimated. This is in line with recent findings from other continents which suggest that in fact more than half of Old World monkey genera use mangroves at least seasonally or opportunistically (Nowak 2012; Rowe & Myers 2011).

Our review suggests that at least some of the observed lemur species use mangrove patches as part of a mosaic habitat where other forest types are also present. Mainly frugivorous lemurids, for example, seem to rely on mangroves opportunistically as refuges for their resting/sleeping or as corridors for travelling between preferred habitats, while feeding in mangrove forests is rarely observed. However, this interpretation may simply be the result of a lack of systematic follows in this habitat type and, therefore, further survey efforts are needed to verify this assumption. Other species are also known to use mangroves or inundated forests as part of an effective defence response to terrestrial predator risk (Galat-Luong & Galat 2005; Matsuda *et al.* 2009b; Matsuda *et al.* 2011; Nowak 2012). The observation that frugivorous *E. collaris* in Mandena spend much of their resting time in the swamp, while most of their feeding occurs in the forest, fits well with an antipredator strategy (G. Donati, unpublished data). Although observational data from mangrove areas are not available, a similar reasoning can be used to explain the opportunistic use of mangroves by other frugivorous lemurids which are unlikely to meet their dietary requirements in this habitat year round. In fact, mangrove habitats have been shown to provide monotonous food supply as well as to have relatively low fruit availability (Matsuda *et al.* 2009b; Nowak 2008).

Folivorous lemurs, such as *Propithecus* and *Hapalemur* species, seem to use mangroves or inundated forests, respectively, more extensively. Phytochemical analyses show similar nutritional values of leaves in forest and freshwater swamps in Mandena, although we found a trend for higher nitrogen content in leaves in upland forest. Keeping other factors constant, this finding would theoretically explain the large amount of time allocated by the resident folivore, *H. meridionalis*, to feeding in the swamp (Eppley *et al.* 2011, 2015), since clear nutritional benefits for feeding in the more exposed/risky upland forest seem to be lacking.

As for mangrove phytochemical characteristics, our sample of leaves was too small to allow statistical comparisons; however, the two species collected show nitrogen content comparable to forest and freshwater swamp trees and lower fibre content. This would make mangrove resources of potentially high value for folivores and it matches with other studies reporting similar nitrogen content in upland forest and mangrove species (Agoramoorthy & Hsu 2005; De Lacerda *et al.* 1986; but see Ellison 2002). However, in our analysis, mangrove leaves also show substantially higher median values of tannins and polyphenolics, which are known to work as a deterrent to folivory (Ganzhorn 1988; Glander 1982). If our preliminary result is confirmed by future analyses of larger datasets, then this would support the hypothesis that mangroves do not represent a food source that lemurs may rely upon extensively or exclusively (Birkinshaw & Colquhoun 2003). High values of tannins and phenolics may induce thirst in mammals as animals need water to flush out salt and secondary plant metabolites (Glander 1982). For example, water consumption was found to be associated with mangrove leaf consumption and mangrove use in Zanzibar red colobus (*Procolobus kirkii*) (Nowak 2008). In Borneo, proboscis monkeys (*Nasalis larvatus*) occupy mangroves, but feed preferentially on non-mangrove leaves even though mangrove and non-mangrove leaves do not differ in protein concentration (Agoramoorthy & Hsu 2005; Yeager, 1989). We may hypothesize that some lemur species, such as *Propithecus* spp., may tolerate higher secondary component intakes, and thus have more continuous access to mangrove habitats. However, this suggestion should be regarded as highly speculative until more data are collected.

Mangrove extent and diversity relative to other forest types in Madagascar are low, and worldwide comparative analyses indicate that species richness in any given mangrove habitat is correlated with its size (Ellison *et al.* 1999). Thus, the limited variety of plants in Malagasy mangrove swamps may be an additional factor which explains the presence of only subsets of lemur communities (or lack thereof) in these areas and lemurs' need to rely on other forest types. That said, mangrove deforestation rate in Madagascar is lower compared with other forested habitats and these swamps may thus represent a refuge in areas where most other types of forest have disappeared (Giri & Muhlhausen 2008; but see also Jones 2013). Given that it is not known to what extent mangroves can be used as dietary complements, it seems crucial to conserve the matrix of mangroves and other forest types in an integrated way. The need to protect these neglected areas should be seriously considered in light of the growing threats to upland forests from expanding agriculture, aquaculture and logging (Jones 2013).

Acknowledgements

We thank the members of the IUCN Lemur Specialist Group for sharing their observations. Lucienne Wilmé kindly provided the figure. Thanks to Benjamin Freed, Katarzyna Nowak, Bruno Simmen, Christoph Schwitzer and Urs Thalmann for sharing their observations and for useful discussions. G.D.'s work on *Eulemur collaris* was funded by the Rufford Grant and the University of Pisa. T.M.E.'s work on *Hapalemur meridionalis* was funded by the American Society of Primatologists, Conservation International's Primate Action Fund, IDEAWILD, Mohamed Bin Zayed Species Conservation Fund (Project Number: 11253008), Primate Conservation Inc., the Primate Society of Great Britain/Knowsley Safari Park. J.Y. and J.U.G. collaborate on lemurs with support from SuLaMa.

Chapter 6

Survey and Study Methods for Flooded Habitat Primatology

Adrian A. Barnett, Joseph E. Hawes, Antonio R. Mendes Pontes, Viviane M. Guedes Layme, Janice Chism, Robert B. Wallace, Nayara de Alcântara Cardoso, Stephen F. Ferrari, Raone Beltrão-Mendes, Barth Wright, Torbjørn Haugaasen, Rose Marie Hoare, Susan M. Cheyne, Bruna M. Bezerra, Ikki Matsuda and Ricardo Rodrigues dos Santos

Introduction

As a number of chapters in this book show, flooded habitats, although often highly productive, can be tough places to live (see Chapters 4, 12 and 25). But, if it is hard for the primates, it can be even harder for the primatologists who wish to study them. One of the reasons why primates in flooded forests are so poorly known is the perceived and actual difficulties in working on them. This chapter delineates what these difficulties are and describes the innovative solutions that have been devised to solve them. Some of the contributions are full-sized methods, while others are more like tips from a colleague. It is hoped that all will serve as means of making fieldwork in flooded habitats, easier, more pleasant, effective and productive.

The chapter is organized into the following sections:

- *General accessibility* – moving in mud; moving in leaves, surveys in swamps; working from canoes; cutting and maintaining trails; clothing and protection
- *Measuring habitat and resources* – marking trees; fruit traps; ground surveys; food resources in mangroves
- *Technological advances: remote detection of primates* – camera traps; acoustic surveys; satellite monitoring; drones.

All of the ideas in this chapter are posted on the Pitheciine Action Group (PAG) website (https://sites.google.com/site/pagprimate/home) (a taxonomic group with many species that use flooded habitats), but readers are welcome to send in new ideas and comments to P.A.G., and so keep the content fresh and up to date.

General Accessibility

Moving in Mud

Beneath a covering of fallen leaves, the floor of unflooded seasonally inundated forest (such as the Amazonian *igapó* and *várzea*) is often consolidated by a mat of fine roots that is resistant (if springy) to the step. This is not the case with such habitats as mangroves and palm swamp, where mud generally dominates the substratum. Often of uncertain depth, but near-guaranteed glutinous consistency, safe passage across it requires skill and experience, especially as the mud often lies beneath a thin layer of concealing water.

The following is advice based on walking through Amazonian palm swamps, but should work for other swamps too: it is essential to always step on something solid, so try to avoid stepping into water, as the bottom is invisible and consists of mud which will suck your foot in; try to get at least one foot on a floating piece of wood, a root or clump of vegetation. Even if the floating piece of wood sinks it gives you more purchase in pulling the other foot out. A strong walking stick is very helpful: it improves balance, can be used to probe murky water landing spots, and assists when freeing yourself from mud (Figure 6.1).

If a standard transect is being followed, consider positioning long strong sticks at intervals into the mud. Linking other sticks or a cord at waist height between these uprights creates a supportive railing across the swamp. Finally, when you do go in, having a companion to help pull you out is extremely helpful.

Moving in Leaves

Once the water levels fall, seasonally flooded forests (like *igapó* and *várzea*) can dry out very quickly, leaving a deep layer of dry leaves which makes it very difficult to move silently through the forest. The solution for transect-based surveys is to sweep the transect with a broom, either during the survey, or on the return from it.

Surveys in Swamps

Suspended trails work well for long-term studies at a fixed locality (e.g. Cheyne 2007, 2010; Cheyne *et al.* 2013a,b; Cheyne & Tuttle 2011; Ehlers-Smith *et al.* 2013a; Harrison *et al.* 2010a). Because they walk on suspended trails, researchers avoid problems of contact with mud, do not damage crustacean and other taxa refuges, and the breeding and feeding places of shellfish. This, therefore, also alleviates impacts on the availability of potential prey species for the primates under study. Use of such trails also minimizes the impact on the growth of

Figure 6.1 Getting support from whatever is safe: partially flooded forest, Borneo. Photo: Ikki Matsuda. (A black and white version of this figure appears in some formats. For the colour version, please refer to the plate section.)

seedlings, reduced substrate compaction (such as in Bornean peat swamps: see Chapter 29) and, once the initial disturbance of construction is past, may be less intrusive to primates, facilitating habituation and study (Figure 6.2). Such methods are, however, restricted to sites where studies are transect based.

Boardwalk lengths and locations clearly depend on study site, but the system should be built above the maximum local daily or seasonal water level (tide or flood pulse), so allowing for continued study during high- and low-water periods. Logging can be avoided by having the trail system kink and curve to accommodate existing trees. In mangroves, the removal of a few props roots during the installation of the system does not compromise tree survival.

While deterioration of wood is less problematic for trails constructed in freshwater swamps and forests, burrowing bivalve molluscs ('shipworms', including the teredo worm (*Teredo navalis*, Teredinidae)) can quickly compromise the structural integrity of a wooden structure in marine swamps. Consequently, the suspended floor of any such trail should be made of either bound nylon ropes or strong netting. The supporting columns can be of wood, if they are clad in metal sheet (any wooden parts of the structure above the maximum water level do not require cladding). If using wooden or plastic planks, these should be constructed to overlap at the joints so as to provide stability. The surface can be covered in mesh to provide grip during the wet season. New staff/researchers need to be made aware that raised boardwalks/trails can be hazardous due to tripping and slipping.

Areas of low scrubby seasonally flooded forest (*chavascal* in the Brazilian Amazon) are problematic to survey when partly flooded as they are too dense to work with in canoes (see below), but lack strong supports for suspended walkways. Long transects (> 5 km) are not generally successful due to the low visability and the noise generated when moving through such habitats. A solution that increases the chance of contact (though not prolonged observation) is to use a large number of short transects across such study areas. Fallen logs can form bridges across most flooded areas (Cardoso 2015). Chest-waders can be used, though progress is often noisy and should weigh-up the risk of attack by snakes and crocodiles. Under such conditions, movement, followed by 5-minute listening and watching pauses ('point counts', a standard method in bird surveys: Bibby *et al.* 1998) is an alternative to standard speed surveys (see Barnett 1995; Ross & Reeve 2011).

Working from Canoes

In mangroves, canoe access is restricted to only those channels navigable during high tide. This makes the use of canoes impractical for studies that require continuous monitoring. However, canoes are extremely useful for studies of more open freshwater flooded forests where water levels change seasonally rather than daily, allowing access to subtle aspects of the ecology of populations inhabiting seasonally flooded forests. For example, in the Turtle Mountain region of Guyana, it has been noted that primates such as *Cebus olivaceus* increase insect and arachnid foraging in flooded areas as the water rises (B. Wright, unpublished data). Such changes in diet occur because such resources become more concentrated (and so more readily accessed) in the trees during flooding since invertebrates leave the forest floor as inundation proceeds (Adis 1984).

A small wooden or fibreglass canoe, 3–4 m in length and some 0.6–0.8 m wide provides a nimble but stable platform with which to negotiate flooded riparian forests (Figure 6.3a–d). Ideally, two people work a canoe, one as the pilot and one as observer/data recorder. While phenology and fruiting tree points can use the sliding tag design mentioned below, if trails are being used, mark guide trees at three different levels (the topmost at a level above the highest historical floodmark). Ensuring each guide tree is visible from two other trees should minimize the risk of getting lost in what can be a very visually confusing habitat.

Larger canoes are good for moving between study sites. Their extra stability means they are also to be preferred when working with fruit traps or collecting botanical specimens (where sampling generally involves pole-cutters). The effort needed to reach the canopy can be minimized by sampling during peak inundation (when many species are either fruiting or flowering: see Chapters 2, 4 and 27). It is best that such canoes have a flat bottom to facilitate specimen storage.

When surveying, remember to keep cameras and other electronic equipment in zip-lock plastic bags (see Figure 6.3d),

Figure 6.2 Short- and long-term walkways in flooded forest. (a) branches in Brazilian *várzea*; (b and c) boardwalks in Bornean flooded forest. Photos: (a) Nayara Cardoso; (b) and (c) OxTrop.

and everything in a kayak-sack (dry-sack) or two when in motion). Putting specimens (in their individual bags) inside a large polystyrene container should ensure they are saved if the canoe capsizes or sinks (a rare occurrence). Notebooks should be of waterproof paper (e.g. rite-in-the-rain™), or hand-made from waterproofed paper. Always tie your pen or pencil onto something (e.g. your notebook, shirt buttonhole, belt), that way, if you drop it in the water, you get it back.

Annotating one version of a photograph so that the subject, location and time are recorded is always a good idea (Barnett 1995). As balance, hands and space are at a premium in canoes, an easy and waterproof way to place such data in the photograph is to write it out on a small stick-mounted whiteboard. Writing on such boards with pens with water-soluble ink means they are instantly wipeable.

Canoes often have little freeboard and leaning over to retrieve floating seeds and flowers for samples can cause them to capsize or sink. To reduce the chance of destabilizing the canoe use a seed spoon to get samples, rather than reaching for them. This consists of a mesh bag sewn to a metal ring and mounted on a lightweight pole (Figure 6.4). These also minimize hand contact with water. This is made all the more desirable when, as in several freshwater systems, synchronized death of freshwater sponges have filled the water with spicules that can lead to sight-threatening eye infections (*cauixi* in Brazil – see Volkmer-Ribeiro *et al.* 2006). Consulting local medical professionals before starting fieldwork can provide information on any water quality issues. This can be important as waterways may have high levels of pollution which can cause skin irritation and infection, and many areas currently or formerly mined will have high levels of mercury (in water, sediment and fish).

A canoe in flooded forest is not as manoeuvrable as a person in unflooded habitat. You cannot simply move a canoe laterally as you yourself would step around a tree. Consequently, manoeuvring for visibility when the study animals move (e.g. out of sight behind branches as sakis constantly do), is not easy. But, if you design a sampling method with intervals that accommodates this difficulty, remember that statistical comparability demands that you use the method in all other habitats that the study embraces.

Both for reasons of safety and to give the observer maximum sampling opportunities, it is normal (indeed very highly recommended) to employ a local person as canoe pilot. However, unless they have worked with ornithologists or primatologists before, remember to explain beforehand that, because the canoe will angle, just because the pilot can see the animals does not mean the observer can see them too. In addition, most pilots have a tendency to try to get right up under the tree where the animals are, but this may not be where the observer has best visibility. Explaining such things in advance can save a lot of potentially lost data when in the field.

Finally, although paddling a canoe may be the surest way of navigating where water is shallow and/or vegetation is dense, in deeper water it is possible to use small electric outboards. Sold as 'trolling motors', they can be run from a car battery and, as they are near-silent, do not frighten animals with noise as do petrol-driven outboards. If such motors are not available or are difficult to import, a 4-stroke engine is much quieter than a 2-stroke (additionally, though initially more expensive, they consume less fuel). A 15 hp or 20 hp motor is appropriate for flooded forest work, as there is rarely a strong current.

Cutting and Maintaining Trails

With rare exceptions (e.g. Barnett 2010; Barnett *et al.* 2012b; Chapter 27), primates of flooded habitats do not forage close to the water surface, and restrict their activities to the canopy. Consequently, when cutting trails it may be wise to avoid the use of a machete-cut grid system and try to use natural

Figure 6.3 (a) Flooded forest workhorse – the wooden canoe; (b) native pilot and two biologists, Brazilian Amazon; (c) moving canoes through dense flooded forest, Peru; (d) a good bailing technique is also essential. Photos: (a, b and d) Adrian Barnett; (c) Janice Chism.

spaces for positioning trails. This minimizes damage to vines and overstory vegetation used by the study animals. For example, Martinez Mollinedo (2014) found that titi monkeys in flooded studies not only fed from vines, but used them as natural bridges. Avoiding damage to these important features, causes less disruption, and speeds the process of study group habituation.

Where researchers do choose to cut and maintain trails (e.g. in a grid system) then trail maintenance will need to track the rising and falling water levels, with rising water levels involving

Figure 6.4 A seed spoon. Very simple, but very effective as a means of sampling from water infested with sponge spicules (or any other unpleasant substance). Photo: Adrian Barnett.

Figure 6.5 'Mud moks', better than rubber boots when working in mangroves. Photo: Ricardo Rodrigues dos Santos.

cutting branches to permit access through the canopy, and falling water levels involving debris accumulated during the flood period. Such a process may involve substantially more effort than in non-flooded forests, especially if the work has to be carried out by canoe. Clearing fallen trees often requires chainsaws or manual saws which can cause disruption to the primates.

Protection and Clothing

For muddy habitats, the best footwear is just-below-the-knee-high rubber boots (the ones you can get in local markets). It is essential to get ones that fit tightly enough that they will not pull off when your feet get stuck in mud. Make sure you bring an extra pair in case of loss. When working in canoes, remove rubber boots on entering as, if the canoe capsizes, they can fill with water and drag you under.

In mangroves, rubber boots can be a liability because of the difficulty imposed by stilt roots. Under such circumstances, the neoprene shoes (Figure 6.5) used by divers combine great flexibility, firm grip and good protection against barnacles. In forests with spiny palms (e.g. the Neotropical genera *Astrocaryum* and *Bactris*), such shoes provide some protection from spines floating in the water.

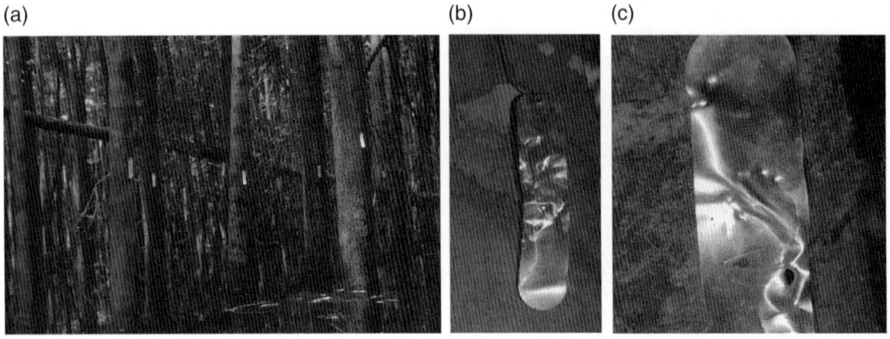

Figure 6.6 Tag vandals: (a) Tags in place in partially flooded *igapó* forest; (b) bites made on submerged tag made by big-headed river turtle (*Peltocephalus dumerilianus*) mistaking tags for fish; (c) bites made by squirrel monkey (*Saimiri sciureus*) on non-submerged tag. Photos: Adrian Barnett.

When walking in flooded forests where water is high, but mud largely absent (e.g. Bornean peat swamps), rubber boots may not be the best option as they can become filled with water at each step, reducing manoeuvrability and creating extra weight to be carried. Trainers (sneakers) which can allow water to drain quickly are good, ankle height boots which also allow water too drain out are recommended for people whose ankles require more support.

In mangroves, waterproof trousers protect legs from moisture, mud and mosquitoes, the latter two being abundant in such areas. Neoprene waders are a fully waterproof, puncture-resistant option that provide protection to waist-level. However, they are less flexible than a combination of thin waterproof trousers and neoprene divers shoes, and some models – designed for trout fishing in cool temperate streams – may become uncomfortably hot when used in the tropics.

In some flooded habitats, commensal ants can be common and often have ferocious stings. In the Amazon, for example, they inhabit the hollow stems of such legume genera as *Tachigali*, *Sclerolobium* and *Macrolobium*, and the bites are sufficient to deter almost all vertebrate seed predators (Barnett *et al.* 2015a). Not only are the bites unpleasant and distracting, they may induce strong allergic reactions in some people. It is worth learning to recognize such trees at a distance, as well as packing allergy-combating medicines as part of your field medicine kit (see Bearder & Nekaris 2011 for further details of field medicine).

Leeches are generally rare in the Neotropics, but can be disconcertingly common in some Paleotropical areas. Leech socks are available (see www.mysabah.com/wordpress/how-to-use-and-where-to-buy-a-anti-leech-sock/) to protect against leeches on the ground and in low vegetation. Leeches that drop from higher vegetation may be encouraged to disengage with a drop of preserving alcohol. Other sanguinovores, such as mosquitoes and sandflies, are often less abundant in blackwater and nutrient-poor ecosystems, but can be abundant in mangroves and nutrient-rich freshwater ecosystems (Junk *et al.* 2011; Knight 2011). For general comfort, preservation of sanity and minimization of the risk of malaria, it is important to reduce mosquito bite rates. One strongly recommended way is to put field clothes in a plastic trash (bin) bag, then spray-in aerosol insecticide so making a balloon; then tie-off the bag, shake and leave for the night. A similar effect can be obtained with the smoke from wood cook fires. Alternatively, clothes (and mosquito nets) can be soaked in pyrethrin-based liquids on a regular basis. These liquids are long lasting, bind well to most fabrics and have a satisfyingly quick knock-down time. Finally, applying lemon directly to affected bite areas will quickly reduce itching. In extremely fly-rich areas a headnet may be advisable.

It is important to note that in any flooded/muddy habitat, feet will be wet/submerged for long periods of time. After every day of fieldwork it is vital that feet are dried and remain dry for as long as possible to prevent such fungal infections as 'trench foot'. Talcum powder is good for helping feet stay dry.

Measuring Habitat and Resources

Marking Trees When Water Heights Vary Over Time

The need to mark trees for phenological studies and identification of feeding trees is a complication in flooded forest fieldwork, as the marked objects can be easily covered by a rise in water level. Putting several metal tags up the vertical length of the trunk risks more damage to the tree and uses a lot of tags. In addition, tags covered by water may not be usable the next season, because predators mistake them for fish and bite them to unreadability (Figure 6.6), while cleaning tags covered in aufwuchs (periphyton) is an expensive use of field time.

Using plastic tape to mark trees, encircling a trunk with it (Figure 6.7a, b), then moving and each circle of tape up or down and retrying it as the water levels change, is also a time-consuming process. Additionally, such tapes are often stolen by birds or squirrels for nesting material, or chewed by insects. A method used with success in Amazonian *igapó* has been a design based on the tighteners of tent guy ropes (Figure 6.8). Run from eye-hooks at lowest and highest water levels, the passage of a cord-under-tension through the tag allows it to be moved easily to a new level at each phenological checking. If water has covered the tag the presence of the cord allows it to be located easily and, since it is looped, raising it is easy and swift. All these advantages easily compensate for the greater investment in set-up time.

Carnivorous turtles often mistake submerged metal tags for fish (Figure 6.6); however, plastic tags, being matte, do not attract their attention. In addition, if used with the guy-based system mentioned above, plastic tags, being thicker, grip the cord better than do metal tags and so are less likely to back-slide.

Figure 6.7 (a) A flooded forest phenology plot with ID tags flagged with marking tape; (b) moving the tape up and down to accommodate water-level changes is a time-consuming task. Photos: Adrian Barnett.

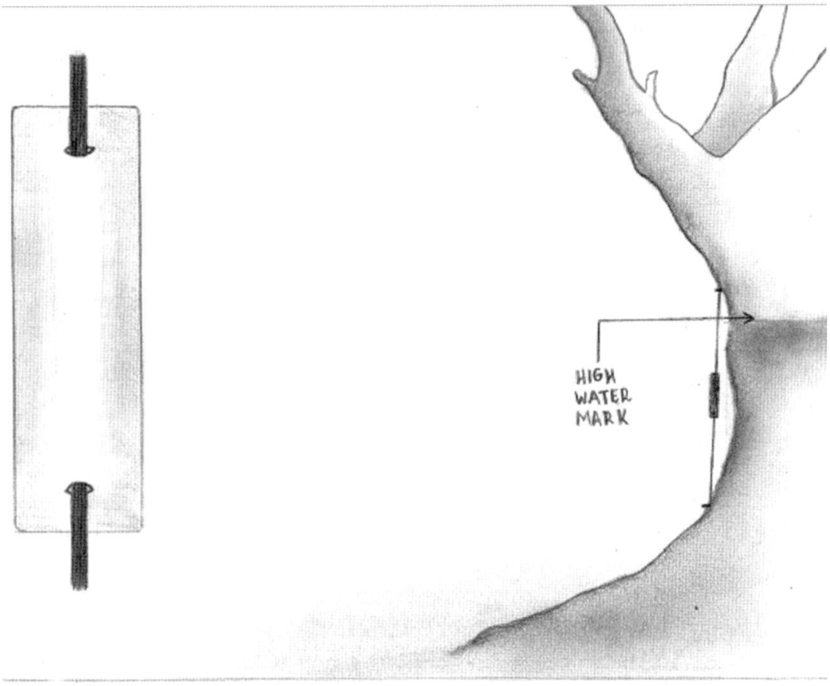

Figure 6.8 Design for tree-marking tag for use in flooded forest based on tent guy rope tighteners. Illustration by Richelly Andrade.

When marking trees in mangroves, tags should be secured with stainless steel nails like those used in sea-going wooden boats. These will not rust when in contact with salt. Tags themselves should be plastic. Plan for plenty of spares as primates that use mangroves are often both smart and curious (e.g. capuchins, macaques) and may remove tags as playthings.

Fruit Traps

To better understand primate abundance, distribution, behaviour and ecology, it is often necessary to quantify production of plant resources (fruits, seeds, flowers and leaves) across both spatial and temporal scales. Primate studies typically use either direct observations of tree and liana crowns (Chapman *et al.* 1992) or fruit traps (Peres 1994a; Stevenson *et al.* 2000).

There are a range of techniques for canopy and visual estimation of phenology and productivity (Barnett 2010, Table 6.1), but difficulties with visibility are a frequently cited problem (Barnett 1995; Ganzhorn *et al.* 2011; Stirling *et al.* 2013). With the use of canoes (see below, and Figure 6.3a, b) canopy observations may actually be easier to conduct in flooded locales, at least when high water levels coincide with fruit production. In contrast, while traps have been widely employed in upland forests to quantify fruit/seed- or flower-fall (Chapman *et al.* 1992; Wright & Calderón 2006), and the fall of leaves and fine litter (Clark *et al.* 2001), their use in flooded forests has proved much more problematic. Despite their potential underestimation of fruit production due to the prior consumption of fruits/seeds by arboreal frugivores in the canopy (Terborgh 1983), and their unsuitability for recording unripe fruit (which represents an important resource for specialist consumers such

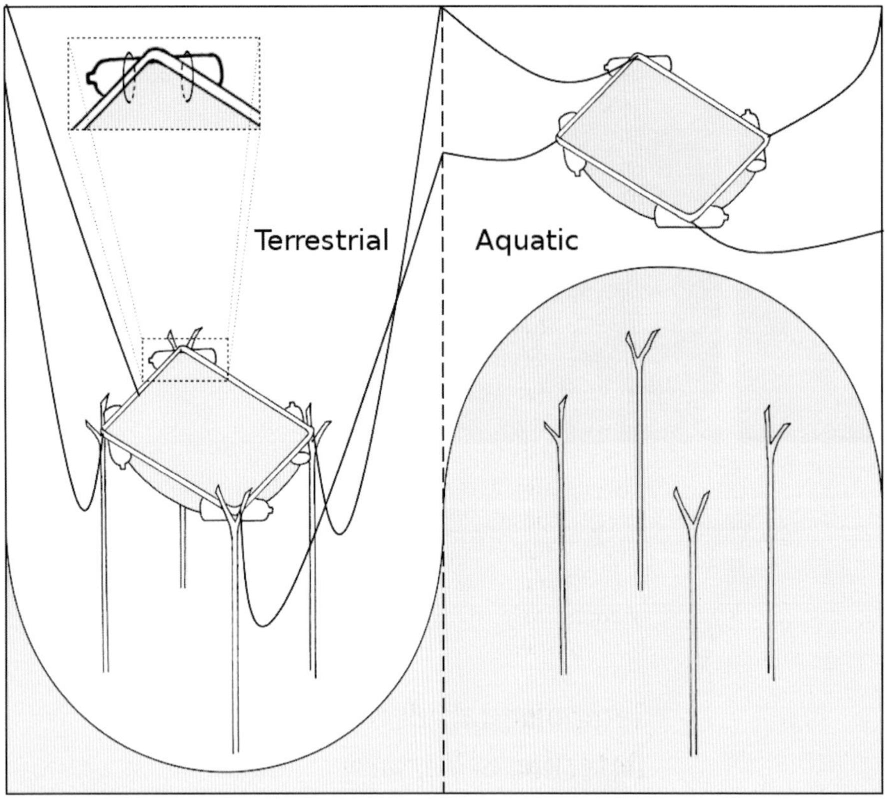

Figure 6.9 Floating fruit/seed trap design used in seasonally flooded forests: supported at a height of 1 m during the terrestrial phase but floating with fluctuating floodwaters during the aquatic phase. Inset shows empty plastic bottle attached to trap frame with wire loops.

as the Pitheciines: Barnett 2010; Pinto *et al.* 2013), traps do not suffer from the high levels of observer biases (Barnett 1995; Ganzhorn *et al.* 2011) found in direct crown observations and represent an important complementary method, which has been under-utilized in flooded forests to date.

Traps have been employed in the tidal forests of the Amazonian estuary (Cattanio *et al.* 2004), and in the upper Amazon, where they were strung between branches during high water-level periods (Nebel *et al.* 2001). But the greater flood depths in the central Amazon provide far more of a practical challenge, and field attempts at trap deployment have been met with variable success (Haugaasen & Peres 2005b; Schöngart *et al.* 2010). Attempting to adapt methods to habitats where flood depths may reach 10–15 m and inundation lasts for up to 210 days (Parolin *et al.* 2004; Chapter 2) is clearly not trivial, especially as the annual water level is variable across the annual flood pulse (Junk *et al.* 1989). Increasing the height of trap supports or suspending traps from tree branches as the water rises may require repeated maintenance and modification. One possible solution is a floating trap that can be constantly adjusted to the current water level in order to track tree phenology and fruit productivity in the seasonally flooded forests. This worked successfully in *várzea* forests, western Brazilian Amazonia (Hawes 2012; Hawes & Peres 2016). It is described below.

The floating trap is a standard square and constructed of fine polyester mesh with PVC tubing support (Stevenson & Vargas 2008). It has a collection area of 0.5 m^2 (0.71 × 0.71 m) and is supported 1 m above the ground. To cope with the seasonal fluctuation of floodwaters in *várzea* forest, buoyancy is added to this basic design with four empty, water-tight 2-litre plastic bottles, one at each corner of the trap, to maintain the polyester mesh above water. Bottles are attached to the frame with wire loops and then, to stabilize the trap's vertical position, the trap is tied loosely with string to the upper branches of surrounding vegetation. If cords are slack, the bottles will rise and fall on their supports with the changes in floodwater level (Figure 6.9). The standard design can be modified for deployment in both flooded and unflooded areas (Figure 6.9), ensuring methodological continuity and, hence statistical comparability, between habitats.

A method reported by Hawes (2012) and Hawes and Peres (2016) placed traps at 100 m intervals along 1 km transects arranged in a grid within two 100 ha plots. Material was collected from traps twice monthly (by canoe during the aquatic phase in *várzea*); dried for 72 h, and separated by plant part into fruits and seeds, flowers, leaves and twigs; each fraction was then dry-weighed separately (using electronic scales with 0.01 g precision). All fruits and seeds were retained for identification and herbarium deposition. This method was employed for 12 consecutive months in a study comparing the plant phenology of adjacent flooded (*várzea*) and unflooded (*terra firma*) forest.

The floating trap method described above succeeded in collecting fallen plant material throughout the year, with minimal adjustment of the ties to branches in surrounding vegetation required as the water level rose or fell. Any occasional damage to traps (e.g. from fallen tree branches) was not more prevalent in *várzea* than using standard traps in *terra firma* forest, and plastic bottles were easily replaced when necessary. Collected plant material was used to assess the fruit and seed resources available to frugivores in both flooded and non-flooded forest sites (Hawes & Peres 2014), and to calculate monthly variation in fruit production and litterfall (Hawes 2012; Hawes & Peres 2016).

Ground Fruit Surveys in Swampy Areas

Seasonally flooded forests may occupy substantial areas (Wittmann *et al.* 2004). However, forests along the margins of watercourses can be difficult to work in because they often occur as short, narrow, highly meandering strips, so that the standard quadrant methods cannot be used, as forest width may be less than required to establish such plots. Assessing productivity and phenology can also be complicated under such conditions, especially in the wet season. For example, in Bornean peat swamp forests, the time required to cover any distance becomes much greater when the forest is fully flooded, so that some areas become effectively inaccessible to sampling. However, a method developed during the study of swamps dominated by the buriti palm (*Mauritia flexuosa*, and known in Brazil as *buritizais*) should also be applicable in other types of swamp forest.

The traditional view in vegetation sampling is that large intervals between sampling stations along an extensive transect (e.g. 10 m intervals on a 100 m transect), increases the chances of adding new species to the sample, and so more effectively capture forest heterogeneity (Muller-Dombois & Ellenberg 1974). But, because *buritizais* are often narrow short strips of forest inserted within other major forest types, it is more effective to establish much shorter transects and sample intervals (e.g. an interval of 4 m along a 40 m transect, so as to total ten selected trees and be consistent with the methodology of Muller-Dombois and Ellenberg (1974)). In addition, instead of a straight line, the transect should follow the elevation contour, so controlling for any variation due to different soil types and/or inundation durations (Magnusson *et al.* 2005).

The use of fruit traps to assess fruit productivity (see above) is inviable when a forest's small size made it hard to position a sufficient number of traps. Under such conditions a raked-ground fruit survey can be applied (Guillotin *et al.* 1994; Sabatier 1985; Zhang & Wang 1995a), where dead leaves and debris are cleared from 40 × 10 m rectangles constructed with string-and-stakes and positioned at one-per-swamp along a 40 × 30 m-wide survey path 10 km in length. Fallen fruit can be removed from the sampled area after each census. Monitoring fallen fruits densities on a weekly basis, reduces risks that deposited fruit will be lost to visiting frugivores or strong rains. In flooded periods, if the water rarely exceeds 1 m in depth, plastic sheeting suspended on stakes hammered into the ground, is a viable modification.

Such modified methods have been successful in assessing community-wide patterns of phenology and fruit productivity in *buritizais*, and assessing the importance of the habitat to local wildlife, including primates (Mendes Pontes & Chivers 2007). It is hoped that they will also be applicable to similar forest types elsewhere in the tropics.

Studying Food Resources in Mangroves

Mangroves possess an abundant and diverse vertebrate and invertebrate fauna, and this is attractive to a variety of foraging primates (see Chapters 7 and 8). However, quantifying availability of such resources is complicated by the presence of mud and above-ground roots. When sampling mud-dwelling potential prey species (notably annelids, crabs and fish), it is possible to conduct hole counts (providing the makers can be distinguished). If it becomes necessary to capture animals for identification, this can be done by inverting a large plastic jar over the animal. With seawater added, this can then be used to store the animal. A screw-top is easier to handle under field conditions.

If field methods call for marking food resources, it is best to use waterproof fluorescent paint, making the marks so they are visible at any relative angle between animal and observer. If a shell is being marked, it is best to make more than one mark, to avoid data loss from natural abrasion. As a downside, such marked animals may be preferentially predated (LeBreton *et al.* 1992). Tape and tags (preferably plastic) can be used to mark animals, but are removed by some species of primates, such as capuchin monkeys. Tape also has the disadvantage, if it falls off, of adding to localized or general marine pollution, and perhaps being mistakenly consumed by marine life.

Technological Advances: Remote Detection of Primates

Camera Traps

With rare exceptions (Schipper 2007), camera trapping has proven highly effective in detecting species hard to find by other methods (Bezerra *et al.* 2014; O'Connell *et al.* 2011; Silveira *et al.* 2003; Srbek-Araujo & Chiarello 2005; Figure 6.10a), and even been responsible for the discovery of new primate species (e.g. *Macaca leucogenys*: Li *et al.* 2015). Left in place near fruiting trees, camera traps can provide a valuable record both of the temporal patterning of resource use and of the suite of species that are visiting it (Jansen & Den Ouden 2005). In surveys, they are especially useful in habitats that are difficult to access, such as mangroves, where human passage can be highly disruptive and noisy (Bezerra *et al.* 2014; Nowak 2012). To further diminish interference caused by human presence, it is possible to use a wireless- (Mainwaring *et al.* 2002; Simasathien *et al.* 2015), or Wi-Fi-based system (O'Connell *et al.* 2011), and access the cameras with a tablet or mobile phone (Nichols *et al.* 2011). In addition, control of cameras using satellite connections allows researchers to remotely control video cameras by tilting the camera and zooming or panning to animal subjects. Remotely operated robotic cameras are another option (Swann *et al.* 2011).

Species accumulation curves indicate that around 400–500 camera days are required to capture the majority of common mammal species in an area, and Tobler *et al.* (2008) obtained images of 86% of the mammal species known from their study area in two months of camera trapping. However, high levels of survey effort may be required, and all habitats should be surveyed, for an inventory to be considered complete (Tobler *et al.* 2008).

Combining the use of camera traps and feeding stations allows species presence to be recorded (see figures in Bezerra

Figure 6.10 Camera traps in mangrove forest, coastal Brazil. Note (a) the heavy armature and monkey-resistant chain, and (b) the use of a flat topped stilt root as a baiting site. Photos: Ricardo Rodrigues dos Santos.

et al. 2014 and Kierulff *et al.* 2013). If the camera traps takes continuous images (rather than just time-limited snapshots), it is possible to record how the animals process different food items, as well as social interactions and other behavioural aspects. Even though provisioning risks altering behaviour, such remote recording mechanisms may be the only way to register primates in some poorly accessible and dense habitats. If there are a network of camera traps and feeding stations, it becomes possible to track where the animals are ranging by the timing of the visits to stations or sequential passage in front of various cameras. When positioning camera traps whether tied to branches or trunks, or atop one or more poles (Figure 6.10a), care should be taken to remember to position them above the highest water level. In mangroves, the strictures against the use of wood in construction of supportive platforms (mentioned above for walkways) will apply. As always, to avoid 'ghost images', make sure the stage in front of the camera is unobstructed by vegetation (Figure 6.10b). Méndez-Carvajal (2014) provides a simple, practical and well-illustrated guide to methods that allow the installation of camera traps in lower and mid-canopies without the need to climb trees. These methods can be adapted for high canopy work.

Remote Sensing with Acoustic Surveys

Playback has long been widely used in bird surveys to establish the presence of particular species (Gregory *et al.* 2004). In primates, playback was initially used in investigations of territoriality (e.g. Chivers & MacKinnon 1977; Raemaekers & Raemaekers 1985), communication studies (Cheyney & Seyfarth 1982; Wich *et al.* 2002; Windfielder 2001; Zuberbühler 2000a, b), predator recognition (Hauser & Wrangham 1990; Macedonia 1990), and diagnosing species/subspecies differences (Davila Ross & Geissmann 2007; Méndez-Cárdenas *et al.* 2008; Zimmermann 1990). However, it can also be used to detect the presence of primate species, either because they are so rare that observation rates are low (e.g. Dacier *et al.* 2011; Peck *et al.* 2011; Marques *et al.* 2013), or because the habitat to be surveyed is difficult logistically. Studies in the latter category have included mangrove and flooded forests (Bastos *et al.* 2015; Bezerra *et al.* 2008, 2010a,b; Hilário & Ferrari 2015; Savage *et al.* 2010). Recording of calls without playback have also been used in flooded forest primate surveys (Buckley *et al.* 2006).

For longer term, passive monitoring and acoustic sensors can be used (see Kalan *et al.* 2015 for primates, Haselmeyer & Quinn 2000; Kays & Alison 2001; Thompson *et al.* 2010a,b; Wimmer *et al.* 2013 for related papers on other taxa, and Marques *et al.* 2013 for a general review). While the disadvantage with these, clearly, is that the resulting soundscapes have to be listened to, the amount of time expended can be limited by setting triggering frequencies in analytical software.

The use of automated bioacoustic recorders associated with call classifying software has proven more efficient than experienced human surveyors for some species (e.g. Zwart *et al.* 2014, for nightjars, a group that is particularly difficult to survey). This technique is still in its infancy for primates, but has shown some promising results for chimpanzees, Diana monkeys, king colobus and western red colobus (Kalan *et al.* 2015; Heinicke *et al.* 2015). The Wildlife Acoustics Company has a range of products and software to facilitate passive acoustic surveys (www.wildlifeacoustics.com). The method looks likely to become a powerful non-invasive and efficient tool to investigate elusive, rare and highly threatened primate species (Figure 6.11).

Remote Sensing with Satellite Tags

The use of Global Positioning System (GPS) collars to remotely report primate positions, and so assess movement patterns and habitat preferences is well established (Greuter *et al.* 2009; Ren *et al.* 2008), though in earlier studies it was conducted only in open habitats (e.g. Markham & Altmann 2008; Pebsworth *et al.* 2012; Sprague *et al.* 2004). However, the technology has improved sufficiently that primates in dense vegetation, including flooded forests can now be investigated by this means. For example, a recent study by Stark *et al.* (2014) used collar-mounted GPS units to track the movements of proboscis monkeys in the Kinabatangan floodplain area of Sabah, Malaysian Borneo. Collars stayed on for up to a year and had an average fix time of 43 seconds.

Remote Sensing with Drones

Drones (Unmanned Aerial Vehicles – UAVs) are small, lightweight, stable, remotely controlled and low cost in comparison to the benefits gained. They provide a mobile image acquisition system providing data that can be viewed and remotely stored in mobile phone, tablet or computer. They can be used to

Figure 6.11 Passive Acoustic Survey unit in coastal Brazilian mangroves (with signs to protect against the curious). Photo: Bruna M. Bezerra.

monitor and measure the speed of animal movements between and within habitats. They can do this using smaller time intervals, more accurately and with lower costs than the data acquired from high-resolution satellite images. The equipment is ideal for monitoring the environment because the images can be acquired at a time under the control of the observer, without interference from clouds and changes can be analysed in near real-time. Depending on the type of environment and the species studied, the equipment can also help locate high canopy-dwelling primates (such as howler monkeys) or their nests (e.g. orangutans and chimpanzees) (van Andel *et al.* 2015), and analyse habitat associations (Chabot & Bird 2014; van Gemert *et al.* 2014). They are also useful for monitoring areas that are difficult or arduous to access by foot or boat (such as flooded habitats) (Chabot & Bird 2014; Vermeulen *et al.* 2013), checking for presence of clearings (Getzin *et al.* 2014), and detecting human actions that jeopardize conservation, such as deforestation, burning (Paneque-Gálvez *et al.* 2014) and poaching (Mulero-Pázmány *et al.* 2014). It is also possible to quantify the vegetation, map land cover known to be favoured by a particular species (Szantoi *et al.* 2017) and to distinguish vegetation types and possibly food resources in the treetops (Koh & Wich 2012). Drones have been used extensively, and the field is one of rapidly expanding technological capacity and methodological possibility (Whitehead & Hugenholtz 2014; Wich 2015; and see http://www.ted.com/talks/lian_pin_koh_a_drone_s_eye_view_of_conservation from conservationdrones.org). However, they (currently) have limited flight time, and are dependent on energy sources available in the field to power and recharge the batteries.

An alternative to an off-the-shelf model is the conservation drone, an autopilot system (APM) developed by diydrones.com, a free-access online community. This combines the APM with open-source mission planning software (APM Planner), making it possible to upgrade off-the-shelf model airplanes to fully functioning conservation survey drones. The cost is further kept down by choosing simple and inexpensive airframes, which also saves on maintenance costs (Koh & Wich 2012).

There appears to be an unspoken assumption that drones are benign and do not influence animal behaviour in any way, possibly because they do not resemble any known predator and do not provoke alarm responses (Mourté & Barnett 2014). However, using cardiac biologgers, Ditmer *et al.* (2015) found that North American black bears (*Ursus americanus*) exhibit a stress response to drone flights, with magnitudes of heart rate spikes correlated to the proximity of drones. Such reactions occur even though the bears rarely display an obvious behavioural response, as measured by GPS collars, to such flights (Ditmer *et al.* 2015). That primates can be in some way bothered by drones (perhaps thinking of them as large flying insects) comes from a report by Van Hoof and Lukkenaar (2015) that captive chimpanzees at Royal Burgers' Zoo in the Netherlands used long sticks to swat from the air a drone that was filming them at close range. These reports have clear implications for primate research, especially for long-term studies.

Final Words

It is hoped this selection of tips, suggestions and methods has been helpful. There are many more ideas out there, both because technology is developing quickly and because primatologists are an innovative crowd who, like their study subjects, are constantly inventing new solutions to the problems they face. While some, such as new developments in drone or camera-trap technology, fit the consensus of what is appropriate for publication, many of the tweaks and tips do not. Accordingly, please share yours at the PAG website (https://sites.google.com/site/pagprimate/).

Acknowledgements

The authors thank: Thomas Defler, Eckhard Heymann, Rose Hores, Shawn Lehman and Marilyn Norconk and an anonymous reviewer for their inputs and collective kindnesss.

Chapter 7

Worldwide Patterns in the Ecology of Mangrove-living Monkeys and Apes

Katarzyna Nowak and Rebecca Coles

Introduction

Our knowledge of primate socioecology is biased towards tropical *terra firma* forests. With limited access and visibility, mangrove-dwelling primates present a challenge for monitoring and research (Chapter 6). Here, we attempt to review some convergent patterns in primates' use of mangroves across the globe using information drawn from chapters in this volume, as well as data collected by means of a short questionnaire sent to this book's contributors and to the IUCN Primate Specialist Group. In addition, we contacted individual published researchers working on mangrove-dwelling primates who provided us with *ad hoc* information. Our chapter is intended to set the scene for other chapters in this section, and provide a gross overview of primate natural history in mangroves highlighting some of the behaviour unique to this habitat.

Questionnaire

Questionnaire participants were asked to list the primate species and site(s) where mangrove habitat is used, and specify whether they observed the species in mangrove habitat firsthand. They were asked about the extent to which and ways in which the primate species used mangrove, including for ranging, refuging, sleeping, travelling, swimming and feeding. Via the questionnaire, we received data for 25 taxa, mostly from Africa (Table 7.1), with more than one questionnaire submitted for nine of these taxa from independent authors. 'Pers. comm.' is used to cite data acquired by means of our questionnaire (Table 7.1). We also accepted data on an *ad hoc* basis from authors outside the context of the questionnaire (seven species indicated with an asterisk in Table 7.1), and incorporated data on five species (i.e. western gorilla, *Gorilla gorilla gorilla*; Bornean orangutan, *Pongo pygmaeus*; southern pig-tailed macaque, *Macaca nemestrina*; blonde capuchin, *Sapajus flavius*; yellow-breasted capuchin, *S. xanthosternos*) described to use mangrove in this volume (Chapters 4, 10, 11 and 12). For lists of species occurrence in mangroves in the Neotropics and Africa, refer to Chapters 9 and 12.

Results

Types of mangrove use by 36 primate taxa are presented in Table 7.1, including data on 17 primate species in Africa (12 genera), 12 in the Americas (6 genera) and 7 in Asia (4 genera). No apparent phylogenetic pattern is evident, as primates across all taxa exploit mangroves from the pygmy mouse lemur (*Microcebus myoxinus*; J. Ganzhorn, pers. comm., 2014) to the western lowland gorilla (Chapter 12). The extent to which mangroves are used varies across locations and species, with seasonal use, opportunistic feeding and refuge-seeking emerging as dominant incentives. We break down types of use into several categories: ranging and refuging, resting, moving and travelling (including swimming), and feeding and foraging (Table 7.1).

Mangrove Use Extent: Ranging and Refuging

There appear to be no truly obligate mangrove users, perhaps with the exception of proboscis monkeys (*Nasalis larvatus*; Salter *et al.* 1985; A. Bennett, pers. comm., 2014; V. Nijman, pers. comm., 2014; Chapter 13) and possibly long-tailed macaques (*Macaca fascicularis*; I. Matsuda, pers. comm., 2014). Some local populations of capuchins (*Cebus* and *Sapajus* spp.) in the Americas, guenons (*Cercopithecus* spp.) and colobus (*Procolobus* spp.) in Africa, and macaques and lutungs (*Macaca* and *Trachypithecus*) in Asia appear to specialize on mangroves, but some of these taxa may still possess preferences for upland *terra firma* forest for forage, and/or be unable to subsist exclusively in mangrove. Some species such as the yellow-breasted capuchin (*Sapajus xanthosternos*; G. Canale, pers. comm., 2014) in Brazil (Sitio do Conde, Bahia; unprotected area) and the pygmy mouse lemur in Besalampy, Madagascar (J. Ganzhorn, pers. comm., 2014) live in mangroves permanently, while others such as the Geoffroy's spider monkey (*Ateles geoffroyi*) in El Salvador make seasonal use of mangrove (K. Morales Hernández, pers. comm., 2014).

In some areas, such as in Balikpapan Bay (S. Lhota, pers. comm., 2014), there is a sharp distinction between mangrove and non-mangrove habitats, while in other areas, such as Lower Kinabatangan or Tanjung Puting, transitional forest makes this gradation less dramatic. Isolation of mangroves and the sharpness of the transition from mangrove to non-mangrove (Chapter 2) may be associated with and help to explain the types and extent of mangrove use by resident primates.

We ask: What patterns emerge for areas where groups or populations specialize or spend a disproportionate amount of time in mangroves that help explain mangrove use at these sites (e.g. some groups of the Zanzibar red colobus (*Procolobus kirkii*) spend up to 85% of their time in the mangroves; Nowak 2008)?

Table 7.1 Thirty-six mangrove-using primate species for which we were able to gather data on ranging or on at least one type of activity/behaviour in mangrove habitat.

Region	Species	No. Authors reporting sp. in mangrove	No. Sites at which reported	Ranging (core, peripheral, both)	Refuging (natural and/or human risk)	Resting (day, night, both)	Moving/ Travelling (strata if specified)	Swimming (context if specified)	Feeding (diet items if specified)	Reference[a]
Africa	*Cercocebus torquatus* **Red-capped mangabey**	1	1	Both	NA	NA	Upper and lower strata	Swims when fleeing	Shellfish/crabs, insects, leaves, fruit	J. Head
	Cercopithecus mitis **Blue monkey**	3	> 5	Both	Yes	Yes	All strata	No	Yes	T. Butynski & Y. de Jong; Nowak, pers. obs.; W. Jubber & J. Trindade
	Cercopithecus mona **Mona monkey**	1	1	Core	NA	NA	NA	Crossing a 4-m wide creek in Cameroon mangrove swamp	NA	Gartlan and Strusaker 1972; T. Struhsaker
	Chlorocebus sabaeus **Green monkey**	1	2	Both	Yes	Yes	Yes	Yes, up to 100m at high tide	Fiddler crabs (*Uca tangeri*), *Rhizophora* sp.	A. Galat-Luong & G. Galat (also Chapter 39)
	Chlorocebus pygerythrus **Vervet monkey**	2	4	Peripheral	NA	NA	Upper and lower strata	NA	Yes	T. Butynski & Y. de Jong; M.J.F. de Silva
	Colobus angolensis palliatus **Tanzanian black-and-white colobus**	1	4	Peripheral	NA	NA	Yes	No	*Rhizopora mucronata, Heritiera littoralis, Ceriops tagal* young leaves, leaf buds	Anderson *et al.* 2006; J. Anderson
	Erythrocebus patas **Patas monkey**	2	4	Peripheral	NA	NA	Lower strata	No	Fiddler crabs (*Uca tangeri*)	A. Galat-Luong & G. Galat (also Chapter 39)
	Gorilla gorilla gorilla **Western gorilla**	1	1	Peripheral	Yes, from humans	NA	Lower strata	NA	NA	Chapter 12
	Lemur catta **Ring-tailed lemur**	1	>1	Core	Yes	No	Upper and lower strata	No	*Avicennia marina* leaves and fruits	Y. Jacky
	Microcebus myoxinus **Pygmy mouse lemur**	1	1	Core	NA	NA	NA	NA	Insects	J. Ganzhorn
	Otolemur garnettii **Northern greater galago**	2	1	NA	NA	NA	Yes	No	NA	T. Butynski & Y. de Jong; A. Perkin
	Papio cynocephalus **Yellow baboon**	1	3	Both	NA	NA	NA	NA	NA	T. Butynski & Y. de Jong

	Papio papio **Guinea baboon**	2	>1	Both	Yes	Yes	Yes	No	Crabs and mangrove flowers	A. Galat-Luong & G. Galat (also Chapter 39); M.J.F. de Silva & C. Casanova
	Procolobus badius temmincki **Temminck's red colobus**	1	1	Peripheral	Yes	NA	Lower strata	No	NA	A. Galat-Luong & G. Galat (also Chapter 39)
	Procolobus kirkii **Zanzibar red colobus**	2	3	Both	Yes	Yes	Upper and lower strata	No	*Rhizophora mucronata, Sonneratia alba, Ceriops tagal, Bruguiera gymnorhiza, Avicennia marina* young leaves, mature leaves, flowers, fruits	Nowak 2007, 2008; T. Butynski & Y. de Jong
	Propithecus coquereli **Coquerel's sifaka**	1	1	NA	NA	Yes	Yes	No	NA	M. Markolf
	Propithecus coronatus **Crowned sifaka**	1	1	NA	NA	Yes	Yes	No	NA	M. Markolf
Americas	*Alouatta palliata* **Mantled howler monkey**	1	1	Core	Yes	Yes	Upper and lower strata	No	Yes	L. L. Bridgeman
	Alouatta pigra **Yucatán black howler**	1	1	Core	Yes	Yes	Upper and lower strata	No	All food items consumed are in mangrove, e.g. *Lonchocarpus hondurensis* leaves and flowers	L. L. Bridgeman
	Alouatta seniculus **Red howler monkey**	1	1	Peripheral	NA	NA	Yes	No	Yes	O. J. Linares & B. A. Rivas
	Alouatta ululata **Maranhão red-handed howler**	1	2	NA	Yes	NA	Yes	NA	Yes	T. Pinto
	Ateles geoffroyi **Geoffroy's spider monkey**	1	2	Peripheral	NA	No	Yes	NA	Yes, seasonally	K. Morales Hernández
	Cebus capucinus **White-throated capuchin**	1	1	Both	No	Yes	Yes	No	Insects	M. Baker

(*continued*)

Table 7.1 (cont.)

Region	Species	No. Authors reporting sp. in mangrove	No. Sites at which reported	Ranging (core, peripheral, both)	Refuging (natural and/or human risk)	Resting (day, night, both)	Moving/ Travelling (strata if specified)	Swimming (context if specified)	Feeding (diet items if specified)	Reference[a]
	Cebus olivaceus **Wedge-capped capuchin**	1	1	Peripheral	Yes	NA	NA	NA	Insects, spiders and small vertebrates	O. J. Linares & B. A. Rivas
	Macaca mulatta **Rhesus macaque**	1	1	Core	Yes	NA	Yes	Yes	NA	J. Fellowes
	Sapajus apella **Tufted capuchin**	1	>1	Both (usually peripheral)	NA	Yes	Upper and lower strata	No	Main items are molluscs, but also crabs (*Ucides cordatu*)	R. R. Santos
	Sapajus flavius **Blonde capuchin**	1	1	Peripheral	NA	NA	Yes	No	Lizards, flowers (e.g. *Calophyllum brasiliense*), leaves (e.g. *Tapirira guianenses*), fruits (e.g. *Chrysobalanus icaco*)	Chapter 10
	Sapajus libidinosus **Bearded capuchin**	1	2	Core	No	Yes	Upper and lower strata	No	Yes, main items are molluscs and crabs	R. R. Santos
	Sapajus xanthosternos **Yellow-breasted capuchin**	2	>1	Both	Yes	Yes	Yes	NA	Crabs (from traps)/ oysters, shipworm (*Neoteredo reyni*), social insects, red mangrove propagules and coconuts (*Cocos nucifera*)	Chapter 11
Asia	*Macaca fascicularis* **Long-tailed macaque**	3	12	Both	No	Yes	All strata	Yes; crosses rivers and goes far into sea for crabs	Yes; fruit of *Sonneratia caseolaris*; fruit of *Nypa frutican*; fruit of *Ficus microcarpa*	S. Lhota; V. Nijman; I. Matsuda
	M. mulatta **Rhesus macaque**	1	5	Both	Yes	Yes	Yes	Yes	Main mangrove spp.; honey; eggs of birds, turtle and crocodile; insects, molluscs, crabs and fish	J. Kumar- Mallick
	M. nemestrina **Southern pig-tailed macaque**	1	1	Peripheral	No	NA	All strata	NA	NA	I. Matsuda

Nasalis larvatus **Proboscis monkey**	4	> 6	Both	Yes	Yes	All strata	Yes, frequently	*Rhizophora* young leaves a key seasonal fallback; young leaves of *Xylocarpus granatum* and *Sonneratia alba*; young leaves, fruit, flowers of *S. caseolaris*; propagules of *Bruguiera* sp.	E. Bennett; S. Lhota; V. Nijman; Chapter 4
Pongo pygmaeus **Bornean orangutan**	1	1	Both	No	Yes	All strata	No	Shoots of *Nypa fruticans*	I. Matsuda
Trachypithecus auratus **Ebony leaf monkey**	1	5	Core	No	Yes	Mostly upper strata	No	Yes	V. Nijman
T. cristatus **Silvery leaf monkey**	3	6	Both	No	Yes	Mostly upper strata	Yes; for crossing rivers/ creeks; not long-distance	Yes	S. Lhota; V. Nijman; I. Matsuda

[a] Unless the year is provided, the source of information is a personal communication via our questionnaire or *ad hoc*.

Habitat disturbance and hunting are two anthropogenic threats that drive mangrove usage in the form of 'refuging' (Chapter 39). Mangroves are used by red colobus monkeys (*Procolobus* spp.) in Senegal and Zanzibar (Galat-Luong & Galat 2005; Nowak 2012), and by ebony leaf monkeys (*Trachypithecus auratus*) in Java, Bali and Lombok (Chapter 14) because of terrestrial forest disturbance. The Yucatán black howler and mantled howler monkeys (*Alouatta pigra* and *A. palliata*) are now restricted to the mangroves of the Mexico Pantanos de Centla Biosphere Reserve because lowland forest in the area was cut down to create more pastureland (Bridgeman 2012). A similar scenario is observed in southern Bahia where yellow-breasted capuchins have become permanent residents of mangroves due to the loss of remnant *terra firma* forests (Chapter 11).

In Brazil, high hunting pressure has driven the Maranhão red-handed howler (*Alouatta ululata*) into mangrove (T. Pinto, pers. comm., 2014). According to E. Bennett (pers. comm., 2014), a significant part of proboscis monkey range today includes mangroves because of hunting in other habitats. Guinea baboons (*Papio papio*) retreat into mangroves in Cantanhez Woodlands National Park, Guinea-Bissau because of bushmeat hunting and people's use of their skins for traditional medicine (M.J.F. da Silva, pers. comm., 2014). While these baboons spend most of their time in mangroves, where they become quiet upon detecting humans, they do occasionally venture out to raid crops (M.J.F. da Silva, pers. comm., 2014). Similarly, gorillas retreat into mangroves to escape humans, but in this case, the humans are researchers in Loango NP, Gabon (Chapter 12).

Proboscis monkeys and sympatric primates use riverine and mangrove habitat for night-time roosting providing long-range visibility against feline predators, such as leopards (Feilen & Marshall 2014; Matsuda *et al.* 2011). The rhesus macaque (*Macaca mulatta*) in the Tiger Reserve, Sundarbans use mangrove to escape tigers, their natural predator, and take shelter high up in mangrove trees such as *Sonneratia apetala* (Lythraceae), an emergent in the area (J. Mallick, pers. comm., 2014). However, predation – aerial predation in particular – has also been proposed as one hypothesis explaining the lack of lemurs that inhabit or specialize on mangroves (Chapter 5), suggesting that mangroves are not always 'safe havens' from human or natural predation risk. Interestingly, the dense vegetation of mangroves may provide a suitable niche for conspecific avoidance. For example, rhesus macaques in Cayo Santiago, Puerto Rico, notably young solitary males, have been observed to enter mangroves to avoid conflict, and moribund individuals may withdraw to the dense vegetation afforded by mangroves to die (J. Fellowes, pers. comm., 2014).

Finally, some species, such as the green monkey (*Chlorocebus sabaeus*; G. Galat & A. Galat-Luong, pers. comm., 2014), use mangroves to escape biting insects. The ring-tailed lemur (*Lemur catta*) in Tsinjoriake, southwestern Madagascar also rest in mangroves because they provide cool shelter from heat (Y. Jacky, pers. comm., 2014). Thus, thermoregulatory benefits can also spur mangrove use despite other constraints, such as those on movement and diet, discussed next.

Movement: Navigating Available Strata

The prop roots of the mangrove tree genus *Rhizophora* (Rhizophoraceae) offer a tangled lower stratum that some arboreal primates exploit. For example, the Senegal and Zanzibar red colobus become more quadrapedal when moving along these prop roots at low tide, ascending into the canopy at high tide or when under threat. Other species including proboscis and Sykes's (*Cercopithecus albogularis*) monkeys move along muddy and sandy mangrove soils at low tide. Others, such as the lutungs, move solely in the crown, while still others use only the lower stratum (see Table 7.1), despite predation and other risks. Mallick describes the adaptive gait by which rhesus macaques avoid sharp pneumatophores in mangroves (J. Mallick, pers. comm., 2014).

Mangroves may be used for long-distance movement and dispersal, and provide connectivity between other habitats. The matrix used by black-and-white colobus (*Colobus guereza*) in Kenya includes mangrove and facilitates colobus movement between other habitats (J. Anderson, pers. comm., 2014). Likewise, Geoffroy's spider monkey (K. Morales Hernández, pers. comm., 2014) uses mangroves as corridors to access other forest fragments; although they forage seasonally in the mangroves, groups never sleep there. Mallick (pers. comm., 2014) observed no connectivity between mangroves and terrestrial habitats at his study site in the Sundarbans, and describes rhesus macaques moving from island to island by swimming and rhythmically responding to tidal fluctuations.

Many primates swim, including to move between forests following the fragmentation of their habitat (e.g. Gonzalez-Socoloske & Snarr 2010). Proboscis monkeys frequently swim (A. Bennett, pers. comm., 2014; Bennett *et al.* 1988; Yeager 1991a; Boonratana 1993; Matsuda *et al.* 2008) and even have partially webbed back feet. Long-tailed and southern pig-tailed macaques have been observed to swim on at least several occasions (I. Matsuda *et al.*, pers. comm., 2014). Red-capped mangabeys (*Cercocebus torquatus*) have also been observed to swim including when fleeing people (J. Head, pers. comm., 2014) although this behaviour is probably infrequent relative to that of the sympatric mona monkey (*Cercopithecus mona*), a frequent swimmer and relatively more abundant in mangrove habitat than the red-capped mangabey (Gartlan & Strusaker 1972; T. Struhsaker, pers. comm., 2014).

Mangrove Foraging Ecology

We naturally expected some convergence in the foraging behaviour of primates inhabiting mangroves, the relatively few genera and species of which occur worldwide (see Table 7.1, column on foraging). For example, *Rhizophora* spp. turn up in the diets of primate species in both Asia and Africa. Protein-rich seafood such as crabs, oysters and molluscs are sought after by patas monkeys (*Erythrocebus patas*) and Guinea baboons in West Africa (G. Galat & A. Galat-Luong, pers. comm., 2014), capuchins in Brazil (R. R. Santos, pers. comm., 2014) and macaques in the Sundarbans (J. Mallick, pers. comm., 2014; Chapter 18) (see 'innovative behaviour').

More generally, we expected more folivores than frugivores to use mangroves and at least some foraging innovations given the limits of mangrove habitat both in terms of plant species diversity and high tannin and salt loading on the diet. Folivores still need to deal with such salt loading by, for example, searching for fresh, or at least brackish, water (Nowak 2008; J. Mallick, pers. comm., 2014; see below on 'drinking') or supplementing mangrove diets with terrestrial forest species, or even human food crops (e.g. Guinea baboons mentioned above) which brings them into conflict with people.

Do more folivores than frugivores use mangroves? For lemurs, this appears to be the case (Chapter 5): leaf availability may exceed fruit availability in mangroves, and folivorous lemurs can deal with tannins and phytochemicals better than frugivorous ones. Bridgeman (2012) found that in mangrove-dwelling populations of howlers, they eat significantly more flowers and seeds and less fruit compared with *terra firma* forest populations; however, their leaf consumption is within the range of other howler diets.

The overall health of primates may not necessarily be limited by low species diversity. Mangrove phenology, while little studied by primate researchers, is suggestive of aseasonal production and year-round, continuous availability of young leaves, buds, flowers and fruits in mangrove forest stands (Matsuda *et al.*, in Borneo, Chapter 4; Nowak in Zanzibar, Nowak 2008). Such availability could counter the poor plant species diversity of mangrove habitat.

A potential outcome of mangrove aseasonality is a high abundance of invertebrates and small animals, providing an alternative, fallback or preferred food source for certain primates, especially the omnivorous ones. For instance, the white-throated capuchin (*Cebus capucinus*) in Curú, Costa Rica eats clams, insects and even small mammals in mangroves (M. Baker, pers. comm., 2014). Based on vocalization surveys, the blonde capuchin (*Sapajus flavius*) is thought to use the mangrove primarily for foraging, feeding on lizards, fruits and flowers (Chapter 10). The red-capped mangabey eats shellfish, crabs and insects in addition to plant matter in mangroves in Loango National Park, Gabon (J. Head, pers. comm., 2014). The rhesus macaque eats the eggs of birds, turtles and crocodiles, as well as insects, molluscs, crabs and fish (J. Mallick, pers. comm., 2014) supplementing a diet of plant matter.

Drinking

As a result of high salt and tannin consumption in mangroves relative to terrestrial forests, primates may need to drink more frequently and even develop innovative water foraging behaviour (Nowak 2008). The Zanzibar red colobus drinks from mangrove treeholes and limestone crevices, and licks dew off of mangrove leaves and rainwater off tree trunks on Uzi Island, Zanzibar, where upland coral rag forest has been severely degraded (Nowak 2007, 2008). Similarly, the ring-tailed lemur drinks from limestone (Y. Jacky, pers. comm., 2014). Guinea baboons drink from springs which are also used by people in the Cantanhez Woodlands NP, Guinea-Bissau (M.J.F. da Silva, pers. comm., 2014). Seasonal changes in water consumption has been observed among rhesus macaques in the Sundarbans (J. Mallick, pers. comm., 2014; Figure 7.1); when rainwater accumulated in small ditches or when sweet water ponds dried up during the dry season, the rhesus macaques drank brackish water or licked dewdrops, when available. During the rainy season, their water requirement was met primarily through their consumption of succulent plant matter (J. Mallick, pers. comm., 2014).

Primate Group Sizes in Mangroves

How do sizes of primate groups in mangroves compare with groups in adjacent *terra firma* habitats? We expected smaller group sizes in mangroves relative to groups in *terra firma* habitats given the challenges of exploiting mangroves: dense vegetation, muddy soils, low plant species diversity, a changing tide table, and salty, tannin-packed leaves. Consistent with this expectation, are observations of small group sizes of mantled howler monkeys in the mangroves of Cuero y Salado Wildlife Refuge, Honduras compared with other sites (Snarr 2006). Small groups (of around 50 individuals) of Guinea baboons inhabit mangroves compared to larger groups of ~150 individuals (A. Barata, pers. obs.) in dry and sub-humid Guinean forest patches (M.J.F. da Silva, pers. comm., 2014).

There are however, exceptions, as per usual in the Primate order, and some primate species' groups are larger in the mangroves than in other habitats, possibly because of the safety this habitat confers or a diet to which some folivores or opportunists are well suited. For example, ebony langurs (Chapter 14) and Zanzibar red colobus (Nowak & Lee 2010) have relatively larger group sizes in mangroves compared with upland forest. *Rhizophora* mangrove supports the highest group densities of long-tailed macaques in Sumatra (Crockett & Wilson 1980). Rhesus macaques in the Sundarbans are found in splinter groups of 30–40 individuals on average with a maximum group size of 80 individuals (J. Mallick, pers. comm., 2014); meanwhile, in the tropical and subtropical forests of northern West Bengal, group sizes do not exceed 50 individuals (J. Mallick, pers. obs.). In *Rhizophora/Laguncularia* mangrove in Mexico, group sizes of the Yucatán black howler monkey are not significantly different from those in other habitats (Bridgeman 2012).

It is possible that mangroves could therefore function as a source rather than a sink habitat for some primate species, and not necessarily just the leaf-eating or omnivorous ones.

Mangrove-Adapted Behaviour and Innovation

Primates, capable of learning and flexing behaviour, should be able to capitalize on mangrove use despite challenges. Such behavioural adaptations may be key factors in the ecological success of some of these species whose *terra firma* habitats may be undergoing accelerated anthropogenic change.

Consumption of animal matter is one such adaptation and, in the mangroves, animal matter comes packed with protein and omega-3 fatty acids. According to Mallick (pers. comm., 2014), rhesus macaques have innovative ways of catching fish at Tiger Reserve, Sundarbans. Yellow-breasted capuchin eat crabs

Figure 7.1 A pair of thirsty rhesus macaques drinking vigilantly at a sweet water hole in tiger habitat in the Jingakhali Forest in the Sundarban Tiger Project area, India. Photo: Subrata Pal Chowdhury. (A black and white version of this figure appears in some formats. For the colour version, please refer to the plate section.)

out of traps set by humans (G. Canale, pers. comm., 2014) and have also been observed to catch crabs for consumption (R. Beltrão-Mendes, pers. comm., 2014). It would therefore appear worthwhile to venture into challenging mangrove habitat for such valuable foods.

Tool use has only been observed in two mangrove-dwelling Neotropical species: the tufted and bearded capuchin monkeys, enabling them to exploit abundant marine invertebrates that are protected by shells and exoskeletons. Male tufted capuchins repeatedly hit oyster shells fixed to mangrove trees using another piece of the oyster colony in Maranhão, Brazil (Fernandes 1991). Various populations of bearded capuchin monkeys frequently use *Rhizophora* as anvils and wooden branches as hammers to eat snails (*Neritina zebra*) and crabs (*Ucides cordatus*). Tool use for the exploitation of these resources is, however, not essential, suggesting that it is not a prerequisite of survival in mangrove habitat and has evolved to maximize the efficiency of exploiting food resources already in the diet (Chapter 9).

Consumption of unusual plant matter, for example mangrove prop roots for moisture and searching out coral rag islands festooned with plants within mangrove forest are other examples of mangrove-adapted behaviour (Nowak 2008). No doubt there are many more examples, the reporting of which this volume may hopefully stimulate.

Conclusion

What do primates that inhabit or exploit mangrove habitats around the globe have in common? It may be easier to say what they do not, including: not necessarily being opportunistic generalists, such as capuchins and guenons, but also being relative specialists including leaf monkeys such as red colobus and langurs, i.e. not only 'swamp specialists'. Mangrove dwellers do not essentially possess broad forest habitat tolerances across their wider geographical range. They are not all arboreal, and therefore presumably, not all strictly foragers of the upper mangrove canopy. It would seem, however, that they are inhabitants of moist forests (e.g. coastal belt, riparian), and adapted to wet and hot conditions. There are, however, exceptions such as the green monkey reported to use mangrove in a number of places including Ébrié Lagoon, Cote d'Ivoire (Bi *et al.* 2009) and Saloum, Senegal, where they have been observed to swim up to 100 m at high tide (G. Galat & A. Galat-Luong, pers. comm., 2014), and patas monkeys observed to enter mangrove in Saloum, Senegal to catch and eat fiddler crabs *Uca tangeri* (G. Galat & A. Galat-Luong, pers. comm., 2014). African apes have been observed in mangroves, for example gorillas in Petit Loango, Gabon (J. Head, pers. obs.; Chapter 12) and chimpanzees in Guinea-Bissau (Hockings, pers. comm., 2014) as well as orangutans (Chapter 4) and gibbons, although Asian apes appear to prefer peat to mangrove swamp (Chapter 29).

Future research could compare diets of primates in and outside mangrove and gather evidence of primates entering mangroves to seek out nutrient-rich food resources. Mangroves support a wide array of non-plant diet items in the form of invertebrates and shellfish, and primates show numerous behavioural adaptations to exploit these resources. Whether terrestrial forest loss is associated with increasing mangrove use and habitat switching, at least in some primate taxa, is yet another potential line of investigation. Mangroves are not necessarily 'ecological traps' – that is, poor quality habitats where primates have been forced to settle – but may represent areas of safety with continuous food supply and thermoregulatory benefits during the hot dry season. Non-occurrence in mangrove in sympatric primate communities would be equally valuable to studying certain species' propensity to exploit this, at first glance, marginal habitat (Chapter 12).

Do we expect primates that have access to – and the ability to exploit – swamp forest to have better chances of persisting under ongoing human-induced change to natural environments? A large number of species (from our questionnaire) were described as using mangroves for 'refuge'. This may be related to the conservation status of the wider area inhabited by a given species where mangroves are protected by extension of designating protected areas that include coastal areas. However, in other places, mangroves may not be well protected, and are visibly under human threat as much as *terra firma* forests (all in this volume: Sundarbans, Chapter 16; Tana River Delta and other deltas, Chapter 30; Niger Delta, Chapter 40; also Chapter 43 on conservation in Africa).

Chapter 8

Mangrove-living Primates in the Neotropics

An Ecological Review

Ricardo Rodrigues dos Santos and LeAndra Luecke Bridgeman

Mangrove Ecology and Distribution in Neotropics

The term 'mangrove' refers to the association of tropical woody trees and shrubs that grow in tidal and saline wetlands (Ball 1988), where land meets the sea. General mangrove ecology is given in Chapter 2 and in Kathiresan and Bingham (2001) and Tomlinson (1986).

By area, 15% of mangroves worldwide are found in North and Central America and 11% in South America (Giri *et al.* 2011). The regional term 'Mesoamerica' used in this chapter includes Mexico, which is part of the North American continent, and Central America, which includes the countries of Belize, Guatemala, Honduras, El Salvador, Nicaragua, Costa Rica and Panama. Together, these eight countries hold 185 designated wetland protected areas of just over 11 million hectares (Ramsar 2001).

South American mangroves occur in Colombia, Venezuela, Ecuador, Suriname, Guyana, French Guyana, Peru and Brazil. The majority of the mangrove forest in the Neotropics occurs in Brazil and Mexico (ranking third and fourth globally, respectively). In Brazil, mangroves cover approximately 962 700 ha or 7.0% of the total global mangrove land cover, and in Mexico, mangroves cover approximately 742 000 ha or 5.4% (Giri *et al.* 2011).

Neotropical Primates in Mangroves

Neotropical forests are home to a high percentage of the world's primates (Alfaro *et al.* 2015; Schneider & Sampaio 2015). Most species are located in South America and mainly in the Amazon rainforest (Boubli *et al.* 2015). Until recently, only brief reports of four species of Mesoamerican primates (Table 8.1) and 13 species of South American primates in mangroves existed (Table 8.2).

Mesoamerica is home to 6 genera, 8 species and 17 subspecies of primates (Rylands *et al.* 2006): Geoffroy's tamarin (*Saguinus geoffroyi*), squirrel monkey (*Saimiri oestedii*), white-faced capuchin monkey (*Cebus capucinus*), night monkey (*Aotus zonalis*), mantled and black howler monkeys (*Alouatta palliata* and *A. pigra*), and Geoffroy's and Colombian black spider monkeys (*Ateles geoffroyi* and *A. fusciceps*). These primates inhabit a wide range of habitats from lowland tropical rainforest to disturbed areas of secondary growth (see Table 8.1). However, only mantled and black howler monkeys (Bridgeman 2012; Luecke & Estrada 2005; Milton & Mittermeier 1977; Serio-Silva *et al.* 2006; Snarr 2006), Geoffroy's spider monkey (Cuarón *et al.* 2008) and white-faced capuchin (Causado *et al.* 2008) have been reported in mangroves in Mesoamerica.

South America is home to 19 genera and 199 species and subspecies of primate (Rylands & Mittermeier 2009). Most records of these taxa in mangrove are single-sighting reports on occurrence or feeding behaviour. There are brief records on the diets in mangroves of red-handed howlers (*Alouatta belzebul*), bearded sakis (*Chiropotes satanas*), common squirrel monkeys (*Saimiri sciureus*), and tufted capuchins (*Sapajus apella*) (Fernandes 1991; Fernandes & Aguiar 1993). There are also poorly documented and limited data on the occurrence in mangroves of mantled howler populations in South America, red howlers (*Alouatta seniculus*), Maranhão red-handed howlers (*Alouatta ululata*), night monkeys (*Aotus azarae infulatus*), and wedge-capped capuchins (*Cebus olivaceus*) (see Tables 8.1 and 8.2 for references).

The common marmoset (*Callithrix jacchus*) is reported to use mangrove, but there is no record of where the sighting was made (M.A.O.M. da Cruz, pers. comm.; see Fernandes & Aguiar 1993). Typically, the common marmoset ranges in dry areas in northeastern Brazil, and forages in coastal vegetation close to mangroves, as has been observed in Rio Preguiças, Maranhão, Brazil, in the northwest part of this species distribution (R.R. Santos, pers. obs.). Because of relatively recent habitat loss over the past two decades, the bearded saki population reported from mangrove in coastal Brazilian Amazonia (at Alcântara, Maranhão state) (Fernandes 2000; Fernandes & Aguiar 1993) may now be extinct (R.R. Santos, pers. obs.).

Although there are no Neotropical primate species endemic to mangrove, some have populations whose individuals may live in this habitat throughout their lives (see Chapter 7 for other examples worldwide). In Mexico, the black howler population at the Pantanos de Centla Biosphere Reserve was forced into the river delta mangrove forests at least 25 years ago, as this was the only suitable habitat left after the region's last remaining lowland forest was destroyed to create livestock pastureland (Bridgeman 2012). In Brazil, Maranhão red-handed howler (R.R. Santos, pers. obs.), tufted capuchin, bearded capuchin (*Sapajus libinosus*) (Santos *et al.* 2016), and yellow-breasted capuchin (*Sapajus xanthosternos*) (Canale: Chapters 7 and 11) have mangrove-living groups distant and isolated from upland populations.

Table 8.1 Details on the four Mesoamerican primate species reported in mangrove forests. The Habit and Diet categories are general for each species.

Species	Common name	IUCN Red List status	Habit	Diet	Geographic range	References
Alouatta palliata	Mantled howler	Least Concern	Arboreal	Folivore	M, B, G, H, N, CR, P	1, 2, 3, 4, 5, 6, 7
Alouatta pigra	Yucatán black howler	Endangered	Arboreal	Folivore	M, B, G	3, 8
Ateles geoffroyi	Geoffroy's spider monkey	Endangered	Arboreal	Frugivore	M, B, G, H, ES, N, CR, P	6, 9
Cebus capucinus	White-throated capuchin	Least Concern	Arboreal/ terrestrial	Omnivore	H, N, CR, P	1, 4, 6, 10, 11

Mesoamerican geographic range given here are derived from Rylands *et al.* (2006): M = Mexico, B = Belize, G = Guatemala, H = Honduras, ES = El Salvador, N = Nicaragua, CR = Costa Rica, P = Panama. 1. Milton and Mittermeier (1977), 2. Reid (1997), 3. Luecke and Estrada (2005), 4. Snarr (2006), 5. Cuarón *et al.* (2008), 6. Bridgeman (2012), 7. World Wildlife Fund (2013), 8. Marsh (2003), 9. Serio-Silva *et al.* (2006), 10. Rose (1998), 11. Causado *et al.* (2008).

Table 8.2 Details of the 13 South American primate species reported in mangrove forests. The Habit and Diet categories are general for each species.

Species	Common name	IUCN Red List status	Habit	Diet	Geographic range in mangrove	References
Alouatta belzebul	Red-handed howler	Vulnerable	Arboreal	Folivore	B	1, 2
Alouatta palliata	Mantled howler	Vulnerable	Arboreal	Folivore	P	3
Alouatta seniculus	Red howler	Least Concern	Arboreal	Folivore	C	4, 5, 6
Alouatta ululata	Maranhão red-handed howler	Endangered	Arboreal	Folivore	B	7, 8
Aotus infulatus	Night monkey	Least Concern	Arboreal	Frugivore	B	9
Callithrix jacchus	Common marmoset	Least Concern	Arboreal	Gum-feeding specialist	B	1, 2
Cebus olivaceus	Wedge-capped capuchin	Least Concern	Arboreal/ terrestrial	Omnivore	–	4
Chiropotes satanas	Cuxiú	Critically Endangered	Arboreal	Frugivore/seed predator	B	1, 2
Saimiri sciureus	Common squirrel monkey	Least Concern	Arboreal	Frugivore/insectivore	B	1, 2
Sapajus apella	Tufted capuchin	Least Concern	Arboreal/ terrestrial	Omnivore	B	1, 10, 11
Sapajus libidinosus	Bearded capuchin	Least Concern	Arboreal/ terrestrial	Omnivore	B	11
Sapajus xanthosternos	Yellow-breasted capuchin	Critically Endangered	Arboreal/ terrestrial	Omnivore	B	12, 13
Sapajus flavius	Blond capuchin	Critically Endangered	Arboreal/ terrestrial	Omnivore	B	14

B = Brazil, C = Colombia, P = Peru. 1. Fernandes (1993), 2. Fernandes (2000), 3. Crocket (1998), 4. Linares and Rivas (2004), 5. Hernandez-Camacho and Defler (1985), 6. Scott *et al.* (1976), 7. Pinto, see Nowak and Coles, Chapter 7, this volume, 8. Santos, pers. comm., 9. Silva Jr and Fernandes (1999), 10. Fernandes (1991), 11. Santos *et al.* (2016), 12. Canale, see Nowak and Coles, this volume, 13. Beltrão-Mendes and Ferrari, Chapter 11, this volume, 14. Bastos *et al.*, Chapter 10, this volume.

Maranhão red-handed howlers and bearded capuchins are found in two protected areas: Rio Preguiças (Área de Proteção Ambiental do Rio Preguiças e Pequenos Lençóis, with mangroves surrounded by sand dunes) and Rio Parnaíba (Área de Proteção Ambiental Delta do Parnaíba, formed by approximately 80 islands) in Maranhão state, Brazil (R.R. Santos, pers. obs.). Tufted capuchins have been observed to use mangrove areas throughout the species' range along the northern Brazilian coast, from east of the Amazon River in Pará through to the Amazon coast of Maranhão state (Santos *et al.* 2016).

To date, only four studies have been undertaken on the two howler species inhabiting Mesoamerican mangrove forests and two capuchin monkey species inhabiting South American mangrove. The results of these studies are detailed below. The behaviour and ecology of other primate species in mangroves remain virtually unknown. Below, we present the ecological and primate population data derived from these studies. For a summary of their comparative behavioural ecology, see Chapter 17.

Studies of Primates in Neotropical Mangroves

There have been two studies of Mesoamerican primate populations in mangroves. LeAndra Luecke Bridgeman (2012) focused her research on Yucatán black howlers on the Pantanos de Centla Biosphere Reserve (PCBR) in Tabasco, Mexico (Figure 8.1), and Karen Snarr (2006) studied a population of mantled howlers on the Cuero y Salado Wildlife Preserve (CSWP) in Honduras.

Bridgeman's research was carried out in riverine mangrove on the PCBR, a wetlands reserve that covers much of the state of Tabasco. PCBR lies on the southern rim of the Gulf of Mexico on the Isthmus of Tehuantepec between the states of Veracruz and Campeche, and covers approximately 300 000 ha. The climate is hot and humid, with a mean annual temperature of 26°C. Annual precipitation ranges from 1500 to 2000 mm (ParksWatch-Mexico 2003).

In the Centla District wetlands, in which much of the PCBR is located, Guadarrama-Olivera and Ortiz-Gil (2000) recorded 110 families, 363 genera and 637 species of trees and other plants. However, the 300 ha mangrove study area of Arroyo Polo had only 17 tree species and was dominated by white mangrove, *Laguncularia racemosa* (Combretaceae), inland and red mangrove, *Rhizophora mangle* (Rhizophoraceae), which form a ring around the island at the water's edge. Other tree species in the mangrove here, found in much lower densities throughout the island, were swamp dogwood, *Lonchocarpus hondurensis* (Fabaceae), black olive, *Bucida buceras* (Combretaceae), and Guianan chestnut, *Pachira aquatica* (Malvaceae) (Bridgeman 2012). White mangrove forest averaged 10 m in height and the diameter of trees averaged 33 cm. Red mangrove forest tended to be shorter at 7.8 m and tree trunks thinner at 19.8 cm. The average canopy cover was 48%.

The primates at the site in Tabasco were Yucatán black howlers, *Alouatta pigra*, hereafter referred to as 'black howlers'. Black howlers, all ages and both sexes, have black fur and black faces (Figure 8.1). Males of this species generally weigh an average of 11 kg and females an average of 6.4 kg (Smith & Jungers 1997). Black howlers tend to live in single-male or multi-male social groups with multiple females. In the mangrove, mean group size was 5.8 individuals. Groups had an average of 1.2 adult males (ADM), 2.4 adult females (ADF), 1.2 juveniles (JUV) and 1 infant (INF). For the mangrove population as a whole, adult male to adult female ratio was 1:2. Adult female to infant ratio was 2.4:1 and adult female to immature (juveniles and infants) ratio was 1.1:1. Primate population density was 15.3 individuals/km^2 and primate biomass was 97 kg/km^2 (Bridgeman 2012).

In this same study, when wet and dry season data were pooled, the mangrove black howlers spent most of their time resting

Figure 8.1 Black howlers eating *Lonchocarpus hondurensis* flowers. Photo: Aubrey Tischer.

(73%), followed by locomotion (nearly 12%), feeding (11.5%) and affiliative social behaviours (3.5%). The study groups fed most frequently on leaves (47% of observations), with mature leaves selected more often than young leaves. Feeding on flowers and seeds was observed just over 41% and 6%, respectively. In the dry season (March, April and May), the mangrove howlers relied heavily on the flowers of *L. hondurensis*. Overall, they were observed to eat from 12 species of tree and liana on the island, the top five being *L. hondurensis, Dalbergia brownei* (Fabaceae: Pap.), *Laguncularia racemosa, Rhabdadenia biflora* (Apocynaceae), and *Rhizophora mangle*.

Snarr's (2006) study of Honduran mantled howlers, *Alouatta palliata*, in mangrove and lowland swamp forest, was also a first for the species in this habitat type. The CSWP is also a floodplain and wetlands reserve of 13 255 ha, and is situated on the northern Caribbean coast of Honduras. Annual precipitation averages 3050 mm and the mean annual temperature is 27.4°C (Snarr 2006).

There were ten tree species in the Honduran mangrove sites which were dominated by red mangrove (*R. mangle*), Guianan chestnut (*Pachira aquatica*, Malvaeae) and dragonsblood trees (*Pterocarpus officianalis*, Fabaceae) that made up 78% of the forest. Snarr's study site also included tropical evergreen broadleaf lowland swamp forest that was inundated for eight or more months of the year. This forest type has 20 species and was dominated by palms and other broadleaved trees. In the study zone, Guianan chestnut, dragonsblood trees, river pear (*Grias cauliflora*, Lecythidaceae) and the cohune palm (*Attalea cohune*, Arecaceae) made up 83% of the forest. The average tree height was 15 m for both forest types. Mean DBH of trees in the eastern portion of the reserve, where the monkeys were found, tended to be bigger at around 33.6 cm (Snarr 2006).

Mantled howlers are blackish or brownish with elongated hair along the flank, which varies in the intensity of its yellow colour (Lawrence 1933). Male mantled howlers are generally around 7 kg. Females are smaller (about 5.4 kg: Smith & Jungers 2007). Mean group size for the mangrove population was

Figure 8.2 Bearded capuchins on aerial roots of *Rhizophora mangle* along the Rio Preguiças, Maranhão, Brazil. Photo: Ricardo Rodrigues dos Santos.

6.5 individuals, with a male to female ratio of approximately 1:1; at the lower end of the range for this species on both counts. Primate population density was estimated to be 78 individuals/km^2, and was within the established range for conspecifics from other sites (Snarr 2006).

Snarr's study groups spent most of their time resting (41–44%) and eating (36–40%), followed by movement (10–17%) and 4–8% of their time engaged in other behaviours. These groups were seen to use 18 different tree and vine/liana species, with the largest percentage eaten from *Pterocarpus officianalis.* Leaves (mature, new and new growth) made up the largest proportion of their diet, followed by flowers and seeds over both the wet and dry seasons.

Studies of primates from South American mangroves focused on capuchin monkeys from the Brazilian north coast. Santos (2010) studied tool-use behaviour in bearded capuchins *S. libidinosus* (for tool-use detail see Chapter 9) and mangrove geographic distribution of *S. apella* and *S. libidinosus* (Santos *et al.* 2016) to verify whether mangroves form peripheral or core areas for capuchins (Figure 8.2). Cutrim (2013) focused her research on one group of bearded capuchins in a study of activity budget and tool-use behaviour. The species are not sympatric in mangrove areas.

The mangroves of the Brazilian north coast are present largely as a continuous forest. This landscape, formed by tidal canals, rivers and islands of different sizes, is one of the largest continuous formations of mangrove in the world (Souza-Filho 2005). The tidal influx ranges from 4 m to 7.5 m (Souza-Filho 2005) and provides an environment of high biological productivity (Menezes *et al.* 2008). The climate is characterized as seasonal tropical, with two marked seasons: a rainy season (January to June) and a dry season (July to December). The temperature is stable throughout the year with an average of 26°C. Six true mangrove species are present in the area: red mangroves (*Rhizophora mangle, R. racemosa* and *R. harrissoni*), black mangroves (*Avicennia germinans* and *A. shaueriana*, Acanthaceae) and white mangrove (*L. racemosa*) (Menezes et al. 2008; Santos 2010).

Santos *et al.* (2016) reported bearded capuchins living in continuous and naturally fragmented areas of coastal mangrove, as well as mangroves on small islands. Bearded capuchins were recorded living in groups of 1.3 ha and 2.0 ha in Rio Novo and 13.0 ha and 37.0 ha in Rio Preguiças, Maranhão state, Brazil (Santos 2010). Such home ranges are smaller than usually recorded for capuchins (Fragaszy *et al.* 2004).

Capuchin monkeys may be the only Neotropical primates that eat crabs and molluscs in mangroves. To date, no other primate species in the Americas has been recorded foraging in mud for such shellfish. Most of these food resources are protected by carapaces or shells and necessitate a forager's ability to break into them (Chapter 9). The success primate species have in colonizing mangrove may depend more on their ability to exploit such animal resources than the relatively low-quality resources offered by fruits, flowers or seeds. This is shown by the Maranhão red-handed howler that has been recorded in the same riverine area (Rio Preguiças) as bearded capuchins, but only occurs in a transitional area to *várzea* flooded forest where a mix of mangrove and non-mangrove plant food is available (R.R. Santos, pers. obs.). In addition to the consumption of crabs *(Ucides cordatus)*, snails (*Neritina zebra*) and shipworms (unidentified), bearded capuchins eat the fruits of black mangrove (*A. germinans*), flowers of white mangroves (*L. racemosa*), and propagules and flowers of red mangrove (*R. mangle*) in this area (Cutrim 2013; Santos 2010).

The activity budget for bearded capuchins was studied in a small 37 ha natural fragment (Cutrim 2013) at the Rio Preguiças site. Activities (resting, travelling, foraging, eating and socializing) carried out in mangrove were apportioned similarly to upland populations, but with a higher proportion of time spent foraging. However, the high density, 31 individuals/37 ha in 2011–2013, combined with the small area available may have operated to increase foraging time and decrease travelling time (Cutrim 2013). Many mangrove fragments in the study area are shrinking naturally due to the movement of sand dunes. Currently, the group size of the study population is approximately 60 animals (R.R. Santos, pers. obs.), however, it is possible that subgroup formation has affected the counts by Santos (2010) and Cutrim (2013), so this number may be inaccurate.

Summary and Future Directions for Research

The sum total of our knowledge of primate ecology in mangroves comes from a series of brief observations and four detailed studies. Studies of additional taxa and geographic areas are urgently needed.

Future research in mangrove and flooded landscapes should involve surveys and censuses for primate populations along the coastlines and river deltas in areas of Neotropical mangroves, especially in Mexico and Brazil, where mangrove land cover is most abundant. Although coastal areas of Panama, Colombia and Ecuador should not be neglected. Locating more populations in mangroves and conducting detailed ecological and behavioural studies on such populations, including nutritional intake and phytochemistry (an important, but often overlooked part of conservation research) would be informative. Adding data on all aspects of primate behavioural ecology, such as population characteristics, group size variation, activity patterns, foraging and feeding behaviour, and primate habitat preferences in this unusual habitat will allow us to better understand the adaptability and flexibility shown by primates and other animals in low diversity habitats such as mangroves.

Chapter 9

The Role of Tools in the Feeding Ecology of Bearded Capuchins Living in Mangroves

Ricardo Rodrigues dos Santos, Arrilton Araújo de Sousa, Dorothy M. Fragaszy and Renata Gonçalves Ferreira

Introduction

Percussive tool use has been observed in three genera of non-human primates (*Sapajus, Pan* and *Macaca*), all of which share a tendency towards terrestriality (Fragaszy *et al.* 2004; Kortlandt 1986; Malaivijitnond *et al.* 2007; Sugiyama 1989). Tool use is one of several behaviours that is of interest to behavioural science as a means of understanding the origin of human dexterity in manipulative behaviour (Matsuzawa 2001; Roux & Bril 2005). It is also relevant to biologists as an aspect of foraging behaviour, with implications for diet composition and survival in harsh habitats, and those are the approaches taken here.

In Neotropical primates, only the tufted capuchin monkeys (*Sapajus*, especially *Sapajus libidinous*, the bearded capuchin; Ottoni & Izar 2008) use percussive tools frequently, and their use is associated mainly with the consumption of encapsulated fruits and nuts. As most populations of *Sapajus* inhabit non-flooded habitats like *Caatinga* (savanna-like vegetation) or *Cerrado* (wood savanna vegetation), most percussive tool use occurs on dry land. *Sapajus* primarily use stone hammers and anvils (Canale *et al.* 2009; Emidio & Ferreira 2012; Fragaszy *et al.* 2004; Waga *et al.* 2006) as do chimpanzees (McGrew 1994) and as (based on remains found in archaeological sites) early hominins are inferred to have done (Beaune 2004; Haslam *et al.* 2009; Visalberghi *et al.* 2013).

Percussive tool use is also observed in *S. libidinosus* groups in mangrove, a distinct saline, cyclically flooded biome dominated by *Rhizophora* vegetation along the north coast of South America (Santos 2010) (see Chapter 2 for a more complete description). Differences exist in the materials used as percussors and anvils, and the food resources exploited through their use as tools, between populations of bearded capuchin monkeys living in these two habitat types.

In non-flooded habitats, tool use by primates has been linked to the tendency to forage on the ground (Visalberghi *et al.* 2005). Currently, it is thought that the discovery of tool use is more likely when resources are abundant (Koops *et al.* 2013; Spagnoletti *et al.* 2012). A third hypothesis is that tool use increases when food resources are scarce (Moura & Lee 2004).

Variation in the ways of using tools allows *Sapajus* to exploit a wide range of plant and animal resources (Mannu & Ottoni 2008). Santos (2010) has proposed that when *Sapajus* live exclusively in mangrove, they occupy a distinct food niche characterized by the exploitation of encapsulated foods such as molluscs and crustaceans, and that this diet is both accompanied by, and possible because of, percussive tool use.

In this chapter, we describe the nature and extent of tool use by *S. libidinosus* in mangrove and in non-flooded habitats in Brazil. Descriptions of the mangrove areas, surveys of anvil sites and monkeys' feeding opportunistic behaviour are taken from Santos (2010). Records of tool use in non-mangrove habitats in *Caatinga, Cerrado* and Atlantic forest are described in the literature. Based on comparative records made by Santos (2010) from two mangrove areas in the north of Brazil, we propose that percussive tool use is not a necessary condition for capuchin monkeys' survival in mangrove.

Mangrove Colonization by Capuchin Monkeys

Several species of Neotropical primates use mangrove forests (Chapters 8, 10 and 11). Along the northern coast of Brazil near the Amazon River mouth, on the northern coast of South America, several populations of tufted capuchin monkeys range primarily in non-flooded forests, but sporadically use the mangroves as a secondary habitat, possibly attracted by specific foods, including crabs, shipworms and oysters (Santos *et al.* 2016). Coastal *S. apella* populations in northern Brazil have been observed to range deeply into mangroves (Fernandes & Aguiar 1993; Santos *et al.* 2016) while in northeast Brazil, *S. xantosthernos*, yellow-breasted capuchins (Chapter 11), and *S. flavius*, blond capuchins (Chapter 10), penetrate these forests less extensively.

In estuaries on the northern Brazilian coast, *S. libidinosus* and *S. apella* are not sympatric. Here, some populations of *S. libidinosus* have no apparent connection to non-flooded forest habitats; there are no records of excursions to dry habitat. Hence, individuals in these populations depend exclusively on mangrove forests for resources. It is not known how isolated such populations are from other *S. libidinosus* colonies.

Like other Neotropical mangrove communities, those in Brazil have low plant diversity (Schaeffer-Novelli 1990). However, unlike non-flooded habitats, edible plant resources are easily accessible year round (Fernandes 1999; Fernandes *et al.* 2005; Menezes *et al.* 2008) and the environment is rich in such animal prey as shellfish and crustaceans (Macintosh & Ashton 2002). In this daily flooded forest system, patterns of food resource availability for capuchin monkeys differ from those in non-flooded areas. In mangroves, high tides submerge the ground and its associated food resources twice a day, but these same tides also contribute to high plant productivity.

Feeding Ecology and Tool Use of Bearded Capuchins Living in Non-flooded Habitats

Many of the foods tufted capuchins exploit are mechanically defended, meaning that they are encased in a tough husk or shell that the monkeys breach by banging, rubbing, biting or some combination of these activities (Chalk 2011; Fragaszy *et al.* 2004; Wright *et al.* 2009). Thus tufted capuchin monkeys can exploit foods that are difficult for other primates to obtain (Fragaszy *et al.* 2004; Wright *et al.* 2009). For example, they eat encapsulated foods, such as tubers that must be excavated from the soil, invertebrates that must be extracted from bark or dead branches, and seeds that must be plucked from the interior of thick husks or other embedding matrices. These monkeys have robust jaws and a strong bite force, so many tough items are opened by biting (Makedonska *et al.* 2012). However, they also eat a variety of encased foods that cannot be bitten open. These are mostly tough-husked fruits and, across their range, bearded capuchins use stone hammers and anvils to crack them, often matching the size of the stone to the size of the fruits to be opened (Canale *et al.* 2009; Ferreira *et al.* 2010b; Fragaszy *et al.* 2004; Ottoni & Izar 2008; Waga *et al.* 2006).

Tufted capuchin monkeys also use tools to access foods for reasons other than that they cannot bite them open. Some preferred foods are defended chemically. For example, the monkeys eat the endosperm (the commonly called nut) of the cashew tree (*Anacardium occidentale*, Anacardiaceae) a species native to the *Cerrado* and *Caatinga* of Brazil. To eat the nut, the monkeys have to extract it from the surrounding inedible husk, which contains irritating latex. If the monkey bites the husk to get to the nut, it risks smearing the caustic latex on its mouth and hands. Skilled foragers do not bite the fruit. Instead, depending on the maturity of the fruit, they either rub it persistently until a hole develops in the husk, and then they extract the edible kernel using one finger or they crack the husk on an anvil using a stone hammer (Sirianni & Visalberghi 2013). Both methods allow them to avoid contact with the latex while accessing the lipid-, protein- and mineral-rich nut. Bearded capuchins use percussive tools to pound cacti, another food that can be damaging to the skin if handled carelessly (Moraes *et al.* 2014). Pounding separates the spines from the fibres, rendering it an easy task to collect the edible fibres without the risk of piercing the fingers. Bearded capuchins in Serra da Capivara National Park, in Piauí State, Brazil, use stones to scrape the soil to expose tubers, and then follow this activity by pounding at the tubers in a way that breaks the long tap root and allows the monkey to free the tuber from the soil (Falótico 2011). Finally, bearded capuchins and blond capuchins (*S. flavius*) use sticks to probe insect nests to collect honey (Souto *et al.* 2011) and to flush vertebrate prey from rock crevices (Falótico 2011).

The best-documented example of tool use in bearded capuchins is cracking tough palm nuts using large hammer stones (Fragaszy *et al.* 2004, 2010a, b; Spagnoletti *et al.* 2011). The behaviour involves skill at several levels, from selection of appropriate tools, selection of nuts on the basis of their resistance to cracking, selection of anvil sites, positioning the nut on the anvil, orienting the hammer stone in the hands and striking with effective force (Fragaszy *et al.* 2010a, b, 2013; Liu *et al.* 2009, 2011; Massaro *et al.* 2012; Visalberghi *et al.* 2009a).

In all *terra firma* habitats where capuchin monkeys use tools for feeding, they do so all year long. However, so far as is known, tool use is not necessary for tufted capuchin survival in any of the habitats in which these primates have been studied. While capuchins at some *terra firma* sites use tools in feeding more often during the drier part of the year, this variation does not always correlate with food abundance. For example, in Serra da Capivara National Park, in Piauí State, Brazil, bearded capuchin monkeys use stone tools more often when food is more abundant (Falótico 2011), while at Fazenda Boa Vista, Gilbués (also in Piauí), bearded capuchin monkeys use stone tools year round, and tool-use frequency correlates only with the abundance of the most common palm nut that they crack (Spagnoletti *et al.* 2012; Izar *et al.* 2018).

Thus, it seems that monkeys use tools in feeding on an opportunistic basis, rather than as a last resort when food is scarce. We can therefore expect that the nature of tool use will vary from site to site and across seasons, depending on which foods are abundant and whether those foods are most easily or comfortably processed with tools versus some other method. The use of tools broadens the range of items that tufted capuchins can feed upon. One can imagine that in times of extreme scarcity, being able to access tough foods with tools could provide a needed means of feeding, but to date we have no evidence that using tools in feeding is essential for survival in any habitat where capuchin monkeys are permanent residents.

Tool-Use Behaviour in Mangroves

In the mangroves of northern Brazil, in the eastern portion of the Amazon River estuary, molluscs and crabs are eaten by both *S. apella* and *S. libidinosus* (Santos 2010). The way these resources are exploited and the level of dependence on marine animals vary with how the mangrove is used and with species of capuchin monkey. Although mangrove-dwelling *S. apella* have not been reported to use tools routinely (there is just a single eye-witness report; Fernandes 1991), this behaviour is known extensively for many *S. libidinosus* groups, and it is this that will be reported here from observations made by Santos (2010). At least seven groups were observed to be present in three continuous forest areas and four natural forest fragments in Rio Preguiças, Barreirinhas, Maranhão. Groups in three fragments were composed of 8 individuals in one area and approximately 30 individuals in each of the other two areas.

The description of the behaviour, food resources and cracking sites is derived from observations of the last two groups. Behavioural descriptions were taken opportunistically of the Vassouras' group and individual animals could not be identified. Cracking sites, hammer and anvil data were taken from Morro do Boi group. Availability of cracking sites and

Figure 9.1 Cracking sites with hammers and food items used by *Sapajus libidinosus* in mangrove forests of Rio Preguiças, Barreirinhas, Maranhão, Brazil. (a) Crabs (*Ucides cordatus*) and hammer on stilt roots from *Rhizophora* sp. tree (suspended cracking site). (b) Fallen mangrove trunk (fallen trunk cracking site), gastropoda mollusc shells (*Neritina zebra*) and four hammers. Details about the average size of hammers and anvils are given in the text.

food (crabs) and the use of cracking sites by capuchins were comparable between Rio Preguiças populations and those 20 km away on the Rio Novo (Santos *et al.* 2016).

S. libidinosus use tools to exploit two species of marine invertebrate: the snail *Neritina zebra* Bruguière, 1792 (Gastropoda, Neritidae), and the crab *Ucides cordatus* Linnaeus, 1763 (Crustacea: Ucididae) (Figure 9.1). *U. cordatus* are the largest crabs in the Brazilian mangroves (N = 8 crabs, average weight: 59.94 g, SD: 16.85 g; carapace average length: 3.88 cm, SD: 0.32; carapace average width: 5.25 cm; SD: 0.51) and, along with *N. zebra*, form part of the benthic fauna of the region's mangroves. Like *S. apella, S. libidinosus* exploit shipworms without using tools. Here we describe tool-use behaviour to exploit two encapsulated food sources: crabs and snails.

The invertebrates are exploited at low tide, when the mangrove is not flooded and the monkeys can catch them on the mud. The capuchins may forage by travelling through the stilt root regions of *Rhizophora* spp. and plucking crabs from their holes with one hand. To immobilize the crabs they use both hands to disarticulate the legs near the cephalothorax (they do not take the legs off at this time), pounding the crabs directly on a nearby trunk or branch before carrying them to a cracking site (anvil). Here, branches as hammers are used to break down both the most rigid parts such as the claws and carapace, as well as more flexible parts, such as the locomotor appendages. The animals feed from all parts of the crab body. The density of the crabs in one area was measured at 5.7 crabs/m^2 (Cutrim 2013), making them an abundant source of food. Snails, in contrast, are not processed prior to cracking at anvils. Here, capuchin monkeys use wooden branches as hammers to crack *N. zebra* shells.

In *terra firma*, capuchins famously use stones as tools (Ottoni & Izar 2008). However, there are no stones in the mangroves surveyed, and the tools used as hammers to crack shells and exoskeletons were pieces of wood from *Rhizophora* mangrove trees. Such material is both abundant and readily available. Those used had an average weigh of 144.5 g (SD: 128.1, N = 29), and an average length of 34.2 cm (SD: 17.3, N = 29).

The wooden hammers are transported from the ground to the cracking sites and often remain on the anvil after use. The monkeys do not break branches from trees to use as hammers. Hammers are characterized by abrasion marks in the middle area of the tools and remains of shells and exoskeletons bonded to their surface by the force of striking (Figure 9.2). Hammers are collected on the ground and are bare of leaves and petioles. There was no evidence of modification or manufacture of tools, such as removal of leaves or petioles, or any activity that could be related to altering the objects to increase their suitability as tools.

Importantly, monkeys in the same groups also eat *U. cordatus* without using tools, grabbing an individual with one or both hands and battering it against the same surface (stilt roots or fallen trunks) repeatedly to extract the meat. In both cases the monkeys extract meat from cephalothoraxes and the legs, using one or both hands. Feet are sometimes used to secure the prey during such processing.

Figure 9.2 Hammer used by *Sapajus libidinosus* and shells of *Neritina zebra* on the anvil and hammer. Details about the average size of hammers and anvils are given in the text.

Capuchin monkeys were also observed cracking food on stilt roots of three species of *Rhizophora* (*R. mangle, R. racemosa* and *R. harrisonii*; referred to here as 'suspended cracking sites') or fallen tree trunks (Figure 9.1). Studying a group of 30 *S. libidinosus* in a natural mangrove fragment, Santos (2010) found that *Rhizophora*-derived cracking sites have narrow anvils (11 cm, $n = 10$, SD: 2.9 cm) and occur at a height of 2.3 m ($n = 12$, SD: 0.9 m) above the substrate (Figure 9.1a). In contrast, anvils at fallen trunks (Figure 9.1b) have average wide bases of 21.3 cm ($n = 13$, SD: 5.6), and are closer to the ground (height: 60.0 cm, SD: 22 cm, $n = 12$).

Over a period of two years (2008 and 2010), Santos (2010) surveyed two regions of northern Brazilian mangrove, using transects to quantify the number of cracking sites used by *S. libidinosus* living exclusively in these forests (see Figure 9.3 for mangroves from Rio Preguiças and Rio Novo). The density of cracking sites and hammers in mangrove is greater than in non-flooded areas. In a longitudinal survey, from 2008 to 2010 at the Rio Preguiças, the densities were 17/ha within the 37 ha home range of one study group. This rose to 33.3 cracking sites/ha in 2010. This means an increase of nearly 200% in 2 years, to an estimated number of cracking sites increased from 629 to 1233 over the entire range. This greatly exceeds densities of nut-cracking anvil in non-flooded areas (< 2/ha, Visalberghi *et al.* 2009*b*; 26.7/ha, Emidio & Ferreira 2012), and makes it clear that at Rio Preguiças, *S. libidinosus* use tools to eat hard-shelled marine invertebrates at a very high frequency.

While this shows that tool use occurs in some mangrove-living *S. libidinosus* populations, this behaviour is not universally observed: tool-using behaviour appeared common only in one (the Rio Preguiças: Figure 9.3a) of the two estuaries analysed. On the other (the Rio Novo: Figure 9.3b), *S. libidinosus* populations consume crabs, but tool use was not observed.

However, a comparison of the availability of fallen trunk cracking sites and suspended cracking sites by capuchins at Rio Preguiças and Rio Novo showed no between-site difference in their availability (Santos 2010). In addition, *U. cordatus* densities were comparable at these two sites and showed no differences. Yet tool use occurred at only one of the two sites. Therefore, tool use may not be essential for the exploitation of crabs in mangrove areas.

S. libidinosus groups living exclusively in mangrove forest seem to have a high proportion of animal prey in their diet. This is supported by the high number of cracking sites in both populations that use areas without connection to *terra firma*. In principle, percussive tool use affords access to encapsulated resources, broadening the diversity of food items (Mannu & Ottoni 2008; McGrew 2007). Tools can also be used to maximize the efficiency of exploiting food resources that are already part of the diet.

In mangrove forests, tool use during the consumption of molluscs and crustaceans appears to produce both these outcomes for bearded capuchin monkeys. For example, bearded capuchins living in mangrove along Rio Preguiças, consumed *N. zebra* snails only after they had been processed with tools, while *U. cordatus* crabs were processed using both tool use and direct percussion (Santos 2010). Thus, tool use was necessary for exploiting one, but not both, of these resources. Therefore, using tools might allow for increased consumption of crustaceans or extend the variety of molluscs in the diet, by adding *N. zebra*, for example.

Conclusions

That an omnivorous primate has developed a diet that includes marine invertebrates is predictable when mangrove is used as a peripheral habitat, and even more so when the monkey in question lives exclusively in mangrove (Santos *et al.* 2016).

Figure 9.3 Northeastern coast of Maranhão state, Brazil. (a) Groups of *S. libidinosus*, Rio Preguiças mangrove, Barreirinhas (2°37′21.7′ S; 42°41′18.5′W) use tools. (b) Groups of *S. libidinosus*, Rio Novo mangrove, Paulino Neves (2°42′52.5′ S; 42°31′38.5′W) do not use tools. Tidal floodplain forest = *Várzea de maré*; Apicum = high salinity area inside or behind the mangrove forest that does not support mangrove or other tree species. (A black and white version of this figure appears in some formats. For the colour version, please refer to the plate section.)

Studies providing estimates of animal protein in the diet of monkeys living in mangrove versus other habitats are needed to determine how far the monkeys shift their diet towards animal-based resources when living exclusively in mangrove. However, bearded capuchin monkeys' survival in the mangrove does not depend upon using tools. Instead, using tools in the mangroves appears to increase the efficiency of consuming animals and/or expands the diet.

Chapter 10

Use of Mangrove Habitats by *Sapajus flavius* Assessed by Vocalization Surveys

Monique Bastos, Karolina Medeiros, Antonio Souto, Gareth Jones and Bruna M. Bezerra

Introduction

Primate vocal signals are usually classified according to behavioural context during signal emission (e.g. Bastos 2013; Bezerra & Souto 2008; Gros-Louis *et al.* 2008; Hammerschmidt & Fischer 1998; Leliveld *et al.* 2011; Winter *et al.* 1966). Investigations on specific calls from *Cebus* and *Sapajus* species have allowed researchers to gain a deeper knowledge on the behavioural context of several vocalizations (e.g. Boinski 1993; Boinski & Campbell 1996; Di Bitetti 2003; Di Bitetti & Janson 2001; Gros-Louis 2002, 2003, 2006). To investigate the vocalizations of species flagged as 'threatened' is a non-invasive manner of acquiring information on their behaviour and ecology. Information on vocalizations is particularly valuable when there is a problem either with habituation or when contact with humans may pose risks to study animals and potentially increases disease transmission across monkey populations (Crockett 1998; Holzmann *et al.* 2010; Vasconcelos *et al.* 2001, 2003).

Animals belonging to the genera *Cebus* and *Sapajus*, known as capuchin monkeys, are largely arboreal and inhabit a range of biomes across Central and South America, including the Atlantic forest, Subtropical forest, Amazon forest, *Caatinga* and *Cerrado* (Alfaro *et al.* 2012; Di Bitetti 2001; Fernandes 1991; Freeser & Oppenheimer 1981; Oliveira & Langguth 2006; Redmond 2008; Rowe 1996; Vilanova *et al.* 2005). They are capable of living in small and highly fragmented areas (Beltrão-Mendes *et al.* 2011; Chagas & Ferrari 2010; Janson & Di Bitetti 1997; Ludwing *et al.* 2006; Souto *et al.* 2011), due, at least in part, to the efficient and versatile foraging strategies they adopt.

Blonde capuchin monkeys (*Sapajus flavius*) inhabit fragments of Atlantic forest in northeast Brazil, with populations concentrated in the states of Alagoas, Paraíba and Pernambuco (Malta & Mendes Pontes 2013; Masseti & Veracini 2010; Oliveira & Langguth 2006; Souto *et al.* 2011). The species has a diverse vocal repertoire with 31 call types (Bastos *et al.* 2015; Bastos *et al.* 2018) and frequents densely vegetated habitats that impose limitations on the transmission of visual signals. Because vocalizations can propagate over long distances (e.g. ~0.5 km the *Lost call by Cebus capucinus* – Gros-Louis *et al.* 2008) even when there are physical barriers (Bezerra *et al.* 2012; Brumm 2004; Sabatini & Ruiz-Miranda 2008), acoustic communication is essential for the coordination of the large groups of *S. flavius* that seem to live in fission-fusion groups (Bezerra *et al.* 2014).

In this chapter, we use vocalizations to investigate the behavioural activity patterns of the Critically Endangered *S. flavius* in dense mangrove habitats. Until not long ago, *S. flavius* were considered to be one of the 25 most endangered primate species in the world (Mittermeier *et al.* 2012). Mangrove habitats impose major challenges for researchers including for approaching and observing owing to the daily flooding (Alongi 2008). Consequently, it is difficult to access mangrove areas, and not always feasible to directly observe the behaviours of the animals living there. Considering the importance and the progressive destruction of mangrove ecosystems across the world (Alongi 2008; Ribeiro *et al.* 2009; Rodrigues *et al.* 2009), understanding how the Critically Endangered *S. flavius* exploits such habitats will aid in the effective management and conservation of this threatened habitat and primate.

Methods

Study Site and Animals

The study was conducted in mangroves within an Atlantic forest fragment, 15 km from Mataraca, Paraíba, Brazil (06°29.423′S – 034°58.847′W). The fragment belongs to the mining company Cristal and has an area of 1050 ha. About 27 ha consists of flooded forest (mangrove and *várzea*) intersected by the Guajú river. In addition to mangrove habitats, the area comprises primary and secondary Atlantic forest, coastal scrubland (*Restinga*) and sand dunes, and *S. flavius* has been observed using all of these habitats (M. Bastos & B. Bezerra, unpubl. data).

The *S. flavius* in the area were semi-habituated, allowing the researchers to approach at ≥ 5 m. The maximum visual count of the group at the time of study was 77 individuals, with animals of all age classes (i.e. adults, juveniles and infants) present. The presence of the animals in the mangrove habitat was initially confirmed by direct visual contact and from hearing diagnostic vocalizations.

This study was part of a long-term conservation research project on *S. flavius*, which started in September 2010. Fieldwork was conducted from September 2010 to February 2011, and from August 2011 to July 2012, and included 856 hours of fieldwork in 91 days (average of 5 days/month). The study group was observed entering the mangrove habitat on 35

Table 10.1 Vocal repertoire of *Sapajus flavius* (adapted from Bastos *et al.* 2015).

Call category	Call name	Call context
1. Agonistic	*Heh*	Low-intensity aggressive situations, with no physical contact between the animals.
2. Fear	*Nhan*	Infants emitted the call when carried by young group members. Facial expression of the infant indicated fear.
3. Contact	*Trill*	Used during close interactions between animals.
4. Locomotion	*Flick* *Snap* *Clack* *Popped*	Locomotion-associated calls.
5. Screams	*Hoot* *Bellow* *Howl* *Yell* *Shout*	Not identified.
6. Calls emitted prior to sleeping	*Cheep* *Woop*	Emitted when preparing to sleep.
7. Invitation calls	*Sleck* *Ghiu* *Clok* *Houh*	After emitting the *sleck*, *ghiu* and *clok* described below, the animals presented alert behaviours. After hearing the calls, the animal tended to look at the caller and approached or followed him. Hough call was a long-distance contact call.
8. Food-associated calls	*Huh* *Nosh* *Huh*-1var *Huh*-2var *Huh*-3var *Huh*-4var	*Huh* and *nosh* were associated with foraging (i.e. looking for food on both the ground and in the trees). *Huh*-1var, *huh*-2var, *huh*-3var and *huh*-4var were associated with feeding (i.e. emitted as soon as the monkeys spotted the food and while eating it). The latter were variants of the *huh* call.
9. Grooming	*Ghrr*	Social grooming.
10. Mangrove specific calls	*Troll* *Tiny* *Phliu* *Kuen*	Not identified.

(38.5%) of the 91 days of fieldwork. On 9 (10%) of those days, 64 min of recordings of the animals vocalizing in the mangrove were made.

We used a Marantz PMD670 digital recorder (Kanagawa, Japan; linear frequency response: 20 Hz–20 kHz), connected to a directional microphone Sennheiser ME67 (Sennheiser, Germany; linear frequency response: 20 Hz–20 kHz) with a K6 module. We used AKG K44 headphones (frequency response: 18 Hz–20 kHz) to monitor the sounds while conducting the recordings. The animals were at a maximum distance of 20 m from the recording equipment.

The recordings of the vocalizations of *S. flavius* were inspected using the software BatSound 3.31 (Pettersson Elektronik, Uppsala, Sweden). Spectrograms were constructed and vocalizations of different types were then quantified (Figure 10.1). The context of vocalizations was ascertained from previous knowledge of the vocal repertoire of the species (Table 10.1 adapted from Bastos 2013 and Bastos *et al.* 2015). The following settings were used when inspecting the spectrograms: FFT size = 1024; threshold = 15; and a Hanning window.

A chi-square test was used to verify whether there was a difference in the frequency of use of different vocalizations in mangrove (i.e. to test whether the number of times a specific vocalization was emitted differed from an even use of vocalizations). The pattern of vocalizations was quantified from the recordings. The behavioural context attributed to each vocalization was used to infer behaviours of *S. flavius* in the mangrove areas (Bastos *et al.* 2015).

Results and Discussion

We extracted 1023 vocalizations from the sound files obtaining a total of 16 call types recorded when the animals were using the mangrove habitat (Figure 10.2). The frequency of use of the different calls varied significantly ($N = 12$; Chi square = 2771.1; $df = 15$; $p < 0.001$). The most frequent calls were *huh*-1var (41.20%), *trill* (15.67%) and *huh* (13.15%). Based on the calls' context, 62% of vocalizations were feeding associated (Figure 10.3; Table 10.1).

Sixteen out of 31 vocalizations of the vocal repertoire of *S. flavius* (Bastos *et al.* 2015; Bastos *et al.* 2018) were recorded when the animals were in the mangrove habitat. The majority of the calls were related to foraging, suggesting that the mangrove is primarily used for feeding. Other species of capuchin monkeys have been reported using the mangrove for foraging. For instance, Fernandes (1991) observed *S. apella* and Santos (2010) observed *S. libidinosus* feeding on oysters in the mangrove. We have seen *S. flavius* eating lizards, flowers (e.g. from *Calophyllum brasiliense*, Calophyllaceae), leaves (e.g. from *Tapirira guianenses,* Anacardiaceae) and some fruits (e.g. from *Chrysobalanus icaco,* Chrysobalanaceae).

The vocalization most commonly used was the foraging-associated call *huh*-1var, which in *Cebus capucinus* has been associated with arthropod predation (Boinsk & Campbell 1996). In *S. flavius*, the call is used for foraging in general, with no apparent specificity between the call and the exploitation of any prey type. Four calls of the vocal repertoire of the species (i.e. *troll, tiny, phliu* and *kuen*) were exclusively produced when the animals were in the mangrove (Bastos *et al.* 2015). The calls could not be associated to a specific behavioural context due to the poor visibility in the area. Further investigation, possibly

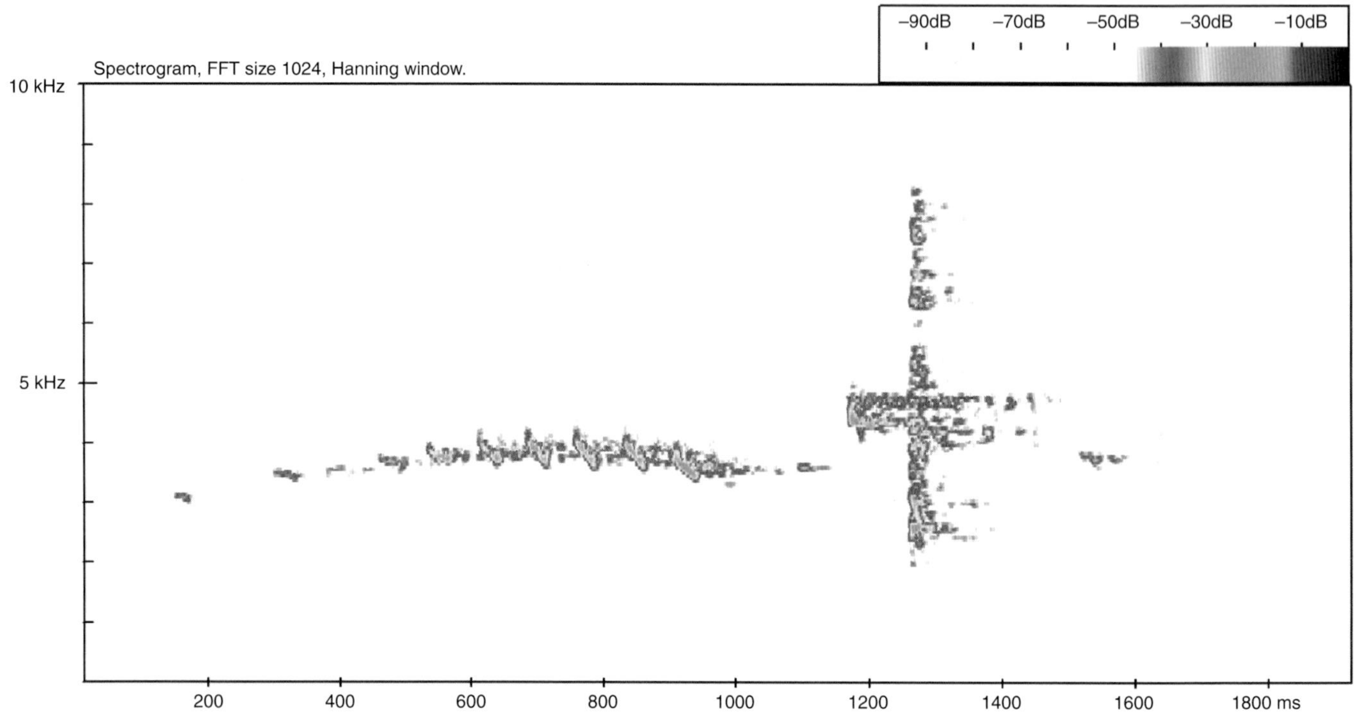

Figure 10.1 Spectrogram showing two different call types of *Sapajus flavius*. Each acoustic signal was counted in the spectrograms. For instance, in the depicted spectrogram, we counted two calls (*trill* call on left and *sleck* call on the right).

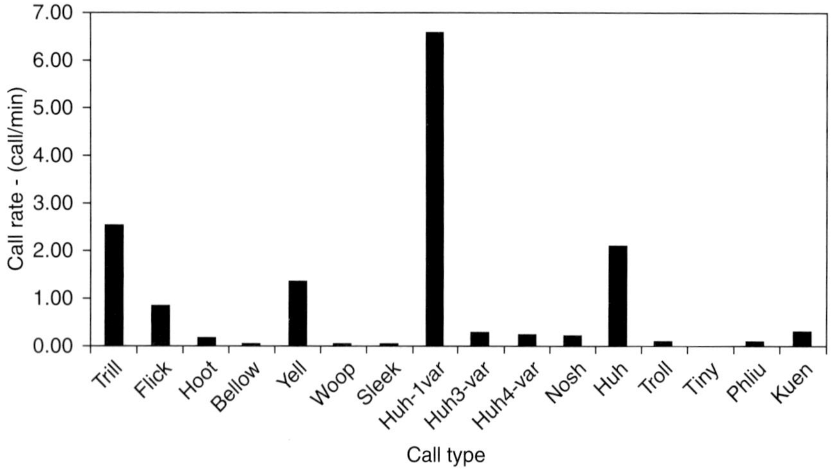

Figure 10.2 Preliminary vocal pattern of *Sapajus flavius* in mangrove habitat.

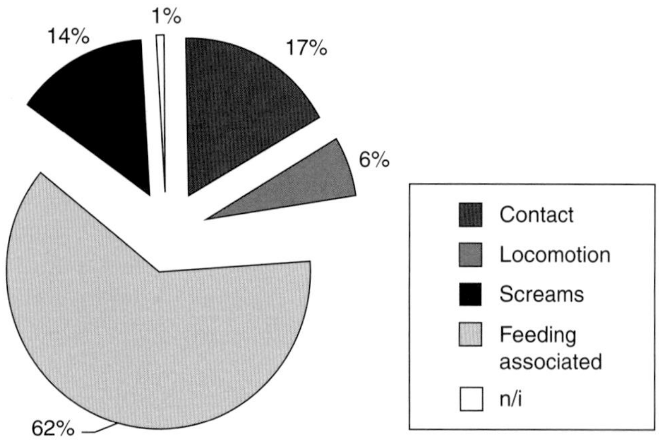

Figure 10.3 Inferred behavioural context of the *Sapajus flavius* calls in mangrove habitat.

using playback experiments, would be necessary to ascertain the exact context of these calls.

The behavioural contexts of vocalizations suggested that *S. flavius* in mangrove habitats behave in similar ways to other capuchin monkeys in general, i.e. that foraging-related behaviours are usually predominant (de Oliveira *et al.* 2014). Even though some of the vocalizations of *S. flavius* could not be attributed to specific behaviours in the recordings, the percentage of those was very small (1.0%). Thus, overall we confirm that recordings of vocalizations could be used as a tool to infer behaviours of *S. flavius* and possibly other primate species. Considering that knowledge on the vocal repertoire of *S. flavius* is available, an automated data collection system (Aide *et al.* 2013) could be developed and used effectively in different study sites for comparative purposes. Other species inhabiting the study sites could also be investigated from the same database of recordings. Common marmosets, for example, inhabit many areas where *S. flavius* exist and their vocal repertoire is fairly well known (Epple 1968; Bezerra & Souto 2008; Bezerra *et al.* 2011). Equipment such as the SM4 for land animals developed by Wildlife Acoustics, Inc. (www.wildlifeacoustics.com), can record over 300 hours of acoustic data and can be programmed to record in different hours to adapt to the activity period of the study species. Like camera traps (Bezerra *et al.* 2014; de Moraes *et al.* 2014; Olson *et al.* 2012; Tan *et al.* 2012), sound traps could become powerful tools for exploring behaviour of non-habituated primates especially in areas where it is difficult to visualize and follow the animals, or where risk of disease transmission occurs.

Acknowledgements

This study is part of the Blonde Capuchin Research Conservation Project/Projeto Galego. Our study was non-invasive and complied with Brazilian law (Licence number: 25727-1, MMA, ICMBio, SISBIO). The project has been supported by Mohamed bin Zayed Species Conservation Fund, Margot Marsh Biodiversity Foundation, Rufford Small Grant, FACEPE (Fundação de Amparo a Ciência e Tecnologia do Estado de Pernambuco) and CNPq (Brazilian National Counsel of Technological and Scientific Development). Logistical support has been kindly provided by the Mineradora Cristal. Monique Bastos was supported by a FACEPE PhD scholarship - Grant no: IBPG-0119-2.04/11). Karolina Medeiros is supported by a FACEPE MSc scholarship (Grant no: IBPG-0225-2.04-15). We are grateful to Sr. Rodrigo Costa and Sr. João Maria for assistance in the field. We also thank Sr. Geraldo Moraes and Virgílio Pinto for allowing access to the study sites.

Chapter 11

Mangrove Forests as a Key Habitat for the Conservation of the Critically Endangered Yellow-breasted Capuchin, *Sapajus xanthosternos*, in the Brazilian Northeast

Raone Beltrão-Mendes and Stephen F. Ferrari

Introduction

Tufted capuchins (*Sapajus* spp., cf. Lynch Alfaro *et al.* 2012a, 2012b) are the most versatile platyrrhine monkeys in terms of their behaviour and ecology. They occupy a wide range of habitats throughout most of tropical South America, including rainforests, savannas, scrub forests, swamps and mangroves (Canale *et al.* 2009; Freese & Oppenheimer 1981; Hill 1960; Jerusalinsky *et al.* 2006; Santos 2010; Terborgh 1983). The ecological success of the tufted capuchins is at least partly due to their physical strength and manipulative abilities (Janson & Boinski 1992; Terborgh 1983; Visalberghi & Antinucci 1986; Wright 2007), which enable them to forage for foods that may be unavailable to most, if not all, other platyrrhine monkeys. Under extreme conditions, tufted capuchins may resort to the use of tools in order to exploit relatively inaccessible resources, such as oysters (Fernandes 1991), coleopteran larvae (Rocha *et al.* 1998), palm kernels (Moura & Lee 2004) and crabs (Jerusalinsky *et al.* 2006; Santos 2010).

Despite their high productivity, mangrove forests are characterized by a greatly reduced plant diversity compared to other tropical forests, which is generally restricted to between one and four tree species adapted to the highly saline conditions of tropical estuaries (Chapters 2 and 4). These forests possibly represent the most extreme type of habitat occupied by tufted capuchins, due primarily to the general lack of palatable plant parts, as well as the scarcity of social insects, which constitute a major component of the diet of these monkeys at most sites (Freese & Oppenheimer 1981).

Until very recently, only anecdotal accounts were available of the occurrence of tufted capuchins in mangrove forests (Fedigan *et al.* 1996; Fernandes 1991; Jerusalinsky *et al.* 2006). These initial observations nevertheless indicated that these monkeys are able to exploit alternative resources available in these habitats, such as oysters (Dampier 1697; Fernandes 1991) or crabs (Jerusalinsky *et al.* 2006). Between 2004 and 2008, Santos (2010) conducted the first systematic study of tufted capuchins (*Sapajus apella* and *S. libidinosus*) in mangrove forest, on the eastern coast of the Brazilian state of Maranhão. Among other results, this study confirmed the use of tools (wooden hammers and anvils) for the predation of crabs, *Ucides cordatus* (see Chapter 9). More recently, Cutrim (2013) also conducted a systematic study of *S. libidinosus* tool use and feeding behaviour in a mangrove.

Perhaps surprisingly, there are few records of the occurrence of untufted capuchins (*Cebus* spp.) in mangrove habitats (Freese 1976; Manson *et al.* 1999). This is despite the fact that these monkeys are probably at least as flexible, in ecological and behavioural terms, as the tufted capuchins, to which they are closely related, having been considered congeners until very recently (Lynch Alfaro *et al.* 2012a, 2012b). In particular, *Cebus capucinus* has been well studied at a number of sites in Central America, including coastal areas (e.g. Fedigan *et al.* 1996; Fedigan & Jack 2001; Manson *et al.* 1999; Rose *et al.* 2003; Santos 2013; Chapter 8), and is the only untufted species to have been recorded in mangroves (Freese 1976). This may represent a sampling artefact, however, rather than an absolute difference in the ecological characteristics of the two genera. In fact, Dampier (1697) described the use of tools by *C. capucinus* to open oysters on Gorgona Island off the Pacific coast of Colombia more than 300 years ago. Similarly, in 1789, the catholic priest Johann Breuer reported the use of tools by a monkey (presumably *S. libidinosus*) in Ceará, northeastern Brazil (Papavero *et al.* 2011). While capuchins have gained increasing notoriety over the past decade for their tool-using abilities in non-flooded habitats (e.g. Spagnoletti *et al.* 2011), this behaviour has been documented since colonial times and appears to be especially relevant to the ecology of these monkeys in the mangrove (Chapter 8).

The yellow-breasted capuchin, *Sapajus xanthosternos*, occurs in eastern Brazil between the Jequitinhonha River in the south and the São Francisco River in the north and west (Figure 11.1). Much of this area is covered in semi-arid *Caatinga* scrub, a relatively harsh environment for primates, in which the occurrence of tufted capuchins may be determined by the availability of suitable habitats, such as gallery or other humid forests, as well as specific resources, such as *Attalea* and *Syagrus* palms, and appropriately sized stones for use as nut-cracking hammers (Canale *et al.* 2009). However, little is known of the occurrence of *S. xanthosternos* in the *Caatinga*,

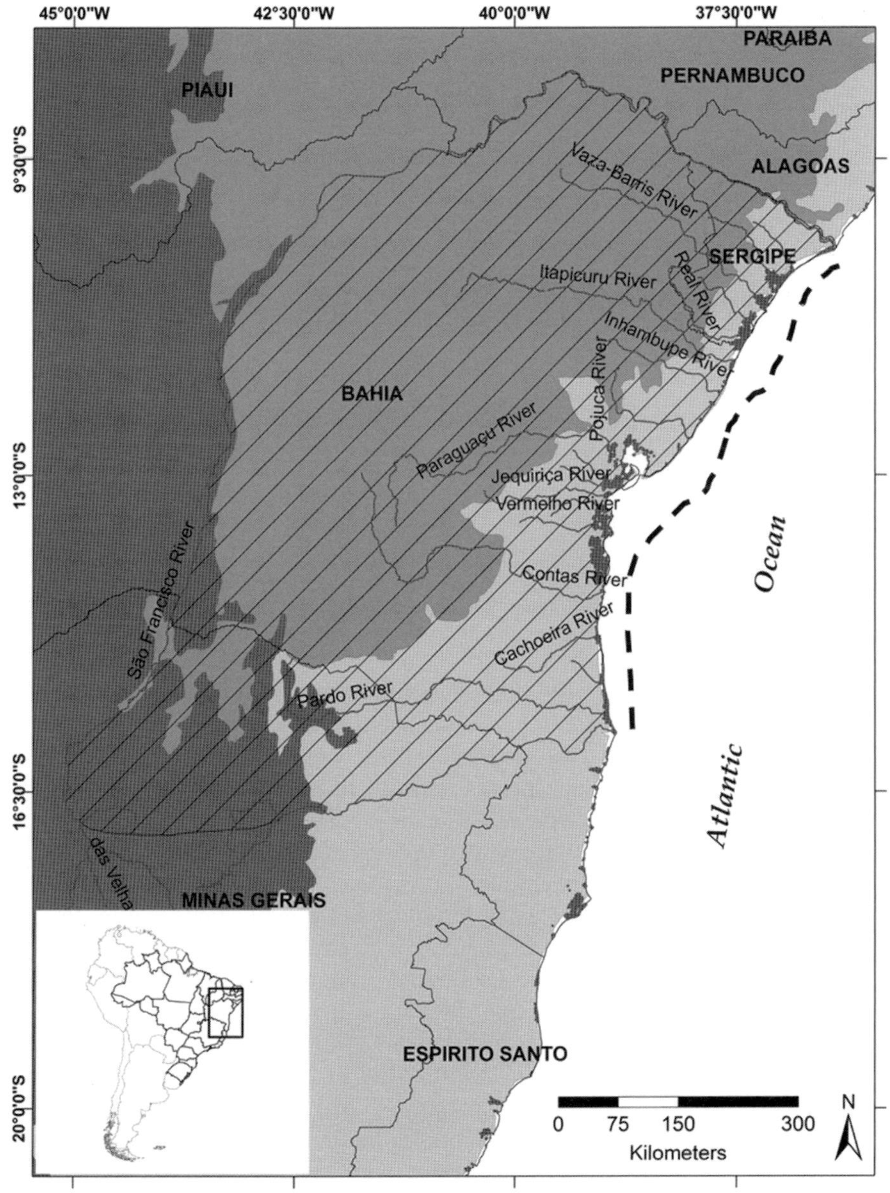

Figure 11.1 Distribution of *Sapajus xanthosternos* (hatched area) limited by the major rivers in the west and north (São Francisco), and south (Jequitinhonha). The species occurs mainly in the coastal Atlantic forest (light grey) and *Caatinga* (grey), and some areas of *cerrado* (dark grey). The survey area is indicated by the dashed line, and the principal areas of mangrove are shown in dark grey in the coastal zone.

except that the species may be patchily distributed and is probably rare at most sites where it does occur (Canale *et al.* 2009; Printes 2007), although it was more abundant and widespread in the past (Oliver and Santos 1991; Santos *et al.* 1987). Since these surveys, extensive deforestation has drastically reduced the available habitat, and most remaining *S. xanthosternos* populations appear vulnerable to hunting pressure (Beltrão-Mendes *et al.* 2011). The species is classified as Critically Endangered by the IUCN (Kierulff *et al.* 2008). Even so, the ecology of *S. xanthosternos* is still poorly known in general, and in particular with regard to its tolerance of anthropogenic impacts and the potential for the conservation of the species over the long term.

Most of the fieldwork on *S. xanthosternos* has focused on the populations in southern Bahia, especially in the Una Biological Reserve and neighbouring areas of forest (Canale 2010; Moreira 2009; Suscke 2014), which represent the largest tracts of Atlantic forest in the region. Further north, this biome is reduced to a narrow coastal strip, which is even more vulnerable to ongoing deforestation and habitat fragmentation (Jerusalinsky 2013) as well as hunting pressure (Beltrão-Mendes *et al.* 2011), although relatively large tracts of well-preserved mangrove forest persist in many areas. As mangrove ecosystems are fully protected under Brazilian legislation, they represent a potentially useful resource for the conservation of habitats and species, such as *S. xanthosternos*, depending on its capacity to survive in this environment over the long term.

Given this potential, the Atlantic coast at the eastern extreme of the geographic range of *S. xanthosternos* was investigated in order to identify mangrove-dwelling populations of this species, and their relationship with this ecosystem. The survey revealed a number of *S. xanthosternos* populations using mangroves, either exclusively or partially. As hunting pressure is negligible in the mangroves, these forests may represent an

Figure 11.2 Typical mangrove channel with coconut plantation and salt flats at Parapuca, at the mouth of the São Francisco River, Sergipe, Brazil.

important ecological refuge for the dwindling *S. xanthosternos* populations in coastal areas.

Methods

Study Area

The present study focused on the mangrove forests of the Atlantic coast of the Brazilian states of Sergipe and Bahia (Figure 11.1), ranging from the mouth of the São Francisco River (10°29′S, 36°25′W) as far south as the mouth of the Jequitinhonha River (15°49′S, 38°52′W), encompassing a total coastline of 750 km. This area represents the eastern extreme of the known geographic distribution of *S. xanthosternos*, and includes 13 major river estuaries and over 580 km^2 of mangrove.

Mangroves (Figure 11.2) represent an important natural resource for coastal communities throughout northern Brazil, especially for artisanal fisheries, in particular the harvesting of the mangrove crab, *Ucides cordatus* (Ucididae). Despite the fact that mangroves have been assigned permanent protection status under current Brazilian legislation, they are still subject to a certain degree of impact, especially where they coincide with urban areas, industrial installations, touristic developments or shrimp farms. In many rural areas, however, local communities are aware of the value of the mangrove for the sustainability of natural resources, especially fishery stocks, and generally do not disturb this habitat. Within the coastal *S. xanthosternos* study area, in addition, a number of sustainable-use protected areas, known as extractive reserves and environmental protected areas, can be found, covering some 11 873 km^2 (IBAMA 2014).

Procedures

The distribution of mangrove forests within the study area was evaluated initially through the combined analysis of the IBGE geographic database of rivers and human settlements (IBGE 2012) with the shapefile of the geographic distribution of *S. xanthosternos* available in the IUCN Red List (Kierulff *et al.* 2008). Based on this analysis, communities located in close proximity to estuaries were identified and subsequently visited for the collection of data on (1) mangroves, (2) fragments of *terra firme* (the term used in Brazil to describe types of never-flooded lowland rainforest as opposed to *terra firma*) forest, (3) the general presence of *S. xanthosternos* populations and (4) the specific occurrence of *S. xanthosternos* in local mangroves.

Data were collected from experienced local long-term residents (crab fishers) using non-directive, unprompted interviews (Chizzotti 2005). Crab fishers were considered to be local experts (Davis & Wagner 2003) on the natural resources found in the mangroves, including the native fauna. The presence of capuchins in the area surrounding the settlements visited was confirmed with the assistance of the coloured plates in Mittermeier *et al.* (2007) and recordings of *Sapajus* vocalizations available in Emmons *et al.* (1998).

Where the occurrence of capuchins in the local mangrove was confirmed, additional information was collected on the characteristics of the species, including its distribution, occurrence and behaviour. The presence of capuchins in the mangrove could be confirmed reliably by most informants due to the monkeys' habit of raiding their crab or fish traps. In a parallel study on Cabeço Island (Site 2, Table 11.1), camera trapping using baited platforms was deployed to obtain complementary information on the occurrence of capuchins (Figure 11.3).

Results

Between July and November 2012, a total of 61 sites were visited for the study (Table 11.1), during which 20 mangrove sites were located close (≤ 2 km) to fragments of *terra firme* forest. Evidence of the occurrence of local *S. xanthosternos* populations was found in 14 of *terra firme* fragments and in four areas of mangrove.

Table 11.1 Mangrove sites surveyed on the coasts of Sergipe and northern Bahia during the present survey.

Ref.	Locality	Geographic coordinates	Capuchins present		Area of mangrove (km²)	Distance to nearest *terra firme* forest (km)	Distance to nearest urban area or settlement (km)
			Mangrove	*Terra firme* forest			
1	Saramém	10°28′35′S, 36°25′40′W	–	Yes	32	Contiguous	Contiguous
2	Cabeço / Costinha	10°30′31′S, 36°25′23′W	Yes	–		–	–
3	Carapitanga	10°30′38′S, 36°29′35′W	Yes	Yes		Contiguous	Contiguous
4	Aracaré / Oitizeiro	10°31′30′S, 36°31′38′W	Yes	–		–	–
5	Ponta dos Mangues	10°33′16′S, 36°34′22′W	–	–		–	–
6	Pirambu	10°43′17′S, 36°51′15′W	Yes	Yes	8	> 2	Contiguous
7	Santo Amaro das Brotas	10°47′21′S, 37°03′23′W	–	Yes	25	> 2	Contiguous
8	Areia Branca	11°03′03′S, 37°07′55′W	–	–	12	–	Contiguous
9	Porto dos Caibros	11°06′33′S, 37°13′40′W	–	Yes	9	> 5	> 10
10	Porto do Mato	11°24′40′S, 37°21′01′W	–	–	38	Contiguous	> 20
11	Pontal (Lado da Bahia)	11°27′50′S, 37°24′55′W	Yes	Yes	38	Contiguous	> 1.0
12	Poças	11°48′54′S, 37°32′48′W	Yes	Yes	19	Contiguous	> 2.0
13	Barra do Itariri	11°56′55′S, 37°37′16′W	–	–	2		Contiguous
14	Povoado Mata	12°04′29′S, 37°43′12′W	–	–	3	> 20	Contiguous
15	Subaúma	12°14′40′S, 37°47′13′W	–	–	> 5	–	Contiguous
16	Massarandupió	12°18′20′S, 37°51′09′W	–	–	Absent	–	–
17	Porto Sauípe	12°22′18′S, 37°53′05′W	–	–	3	–	Contiguous
18	Imbassaí	12°29′36′S, 37°57′33′W	–	–	Absent	–	Contiguous
19	Praia do Forte	12°34′28′S, 38°00′36′W	–	Yes	Absent	> 1	Contiguous
20	Comunidade Beira do Rio	12°35′39′S, 38°02′41′W	–	Yes	> 1	> 1	Contiguous
21	Barra do Jacuípe	12°41′56′S, 38°08′08′W	–	–	5	> 10	Contiguous
22	São Francisco do Paraguaçu – north of Todos os Santos Bay	12°44′37′S, 38°52′24′W	–	Yes	17	> 1	Contiguous
23	Foz do Rio Joanes	12°51′39′S, 38°17′23′W	–	–	> 1	> 10	Contiguous
24	Salinas das Margaridas	12°52′22′S, 38°45′22′W	–	–	2	Contiguous	Contiguous
25	Cações	13°00′19′S, 38°47′30′W		–	–	–	–

(continued)

Table 11.1 *(cont.)*

Ref.	Locality	Geographic coordinates	Capuchins present		Area of mangrove (km²)	Distance to nearest *terra firme* forest (km)	Distance to nearest urban area or settlement (km)
			Mangrove	*Terra firme* forest			
26	Matarandiba	13°00′29′S, 38°45′42′W		–	–	–	–
27	Ilha da Banca	13°02′15′S, 38°48′40′W	–	–	14	–	Contiguous
28	Aratuípe	13°04′17′S, 39°00′13′W	–	–	5	–	Contiguous
29	Jaguaripe	13°16′47′S, 38°58′05′W	–	–	18	–	Contiguous
30	Cacha Pegros / Jeribatuba	13°22′01′S, 39°04′22′W	–	–	5	> 10	Contiguous
31	Camassandi	13°24′57′S, 39°04′52′W	–	–	6	–	Contiguous
32	Guaibim	13°28′43′S, 39°05′23′W	–	–	3	–	> 10
33	Valença	13°29′25′S, 39°02′47′W	–	–	170	–	Contiguous
34	Maricoabo/Cajaíba	13°32′16′S, 39°06′03′W	–	–		–	Contiguous
35	Graciosa (28)	13°35′03′S, 39°00′51′W	–	–		–	Contiguous
36	Cairu	13°36′24′S, 39°06′02′W	–	–		–	Contiguous
37	Taperoá (São Felipe)	13°39′03′S, 38°58′45′W	–	–		–	Contiguous
38	Boitaraca (quilombo)	13°43′29′S, 39°08′53′W	–	–		> 1	Contiguous
39	Nilo Peçanha	13°50′56′S, 39°04′58′W	–	Yes		–	Contiguous
40	Barra dos Carvalhos	13°53′31′S, 39°08′18′W	–	–		> 1	Contiguous
41	Ituberá	13°54′24′S, 38°58′48′W	–	Yes		–	Contiguous
42	Pau d'Óleo (Igrapiúna)	13°56′23′S, 39°06′25′W	–	–		> 1	–
43	Pinaré	13°56′31′S, 38°58′32′W	–	–		–	Contiguous
44	Campinho	13°57′44′S, 39°01′55′W	–	–		> 20	–
45	Camamu	13°57′52′S, 39°01′00′W	–	–		–	Contiguous
46	Taipu de Dentro	14°02′27′S, 39°00′48′W	–	–	25	> 20	Contiguous
47	Cajaíba	14°04′36′S, 39°05′29′W	–	–		> 10	Contiguous
48	Aldeia Velha	14°06′12′S, 39°00′49′W	–	–		> 10	Contiguous
49	Barcelos do Sul	14°16′42′S, 38°59′45′W	–	–		> 10	Contiguous
50	Tapuia	14°41′24′S, 39°05′18′W	–	–		> 2	Contiguous

51	Marau	14°51′08′ S, 39°04′00′W	–	Yes		> 2	Contiguous
52	Itacaré	14°52′46′ S, 39°01′37′W	–	–	1	–	Contiguous
53	Aritaguá	15°04′59′ S, 38°59′55′W	–	–	2	–	Contiguous
54	Rio do Engenho	15°16′23′ S, 39°01′20′W	–	–	8	–	Contiguous
55	Cururupe	15°21′03′ S, 39°00′04′W	–	–	> 1	–	Contiguous
56	Acuípe	15°23′58′ S, 38°59′51′W	–	–	> 1	Contiguous	Contiguous
57	Pedras de Una	15°26′43′ S, 38°59′58′W	–	Yes	14	Contiguous	Contiguous
58	Comandatuba	15°40′19′ S, 38°57′13′W	–	–	90	–	Contiguous
59	Oiticica	13°16′47′ S, 38°58′05′W	–	–		–	Contiguous
60	Poxim do Sul	13°22′01′ S, 39°04′22′W	Yes	–		–	Contiguous
61	Canavieiras (porção central)	13°24′57′ S, 39°04′52′W	Yes	–		–	Contiguous

Figure 11.3 Camera trap images of *Sapajus xanthosternos* from Cabeço Island, in the Parapuca mangrove complex, at the mouth of the São Francisco River, Sergipe, Brazil. (a) An adult individual and (b) a partial view of the baiting platform.

Figure 11.4 A funnel trap type, made of bamboo, used to capture fish and crabs by local fishers in (a) lateral and (b) frontal view.

At the remaining 10 sites, local informants reported that capuchins were very occasional visitors, only coming to the mangrove to exploit specific resources, such as coconuts (*Cocos nucifera*), before returning to the *terra firme* forest. As plantations of coconuts and other crops (e.g. beans, cassava, and watermelon) occur only on salt flats – known locally as the *apicum* – close to the mangroves, it seems likely that the capuchins are using the mangroves primarily as corridors of access to preferred habitats.

At an additional five sites, the presence of capuchin populations was confirmed in mangroves completely isolated from *terra firme* forests, indicating that the monkeys are permanent residents in these mangrove areas. At these sites, local residents reported that the fragments of *terra firme* forest persisted until the recent past (within the last decade), when they were replaced by agricultural plots.

Local informants at two sites (6 and 12) reported that the capuchins have started to frequent the mangroves more intensively during the past three or four years, and rather than just passing through the forest on their way to raid plantations, they have begun to exploit new resources available in the mangrove itself, such as fish and crabs (obtained from traps) and the wood-boring mollusc (*Neoteredo reyni*, Teredinidae – known as 'shipworm'). One informant at Pirambu (Site 6), eastern Sergipe, reported that he had stopped using the traditional funnel traps, known locally as *covos* (Figure 11.4), because capuchins had begun raiding them to remove the fish they contained, destroying the traps in the process. Two other informants reported the same behaviour at Poças (Site 12).

In the Canavieiras mangrove complex (Sites 59–61), a local informant reported that in the past, capuchins used to cross a local village on the ground to raid coconut plantations and to move between the *terra firme* forest and the mangrove. The capuchins stopped doing this, however, when the local road was paved. A second informant at Canavieiras confirmed that the capuchins had become permanent residents of the mangrove due to the loss of the remaining *terra firme* forests.

A similar situation was observed in the Parapuca area (Sites 2–4), where the long-term survival of local capuchin populations may be dependent on the natural dynamics of the

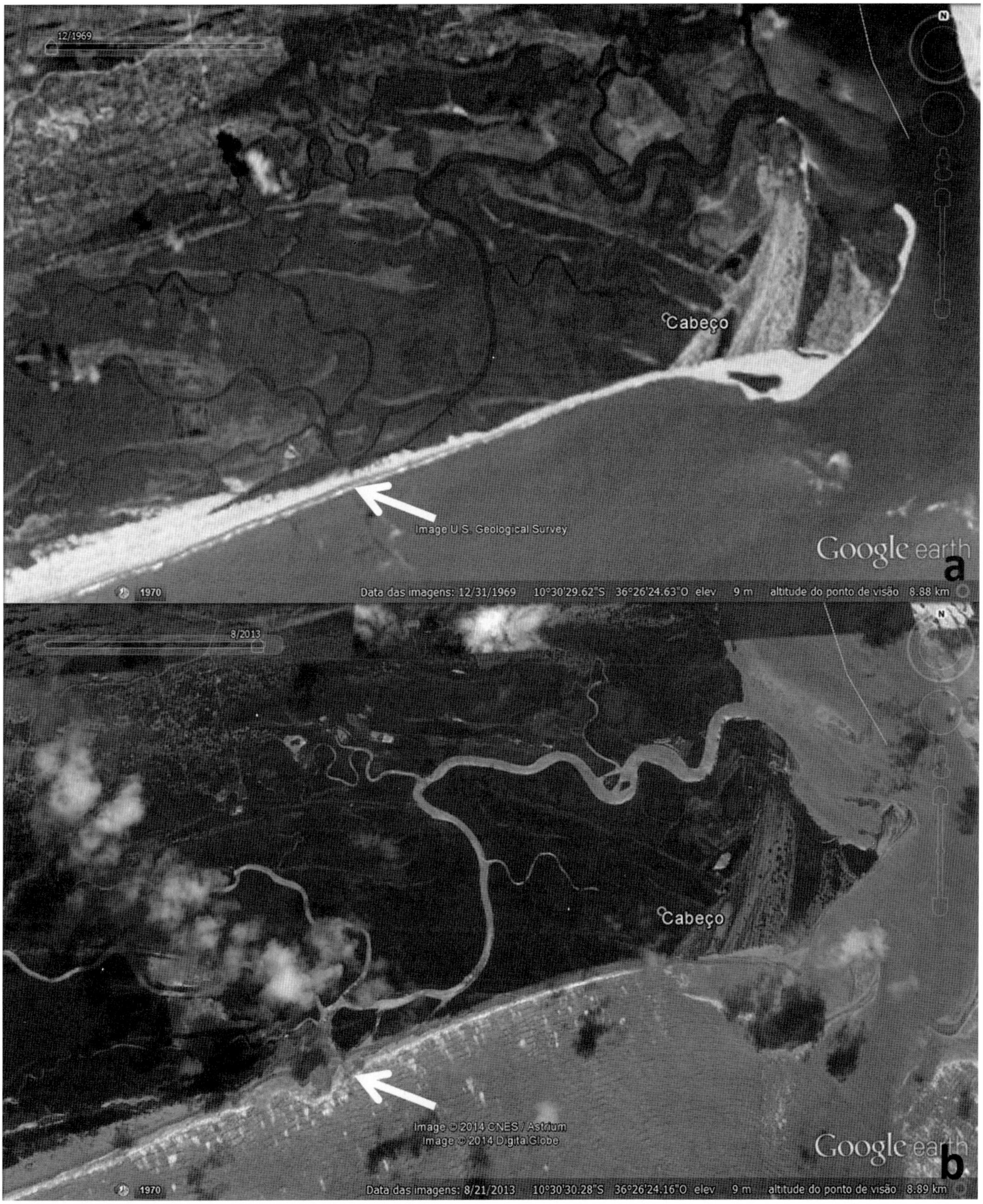

Figure 11.5 Satellite images showing the changes in the geomorphology of the Parapuca mangrove complex at the mouth of the São Francisco River in the eastern extreme of Sergipe. The white arrow indicates (a) the continuous land in 1969 and (b) the secondary mouth of the São Francisco River in 2013, creating a new island since 2010. Map data: Google, U.S. Geological Survey, CNES/Astrium and DigitalGlobe.

mangrove environment, characterized by constant shifts in the configuration of its islands and channels. This process can be observed in the four-decade sequence of satellite images of the area covering the past four decades (Figure 11.5). Recently, a tidal creek formed between Cabeço/Costinha Island and the mainland, effectively isolating this island's *S. xanthosternos* population. In previous years, no such channel was present, even during high tide, and the contiguity of the mangroves almost certainly represented an effective corridor for the dispersal of the capuchins to neighbouring areas.

As no satellite images are available prior to 1969, the longer-term time scale of these changes in the configuration of the habitat is unclear, although the remains of silted creeks and channels can be seen clearly in many images. In addition to guaranteeing access to specific habitats and resources, this process may be important to maintain gene flow and genetic variability in the local populations, which will almost invariably be small in size. Over the long term, the natural dynamics of the mangrove ecosystem may effectively create a local capuchin metapopulation.

In contrast, capuchins were reported to be absent or only occasional visitors at the other 11 sites where fragments of *terra firme* forest persist in the vicinity of the mangroves. In general, the evidence indicates that where *terra firme* forests provide an adequate resource base (e.g. Site 22), the capuchins may ignore the mangrove completely.

Discussion

The geographic distribution of the Critically Endangered yellow-breasted capuchin, *Sapajus xanthosternos*, was surveyed extensively in the 1980s and 1990s (Santos *et al.* 1987; Coimbra-Filho *et al.* 1991; Oliver & Santos 1991), but while a number of new sites were identified in the Atlantic forest, no data were obtained on the occurrence of the species in adjacent mangrove forests. While Santos *et al.* (1987) and Oliver and Santos (1991) recorded *S. xanthosternos* populations in coastal areas in the vicinity of mangrove habitats, Jerusalinsky *et al.* (2006) were the first to confirm the occurrence of this species in mangrove when they encountered a population on Cabeço Island, at the mouth of the São Francisco River, the eastern extreme of the species' distribution.

The present study has shown that the presence of yellow-breasted capuchin populations in mangrove habitats tends to result from the loss of adjacent *terra firme* forests, and is made possible by a combination of the natural and anthropogenic dynamics within these flooded forest habitats, together with the ability of the capuchins to exploit relatively difficult to access resources, such as crabs, oysters, shipworm and coconuts. While the capuchins do appear to be able to survive in mangrove, at least over the short term, they may possibly be dependent on adjacent areas of *terra firme* forest or plantations of crops such as cassava as a complement to their diet, although the exploitation of these resources also implies increasing anthropic pressures and other risks.

In many other areas of Sergipe state, further inland, hunting pressure from neighbouring communities may have contributed to the local extinction of capuchin populations, even where remnant forests are relatively extensive by regional standards (Beltrão-Mendes *et al.* 2011). In most cases, however, the mangrove-dwelling populations of *S. xanthosternos* are found in areas of relatively sparse human populations, where most residents depend on the harvesting of resources such as fish and crustaceans for subsistence, rather than on hunting. The lack of a hunting tradition appears to be an additional positive factor, although this situation may vary considerably, and the lack of mangrove-dwelling capuchin populations in the vicinity of larger urban centres likely reflects increasing anthropic pressures, as seen in other areas further inland.

While no direct measurements have been made, feeding resources for capuchins in the mangrove appeared to be limited primarily to molluscs and crustaceans, as well as some social insects (termite and wasp nests: R. Beltrão-Mendes, pers. obs.), although red mangrove propagules (*Rhizophora mangle*, Rhizophoraceae) and coconut fibres (*Cocos nucifera*, Arecaceae) were found in faecal samples collected at Site 1. Overall, capuchins are clearly rare in the mangroves of the study area, being absent altogether from most sites and occurring at low densities in others. As population density will generally correlate with the availability of resources (Robinson & Janson 1987), large home ranges and low densities would be expected in the mangrove in general, and this ecosystem may impose other limitations, related to both resource availability and population dynamics.

Overall, then, while the capuchins obviously prefer *terra firme* forest, probably because of the richer resource base offered by this type of habitat, mangroves appear to offer a refuge of last resort. It is still unclear whether *S. xanthosternos* populations will be able to survive over the long term in mangrove alone, but in areas such as that of Cabeço Island (Site 1), Pontal (Site 11) and Poças (Site 12), the existing scenario, which encompasses a constantly shifting landscape and a mosaic of resources, may represent a viable refuge for future generations. Given the bleak outlook for the species in areas further inland (Beltrão-Mendes *et al.* 2011), these mangrove systems appear to represent an important resource for the conservation of *S. xanthosternos* in years to come.

Acknowledgements

This study was financed by CAPES through a postgraduate stipend to RB-M, CNPq funding to RB-M (503372/2014-5 and 150123/2018-3) and SFF (Projects 476064/2008-2 and 303994/2011-8), Mohamed bin Zayed Species Conservation Fund (Project 12055114), Primate Conservation Inc. (Project 1158) and Primate Action Fund (Project 1001257), PROBIO II. We also thank P. Rocha, J. Ruiz-Esparza, S. Silvestre, L. Jerusalinsky and G. Giné for their help and support during the collection of data.

Chapter 12

Primates of African Mangroves

Josephine Head, Aoife Healy and Katarzyna Nowak

African Mangroves

Mangrove forests provide important habitat for a wide variety of fauna and can be rich in biodiversity. To date, 70 species of mangrove have been recorded worldwide (Spalding *et al.* 1997 in Corcoran *et al.* 2007). While mangroves represent important communities across the globe, this chapter is focused on discussing the use of mangroves by anthropoid primates in Africa. African mangroves represent almost one-fifth of the world's mangroves, and are found across 26 countries of sub-Saharan Africa (Corcoran *et al.* 2007). There is variation in the phytogeographical distribution of mangrove species across Africa, with an estimated 70% of Africa's mangroves located in the equatorial regions of West and Central Africa, the same region in which African primate species richness is highest (Eeley & Foley 1999). Mangroves also occur locally in East Africa, where species composition resembles other mangroves of the Indian Ocean, while species composition in West African mangroves is more similar to mangroves of the Americas (WWF 2001).

In West and Central Africa, six mangrove species from three families occur including Acanthaceae (*Avicennia germinans*), Combretaceae (*Laguncularia racemosa, Conocarpus erectus*) and Rhizophoraceae (*Rhizophora harrisonii, R. mangle, R. racemosa*); *R. racemosa*, characterized by prop roots, is particularly dominant in this region. In Eastern Africa, ten species of mangroves are found, with three dominant species (*R. mucronata, Ceriops tagal* (Rhizophoraceae) and *A. marina* (Semesi 1998)) occupying a total of 1.1 million hectares (Spalding *et al.* 1997). Mangroves are restricted to coastal areas at river mouths or tidal lagoons and prefer warmer seas in humid tropical climates, although in parts of Mozambique and Tanzania, mangroves extend up to 50 km inland (WWF 2013).

Increasing human populations, demand for resources, and agricultural development all threaten the future of African mangroves and the species that inhabit them (see Chapter 43). Given increasing fragmentation of remaining forests, improving our understanding of the importance of mangroves as a refuge for species under intense human pressure and establishing what level of biodiversity mangroves can support is relevant for the conservation of primate populations (Galat-Luong & Galat 2005, 2007; Nowak 2012). This is particularly important in areas with high primate species richness, as species richness is associated with reduced habitat and dietary breadth among primates, or greater specialization in habitat use and diet (Eeley & Foley 1999). As explored in this chapter, some primates have become highly specialized mangrove users, and the disappearance of mangrove habitat could have dire consequences for their survival in areas where other niches are already occupied by sympatric primates. This chapter aims to explore African primates' different occupancy types in mangroves, examine how widespread primate use of mangroves is across Africa, and in doing so, review the importance of mangroves for African anthropoid primates.

Literature Review

Only studies or observations that made explicit reference to mangrove use by primates were included in this chapter. Mention by *The Directory of African Wetlands, All the World's Primates* website, *Ramsar Site Information Service – African Wetland* or *The IUCN Red List* of mangrove use (though essential as literature search tools) did not qualify as sufficient for inclusion here. Where the above online resources listed mangrove use by certain primate species, such occurrences were investigated further and relevant primary sources cited. Additional sources included *Field Guide to Primates of West Africa* (Oates 2011) and *Histoire Naturelle des Primates d'Afrique Centrale* (Gautier-Hion *et al.* 1999). If no further information was found, the data were excluded from this chapter. Table 12.1, organized according to subregion, taxon, area of study and primary occupancy type (as and when sufficient information was available), is not a definitive or complete table, given that mangrove use by primates is a relatively data-deficient area of study. More comprehensive, but still preliminary lists, are given in Nowak and Coles (Chapter 6), Nowak *et al.* (Chapter 43) and Nowak (2012).

Mangrove Distribution and Primate Species Richness

A map of mangrove sites used by primates was created (using ArcMAP v. 9.2) indicating number of mangrove-using primates at each site (Figure 12.1). Our definition of mangroves is consistent with Bennett *et al.* (Chapter 2).

In Table 12.1, extent of study (1–3) is given, with '3' representing well-detailed studies, for example that in the Saloum Delta National Park, Senegal, specifically aimed at

Table 12.1 Primate species, their location and primary occupancy type in mangroves.

Subregion	Country	Area	Wetland status	Common name	IUCN Red List status	Genus	Species	Subspecies	Level of detail	Occupancy type	Reference
W	Senegal	Niokolo-Koba	National Park	Guinea baboon	Near Threatened	*Papio*	*papio*		1	1	Galat-Luong *et al.* 2006
W	Senegal	Saloum Delta	National Park & Ramsar	Green monkey	Least Concern	*Chlorocebus*	*sabaeus*		3	1	Galat & Galat-Luong 1976
W	Senegal	Saloum Delta	National Park & Ramsar	Green monkey	Least Concern	*Chlorocebus*	*sabaeus*		2	NEI	Galat-Luong & Galat 2005
W	Senegal	Saloum Delta	National Park & Ramsar	Patas monkey	Least Concern	*Erythrocebus*	*patas*		2	3	Galat-Luong & Galat 2005
W	Senegal	Toubacouta-Sangalo area		Guinea baboon	Near Threatened	*Papio*	*papio*		2	NEI	Galat-Luong & Galat 2013
W	Senegal	Saloum Delta	National Park & Ramsar	Temminck's red colobus	Endangered	*Piliocolobus*	*badius*	*temmincki*	3	3	Galat-Luong & Galat 2005
W	The Gambia	Saloum Delta	National Park & Ramsar	Green monkey	Least Concern	*Chlorocebus*	*sabaeus*		1	NEI	Pourrut *et al.* 1996
W	Guinea-Bissau	Cantanhez Forest	National Park	Green monkey	Least Concern	*Chlorocebus*	*sabaeus*		2	1	Gippoliti & Del'Omo 1996
W	Guinea-Bissau	Cantanhez Forest	National Park	Western chimpanzee	Endangered	*Pan*	*troglodytes*	*verus*	2	NEI	K. Hockings, pers. comm.
W	Sierra Leone			Campbell's monkey	Least Concern	*Cercopithecus*	*campbelli*		2	NEI	Grubb *et al.* 1998
W	Côte d'Ivoire	Ébrié Lagoon		Green monkey	Least Concern	*Chlorocebus*	*sabaeus*		2	NEI	Galat 1983
W	Côte d'Ivoire	Ébrié Lagoon		Eastern lesser spot-nosed monkey	Least Concern	*Cercopithecus*	*petaurista*	*petaurista*	2	NEI	Galat & Galat-Luong, pers. comm.
W	Côte d'Ivoire	Ébrié Lagoon		Olive colobus	Near Threatened	*Procolobus*	*verus*		1	?	Galat & Galat-Luong pers. comm.
W	Côte d'Ivoire	Iles Ehotilé	National Park & Ramsar	Green monkey	Least Concern	*Chlorocebus*	*sabaeus*		3	3	Bi *et al.* 2009
W	Nigeria	Niger Delta	UNESCO World Heritage Site (tentative list)	Mona monkey	Least Concern	*Cercopithecus*	*mona*		1	3	Werre 2001a
C	Cameroon	Southern Bakundu	Forest Reserve	Mona monkey	Least Concern	*Cercopithecus*	*mona*		2	3	Gartlan & Struhsaker 1972
C	Equatorial Guinea	Rio Muni	Ramsar	Red-capped mangabey	Vulnerable	*Cercocebus*	*torquatus*		3	2	Jones & Sabater Pi 1968

C	Gabon	Sette Cama	On edge of Loango National Park, Ramsar	Red-capped mangabey	Vulnerable	*Cercocebus*	*torquatus*		2	NEI	Cooke 2005
C	Gabon	Loango	National Park & Ramsar	Red-capped mangabey	Vulnerable	*Cercocebus*	*torquatus*		2	1	Head, pers. obs.
C	Gabon	Loango	National Park & Ramsar	Northern talapoin	Least Concern	*Miopithecus*	*ogouensis*		1	NEI	Head, pers. obs.
C	Gabon	Loango	National Park & Ramsar	Western lowland gorilla	Critically Endangered	*Gorilla*	*gorilla*	*gorilla*	3	2	Head, pers. obs.
C	Gabon	Loango	National Park & Ramsar	Central chimpanzee	Endangered	*Pan*	*troglodytes*	*troglodytes*	3	2	Head, pers. obs.
C	Gabon	Loango	National Park & Ramsar	Grey-cheeked mangabey	Least Concern	*Lophocebus*	*albigena*		2	2	Head pers. obs.
C	Gabon	Loango	National Park & Ramsar	Crowned monkey	Least Concern	*Cercopithecus*	*pogonias*		2	2	Head, pers. obs.
C	Gabon	Loango	National Park & Ramsar	Moustached monkey	Least Concern	*Cercopithecus*	*cephus*		2	2	Head, pers. obs.
C	Gabon	Loango	National Park & Ramsar	Putty-nosed monkey	Least Concern	*Cercopithecus*	*nictitans*		2	2	Head, pers. obs.
E	Mozambique	Vamizi Island		Sykes' monkey	Least Concern	*Cercopithecus*	*mitis*	*erythrarchus?*	1	NEI	Jubber & Trindade, pers. comm.
E	Mozambique	Zambezi Delta		Sykes' monkey	Least Concern	*Cercopithecus*	*mitis*	*albogularis*			Nowak, pers. comm.
E	Tanzania/ Mozambique	Ruvuma River		Sykes' monkey	Least Concern	*Cercopithecus*	*mitis*	*albogularis*			Nowak, pers. comm.
E	Tanzania	Saadani	National Park	Sykes' monkey	Least Concern	*Cercopithecus*	*mitis*	*albogularis*			Nowak, pers. comm.
E	Zanzibar	Uzi Island		Sykes' monkey	Least Concern	*Cercopithecus*	*mitis*	*albogularis*	1	2	Nowak & Lee 2010
E	Tanzania	Pangani		Vervet monkey	Least Concern	*Chlorocebus*	*pygerythrus*	*nesiotes*	2	2	de Jong & Butyksni, pers. comm.
E	Tanzania	Saadani	National Park	Vervet monkey	Least Concern	*Chlorocebus*	*pygerythrus*	*nesiotes*	2	2	de Jong & Butyksni pers. comm.
E	Zanzibar	Pemba Island		Vervet monkey	Least Concern	*Chlorocebus*	*pygerythrus*	*nesiotes*	2	2	de Jong & Butyksni pers. comm.

(continued)

Table 12.1 *(cont.)*

Subregion	Country	Area	Wetland status	Common name	IUCN Red List status	Genus	Species	Subspecies	Level of detail	Occupancy type	Reference
E	Zanzibar	Uzi Island		Zanzibar red colobus	Endangered	*Piliocolobus*	*kirkii*		3	3	Nowak 2008; Nowak & Lee 2010
E	Zanzibar	Uzi Island		Garnett's great galago	Least Concern	*Otolemur*	*crassicaudatus*		1	3?	Nowak *et al.* 2011
E	Zanzibar	Uzi Island		Zanzibar galago	Least Concern	*Galagoides*	*zanzibaricus*		1	3?	Nowak *et al.* 2011
E	Kenya	Kiunga		Vervet monkey	Least Concern	*Chlorocebus*	*pygerythrus*	*hilgerti*	2	2	de Jong & Butyksni pers. comm.
E	Kenya	Lamu Archipelago	Lamu town is a UNESCO World Heritage Site	Vervet monkey	Least Concern	*Chlorocebus*	*pygerythrus*	*excubitor??*	2	2	de Jong & Butyksni, pers. comm.
E	Kenya	Lamu Archipelago	Lamu town is a UNESCO World Heritage Site	Yellow baboon	Least Concern	*Papio*	*cynocephalus*	*ibeanus*	2	2	Butynski & De Jong, pers. comm.
E	Kenya	Lamu Archipelago – N Lamu and NW Manda Is.	Lamu town is a UNESCO World Heritage Site	Pousargues's white collared monkey	Vulnerable	*Cercopithecus*	*mitis*	*albotorquatus*	1	NEI	De Jong & Butynski 2009
E	Kenya	Kwale District		Angolan black-and-white colobus	Least Concern	*Colobus*	*angolensis*	*palliatus*	1	2	Anderson *et al.* 2007

Level of study 1–3 (1 = understudied, 2 = inventories/surveys, 3 = detailed study; NEI = not enough information).

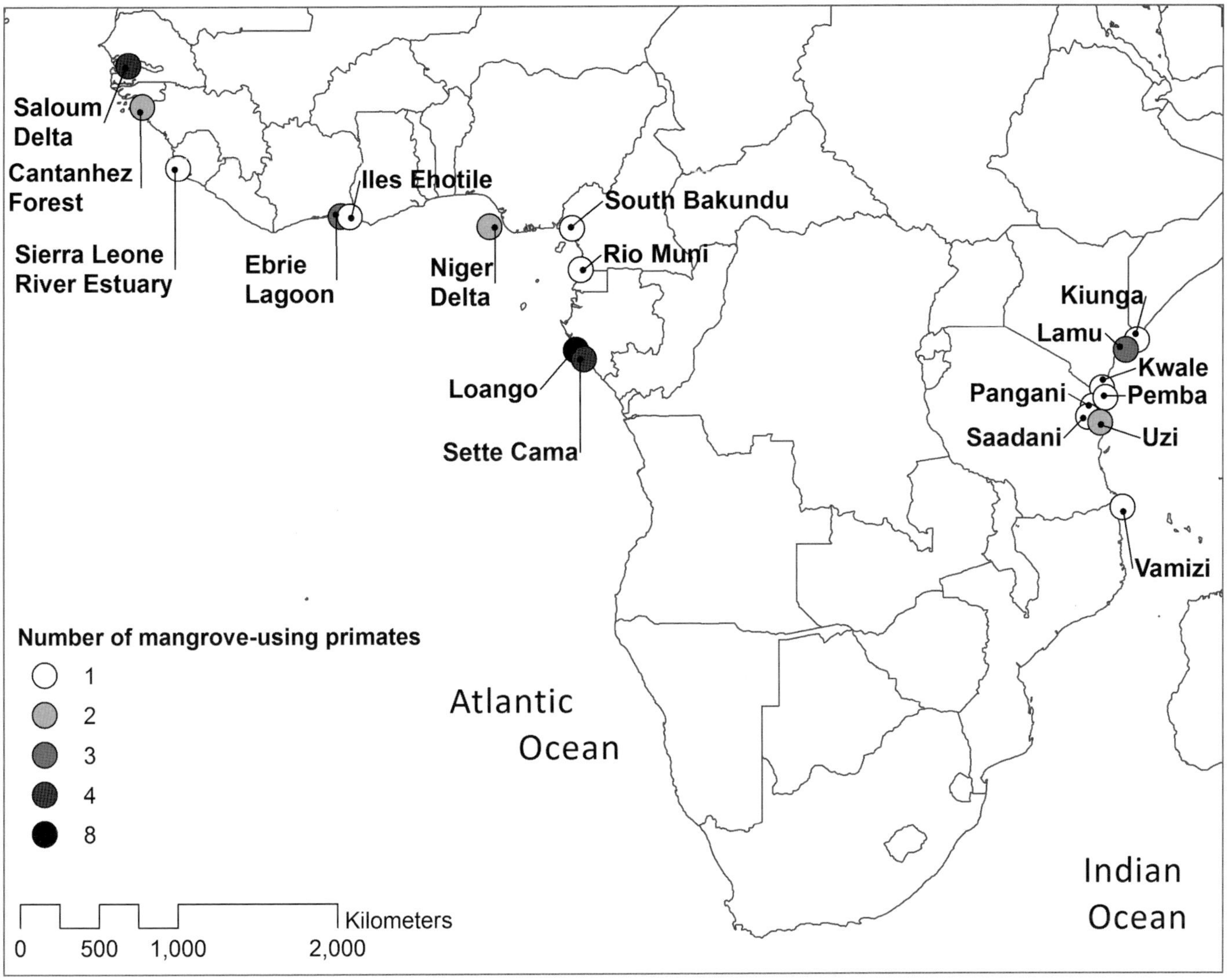

Figure 12.1 Number of mangrove-using primates at 18 sites included in this review.

investigating mangrove use by green monkeys *Chlorocebus sabaeus*. Sites at which only an observation (such as a sighting or vocalization) was used to confirm presence of a species in mangroves are ranked '1'. The purpose of the map and table is to provide visual representation of mangrove areas used by primates and to highlight those in need of further study, respectively.

Occupancy Types

Occupancy was defined as one of three types:

1. Historical occupancy:
   Mangrove use is frequent, and mangrove habitat plays an important role in daily activities. Species exhibit morphological or behavioural adaptations for mangrove use, such as dietary specialization on mangrove plants.
2. Seasonal, occasional and opportunistic occupancy:
   Occupancy of mangroves is temporary or seasonal for feeding or shelter, or associated with a specific function such as dispersal or movement between adjacent habitats.
3. Potentially novel extension of range or habitat shift:
   Novel uses of mangroves resulting from a habitat shift, an increase in frequency or duration of mangrove use, or a change of occupancy type in mangroves (e.g. from use of mangroves as a corridor to feeding in mangroves). Such changes may occur in response to human pressure, interspecific competition, or environmental change.

Results

Twenty-four mangrove-using primate species in 23 locations and their primary occupancy types are given in Table 12.1. This table is followed by more detailed case studies describing the nature of mangrove use by some of these specific species, for which relatively rich data were available.

Of the 42 studies in Table 12.1, five give a detailed report of mangrove use by the study species. The remainder of the studies include minimal data, sightings or vocalizations to confirm presence, often as part of a survey/census. We know little about the nature of mangrove use by many of the species included here including *Pan troglodytes, Lophocebus albigena, Miopithecus ogouensis, Cercopithecus campbelli, C. pogonias, C. cephus* and *C. nictitans*.

Mangrove sites included here supported between one and eight species with mangroves of Loango National Park in Gabon supporting the most at eight species.

Case Studies of Mangrove Use

Historical Occupancy

Chlorocebus sabaeus, West African Mangroves

Chlorocebus sabaeus is an opportunistic generalist exploiting a broad habitat range from savanna woodland and dry forest to gallery forest and coastal scrub. It adapts relatively well to disturbed, secondary growth areas, and will exploit tourist lodges and cultivated land (Boulton *et al.* 1995; Brennan *et al.* 1985; Horrocks & Hunte 1986; Kavanagh 1980). *C. sabaeus* also colonizes mangroves, an ability attributed to its semi-terrestrial travel and capacity for dietary adjustments (Galat & Galat-Luong 1976).

Mangrove use by *C. sabaeus* has been reported in several West African countries including Guinea-Bissau, Ivory Coast, The Gambia (Figure 12.2) and Senegal with occupancy types varying across study sites. A study of *C. sabaeus* in mangroves in Senegal demonstrated the central role which mangroves may play in the lives of some primate species (Galat & Galat-Luong 1976). The home range of the study group along the mangrove – *terra firma* border in the estuary of the Saloum River was characterized by the presence of *Rhizophora mangle, R. racemosa* (Rhizophoraceae) and *Avicennia nitida* (Acanthaceae), and the study site was described as one of the best preserved mangroves in Senegal. Mangroves were used by *C. sabaeus* for all activities including moving, resting, feeding, sheltering from predators, and even establishing sleeping sites with up to 80% of the group's time spent in mangroves. Mangroves were frequented most often during the hotter parts of the day for rest, shade, and water when monkeys were often seen with their bodies flush with the water's surface, presumably for thermoregulatory purposes. Movement through the mangroves was generally from crown to crown, and less frequently by wading through the water (in contrast to the use of mangroves' prop roots for travel by *Cercocebus torquatus* and *Piliocolobus kirkii*, see below).

Mangroves were also important for safety. When in mangroves, *C. sabaeus* were observed eating in a measured manner throughout the day, whereas when on the ground, eating was rushed and vigilant, with monkeys filling cheek-pouches and retreating to the relative safety of the mangroves. *C. sabaeus* inevitably fled into the mangroves for shelter in the presence of perceived predation risk. Seventy-five per cent of feeding took place in the mangroves with mangroves themselves constituting an important part of the diet including fruits of *A. nitida*, and the fruit, flowers, young shoots and leaves, seeds, twigs and spinal roots of *Rhizophora sp.* Fiddler crabs (*Uca tangeri*) were hunted daily and formed an apparently significant part of the diet, an observation supported by much earlier reports that 'mangrove monkeys were fond of *U. tangeri*' (MacLaud 1906). Small unidentified crabs and oysters were also consumed, and local people reported *C. sabaeus* eating fish. Evidence of consumption of other crustaceans (molluscs) was evident from faeces, highlighting the importance of mangrove habitat for this species.

Papio papio, Senegal

P. papio inhabits a wide range of habitats from coastal mangrove to Sahelian steppe within reach of water (Oates 2011), also occurring in shrub, woodland savanna, gallery and secondary forests in the south of its range (Oates *et al.* 2008). While *P. papio* is the least studied of the five *Papio* species (Oates 2011), there is evidence of mangrove use. In one case, a released group of *P. papio* in West Senegal in the region of Toubacouta, was observed entering mangroves to feed on *Uca tangeri* (Galat-Luong & Galat 2013). These baboons were released over a period of several years (at least 1989–1992) by a monkey dealer who was disposing of excess 'stock' (that was no longer 'useful').

Cercopithecus mitis albotorquatus, Kenya

Cercopithecus mitis albotorquatus is reported at apparently low densities in the extensive mangrove forests of North Lamu Island, and both northeastern and northwestern Manda Island, Kenya (De Jong & Butynski 2009; Figure 12.3). In the mangroves of North Lamu Island, calls were heard from at least one group but on account of poor weather, no monkeys were actually sighted. Very low densities, likely due to a lack of year-round access to fresh water, were also found on Manda Island where vocalizations were heard from the mangrove forest in the northwest of the island. In the southeast of the island an individual was seen in the coastal shrub on coral rag on the edge of the mangroves. Local people of both islands are familiar with the monkeys and on Manda Island report that *C. m. albotorquatus* occurs mainly in the mangrove forests. These are the first reports of *C. mitis* on these islands.

The '*mitis*' group is highly polytypic and its taxonomic organization is as yet unsettled (Dalton *et al.* 2015). It is not definitively known which subspecies was observed on Lamu and Manda Islands as it could be either *C. m. albotorquatus* or *C. m. phylax*. The distribution of *C. m. phylax* is restricted to the Lamu Archipelago (Groves 2001). However, Hill (1966) notes that it is known only from its type locality, Patta Island. No published locality records of *C. mitis* for Manda and Lamu Islands have been found to clarify this inconsistency (De Jong & Butynski 2009). *C. m. phylax* may be a synonym for *C. m. albotorquatus*. If Hill (1966) is correct in that *C. m. phylax* is restricted to Patta Island, then this survey reporting the presence of the species in mangroves is the first to confirm the presence of *C. mitis* on Lamu and Manda Islands.

Seasonal, Occasional or Opportunistic Occupancy

Cercocebus torquatus, Central African Mangroves

Cercocebus torquatus is limited to the Atlantic coastal basins of West and Central Africa (Gautier-Hion *et al.* 1999; Kingdon 1997; Lee *et al.* 1988; Malbrant & Maclatch 1949; Oates 1996),

Figure 12.2 *Chlorocebus sabaeus* on the edge of a mangrove stand, Gambia. Photo: Simon Bearder.

Figure 12.3 Adult male Zanzibar Sykes's monkey *Cercopithecus mitis albogularis* in mangrove at Vanga, Kenya. Photo: Y.A. de Jong and T.M. Butynski (www.wildsolutions.nl). (A black and white version of this figure appears in some formats. For the colour version, please refer to the plate section.)

its range apparently restricted to coastal areas, extending to 80–100 km inland (Matthews & Matthews 2002). Almost all known populations of *C. torquatus* occur within 300 km of the coast with a decrease in abundance from west to east in Cameroon and Gabon; while further inland it is ecologically replaced by *Mandrillus leucophaeus* (Oates 2011). *C. torquatus* is rarely seen in dry forest, frequently colonizing waterfronts and wetlands where it is known to inhabit mangrove forest (Gautier-Hion *et al.* 1999). *C. torquatus* typically occupies the lower forest strata and most observations of the species occur at 3–10 m in the canopy (Astaras *et al.* 2011), on the ground, and on mangrove roots (Jones & Sabater Pi 1968).

Abstracted information from a long-term study in Sette Cama (Cooke 2005) refers to mangrove use by *C. torquatus*, reporting that although primarily observed in dryland forest (70%), *C. torquatus* was frequently observed in mangrove forest (24%), and occasionally in beach forest (5%), and was exclusively observed on the ground. More detailed information on mangrove use by *C. torquatus* comes from an earlier study in Equatorial Guinea by Jones and Sabater Pi (1968). The study area in Rio Muni was a mangrove swamp dominated by *Rhizophora mangle* and surrounded by well-drained soils supporting adjacent primary and secondary forests. In this area, the authors carried out a comparative ecological study of *C. torquatus* and *Lophocebus albigena*, and reported that the mangrove swamp (its lower strata often consisting entirely of mangrove roots) was the typical habitat of *C. torquatus*, with 11 of the 23 individuals collected during the study period captured in these swamp forests. Jones and Sabater Pi also report that mangrove use by *C. torquatus* was relatively seasonal, with the species frequenting the adjacent primary and secondary forest in the dry season and the mangrove swamps during the rainy season. In addition, on all occasions when *C. torquatus* was disturbed by humans they sought refuge in the mangrove swamp forests, suggesting that this habitat may serve as a shelter for species under increased human pressure.

More recently, Maisels and colleagues (2007) confirmed the presence of *C. torquatus* in Mayumba National Park, southern Gabon. Here it was commonly observed several hundred metres behind the beach, in the littoral forests along the shore and the marsh forests along the lagoon. In addition, *C. torquatus* was frequently observed feeding, resting and travelling in the mangroves overlooking the beach in Loango National Park, Gabon and spent a large proportion of time in this habitat (J. Head, pers. obs., 2005–2012). Conversely, Maisels *et al.* (2007) report that during surveys in Conkouati National Park, Republic of Congo, *C. torquatus* was never seen close to the coast but observed further inland, and the authors attribute this difference to site-level habitat preferences. While there is apparent variation in mangrove use by *C. torquatus* (probably as a result of a combination of ecological site-specific factors), the species' restricted range in coastal areas of Central Africa support the notion that mangroves are clearly an important habitat for this species.

Chlorocebus pygerythrus, from Senegal to Kenya

The eastern form of the *Chlorocebus* genus (which includes *C. sabaeus*) is *C. pygerythrus*, and this species has also been observed in mangroves in the Wittu Islands of the Lamu Archipelago (*C. p. excubitor*), in Kiunga on mainland Kenya (*C. p. hilgerti*), in Saadani National Park and Pangani, Tanzania; and on Pemba Island, Zanzibar (*C. p. nesiotes*) (T.M. Butynski & Y.A. de Jong, pers. comm., 2014). However, while mangrove use has been observed, *C. pygerythrus* appears unable to survive exclusively in this habitat, unlike *C. sabaeus* which can. This unsuitability has been attributed to a lack of fresh water, absence of tall sleeping trees and insufficient food sources (T.M. Butynski, pers. comm., 2013). It is noteworthy that annual rainfall in the Cantanhez Forest and the Saloum Delta (approx. 1600 mm per annum) is substantially higher than on the coast of East Africa (889 mm per annum in the Lamu Archipelago), making drinking water available for *C. sabaeus* in the form of rainwater. As for the establishment of safe sleeping sites, it may be that the necessity for taller trees is negated by a lack of predators such as leopards in West African mangroves, unlike the eastern coast where predation avoidance remains a priority (with the exception of islands, such as Zanzibar, where the leopard is long extinct).

Papio cynocephalus, Kenyan Coast

Papio cynocephalus is a highly adaptable species that persists in secondary and heavily fragmented vegetation including cultivated land close to human settlement (De Jong & Butynski 2009). *P. cynocephalus* uses a range of habitat types, avoiding forest but successfully occupying forest edge. Over a large part of its range, *P. cynocephalus* is specific to fire-climax miombo (*Brachystegia*) woodland (Kingdon *et al.* 2008), but within this zone it also occupies dry bushland, open woodland, forest-grassland mosaic, thickets, steppes and the coastal littoral forests, including mangroves.

De Jong and Butynski (2009) observed *P. cynocephalus* in the coastal forests of Kenya where the subspecies *P. c. ibeanus* is locally common. It was observed moving along the edges of the mangroves as well as moving through mangroves to cross channels in the Lamu Archipelago. It has been suggested that *P. c. ibeanus* uses mangrove forest for foraging (although individuals were not observed within the mangroves themselves) but that it could not live solely in mangroves (T.M. Butynski & Y.A. de Jong, pers. comm., 2013). The IUCN Red List (Kingdon *et al.* 2008) and Kingdon (1997) also list mangroves as one of many habitats exploited by *P. cynocephalus*, but no further detail is provided.

Potential Novel Extension of Range or Habitat Shift

Cercopithecus mona, Niger Delta and Cameroonian Creeks

Cercopithecus mona is a generalist and versatile lowland forest species, relatively abundant close to rivers and in gallery forest and extending into savanna. It is highly adaptable and remains relatively common, even with the fragmentation and degradation of habitat throughout much of its geographic range (Oates *et al.* 2008). In some parts of eastern Nigeria where there is little remaining forest and the majority of anthropoid forest primates are now extinct, *C. mona* is often the only monkey species left (Oates 2011), and is relatively common in marginal habitats including mangrove forest where other primate species can be rare or absent. In the mangrove zone of the Niger Delta, *C. mona* appears to be the only primate species present (Were 2001) while in Cameroon it has been observed swimming across a wide creek in a mangrove swamp (Gartlan & Struhsaker 1972). Oates (1988) describes the species as 'particularly frequent in mangroves'. Despite the ubiquity of the species in West Africa, there are few published data of field studies and the observations outlined here highlight the potential importance of mangroves as a refuge habitat for this adaptable species.

Chlorocebus sabaeus, 'Mangrove Monkey', West African Mangroves

A study in Côte d'Ivoire (Bi *et al.* 2009) observed a population of *Chlorocebus sabaeus* outside its previously described range in the littoral forest of Iles Ehotilé National Park, restricted to the swamp and mangrove forests. The presence of *C. sabaeus* in the southern part of the country is not well known and its distribution in Côte d'Ivoire is discontinuous, with the southern and northern populations separated by a distance of approximately 300 km of adjacent forest zone. Two hypotheses are posed to explain this disconnected occurrence. The first suggests this population was descended from reintroduced pets released by foreign tourists upon leaving the country, since the lagoon forests where *C. sabaeus* occurs are along the former north–south road to Abidjan or near points of tourist interest (Bi *et al.* 2009). The adaptability of the species suggests that they would be capable of successfully colonizing the mangroves and surviving to reproduce in these areas. The second hypothesis is that the colonization of this coastal belt is the outcome of pressure from expanding agriculture and conversion of rainforest to a forest-agricultural mosaic, and a habitat shift that *C. sabaeus* has undergone elsewhere (Kavanah 1980). It is also possible that these are relict populations from a former continuous distribution, since there is some evidence for climatic fluctuations during the Pleistocene that caused several retreats and expansions of rainforest. If *C. sabaeus* subsisted off mangrove habitat when rainforest regrowth isolated them from the northern populations, this would represent evidence of the importance of mangroves to some primates during periods of environmental change.

In Guinea-Bissau *C. sabaeus* was observed in mangroves of the Cantanhez Forest in the Cacine Basin (Gippoliti & Dell'Omo 1996, 2003). The local name for *sabaeus* in the area is '*macaco de terrafe*' which translates literally to 'mangrove monkey'. This suggests that the occurrence of *C. sabaeus* in this area may be limited to mangroves. This narrow habitat preference of *C. sabaeus* may be the result of competition with *Cercopithecus campbelli campbelli*. *C. campbelli campbelli* was observed in closed forest and woodland savanna where one would typically expect to observe *C. sabaeus*, indicating that mangroves may facilitate niche separation in areas where interspecific competition is intense. Similar behaviour was reported for *C. sabaeus* in the mangroves of the Pirang Forest in The Gambia (Pourrut *et al.* 1996), where in the event of polyspecific associations with other primates, *C. sabaeus* increased its frequency of mangrove use compared to the use of the adjacent *terra firma* forest block.

Gorilla gorilla gorilla, Gabon

Loango National Park is one of 13 national parks created in Gabon in 2002, and its 1550 km^2 area protects diverse coastal habitats including mangrove, salt marsh, coastal lagoon, beach and primary forest. Loango's coastline runs for more than 100 km and exhibits mangroves along most of its length, providing a unique refuge for many terrestrial, arboreal and aquatic species. For example, *Crocodylus niloticus* uses the protective roots to lay its eggs, several bird species nest within the dense upper canopy, and many mammal and reptile species use the mangroves for foraging, travel or refuge. Eight species of diurnal primates live sympatrically in Loango National Park: *Gorilla gorilla gorilla, Pan troglodytes troglodytes, Cercocebus torquatus, Lophocebus albigena, Cercopithecus nictitans, C. cephus, C. pogonias* and *Miopithecus ogouensis*. All of them use mangroves to some degree, but there is much variation in their frequency of use and occupancy type, with *C. torquatus, G. g. gorilla* and *M. ogouensis* most often observed in the Loango mangroves (J. Head, pers. obs).

The Max Planck Institute of Evolutionary Anthropology set up a research site in Loango National Park in 2005 in order to carry out ecological and behavioural research on *P. t. troglodytes* and *G. g. gorilla* within a 100 km^2 area in the centre of the national park (Figure 12.4). Research on the feeding ecology of the two species revealed niche separation in diet and habitat use (Head *et al.* 2011, 2012), in addition to variation in the use of mangroves for transport and feeding. While *P. t. troglodytes* used mangroves primarily as a corridor for travel between neighbouring forest blocks, *G. g. gorilla* used the mangroves for both travelling and nesting, in addition to consuming the pith of young aerial mangrove roots.

One of the goals of the project was to habituate a group of *G. g. gorilla* to human presence to enable detailed observations of their behaviour. Habituating gorillas is a process which takes many years and involves tracking and contacting the group daily until individuals gradually lose their fear of humans. One group of *G. g. gorilla* was followed in Loango between 2009 and 2012 in an attempt to habituate them to human presence. In the first few months, the group exhibited a fear response and fled each time observers approached, but after two years they responded increasingly calmly such that it became possible to observe them for long periods of time and to follow behind them as they travelled.

While no data were specifically collected on the frequency of mangrove use by *G. g. gorilla* in Loango, their behavioural response during the habituation process was particularly interesting and may offer some insight into the possible role of mangroves as a refuge for species under increased pressure from humans. Comparisons of mangrove use by the gorillas were made from two months of group tracking data in 2009 and 2011. Data were collected at the same time of year and for the same duration to avoid introducing bias from possible seasonal variation in ranging patterns. The particular month was chosen because in both years gorillas were contacted nearly every day, making the data more comparable. Specifically, observers recorded whether or not the group travelled into mangroves after a contact with human observers on each day. No data were systematically collected on how much time was spent in the mangroves, their behaviour while in the mangroves, nor on how quickly the group moved to the mangroves after a contact.

Table 12.2 compares the response of the group under the pre- and post-habituation conditions (noting that 'post' here does not imply complete habituation, but simply an increase

Table 12.2 Changing response of gorillas to human contact in the Max Planck Institute research area.

	Pre-habituation	Post- habituation
Period	May 2009	May 2011
Number of days followed	30	30
Number of days contacted	29	28
Mean bi-directional[a] contact duration (mins)	2	21
Mean total[b] contact duration (mins)	32	61
Number of days refuged in mangroves post contact	18	9
Percentage of days refuged in mangroves post contact	62%	32%

[a] 'bi-directional' refers to gorillas being clearly visible to human observers and vice versa.
[b] 'total' refers to gorillas being with 50 m proximity of human observers.

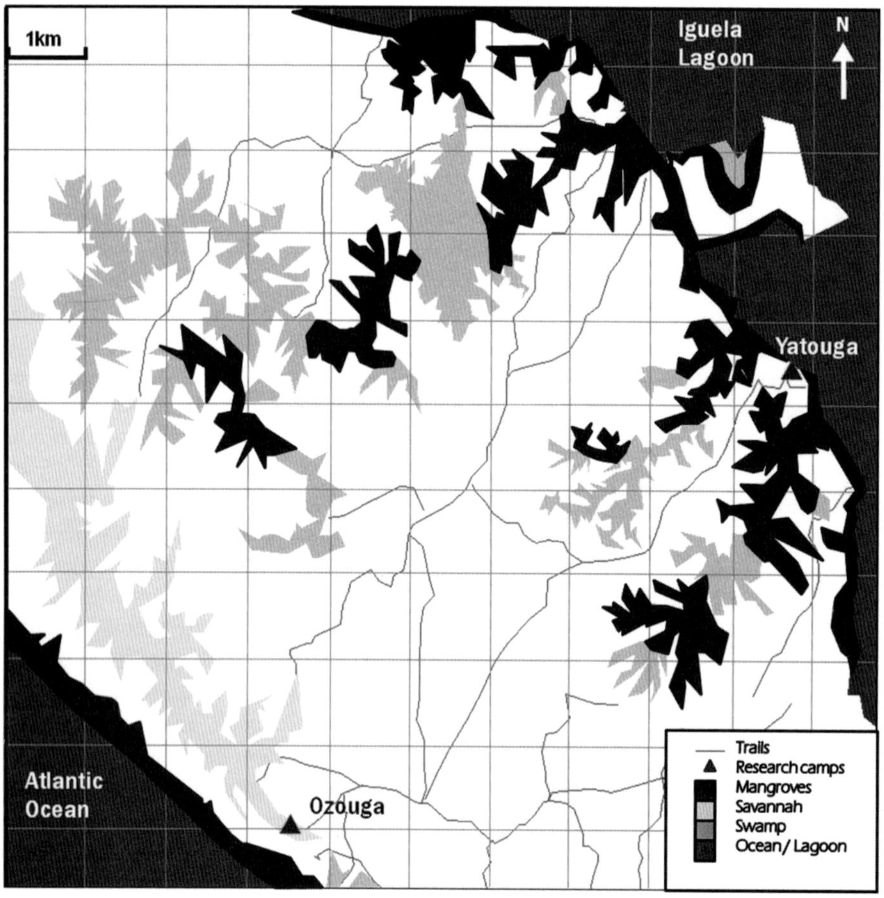

Figure 12.4 Max Planck Institute Research area in Loango National Park.

in the tolerance of *G. g. gorilla* to human presence as indicated by mean contact duration). It shows that at the beginning of the habituation process the focal group appeared to seek refuge in the mangroves after 62% of contacts with human observers, while two years later when the habituation had progressed substantially the focal group went in the mangroves after only 32% of contacts. While tentative and based on short-term data, these findings suggest that gorillas may feel more secure in mangroves when humans are nearby, and mangroves may be an important refuge habitat for species which are forced to shift habitat use patterns as a result of increased human pressure. However, given the small dataset used, we cannot rule out the possibility that the difference in mangrove use between years was influenced by ecological factors such as variability in food availability, which has been shown to vary inter-annually at this location (Head *et al.* 2011).

Piliocolobus kirkii, Zanzibar Archipelago, Tanzania

Arguably, members of the genus *Cercopithecus* are more behaviourally flexible than members of *Colobus* and *Piliocolobus*. But given the high tannin and folivorous diets tolerated by colobines, they, like many flexible and opportunistic cercopithecines, can

Figure 12.5 Adult female *Piliocolobus kirkii* avoiding the mid-rib of mature leaves of *Rhizophora mucronata*. Photo: K. Nowak.

also exploit mangrove habitat. At several East African mangrove sites, *Cercopithecus mitis* is sympatric with either *Colobus angolensis* or one of two species of *Piliocolobus* in coastal forests.

On Uzi Island, just south of the southern and main island of Zanzibar called Unguja, *Piliocolobus kirkii* and *Cercopithecus mitis albogularis* inhabit patches of mangrove forest. These mangrove patches now represent Uzi Island's last remaining forest with an intact canopy as the coral rag forest has been decimated, cleared for charcoal making and agriculture. The Uzi Island *P. kirkii* can spend > 80% of their day in the species-poor mangroves (Figure 12.5), where water foraging and drinking increases with proportion of time monkeys spend in mangroves and with the proportion of diet made up of mangrove leaves (Nowak 2008). While the consequences of mangrove herbivory for large mammals are not well studied (Tomlinson 1986), it appears that in *P. kirkii*, mangrove leaf consumption induces thirst making it unlikely that this species can subsist exclusively off a mangrove diet or survive a dry year while living mainly in mangroves.

Mangrove-dwelling groups of *P. kirkii* aggressively defend their home range, unlike their coral rag living counterparts in other forests in Zanzibar, and this defence may have more to do with defence of water-collecting surfaces and tree holes than food resources. Mangrove groups are also significantly larger in size and more cohesive than coral rag groups, and, it would appear, infant survival is higher in mangroves than in disturbed coral rag (Nowak & Lee 2010). This suggests that mangroves are possible source rather than sink habitats possibly because they make good refuges for *P. kirkii*, despite constraints on locomotion due to the tides and high salt loading on the diet. However, increasing pressure on mangrove-adjacent coastal thicket means that *P. kirkii* and sympatric *C. m. albogularis* on Uzi Island are increasingly limited to and reliant on mangroves. As a result, they come into conflict with people as they are forced to supplement their mangrove diets by raiding human food crops (e.g. cassava, papaya) which now grow immediately next to mangrove forest patches where there once were coral rag species such as *Terminalia boivinii* (Combretaceae), *Diospyros consolatae* (Ebenaceae), *Sorindeia madagascariensis* (Anacardiaceae) and *Grewia bicolor* (Malvaceae) – all exploited by *P. kirkii* for food. Without these *terra firma* species, *P. kirkii* is destined to decline in these areas given that its tolerance of mangroves is not absolute.

Discussion

Widespread observations of primates in mangrove habitat across Africa suggest its important and neglected role in African primate ecology, evolution and persistence. Across their range, mangroves are used by many different primate species for feeding and socializing, as means of moving between adjacent habitats and for refuge from humans and likely other potential threats. While there is clear variability in mangrove occupancy across primate species, the behavioural flexibility and variability facilitated by mangroves highlights the importance of conserving this habitat type across Africa and worldwide.

The role of mangroves as a secure refuge deserves particular mention, since this has been observed for *Chlorocebus sabaeus, Gorilla gorilla gorilla, Cercocebus torquatus* and *Piliocolobus kirkii* (Chapter 39). It is likely that more in-depth studies will reveal similar patterns for other primate species. Furthermore, observations of *Cercopithecus mona* across West Africa indicate that for species with high adaptability, mangroves may become one of the only remaining shelters in areas under increased human pressure, or areas undergoing a reduction in terrestrial forest habitat.

Further evidence that mangroves may become one of the only remaining refuges available to primates is apparent in one study that reports *Erythrocebus patas* entering mangroves and catching and eating *Uca tangeri* (fiddler crabs), in the Pirang Forest, Senegal (Galat-Luong and Galat 2005). Galat-Luong and Galat (2005) analysed data from Gatinot (1975), Lykke

(1993), Galat-Luong *et al.* (1998) and aerial photographs of the area. They showed that human encroachment resulted in a significant decline in tree coverage from 88% to 30% in the plateaus and a drastic decline from 98% to 23% in the gallery forest. Diversity of woody species and overall tree density also decreased in both habitats – density of the most abundant woody species decreasing by around 50%.

Because current patterns of primate distribution and abundance are so strongly influenced by human behaviour, the impact primates can have on one another can sometimes be neglected. However, competitive exclusion is commonplace among many different species, and this chapter provides some preliminary evidence of the evolutionary role mangroves may have played in facilitating niche separation and thus enabling the coexistence of sympatric primates. For example, observations of *Cercocebus torquatus* in Central Africa indicate that this species may have adapted to exploit mangrove habitat in order to avoid competition with both *Lophocebus albigena* and *Mandrillus leucophaeus*. In addition, a study in Gabon showed clear evidence of niche separation between sympatric *P. t. troglodytes* and *G. g. gorilla* in the use of swamp habitat (including, but not restricted to mangroves; Head *et al.* 2012).

Further study of many of the species mentioned in this chapter, several of which are considered Endangered or Vulnerable (Table 12.1; IUCN Red List), would therefore be valuable for increasing our understanding of the importance of mangroves to primates, both in terms of influencing speciation and niche separation, and in ensuring adequate conservation measures are put in place in time to secure their long-term survival and persistence.

Chapter 13

Feeding Ecology of the Proboscis Monkey in Sabah, Malaysia, with Special Reference to Plant Species-Poor Forests

Henry Bernard, Ikki Matsuda, Goro Hanya, Mui-How Phua, Felicity Oram and Abdul Hamid Ahmad

Introduction

Annual and/or seasonal activities, especially feeding behaviour, are largely affected by availability, distribution and quality of food sources in relation to primates' strategies for energy conservation (e.g. Dasilva 1992; Fan *et al.* 2008; Hanya & Bernard 2012; Matsuda *et al.* 2013; Oates 1977; Sha & Hanya 2013; Stanford 1991; Tsuji *et al.* 2008). To understand how primates interact with their environment and allocate energy and time, it is crucial to study their feeding behaviour. Colobines, including proboscis monkeys (*Nasalis larvatus*), are characterized by an enlarged and sacculated forestomach which enable them to exploit a diet of leaves in greater quantities than other sympatric primates (Chivers 1994). It has therefore been assumed that colobines mostly exploit ubiquitous food sources such as leaves, though some studies provide evidence for their dietary flexibility, i.e. they show high levels of fruit and/or seed consumption in response to local conditions with significant effect of fruit eating on their wider behavioural repertoire (e.g. Davies 1991; Dela 2007; Hanya & Bernard 2012; Kool 1993; Matsuda *et al.* 2009a; Stanford 1991; Yeager 1989). Proboscis monkeys may be one of the best colobine species for studying dietary flexibility and adaptation as the monkeys inhabit various forest types, such as riverine, peat swamp and mangrove forests, and possibly change their diets depending on the habitat types in which they occur; for example, higher dietary diversity in riverine and peat swamp forests, but lower dietary diversity in mangrove forests (Bennett & Sebastian 1988; Boonratana 2003; Salter *et al.* 1985; Yeager 1989; Hayakawa *et al.* 2018; Chapter 4).

In this study, we investigated the monthly distribution patterns and feeding behaviour of proboscis monkey groups inhabiting riverine, mangrove and mixed mangrove–riverine forests that are characteristically plant species poor. Although previous studies investigating the feeding ecology of proboscis monkeys have revealed their high dietary diversity (36–188 plant species eaten) with particular preference for unripe fruits or seeds in riverine and peat swamp forests (Boonratana 2003; Matsuda *et al.* 2009a; Yeager 1989), knowledge of the feeding ecology of proboscis monkeys inhabiting poor vegetation habitats like mangroves and mixed mangrove–riverine forests, is still incomplete. Bennett and Sebastian (1988), who made observations of proboscis monkeys in mangrove forest by boat, described that young leaves and fruits accounted for the majority of feeding observations. However, the sample sizes in their study were rather small (total 34 feeding events). Similarly, Boonratana (2003) documented the feeding ecology of proboscis monkeys, also in a mangrove forest, and revealed low dietary diversity (18 plant species eaten) with a strong preference for fruits (< 30% of their feedings). But, as with Bennett and Sebastian (1988), the sample size obtained was small (total 15-min scans over 8 months = 188 feeding events) and only limited analytical interpretation was possible. Hence, an intensive study on the feeding ecology of proboscis monkeys inhabiting a low plant diversity forest will provide essential information for understanding their foraging strategies in this type of forest. Moreover, given that a large extent of the remaining habitats of proboscis monkeys consists typically of species-poor forests that are increasingly disturbed and fragmented, much baseline ecological information about the monkey inhabiting such forests is needed to support conservation efforts and management decisions (Boonratana 2013; Meijaard & Nijman 2000a; Sha *et al.* 2008; Stark *et al.* 2012). Information obtained from such studies may also generally contribute to an improved understanding of behavioural adaptations and resilience of colobine monkeys across their greater habitat range.

Unlike studies on proboscis monkeys in riverine forests which are floored with relatively hard soil, making observations of monkeys on foot possible (Boonratana 2000; Matsuda *et al.* 2009b), studies of proboscis monkeys inhabiting mangrove and mixed mangrove–riverine forests, that are constantly or periodically inundated, are challenging and fraught with difficulties, particularly with respect to tracking the monkeys inland (Bennett & Sebastian 1988; Salter *et al.* 1985). Therefore, most studies on proboscis monkeys in such swampy habitats have been conducted either exclusively or primarily using the boat-based approach, i.e. observing the monkeys from a boat while the monkeys are on the riverbank in the early morning and late afternoon. Even though the boat-based method for studying proboscis monkeys has its limitations, for example it prevents the observation of activity patterns inland, this method is, nevertheless, the most pragmatic way of studying this monkey in swampy habitats on Borneo (Bennett & Sebastian 1988; Kawabe & Mano 1972; Salter *et al.* 1985; Yeager 1989).

In this chapter, we describe the seasonal habitat preferences of proboscis monkeys over 23 months and their feeding ecology over 19 months based on boat-based surveys carried out in increasingly fragmented and isolated forests characterized by poor plant species diversity in the Klias Peninsula, Sabah, Malaysian Borneo. Previous studies estimated that between 439 and 578 proboscis monkeys inhabit the study area (Bernard & Zulhazman 2006; Sha *et al.* 2008), making it the third largest proboscis monkey population in the state of Sabah, and likely the only viable populations of proboscis monkeys in the west coast of this state. The proboscis monkey population at this site is isolated from the nearest proboscis monkey populations located > 20 km in the south of the Klias Peninsula. We describe the seasonal distribution patterns of the proboscis monkey population in three different habitat types that are contiguous with each other and investigate the monthly dietary variation, as well as the relationship between food habits and food resource availability. Lastly, we discuss the relevance of our results for conservation management of the proboscis monkey.

Methods

Study Site

We conducted this study in the riverine, mangrove and mixed mangrove–riverine forests along the Klias river, its tributary (Garama River) and a small channel (Terusan Pura: approximately 3 m wide and 4.2 km long) in Padas Damit Forest Reserve (PDFR) (*c.* 9000 ha; 115°30′E, 5°21′N) located in the central part of the Klias Peninsula in western Sabah, Malaysia, northern Borneo (Figure 13.1). As the name suggests, riverine forest in this study generally refers to the type of vegetation that is found in narrow strips along the river where regular inundation occurs. The vegetation composition of this forest is clearly different from the mangroves which are found mainly further down river. Whereas, mixed mangrove–riverine forest basically refers to the transition zone between mangroves and riverine vegetation along the river banks. In PDFR, the mixed mangrove–riverine zone is characterized by very sparse trees and dense undergrowth. The predominant vegetation types in the general study site region consist of a complex mixture of nipah forest, freshwater swamp forest and degraded peat swamp forest interspersed with bare lands and wet grasslands (Phua *et al.* 2007). The terrain of PDFR is generally flat with elevation 0–10 m asl. The highest point is only 150 m asl. A large proportion of natural vegetation outside of the PDFR has been either heavily disturbed or cleared to make way for human settlements and subsistence agriculture, mainly for rubber tree plantations and, more recently, the establishment of small- and medium-scale oil palm plantations (Phua *et al.* 2007). This area is prone to forest fires, particularly during the long dry period associated with the El Niño Southern Oscillation event (Phua *et al.* 2007). During the study period, the average annual precipitation in the study site was 3540 mm and average monthly minimum and maximum temperatures were 22°C and 35°C. The study site was influenced by tidal conditions. Gazetted in 1984, PDFR is managed as an Amenity Forest Reserve by the Sabah Forestry Department (Sabah Forestry Enactment 1968).

Vegetation and Phenology Surveys

To study the phenology of the plants in the study site, we set up eight botanical plots measuring 10 m wide and 80–280 m long perpendicular to the Garama river banks on both sides of the banks (Figure 13.1). The total combined length of the botanical plots was 1 km (i.e. survey area: 1.0 ha). All trees ≥ 30 cm DBH located within the plots were marked using uniquely numbered aluminium tags. All labelled trees were taxonomically identified with the support of a botanist, D. Sundaling, from the Forest Research Center of the Sabah Forestry Department, Sabah, Malaysia. At the end of every month during the study period, we recorded the phenology of all labelled trees by examining each tree for the presence or absence of fruits, flowers or mature/young leaves. Young leaves were observed using binoculars and were detected based on the presence of clusters of shoot flush that could be distinguished from mature leaves by their pale yellowish-green or reddish-green colour. We used the monthly proportion of trees with mature/young leaves, fruits and flowers to show the monthly variations in the phenology of plants from the vegetation plots.

Study Animals

Proboscis monkeys are large, sexually dimorphic (Bismark 2010: male, *c.* 20 kg; female, *c.* 10 kg), diurnal, arboreal colobines, endemic to the island of Borneo in Southeast Asia. Their typical social unit is a unimale, multi-female group, consisting of an adult male, several adult females and immatures, or all-male groups, consisting mainly of young males (Bennett & Sebastian 1988; Matsuda *et al.* 2012; Murai 2004; Yeager 1990). Proboscis monkeys inhabit mainly mangrove, peat swamp and riverine forests (Sha *et al.* 2008), where they are closely associated with waterways. They travel inland to forage during the day (generally up to 1 km) and return to their sleeping sites (trees) along the river edge every evening (Bernard *et al.* 2011a; Boonratana 2000; Matsuda *et al.* 2009b). The sleeping sites of the groups are distributed over a wide area and overlap with those of other unimale, multi-female and all-male groups (Bennett & Sebastian 1988; Bernard *et al.* 2011a; Matsuda *et al.* 2010a; Yeager 1991b).

Population Distribution Survey

We collected data on the monthly distribution pattern of the proboscis monkey groups in the late afternoon (16:00–19:00) while the monkeys were at the riverside sleeping trees (see Bernard *et al.* 2011a for details). We determined the variations in their monthly distribution patterns based on monthly changes in their sleeping tree locations. Locations of sleeping trees were determined by cruising a looping survey route starting from the main jetty at Kampung Garama going down stream along the Garama River to Klias river to Terusan Pura and back to Garama River (total 23.6 km, Figure 13.1), or going

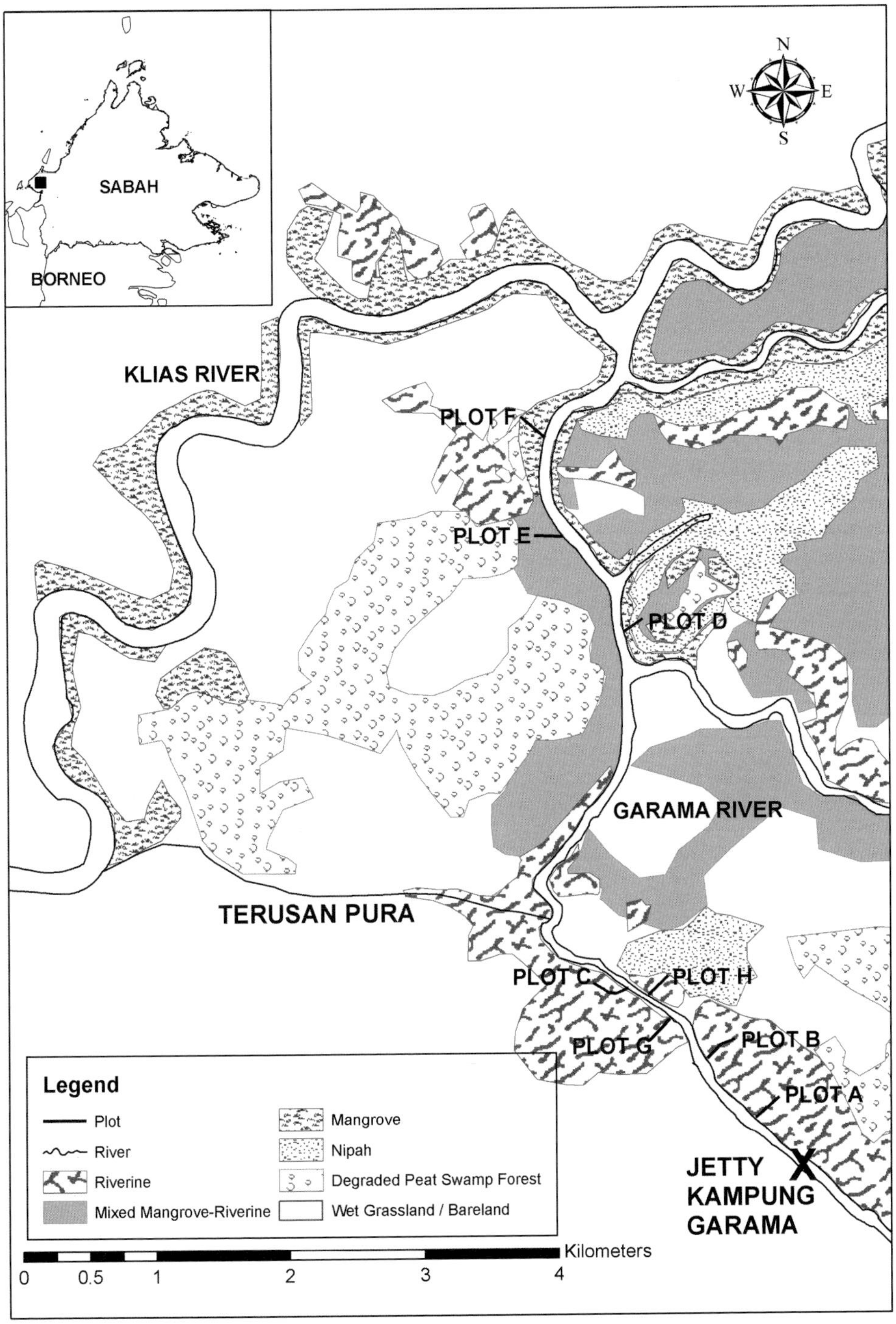

Figure 13.1 The central part of the Klias Peninsula in the west of Sabah, Malaysia, northern Borneo (inset), showing the general vegetation types of the study site within the Padas Damit Forest Reserve. Locations of the vegetation plots are indicated as A–H.

the other way around (Garama River > Terusan Pura > Klias River > Garama River). This was conducted monthly during the third week of the month from January 2008 to December 2009, except in August 2008 where only part of the usual survey route was inspected. Two spotters were normally involved in the survey to observe monkeys from both riverbanks. Binoculars were used for observations from a boat cruising at a speed of about 5–10 km per hour. Whenever we spotted proboscis monkeys in the tree, we stopped the boat and recorded their location (using a GPS unit) and the habitat type (riverine, mangrove or mixed mangrove–riverine forests) occupied. When individual monkeys were spread over several trees along the riverbank, the GPS coordinates of the tree located approximately at the group's centre was taken to represent the location of the monkey group. We also counted the monkeys in the group and recorded the group type (i.e. either unimale, multi-female group or all-male group) and the structure using age/sex categories (adult male/female, subadult, juvenile and infant) from Bennett and Sebastian (1988). Individuals that were partially hidden behind trees were classed as 'unidentified' animals, the minimum number of which was recorded as one individual. We are aware that separate proboscis monkey groups regularly sleep within 100 m of each other (Bennett & Sebastian 1988; Matsuda *et al.* 2010a; Yeager 1991b). However, to enable data comparisons with previous studies, we generally regarded all individual monkeys within approximately 50–100 m of each other along the riverbanks as belonging to the same group, except on opposite sides of the river where they were considered to be in separate groups, i.e. following the operational definition adopted by Kren (1964), Kawabe and Mano (1972), Macdonald (1982) and Salter *et al.* (1985).

Behavioural Data Collection

Earlier studies suggest that proboscis monkeys typically set up their sleeping sites in riverside trees (Bennett & Sebastian 1988; Bernard *et al.* 2011b; Matsuda *et al.* 2010b; Salter *et al.* 1985; Yeager 1991a). Therefore, we conducted our boat-based observations to collect behavioural data on proboscis monkeys, i.e. by approaching the monkeys by boat and observing their behaviour using a pair of binoculars when the monkeys were in the trees located by the riverbanks up to a maximum perpendicular distance of *c.* 50 m away from the riverbanks. We mainly collected behavioural data in riverine and mixed mangrove–riverine forests along an 8 km stretch of the Garama River. The monkeys inhabiting such areas were undeterred by the presence of humans in boats, so we could approach the animals up to 15 m by the riverbanks without agitating the monkeys or affecting their natural behaviour. The frequent visits of tourists to the study area by boats for > 10 years may have habituated the monkeys to the presence of humans in boats in the river, though animals are rarely disturbed by tourists due to strict rules imposed by the Sabah Forestry Department (H. Bernard, pers. obs.). Although we also attempted as much as possible to collect behavioural data on the proboscis monkey on foot, i.e. following monkeys on foot into and inside inland forest, we did not succeed in doing so due to the shyness of the monkeys, the swampy terrain and the dense undergrowth of the study area, which prevented us from following the monkey groups quietly and efficiently.

We used the instantaneous group scan sampling method (Altmann 1974) to collect behavioural data. Scans were conducted at every five minutes. We started observing monkeys in the early morning (06:00–06:30) at the monkeys' night-sleeping trees which we had located during the previous evening. We observed the monkeys' behaviour for as long as possible along the riverbanks until the monkeys moved into the forest's interior. We conducted behavioural data collection throughout the day (6:00–18:00) on some days when it was possible to do so even from the boat due to the narrow strips of forest alongside riverbanks (≤ 35 m), though on most days, we could only observe monkeys from the boat intermittently for brief periods and at different times of the day (i.e. opportunistically). When the monkeys disappeared from our sight, we terminated the observation period immediately and moved on to search for a new group along the riverbanks. When we could not locate any groups within 3 h in the morning or afternoon we terminated our observations. We returned to the study area again in the late afternoon from 16:00 and continued searching for proboscis monkey groups until 18:00, except in the event of heavy rain. Proboscis monkey groups were identified using individually distinctive animals as markers, usually the alpha males in unimale, multi-female groups and the largest males in all-male groups.

Behavioural data collection took place for 19 months between February 2008 and December 2009, with a four-month interruption between September and December 2008 because of logistical difficulties associated with monitoring proboscis monkeys. We (H.B. and two trained research assistants) recorded behavioural activities, i.e. feeding, moving, resting and others (such as vocalizing, agonistic behaviours, etc.), on all visible individuals, except dependent infants. Definitions of different behaviour categories followed Matsuda *et al.* (2009a). When animals were feeding, we recorded their age-sex categories (adult male/female, subadult and juvenile), plant species and the food item eaten (i.e. mature/young leaves, fruits and flowers). We taxonomically identified food plant species *in situ*. If food plant species were unknown, we collected plant samples for later taxonomic identification.

Statistical Analysis

To compare the monthly proboscis monkey group encounter rates, i.e. number of groups encountered per kilometre of rivers surveyed, in different forest types (riverine, mangrove and mixed mangrove–riverine forest), we used the non-parametric Kruskal–Wallis analysis of variance (ANOVA) test as the data did not conform to a normal distribution (Klomogrov–Smirnov test, $p < 0.05$). Detailed pairwise comparisons of encounter rates with groups between habitat types were made using a non-parametric Mann–Whitney *U*-test with a Bonferroni correction ($0.05/N$, i.e. $p = 0.017$) of the p values for these tests (Rice 1989). To test the effects of food availability (fruits, flowers and young leaves expressed as proportions using phenology data) on distribution patterns of proboscis monkeys, i.e. based on the monthly numbers of detected proboscis monkey groups in riverine, mangrove and mixed mangrove–riverine forests (where all eight of the botanical plots were located), we used a generalized linear model (GLM), applying the Poisson family and log link function to calculate the Akaike's information criterion (AIC; dispersion parameter, 1.03). We treated the sum of the groups detected in the riverine and mixed mangrove–riverine forests from the population distribution surveys for each month as a dependent variable. The sum of all groups detected in all forest types from the population distribution surveys in each month was used as an offset term.

We used both unweighted and weighted scans to analyse feeding data. Unweighted scans were used only to summarize information regarding the general sampling intensity. To compare data with different sampling effort in each month, we used weighted scans. In this study, we defined weighted scans as an observation of one animal's behaviour recorded in a scan divided by the total number of individuals observed within that scan. As such, the monthly variations in behavioural activities based on weighted scan were expressed on the basis of an individual monkey. As individuals between different proboscis monkey groups could not always be identified with high certainty, we pooled data across groups. Observations made on different days of all proboscis monkey groups in a given month were pooled for that month in order to increase sample sizes. Pooling data also gave a better representation of behavioural activities from 06:00 to 18:00 across months; note that this analytical protocol may have introduced some degree of pseudoreplication or non-independence into the data set, though we believe that its effect on the analysis was negligible.

We examined the effect of monthly proportions of trees with young leaves, fruits and flowers (independent factors) in the phenology survey on the monthly feeding frequencies of the proboscis monkey based on weighted scans (dependent variable) using a GLM, applying the gamma family (link function = inverse, i.e. the calculated coefficient value reflects the inverse effect) to calculate the AIC (dispersion parameter for gamma family, 0.11). To select the best model, we assessed the models with the AIC using the dredge function in the MuMIn package version 1.9.13 (Bartoń 2013) in R version 3.1.0 (R Development Core Team 2014). The variance inflation factors were 1.21 for flowering trees, 1.18 for fruiting trees and 1.04 for flushing trees, which were less than the cut-off value (5), therefore collinearity among independent factors did not affect the results. We chose the model with the smallest AIC among all possible combinations of independent variables, including the null model.

Finally, we also analysed the correlation between monthly availability of each plant part (young leaves, fruits and flowers) and feeding frequency on that plant part, as well as between the feeding frequencies of each plant part with the monthly dietary diversity. Because data did not fit a normal distribution (Klomogrov–Smirnov test, $p < 0.05$), we used the non-parametric Spearman's rank correlation analysis with a Bonferroni correction ($0.05/N$, i.e. 0.007) of the p values for these tests (Rice 1989). We used proportions data for availability of each plant part and weighted scans for feeding frequencies of all plant parts. Monthly dietary diversity was calculated using the Shannon–Weiner index of diversity (H') (Pielou 1966). We calculated dietary diversity (H') in the program Species Diversity and Richness (PISCES Conservation, Oxford, United Kingdom: www.pices-conservation.com/pdf/SDRInstruction.pdf).

Results

Vegetation and Phenology

A total of 248 trees (≥ 30 cm DBH) were recorded from the botanical plots representing only 16 species, 16 genera and 14 families (Table 13.1). The five most abundant species were *Excoecaria indica* (Euphorbiaceae), *Cerbera odollam* (Apocynaceae), *Bruguiera gymnorrhiza* (Rhizophoraceae), *Psydrax* sp. (Rubiaceae) and *Ficus binnendjikii* (Moraceae), which accounted for 70% of all marked trees. The cumulative number of tree species, estimated using abundance-based rarefaction techniques with number of individual trees as sampling effort and constructed in EstimateS (Colwell 2013; Colwell & Coddington 1994), appeared to have reached an approximate asymptote after about 250 trees were sampled (95% CI: 14.11–17.89 species). This indicates that trees in the botanical plots sufficiently represented the general vegetation patterns of the study area (Figure 13.2).

Table 13.1 The frequency distribution of tree species (> 30 cm DBH) from eight botanical plots (1.0 ha) in Padas Damit Forest Reserve.

Family	Species	Number of trees
Euphorbiaceae	*Excoecaria indica*	71
Apocynaceae	*Cerbera odolam*	33
Rhizophoraceae	*Bruguiera gymnorrhiza*	24
Rubiaceae	*Psydrax* sp.	24
Moraceae	*Ficus binnendijkii*	23
Symplocaceae	*Symplocos celastrifolia*	13
Myrsinaceae	*Rapanea avenis*	13
Meliaceae	*Dysoxylum cyrtobotryum*	13
Myrsinaceae	*Ardisia elliptica*	10
Sterculiaceae	*Heritiera littoralis*	8
Fabaceae	*Pongamia pinnata*	6
Rhizophoraceae	*Rhizophora apiculata*	6
Palmae	*Oncosperma tigillarium*	4
Flacourtiaceae	*Casearia grewiaefolia*	1
Malvaceae	*Hibiscus tiliaceus*	1
Rubiaceae	*Nauclea orientalis*	1
	Total	248

In terms of phenology, trees with mature leaves dominated the botanical plots throughout the study period (mean percentage: 64.0%; range: 44.8–91.4% of total number of trees) (Figure 13.3a). Trees with young leaves, fruits and flowers averaged 13.0%, 9.4% and 10.8%, respectively. The number of trees with young leaves peaked in February (2008), October (2009) and December (2009) (range: 20.4–26.2% of the trees: Figure 13.3a). Number of trees with fruits peaked in January (2008), May to September (2008) and July (2009) (range: 14.7–28.3%). The number of trees with flowers peaked in October–December (2008) and April (2009) (range: 14.3–17.6%).

Proboscis Monkey Distribution Patterns

A cumulative total of 305 proboscis monkey groups were detected during the 23 months distributional survey. On average 13.3 groups were detected per month (range: 5–22 groups). The maximum count of proboscis monkey individuals and groups detected during the monthly population distribution survey were both recorded in December 2009 with 218 individuals and 22 groups (unimale, multi-female groups: 18; all-male groups: 4), respectively. Sleeping sites of the proboscis monkey groups were widely distributed and overlapped with each other across different months. The age–sex compositions were 32 adult males, 74 adult females, 35 subadults, 38 juveniles, 31 infants and 8 'unidentified' individuals.

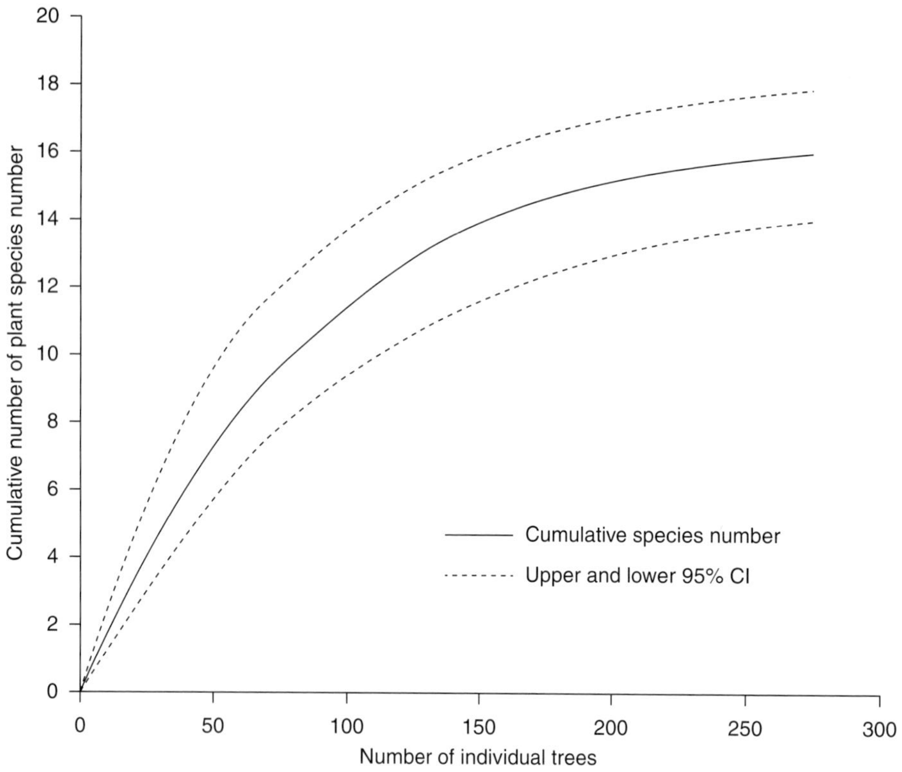

Figure 13.2 Cumulative number of plant species with 95% confidence intervals constructed using an abundance-based (i.e. number of individual trees) rarefaction approach with 100 iterations in Estimates (Colwell 2013).

Group encounter rates (group/km) varied significantly between riverine forest, mangrove forest and mixed mangrove–riverine forest (Kruskal–Wallis ANOVA test: $X^2 =$ 18.6, $df = 2$, p-value = 0.001). The highest groups encounter rate was recorded in riverine forest, i.e. mean = 0.87 ± 0.34 (SD) (riverine versus mangrove forests: Mann–Whitney U-test: $U = 93$, p-value < 0.001; riverine versus mixed forests: $U = 95$, p-value < 0.001, see Figure 13.4). Groups encounter rates were not significantly different between mangrove (mean: 0.45; SD: 0.36) and mixed forest (mean: 0.42; SD: 0.49) (mangrove versus mixed forests: $U = 223$, p-value = 0.36).

In the GLM, the monthly numbers of detected groups in the riverine and mixed forest were not significantly influenced by any inserted variables as shown in Table 13.2; but there was a tendency for fruit availability to be associated with higher monthly numbers of detected groups, i.e. more groups were found in the riverine and mixed forest in months when fruit availability was relatively higher.

Feeding Ecology

Overall Observations of Proboscis Monkey Groups

During the study period, a total of 14 different proboscis monkey groups (unimale, multi-female groups: 13; all-male groups: 1) consisting of 169 individuals, were followed on an opportunistic basis. The age–sex compositions of all groups combined included 22 adult males, 56 adult females, 28 subadults, 34 juveniles, 21 infants and 8 unidentified individuals (mean group size for unimale groups: 12.38 ± 10.26 (SD) individuals; group size for all-male groups: 8 individuals).

The cumulative frequency of observations of different age–sex classes of the proboscis monkey groups during behavioural observations did not differ significantly from the expected proportion of the different age-sex classes of the proboscis monkey population in PDFR i.e. based on the age-sex proportion data collected from the monthly distributional survey (Chi-square goodness-of-fit test: $X^2 =$ 6.26, $df = 4$, p-value = 0.18). This indicated that the proboscis monkey groups observed for behavioural sampling were representative of the wider population of proboscis monkeys in PDFR.

Overall, we observed the proboscis monkeys for a total of 70 days over 19 months (range: 2–7 days/month). Total observation time was 192 h, and the monthly observation time was 2–21 h (mean: 10.1 h). Although we attempted to observe the monkeys from 06:00 to 18:00 as much as possible, because of the limitations of our boat-based survey method the majority of our observation time took place between two time periods, i.e. 06:00–10:00 (110 h) and 16:00–18:00 (56 h). Observation time in the late morning till early afternoon, between 10:00 and16:00, was represented by 26 h. This resulted in a total number of 2304 scans and 10 214 individual observations of monkey behaviour, excluding that of infants. The average number of groups observed per month was 2.0 (range: 1–5 groups) and the average number of individuals observed per scan was 4.36 (range: 1–15 individuals).

Food Habits

Throughout the study period, the three major behavioural activities recorded were resting, feeding and travelling accounting for 53.6% (n = 5475), 17.7% (n = 1811) and 15.7% (n = 1598) from the total number of observations made, respectively. A total of 19 plant species were identified as food

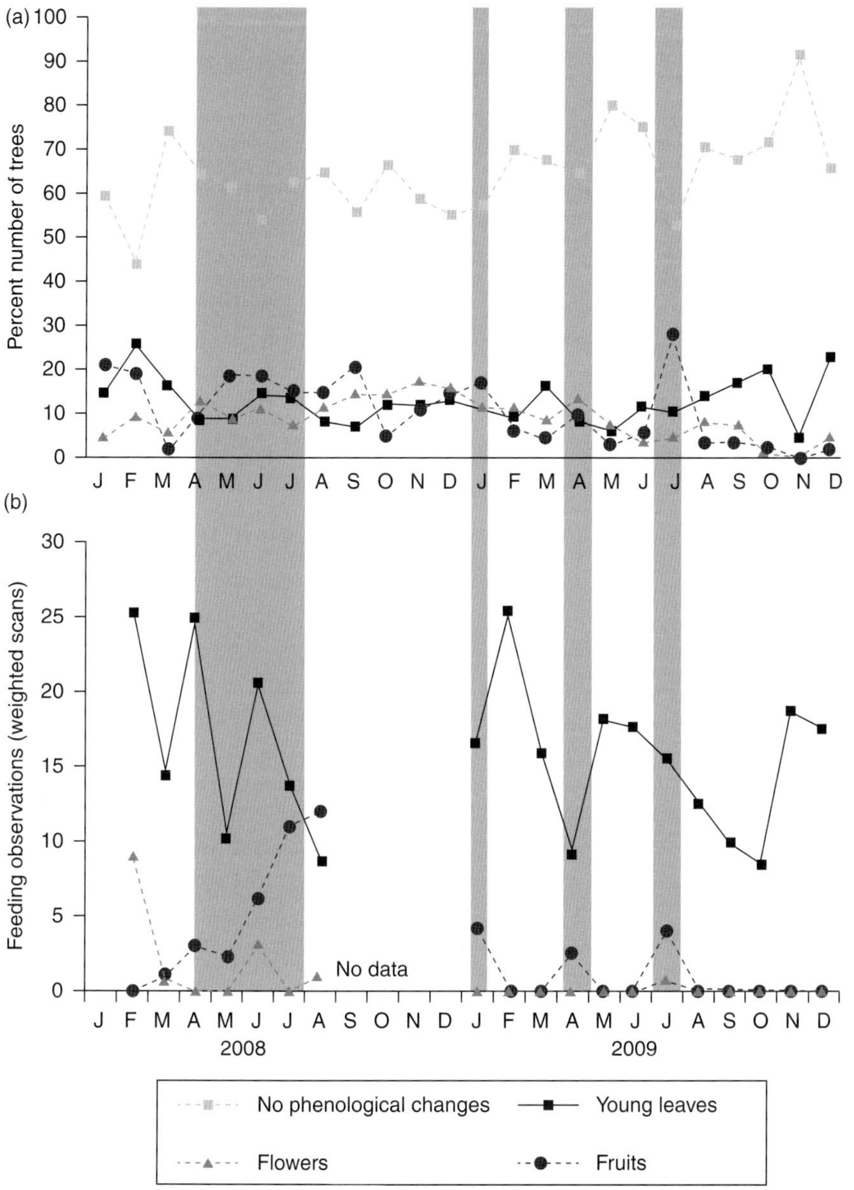

Figure 13.3 (a) Monthly availability of mature leaves, young leaves, flowers and fruits and (b) monthly changes in young leaves, flowers and fruits in the diet of proboscis monkeys in Padas Damit Forest Reserve, Sabah.

plants, with two species, *B. gymnorrhiza* and *F. binnendjikii*, dominating 80.4% of feeding records (Table 13.3). The average number of plant species consumed per month was 5.16 ± 2.22 (SD) (range: 2–10 species). The monthly plant diversity (*H'*) in the diet averaged 1.03 ± 0.30 (range: 0.59–1.75).

The main food items were young leaves accounting for 91.6% of total records of feeding frequency. Fruits represented 3.13% and flowers 3.45%. Other food items represented small proportions in the monkeys' diet and included mature leaves, seeds and bark (totalled 0.28%) and 'unknown foods' (1.53%). The proboscis monkeys ate young leaves from a total of 16 plant species with two species, *B. gymnorrhiza* and *F. binnendjikii*, accounting for 82.5% of feeding on this plant part. Fruits (including inflorescence of human planted coconut trees, *Cocos nucifera*) were taken from six plant species with two species, *H. littoralis* and *Rhizophora apiculata*, together accounting for 68.0% of total feeding on this plant part. The hypocotyl (the new leaves) of the viviparous fruits (*R. apiculata*) and the cotiledone (the proper fruit body) were consumed, though the epigeal (embryonic root) was frequently discarded. Flowers of only two plant species, *H. tilliaceus* and *B. gymnorhiza*, were consumed by the proboscis monkeys.

Monthly Variation in Diet

The variation in monthly feeding frequencies (based on weighted scans) of the three main plant parts in the diet of the proboscis monkeys showed that leaves formed the major component of their monthly diet throughout the study period (Figure 13.3b). Fruits and flowers formed only minor components, though during two months – July and August 2008 – fruits also contributed a major proportion to the monkeys' diet in addition to young leaves.

The best-fit model predicting effects of food availability on the feeding frequencies of proboscis monkeys included only

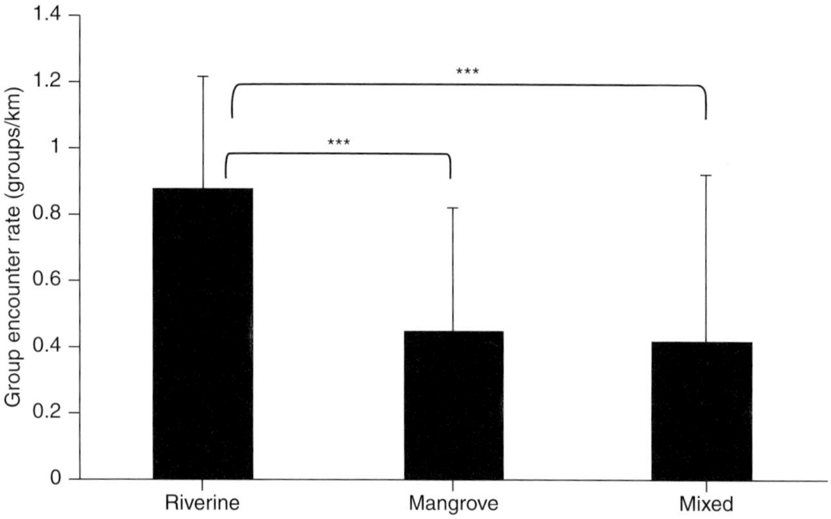

Figure 13.4 Proboscis monkey average group encounter rates ($n = 23$) in three different habitat types in the Padas Damit Forest Reserve: riverine (range: 0.25–2.00 groups/km); mangrove (0.0–1.36 groups/km); mixed mangrove–riverine (0.0–1.75 groups/km). Error bars indicate standard deviation with asterisks (***) indicate significant differences at p-value = 0.001.

Table 13.2 GLM showing the effect of food items availability (proportion data) on the monthly number of detected proboscis monkey groups in the riverine, mangrove and mixed mangrove–riverine forests.

	Coefficients	SE	Z	*P*-value
Intercept	−1.022	0.289	−3.517	0.000
Flowers availability	0.006	0.007	0.895	0.370
Fruits availability	0.007	0.004	1.884	0.059
Young leaves availability	0.001	0.005	0.256	0.797

fruit availability with a negative effect (Table 13.4). The negative linear term, however, indicated that the monkeys spent more time feeding when fruits were more common (see Methods). In addition, monthly fruit availability was significantly positively correlated with fruit eating (Spearman's rank correlation coefficient $r = 0.674$, $n = 19$, p-value = 0.002), but no correlation was found between monthly young leaf availability and young leaf consumption ($r = 0.068$, $n = 19$, p-value = 0.782), and between monthly flower availability and flower-eating ($r = 0.175$, $n = 19$, p-value = 0.473). Similarly, no correlation was found between fruit availability and young leaf-eating (r = - 0.219, $n = 19$, p-value = 0.367), and between fruit-eating and dietary diversity (H') ($r = 0.146$, $n = 19$, p-value = 0.552).

Discussion

We found that encounter rates with proboscis monkey groups in PDFR were comparatively higher in riverine forest compared to mangrove forest and mixed mangrove–riverine forest. Echoing observations of higher encounter rates of proboscis monkey groups in this study, other studies have found proboscis monkeys to be more common in riverine forest (Sha *et al.* 2008), suggesting that proboscis monkeys almost universally prefer riverine to other forest types. The higher preference of proboscis monkeys for riverine forest may be due to the availability of more diverse and higher-quality food types including young leaves and fruits in such habitats compared with other forest types (Boonratana 2003; Matsuda *et al.* 2013; Matsuda *et al.* 2009a; Yeager 1989). In fact, other Bornean primates that are more frugivores in their feeding habits, such as the long-tailed macaque (*Macaca fascicularis*), pig-tailed macaque (*Macaca nemestrina*) and orangutan (*Pongo pygmaeus*), have been reported to occur more frequently in riverine forests than other forest types (Lackman-Ancrenaz *et al.* 2001; Sha 2006), suggesting that more food resources, especially fruits, are available in riverine forests. Although the relationship between proboscis monkeys' distribution patterns and fruit availability in riverine forest was not statistically significant in our study, our results potentially support a link between fruit availability and monthly variation in the habitat use of this monkey. Indeed, we found that proboscis monkeys in our study site preferred fruits to leaves when fruits were available, and fruit availability in the riverine forest, mangrove forest and mixed mangrove–riverine forest was an important factor affecting their feeding behaviour.

Although riverine forest seems to contain more food plant species than mangrove or mixed mangrove–riverine forests in PDFR, this forest type contains much less food plant sources in comparison with other comparable habitats or different habitats elsewhere in Borneo. For example, even when the numbers of potential food tree species at all three forest types in PDFR were considered, the recorded species density in a 1.0 ha vegetation survey area in PDFR was only 16 species compared to 46 species in peat swamp in Kalimantan (Yeager 1989) and 84 species in riverine forest in Sukau (Chapter 4). The number of food plant species recorded for proboscis monkey in PDFR was 19 (including human planted coconut trees not found in the vegetation plots and a species of fern and vine not enumerated in the vegetation plots). Whereas, in the previous studies, the number of food plant species recorded for proboscis monkeys was 55 (Yeager 1989) and 188 species (including vines) (Matsuda *et al.* 2009a). Therefore, the low food plant species diversity in the diet of the proboscis monkey in PDFR appeared to be related to the fact that PDFR contains far less food sources available for the monkeys compared to sites elsewhere. Although detailed information is limited, previous studies on proboscis monkey feeding habits conducted in low plant species diversity forests

Table 13.3 Food plant and plant parts eaten by proboscis monkey groups from February 2008 to December 2009 in the Padas Damit Forest Reserve.

Family	Species	Parts eaten	Percentage of feeding frequency
Rhizophoraceae	*Bruguiera gymnorrhiza*	YL/ML/FL	42.4
Moraceae	*Ficus binnendijkii*	YL/FR	38.0
Leguminosae	*Intsia* sp.	YL	3.81
Vitaceae	*Cayratia trifolia*	YL/ML	2.87
Euphorbiaceae	*Excoecaria indica*	YL	2.60
Malvaceae	*Hibiscus tilliaceus*	YL/FL	2.32
Dennstaedtiaceae	*Acrostrichum aureum*	YL/ML	2.32
Rhizophoraceae	*Rhizophora apiculata*	YL/FR	1.77
Sterculiaceae	*Heritiera littoralis*	YL/FR	1.77
Una (local name)	*XX*	YL/ML/FR	0.55
Meliaceae	*Dysoxylum cyrtobotryum*	YL	0.44
Fabaceae	*Pongamia pinnata*	FL/FR	0.33
Acanthaceae	*Acanthus ilicifolius*	YL	0.28
Apocynaceae	*Cerbera odollam*	YL	0.17
Rubiaceae	*Psydrax* sp.	YL	0.11
Jaur (local name)	*XX*	YL	0.11
Arecaceae	*Nypa fruticans*	IF	0.06
Arecaceae	*Cocos nucifera*	IF	0.06
Jalungan (local name)	*XX*	YL	0.06

YL = young leaves; ML = mature leaves; FL = flowers; FR = fruits; IF = inflorescence; *XX* = unknown.

Table 13.4 Best-fit GLM on the effect of availability of food items (proportion data) on the feeding frequencies (weighted scans) of the proboscis monkeys.

	Coefficients	SE	T	*P*-value
Feeding frequencies (weighted scans) (AIC = 126.06)				
Intercept	0.064	0.007	9.59	0.000
Fruits availability	−0.000	0.000	−2.40	0.028

such as in pure mangrove and mixed mangrove-nipah forests also supported the trend found in our study, i.e. low plant species diversity in forests were reflected in the low dietary diversity of proboscis monkeys (Kawabe & Mano 1972; Kren 1964; Macdonald 1982). Findings from these earlier studies and those from ours show that proboscis monkeys are capable of adapting to living in forests that are characterized by poor plant species diversity, thus suggesting their behavioural flexibility i.e. ability to change feeding behaviour in space and/or time, with regard to dietary diversity across its geographic range. This dietary diversity associated with habitat types may be one of the reasons why colobines in general are able to utilize a wide range of habitats (Fashing 2011; Kirkpatrick 2011).

Despite the much poorer plant species diversity in our study site, we found that proboscis monkeys in PDFR had similar tendencies regarding fruit availability, feeding activity and fruit-eating patterns, as is the case with proboscis monkey populations elsewhere and several other Asian and African colobines in general. For example, the proboscis monkey (Matsuda *et al.* 2014; Matsuda *et al.* 2009*a*; Yeager 1989), silver langur, *Trachypithecus auratus sondaicus* (Kool 1993), maroon langur, *Presbytis rubicunda* (Hanya & Bernard 2012) and purple-faced langur, *Semnopithecus vetulus nestor* (Dela 2007), all have been reported to adjust their feeding activity to fruit availability, and for some species, monthly availability of fruits is positively related to monthly feeding activity and extent of frugivory.

An explanation for the tendency to eat more fruits with increasing fruit availability is that as compared to leaves, fruits (as well as seeds) are composed of higher and more easily digestible carbohydrates and probably also contain fewer toxins (Davies *et al.* 1988). Therefore, fruits may provide a better energy source than leaves. Although leaves may contain higher levels of digestible protein (Davies *et al.* 1988; Hanya & Bernard 2012; Matsuda *et al.* 2013), the preference for fruits over leaves in response to higher availability of fruits, suggests that fruits are likely mainly exploited as energy food sources, whereas leaves are exploited mainly as protein sources.

Although fruit availability seems to play an important role in the feeding activity of proboscis monkeys in this study, there was no correlation between monthly fruit availability and young leaf-eating. In other words, when fruits were available, the proboscis monkey was engaged in higher frequency of fruits-eating, but they did not reduce their frequency of young leaf-eating to a significant level. This suggests that proboscis monkeys in PDFR are more folivorous (young leaves constituted 92% of total feeding observations) than those in other study sites on Borneo, e.g. 52% (Yeager 1989), 38% (Bennett & Sebastian 1988), 66% (Matsuda *et al.* 2009*a*) and 50–73% (Boonratana 2003). Interestingly, although the average monthly availability of young leaves (13%) at PDFR was not much higher than the monthly availability of fruits (9%; i.e. including species that were not recorded to be eaten by proboscis monkeys during the study), this food item (fruits) formed only a minor (3%) dietary constituent of the proboscis monkey's diet throughout the study period. In contrast, even though using different methods from that of our study for calculating feeding intensity on different food items, Yeager (1989), Bennett and Sebastian (1988), Matsuda *et al.* (2009*a*) and Boonratana (2003) reported proboscis monkeys appropriated a much higher proportion of their total feeding time to fruits, i.e. 40%, 50%, 26% and 11–39%, respectively. It is probable that the different feeding habits exhibited by the proboscis monkeys in PDFR as opposed to elsewhere was due to a sampling artefact resulting from the limitations associated with the boat-based study method employed in our study out of necessity. However, similar limitations also have been noted in other proboscis monkey studies, yet higher feeding intensity on fruits was observed (e.g. Bennett & Sebastian 1988; Yeager 1989).

We propose that the high folivory, low frugivory feeding habit of the proboscis monkey in PDFR was due to a combination of low fruit productivity and likely clumped distribution of the fruit resources in this area. It might not be highly advantageous for the proboscis monkey to expend their energy to have access to small, isolated, patches of fruit resource and rely chiefly on them due to energetic losses. As a result, in order to maximize the energy gain per unit of time travelling and foraging, the proboscis monkeys at PDFR consume more evenly distributed and easily accessible young leaves. Though further studies are needed, this feeding strategy appears to be in general conformity with the classic optimal foraging strategy model (Krebs & Davies 1993) and has been shown in the maroon langur (Hanya & Bernard 2012) and proboscis monkey in Eastern Sabah (Matsuda *et al.* 2013).

Lastly, we have to note that the preference of proboscis monkeys for riverine forest in PDFR stresses the need for this habitat to be protected. However, this does not necessarily imply that riverine forest is more important habitat than mangrove or mixed mangrove–riverine forests. A study at Samunsam in Sarawak, where mangrove and other habitats are contiguous with one another, showed proboscis monkeys moving in and out of mangrove forest at different times of the year indicating seasonal changes in habitat preferences (Bennett 1986; Rajanathan & Bennett 1990). Therefore, during certain parts of the year when preferred food sources in riverine forest are scarce, proboscis monkeys may use mangrove forest or other habitats for feeding. In addition, mangroves and mixed mangrove–riverine forests in our study were used for roosting on some occasions, even though it was less frequent than in riverine forest. Principally, however, as young leaves and preferred fruiting trees affecting proboscis monkey behaviours were distributed not only in riverine forest but also in other forest types, especially for several of their important plant species, e.g. *B. gymnorrhiza* (10% of sampled trees from the vegetation plots), *F. binnendjikii* (9%), *H. littoralist* (3%) and *R. apiculata* (2%), the protection of all habitat types in PDFR is considered necessary for the proboscis monkeys in PDFR. While it is a common practice to allocate the mangroves for conservation, as the economic values of the intact habitat vastly outweigh those of cutting the trees (Bennett & Reynolds 1993), the importance of non-mangrove forests are usually overlooked, particularly with respect to proboscis monkey conservation. With the main ongoing threats to the populations of proboscis monkey in Klias Peninsula identified as habitat loss and fragmentation due primarily to anthropogenic activities and also forest fires associated with the El Niño event (Bernard & Zulhazman 2006; Sha *et al.* 2008), the protection of all useful remaining habitats that are contiguous with each other and maintenance of their connectivity is vital for the continued survival of the proboscis monkey in this region of Borneo.

Acknowledgements

We thank our research assistants, Lee Shan Khee and Gilmoore Bolongan, and field enumerators, Rozan, Awang Masis and family, for assistance rendered during the field work. Data entry was conducted by Julia George, Pius Pansang and Lucy Wong. This study was partly financed by the Ministry of Higher Education Malaysia through its Fundamental Research Grant Scheme (FRGS0085-BD-1/2006) and ProNatura Foundation Japan 2008/2009 awarded to H. Bernard. An earlier draft of this chapter was prepared at KUPRI, through funding provided by the Daiko Foundation, Japan and Universiti Malaysia Sabah. I. Matsuda gratefully acknowledges the support of HOPE and Human Evolution Project of KUPRI, the National Geographic Society (9254-13), the Wildlife Reserves Singapore, the Inamori Foundation and JSPS Grants-in-Aid for Challenging Exploratory Research (24657170 & 15K14605), for Young Scientist (26711027 & 21770261) and Strategic Young Overseas Visits Program for Accelerating Brain Circulation (S2508).

Chapter 14

Ebony Langurs in Mangrove and Beach Forests of Java, Bali and Lombok

Vincent Nijman

Introduction

> Seen from a distance the mangrove makes the impression of a dark-green more-or-less monotonous type of forest. On entering it on foot with ebb its eerie aspect appears at once from the oppressing heat, the damp atmosphere, the bare, stinking mud, covered with stilt-rooted trees ... (van Steenis 1958, p. 431)

Giesen *et al.* (2006, pp. 11–12), reflecting on the views of van Steenis partially quoted above, commented that (in Southeast Asia) mangroves are very interesting, but not the place you would go for a picnic. They noted, from experience, that mangroves are not easy environments to work in but that they can be very rewarding. From a botanical perspective, their structure is generally straightforward and simple, and the number of species is limited. However, the species that do occur may be very abundant, and as long-term studies show, they are often highly productive. The same may be said about mangrove faunal communities, and in particular primate communities, which tend to be 'straightforward and simple'. In Southeast Asia, there are a number of primates that can be found in mangroves, including several species of macaque and langur, but no species are confined to it. Proboscis monkeys (*Nasalis larvatus*) are perhaps viewed as a typical mangrove primate (Kawabe & Mano 1972; Chapter 4) as is, to a lesser extent, the long-tailed or crab-eating macaque (*Macaca fascicularis*); however, both species are found in a range of other habitats as well. Ebony langurs (*Trachypithecus auratus*) have been characterized as being largely restricted to coastal and riverine habitats (Bennett & Davies 1994) but in fact they occur in a wide range of habitats (Nijman 2012).

Colobines in Asia including the proboscis monkey and langurs of the genus *Trachypithecus*, show adaptations that make it possible to live in the 'dark-green monotonous' mangroves. They have an enlarged and sacculated, ruminant-like stomach, an adaptation to a folivorous diet. This comes either in the form of a tripartite stomach with a saccus, tubus gastricus and pars pylorica as seen in, for instance, *Trachypithecus* or as a quatropartite stomach through the addition of a presaccus as seen in *Nasalis* (Caton 1999; Matsuda *et al.* 2015). The saccus is large and contains bacteria for fermentation of plants. The sacculated part of the stomach and the proximal two-thirds of the gastric tube are lined with cardial glands. These structures allows the separation of ingesta between proximal and distal parts, and thus between alkaline and acidic environments (Bauchop 1978). Colobines have large salivary glands that act to neutralize stomach acid. Due to these adaptations, colobines can persist on a seemingly monotonous diet of leaves, comprising typically 60–80% of their diet (Kirkpatrick 2011).

This chapter focuses on the ebony langur in the mangroves and beach forests of the Indonesian islands of Java, Bali and Lombok. The ebony langur is endemic to this region and shares these three islands with almost 145 million people (135 million on Java, 4 million on Bali and 3 million on Lombok; BPS 2004). As detailed below, most of the mangroves and beach forest have been converted for other land uses and only small isolated pockets remain. Mangroves and beach forests are the two dominant coastal forest types found in Java (and indeed elsewhere in Southeast Asia). While mangroves are generally restricted to the tidal zone, the strip of coast starting from the lowest water level up to the highest water level, beach forests generally occur along exposed, sandy or coral coasts, root in freshwater, and are almost never inundated by seawater. Clearly, both mangroves and beach forests, are strongly influenced by the sea.

Here, an overview of the occurrence of ebony langurs in mangroves and beach forests on the islands of Java, Bali and Lombok is given. First, a physical description of these forest types is provided, including their predominant tree species, followed by a summary of the main sites at which ebony langurs have been recorded. Second, a concise overview of the ecology of langurs in coastal forest focusing on densities and diets is given. Finally, data on the demise of mangroves and beach forests and the langur populations they support are presented, alongside a description of some recent mangrove restoration initiatives and their relevance for ebony langur conservation.

Methods

Data on ebony langurs were collected by conducting surveys throughout the species' range on Java, Bali and Lombok, between 1994 and 2012, details of which can be found in earlier publications (Nijman 2000, 2012). Ebony langurs occur in a range of forest types, from coastal to lowland forest, both dry deciduous and perhumid rainforest, all the way up to the highest mountains above 3000 m asl (Nijman 2012). Surveys included several areas with mangroves and beach forest,[1] lasting several days, within area multiple sites were visited and most sites were revisited over the study period. Data were collected on the occurrence and distribution of ebony langurs; upon each sighting of individuals or groups, an estimate of group size and general description of habitat and level of disturbance was made.

In addition to surveys, data on the distribution and socioecology of ebony langurs throughout Java were compiled from the literature. While hitherto only one study has been conducted specifically on the ecology of ebony langurs in a mangrove forest (Supriatna *et al.* 1988) and one in beach forest (Kartikasari 1986), several (botanical or ecological) studies in coastal forest have noted the presence of ebony langurs (e.g. Appelman 1939; Hoogerwerf 1972; Erftemeijer *et al.* 1988; Gurmaya *et al.* 1994; Supardjo 2008). In addition, several studies on ebony langurs have noted their presence in mangrove and beach forests. Combining these data allowed for a detailed, albeit incomplete overview of the importance and future of mangroves for ebony langurs.

Results and Discussion

Mangroves, Beach Forests and the Distribution of Langurs

The most commonly recorded plant families in Southeast Asian mangroves include the Leguminosae or legumes (22 species), Rhizophoraceae – usually regarded as the family of mangrove trees, many with stilt roots and other adaptations (12 species), Combretaceae family or *Terminalia* genus (six species), Avicenniaceae, another family of true mangrove trees, characterized by pneumatophores, i.e. roots that emerge, peg-like, from the mangrove soil (five species) and Sonneratiaceae, another family consisting predominantly of mangrove tree species (five species) (Giesen *et al.* 2006). Abundant and characteristic of the mangroves of Java, Bali and Lombok are species such as *Rhizophora apiculata, Bruguiera sexangula* and *B. gymnorrhiza, Ceriops decandra, Avicennia alba* and *Sonneratia alba* (Whitten *et al.* 1996). The other coastal forest is beach forest, which in Southeast Asia is usually identical to the '*Barringtonia* formation' (e.g. van Steenis 1958). Beach forests generally occur along exposed, sandy or coral coasts. Unlike mangroves, beach forests are almost never inundated by seawater, but root in freshwater, with influence from the sea. Many typical beach forest species such as *Barringtonia* species, *Pemphis acidula, Terminalia catappa, Calophyllum inophyllum* and *Thespesia populnea* can often be found in the landward fringe of mangroves as well (Giesen *et al.* 2006).

Ebony langurs occur in both mangroves and beach forest. In some areas with extensive mangroves or in uninterrupted stretches of beach forest, individual groups or whole populations are confined to such coastal forests. In many areas, however, mangroves and/or beach forest cover relatively small areas and are continuous with other forest types including backswamps and lowland forest. In these areas, individual ebony langur groups may use mangroves or beach forest as part of their range.

Based on surveys and review of the literature, it seems that at present ebony langurs are found in just 14 mangrove and beach forests on Java, one on Bali and possibly one on Lombok. Some areas are very small and cover only a small stretch of the coast whereas others are extensive and cover large areas (Table 14.1; Figure 14.1).

Java's West Point

Ujung Kulon is a national park on the westernmost tip of Java best known as the home of the last Javan rhinoceros, *Rhinoceros sondaicus*. The park includes a mainland section (Mt Honje), a peninsula and two islands, Panaitan and Peucang. On Penaitan Island, there is a well-developed beach forest covering most of the 17 km^2 of the island's area and also approximately 1 km^2 of mangroves (Giesen *et al.* 2006). Beach forest is found on the southern part of the Peninsula, covering several kilometres of coastline. Approximately 10 km^2 of mangroves occur on the northern and northeastern shores with species *S. alba, Lumnitzera racemosa, N. fruticans*, and species of *Avicennia, Rhizophora* and *Bruguiera* dominating (Hommel 1987). Based on the author's data, ebony langurs do not occur on the islands of Panaitan or Peucang (cf. Hoogerwerf 1970), but are found throughout the remainder of the reserve, including the areas of mangrove and beach forest (cf. Gurmaya *et al.* 1994). The ebony langur population in Ujung Kulon is likely the second largest remaining and among the best protected.

Jakarta Bay

The entire area where the current Indonesian capital Jakarta is situated was once covered in coastal and swamp forest. At present, little remains, and only three coastal areas in the Jakarta Bay may still contain ebony langurs, Muara Angke-Kapuk, Muara Gembong and Tanjung Sedari. Muara Angke-Kapuk comprises three gazetted areas, the Muara Angke wildlife sanctuary (25 ha), a nature tourism forest (100 ha) and the Muara Angke strict nature reserve (44 ha). The areas are under constant threat from the ever expanding capital and, while in 2009 the mangroves fringing Jakarta were declared as an important 'green belt', poor enforcement of environmental laws and regulations in Jakarta, and indeed Indonesia as a whole, mean that these last vestiges of mangroves are not safe. Muara Gembong is the one area where ebony langurs have been studied in mangroves intermittently over 8 months in 1981, 1984 and 1987 (Supriatna *et al.* 1988). At the start of the study in 1981, at least 10 km^2 of mangroves remained standing and the density of ebony langurs was estimated at 20 individuals/km^2, suggesting a population of some 100 langurs. Midway through the study, parts of the area were converted into rice fields, fish ponds and settlements and, in the final year of the study, nearly all the mangrove forest of Muara Gembong had been destroyed. At the easternmost part of Jakarta Bay, a population of ebony langurs is still present in the mangroves of Tanjung Sedari (Nijman 2000). Like other mangroves along the north coast, the area has suffered heavily from ongoing encroachment, reclamation of land and conversion into fish and shrimp ponds. No recent data on the ebony langurs are available.

West Java's South Coast

Two populations of ebony langurs are found in the coastal forest of the south coast of West Java, i.e. Leuweung Sancang and

Table 14.1 Overview of ebony langurs *Trachypithecus auratus* in beach forest and mangroves on Java, Bali and Lombok, listed in a west-to-east sequence, with estimates of the amount of available habitat.

Site	Status	Beach forest	Mangrove forest	Coast line (km)	Comments
Ujung Kulon	NP	B	C	90	
Cibanteng	NR	A		13	degraded; adjacent to Cikepuh
Cikepuh	WR	A		17	degraded
Muara Angke	various		A	3	severely degraded; langurs possibly extinct
Muara Gembong			A		severely degraded; langurs possibly extinct
Tanjung Sedari			A		severely degraded; langurs possibly extinct
Leuweung Sancang	WR	A	A	7	degraded
Pangandaran	TW/NR	B	A	8	
Segara Anakan		B	D	40	degraded
Balekambang		A		2	
Pasir Putih			A	1	severely degraded; langurs in adjacent inland forest
Meru Beteri	NP	B		60	
Baluran	NP	B	B	30	
Alas Purwo	NP	B	C	80	
Bali Barat	NP	B	A	15	Prapat Agung Peninsula only
Kerandangan	TP	?		?	langurs possibly in inland section only

Status: NP = National Park (*taman nasional*); NR = Strict Nature Reserve (*cagar alam*); WR = Wildlife Reserve (*suaka margasatwa*); TP = Nature Tourism Park (*taman wisata alam*); Muara Angke is a patchwork of protected areas; blanks indicate unprotected areas.
Key: A = < 1 km^2; B = < 10 km^2; C = < 30 km^2; D > 30 km^2

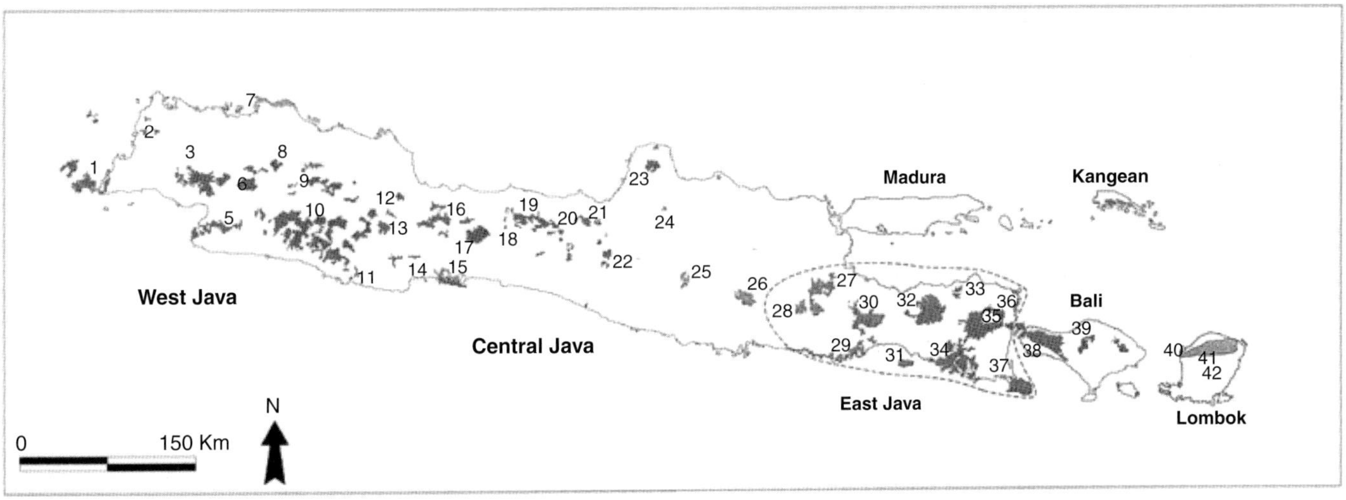

Figure 14.1 Map of langur distribution from Nijman (2000) – coastal sites are numbers 1 (Ujung Kulon), 5 (Cibanteng), 7 (Jakarta Bay), 11 (Leuwang Sancang), 14 (Pangandaran), 15 (Segara Anakan), 34 (Meru Beteri), 37 (Alas Purwo), 36 (Baluran), 33 (Pasir Putih), 38 (Bali Barat), 40 (Kerandangan).

Pangandaran. Unlike the north coast, the south coast has few shallow areas, and mangroves are found in shallow sheltered bays and inlets. In Leuweung Sancang, a small extent of mangrove is present at the mouth of the Cibako River, and, adjacent to it, are some stretches of beach forest. Three groups of ebony langurs were observed in the beach forests of Leuweung Sancang and one in the mangroves by a team of the University of Bandung in July 2011 (Anonymous 2011). Pangandaran is a well-known strict nature reserve containing rainforest on the south coast of West Java. The reserve comprises an uplifted limestone peninsula and small sections of beach forest particularly along the northern part of the peninsula, totalling probably < 1 km^2. The beach forest is characterized by *Pterospermum javanicum, Pilea trinervia, Lagerstroemia speciosa* and *Barringtonia asiatica* (Nugraha 2007). Ebony langurs are especially abundant in the northern part of Pangandaran where they have been studied by, among others, Kool (1989, 1993), Brotoisworo (1983) and Brotoisworo and Dirgayusa (1991).

Central Java's Inland Sea

The Segara Anakan lagoon, located on the south coast of central Java consists of a central lagoon surrounded by mangrove swamps and recently accreted intertidal land that has been partially converted into rice fields. The central lagoon has a remaining surface area of approximately 17 km^2, and there are about 122 km^2 of mangrove forests of which about 56 km^2 remains in slightly to moderately disturbed conditions (Abubakar *et al.* 2001). Mangroves here are dominated by *Avicennia, Rhizophora* and *Sonneratia* (especially *S. alba*). This area was proposed as a reserve in the early 1980s, but a combination of conservation and sustainable use is now considered the best option because of heavy development pressures (White *et al.* 1989). No specific studies have been conducted on the ebony langurs of Segara Anakan, but the author has observed numerous groups during boat surveys in the 1990s as have Erfemeijer *et al.* (1988) who reported their presence in the central mangrove region. Based on the large extent of mangrove forest and the distribution of ebony langurs within them (essentially everywhere where good forest remains), without a doubt Segara Anakan represents/supports the largest remaining population of ebony langurs in mangrove forest.

Java's Easternmost Corner

Ebony langurs have been recorded in the three large national parks situated along the coast in this part of the island, i.e. Meru Beteri in south, Alas Purwo in the southeast and Baluran in the north. Ebony langurs are present in the beach forests in all three protected areas as well as in the fringes of mangrove forests in Baluran (in particular in Kelor, Bilik, Lambuyan, and Tanjung Sedano) and Alas Purwo (Bedul and Segoro Anak) (Supardjo 2008). In particular, the mangroves of the Segoro Anak are still extensive covering some 23 km^2 and, albeit intensively used (Satyasari 2010), appear to harbour a substantial population of ebony langurs. In other areas, most of the ebony langurs are not confined to coastal forest and range also in the adjacent rainforest (Meru Beteri) or deciduous forests (Alas Purwo, Baluran) (Appelman 1939; Hoogerwerf 1972; Kartikasari 1986; Pfeffer 1965). West of Baluran, along Java's north coast, a small population of ebony langurs is found in the mangroves near Pasir Putih (Nijman 2000).

West Bali

Ebony langurs are found predominantly in the western part of the island, where much remaining forest is contained within Bali Barat National Park. Here, langurs are found throughout the reserve, including in small numbers in the fringe mangroves along the Prapat Agung Peninsula. Vogt (2003) conducted surveys in the Peninsula and studied two groups of langurs showing them to be present in the mangroves as well as in the adjacent deciduous forests.

Northwest Lombok

Ebony langurs on Lombok were thought to occur in the mountainous interior only (Kitchener *et al.* 1990; Nijman 2000); however, a booklet from the provincial conservation agency (Wahyuni & Mildranaya 2010) specifically mentions the presence of ebony langurs in the newly established (1992) Kerandangan Nature Tourism Park near the island's northwest coast. The forests of Kerandangan are continuous with those near Pucuk where ebony langurs were reported by Nijman (2000). It is unclear whether Kerandangan itself borders the coast, but aerial photographs suggest that it does, at least in part.

Ecology of Langurs in Coastal Forests

While there are some estimates of the densities of ebony langurs in inland forest (Nijman 2000, 2012), it is often difficult to systematically survey primates in coastal forest. Mangroves frequently have to be surveyed by boat, and langurs in mangroves and beach forest have a more or less 'linear' distribution. Both make it problematic to estimate densities per area (as in individuals per square kilometre). Yet the data we have from several studies suggest that ebony langurs can be particularly abundant in coastal forest. Densities in Pangandaran are in the order of 185–195 individuals/km^2 (Brotoisworo 1986; Kool 1989) and numbers appear to be higher in the coastal regions than in the inland areas (Kool 1989, pers. obs.). Likewise, langur densities in the coastal forests of Baluran appear to be higher than in the inland parts of the reserve. Finally, data from Supriatna *et al.* (1988) suggest particularly high densities of ebony langurs in the mangroves of Muara Gembong, at least in the early 1980s when their first surveys were conducted. Group sizes of ebony langurs correlate with rainfall seasonality and elevation, such that the largest groups are found at sea level in the highly seasonal parts of the islands (Nijman 2012). While group sizes in montane areas and lowland rainforest are typically on the order of six or seven individuals, group sizes in coastal forests typically comprise between 10 and 11 individuals. Moreover, very large groups of 20 to 30 individuals are regularly recorded in coastal forest but not in montane or inland rainforest (Kool 1989; Kartikasari 1986; Nijman 2012; Supriatna *et al.* 1988).

The number of food plants that ebony langurs use in coastal forest is significantly lower than in non-coastal forest (Figure 14.2). In mangroves, langurs fed on merely nine species of, by and large, abundant mangrove species and in beach forest the langurs were recorded to feed on just 14 species. In the beach forest of Baluran, the top nine food plant species comprised 93% of the langur's diet, in Bali Barat this was less at 68%, in Pangandaran at 57%, whereas on Mt Gede the nine most commonly used food plants made up a mere 39% of the langur's annual diet. It is important to note that this is not an artefact of sampling as, for instance, Kool (1989) reports high dietary diversity even within a single month; in several months,

Table 14.2 Chemical composition of food plants eaten by ebony langurs *Trachypithecus auratus* and proboscis monkeys *Nasalis larvatus* in three coastal sites. Presented are means ± standard deviation and sample size.

Species	Item	Water	Ash	Crude proteins	Crude fat	Site
T. auratus	leaves	64.5 + 14.0 (5)	4.0 + 2.3 (5)	4.9 + 1.2 (5)	2.1 + 2.2 (5)	1
N. larvatus	leaves	31.7 + 19.0 (13)		11.7 + 5.5 (13)		2
N. larvatus	leaves		6.3 + 2.8 (23)	9.9 + 3.4 (20)		3
T. auratus	fruit	65.7 + 8.6 (4)	1.8 + 1.7 (4)	3.7 +1.1 (4)	0.7 + 0.5 (4)	1
N. larvatus	fruit	78.3 + 1.7 (10)	3.5 + 1.4 (17)	5.2 + 1.6 (22)	7.4 + 6.9 (10)	3

Key for sites 1 = Muara Gembong, West Java; 2 = Samunsam, Sarawak, Borneo; 3 = Tanjung Puting, Central Kalimantan, Borneo. Sources: Based on data from Supriatna *et al.* (1988) and Nijboer (2006).

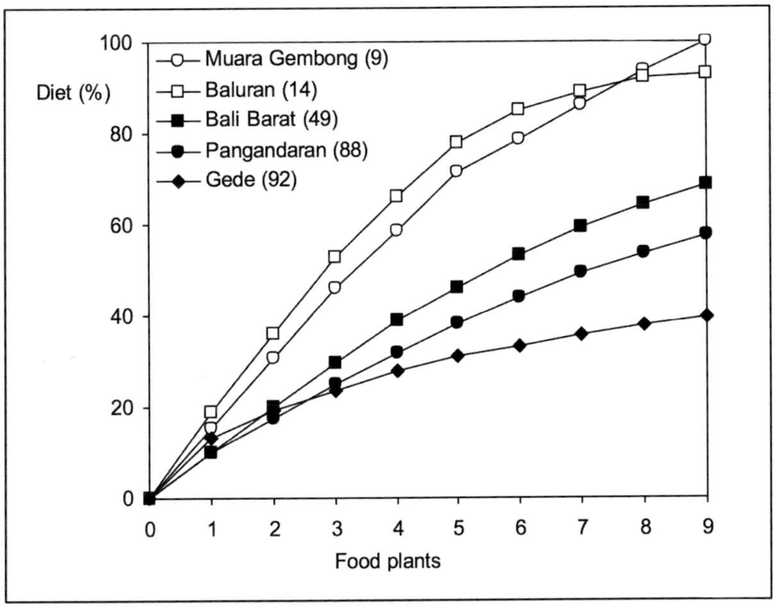

Figure 14.2 Contribution of food plants to the diet of ebony langurs *Trachypithecus auratus* in five study areas in Java and Bali, Indonesia comparing two coastal forests (open symbols: Muara Gembong, mangroves and Baluran, beach forest) with three non-coastal forests (closed symbols). Numbers within brackets indicate the total number of food plants used. Sources: Muara Gembong: Supriatna *et al.* (1988); Baluran: Kartikasari 1986; Bali Barat: Vogt (2003); Pangandaran: Kool (1989); Gede: Beckwith (1995).

even the most commonly used food plant contributes < 15% of that month's diet. Hence the above data suggest that the diet of ebony langurs in mangroves and beach forest is simple with only a few plant species contributing to the langurs' diet. Both in beach forest and mangrove forest, floristic composition is less diverse than in rainforest or deciduous forest and this explains the small potential dietary breadth of mangrove- and beach forest-dwelling langurs.

Typical mangrove and coastal forest plant species consumed by ebony langurs include *Sonneratia alba, Aegiceras corniculatum* (Muara Gembong, Segara Anakan), *Avicennia alba, A. marina, A. officinalis, S. acida, Rhizophora apiculata, Nypa fructans, Bruguiera gymnorrhiza* (all in Muara Gembong), *Sonneratia caseolaris* (Segara Anakan), *Bruguiera cylindrica* (Baluran), and *Terminalia microcarpa* (Bali Barat). In terms of chemical composition, and comparing with data from proboscis monkeys (Table 14.2), ebony langurs' diet in the mangroves of Muara Gembong appears to be low in crude proteins (both from leaves and fruits), ash (leaves and fruit) and crude fat (fruit). However, the variation around the means is large for both species suggesting these differences are not statistically significant. The major selective driver in leaf selection is proteins and the low values observed in the diet of the ebony langurs suggest a very low-quality diet.

Conservation and Restoration of Mangroves

In the past, ebony langurs almost certainly were present along the entire coastal strips of all three islands of Java, Bali and Lombok. With many coastal villagers being dependent on the sea and little incentive for converting mangrove or beach forest into other land uses, the langurs must have lived more or less in sympatry with humans in these areas. Probably from the early twentieth century onwards, mangroves were increasingly converted into fishponds (or *tambak* in Indonesian). Previously, these fishponds were established within mangrove forest, and trees were retained on pond dykes or on islands in the *tambak*. Later, however, clear-felling was carried out prior to the construction of a *tambak*, leaving these more recent fishponds with a tree cover of almost zero (Giesen *et al.* 2006). At present, very large parts of the coastal region of especially Java are covered in fishponds with very limited opportunity for species such as langurs to coexist (with people and their farms).

A rather recent development is land reclamation of former mangrove areas for housing and recreation estates. In Java, this started with the successful reclamation of the Ancol marshlands, on the outskirts of Jakarta, in the early 1970s (Giesen *et al.* 2006). Reclamation of Segara Anakan was proposed as early as the 1930s when Indonesia was still under Dutch colonial rule. Similar to how land reclamation has been done on a grand scale in the Netherlands, the Segara Anakan lagoon was to be closed off by sea dykes, diverting the Citanduy River into the Indian Ocean and draining the resulting polder by pumping. Local rainfall runoff then flushes out the remaining salt water allowing the area to be converted into productive agricultural land (White *et al.* 1989). It seems that the eruption of World War II put these plans on permanent hold, and subsequently Segara Anakan is one of the few areas in Java where mangrove forest of any significance still remains. Despite the conversion and reclamation, there are numerous initiatives to replant mangroves on Java and Bali (the author is not aware of any on Lombok). While these initiatives are to be applauded and will clearly benefit mangrove ecosystems and the animal communities in them, they are often carried out at too small a scale to be of benefit to ebony langur groups. Mangroves are frequently planted in strips, or in small patches, with relatively few species included in the planting scheme. Langurs have home ranges in the order of 20–30 ha at least and few, if any, mangrove restoration projects on Java or Bali reach this size.

While undoubtedly the outlook for ebony langurs in Java, Bali and Lombok is bleak (cf. Nijman & Supriatna 2008), there is still much to learn about langurs in coastal ecosystems. It would be worthwhile to conduct a long-term (at least one year) study on the ecology of ebony langurs in mangrove and/or beach forest, and to properly examine the adaptations (behavioural and physiological) exhibited by langur groups inhabiting such extreme environments. The two best areas to initiate such studies are Ujung Kulon and Segara Anakan (Figure 14.1). A study in Ujung Kulon would preferably be conducted along the northern part of the Peninsula where mangroves and beach forest are found. The study in Segara Anakan could focus on groups living in the central part of the lagoon, where langurs are confined to mangroves; or, the study could focus on the area closer to Nusa Kembangan, where they occur in the beach forest. Raising attention to the occurrence of ebony langurs in coastal forest may also bring some hope of allowing them to persist in these 'dark-green monotonous forests' into the future.

Acknowledgements

I would like to thank the Indonesian Institute of Sciences (LIPI) and the Directorate General for Forest Conservation and Nature Protection (PHKA) for allowing me to conduct my research. Financial support for my studies was provided by the Netherlands Foundation for International Nature Protection, Society for the Advancement of Research in the Tropics, and Martine de Beukelaar Foundation. Contributing to this volume gave me an opportunity to put the data I had collected for over a decade into a clearly defined contextual framework: I thank Katarzyna Nowak for the invitation to contribute. Nicola Thurley helped with compiling information on langur diets, and with locating literature, and Giuseppe Donati helped with interpreting aspects related to diet and feeding behaviour.

Note

1 Within the Indonesian names of areas surveyed, several words appear regularly. Provided are translations of these words from Bahasa Indonesia, Bahasa Sunda or Bahasa Jawa to English. *Ci-*: water or river; *muara*: estuary or river mouth; *nusa*: island; *pasir putih*: white sand; *segara anakan*/*segoro anak*: child [of the] sea (used to describe a lagoon); *tanjung*: cape; *ujung*: end or cape.

Chapter 15

Mangrove

A Possible Vector for Tarsier Dispersal Across Open Ocean

Myron Shekelle, Joan Stevenson, Blair Kaufer, Steven Stilwell and Agus Salim

Introduction

Phylogenetically, tarsiers are an ancient branch of the primate tree, having split off from other primates perhaps 63–71 Ma (Springer *et al.* 2012). This long separation from both strepsirrhine and anthropoid primates makes tarsiers a third, nearly independent, evolutionary branch with which to test hypotheses about primate evolution. Crown tarsiers date to the Miocene, minimally, with estimates varying from 20.6 Ma (10.6–32.1) (Shekelle *et al.* 2010) to 18.64 Ma (8.75–37.19) (Springer *et al.* 2012). Groves and Shekelle (2010) classified tarsiers into three genera, *Tarsius* from Sulawesi, *Cephalopachus* from Borneo and Sumatra, and *Carlito* from the southern Philippine islands, each with three or more species or subspecies. The latter two genera form a clade supported by genetic (Shekelle *et al.* 2010, Springer *et al.* 2012) and morphological data (Groves 1998). To come to this distribution, tarsiers must have crossed the Wallace/Huxley line at least twice (Figure 15.1).

Ecologically, tarsiers are small-bodied nocturnal primates, obligate faunivores, and vertical clinging and leaping primates that primarily use the forest understory (Shekelle *et al.* 2013). Most tarsier species average about 120 g for males, with average female body weight being about 90% that of males (Sussman 1999). There is one montane form, sometimes called a pygmy tarsier, which weighs about half as much as the lowland forms (Grow & Gursky 2010). Tarsiers are unique among primates in that they eat only live-caught animal prey (Shekelle *et al.* 2013). Although their diet is highly unique for primates, some might say specialized, they are in fact very generalized carnivores, eating just about anything they can catch and kill (Shekelle *et al.* 2013). Skeletally, they have the lowest intermembral index among primates, with forelimbs only about 52–58% as long as their hindlimbs (Fleagle 2013). This extreme skeletal adaptation corresponds to their propensity for using vertical clinging and leaping as a means to move among the saplings and fine branches of the understory (Niemitz 1984). Tarsiers spend most of their time within 2 m of the forest floor, occasionally coming to the ground to feed or cross open spaces, and occasionally ranging higher to forage, cross gaps, avoid predators and for various social activities (Shekelle *et al.* 2013). Tarsiers are documented in a great variety of primary and secondary habitats, including: 'mossy, microphyll, montane, bush, gallery, deciduous rain, and mangrove forests; thorn scrub, shrubland, swamps, riparian, palm, and bamboo habitats; seashore scrub; and urban gardens and villages ... [e]ven grassland areas' (Shekelle *et al.* 2013).

The systematic study of tarsiers in the wild is relatively recent and rare compared with many of the better-known primates (Niemitz 2010). Much of the effort to study them has focused on the behavioural ecology of a given population, with only one study explicitly examining tarsiers in different habitats (Merker *et al.* 2005). Never has a study explicitly set out to study tarsiers in mangrove. Nevertheless, some records of tarsiers in mangrove exist, proving that they are there. Given that, questions that beg answers include:

1. Are tarsiers native to mangrove, or have they been pushed there by anthropogenic processes?
2. When tarsiers are found in mangrove, does mangrove constitute the totality of those tarsiers' home range, or is mangrove only a portion of their home range?
3. Is mangrove in any way critical for tarsiers? Are tarsiers in any way critical for mangrove?

This last set of questions is particularly intriguing when it comes to considering how tarsiers disperse across open oceans.

Mangrove Habitat within the Extent of Occurrence of Tarsiers

Following the methods of Shekelle and Salim (2009), i.e. using the Southeast Asian Mammals Database to estimate tarsier distributions and the US Geological Survey (USGS) 2011 to estimate habitat cover, we calculated the size of the extent of occurrence (EOO) and the amount of mangrove within the EOO for each tarsier taxon. From that, we calculated the percentage of each EOO that is covered by mangrove (Figure 15.2).

Globally, mangrove accounts for only about 1.4% of the EOO of tarsiers (Tables 15.1 and 15.2). The tarsier species with the largest percentage of mangrove within its EOO is *Cephalopachus bancanus*, with 1.7% of its EOO covered by mangrove. Two species, *Tarsius lariang* and *T. pumilus* have no mangrove at all in the EOO. All other known species vary in the percentage of their EOO covered by mangrove from 0.22 to 0.57%.

Direct Evidence of Tarsiers in Mangrove

Published reports of tarsiers in mangrove include Niemitz (1979), who reported setting nets in mangrove forest at Bako

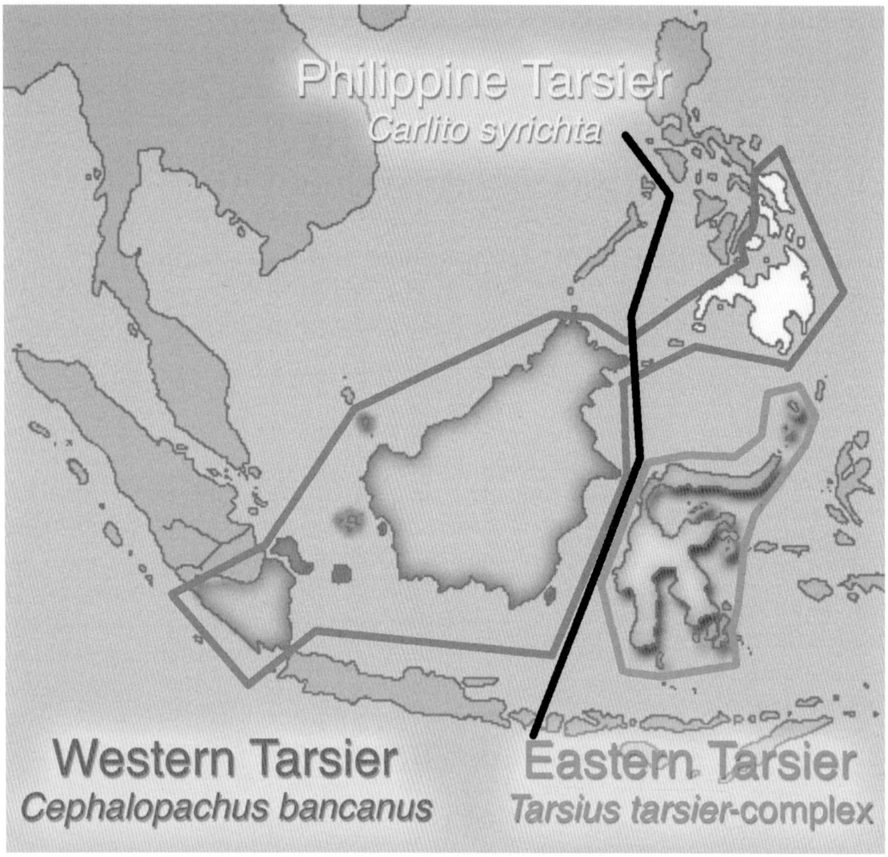

Figure 15.1 Tarsier taxonomy and biogeography. (A black and white version of this figure appears in some formats. For the colour version, please refer to the plate section.)

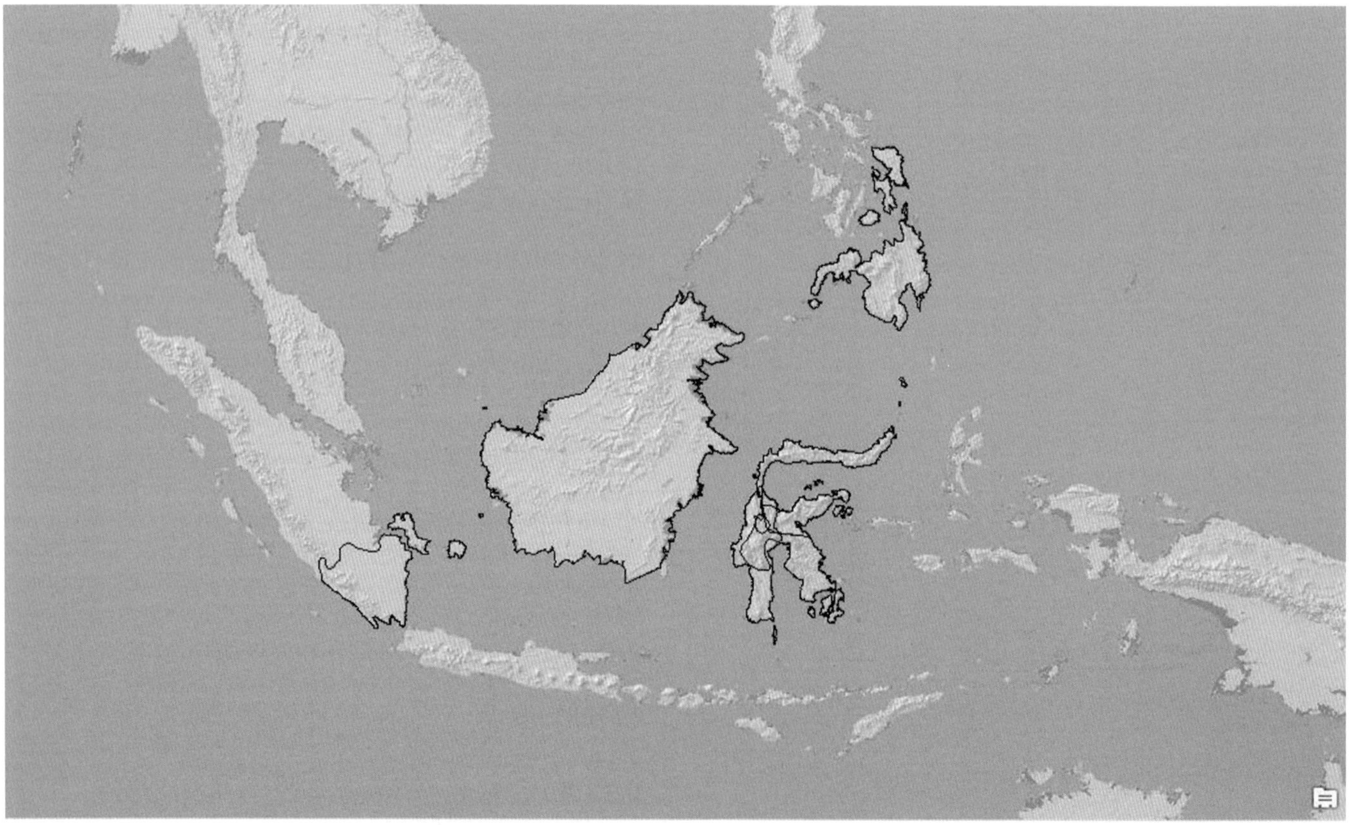

Figure 15.2 Distribution of mangrove in Southeast Asia: mangrove in dark grey outline represents tarsier distributions.

Table 15.1 Mangrove habitat within EOO of tarsiers.

EOO Tarsiers	Mangrove (ha)
Eastern Tarsiers	**131 031.62**
Philippine Tarsiers	**80 886.51**
Carlito syrichta	**80 886.51**
C. s. carbonarius	47 587.34
C. s. fraterculus	8 068.17
C. s. syrichta	25 231.00
Western Tarsiers	**1 655 966.16**
Cephalopachus bancanus	**1 655 966.16**
C. b. bancanus	110 937.60
C. b. borneanus	1 537 738.34
C. b. natunensis	233.05
C. b. saltator	7 057.16
Grand total	**1 867 884.29**

National Park, north of Kuching, Sarawak, and though 48 net nights yielded no captures, 'the unmistakable smell of tarsiers' (p. 180) was detected. In refuting the commonly held assumption that tarsiers are exclusive to, or prefer, secondary habitats, Niemitz (1979) noted that tarsiers 'may appear … in mangrove vegetation' (p. 223). Niemitz (1984) reiterated this view, stating tarsiers 'may also penetrate mangrove areas on the coast' (p. 87). MacKinnon and MacKinnon (1980) reported a density of four tarsiers per hectare in mangrove near Tanjong Panjang, Gorontalo.

There are a few unpublished reports of tarsier in mangrove. Sharon Gurksy (pers. comm.) commented that she has little knowledge of tarsiers in mangrove 'beyond that they do inhabit them'. A report by Shekelle *et al.* (1996, unpublished data) documented fieldwork conducted in 1995 and 1996 including at Libuo (equivalent to MacKinnon and MacKinnons's site Tanjong Panjang) where: 'We caught 15 tarsiers from five families. Tarsiers were observed in the short dense spiked forest around mangrove.' (p. 4).

During the course of Shekelle's phylogeographic studies of *Tarsius*, tarsiers were encountered in mangrove on multiple occasions. Shekelle (2003) documents results of fieldwork in 1995 and 1996, and includes reports of tarsiers in mangrove, and on the border of mangrove and other habitats. Of ten surveyed tarsier sleeping sites on Sangihe Island, midway between the northern tip of Sulawesi and the Philippine Island of Mindanao, one site was relatively high in a mangrove tree (Shekelle 2013). The tree itself was on the border between a thin strip of mangrove and a relatively clean coconut grove. A second site on Sangihe Island was in a tall tree in a village, which was near a thin strip of mangrove. On Malenge Island, in the Togian Islands of Central Sulawesi, one of the three surveyed sleeping sites was a vine-covered coconut tree at the meeting of a mixed garden (similar to agroforestry) and mangrove. The group composed of a minimum of four tarsiers, two of which were captured, an adult male and an adult female. In the tiny village of Basaan, near Ratatotok, in North Sulawesi a pair of tarsiers was heard dueting at the junction of coconut grove and mangrove; the adult male was captured. A second sleeping site at Basaan was built in matted vegetation in a large tree in a mangrove swamp. The group had a minimum size of four tarsiers and an adult female was captured.

Indirect Evidence that Tarsiers Are Suited to Mangrove Habitat

Cuming (1838) noted that tarsiers will eat shrimp when 'extremely hungry' (p. 67). Similarly, Cook (1939) reported on captive *Tarsius syrichta carbonarius*, 'fish and shrimp were readily eaten, so it may be concluded that they were not new foods; but from the fact that they were the least preferred, it is believed that the tarsiers only resort to fishing when better liked food is unavailable' (p. 176). Cook discusses the diet and feeding preferences of these captive tarsiers further. Both shrimp and fish – those under 3.8 cm – were taken from the water and consumed. Shrimp were consumed even after they had been dead for several hours, whereas fish were typically eaten only when fresh. Small land crabs, up to 3.8 cm, were also consumed, but hermit crabs were not, nor was crab roe.

Luther and Greenberg (2009) reviewed mangrove and the terrestrial vertebrates found there. They found that crabs formed 'a large part of the diet of mangrove-restricted lizards, which are otherwise omnivorous and have a highly opportunistic diet' (p. 608). Furthermore, of six mammal species restricted to mangrove, 'four eat primarily leaves and two are insectivorous bats'. Furthermore, they found that fully '51% of mangrove-restricted birds feed primarily on insects, followed by smaller proportion that feed on crabs (27%), nectar (16%), and fish (4%)'. Thus, the faunivorous diet of tarsiers, consisting mostly of insects, but also including crabs and fish is consistent with the diet of mangrove-restricted lizards and birds.

Morphometric Comparisons of Tarsiers in Mangrove and Non-mangrove Habitat

In two species, *Tarsius sangirensis* and *Tarsius sp.* ('Manado form') (see Groves & Shekelle 2010) we were able to compare morphometric data from wild caught tarsiers in both mangrove and non-mangrove habitats. The variables in question were body weight, head size, and tail length. Samples sizes were small, but the available evidence indicated that tarsiers within mangrove have measurements that are within the expected range of variation for those from non-mangrove habitats (Table 15.3).

Discussion

Although the data are scant, we can make some progress towards answering our research questions. First, the available evidence provides nothing whatsoever to refute the hypothesis that tarsiers are native to mangrove. Thus, for now, the safe assumption is that tarsiers exist naturally within mangrove, and

Table 15.2 Mangrove as a percentage of EOO for several species of tarsiers.

SAMD 2006	Mangrove	Non-mangrove	Total (ha)	% mangrove
Carlito syrichta	67 156	12 263 221	12 330 377	0.544641031
Cephalopachus bancanus	1 480 858	84 304 989	85 785 847	1.72622612
Tarsius dentatus	7 871	3 514 300	3 522 171	0.223466391
Tarsius lariang		366 185	366 185	0
Tarsius pelengensis	641	235 463	236 105	0.271600229
Tarsius pumilus		2 597 861	2 597 861	0
Tarsius sangirensis	168	67 529	67 697	0.247681112
Tarsius tarsier	81 649	14 239 440	14 321 089	0.570128204
Grand Total	**1 638 342**	**117 588 988**	**119 227 330**	**1.374133221**

Table 15.3 Average body weight, head length, and tail length for *Tarsius sangirensis* and *T.* sp. ('Manado form'), by sex and habitat.

	Tarsius sangirensis				***Tarsius* sp. ('Manado form')**			
	Not in mangrove		**In mangrove**		**Not in mangrove**		**In mangrove**	
	Male	**Female**	**Male**	**Female**	**Male**	**Female**	**Male**	**Female**
	$n = 1$	$n = 2$	$n = 1$	$n = 1$	$n = 21$	$n = 10$	$n = 1$	$n = 1$
Mean Body Weight	110	134	104	130	106.2	115.5	110	110
Stand. Error		4			1.90	2.51		
Mean Head Length	38	40	40	40	39.10	39.5	39	39
Stand. Error		1			0.28	3.24		
Mean Tail Length	245	252	261	247	231.52	233.44	236	258
Stand. Error		6			6.30	3.24		

have existed there since the time before any human pressure might have pushed them there.

There are not enough data to answer the second question at this time. No one has systematically studied tarsiers in mangrove, thus we do not know for certain if the home ranges of some tarsiers lie entirely within mangrove. It follows, therefore, that neither do we know if mangrove is an ecological sink, where mortality exceeds reproduction, or if mangrove alone can sustain viable tarsier populations. What we do know is that mangrove meets some of the basic needs of tarsiers, including sleeping sites and suitable prey.

Regarding the third set of questions, mangrove forms a very small percentage of the EOO of tarsiers, and in some cases, there is no mangrove at all within the EOO of a given tarsier species. Thus, it is very unlikely that mangrove is critical to the survival of tarsiers. Likewise, mangrove exists outside of the global EOO of Tarsiidae, making it unlikely that tarsiers form a critical part of the mangrove ecocommunity. These conclusions would have to be revisited if it turned out that future research showed there to be finer distinctions within tarsier taxonomy or among mangrove ecocommunities, such that a hypothetical tarsier population existed largely within mangrove, or the mangrove ecocommunity within the EOO of Tarsiidae differed substantially from those outside of that EOO.

There remain at least two interesting possibilities concerning tarsiers and mangrove. The first of these concerns the historical biogeography of tarsiers. Tarsiers are quite unusual in that they must have dispersed across Wallace's Line at least twice, and all available evidence is that they did so over open water. Thus, there is something peculiar to the ecology of tarsiers that seemingly makes them comparatively successful rafters. Does their existence in mangrove have something to do with this? Interestingly, Lo *et al.* (2014) found the global distribution of mangrove to be best explained by a combination of vicariance and open ocean dispersal. This combination is widely assumed to also explain the global distribution of primates, although the seeming incongruity of primate dispersal across open oceans remains perplexing. Lo *et al.* (2014) also found open ocean dispersal in the early Eocene to be the best explanation for New World mangroves. If we assume reasonably large error brackets around the estimated time of dispersal, a large raft of floating mangrove might have provided the vector to get anthropoid primates from Africa to the New World. Anthropoid primates of that time period were small and tarsier-like, and thus, tarsiers, with their demonstrated rafting ability, might serve as good models for early anthropoid dispersal.

Finally, if mangrove is shown to be suitable for maintaining viable tarsier populations, then might reserves consisting entirely of mangrove become a component of a global conservation plan for tarsiers? This possibility is increasingly intriguing in the aftermath of the tsunami of 2004, which has seen a drive to restore mangrove throughout the region as a protection against future tsunamis.

Future Directions

There are two studies that could prove quite useful for both the study of tarsiers and that of mangrove. One study would be to assess the suitability of mangrove for sustaining tarsier populations. A second study would be to examine the Southeast Asian taxa that tend to raft as a community, and compare these to the community of organisms in mangrove and other habitats in order to see if floating islands of mangrove might be the vector for rafting communities of organisms across open ocean. Duke *et al.* (2007) estimate that at the current rate of loss of mangrove habitat, i.e. about 2.7% per year between the early 1980s and 2001, mangroves could be extinct within 100 years. Thus, there is an imperative to conduct these studies soon.

Acknowledgements

M.S. acknowledges that the work was supported by the Ewha Global Top 5 Grant 2013 of Ewha Womans University.

Chapter 16

Primates in the Sundarbans of India and Bangladesh

Jayanta Kumar Mallick

Introduction

Although the biological system in the Sundarban mangroves is tough, resilient, complex, interrelated and interacting, it is adaptive, designed for survival and the preferred natural abode of primates. The Sundarban mangroves provide diverse and abundant floral and faunal food resources with high caloric concentrations, absence of human settlements or infrastructural development within the forest reserves and restricted biotic interference.

The endangered Sundarbans Mangroves (Mangals) Ecoregion (N° 33), the world's largest estuarine wetland (> 10 000 km^2) in India and Bangladesh, was formed at the fertile delta of Ganges–Brahmaputra–Meghna (Bengal Basin) approximately 31 750 (± 2030) years before present (BP) and was managed as a single unit before partition in 1947 (Mandal *et al.* 2010; Mallick 2011). Nourishment of this rarest, unfragmented, dynamic and most productive 'Bengalian Rainforest' is derived from both land and water, harbouring a very rich and healthy population of non-human primates (Cercopithecidae: *Macaca*), that have not been studied rigorously. This review intends to bridge the persisting knowledge gap.

Methods

The Great Sundarban (about 266 km east–west and 100 km north–south; 21°27'–22°40′N, 88°05′–90°18′E) is bounded by the River Muriganga (west), River Baleswar-Haringhata (east), human-dominated landscapes (north) and the Bay of Bengal (south) (Figure 16.1). The River Raimangal-Harinbhanga flows between the two countries. The western (Indian) Sundarban covers 4264 km^2: 2483 km^2 forest and 1781 km^2 aquatic sub-ecosystems; the eastern part (2585 km^2) under Sundarbans Tiger Reserve (STR) and the western part (1680 km^2) under South 24-Parganas Forest Division, separated by the River Matla. The eastern (Bangladesh) Sundarban extends over 6016 km^2: 4143 km^2 forest and 1873 km^2 hydrological regime under East (Bagerhat) South and West (Khulna) Forest Divisions.

The subtropical moist climate is characterized by high humidity (60–90%). The average temperature/kilometre per hour wind speed during post-monsoon (November–February), monsoon (July–October) and pre-monsoon (March–June) is 20°C/11.5 km/h, 29°C/11.1 km/h and 34°C/60–65 km/h (maximum 43°C/120 km/h), respectively. The average annual rainfall is 1920 mm (65.79/20.35/14.81 days during monsoon/pre-monsoon/post-monsoon), increasing west to east and decreasing southeast to northwest. The area is prone to mid-year monsoon flooding of > 50% forest area and disastrous pre- or post-monsoon storms and super cyclones, e.g. Sidr (November 2007) in Bangladesh and Aila (May 2009) in India.

Owing to the ramifications of the riverine systems (> 250 islands separated by two dozen major tributaries (north–south) and > 500 interconnected tidal estuaries, creeks and canals) and complex ecological conditions, growth of diverse mangrove species (34 true and 62 associates) and community zonations are classified under two forest groups: 4A. littoral (L_1), facing the sea, and 4B. tidal swamp (TS_1 mangrove scrub, TS_2 mangrove, TS_3 salt water (mixed) – *Heritiera* (Sterculiaceae) and associates, TS_4 brackish water (mixed) – *Heritiera–Rhizophora* (Rhizophoraceae) and E_1 palm swamp – *Phoenix* (Arecaceae)). The northern part, new depositions and intertidal mudflats are characterized by *Avicennia* (Avicenniaceae) and *Sonneratia* (Sonneratiaceae), flanked by foreshore grassland (*Oryza coarctata* (Poaceae)), gradually replaced by *Excoecaria* (Euphorbiaceae) and *Ceriops* (Rhizophoraceae). The southern and eastern associates are composed of *Rhizophora* (Rhizophoraceae), *Bruguiera* (Rhizophoraceae) and a few patches of *Heritiera.* Occurrence of *Phoenix, Xylocarpus* (Meliaceae) and *Nypa* (Arecaceae) is extremely limited. Depending on the Practical Salinity Scale expressed as permille (parts per thousand, ppt), greater tree growth, height and canopy closure were observed in the east and vice versa towards the west.

Since transect surveys were not possible due to hostile habitat, lack of logistic infrastructure, stilt root/pneumatophores and muddy forest floor, this study (1 January 2011 to 31 December 2012) was based on a variety of methods including (1) a literature review of museum records and camera trap data (undertaken by WWF-India in collaboration with the Sundarban Biosphere Reserve); (2) a questionnaire survey of 102 resource persons; and (3) up- and downstream monitoring during high and low tide with the help of stationed and mobile forest staff by using a mechanized boat (speed 1.5–3 km/h) from 06:00 h to 18:00 h and also from the fenced watch tower locations.

The field activities included a reconnaissance survey, preliminary and final location surveys. Most of the channels were surveyed from the boat with the motor running, but in some places, where many monkeys aggregated, the motor was

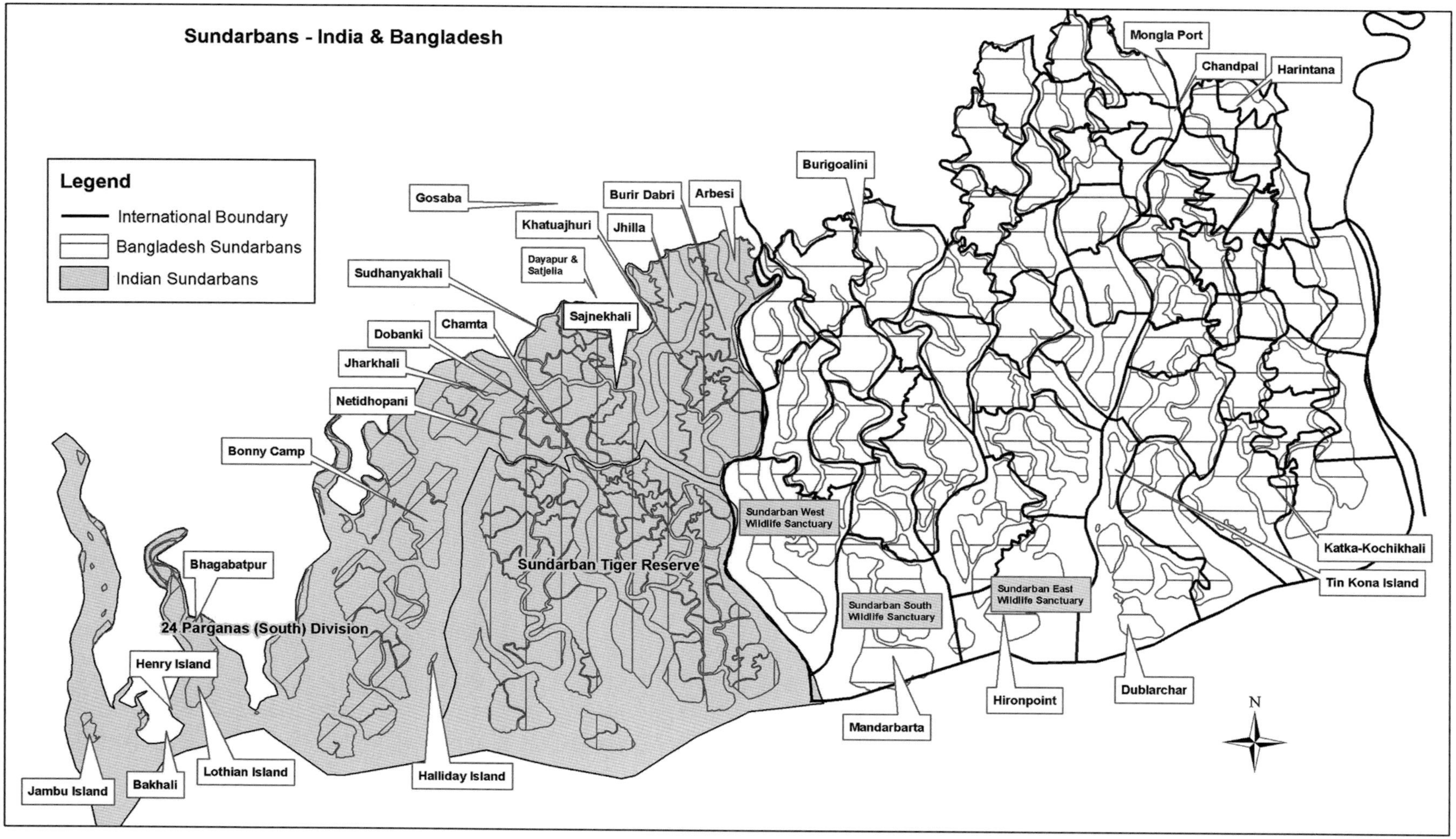

Figure 16.1 Map of Sundarbans: India and Bangladesh.

stopped to complete the count and observations. The surveys were mainly carried out at high tide, but some important areas had to be surveyed at low tide. Several areas were also surveyed two or three times and at different stages of the tidal cycle to assess tidal differences in numbers and distribution.

Results and Discussion

Primate Species Diversity

Among the primates of India (Mitra 2011), the monotypic (as revised by Fooden (2000)) rhesus macaque (= Common Bengal or 'mangrove monkey') *M. mulatta* Zimmerman, 1780, is the only extant primate species in the Sundarbans (Hasan *et al.* 2013). Six type-specimens (five males and one female (N° 11905–11907, 11925, 11931, 11985)) and one skin (N° 11984) in the National Collections of Zoological Survey of India (ZSI) were hunted in the Eastern Sundarbans on 26 April 1870 (Agrawal *et al.* 1992).

The only historical record of the Assamese macaque *M. assamensis* McClelland, 1839, in the Sundarbans (Anderson 1872) is considered peculiar due to isolation and disjunction (broken by > 700 km from the sub-Himalayan north Bengal), but a mistake over locality was declined (Md. Anwarul Islam and P.M. Thompson, pers. comm., 2010 (Timmins & Duckworth 2013)). Having heard a description of a sympatric occurrence of two primate species from the local inhabitants, which fitted the rhesus and Assamese macaques, a collector was sent by Anderson (then Curator of the Indian Museum) to the area and undoubted examples of both species were brought back. Pocock (1939, p. 54) observed:

> Further information is also required about the representatives of this species that occur in the Sundarbans, whence Anderson had at least one specimen, as well as several examples of *M. mulatta*. In this low-lying district, the exact spot being about 50 miles [= 80 km] east of Calcutta, might well occur a race distinct from typical *assamensis*, which is found at tolerably high altitudes.

Species identification was revalidated by Fooden (1982) based on (1) two museum specimens hunted in 1870 (one skin (ZSI No.11999)), with head–body and tail length of 559 mm and 318 mm, respectively, having no red about the face or on the hindquarters and (2) no contradictory evidence. This allied plain-form was referred to as a 'local race' (Finn, 1929) or 'zoogeographic relict' (Fooden, 1982). Whereas *M. a. pelops* was not known to inhabit the western Sundarbans (Khajuria 1954; Mohnot 1980), the site of occurrence was assumed to be Satkhira (*c.* 22°35′N, 89°15′E) in the eastern Sundarbans (Fooden 1982). But, during the last 140 years, further sightings of this race were not reported from the Sundarbans (and presumed extinct).

Two races of the crab-eating or long-tailed macaque *M. fascicularis* Raffles, 1821, are found in the mangrove forests of both India and Bangladesh, but neither is reported from the Great Sundarbans. Whereas *umbrosa* Miller, 1902, is endemic to the Nicobar Islands (India), *aurea* I. Geoffroy Saint-Hilaire, 1830, is recorded from the coastal mangrove forests of Whykheong-Keruntoli along the River Naf, Jhaliar Island and Fashiakhali of Chakaria Sundarbans (Cox's Bazar, Bangladesh).

Distribution and Locality Records

M. mulatta is widely distributed in the entire landscape of the Sundarbans (Table 16.1). In the western Sundarbans, the highest concentration (based on personal observations) was recorded in the protected areas and surrounding reserve forests of Sajnekhali Sanctuary (362.40 km^2) between the rivers Peechkhali and Gomdi, Halliday (5.95 km^2) on the River Matla and Lothian Islands (38 km^2) at the confluence of the River Saptamukhi and the bay. Troops were sighted at the popular tourist spots-cum-supplementary feeding sites with well-maintained sweetwater ponds, e.g. Sajnekhali, Sudhanyakhali, Choragazi, Dobanki and Netidhopani (STR). A small troop was camera-trapped in South 24-Parganas Division (Das *et al.* 2012). The monkeys also inhabit the northern anthropogenic (civil) areas (e.g. Gosaba).

M. mulatta was recorded in the eastern Sundarbans by Hendrichs (1975) (29 January–21 April 1971), Green (1978) (July–November 1976), Gittins and Akonda (1982) (early 1980), Khan (1985) (1980 and 1982) and Feeroz *et al.* (1995) (February 1990–June 1993). Sightings were most common in the East Sanctuary (312.26 km^2) than the West (715.02 km^2) and South (369.70 km^2) Sanctuaries. Ten individuals were also camera-trapped in the East Sanctuary during 6 September–4 December 2006 (Khan 2012).

Adaptations

The biology and behaviour of *M. mulatta* has been modified due to ecological stresses (Mukherjee 2006). They adapt efficiently to various elements of the complex habitat, such as anomalous tract, swampy/muddy terrain, seasonal, lunar, dial or daily phenomenon, thermal dynamics, strong winds and rains, increasing salinity in vertical and horizontal planes, fluctuating tidal current (opposite, unidirectional and oscillating) and flooding.

Morphology

Seasonal variation in pelage was difficult to interpret from seven skins of *M. mulatta*, collected in April 1870 (Fooden 2000). The unusual pelage condition may be a result of seasonal fading (Agrawal *et al.* 1992), deterioration in storage or both (Fooden 2000). Anteriorly, greyish brown to golden brown pelage, variably washed with burnt orange posteriorly, was generally observed. In two museum specimens, the orange-red hue was extended to almost the entire dorsum, whereas in the third it was limited to extreme posterior part. The winter coat was found to be thick and soft, whereas during the summer it was decidedly harsh and shaggy. Sexual dimorphism was also reported (♂ Length of head and body (H & B) 508 mm, length of tail (Tl) 203 mm, length of hind foot (Hf) 127 mm; ♀ H & B 491 mm, Tl 193 mm, Hf 120 mm) (Agrawal *et al.* 1992).

The islanders appeared to be smaller than the mainlanders (H & B up to 590 mm and Tl 280 mm). An average weight of

Table 16.1 Distribution records of rhesus monkeys in the Great Sundarbans.

Location	Coordinates	Administrative jurisdiction	Reference
India			
Khatuajhuri	21°59′N, 89°01′E	Reserve Forest	Mukherjee & Gupta 1965; present study
Harinbhanga	21°45′N, 89°E	Reserve Forest	
Jhilla	22°N, 89°E	Sanctuary	
Arbesi	22°40′N, 88°53′E	Reserve Forest	
Sajnekhali	22°07′N, 88°49′E	Sanctuary	Chaudhuri & Choudhury 1994; present study
Sudhanyakhali	22°06′N, 88°48′E	Sanctuary	Present study
Dobanki	22°59′N, 88°46′E	Sanctuary	
Burirdabri	21°59′N, 88°45′E	Reserve Forest	
Netidhiopani	21°55′N, 88°44′E	National Park	
Dayapur Island	22°07′N, 88°50′E	Fringe villages (north of STR)	
Satjelia Island	21°92′N, 88°80′E		
Halliday Island	21°39′N, 88°37′E	Sanctuary	Chaudhuri & Choudhury 1994; present study
Lothian Island	21°39′N, 88°19′E		
Bakkhali	21°33′N, 88°15′E	Reserve Forest	Present study
Jharkhali	22°02′N, 88°41′E	Reserve Forest	
Henry Island	21°34′N, 88°17′E	Fringe villages (24-Parganas(S))	
Bangladesh			
Katka-Kachikhali (Tiger Point)	21°49′-21°57′N, 89°43′-89°51′E	East Sanctuary	Khan 2012; present study
Hiron Point (Nilkamal)	21°45′-21°52′N, 89°21′-89°29′E	South Sanctuary	
Mandarberia	21°38′-21°47′N; 89°12′-89°18′E	South Sanctuary	
Harintana	22°04′-22°11′N, 89°42′-89°49′E	Reserve Forest	
Chandpai	21°18′-22°25′N, 89°38′-89°47′E		
Burigoalini	22°07′-22°15′N, 89°07′-89°15′E		
Karamjal (monkey trail)	22°25′N, 89°35′E		
Tin Kona Island	21°49′N, 89°46′E	East Sanctuary	
Dublar char	21°50′N, 89°46′E		

4 kg recorded in the Sundarban mangroves (Khan 2012) is also markedly lower than in non-mangroves (7–8 kg) (Feeroz *et al.* 2010). The advantages of this insular dwarfism are assumed to be lower food requirement, ability to move quickly through the muddy terrain while foraging or negotiating waterways with minimum loss of energy and quick escape from sudden predatory attacks (Mallick 2011). These reasons may explain dwarfism in other mangrove-using primates (see Nowak *et al.* 2008, who sourced most of the skulls in her sample from a mangrove-dwelling population).

Ecology

The mangrove habitat, intersected by anastomotic tidal waterways, mudflats and alluvial islands of various sizes and shapes, has changed as a result of ongoing ecological processes (monsoon rains, tidal flooding, salinities and climax vegetation). The monkey distribution in the substratum between the high and low tides is determined by the extent and depth of flooding. Six inundation classes in different tidal zones, leading to constant remodelling of the landscape, were recorded (Table 16.2). A rhythmic response of the macaques to the different levels of tidal fluctuations was observed. For example, they take the opportunity to feed on the river banks and in low water at the time of low ebb and prefer arboreal feeding and shelter, when the high tide floods this feeding ground (Figure 16.2).

The sequence of semi-diurnal tidal amplitude (two high water and two low water levels in a cycle of 12 h, 20 m) in the lowland habitats is directly connected with the lunar position. Each lunar month is broadly divided into two phases: the

Table 16.2 Inundation classes of monkey habitats in the tidal zones.

Zone	Inundation conditions	Habitat	Periodicity	Mangrove elements
I	All the high tides up to 7.5 m	River flats/slopes	30 days/month	Major and minor elements (32 + 69 species) of mangrove and mangrove associates
II	Medium high tides	River banks/ridge forests	20 days/month	
III	Normal high tides	Ridge forests (flat land and dense vegetation)	15 days/month	Minor elements, mangrove associates and back mangrove
IV	Spring tides	Ridge forests (flat lands with dense/sparse vegetation and salty patches)	10 days/month	
V	Abnormal/equinoctial tides (monsoon/summer)	Ridge forests, sparse vegetation and reclaimed/naked areas	5 days/month	Back *mangal*
VI	Above the tidal reaches	Deltaic region	–	Non-littoral plants

Figure 16.2 Arboreal perch of a rhesus monkey in mangroves. (A black and white version of this figure appears in some formats. For the colour version, please refer to the plate section.)

darker new moon period (male tide) and the brighter full moon period (female tide). The thirteenth to seventh and the eighth to twelfth day of each phase are locally called *bhara kotal* (peak period) and *mara kotal* (lean period), respectively. The impact of high tide along the inland boundaries was observed approximately 2.5 hours later and that of the ebb flow after two hours of commencement of the tide. The higher tides (5–8 m) were seen during the spring tides of vernal (March–April) and autumnal (August–September) equinox and lower (3–4 m) in the winter months (lowest in February), but the highest tide or *sarasari baan* (10 m) was observed in September, particularly during the cyclonic storms. Following cyclones such as Aila, a major portion of the forest remained inundated for a few days, the monkeys could not forage on the forest floor; even the leaves under water became unpalatable, posing a threat to their food source. Most of the sweet water ponds were inundated with saline water. It was reported from Netidhopani, Dobanki, Sudhanyakhali and other areas that the animals refrained from licking the saline water from those ponds.

A higher tidal range was observed in the northern fringe of the forest than the southern bay. Higher rise and fall was faced in the western rivers than in the east. During the summer and winter, the rivers have become rough at low and high tides respectively. The velocity of current in the large rivers is variable from 3 km/h near the sea surface to 6 km/h higher up in the forests on account of gradual constriction of the waterways, each flow taking a different pathway through the maze of minor channels. The tidal change was seen earlier in the west than in the east. Inundation of forest area for longer duration is recorded during the spring tide in monsoon (3–4 h), followed by the pre-monsoon season. The short flood-tide (2–3 h) and long ebb-tide (8–9 h) during the dry season, when the higher land does not get flooded, is advantageous to the foraging monkeys.

The preferred biotope or permanent abode of the monkeys is the well-drained or less predation-prone arboreal habitat, but the treeless reed swamp does not support these animals (Green 1978). Increase in the numbers of monkeys per square kilometre was observed in the areas above general tide level compared to those frequently inundated and below tide levels. Ecologically, the scattered small patches of Keora *S. apetala* (2% of the mangrove forests) are most important in the mangrove food chain (Chaudhuri & Choudhuy 1994; Chowdhury *et al.* 2001). The monkey troops were primarily found on this quick-growing, tall emergent (10–20 m) and spacious tree with heavy foliage relative to other natural mangrove genera

Figure 16.3 Rhesus monkey foraging in the mud flats during low tide. (A black and white version of this figure appears in some formats. For the colour version, please refer to the plate section.)

of lower height and foliage. Keora is common along the border of islands near estuaries, intertidal zones of several creeks and channels towards upstream swamps and invariably in places affected with fresh and brackish water mixture in association with *Avicennia, Ceriops, Excoecaria* and *Xylocarpus*.

When the floodwater is drained, a rich foraging and hunting ground is exposed on the intertidal mudflat, sparsely vegetated or with no vegetation and exploited by the monkey troops (Figure 16.3). The third biotope, i.e. the estuary or creek with characteristically low-saline brackish water system, is foraged during the low tide for fish, crab, floating tender leaves or ripe fruits.

The brackish water is also taken in when sweet water in the forest is scarce in the vicinity (out of 41 sweet water ponds dug in STR for rainwater-harvesting, only 12 are well maintained). During the dry season, the monkeys were also seen to negotiate the narrow channels in quest for new feeding sites.

Conservation Value

STR is the most resourceful habitat in the western Sundarbans, but most of the forest blocks in South 24-Parganas Division are heavily degraded (250 km^2). The habitat quality in East Sanctuary of Bangladesh is comparatively better than those of South and West Sanctuaries.

The ecological status of the original mangrove forest is declining. The upper and lower Sundarbans were divided into freshwater and moderately saline zones by Curtis (1933), but after the construction of the Farakka Barrage and reduced freshwater flow since 1975, the Sundarbans were classified into freshwater (northern part), moderately saline (central part) and saline (southern part) zones in more-or-less equal proportions. The western part has become more saline, the central part moderately saline and the eastern part freshwater dominated. From west to east of Bangladesh, 60% is polyhaline (e.g. Burigoalini), 35% mesohaline (Chandpai) and 5% oligohaline (Sharonkhola) with indicative features (Islam & Gnauck 2009).

High salinity was observed during February–April due to depletion of soil moisture and reduced upstream freshwater flow from the Ganges–Brahmaputra river system (Bidyadhari in the western Sundarbans, after cut off from the old Ganges, dried up completely and Gorai in the eastern Sundarbans seasonally (January–May)). Salinity in the eastern Sundarbans is now influenced by the River Baleswar–Haringhata, where almost zero salinity was observed throughout the monsoon due to large discharge of the River Sibsa, but increased during the post-monsoon period. The northern and northcentral parts are also fed by the freshwater flow of the River Pasur. Freshwater flow of some significance is carried by the Raimangal feeding the northeastern STR. Salinity in the southern part was < 5 ppt during monsoon and 15 ppt during the dry season, whereas in the western part it is not reduced to low range even during the monsoon periods, but increased during the dry season. Peak salinity is recorded 26 ppt in summer.

It was observed that the combined seasonal (above and below ground) biomass of three dominant mangrove species (*S. apetala, A. alba* and *E. agallocha*) in the western Sundarbans is correlated with the volume of upstream freshwater flow (Table 16.3). The biomass produced in the low-saline zone was 17.63% more than that in the moderate saline zone. Salinity-oriented negative impact was marked more on the growth and survival of *S. apetala* than *A. alba* and *E. agallocha*. Naturally, the monkeys prefer the low salinity zone for its higher production of food resources.

The adverse impact of the rise in water level and increasing salinity in the upstream was marked in the mangrove zonations: succession of the common southern species in the

Table 16.3 Comparative biomass production of three dominant mangrove species (*S. apetala, A. alba* and *E. agallocha*) in the low and moderate saline zones of western Sundarbans.

Salinity	Season	Above-ground biomass (tha^{-1})			Below ground biomass (tha^{-1})		
		2010	2011	2012	2010	2011	2012
Low	Pre-monsoon	355.41	414.39	475.55	82.04	104.42	120.94
	Monsoon	408.46	469.05	535.66	101.68	126.63	144.25
	Post-monsoon	452.95	514.24	574.51	118.13	144.98	162.20
Moderate	Pre-monsoon	255.12	338.58	414.38	58.95	84.4	106.50
	Monsoon	314.82	399.59	491.47	76.51	104.91	131.96
	Post-monsoon	364.56	450.47	541.55	94.74	125.62	154.57

Source: Records of Forest Department.

Figure 16.4 A juvenile rhesus monkey searches for crabs. (A black and white version of this figure appears in some formats. For the colour version, please refer to the plate section.)

northern part, increasing tree mortality rate, dwarfing, poor development, lower production of leaves, reduced leaf longevity and food availability. This resulted in the macaques migrating north. They are now more abundant in the interior northern zone of STR with comparatively lower salinity (Dobandi–Sudhanyakhali–Sajnekhali (average 18.03 ppt/year), Burirdabri of Arbesi (17.70), Khatuajhuri (16.66) and Harinbhanga–Jhilla (18.6)) than the higher salinity zones (central and southern) including Netidhopani (21.38), Chamta (20.62), Matla (21.13) and Gosaba (21.71) forest blocks.

Behaviour

Under the harsh conditions of a difficult forest floor, sharp pneumatophores, muddy substratum, fluctuating tidal rhythms, extensive flooding, variable salinity and lack of fresh-water sources, the rhesus monkey has adapted by modifying its habits (Mukherjee 1980). It was equally at home in the trees, spike roots-infested ground and mudflats (Chowdhury *et al.* 2001). It behaves like a leaf-eating monkey, as well as a crab-eating monkey (Figure 16.4) (Mukherjee 1980).

Its activities were mostly visible in the morning (06:00-10:00 h) and afternoon (15:00-19:00 h). Long-range foraging or inter-island migration is restricted during the high tide and monsoon. When the habitat is partially submerged, the monkeys are mostly confined to their arboreal abode, rarely descending from the trees (Mandal 1964; Mukherjee & Gupta 1965). The high trees are their nighttime sleeping sites (Mandal 1964), but during storms and rains they were seen coming down to the dense canopy of lower branches for protection. When the tidal water receded back exposing the clay bed, they were observed to come down the shoreline in small troops for ground and semi-aquatic foraging, but always remained alert against sudden attacks by tigers. In response to such danger, the macaques flee into the trees (Mandal 1964). While foraging at

Sudhanyakhali, the alpha male, sensing a tiger lurking around, became tense and emitted a shrill bark (alarm call), thumping his feet a few times on the ground to warn other monkeys of the approaching predator, running a short distance and then climbing the nearest available tree at lightning speed. All the female monkeys, along with the subadults and juveniles, followed him.

During leisure time, pairs were observed to spend several hours grooming (taking care of the fleas and ticks) quietly both on branches and the forest floor. They were also seen to take refuge and groom in the dense thickets formed by the stilt roots.

The intertidal mudflats are essentially semi-terrestrial or semi-aquatic habitats. The monkeys are inclined to follow the alternate tidal flooding and either restrict or increase their foraging movements within the home range. Besides gliding over the soft and sticky mudflats, they can easily negotiate the tricky mangrove roots. Foraging on the banks of smaller or narrower channels close to the thickets was not usually observed because they are prone to sudden tiger predation there. Moreover, the rhesus macaque turns out to be an agile swimmer too (Mukherjee 2006). Aquatic locomotion is an adaptation for negotiating the creeks in search of food (Mukherjee & Gupta 1965) and also escaping predators. It has also evolved a unique hunting and feeding mechanism, i.e. standing upright while fishing in the shallow river water.

Diet

As a primary consumer in the Sundarban, the rhesus monkey has the capacity to consume 250 g/day. This species is able to feed on a variety of food showing a high degree of adaptability to arboreal, terrestrial and aquatic food sources. Basically, they are reported to be frugivores and folivores (Mandal 1964). They are dependent on both mangroves and mangrove associates. Most of the mangrove leaves are suitable fodder for rhesus (Mandal *et al.* 2010). The succulent leathery leaves of Keora *Sonneratia*, Baen *Avicennia*, Genwa *Excoecaria*, Garjan *Rhizophora*, Kankra *Bruguiera, Ficus* (Moraceae), etc., are taken (Mallick 2011; Pandit & Mukherjee 2011). They were seen to pluck the tender leaves from top branches. Different types of creepers are also consumed. Feeding on the flowers and flower buds, petioles, and other parts of a variety of trees, shrubs and climbers was observed. The monkeys were observed to feed on the rhizomes of riverside grasses and sedges stabilizing sand dunes and uncompacted sediments (Naskar & Guha Bakshi 1987). They were also seen to graze on the mangrove seedlings and grasses *(P. coarctata)* in STR.

At Hiron Point or Nilkamal of Bangladesh Sundarbans, a troop of monkeys was observed to eat grass and seeds. They were also observed to eat mushrooms (Mukherjee 1980).

The dominant individuals prefer the good quality ripe fruits (Mukherjee 2006). The fruit of *S. apetala* is preferred (Pandit & Mukherjee 2011; Sanyal 1992). Fruits of Sundari *Heritiera*, Hental *Phoenix* and *Excoecaria* are also favoured by the macaques (Pandit & Mukherjee 2011). At low tide, the monkeys come to the edge of the water to pick through the flowers, fruits and leaves that the mangroves drop into the water, which the tide then brings to land (Montgomery 2009).

During August–October, when large amounts of fruits are found floating on the river water, the monkeys congregate near the bank, pick up and devour them voraciously. These vegetable foodstuffs are responsible for fat deposition and growth of these monkeys (Milton 1999). Flowering and fruiting of most of the mangrove species is reported during May–October (Table 16.4). Monkeys were reported to frequently kill off the young golpata (*N. fruticans*) regeneration by pulling up the plants, nibbling the young shoots and throwing them away (Curtis 1933). At Sudhanyakhali, a large male (alpha) was observed to pluck and consume every single flower from *Hibiscus* (Malvaceae), whose tiny branches could barely hold its weight, thereafter coming down to forage for the flowers fallen to the ground.

It was estimated that *Aviccenia* and *Sonneratia*-dominated forests may produce about 212 tons/ha of biomass (Mandal *et al.* 2010). There is some seasonality in flowering/fruiting in the Sundarbans (Brahma *et al.* 2008; Sarkar *et al.* 1993). The higher stands of *S. apetala* are most important from the monkey's ecological point of view because of the species' longer fruiting period. It produces more fruits per mature tree and more seeds per fruit in autumn than spring. The monkeys were reported to move from one Keora grove to another for feeding (Gupta 1966).

It was reported that seasonal honey is much liked by the rhesus macaques and they break the low-hanging (1.5–2.5 m) combs to drink the honey. It was found that combs of one cubic foot (0.03 m^3) volume yield about 3 kg of honey and serve a good number of monkeys. The monkeys are said to smear their body with a thick layer of silt before approaching the combs to protect themselves from bee stings.

An unfamiliar feeding behaviour of *M. mulatta* was observed in the Indian Sundarbans. An adult male rhesus was plunging a small dead bird in the salt water of a muddy tidal creek repeatedly, washing it meticulously to enhance its taste and then consuming it bit by bit. A similar practice was reported before in case of a natural troop of *M. fuscata* (Kawai 1965).

M. mulatta is a predator of the eggs of birds, turtles and crocodiles in the Sundarbans. Resting sites are dominated primarily by *Avicennia* sp. followed by *Excoecaria, Xylocarpus, Aegiceros* and *Rhizophora*, particularly in Sajnekhali. It was reported that while passing by a tree-nest (24 August 2004), one monkey spotted the tree-nest of an incubating masked finfoot *Heliopais personatus* (Heliornithidae), but kept away from the defensive mother and moved on; but, after the mother and the hatchlings had left the nest (26 August), presumably the same monkey approached the nest, grabbed one of the two abandoned eggs and dropped it into the water (Neumann-Denzau *et al.* 2008). They were also observed catching various insects.

The rhesus macaques are also reported to prey on molluscs and crabs (Mandal 1964; Mukherjee 1980; Sterndale 1884). The behavioural and food habit patterns of the monkey are related to the abundance of oysters, crabs and fish (Chaudhuri 2007). Rhesus monkeys have been known to forage for crabs (fiddler and mud crabs), which account for the largest portion of the animal biomass in the area. The mud

Table 16.4 Phenological records of the mangrove species in the primate habitat based on random sampling and accessibility during the present study (2011–2012).

Family	Species (true mangroves* and mangrove associates**)	Flowering period	Fruiting period
Acanthaceae	***Acanthus ilicifolius*	April–June	June–August
Arecaceae	**Nypa fruticans*	February–June	June–September
	**Phoenix paludosa*	January–April	May–August
Avicenniaceae	**Avicennia alba*	March–July	July–October
	**A. marina*	April–June	June–August
	**A. officinalis*	April–August	August–October
Boraginaceae	***Heliotropium curassavicum*	March–August	December–January
Caesalpiniaceae	***Caesalpinia bonduc*	August–January	August–January
	***C. crista*	October–January	December–May
Chenopodiaceae	***Salicornia brachiata*	November–January	November–January
	***Suaeda maritima*	December–March	December–March
	***S. nudiflora*	June–October	June–October
Convolvulaceae	***Ipomoea marginata*	November–March	December–January
	***Ipomoeapes-caprae*	October–March	October–March
Euphorbiaceae	**Excoecaria agallocha*	March–June	June–August
Malvaceae	***Hibiscus tiliaceous*	February–April	August–September
	***Thespesia populnea*	November–January	January–March
Meliaceae	**Xylocarpus granatum*	February–December	February–December
	**X. moluccensis*	March–May	September–October
Myrsinaceae	**Aegiceras corniculatum*	February–April	June–September
Papilionaceae	***Denis scandens*	June–July	August–November
	***Derris trifoliate*	April–July	September–October
Plumbaginaceae	**Aegialitis rotundifolia*	February–April	April–June
Poaceae	***Myriostachya wightiana*	March–April	March–April
	***Porteresia coarctata*	July–October	July–October
Rhizophoraceae	**Bruguiera gymmorrhiza*	March–June	June–August
	**B. parviflora*	April–July	July–September
	**Ceriops decandra*	March–July	July–October
	**C. tagal*	March–July	July–October
	**Rhizophora apiculata*	May–July	July–September
Sonneratiaceae	**Sonneratia apetala*	May–June	June–July
	**S. griffithii*	March–July	July–September
Sterculiaceae	**Heritiera fomes*	February–March	May–June
Tamaricaceae	***Tamarix dioica*	April–December	July–January
Verbenaceae	***Clerodendrum inerme*	December–April	December–April

flats, sands dunes and sand flats are home to a huge variety of fish and crabs. The low mud flat on both sides of the forest canals (*Khanri*) are potentially the best areas to find estuarine crabs. Gupta (2001–2) recorded sighting of a troop feeding on the hermit crabs, *Clibanarius longitarsus* (Diogenidae) exposed by the low tide on the Gomar riverbank at Sajnekhali. To do this, the macaques dextrously manoeuvred themselves to avoid the needle-like pneumatophores.

The unique fishing behaviour of the rhesus monkeys in the Sundarbans is prompted by the onset of the ebb resulting in lower water levels that make it easier to catch fish. Troops were observed many times scooping up nutritious small fish along the rivers, canals and creeks. During a study by Tripura University in February 2011, an adult male rhesus macaque was photographed at Sajnekhali (Majumder *et al.* 2012).

The alpha male was observed walking to the bank of the river and catching live fishes (the prey species could not be identified on the spot (J. Majumder, pers. comm.)), holding and eating them using both hands jointly or alternately. This unique hunting strategy was followed by the rest of the troop members ($n = 7$), after watching the alpha from a distance for about ten minutes. In fact, small (in terms of length and weight) fish form a secondary food item for the monkeys. Fish are abundant in the period mid-June to mid-August, when the salinity of river water is low.

The monkeys were seen to drink the brackish water (de Poncins 1935). They were also observed to lick the dewdrops, when rainwater accumulated in small ditches (during summer) (Mukherjee 1980). During the rainy season, water requirements of *M. mulatta* are met primarily by consumption of succulent plant foods (Mukherjee & Gupta 1965).

Rhesus macaques were also noted for their tendency to seasonally move from the forest to ecotourism locations in search of handouts or refuse from the tourists. Sometimes the tourists throw away half-eaten lunches or dinners, including bread, meat and other food. This has changed the monkeys' feeding habits and behaviour. While feeding on the food wrapped in polythene bags, or just out of curiosity, they sometimes ingest the polythene bag, which can jam their intestines and cause death. These incidents may negatively impact the conservation of the species.

Social Structure

M. mulatta lives within a rigid matrilineal social structure (Bahuguna & Mallick 2011). It is highly socialized, gregarious and found in scattered groups of 30–40 individuals (Mallick 2011). The social composition of the troops encountered at various locations during the present study (2011–2012) is recorded in Table 16.5.

Earlier, de Poncins (1935) found them in packs of 50. Mandal (1964) recorded 47 groups (size: mean 24.1; minimum 2; maximum 100). Troops consisting of 20–30 individuals were seen along the Indo-Bangladesh border and sea-facing islands (Mukherjee 1980). A number of troops were reportedly observed during surveys in South 24-Parganas district (Agrawal *et al.* 1992). During field surveys between 2005 and 2010 in the Bangladesh Sundarbans using a point sampling method (because a transect survey was not possible), Hasan *et al.* (2013) recorded 41 groups with a single population size of 966 found in distinct locations more than 40 km from the nearest group with a mean group size of 23.6 ± 5.2 (23.9, $n = 48$), range 14–31, ratio of adult male:female 1:2.84 and adult:immature (= subadult male, subadult female, juvenile and infant) 1:1.64. Each group was found to have a typically discrete home range of 1–10 km^2, depending on the group size and availability of food resources.

At Sudhanyakhali, a large troop was observed in the afternoon during the peak tourist season in winter. The group size was observed to decline at the end of the peak tourist season in April, most of them migrating to the interior forests in search of natural foods and only a few monkeys were seen in the off-season. The same large troop was again sighted after the monsoon from October onwards. The monkeys (daytime associations) were observed to take shelter in the high trees close to the favourite feeding areas usually before sunset but not after evening. They were seen to sleep in clusters of one (alpha male) or two (breeding pair) to four (mother-yearlings and siblings).

Reproduction

Mating in rhesus monkeys is not limited to a definite season (Mukherjee 2006), but they usually do not breed during the monsoon. Only the dominant males are able to mate and the subordinate males are opportunistic breeders, particularly when the dominant male is away. Two seasonal birthing periods (pre- and post-monsoon, i.e. April–May (major) and September–October (minor)) were recorded (Mandal 1964).

Commensalism

A commensal relationship between *M. mulatta* and *Axis axis* (Cervidae) has frequently been reported (de Poncins 1935; Gupta 1966; Mandal 1964; Mukherjee & Gupta 1965; Sanyal 1983). The deer were seen to move in close association with *M. mulatta* feeding on fruits, leaves, and twigs dislodged by the monkeys from upper branches (Bahuguna & Mallick 2011; Das *et al.* 2012; Mukherjee 1980) or standing upright on their hind legs to feed on the branches pushed down by the arboreal monkeys. A pronounced seasonal variation in the deer–monkey association (common during dry weather but rare in the monsoon) was reported (Bahuguna & Mallick 2011). The deer were also observed to respond to the monkeys' repetitive alarm calls (*kech-kech)* to escape predators (Bahuguna & Mallick 2011; Mukherjee & Gupta 1965; Sanyal 1983). *M. mulatta* was reported to leap down from the branch to ride on the back of deer (Mandal 1964). Occasionally, a deer with lowered antlers was seen to threaten a monkey or a monkey threaten a deer.

Predators

Unlike in other habitats, the tigers in Sundarban hunt scarce prey, including the rhesus macaque (Wikramanayake *et al.* 2002). Scat analysis has revealed that the monkey is the third highest component (6.89% ($n = 10$) to 7.19% ($n = 13.57$)) in the tiger's diet (Mukherjee 2004; Mukherjee & Sen Sarkar 2013; Reza *et al.* 2001). Predation by the tigers, alarm responses and apparent mobbing of a tiger by *M. mulatta* have also been reported (Mandal 1964). Other predators include sharks (Carcharhinidae, Sphyrnidae) and estuarine crocodiles, *Crocodilus porosus* (Crocodylidae) in the rivers (where monkeys swim to cross areas), and terrestrial poisonous snakes (Elapidae and Viperidae) (Hendrichs 1975; Mukherjee & Gupta 1965).

Table 16.5 Social composition of the rhesus troops encountered at various locations during the present study (2011-2012).

Conservation zone	Group	Adult male	Adult female	Juvenile	Infant	Total
STR						
Sajnekhali Wildlife Sanctuary: Panchamukhani & Pirkhali Blocks (362.42 km²)	1	5	17	53	15	80
	2	8	27	30	9	74
	3	4	15	37	12	68
	4	7	13	36	7	63
	5	9	15	27	12	63
	6	6	31	13	9	59
	7	10	26	10	9	55
	8	9	24	10	11	54
	9	6	17	17	13	53
	10	7	13	25	6	51
	11	10	26	9	4	49
	12	8	23	10	8	49
Buffer area: Arbesi, Jhilla, Khatuajhuri & Harinbhanga Blocks (885.27 km²)	1	9	24	8	7	48
	2	10	12	19	6	47
	3	9	22	5	9	45
	4	8	23	10	4	45
	5	3	8	27	6	44
	6	6	10	20	6	42
	7	9	22	5	6	42
	8	4	14	15	6	39
	9	8	10	21	0	39
	10	9	5	20	5	39
	11	6	19	6	7	38
	12	9	12	11	4	36
	13	7	8	15	6	36
	14	6	19	6	4	35
	15	5	8	15	7	35
	16	5	6	19	6	35
	17	10	7	11	7	35
Critical Tiger Habitat: National Park & extended Reserve Forests: Matla, Chamta, Chhotohardi, Goasaba, Gona, Bagmara, Mayadwip, Netidhopani and Chandkhali Blocks (1330.12 km² + 369.50 km² = 1699.62 km²)	1	8	15	10	6	39
	2	5	10	10	10	35
	3	3	7	15	6	31
	4	5	7	13	5	30
	5	2	6	17	5	30
	6	5	6	14	5	30
	7	5	5	15	5	30
	8	6	9	7	8	30
	9	5	12	4	8	29
	10	4	7	11	7	29
	11	7	13	5	2	27

Table 16.5 *(cont.)*

Conservation zone	Group	Adult male	Adult female	Juvenile	Infant	Total
	12	5	7	11	4	27
	13	5	12	4	4	25
	14	4	7	8	4	23
	15	2	4	15	2	23
	16	5	5	7	5	22
	17	4	7	5	5	21
	18	5	6	3	6	20
	19	5	6	4	5	20
	20	3	7	5	6	20
	21	3	6	5	6	20
	22	3	6	6	5	20
	23	3	2	12	2	19
	24	3	6	5	4	18
	25	3	6	3	6	18
	26	2	3	9	4	18
	27	5	9	1	2	17
	28	1	4	8	4	17
	29	3	4	4	4	15
	30	3	3	6	3	15
	31	3	4	8	2	15
	32	2	3	6	3	14
	33	4	5	2	2	13
	34	3	7	2	0	12
	35	4	6	2	0	12
	36	2	4	4	2	12
	37	2	3	4	3	12
	38	3	5	3	0	11
	39	3	5	0	1	9
	40	2	2	3	1	8
	41	2	1	3	1	7
	42	1	4	1	0	6
24-Parganas (South) Division						
Sanctuaries: Lothian (38 km²) & Halliday Islands (5.95 km²)	1	10	25	25	10	70
	2	3	8	14	6	36
Reserve Forests:Ajmalmari, Dhulibhasani, Chulkati, Thakuran, Herobhanga, Saptamukhi & Muriganga blocks	1	7	14	11	8	40
	2	3	7	15	6	31
	3	2	6	17	5	30
	4	2	4	15	2	23
	5	5	12	14	6	37
	6	3	6	10	5	24
	7	1	2	1	0	4

Table 16.6 Rhesus population density in various vegetation types during the present study (2011–2012).

Major pure and mixed genera (Number of species)	Tidal zone* and range (in metre)	Salinity tolerance (in ppt)	Forest zone	Area (in km²)	Rhesus population density (%)
Sonneratia–Excoecaria–Oryza (5)	MLWP–MLWN, 0–4	3–18	River flat	82.86	60
Ceriops–Excoecaria–Sonneratia (6)	ML–MLWN–MLWP, 0–5			648.07	15
Excoecaria–Heritiera (2)	MLWN–MLWS, 1–4		River flat – Ridge forest	1816.76	6
Excoecaria–Ceriops–Xylocarpus (5)	MLWN–ML, 2–5	3–20		346.04	5
Xylocarpus–Bruguiera–Avicennia (9)	MLWN–MTLL–MLWP, 0–6	6–30	Ridge forest – River flat slope – River flat	40.30	5
Pure *Excoecaria* (1)	MLWN, 2–4	3–18	River flat	215.20	4
Heritiera–Xylocarpus–Bruguiera (7)	MLWS–MLWN–MTLL, 1–6	5–20	Ridge forest – River flat slope	95.56	4
Pure *Heritiera* (1)	MLWS, 1–2	5–15	Ridge forest	749.92	1
Total				3994.71	100

* MLWP = mean low water spring; MTLL = mid-tide lower limit; ML = mean level; MLWN = mean low water neap; MLWS = mean low water spring.

Table 16.7 Rhesus densities (extrapolated) and indices of abundance in Bangladesh Sundarbans. Based on data from Khan (2012).

Area	Location	Water salinity (ppt)		Relative density	Absolute density
		Dry season	Wet season		
Katka-Kochikhal	Southeast	5–10	0–5	2.42 (± 0.77)	6.5
Hironpoint	South	20–25	15–20	2.36 (± 0.81)	6.3
Mandarbaria	Southwest	25–30	20–25	2.40 (± 0.69)	6.4
Harintana	Eastcentral	5–10	0–5	1.18 (± 0.59)	3.2
Chandpai	Northeast	0–5	0–5	0.83 (± 0.45)	2.2
Burigoalini	Northwest	20–25	5–10	0.85 (± 0.47)	2.3

Human–Monkey Conflict

Though the rhesus macaques have specific habitat preferences, they also coexist with humans and this commensal adaptation, including seasonal crop depredation, is also a part of their survival strategy in the fringe villages. Monkey depredation is also a serious problem in the populated areas such as Bakkhali, where a number of stray groups operate. But there are no reports of monkey killing either for food or as a crop pest.

Rehabilitation

Introduction of rhesus monkeys to the mangrove plantations of the central coastal islands of Bangladesh Sundarbans was successful (Iftekhar & Islam 2004). Similarly, during the twenty-first century, relocation of seized or rescued rhesus monkeys was also frequently tried by the forest department in the Indian Sundarbans (Mukherjee 2006). Except for health check-ups and acclimatization before release, no other protocol was followed in these cases. Most monkeys were observed to stay at the release site, which is close to a sweet water pond. But some of them were reported either to become easy prey for tigers, go to the fringe village or stay permanently with the forest staff.

Population Estimates

The rhesus monkey was one of the main wild animals (Hunter 1875), common in the Sundarbans (Chaudhuri & Choudhury 1994; de Poncins 1935; Naskar & Guha Bakshi 1987; O'Malley 1914), but its population has gradually declined (Gittins 1981). The estimated population in STR was about 38 000 in the 1990s (Mallick 2011). No such estimate for the South 24-Parganas Forest Division is available, but the population appears to be comparatively low (about 10 000) on the basis of counting the population at Bakkhali in January 2001 and then extrapolating the figure for the entire division (Mallick in Bahuguna & Mallick 2011). The Bangladesh population was between 40 000 and 68 200 (Green 1978; Hendrichs 1975; Khan 1986) against the higher estimate of 88 000 to 126 220 (Eudey 1987; Gittins 1981), but the current estimate is 40 000 to 50 000. Due to shrinkage of habitat, the population has declined over the last two decades.

The rhesus population density, which is determined by the species composition (pure or mixed forests), tidal range and salinity in different forest zones (Table 16.6), was assessed to be 6.5/km² (Table 16.7) in six sites of the East Sanctuary

(Bangladesh) (Khan 2012) and 1.2/km^2 in STR (Mallick 2011). So the density of the Bangladesh population appears to be much higher than India. The occupied area was found in the river flat slope and the ridge forests (tidal mangroves). The mean density was lowest in pure *Heritiera* forest and highest in the *Sonneratia–Excoecaria–Oryza* dominated habitats.

Conservation

M. mulatta is listed in Part I of Schedule II (= non-endangered) of the Indian Wildlife (Protection) Act, 1972 and Part IV of Schedule I of the Bangladesh Wildlife (Preservation) Amendment Act, 1974. It is assigned the 'Least Concern' category in the IUCN Red List in view of its wide distribution, presumed large population and its tolerance of a broad range of habitats (Timmins *et al.* 2008). However, its status should be reassessed because of population declines in the largest conservation area of this species in the world, the Sundarbans. This species is listed on CITES Appendix II.

Threats

1. Change in the ecology of the mangrove forest as a result of reduced inflow of freshwater, flood intensity, rapid erosion, siltation, sedimentations (7 mm/year), sea-level rise. (8 mm/year), marked land subsidence (5 mm/year), rise of salinity and pollution loads.
2. Overexploitation of forest resources (timber, fish, prawns, honey and fodder).
3. Increasing temperature by 0.04°C/year (0.34%) in the western sector, 0.14°C/year (1.34%) in the central sector and 0.9°C/year in the southern sector (near Sagar Island).
4. Erratic monsoon, declining precipitation and 26% increase in severe cyclones over the past 200 years.
5. Soil salinity in the northern Sundarbans at up to 8 ppt and in the south between 8 and 20 ppt (the safety limit being 4–6 ppt).
6. Reclamation of 58% forests during the last two centuries and degradation of the habitat.
7. Changes in vegetation structure, infectious top-dying disease, decrease in tree height (0.4% per year), decline of habitat and nutrient replenishment.
8. Overexploitation and poor regeneration of the mangrove species extensively used by the monkeys leading to rapid decline of their population.
9. Provisioning and continued access to human food is also posing a threat to monkeys and to humans, and contributing to changes in behavioural patterns of wild monkey groups.

Chapter 17

Behavioural Ecology of Mangrove Primates and Their Neighbours

Ricardo Rodrigues dos Santos, LeAndra Luecke Bridgeman, Jatna Supriatna, Rondang Siregar, Nurul Winarni and Roberta Salmi

Introduction

Most studies of primates take place in rich tropical non-flooded forests, with occasional exceptions. One such exception is the mangrove forest habitat, which offers its inhabitants low plant diversity, leaves with high salinity and plants with high carbon to nitrogen ratios and toxins acting as defences against herbivory (Feller 1995; Hogarth 2007; Kandil *et al.* 2004; Tomlinson 1986). Furthermore, mangroves are characterized by high temperature, extreme tidal inundation, high sedimentation, with muddy and unconsolidated anaerobic soils (Macintosh & Ashton 2002; for more detail see Chapter 6).

Mangroves are usually characterized by high productivity, but offer low availability of plant foods, such as fleshy fruits, and fewer drinking possibilities for primates than other types of forest (Bridgeman 2012; Matsuda *et al.* 2009a; Nowak 2008; Chapter 4). Although plant foods may be limited, seafood such as shellfish (molluscs and crustaceans), and insects are abundant in mangroves and available to omnivorous primates. However, many species of shellfish are protected by shells or exoskeletons which may limit their edibility (see Chapter 9) based on the cost–benefit of capture, processing and nutritional return. Additionally, the mangroves offer tiered substrates for primates to travel and forage for animal protein, including both the treetops and the extensive stilt root systems.

In primate populations, the distribution, availability and quality of foods are the key variables affecting morphological and behavioural adaptations, feeding and reproductive strategies, population health and social organization (Strier 2010). Therefore, studies of primates in mangroves provide an opportunity to examine behavioural and ecological aspects under a different scenario. In this chapter, we compile a worldwide review of mangrove monkey studies to help understand the ecology and behavioural flexibility of primates from non-mangrove forests.

Mangrove Distribution and Diversity

Mangrove makes up 3% of the world's tropical forest and occurs in both the tropics and subtropics in 118 countries. In 2000, the total worldwide mangrove area was 137 760 km^2, mostly distributed in Asia (42%), followed by Africa (20%), North and Central America (15%), Oceania (12%) and South America (11%) (Giri *et al.* 2011). However, only 6.9% of these are protected (Giri *et al.* 2011).

In the Americas, mangrove occurs in all Caribbean and Latin American maritime countries, except Argentina, Uruguay and Chile (Lacerda *et al.* 1993), and along the Gulf coast states of the United States. In Africa, mangroves are found along the west coast, from Mauritania to Angola, and along the east coast, from Somalia to South Africa, including the west coast of Madagascar (Spalding *et al.* 2010). Among the countries with the largest coverage of mangroves are Nigeria, Guinea-Bissau, Guinea, Cameroon and Gabon in the west and Mozambique and Madagascar in the east (Spalding *et al.* 2010; Taylor *et al.* 2003, 2007). Southeast Asian mangroves are found mainly in Indonesia (60%), Malaysia (11.7%), Myanmar (8.8%), Papua New Guinea (8.7%) and Thailand (5.0%) (Giesen *et al.* 2006).

Eastern area mangroves, made up of East Africa, Southeast Asia and Australia, tend to have higher diversity of plant species than western area mangroves. The western area is composed of West Africa, the Atlantic and Pacific coasts of North and South America, Caribbean and Central America. In the western area, mangrove diversity is usually no more than 11 plant species (Lacerda 2002), while in the eastern area, particularly in Southeast Asia (excluding East Africa and Australia), there are at least 52 plant species of 'true mangrove species', i.e. those which can only be found in mangrove habitat (Giesen *et al.* 2006). Tomlinson (1986) suggested that there were five plant species widespread across the eastern area: *Avicennia marina* (Acanthaceae), *Bruguiera gymnorrhiza* (Rhizophoraceae), *Excocaeria agallocha* (Euphorbiaceae), *Heritiera littoralis* (Malvaceae) and *Sonneratia alba* (Sonneratiaceae).

Food Resources for Primates

Mangrove forests provide shelter and food for many animals, such as molluscs, arthropods, fish, lizards, snakes, birds and mammals (Giesen *et al.* 2006). For primates, mangroves can work as a refuge (Nowak 2012) or as main (Boonratana 2003; Hasan *et al.* 2013) or alternative food resource (Santos 2010; Santos *et al.* 2016). However, whether primate species are able to colonize mangroves strongly depends on the diversity of available foods.

Accordingly, the high species diversity and richness in Southeast Asian mangroves provides more food for mangrove-dwelling primates than for those using this habitat in other

Table 17.1 Mangrove plant species identified as food items of Asian primates living in mangrove.

Family	Plant species	Primate species	Source
Arecales	*Nypa fruticans*	Long-tailed monkey, ebony leaf monkey	Supriatna *et al.* 1989
Myrsinaceae	*Aegiceras cornicula*	Long-tailed monkey	Supriatna *et al.* 1989
Rhizoporaceae	*Bruguiera gymnorrhiza*	Proboscis monkey, long-tailed monkey, ebony leaf monkey	Salter *et al.* 1985; Supriatna *et al.* 1989
	Bruguiera sexangula	Proboscis monkey	Boonratana 2003
	Rhizophora spp.	Proboscis monkey	Salter *et al.* 1985
	Rhizophora apiculata	Long-tailed monkey, ebony leaf monkey	Supriatna *et al.* 1989; Son 2003
	Rhizophora mucronata	Long-tailed monkey, ebony leaf monkey	Supriatna *et al.* 1989; Son 2003
Sonneratiaceae	*Sonneratia alba*	Proboscis monkey, long-tailed macaque, ebony leaf monkey	Boonratana 2003; Salter *et al.* 1985; Supriatna *et al.* 1989
	Sonneratia asida	Long-tailed monkey, ebony leaf monkey	Supriatna *et al.* 1989
Verbenaceae	*Avicennia alba*	Proboscis monkey, long-tailed monkey, ebony leaf monkey	Salter *et al.* 1985; Son 2003; Supriatna *et al.* 1989
	Avicennia lanata	Long-tailed monkey	Son 2003
	Avicennia officinalis	Long-tailed monkey, ebony leaf monkey	Son 2003; Supriatna *et al.* 1989
	Avicennia marina	Long-tailed monkey, ebony leaf monkey	Supriatna *et al.* 1989

regions. For instance, proboscis monkeys (*Nasalis lavartus*), silvery leaf monkeys (*Trachypithecus cristatus*), and long-tailed macaques (*Macaca fascicularis*) consume leaves, young shoots, fruits, seeds, flowers and buds from the mangroves plants (Matsuda *et al.*, Chapter 4, this volume); the last species is also known to feed on shellfish and crabs that live in mangrove (Gumert *et al.*, Chapter 19, this volume). This high diversity of potential foods is a key factor in allowing some primate species to live exclusively in mangrove. However, there is little information on food resources for primates in mangrove habitats around the world. Data published on primate foods in mangrove described here includes both east and west areas.

Salter *et al.* (1985), based on their study in Serawak (Malaysia), suggested that *Sonneratia alba, Avicennia alba, Bruguiera gymnorrhiza*, and *Rhizophora* spp. (Rhizophoraceae) were the most important food plants for proboscis monkeys. Boonratana (2003) showed that the proboscis monkey was able to eat all parts of *Sonneratia alba* trees and suggested that fruits of *S. alba* were more abundant in mangroves than in riverine forest. Mangroves also provided food for long-tailed macaques and ebony leaf monkeys (*Trachypithecus auratus*) (Son 2003; Supriatna *et al.* 1989). *Rhizophora* spp. and *Avicennia* spp. were two of the main diet items of long-tailed macaques (Son 2003). Important mangrove species as food items of Asian primates are listed in Table 17.1.

The same species (*Sonneratia alba, Avicennia marina, Rhizophora mucronata*, and *Bruguiera gymnorrhiza*), with the addition of *Ceriops tagal* (Rhizophoraceae), were also described by Nowak (2008) as important food items in the diet of the Zanzibar red colobus (*Piliocolobus kirkii*) inhabiting the mangrove forests of Uzi and Vundwe Islands. These colobines prefer the fruits of *A. marina* and the young leaves of *S. alba*, but often feed on the mature leaves and other plant parts of the abundant *R. mucronata*. This last species (especially its flowers and other young plant parts) was also part of the diet of the Temminck's red colobus (*Piliocolobus badius temminckii*) of the Saloum Delta in Senegal (Galat-Luong & Galat 2005) and the Angolan black-and-white colobus (*Colobus angolensis palliatus*) in southern Kenya (Anderson *et al.* 2007). Other *Rhizophora* and *Avicennia* spp. compose part of the diet of green monkey (*Chlorocebus sabaeus*) populations inhabiting the coastal area of Senegal. They feed on fruits, flowers, roots and young shoots and leaves of *R. mangle* and *R. racemosa* and the flowers and fruits of *A. nitida* (Galat & Galat-Luong 1976).

In the delta region of the Usumacinta river system in southeastern Mexico, Bridgeman (2012) found that Yucatan black howlers (*Alouatta pigra*) ate from only a handful of plants in the mangrove, and relied heavily on flowers of the gusano tree (*Lonchocarpus hondurensis*: Fabaceae) as a keystone resource in the dry season.

The fauna of mangroves provide high-quality food resources for omnivorous primates, as opposed to plants, which are considered a low-quality food. Macrobenthos such as insect larvae, crustaceans, and molluscs are examples of mangrove fauna. Long-tailed macaques spend time foraging in mangrove tree canopies to find insects, such as the wood-boring caterpillar of a carpenter moth (*Zeuzera* sp.: Cossidae), and foraging at ground level for macrobenthos such as crustaceans and gastropod shellfish (Son 2003). Capuchin monkeys (*Sapajus apella* and *S. libidinosus*) widely exploit these same kinds of food resources, such as crabs (*Ucides cordatus*) and molluscs (shipworms) (Cutrim 2013; Santos 2010). Together, crabs from

Figure 17.1 Black howlers (*Alouatta pigra*) resting in red mangrove trees (*Rhizophora mangle*), Pantanos de Centla Biosphere Reserve, Tabasco, Mexico. Photo: LeAndra Luecke Bridgeman.

two families (Grapsidae and Ocypodiae) were the most commonly eaten animal foods for these primates (Santos 2010; Son 2003). Crabs, especially *Uca tangeri*, also represent a frequent food source for green monkeys (*Chlorocebus sabaeus*; Galat & Galat-Luong 1976; Head *et al.*, Chapter 12, this volume) and at least one population of Guinea baboons living in West Senegal (*Papio papio*: Galat-Luong & Galat 2013).

Studies of primate ecology and behaviour in mangroves have not received the same attention from primatologists as non-flooded forests. Although mangrove colonization by primates could reveal new insights into the socioecology of these animals, little information and details are currently available.

Mangrove Primates

There are some primate species associated with mangroves, ranging from 'Vulnerable' to 'Critically Endangered' (IUCN 2015), but few data are available on these primate populations. Hunting, habitat loss and degradation, and climate change are the three biggest threats to tropical habitats. These factors are contributing to rapidly decreasing populations of primates. Although mangroves are highly threatened by human development (Kathiresan & Bingham 2001), beside the important ecosystem functions they provide, growing evidence shows that they provide refuge for primates from habitat disturbance and hunting by humans (Bi *et al.* 2013; Galat-Luong & Galat 2005; Nowak 2012). In addition, mangroves may be refugia for primates for natural reasons, such as natural changes in landscape configuration over time on the Brazilian north coast (Santos *et al.* 2016). Consequently, primates may rely on mangroves for either all or part of their subsistence.

There are 67 primate taxa (17 in the Americas, 39 in Africa and at least 11 in Southeast Asia) known to use mangroves (Supriatna & Wahono 2000). In general, primate species use mangroves as a secondary habitat; however, some exclusively live in mangrove, although none are endemic. In the next three subsections we compile data on primate ecology and behaviour and their neighbours from America, Africa and Asia.

American Primates

Although four primate species are reported in Mesoamerican mangrove (see Chapter 8), currently, there are only two completed studies focusing on ecology and behaviour of primates in the region's mangrove. Bridgeman (2012) conducted a study of Yucatán black howler monkeys (*Alouatta pigra*) in a Mexican mangrove swamp (Figure 17.1) and Snarr (2006) studied mantled howlers (*A. palliata*) on a reserve in northern Honduras. Data were collected on habitat characteristics, population attributes, activity patterns, diet, and in the case of the research in Mexico, nutritional components of foods eaten by the howlers in the mangrove were also investigated.

In the mangrove, black howler group sizes, group demographics and population density do not vary significantly from black howlers in other habitats. For mantled howlers, Snarr (2006) found smaller group sizes and lower adult male to adult female ratios than are seen for the species in other sites. Population density at Snarr's site is comparable to other reports for the species (Bridgeman 2012). Activity patterns of the mangrove black and mantled howlers are not significantly different from the means of conspecific populations in non-flooded forests. However, Bridgeman's study (2012) did show that the mangrove black howlers spend overall less time feeding and more time engaged in social behaviours than is typical for the species.

The mangrove howler population in Mexico adjusts its behaviour in response to food availability in general, feeding more frequently and engaging in more affiliative social behaviour when seasonal resources are available, and generally resting more during some months when only leaves are available. A similar seasonal pattern is seen in the activity of the Honduran mangrove howler (Snarr 2006).

Although there is some overlap in the home ranges of the mangrove black howlers, intergroup contacts were rare and resulted in vocalizations by both groups, but without other visible aggressive behaviours. The low aggression rate in this population, both within and between groups, is indicative of low competition for resources when it comes to either food or mating opportunities. The ranging patterns for Honduran mantled howlers in the mangroves are similar to the average for the species (Snarr 2006).

Both howler species in the mangrove ate from fewer plant species (12 and 18) than their counterparts in other habitats, a reflection of the low diversity of plants in the mangrove environment. The Mexican mangrove black howlers rely heavily on flowers and seeds (nearly 50%) during the dry season, much more so than is reported for other black howler populations and other howler species (see Bridgeman 2012). Consumption of leaves was within the range for howlers in general, but was slightly lower than is reported for black howlers at several sites. The consumption of fleshy or propagule fruits in the Mexican mangrove howlers is non-existent and thus is significantly lower than fruit consumption for other howlers. In Honduras, the diet of mangrove mantled howlers is very similar, with vines and lianas providing a significant portion of their diet.

Overall, howlers in mangroves display the behavioural flexibility for which this genus is known (see Bridgeman 2012 and Snarr 2006 for detailed comparative analyses). The effects of the mangrove habitat may be more evident for primates that rely heavily on fruits and do not incorporate significant animal or leafy materials in their diets.

For most South American mangrove primates, there are only brief reports from papers and personal communications (Chapter 8). These species use the mangroves sporadically as a secondary habitat. However, four species are currently being studied and have groups that live exclusively in mangrove forest: the bearded capuchin, *Sapajus libidinosus* (Chapter 8), the yellow-breasted capuchin, *S. xanthosternos* (Chapter 11), the tufted capuchin, *S. apella* (Santos *et al.* 2016; Chapter 8), and the Maranhão red-handed howler monkey, *Alouatta ululata* (Santos, pers. obs.).

Foods from mangroves may not be as important for neotropical primate species that use mangrove occasionally as are food resources from non-flooded areas. For example, Fernandes and Aguiar (1993) reported that *Saimiri sciureus* and *Chiropotes satanas* foraged for insects in red mangrove trees (*Rhizophora mangle*), not far from non-flooded forest. Reports of primate visitor species in mangrove are rare. Thus, ecological and behavioural studies of South American mangrove primates are lacking, making any conclusions about their behaviour and ecology speculative at this point.

Capuchin monkeys have the widest geographical distribution in the mangroves of South America and the number of species increases if we extrapolate using the number and distribution of capuchin species that overlaps with mangrove in the Neotropics. Because of their more omnivorous diet than howler monkeys and other specialized primates, they can exploit the abundance of shellfish and insects that inhabit the mangrove, in addition to the low diversity of potential plant food resources. For example, *Sapajus apella* eat spiders (Araneae) and insects from the Saturnidae, Pentatomidae, Formicidae and Mantidae families in the mangrove (Fernandes 2000; Fernandes & Aguiar 1993), in addition to marine invertebrates such as oysters (*Crassostra rizophorae*) (Fernandes 1991), bivalves (Teredinidae), shrimp (Caridae), and crabs (Leucosiidae and Ocypodidae) (Fernandes 2000; Fernandes & Aguiar 1993; Santos 2010). Because of the omnivore diet and arboreal/terrestrial habit in comparison to others species in the American mangroves (Chapter 8), capuchin monkeys seem to be the only neotropical primates to eat sea animals in the mangroves. Because they increase their foraging area by travelling in the stilt roots of mangroves and eat animal foods from the lower level of the forest, they are more likely to be found in American mangroves than are other primate species.

Although most work on capuchin monkeys has been conducted in *terra firma* forests, there are two studies in mangrove that provide brief comparisons with their non-flooded conspecifics. These studies were conducted by Santos (2010) and Cutrim (2013) and focused on the distribution, diet, and activity patterns of bearded capuchins (*S. libidinosus*).

Bearded capuchins are from the dry habitats of *Caatinga* and *Cerrado* in Brazil (Alfaro *et al.* 2012) but live in mangrove areas of the north coast (Santos 2010). Some groups live exclusively in mangroves and can survive in smaller areas than conspecifics living in non-flooded habitats (Santos 2010; Figure 17.2).

In mangrove forest, capuchins eat encapsulated foods protected by shells and exoskeletons. An important observation is that some bearded capuchins will use tools to crack and open the protective coverings of these food items, which can help them survive in even small patches of mangrove. Such tool use is not widespread throughout the distribution of *S. libidinosus* mangrove, nor in populations from non-flooded habitats. Tool-use behaviour by bearded capuchins in mangrove and non-flooded habitats is detailed by Santos *et al.* (Chapter 9).

Santos (2010) recorded crabs (*Ucides cordatus*) and molluscs (Bivalvia: Teredinidae) in the diet of *S. libidinosus* and *S. apella* in the Brazilian mangrove. However, consumption of *Neritina zebra* (Gastropoda) by *S. libidinosus* in mangrove has only been seen in one estuary. These resources are ground-living and are available only during low tide.

In his study, Santos (2010) recorded *S. libidinosus* eating plant foods from three different species: fruits from black mangrove (*Avicennia germinans*), flowers from white mangrove (*Laguncularia racemosa*) and flowers and propagules from red mangrove (*Rhizophora mangle*). Cutrim (2013) recorded capuchin consumption of root tissue in the stem projections of *Rhizophora mangle*. Propagules from *Rizophora* spp. are an important food in the bearded capuchin diet (Cutrim 2013; Santos 2010). In comparison with primate plant foods in non-flooded areas, the leaves have high levels of salt (Cram *et al.* 2002), and the fruits and propagules from mangrove are drier.

Figure 17.2 Bearded capuchin (*Sapajus libidinosus*) travelling in stilt roots of *Rhizophora* spp. to looking for shellfish, Maranhão, Brazil. Photo: Ricardo Rodrigues dos Santos.

Sapajus libidinosus display fission-fusion dynamics and activity patterns (foraging, eating, resting, travelling, and social interaction) in mangrove similar to some capuchin species inhabiting non-flooded habitats (Cutrim 2013). Cutrim highlights the formation of foraging units (sub-groups) due to the increase in group size. Nakai (2007) and Rímole *et al.* (2008) reported the same behaviour for capuchin monkeys in non-mangrove areas. In both cases, agonistic behaviour increases during encounters of sub-groups. Foraging in mangrove is the main activity and the use of the ground for activities is rare, occurring in < 1% of the time spent in motion (Cutrim 2013).

Capuchin monkeys have a wider distribution in South American mangrove than other primate species. Overall, mangroves appear to provide conditions for survival in small areas and a complementary diet based on shellfish.

African Primates

Although little is known about African primates' use of mangroves, Nowak (2012) recently compiled a preliminary list of primate species reported to use mangroves regularly, seasonally or opportunistically. Thirty-nine primates from 19 genera were found to use mangroves in 21 sites on the coast of Africa. A more conservative list presented in this volume by Head *et al.* (Chapter 12) confirms the use of mangroves for at least 24 species (12 genera) from 21 different sites in mainland Africa. Both studies suggest that no African primates are mangrove specialists and that *Cercopithecus* spp., especially Sykes' monkeys (*Cercopithecus mitis*) inhabiting mainly the coast of Kenya and Tanzania, are among the most adapted species to life in mangroves (Figure 17.3). Green monkeys (*Chlorocebus sabeus*) also appear to do very well in mangrove forests. Although they generally prefer other habitats (savanna, grassland, forest edges), populations living in Senegal and Ivory Coast have been reported to frequently use mangroves. For instance, Galat and Gulat-Luong (1976) reported that in southern Senegal *Chlorocebus sabeus* spent up to 80% of their time in mangrove forest (versus only 13% on *terra firma*) eating *Rhizophora* plant parts, hunting crabs and sleeping overnight in the mangrove.

Other species have been found to use mangroves mostly as refuge from human disturbance (deforestation and hunting) such as red colobus (*Piliocolobus* spp.), Lowe's guenon (*Cercopithecus campbelli lowei*) and the olive colobus (*Procolobus verus*). This finding was also confirmed by the recent survey of Bi *et al.* (2013) in Tanoé Forest, Ivory Coast, where both roloway guenons (*Cercopithecus roloway*) and white-naped mangabeys (*Cercocebus atys lunulatus*) were found to prefer mangrove habitats to non-flooded forests. The authors suggest that the higher encounter rate in mangroves of these two primate species might reflect an adaptive strategy to avoid hunting pressure (Bi *et al.* 2013), which is intense in the region. Similarly, western gorillas in Loango National Park, Gabon, used mangroves as refuge habitat to escape humans during the beginning of the habituation process (Chapter 12).

Although Nowak (2012) included 11 lemur species in the list of African primates using mangrove forests, the information on Malagasy primates is based largely on anecdotal or personal communication (for updated information see Chapters 5 and 7). To our knowledge, only one study conducted in the Baly Bay area of northwestern Madagascar has been published confirming the use of mangroves by *Microcebus* spp. (Hawkins *et al.* 1998). Unfortunately, no information on their ecology and behaviour was provided and, therefore, no comparison can be made with other populations living in different environments.

Figure 17.3 Angola Pluto monkey *Cercopthecus mitis mitis* eating aquatic plant leaf in Kwanza River mangroves, Angola. Photo: Tommy Pedersen. (A black and white version of this figure appears in some formats. For the colour version, please refer to the plate section.)

Only a few studies have compared African primate populations that use mangroves extensively with those living in other habitats and are generally restricted to red colobus species (*Piliocolobus* spp.). Nowak and colleagues (Nowak 2008; Nowak & Lee 2011) have described the demographic structure, ranging patterns, and drinking and feeding behaviours of the Zanzibar red colobus (*Piliocolobus kirkii*) in unprotected coral rag and mangrove forests and Galat-Luong & Galat (2005) conducted a longitudinal study on one population of Temminck's red colobus (*Piliocolobus badius temminckii*) in Senegal whose use of mangrove habitats increased over time.

Zanzibar red colobus groups that use mangroves extensively have been shown to have more heterogeneous age structures, and to be larger, more stable, and more cohesive than those living in more human-disturbed coral rag forests (Nowak & Lee 2011). These researchers also found a relatively higher infant survival probability, more seasonal food availability, and more common play behaviour, which they consider to be an indicator of high habitat quality, in mangrove than on *terra firma* (Figure 17.4).

Differences in behavioural adaptations have also been identified among African primate populations using mangroves and those in other habitats. Primates using mangroves display a diverse activity budget due to more energetically costly travel and less diverse food items, although rich in aquatic plants and water-related animal items (e.g. crabs, molluscs, etc.). In addition, they have been observed to dedicate a greater amount of time to drinking fresh water, possibly to contrast the higher ingestion of both salt and tannins when feeding on mangrove plant parts (Nowak 2007, 2008). In contrast, groups of Zanzibar red colobus living in coral rag forests had smaller group sizes, larger and overlapping ranges, and have been observed exhibiting fission–fusion, which has been interpreted as a response to increased human disturbance occurring in those areas.

Galat-Luong and Galat (2005) reported changes within the Temminck's red colobus population in the non-flooded Fathala Forest (Saloum Delta National Park, Biosphere Reserve of Salouma Delta, Senegal) following drastic habitat degradation. From 638 individuals estimated in 1973, the population decreased to 543 individuals in 1998. Several changes in their ecology and behaviour seemed to be adaptations that enabled the population to survive in a degraded habitat. For example, while Temminck's red colobus were never observed in mangrove and open habitats in the early years (1972–1982), since 1991 they have repeatedly been seen using these habitats while associating with green monkeys. Group size has decreased from 29 in 1973 to 17 in 1993, a size similar to that of groups living in other small *terra firma* forests (e.g. Pirang Forest, Gambia: Galat-Luong 1988). Their diet included fewer species than before with increased consumption of young leaves, flowers, fruits, and especially different parts of *Rhizophora* sp. Finally, they engaged in terrestrial locomotion despite being considered exclusively arboreal. Although their *terra firma* habitat has undergone dramatic and rapid change during the last 30 years, Temminck's red colobus were able to modify their behavioural strategies, enabling them to extend their ecological niche (using new habitat such as mangroves) and overcome the degradation of their historic habitat.

Overall, red colobus species although less adapted to live in mangroves than *Cercopithecus mitis* and *Chlorocebus sabeus* seem to adapt well to a less diverse diet in exchange of a better 'escape cover'.

Asian Primates

In contrast to other continents, in Asia, some species have historically included mangroves in their ecological niche; therefore, we will present most of the available information on those populations using mangroves. To date, the proboscis monkeys (*Nasalis larvatus*), long-tailed macaques (*Macaca fascicularis*), rhesus monkeys (*M. mulatta*), ebony leaf monkeys (*Trachypithecus auratus*), silvered leaf monkeys (*T. cristatus* or *Presbytis cristatus*) and Bornean orangutans (*Pongo pygmaeus*)

Figure 17.4 Zanzibar red colobus (*Piliocolobus kirkii*) adult female in Uzi Island mangroves, Zanzibar, Tanzania. Photo: Katarzyna Nowak.

are known to use mangrove habitats in Asia. Other primates that live in mangrove habitats include: the Asian slow lorises (*Nycticebus* spp.), spectral tarsiers (*Tarsius tarsier*), Mentawai macaques (*M. pagensis*) and white-thigh surili (*Presbytis siamensis*), although the information on the use of mangrove by these animals is minimal (Supriatna & Wahono 2000; Thorn *et al.* 2009). For most of these species, however, mangrove is not the preferred habitat. For instance, proboscis monkeys, pig-tailed macaques and leaf monkeys have been described to prefer other flooded forests, such as riverine forest, when available (Matsuda *et al.* 2011; Salter *et al.* 1985). Similarly the Bornean gibbon (*Hylobates muelleri*), the orangutan (*Pongo pygmaeus*), and again the proboscis monkey were described to prefer riverine habitats for sleeping sites (Cheyne *et. al.* 2013; Matsuda *et al.* 2011; Sha *et al.* 2008). Details of existing mangrove research on proboscis monkeys, long-tailed macaques, rhesus macaques, ebony leaf monkeys, silvered leaf monkeys and orangutans are discussed below.

Proboscis monkeys are found only on the island of Borneo. They prefer coastal regions, where they use a variety of habitat types, including mangrove forest, riverine forest, swamps and lowland rainforest (Kawabe & Mano 1972). Mangrove forest provides young leaves, unripe fruits, flowers and seeds for the proboscis monkey diet (Boonratana 2003). At least 55 different mangrove plant species are listed, with a marked preference for *Eugenia* sp. (ubah or gelam tikus), *Ganua motleyana* (arupa or nyatoh) and *Lophopetalum javanicum* (abuab/Philippines, perupuk/Malay). Young leaves are preferred over mature leaves and unripe fruits are preferred over ripe fruit, most likely because the high sugar content of ripe fruit can lead to gastrointestinal acidosis (Boonratana 2003).

Proboscis monkeys tend to be more frugivorous from January through May and more folivorous from June through December in Tanjung Puting National Park (Yeager 1989). Boonratana (2003) found that young leaves are the preferred dietary items, particularly in Sukau, the riverine forest, compared to Abai, the mangrove forest, where flowers, fruits and seeds contribute significantly to their diets. This finding might be explained by the fact that fruit and flower production is higher in Abai than Sukau (Boonratana 2003). Proboscis monkeys may also eat some invertebrates, including mosquitoes, caterpillars and insect larvae (Yeager 1989).

Availability, distribution and size of food patches affect home range, group size and population dynamics of proboscis monkeys. Mangrove in Abai with extensive *Nypa fruticans* stands (Boonratana 2000b) provided more clumped food resources. At Samunsam, the food resources in mixed-riverine mangrove forest are more seasonal and scarce, likely causing the groups to migrate in and out of mangroves to supplement their diets (Boonratana 2000b).

On a daily basis for proboscis monkeys, feeding activity peaks in the morning and at dusk (Boonratana 1993). The majority of the day is spent feeding, resting, and travelling and at dusk, the groups will move back to the riverside and feed before dark (Bennett & Gombek 1993). Sexual behaviour of proboscis monkeys can be noisy and conspicuous but is rarely observed in mangrove habitats due to the difficult conditions (soft and muddy substrate, entangled roots, and tidal inundations) encountered by researchers in this habitat (Boonratana 2000b, 2011; Chapter 6).

Long-tailed macaques (*M. fascicularis*) prefer forested areas near water and are found at higher densities near riverbanks, lakeshores, or along the seacoast (van Schaik *et al.* 1996). Population densities in mangrove forest were reported at 6.47 groups/km^2 and between 62 animals/km^2 in Vietnam (Son 2004) and 120.3 individuals/km^2 in Sumatra (Crockett & Wilson 1980; see Yanuar *et al.* 2009). Population sizes in mangrove forest are smaller than in non-mangrove forest, due to the limited food availability compared to lowland or secondary forest (Iskandar, pers. comm.).

These primates are able to swim and use mangroves as foraging sites (Son 2004). Living in mangroves made long-tailed macaques adjust their time budgets to tides, spending more time resting during spring tide and less time moving. During high tide, they move inland to drier areas (Son 2004). They seem to prefer foraging and moving in riverine habitats, and the amount of time spent foraging decreases as they move further from the riverbank. Most of their daily activity occurs within 100 m of the river, where resource density is much higher than areas further inland (van Schaik *et al.* 1996). In mangrove, feeding peaks in the morning after sleeping trees are vacated, followed by a rest around mid-day, with another feeding peak in the afternoon before entering the sleeping trees in the early evening (Gurmaya *et al.* 1994; Son 2004). They tend to sleep in the branches of trees that overhang water, which may be a behaviour to avoid predators such as clouded leopards and pythons.

Long-tailed macaques are opportunistic omnivores, eating a variety of animals and plants. In Kalimantan (Borneo), 66.7% of their diet consists of ripe, fleshy fruits, while for macaques on Sumatra, fruit comprises an even higher percentage of their diets (82%) (Yeager 1996; Wich *et al.* 2002). Although fruits and seeds make up 60–90% of their diet, they also eat young and mature leaves, stems, flowers, grass, mushrooms, roots, clay and bark. They sometimes prey on vertebrates (including bird chicks, nesting female birds, lizards, frogs and fish), invertebrates and bird eggs. In Indonesia, the species has become a proficient swimmer and diver for crabs and other crustaceans in mangrove swamps and in rivers (Son 2003; Sussman & Tattersall 1986). Macrobenthos were also considered a high-quality food resource (Son 2003). Son showed that 31% of total feeding time was devoted to mangrove plant species such as *Rhizophora* spp., *Avicennia* spp. and *Ceriops* spp. An important note is that these macaques are able to use stones to process foods such as molluscs (Gumert & Malaivijitnond 2012; Chapter 19).

The long-tailed macaques show behavioural adaptation to living in mangroves. In mangrove habitat, the long-tailed macaques walk on the soft ground, jump on the stilt roots, and are great swimmers (Figure 17.5). Additionally, Son (2003) showed that females living in mangroves in Vietnam were smaller than in other habitats (Son 2003). The difference in body size however was attributed to Bergmann's rule (body size–latitude relationship) and not by a specific adaptation to this habitat (Son 2003).

Another macaque species that uses mangrove is *Macaca mulatta* (rhesus macaque) the natural distribution of which includes India, Bangladesh, Nepal, Burma, Thailand, Afghanistan, Vietnam and Southern China. In Bangladesh, the species inhabits the Sundarban, an important mangrove area of the southwestern part of the country (see Chapter 16). Mean group size in mangrove is lower (23.9 individuals) than other habitat types (24–40 individuals, covering semi-evergreen, evergreen, plantation, scrub forest) but larger than in deciduous forest (19.3 individuals) (Hasan *et al.* 2013). In Sundarban, the rhesus macaque feeds on mangrove leaves and fruits, molluscs and crabs, as well as fish (Majumder *et al.* 2012).

Ebony leaf monkeys (*Trachypithecus auratus*) are found on the island of Java and on the smaller islands of Bali and Lombok, Indonesia (Leca *et al.* 2013; Weitzel & Groves 1985). Mostly studied in the Pangandaran Nature Reserve, West Java, the ebony leaf monkeys live in coastal forest on the eastern side of the park in small groups, avoiding the teak plantations (Watanabe *et al.* 1996). These primates have also been recorded in mangrove, primary and secondary forest, mixed lowland forest of secondary growth containing plant species such as *Tectona grandis* and *Swietenia macrophylla*, and in *Acacia auriculiformis* plantations (Kool 1989; Nijman & van Balen 1998).

Ebony leaf monkeys are found in mangrove in West Java at Muara Gembong, Muara Angke-Kapuk, and Tanjung Sedari and in the coastal area in central Java up to east Java, Bali and Lombok (Leca *et al.* 2013; Nijman 2000; Weitzel & Groves 1985; Chapter 14). In Muara Gembong (Figure 17.6), the ebony leaf monkey lived in groups of up to 30 individuals, at a density of 20 individuals/km^2, but currently there is little mangrove forest remaining (Supriatna *et al.* 1989). The Muara Angke-Kapuk mangrove supports small populations of ebony leaf monkeys and long-tailed macaques, *Macaca fascicularis* (MacKinnon *et al.* 1982).

Ebony leaf monkeys are primarily herbivorous, eating leaves, fruit, flowers, flower buds, and insect larvae. Young leaves of the teak tree, *Tectona grandis*, are an important food source in non-mangrove areas for this species when favoured foods are scarce (Kool 1992, 1993). Although information is limited, *T. auratus* included many mangrove species in their diet such as *Avicennia* spp., *Rhizophora* spp., *Bruguiera* spp., *Sonneratia* spp. and *Nypa fruticans* (Supriatna *et al.* 1989).

Silvered leaf monkeys (*Trachypithecus auratus*) live in a variety of forest types including, primary and secondary growth forests, riverine forests, mangrove forests, swamp forests, and coastal forests in Peninsular Malaysia, Sumatra, and Borneo (Fleagle 1988; Furuya 1961; Harding 2010; MacKinnon & MacKinnon 1987). They primarily inhabit dense forests, but their habitat can vary somewhat depending on the region. In Java and Sumatra, they live in the trees of inland forests,

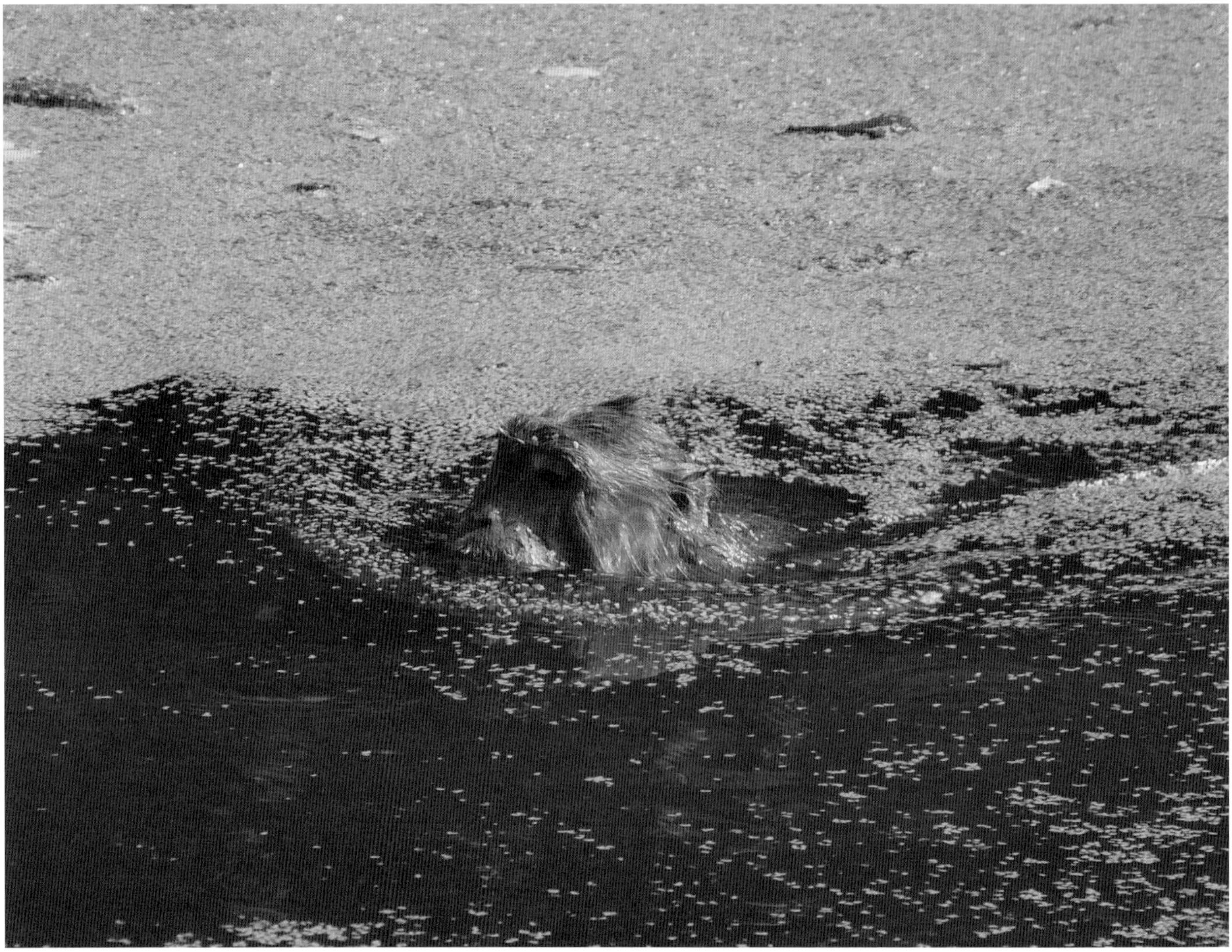

Figure 17.5 Long-tailed macaque *Macaca fascicularis* swimming in Muara Angke wildlife reserve in North Jakarta, the last remaining mangrove in Jakarta, Indonesia. Photo: Nurul Winarni. (A black and white version of this figure appears in some formats. For the colour version, please refer to the plate section.)

whereas on the Malaysian Peninsula, they live in the mangrove and sub-coastal forests. In mangroves, they inhabit areas with *Acanthus illicifolius, Rhizophora conjugata* and *Nypa fruticans* (nipah palm) adjacent to upland forest (Harding 2010). Although the species is found in mangrove and swamp forest, very little is known about the behaviour of silvered leaf monkeys in mangrove here.

Once distributed in Sundaland, now orangutans occur only in Sumatra and Borneo and are divided into two species: *Pongo abelii*, the Sumatran orangutan, and *Pongo pygmaeus*, the Bornean orangutan. Both live in primary tropical lowland forests, including mangrove, peat swamp forests, and riparian forests. Only the Bornean orangutans have been reported to use mangrove as habitat in Sabah, and have been found ranging in the mangrove area of Tanjung Puting, but this association has not been reported in detail (Ancrenaz *et al.* 2004).

Bornean orangutans are mainly frugivorous, with fruits comprising about 61% of their diet. Depending on which fruit it is, they may eat the skin, flesh, juice and/or the seeds. Fruit is always preferentially eaten, but when fruits are in short supply, orangutans forage opportunistically and depend more heavily on other plant foods such as leaves and bark. Part of the orangutan diet also consists of insects, especially termites and ants (Rodman 1988). At least one genus of Rhizophoraceae, a mangrove plant, is listed as part of the orangutan diet in mangrove (Rodman 1988). Additionally, Ancrenaz *et al.* (2004, see Husson *et al.* 2009) indicated that some Bornean orangutans eat the juicy core of young *Nypa* palm found in mangrove habitat.

Overall, Asian primates in mangrove tend to have smaller group sizes and lower densities. There are only four genera of Asian primates that are confirmed to use mangrove habitats with any great frequency: *Nasalis, Macaque, Trachypithecus* and *Pongo*. Mangroves with their entangled roots, muddy substrate and tidal inundations, are difficult to survey which limit the available information on primates.

Conclusion

Although the available information on primate species inhabiting mangroves is limited, three major points emerge from our review. First, the plant diversity of mangroves,

Figure 17.6 An ebony leaf monkey (*Trachypithecus auratus*) in a mangrove habitat from Muara Gembong, Bekasi, West Java, Indonesia. Photo: Ade Wijaya.

which differs especially between Asia and both Africa and the Americas, may strongly influence the ability of primates to colonize this habitat and specialize on it. For instance, only in Asia are there mangrove-obligate primates (e.g. *Nasalis narvatus*). Second, the high productivity of plants (flowers, seeds, leaves, etc.) and shellfish found in this habitat can support healthy populations of folivorous and/or omnivorous primates. For instance, some of the species described here, although they live in other habitat types, seem to do better in mangroves than in other habitats. Some mangrove-dwelling primates dedicate more time to social activities (e.g. *Alouatta pigra*), use smaller areas (e.g. *Sapajus libidinosus*) and have larger and more stable groups (e.g. *Piliocolobus kirkii*) than groups of the same species living in other environments, all an indication of high habitat quality for the species concerned. Third, the high resilience of primate species and their ability to adapt to human disturbance, including both deforestation and hunting pressure, through the use of new and less disturbed habitats such as mangroves underline the critical importance of these habitats for the current and future conservation of primates. Now more than ever, we need to better understand the role mangroves play in the ecology and behaviour of non-human primates. Further investigation into the primates in this particular habitat is important to show how ecological variables may be related to primate biology, behaviour and ecology.

Chapter 18

Maritime Macaques

Ecological Background of Seafood Eating by Wild Japanese Macaques (*Macaca fuscata*)

Yamato Tsuji and Nobuko Kazahari

Introduction

The main diets of non-human primates (hereafter, primates) consist of parts of various plants in the forest, while many species also depend on terrestrial animals, such as insects. Several primate species occasionally include marine organisms in their diet (reviewed by Kempf 2009) sometimes with using stone tools, e.g. long-tailed macaques (*Macaca fascicularis*) and tufted capuchins (*Sapajus apella*) (Fernandes 1991; Malaivijitnond *et al.* 2007). More specifically, long-tailed macaques living in a mangrove forest in Vietnam fed on shrimp, octopus, several species of crabs (Brachyura), shipworm (*Bankia saulii*), razor clam (*Solen gouldii*) and peanut worm (*Phascolosoma arcuatum*) (Son 2003). Long-tailed macaques in Pangandaran Nature Reserve, western Java, Indonesia, frequently fed on dead fish and crustaceans thrown away by fishermen (Hadi 2013). Chacma baboons (*Papio ursinus*) in South Africa capture and consume marine shellfish and crustaceans (Avery & Siegfried 1980). The main diet of savanna monkeys (*Chlorocebus aethiops*) in Saloum Delta, north Senegal was fiddler crabs (Fedigan & Fedigan 1988).

For primates, accessing the seashore, which is not their original habitat, and eating seafoods would seem to have a cost in terms of time and energy. Why do they eat seafoods? Carlton and Hodder (2003) suggested the following three possibilities:

1. Perennial trophic subsidies. Mammals living in perennially impoverished near-coastal ecosystems may seek out seafoods.
2. Seasonal trophic subsidies. Animals living in seasonally or periodically impoverished near-coastal ecosystems may seek out seafoods.
3. Opportunistic use. Animals simply eat seafoods opportunistically.

Japanese macaques (*Macaca fuscata*) are cercopithecine primates endemic to the Japanese archipelago. Since the 1960s, it has been known that wild Japanese macaques come down to the seashore and eat marine organisms, such as seaweeds, shellfish and arthropods (Figure 18.1). Cases of seafood-eating behaviour by Japanese macaques have been reported at seven study sites (Figure 18.2a, Table 18.1). It is known that the habitats of Japanese macaques show clear seasonality in physical condition and plant phenology (Hanya 2005; Tsuji 2010): leaves develop in the spring (March to May) and become harder in the hot summer (June to August), fruits become ripe in the autumn (September to November), and many plants are dormant in the colder winter (December to February). Thus, forest foods are more abundant in the spring and autumn, moderately so in the summer and less available in the winter. Therefore, Japanese macaques may provide a good model for testing the aforementioned possibilities, especially the second one; seafood eating by Japanese macaques may compensate for the seasonal lack of foraging success in forests.

If seafood subsidizes the diets of Japanese macaques, they should be able to use the seashore effectively, because feeding success could affect their survival. For example, seafood should be eaten during seasons, i.e. winter, when nutritional intake from forest diets is scarce. From a nutritional perspective, on the other hand, seafoods might be more nutritious than plant foods in the forest, in terms of energy and protein. In addition, seafood-eating behaviour might occur when the tide is lower because in such situations, seafood can be obtained more effectively. However, most previous reports on primates, including Japanese macaques, eating seafood have been descriptive and lacking analytical interpretation, possibly due to the small samples, and neither the adaptive meaning of this behaviour nor its ecological background has been discussed.

In this study, we tested whether seafood-eating by Japanese macaques occurred in food-poor seasons and whether the seafood items were more nutritious than forest diets. We also preliminarily tested whether the frequency and duration of eating seafood correlated with tidal state. We sought to answer to these questions using long-term behavioural data collected on Kinkazan Island, northern Japan, with a comparison of data on seafood-eating between two troops.

Methods

Subjects and Data Collection

We made behavioural observations of Japanese macaques on Kinkazan Island (141°35′E, 38°16′N), northern Japan. The island is 5.1 km long and 3.7 km wide, undulating over a total area of approximately 9.6 km^2, with a highest peak at 445 m. The mean annual rainfall is about 1500 mm. The mean annual temperature is 11.4°C, being highest in summer (23.5°C in August)

Figure 18.1 Japanese macaques on Kinkazan Island feeding on seaweed (*Gloiopeltis fucata*) in March.

and lowest in winter (0.5°C in January). Tourists come to the island mainly to visit a shrine located in the northwestern part, and few go into the forest.

We followed two troops. The A troop was located in the northwestern part (Tsuji & Takatsuki 2009), and the B_1 troop was located in the western part of the island (Kazahari *et al.* 2013) (Figure 18.2b). The sizes of the A and B_1 troops during the study period were 29–39 (Tsuji & Takatsuki 2012) and 21–30 (Kazahari *et al.* 2013), respectively. The home ranges of both troops consisted of deciduous forest and coniferous forest (Kazahari *et al.* 2013), and part of the home range of the A troop also included a large grassland created by sympatric sika deer (*Cervus nippon*) (Takatsuki 2009; Tsuji & Takatsuki 2004). The A and B_1 troops have been habituated since 1983 and 1985, respectively, and direct observations were possible without disturbance.

We collected behavioural data on the A troop between 2000 and 2013 (observation time: 358 days or 2780 h), and on the B_1 troop between 2002 and 2011 (observation time: 212 days or 1814 h). During the study, whenever we noticed, times when the troop members went down to the seashore and started eating seafood (starting time) and times when they went up from the seashore (ending time) were recorded. The starting/ending times for the A troop were recorded using a scan sampling method in 10-min intervals while the times for the B_1 troop were recorded by the focal animal sampling method. We were unable to record the times at which members of the B_1 troop ended eating seafood in some cases (3/10 cases) due to poor observational conditions. Whenever possible, we also recorded the item(s) on which the macaques fed during the event, but identification of the seafood was incomplete because of their rapid eating and because we sometimes could not get down to the seashore due to the steep topography.

Nutritional Analyses

Data on the nutritional composition, in terms of crude protein (%CP), crude lipid (%CL), and crude ash (%CA), of the forest plant foods (leaves, fruits, and seeds) were taken from previous studies conducted on Kinkazan Island (Tsuji & Takatsuki 2012). From these articles, we obtained nutritional data for food items for which the macaques spent more than 1% of their feeding time: eight fruits, four leaves and six seeds. We averaged these values for later comparison. We collected corresponding

Table 18.1 List of seafoods eaten by Japanese macaques.

Family/genus/species	Study sites			References
(a) Seaweeds				
Seaweed	1	2	3	Izawa & Nishida 1963; Izawa 2009; Kawai 1964b
Laminariaceae				
Laminariales sp.	1	2		Izawa & Nishida 1963; Izawa 2009
Costaria costata	2			Izawa 2009
Sargassaceae				
Sargassum fulvellum	1	2		Izawa & Nishida 1963; Izawa 2009
Hizikia fusiformis	2			Izawa 2009
Ralfsiales				
Analipus japonicus	1	2		Izawa & Nishida 1963; Izawa 2009
Alariaceae				
Undaria pinnatifida	1	2		Izawa & Nishida 1963; Izawa 2009
Alaria crassifolia	2			Izawa 2009
Scytosiphonaceae				
Scytosiphon lomentaria	2			Izawa 2009
Colpomenia sinuosa	2			Izawa 2009
Bangiaceae				
Porphyra tenera	1	5		Izawa & Nishida 1963; Kawai 1964b
Porphyra sp.	1	2		Izawa & Nishida 1963; Izawa 2009
Endocladiaceae				
Gloiopeltis fucata	2	6		Izawa 2009; Iwamoto 1982
Monostromataceae				
Monostroma nitidum	2			Izawa 2009
(b) Arthropoda				
Maxillopoda sp.	2			Izawa 2009
Sessilia sp.	2			Izawa 2009
(c) Mollusca				
Mollusca sp.	1	2	4	Izawa & Nishida 1963; Izawa 2009; Kawai 1964a
Veneroida				
Ruditapes philippinarum	1			Izawa & Nishida 1963
Mytilida				
Volsalla diffcilis	1			Izawa & Nishida 1963
Octopoda				
Octopus sp.	6			Watanabe 1989
Docoglossa				
Cellana sp.	2	7		Kawai 1964a; Watanabe 1989; Izawa 2009
Cellana mazatlandica	1			Izawa & Nishida 1963
Patelloida saccharina	1			Izawa & Nishida 1963
Collisella heroldi	1			Izawa & Nishida 1963
Nipponacmea schrenckii	1			Izawa & Nishida 1963
Cellana dorsuosa	1			Izawa & Nishida 1963
Cellana toreuma	1	6		Izawa & Nishida 1963; Kawai 1964a; Iwamoto 1982
Polyplacophora	6			Kawai 1964a
Discopoda				
Littorina brevicula	1			Izawa & Nishida 1963
Sorbeoconcha	1			Izawa & Nishida 1963
Vetigastropoda				
Trochidae sp.	2			Izawa 2009
(d) Chordata				
Actinopterygii	6			Watanabe 1989
Perciformes				
Lateolabrax japonicus	6			Leca *et al.* 2007
Clupeiformes				
Clupeidae sp.	1			Izawa & Nishida 1963

1: Shimokita Peninsula, 2: Kinkazan Island, 3: Hagachi, 4: Kusuyadake, 5: Tsubaki, 6: Koshima Island, 7: Cape Toi

nutritional data for seafood items that had been eaten by the Japanese macaques on Kinkazan Island (ten seaweeds, two shellfish) from comparable articles (Kirimura 2007; Ministry of Education, Culture, Sports, Science and Technology, Japan 2010; Shizukuishi & Narita 2004; Tsuji 2007; Tsuji & Takatsuki 2012). To calculate the caloric contents (kcal g^{-1}), we performed widely applied method for primates (Maynard *et al.* 1979; Tsuji

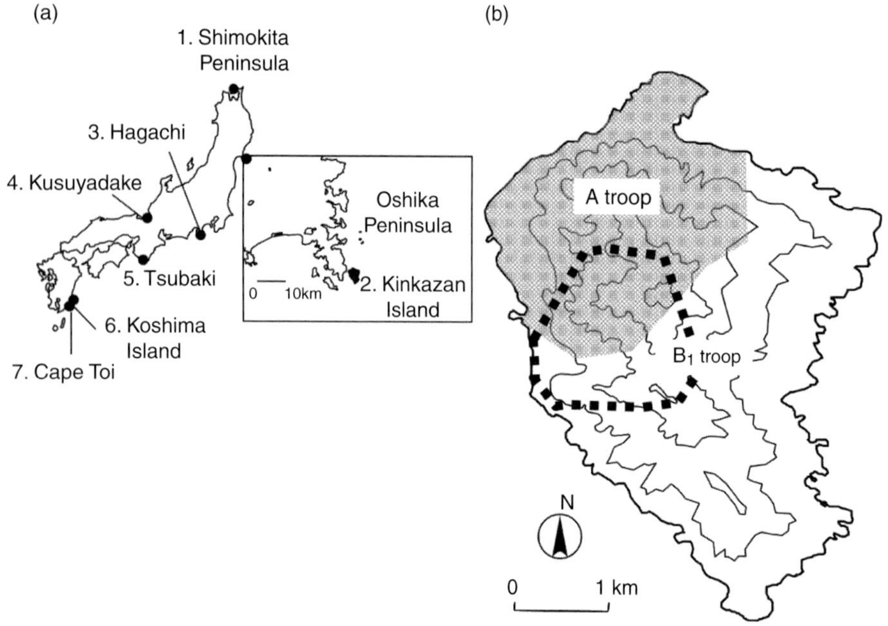

Figure 18.2 Maps of (a) study sites referenced in this chapter, and (b) Kinkazan Island. Annual home ranges of A troop (from Tsuji & Takatsuki 2009; grey colour) and of B_1 troop (from Kazahari *et al.* 2013; dotted line) are also shown. Contours show 100 m intervals.

et al. 2008); the caloric contents of both forest and seafood items were calculated by multiplying each weight (g) of carbohydrate (calculated as total sample weight – [weight of crude protein + weight of crude lipid + weight of crude ash]), protein and lipid by 0.0415, 0.0565 and 0.0940, respectively.

Foraging Success

We calculated daily gross energy (kcal day^{-1}) and gross protein intakes (g day^{-1}) for adult females of the A troop by combining behavioural observation and nutritional analyses from June 2004 to May 2005 (see Tsuji *et al.* 2008 for details). Daily digestible energy intake and daily digestible protein intake were estimated by multiplying gross energy and gross protein intakes by food digestibility (55%), determined from two previous reports for Japanese macaques (Iwamoto 1978; Mori 1979).

Data Analyses

We compared the frequency of eating seafood among different months for each troop and the annual mean frequency of eating seafood between the A and B_1 troops (times/hour) using chi-squared tests for independence. To measure the macaques' feeding activity at the seashore in relation to the tidal cycle, we plotted the time when the A troop started/ended eating seafood against the hourly tidal height at the Oshika Peninsula weather station (141°30′E, 38°18′N) near Kinkazan Island (Japan Meteorological Agency: http://www.data.kishou.go.jp) (Figure 18.2a). To analyse data collected in different months when the intervals between high tides differed, we converted the raw data of day times into times relative to the tidal cycle so that '0.0' referred to low tide, and '1.0' referred to high tide. Due to the small number of cases of eating seafood (n = 10), we did not conduct correspondence analysis for the B_1 troop. We compared the homogeneity of the variances in the starting/ending times of eating seafood relative to the tidal cycle using Levene's tests. To test differences in nutritional composition (gross energy, crude protein, crude lipid and crude ash) between types of seafood and forest foods, we used Kruskal-Wallis tests and post hoc Steel–Dwass tests (Hush 1996). Finally, to test the effect of foraging success of adult females of the A troop in their forest diets on the frequency and duration of seafood eating by this troop, we used Pearson's correlation tests. All statistical analyses were performed using the R.2.15 statistical software (R Developmental Core Team 2012). The significance level (α) was set at 0.05 for all analyses.

Results

During the study, we observed 33 and 10 cases of eating seafood by the A and B_1 troops, respectively, and they fed on seaweeds and shellfish. Due to steep topography near the seashore and the monkeys' quick eating, we could only identify the species of five seaweeds (*Hizikia fusiformis, Undaria pinnatifida, Alaria crassifolia, Analipus japonicas* and *Porphyra tenera*) during the study period. For the A troop, the time spent feeding on seafood comprised up to 6.6% of their daily feeding time, on the basis of a monthly mean (Tsuji, unpublished data). The A troop fed more frequently on seaweed (22 cases) than on shellfish (6 cases). In five cases, monkeys in the A troop ate both seaweed and shellfish. On the other hand, for the B_1 troop, the time spent feeding on seafood reached up to 2.2% of their daily feeding time, on the basis of a monthly mean (Kazahari, unpublished data). They fed primarily on shellfish (six cases) rather than seaweed (one case). In two cases, monkeys in B_1 troop fed on both shellfish and seaweed at same time.

Seasonality in Seafood Eating

Seasonal changes in eating seafood by the two troops are shown in Figure 18.3. The annual mean frequency (times/hour) of eating seafood by the A and B_1 troops were 0.012 and 0.006, respectively, and showed a significant difference between

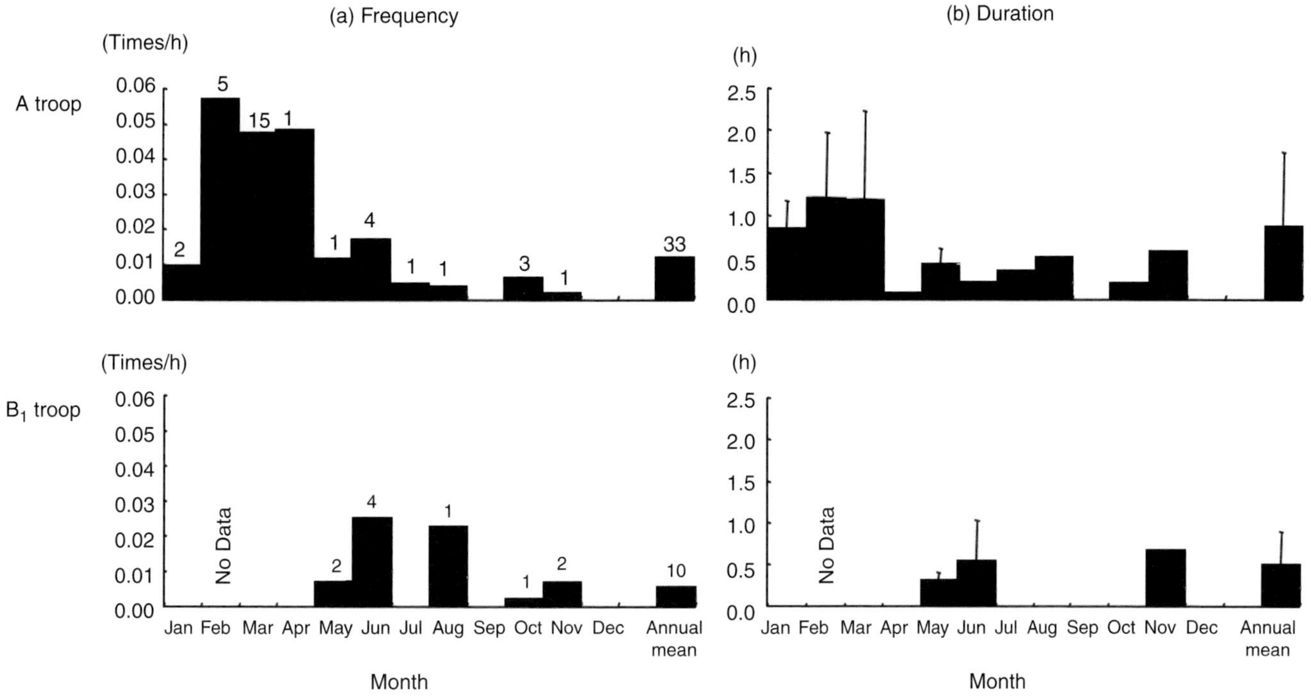

Figure 18.3 Monthly changes in (a) frequency of seafood eating (times/h), and (b) mean (± SD) duration of a single seafood-eating event (h) performed by the macaques in A troop (top) and B_1 troop (bottom). The figures on the top of the histogram represent numbers of seafood-eating events.

the troops (chi-squared test of independence: $\chi^2 = 4.0$, $df = 1$, $p = 0.044$). Seafood eating occurred frequently in February and March in the A troop, while it was common in June in the B_1 troop. The frequency of eating seafood changed significantly on a monthly basis in the A troop ($\chi^2 = 57.8$, $df = 11$, $p < 0.001$), whereas the significance of such changes was marginal for B_1 troop ($\chi^2 = 17.1$, $df = 10$, $p = 0.073$; Figure 18.3a).

The mean (± SD) durations of a single episode of eating seafood (in h) by the A and B_1 troops were 0.86 ± 0.87 ($n = 33$) and 0.49 ± 0.39 ($n = 10$), respectively, but this difference between troops was not statistically significant (Wilcoxon test, $W = 147$, $p = 0.269$; Figure 18.3b). For the A troop, the duration of a single seafood-eating event between January and March was longer than in other months (Figure 18.3b), and this difference was statistically significant (Wilcoxon test, $W = 33$, $p < 0.001$). Thus, we separated January–March and other months in later analyses.

Nutritional Composition of the Seafood

The nutritional composition of the seafood is summarized in Table 18.2. Then we compared the nutritional values of the seafood with those of forest diets (fruits, leaves and seeds; Figure 18.4). Significant differences between the diets were found for all nutritional contents (Kruskal–Wallis test, gross energy: $\chi^2 = 19.3$, $df = 4$, $p < 0.001$; crude protein: $\chi^2 = 18.1$, $df = 4$, $p = 0.001$; crude lipid: $\chi^2 = 19.2$, $df = 4$, $p < 0.001$; crude ash: $\chi^2 = 21.3$, $df = 4$, $p < 0.001$). Post hoc Steel–Dwass tests showed that the gross energy of seaweed was significantly lower than that of fruit and seeds ($p < 0.05$), the crude protein of seaweed was significantly greater than that of fruit ($p < 0.05$), the crude lipid of seaweed was significantly lower than that of fruits and seeds ($p < 0.05$), and the crude ash of seaweed was significantly greater than that of fruit, leaves and seeds ($p < 0.05$). On the other hand, we could not test for any differences between the nutritional composition of shellfish and the forest diets due to the small number of cases ($n = 2$; Figure 18.4).

Relationships Between Seafood Eating and Foraging Success

We found no significant correlation between monthly daily nutritional intake and duration of eating seafood (Pearson's correlation test: $r = -0.294$, $df = 10$, $p = 0.354$ for digestible energy intake (kcal), $r = -0.197$, $df = 10$, $p = 0.583$ for daily digestible protein intake). We also found no significant correlation between frequency of eating seafood and nutritional intake ($r = 0.009$, $df = 10$, $p = 0.979$) or protein intake ($r = 0.354$, $df = 10$, $p = 0.771$). That is, the macaques of the A troop neither increased the frequency nor prolonged the duration of seafood-eating events according to their foraging success.

Relationship Between Seafood Eating and the Tidal Cycle

The time at which the macaques in A troop started eating seafood relative to the tidal cycle ranged between 0.009 and 0.952 (Figure 18.5a). The range of the starting times between January and March (median: 0.184, range: 0.009–0.701) was marginally narrower than in other months (median: 0.281, range: 0.086–0.952; Levene's test: $F = 3.84$, $df = 1$, $p = 0.059$). Between January and March, the macaques in the A troop had a tendency to came down to the seashore and started eating seafood at low tide, whereas the relationship between these factors was unclear in other seasons (Figure 18.5a).

Table 18.2 Nutritional compositions (mean value) of the seafoods eaten by Japanese macaques on Kinkazan Island.

	Nutritional compositions (mean value)				Reference	Notes
	GE	%CP	%CL	%CA		
Seaweeds						
Porphyra sp.	4.53	39.4	3.7	9.8	A	
Porphyra tenera	4.00	34.8	0.7	17.0	A	
Laminariales sp.	3.52	8.2	1.2	19.6	A	*L. japonica*
Costaria costata	3.62	18.3	0.1	19.6	C	
Alaria crassifolia	3.71	14.4	0.1	16.0	C	
Hizikia fusiformis	3.62	10.6	1.3	18.3	A	
Monostroma nitidum	3.62	16.6	1.0	20.1	A	
Sargassum sp.	3.88	18.8	0.4	13.8	B	*S. fulvellum*
Analipus japonicus	4.25	27.9	4.9	13.8	A	
Undaria pinnatifida	3.16	13.6	1.6	30.8	A	
Mollusca						
Trochidae spp.	4.98	53.1	7.2	8.3	D	
Cellana spp.	4.99	64.4	3.3	7.1	E	

GE: gross energy, *CP*: crude protein, *CL*: crude lipid, *CA*: crude ash.
Fibres for seaweeds represent total indigestible fibre, whereas for mollusca represent neutral detergent fibre (NDF).
GE = 0.0415 × (100 − [%CP + %CL + %CA]) + 0.0565 × %CP + 0.0940 × %Cl) (Maynard *et al.* 1979).
A Standard tables of food composition in Japan (2012); B Kirimura (2007); C Shizukuishi and Narita (2004)

The time at which macaques in the A troop stopped eating seafood relative to the tidal cycle ranged between 0.05 and 0.99 (Figure 18.5b). The range of ending times between January and March (median: 0.341, range: 0.070–0.650) was significantly narrower than in other months (median: 0.281, range: 0.053–0.994; Levene's test: $F = 7.69$, $df = 1$, $p = 0.009$). Between January and March, macaques stopped eating seafood and went into the forest before/when the tidal height increased (Figure 18.5b).

Discussion

Forty-three cases of eating seafood were observed during our observation of the two troops, the majority of which were observed in the winter season. As mentioned at the beginning of this chapter, seafood-eating by Japanese macaques has been reported at seven study sites (Table 18.1). Among these sites, the behaviour is frequently observed in northern Japan: in the Shimokita Peninsula and Kinkazan Island, the macaques come down to the seashore, take shellfish and seaweed from rocks, and feed in the daytime during the winter (Izawa & Nishida 1963; Izawa 1999), as we showed in this study. In contrast, on Yakushima Island, southern Japan, despite the long-term history of field study of the macaques there (Yamagiwa 2008), there have been no reports of eating seafood. One of the reasons for the regional variation in degree of seafood eaten is forest productivity. In general, the food environment in northern areas of Japan, especially during winter, is poorer than that in lower latitudes because of the lower productivity of the forests and the concealment of terrestrial diet items by snow (Maruhashi *et al.* 1998; Rosenzweig 1968; Suzuki 1965; Tsuji 2010; Tsuji *et al.* 2013). Thus, frequent seafood eating by the macaques in northern Japan in the winter may reflect the environmental severity of these habitats. Another possibility leading to regional variation in the frequency of eating seafood is differences in the distribution/abundance of seafood. For example, it is known that brown algae (such as Laminariaceae, Sargassaceae and Alariaceae) are distributed mainly in northern Japan, while gleen algae (such as Bangiaceae) occur principally in southern Japan (Imada 2006). Though there is no information on regional variation in arthropod abundance in Japan, a latitudinal cline might exist.

Thus, lower availability of seafood-eating might cause a regional difference in the degree of seafood eating by the macaques across the Japanese archipelago. Studying the availability of both forest and seafood items at various study sites will be necessary to confirm this.

Monkeys of A troop fed more frequently on seafoods than B troop. The inter-troop variation in eating seafood might be explained by the difference in habitat quality between the troops. We found that seafood eating by A troop was more frequent in winter, and that the duration of a single eating event was longer in winter. On the other hand, no clear seasonality was evident in the seafood eating by B_1 troop. The inter-troop difference in seafood-eating might be attributed to difference

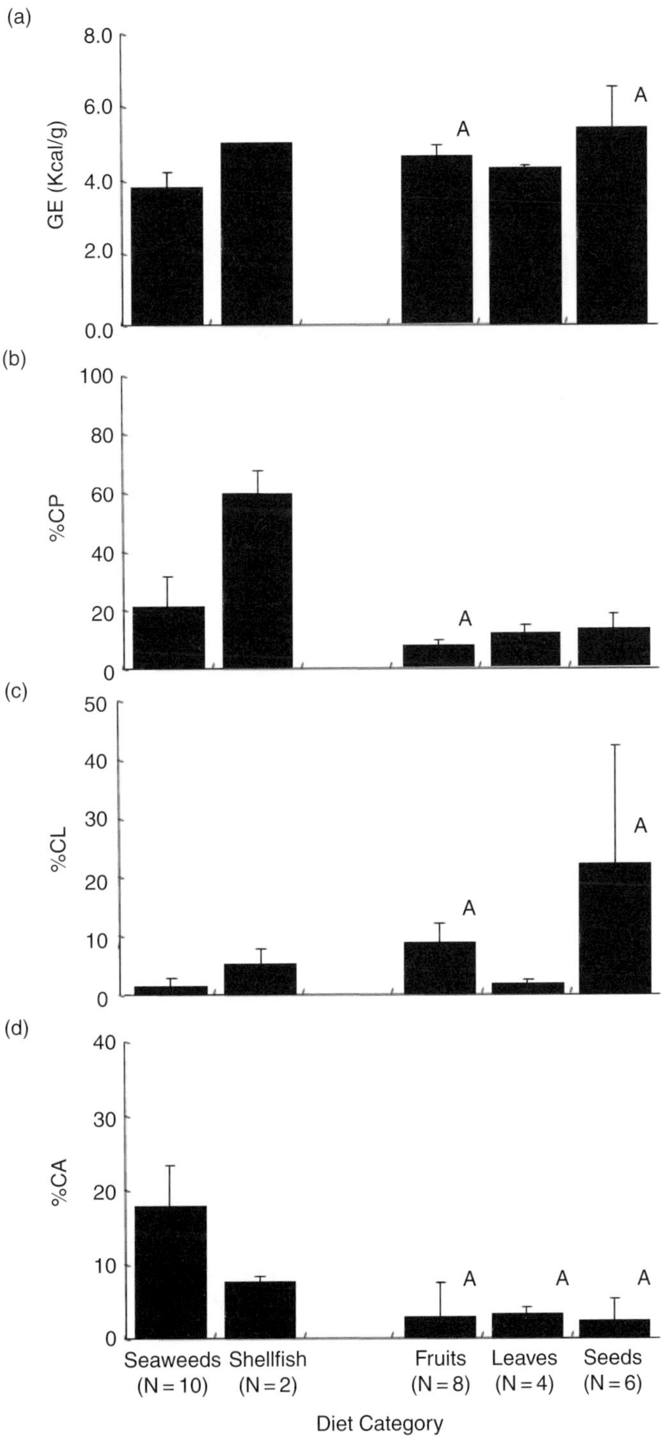

Figure 18.4 Nutritional composition (mean and SD) of the macaque diets on Kinkazan Island: a) gross energy (kcal/g), b) crude protein (%CP), c) crude lipid (%CL), and d) crude ash (%CA). Data sources: Table 18.2 for seaweeds. Tsuji & Takatsuki (2012) for shellfish, fruit, leaves, animals, and seeds. Nakagawa (1989) and Tsuji & Takatsuki (2012) for buds and bark. Tsuji (2007) for shellfish. The letter 'A' indicates a significant difference ($p < 0.05$) between seaweed and given categories (post hoc Steel-Dwass tests).

in habitat quality, but we do not have enough data to address it. To understand the inter-troop difference in eating seafood, it is also important to assess the degree of contact with tourists/ships and tidal fauna in future.

Although the frequency of eating seafood by the macaques might be related to the habitat quality on the island, we found no significant correlation between nutritional intake of A troop from forest diets and frequency or duration of eating seafood. Furthermore, we found that the seafoods did not have prominent nutritional characteristics: the gross energy and crude lipid content of seaweed were significantly lower than forest foods, such as fruits, seeds, and leaves. In

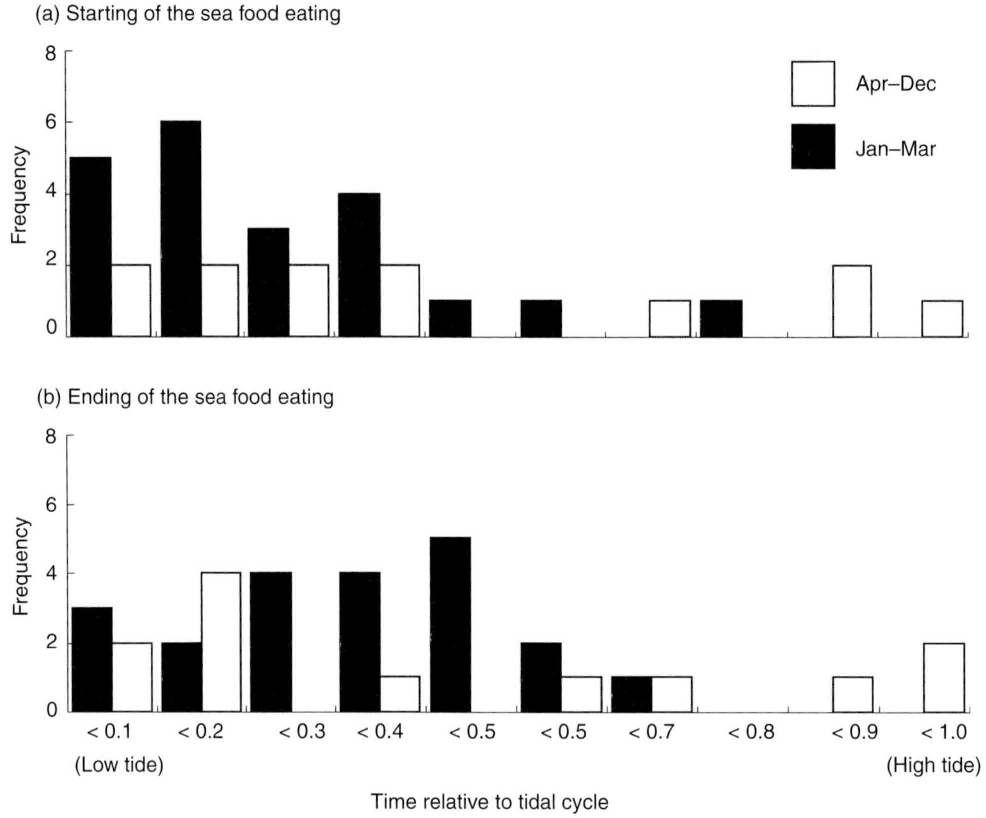

Figure 18.5 Histograms of (a) starting, and (b) ending of seafood eating at different times relative to the tidal cycle (0.0 and 1.0: maximum high tide, 0.5: maximum low tide) at the Oshika Peninsula weather station (141°30′E, 38°18′N) near Kinkazan Island. The seafood-eating events occurring in January–March and April–December are indicated by the filled and open bars, respectively.

fact, the nutritional characteristics of seafoods were similar to those of winter forest foods, i.e. buds and bark, suggesting that eating seafood is not necessarily an adaptive behaviour for the macaques in terms of food nutrition. Nonetheless, availability of seafood, rather than its nutritional value, might be a proximate factor causing the macaques to exploit this resource. Several reports have noted that the amount of shellfish (Kijima *et al.* 2004) and seaweed (Agatsuma *et al.* 2000) around the Oshika Peninsula, near Kinkazan Island, do not change seasonally, indicating that Kinkazan macaques can use seafoods whenever forest diet items are scarce. Thus, seafood seems to play the role of a diet supplement for macaques on the island when forest foods are lacking. It is noteworthy that the crude ash content of seaweed is greater than that of forest foods. It is possible that macaques use seafood to get minerals, but we do not have enough data to confirm this.

We also found that the macaques had a tendency to start eating seafood during low tide and to end when the tide came in again between January and March, when forest diets were poor. This tendency was not detected in other seasons, when forest diets were richer. The seasonal difference in the relationship between tidal height and eating seafood indicates that the macaques in food-poor seasons may adjust their foraging at the seashore in response to the tidal cycle to increase their foraging success for seafood items.

Macaques are one of the non-human primate clades that have succeeded in expanding their habitat widely in the Asian region (Thierry 2011), and one reason for their wider distribution may relate to their dietary plasticity in harsh environments (Tsuji *et al.* 2013). Seafood-eating behaviour by macaques has also been reported in tropical regions, such as Vietnam, Indonesia and Thailand (Son 2003; Malaivijitnond *et al.* 2007; Hadi 2013), but the ecological background of this behaviour has not been discussed. If we apply a similar approach to these populations in a future study, we can generalize the adaptive meaning of this unique behaviour performed to macaques inhabiting different regions.

From an ecological perspective, it may also be worth considering the possibility that Japanese macaques transport nutrients, such as carbon, nitrogen and minerals, from the marine ecosystem to the terrestrial one through eating seafood and subsequently defecating in the forest, as reported in several terrestrial mammals, such as bears (Helfield & Naiman 2006). Mean (± SD) home range size of the macaques is 24 ± 28 km^2 (n = 9), which is same to those of the female Japanese black bears (*Ursus tibetanus*) (19 ± 21 km^2, n = 28) (Oi 2013), and therefore the macaques potentially transport marine nutrients to the forest. This may ultimately affect plant growth and productivity, as in the case of the Pacific salmon, which was transported to terrestrial ecosystem by bears, affecting the riparian forest ecosystem in Alaska (Helfield & Naiman 2006). If so, we should pay more attention to the ecological role of primates as vectors of organic/inorganic matter. We can evaluate the degree of material transportation by macaques using radioisotope techniques (Nakashita *et al.* 2013), and detailed studies on this topic would be interesting.

Acknowledgements

Dr K. Izawa of the Miyagi Prefecture Monkey Research Group and his colleagues offered valuable help and information to our study. The staff of the Kinkazan Koganeyama Shrine provided much hospitality and support during our fieldwork. We are very grateful to Drs I. Matsuda and H. Sugiura, and staff of the Department of Ecology and the social section of the Primate Research Institute, Kyoto University, for their constructive comments on an earlier version of this manuscript. This study was funded by a Grant-in-Aid from the Department of Academics and Technology of Japan (No. 23780160) and the Cooperative Research Fund of the Primate Research Institute, Kyoto University.

Chapter 19

Long-tailed Macaque Stone Tool Use in Intertidal Habitats

Michael D. Gumert, Amanda Tan and Suchinda Malaivijitnond

Introduction

Stone tool use is an uncommon behaviour in non-human primates and has been found only at a few sites in chimpanzees (*Pan troglodytes*), capuchins (*Sapajus libidinosus* and *S. xanthrosternos*) and macaques (*Macaca fascicularis*) (Haslam *et al.* 2009; Wynn *et al.* 2011). Recent studies of macaque tool use have shown that wild Burmese long-tailed macaques (*M. f. aurea*) in Laem Son National Park, Ranong, Thailand, use stones as tools (Gumert *et al.* 2009; Malaivijitnond *et al.* 2007), which was surprising because macaques have generally been considered unskilled tool users (Macellini *et al.* 2012; Panger 2007). We are now challenging this perspective, by presenting our work on the highly proficient and frequent stone tool use that macaques employ for processing encased marine invertebrate prey and coastal plant matter in intertidal habitats (Gumert & Malaivijitnond 2012).

Despite the wide geographical distribution of long-tailed macaques in the Southeast Asian region (Fooden 1995; Gumert 2011), we know of their use of stone tools in only the Isthmus of Kra region of peninsular Thailand and the Mergui (or Myeik) Archipelago of Myanmar (Gumert *et al.* 2009, 2013; Figure 19.1). Carpenter (1887) briefly reported seeing monkeys using tools to process shellfish throughout the Mergui Archipelago, but it was not investigated further until over a hundred years later, when the behaviour was rediscovered just south of Mergui, in Ranong Province, Thailand, at Laem Son National Park (Malaivijitnond *et al.* 2007). At of time of writing (2013), we have found three islands (Piak Nam Yai, Thao and Phayam) and two mainland coastal regions where tool use occurs. In addition, we have found another tool-using population on Koram Island in Khao Sam Roi Yot National Park, Prachuap Khiri Khan Province, on the eastern coast of the Thai portion of the Malay Peninsula, beside the Gulf of Thailand (Gumert *et al.* 2013). At all of these sites, tool use by macaques appears to be specific to intertidal regions of coastal areas and islands, suggesting that the interface with coasts played a significant role in macaque stone tool use.

A few other primates forage in intertidal regions for marine prey (see Chapters 9, 18 and 20), however, reports of tool use in these conditions are almost non-existent. Only one report described the use of detached oyster shells by tufted capuchins (*Sapajus apella*) to strike open oysters (*Crassostrea rhizophorae*, Ostreidae) attached to trees in Brazilian mangroves (Fernandes 1991). Cape baboons (*Papio ursinus*) also forage in coastal habitats for marine prey (e.g. black mussels, *Choromytilus meridionalis*, Mytilidae, and limpets, *Patella barbara* [=*Scutellastra barbara*], Patellidae) (DeVore & Hall 1965; Chapter 19). However, in the wild, these baboons have not been reported to use tools for processing shellfish, despite being tool users in captivity. Even in a detailed investigation of baboon coastal foraging, no evidence of tool use was found (Chapter 19).

The intertidal habitats in which macaques use stone tools are primarily rocky shore and mangrove habitats, and to a much lesser extent, sandy beaches with small stones. Rocky shore habitats are rock-rich environments where thousands of stones are available for use as tools. Also, a continuous distribution of large, flat boulders, which can span up to 5 m across, are available for use as anvils. In the lower littoral zone, macaques forage through oyster beds during low tide, processing oysters with stones (Figure 19.2). In the mangroves, macaques visit areas littered with patches of stones and boulders. Here the intertidal region is not a continuous layer of boulders, as on the rocky shore, but there are still many flat rock surfaces available as potential anvils, generally ranging from 0.5 to 1 m across.

The types of food that macaques process with stones at Piak Nam Yai Island in Laem Son National Park are primarily molluscs and, to a lesser degree, crustaceans and encased plant foods (Gumert & Malaivijitnond 2012). Hooded rock oysters (*Saccostrea cucculatta*, Ostreidae) and nerite snails (*Nerita* spp.) are the most commonly processed invertebrates, but macaques also regularly process trochid snails (predominantly *Monodonta labio*) and muricid snails (predominantly *Thais bitubercularis*). Sea almonds (*Terminalia catappa*, Combretaceae) are the most commonly processed plant matter. Overall, macaques process at least 47 species, consisting of 30 mollusc species, several crustaceans, four other animals (i.e. a gecko *Hemidactylus* spp.; sea cucumber *Holothuria leucospilota*; fish and chiton – both unidentifiable), and 4 plant species. Macaques mainly process food on exposed shores during low tide, but some wade in shallow waters in the mangroves, often bipedally, in search of fish, crabs and other invertebrates.

Macaques use stone tools in a variety of ways and we have discriminated 17 different tool-use action patterns that vary in their combination of material and behavioural elements (Tan *et al.* 2015). Broadly, tool use can be categorized by the stone surface used; point hammering, face hammering or edge

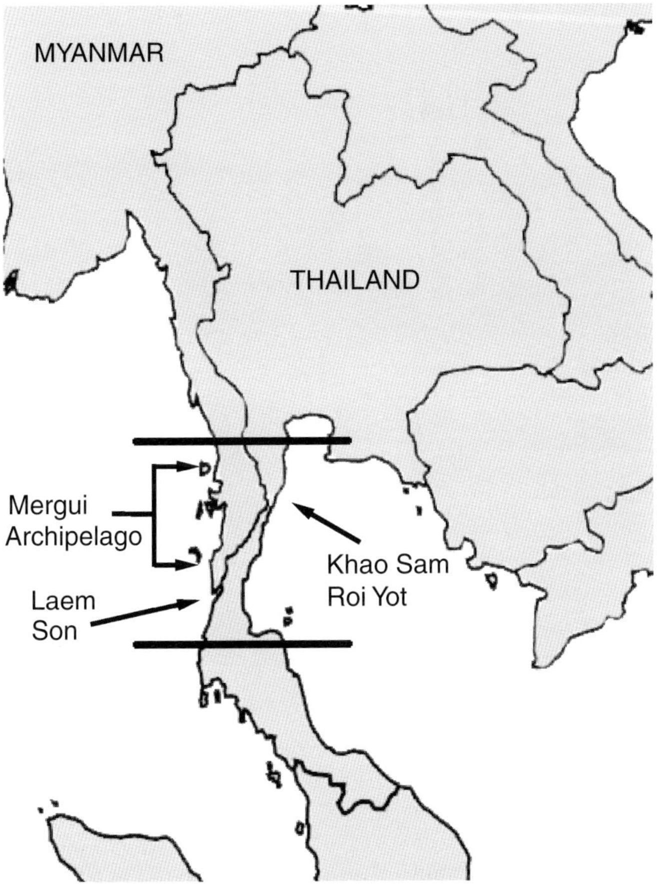

Figure 19.1 Map of Thailand and Myanmar showing the general location where long-tailed macaques are known to use stone tools as of 2013. Arrows point to Laem Son and Khao Sam Roi Yot National Parks, and to the Mergui Archipelago.

hammering. Tools vary in size and mass, with smaller stones being used for sessile and small motile species and heavier stones for larger, tougher motile prey and nuts (Gumert & Malaivijitnond 2013). The type of material also varies, and macaques will use shells as tools (e.g. auger shells, *Turritella attenuata*, Turritellidae; Gumert *et al.* 2009). The behavioural elements of macaque tool use vary in terms of which hands are used, hand synchronization, action of the non-tool hand, how the tool is lifted, striking motion and posture (Tan *et al.* 2015).

The type of actions macaques use during tool use are strongly related to whether the food is sessile or unattached (Tan *et al.* 2015), and these two hammering types have been referred to as axe and pound hammering, respectively (Gumert *et al.* 2009). Macaques generally use axe hammering, which is strongly associated with the use of tool points, to process sessile oysters attached to boulders, tree trunks or roots, of which the most common type is uni-manual point hammering, using a small, hand-sized hammer (Figure 19.3). When cracking motile organisms and nuts, macaques generally use pound hammering, associated with the use of tool faces, of which the most common forms are uni-manual or bimanual face hammering with larger, heavier hammers. Edge hammering tends to be more versatile and is used in similar frequencies for processing both sessile and unattached foods, but is rarer than point or face hammering. Idiosyncratic action patterns of tool use also exist. For example, jabbing, a form of bimanual point hammering using the point of a large tool, is the only action pattern featuring an underhand striking motion. Only one individual has been observed to use jabbing and uses this action pattern predominantly. Other uncommon action patterns are clapping, where the macaque hits together the face of a stone held in one hand and an encased food item held in the other, and fulcrum face hammering, where only one end of the tool is lifted and lowered off the anvil to crush food items.

On Piak Nam Yai Island, we have studied the coastal activity budgets of long-tailed macaques by collecting scan samples from 142 individuals, across eight groups. On average, the macaques used stones for 19% of their total activity and 41% of their coastal foraging activity. These numbers highlight how regularly macaques use stones when in intertidal habitats, with 95 of 107 (88.7%) of all adult and adolescent macaques on the island using tools (Gumert *et al.* 2013). Macaque tool use is a customary activity (McGrew 1992; Shumaker *et al.* 2011; Whiten *et al.* 1999).

Conditions for Macaque Stone Tool-use

Animal stone tool use is rare enough that we can conclude only special conditions favour its emergence and present distribution (Haslam *et al.* 2009). Under the right conditions, macaques are stone-users, and this should not be surprising given their intelligence, dietary flexibility and manual dexterity. At some point in time, the conditions to start using stones as tools were right for macaques living in intertidal habitats. They used this innovation to better process heavily defended food resources in intertidal habitats, which opened a new and highly productive foraging environment for long-tailed macaques living near coasts. The increased access to these new foods would have benefited macaque foraging and influenced survival and reproduction.

What were the critical factors that contributed to macaques using stone tools? Macaques have predispositions that would have made a shift to stone-aided feeding in intertidal habitats a small adjustment. First, macaques are dietary generalists (Richard *et al.* 1989), and can therefore adjust easily to a diet higher in animal prey. For example, in mangrove environments in Vietnam, long-tailed macaques have shifted from the frugivory typical of macaques to consuming greater levels of animal matter, even without the use of tools (Son 2003). Second, several species of macaques handle stones in non-functional ways, suggesting an underlying propensity to manipulate stones (Huffman & Hirata 2003; Huffman & Quiatt 1986; Nahallage & Huffman 2008; 2012). In terms of the mechanical requirements for finding and capturing food, marine prey in intertidal habitats do not present radically different challenges from dietary items foraged in forested environments (i.e. fruits, vegetation and insects) (Verhaegen *et al.* 2007). Shelled marine organisms are often attached or buried, as are fruits and other plant matter (e.g. tubers and roots), while some motile marine prey, such as crabs, move similarly

Figure 19.2 Black arrow points to a female long-tailed macaque foraging on the rocky shores of Piak Nam Yai Island, Laem Son National Park. She carries a hand-sized axe hammer through an oyster bed. Photo: Michael D. Gumert.

Figure 19.3 Black arrow points to an axe hammer being used by a female long-tailed macaque to strike sessile oysters on a boulder in front of her. She is holding the stone by its wider base, and striking with its smaller, pointed end. The stone in the photograph is quartz, although most stones used are basalt. Photo: Michael D. Gumert.

to insects, which are a regular component of macaque diets. Given these predispositions, once macaques interfaced with intertidal habitats containing thousands of stones and an abundance of easily captured, encased food sources, the advent of macaque tool use was just 'a stone's throw away' (or rather, a stone's strike).

Tool Use in the Andaman Sea Region

The region where long-tailed macaques use stone tools is relatively limited when compared to the expansive geographical range of *M. fascicularis*. Therefore, it is important to consider why macaque tool use currently exists in this region but not others (Gumert *et al.* 2009). We suggest three non-mutually exclusive hypothetical scenarios for why macaque tool use occurs here in the Andaman Sea regions, and these relate to previously hypothesized factors for primate tool use, such as necessity, opportunity and relative profitability of exploiting food sources with tools (Fox *et al.* 1999). Last, we do not want to exclude the possibility that macaque stone tool use has simply just not yet been discovered in other regions, as macaques have an enormous geographical range and thus many unstudied populations.

The first hypothesis is that the region's insular geography and history of sea-level change drove the need for macaque tool use (*Insular Isolation Hypothesis*). The second hypothesis is that long-tailed macaques opportunistically exploited intertidal food resources with stones in wild conditions, but since most coastal regions of Southeast Asia are disturbed by human activity, macaques with natural coastal foraging behaviour remain only in the regions around Myanmar, as, along with some Thai islands, these are still little-disturbed by human activity (*Opportunism & Disturbance Hypothesis*). The third hypothesis is that Burmese long-tailed macaques (*M. f. aurea*) are different from the common long-tailed macaque (*M. f. fascicularis*) and other macaques, by having minor adaptations allowing them to recognize and use stones as tools (*Adaptation and Specialization Hypothesis*) (Gumert et al. in review).

Insular Isolation Hypothesis. Hundreds of small islands occur off of the western coast of Thailand and Myanmar, reaching the highest density in the Mergui Archipelago. Historical changes in the sea level have connected and separated these islands from each other and the mainland (Voris 2000), trapping macaques on small islands during periods of high sea level, as we see today (there is not the space here to consider the consequences of occasional longer distance dispersal events). On these small islands, the probability of extinction is increased due to limited variation in food resources (MacArthur & Wilson 1967). Unlike continental macaques, island-dwelling macaques are limited in expanding their range in response to food scarcity but could move into the intertidal habitats. These habitats could have favoured stone tool use, as macaques that could overcome the heavy defences of food items would have benefited by using the new food resources. Indeed, tool-using macaques on Piak Nam Yai show a population density of around 112 individual/km^2 (Gumert *et al.* 2013), which is unusually high for wild long-tailed macaques. This suggests that coastal foraging potentially might increase reproductive success and population carrying capacity, supporting the persistence of larger populations of macaques on small islands.

Opportunism & Disturbance Hypothesis. Macaques are opportunistic feeders and, as described above, could easily co-opt aspects of their typical behaviour and diet for exploiting foods in intertidal habitats. Other long-tailed macaques across Southeast Asia may therefore have been tool users, readily exploiting available coastal biomes when they were encountered. Coastal environments, however, are extremely important to Southeast Asian societies and, thus, Southeast Asia has some of the world's most densely human-populated coastal areas (Small & Nicholls 2003), with most coasts in the region inhabited and disturbed (Szabó & Amesbury 2011). Human disturbance can extinguish animal traditions (van Schaik 2002) and on Piak Nam Yai groups of long-tailed macaques that are affected by human disturbance have lower reproductive success and are easily driven from the open coasts by activity from humans and their familiars (e.g. domestic dogs; Gumert *et al.* 2013). Consequently, we can hypothesize that, historically, long-tailed macaques may have commonly used tools in intertidal habitats, but their lithic traditions have been extinguished by the impacts of human coastal activity.

Adaptation and Specialization Hypothesis. It is puzzling that current findings show stone tool-use to be restricted to only the Burmese subspecies of long-tailed macaque. We therefore question whether this subspecies has any specific adaptations that affect their ability to recognize and use tools, rather than only having undergone a cultural shift in behaviour, of which all macaques would be equally capable. It is even possible, that a historical cultural shift towards tool use driven by insular isolation, could have led to adaptations affecting tool-use development through directional selection across generations of tool users: favouring, for example, individuals with development biases that allowed tool use to be more easily acquired. Research on New Caledonian crows has suggested a genetic relationship between species action patterns and tool-using abilities (Kenward *et al.* 2012), and this case give us a basis by which to query whether any biological difference could be easing the development of stone tool use in Burmese long-tailed macaques.

Determining why long-tailed macaques use stone tools in the intertidal regions of Thailand and Myanmar will require accurate historical information. Perhaps with archaeological (Haslam 2012; Haslam *et al.* 2009) and genetic tools we may be able to uncover information on the history and selection of macaque stone tool use. We have already demonstrated an archaeological approach can accurately reconstruct how macaque tools were used in recent history (Haslam *et al.* 2013); however, using archaeological approaches on macaque tools ultimately will help determine the pre-historical occurrence of their tool use. Genetic research will contribute to our understanding of the speciation of Burmese long-tailed macaques and its possible influence on tool use. By knowing the historical context of macaque tool use, we can examine whether changes in sea levels and the separation of islands in the Mergui Archipelago played a selective role in its emergence. It will be important to continue investigating how intertidal habitats may have affected primate lithic culture and its origins.

Chapter 20

The Ecology of Chacma Baboon Foraging in the Marine Intertidal Zone of the Cape Peninsula, South Africa

Matthew C. Lewis and M. Justin O'Riain

Introduction

The Cape Peninsula is a narrow strip of land that juts into the Atlantic Ocean on the southwestern corner of South Africa. The peninsula's largely oligotrophic terrestrial habitats (Cowling *et al.* 1996) are juxtaposed with productive marine habitats that support abundant invertebrates within the intertidal zone of rocky shores (Robinson *et al.* 2005; Van Erkom Schurink & Griffiths 1991). The tissues of marine invertebrates represent protein-rich food sources that are harvested by maritime mammals (*sensu* Carlton & Hodder 2003), including primates (Galat-Luong & Galat 2013; Malaivijitnond *et al.* 2007; Son 2003; Chapters 7 and 18). It is, therefore, perhaps unsurprising that the Peninsula's chacma baboons (*Papio ursinus*) are known to feed on marine organisms in the intertidal zone (Hall 1962, 1963; Davidge 1978; Lewis and O'Riain 2017).

The marine organisms most frequently preyed upon by peninsula baboons are mussels (*Mytilus galloprovincialis*, Mytilidae) and limpets (*Cymbula* spp. and *Scutellastra* spp., both Patellidae), but crabs (*Cyclograpsus punctatus* Varunidae and *Guinusia chabrus* [= *Plagusia chabrus*], Plagusiidae), sealice (Isopoda) and even shark eggs are also exploited on occasion (Davidge 1978; Hall 1962, 1963; Peschak 2005; M.C. Lewis, unpublished data). Unlike Burmese long-tailed macaques (*Macaca fascicularis aurea*) in Thailand (Gumert & Malaivijitnond 2012; Gumert *et al.* 2009; Chapter 19), the chacma baboons of the peninsula do not use tools to process marine invertebrates with heavy shells. Rather, they pull mussels from the rocks by hand and then crack them using their molars, and lift limpets from the substrate using their teeth or bite off the top of the conical shell and then scoop out the gonad and digestive gland with their fingers (DeVore & Hall 1965; Peschak 2005; Matthew C. Lewis, pers. obs.).

Although marine foods may represent important sources of protein, they are often difficult to acquire as the intertidal zone is typically subject to high wave energy and large, rapid fluctuations in water levels (Bally 1987; Branch *et al.* 1998; Palmer 1995; Chapter 2). Foraging in this environment therefore presents unusual challenges to primates that wish to exploit these foods, and may even involve risk of death by drowning (Hall 1962). Marine foods might therefore only be exploited, or may be far more important dietary items, when nutritional requirements cannot easily be fulfilled by feeding exclusively on terrestrial foods (Chapter 18). We conducted a study of the behavioural ecology of peninsula chacma baboons to determine the importance of marine foods in their diet and to examine how environmental factors (e.g. tide heights and wave action) affect their exploitation of these foods.

Methods

We observed the Kanonkop (KK) troop of chacma baboons, the last remaining troop on the Cape Peninsula, South Africa (34°09′W, 18°23′S) that does not consume anthropogenic foods, from June 2009 to June 2010. A comprehensive description of the study site has been published elsewhere (Lewis and O'Riain 2017). For the purposes of data collection, we followed the study troop, consisting of 56 individuals (five adult males, 17 adult females, 27 juveniles and seven infants) on foot at a distance of ~20 m over 30 days (consecutive days where possible) during each season of a full calendar year. We collected general behavioural and foraging profile data following scan sampling protocols (Altmann 1974), and troop movement data using a hand-held GPS unit for a companion study (Lewis and O'Riain 2017). It was stipulated in our research permit that we should not habituate the baboons to close human presence (< 20 m), and we were therefore unable to conduct focal sampling.

When the troop fed on marine foods, we counted the number of baboons foraging in the intertidal zone each minute during that foraging bout. We inferred consumption of different food categories based on proportional composition of feeding observations in scan data collected for the companion study (Lewis and O'Riain 2017). To examine which environmental factors affected foraging in the intertidal zone, we modelled probability and intensity of marine foraging using two different sets of generalized additive models (GAM).

Results and Discussion

In total, we followed KK troop over 1330 h over five discrete observation periods. KK troop fed on marine foods across all four seasons, but the frequency (per day and per hour) of foraging in the intertidal zone varied with season (Figure 20.1). The baboons of this troop exploited marine foods most frequently in winter, and least frequently in summer. Although the frequency of foraging in the intertidal zone was high when considered at the above-mentioned temporal scales, marine foraging bouts were short (mean ± SD = 23.7 min ± 25.3 min, n = 224), and the number of animals feeding on marine foods during these bouts was relatively low (mean ± SD = 8.8 ± 8.4). Thus, marine foods comprised only a very small proportion of KK baboons' total diet (≤ 0.03 across all four seasons; Figure 20.2). This could be indicative of KK baboons being able to fulfil their nutritional requirements with terrestrial plant foods such as the bulbs of *Watsonia* sp. (Iridaceae), the seeds of *Leucodendron* spp. (Proteaceae) or the leaves of *Pterocelastrus tricuspidatus* (Celastraceae).

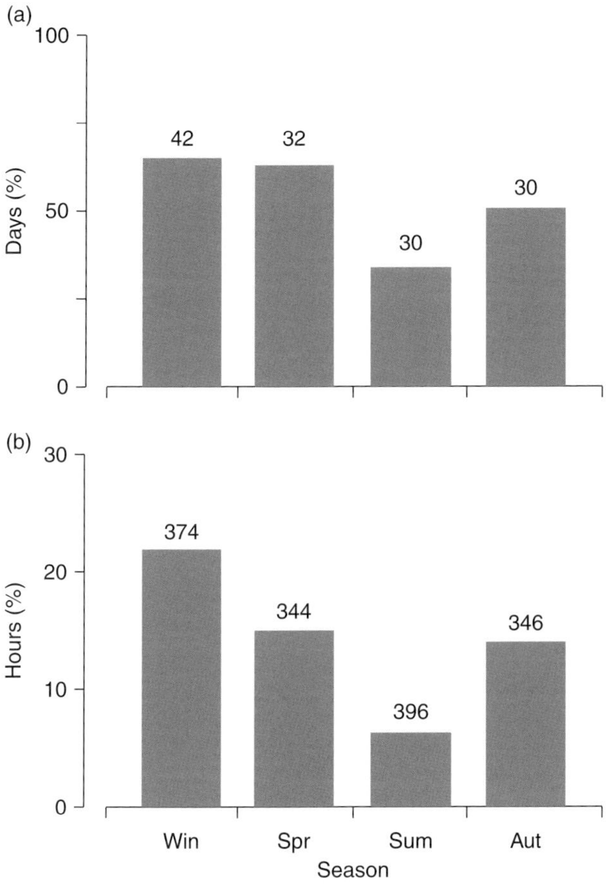

Figure 20.1 Percentages of days (a) and observation hours (b) in different seasons during which marine foraging occurred. Numbers above bars indicate sample sizes. Win = winter, Spr = spring, Sum = summer, Aut = autumn.

The home range occupied by the troop over the entire study period covered 45.3 km^2 in the southerly part of the Cape Peninsula, meaning that the troop's per capita area was 0.81 km^2/baboon (Lewis and O'Riain 2017). The troop's mean (± SE) daily path length was 6.04 (± 0.18) km (Lewis and O'Riain 2017). Given previously observed relationships between aspects of ranging behaviour and resource availability (Barton *et al.* 1992; Henzi *et al.* 1992; Johnson *et al.* 2015), comparison of these data with those for conspecific troops on the peninsula and elsewhere (Lewis and O'Riain 2017) suggests that food is relatively scarce in KK troop's home range.

The dominant activities in the activity budgets of KK troop were feeding (range of median proportions across seasons: 0.34–0.44), and walking (range of median proportions across seasons: 0.29–0.39) (Lewis and O'Riain 2017). Due to the role of locomotion in food acquisition, the sum of time allocated to these two activities has been defined as 'foraging time' (Bronikowski & Altmann 1996). KK troop's allocation to foraging time across the year was similar to that observed in some natural-foraging congenerics elsewhere (Bronikowski & Altmann 1996; Post 1981), but was higher (comparable to that of baboons in harsh, high altitude environments; Whiten *et al.* 1987) during the winter and summer months, indicating that meeting nutritional requirements is more difficult during these periods. This leads us to ask: if marine foods are rich in macronutrients, such as protein, why do the baboons not exploit these foods to a greater extent?

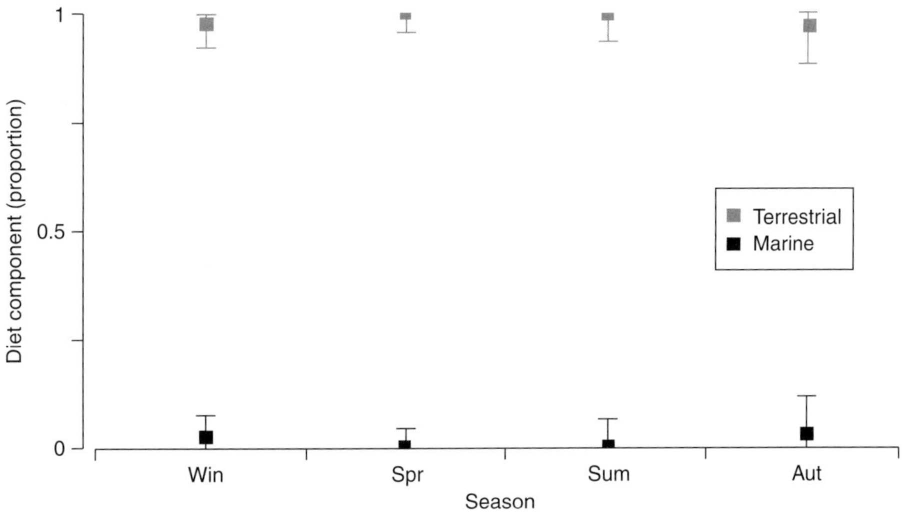

Figure 20.2 Proportions of KK troop's diet comprised of terrestrial and marine food items (derived from scan data) during different seasons. Points indicate median values, and error bars denote inter-quartile ranges. Win = winter, Spr = spring, Sum = summer, Aut = autumn.

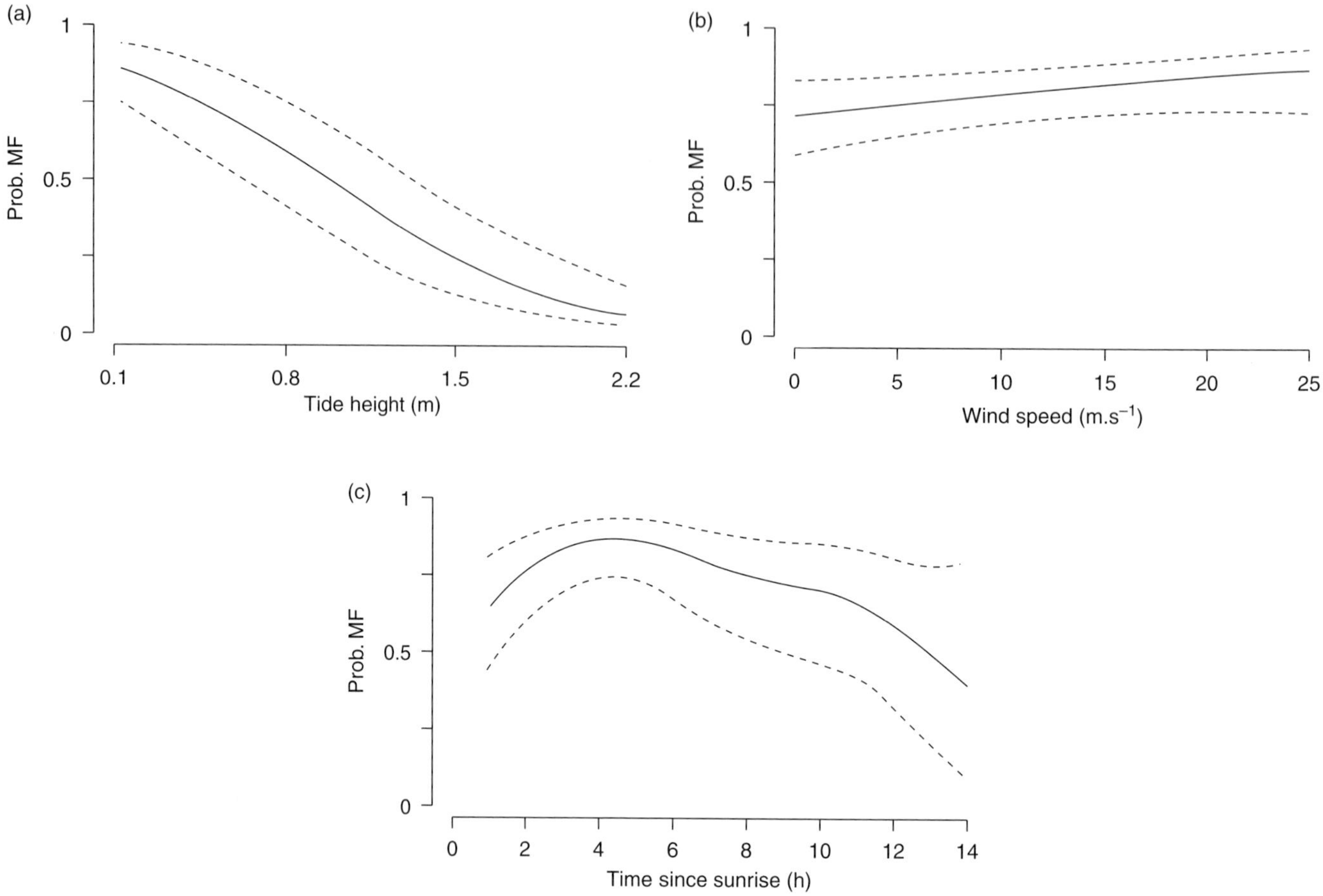

Figure 20.3 Relationships between predicted probabilities of marine foraging (across both coasts) and tide height (a), offshore wind speed (b) and time elapsed since sunrise (c) under optimal marine foraging conditions, based on results of a GAM. Dashed lines indicate 95% confidence intervals.

A possible explanation for KK baboons' tendency to ingest only small amounts of marine food is the difficulty implicit in accessing these foods due to frequent submersion. Indeed, tidal cycles affect foraging in the intertidal zone in other maritime mammals, including primates (Conradt 2000; Hansen *et al.* 2003; Nielsen 1991; Chapter 18). The models of probability and intensity of marine foraging during a given hour, confirmed that this is also true in the case of baboons foraging in the intertidal zone on the Cape Peninsula (Tables 20.1 and 20.2); the probability of marine foraging during a given hour declined with increasing tide and wave height (Figures 20.3 and 20.4). The effect of offshore wind speed on the probability of marine foraging was also significant (Figure 20.4), and is likely a result of the effect of offshore wind on wave action. Finally, probability of marine foraging increased initially and then decreased with increasing time since sunrise (Figures 20.3 and 20.4). This temporal pattern may be a function of the location of most of KK troop's sleeping sites, which allow visual assessment of intertidal conditions in the morning, immediately prior to commencement of foraging.

The predicted intensity of marine foraging declined with increasing tide height, and varied across seasons and between coasts (Figure 20.5). KK baboons exploited marine foods less in summer, when daily journeys were longer and the area used by the troop was larger, than during other seasons. This is surprising as long daily journeys and use of larger areas are associated with declines in food availability within primates' habitats (Henzi *et al.* 1992; Ganas & Robbins 2005; Hanya *et al.* 2005; Riley 2008). The above finding therefore suggests that the use of marine foods is not directly linked to general food availability, but may rather be influenced by as yet unknown factors. These could include nutritional reward (mass of soft tissue) on offer, or the concentration of secondary metabolites (e.g. alkaloids), both of which are known to vary through time in marine invertebrates (López-Legentil *et al.* 2007; Sacristán-Soriano *et al.* 2012). Alkaloids are found in the tissues of the mussels and limpets consumed by peninsula baboons (Lewis, unpublished data) and some alkaloids have been shown to deter fish predators of ascidians and sponges (Lindquist *et al.* 1992; Thoms *et al.* 2004; Thoms and Schupp 2008). Given that baboons tend to favour terrestrial foods that are low in alkaloids (Hamilton *et al.* 1978; Whiten *et al.* 1991), it seems likely that alkaloids in marine invertebrate tissues might also act as feeding deterrents in this context.

In summary, it is clear that marine foraging by baboons is, at least in part, curtailed by tidal fluctuations and the additive

Table 20.1 Summary statistics for predictor variables assessed for significance in GAMs fitted to probabilities of marine foraging.[1]

Both coasts			West coast		
Predictor variable	χ^2	*p*-value	**Predictor variable**	χ^2	*p*-value
te(tide height)	**75.843**	< 0.001	te(tide height)	**60.386**	< 0.001
te(wind speed)	3.941	0.047	te(swell height)	**8.588**	< 0.01
te(time since SR)	**29.305**	< 0.001	te(Wind speed)	0.000	0.999
Wind direction	2.698	0.100	te(time since SR)	**13.966**	< 0.01
te(WSP: Off)	**5.862**	< 0.05	Wind direction	1.892	0.169
te(WSP: On)	0.056	0.813	te(WSP: Off)	0.085	0.771
			te(WSP: On)	3.623	0.298
			ti(TDH × SWH)	**5.299**	< 0.05

Notes: te = tensor spline smooth; Time since SR = time (in hours) since sunrise; WSP = wind speed; TDH = tide height; SWH = swell height; Off = offshore wind; On = on-shore wind; significant values ($p < 0.05$) are in boldface.

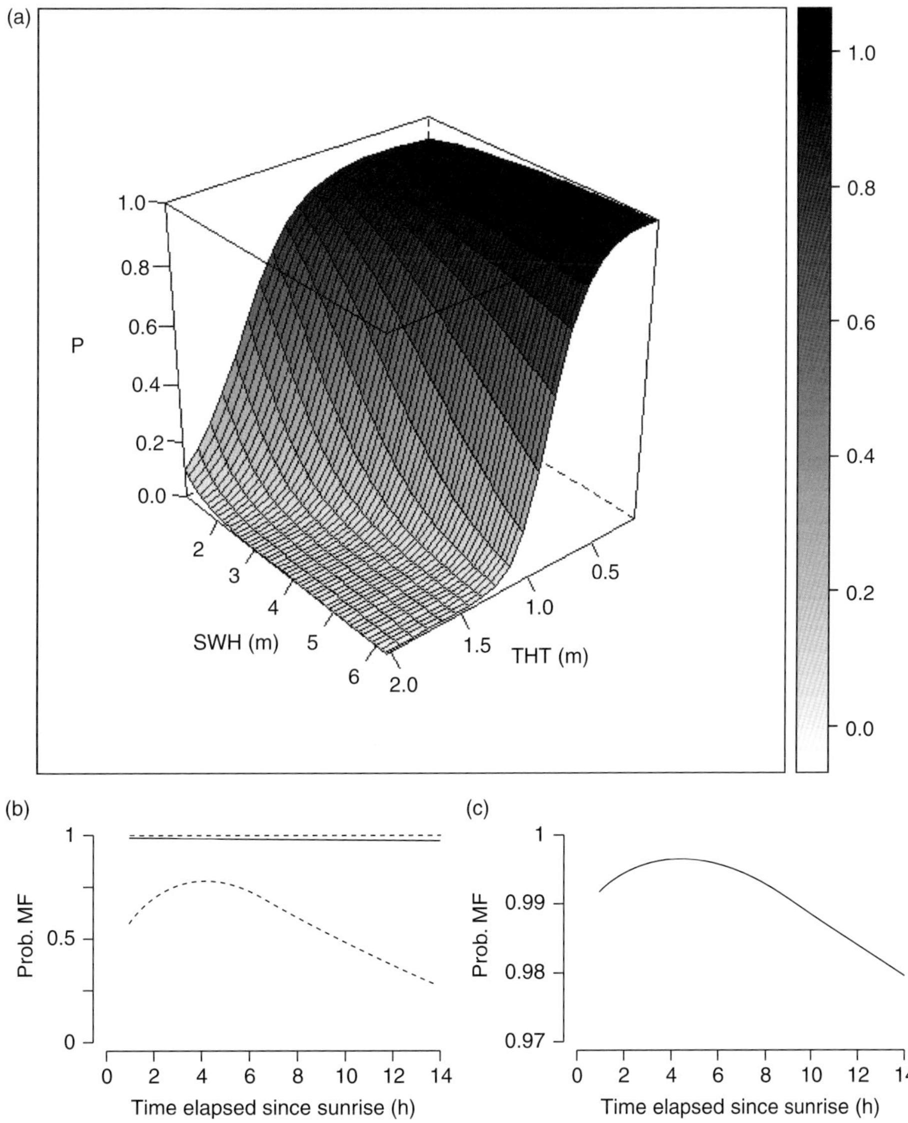

Figure 20.4 Relationships between predicted probabilities of marine foraging (on the west coast) and tide- and wave height (a), and time elapsed since sunrise showing the full range of probabilities (b) and a smaller range of probabilities (c), based on results of a GAM. P = probability, SWH = swell height, THT = tide height; dashed lines indicate 95% confidence intervals; note difference in scale on vertical axes.

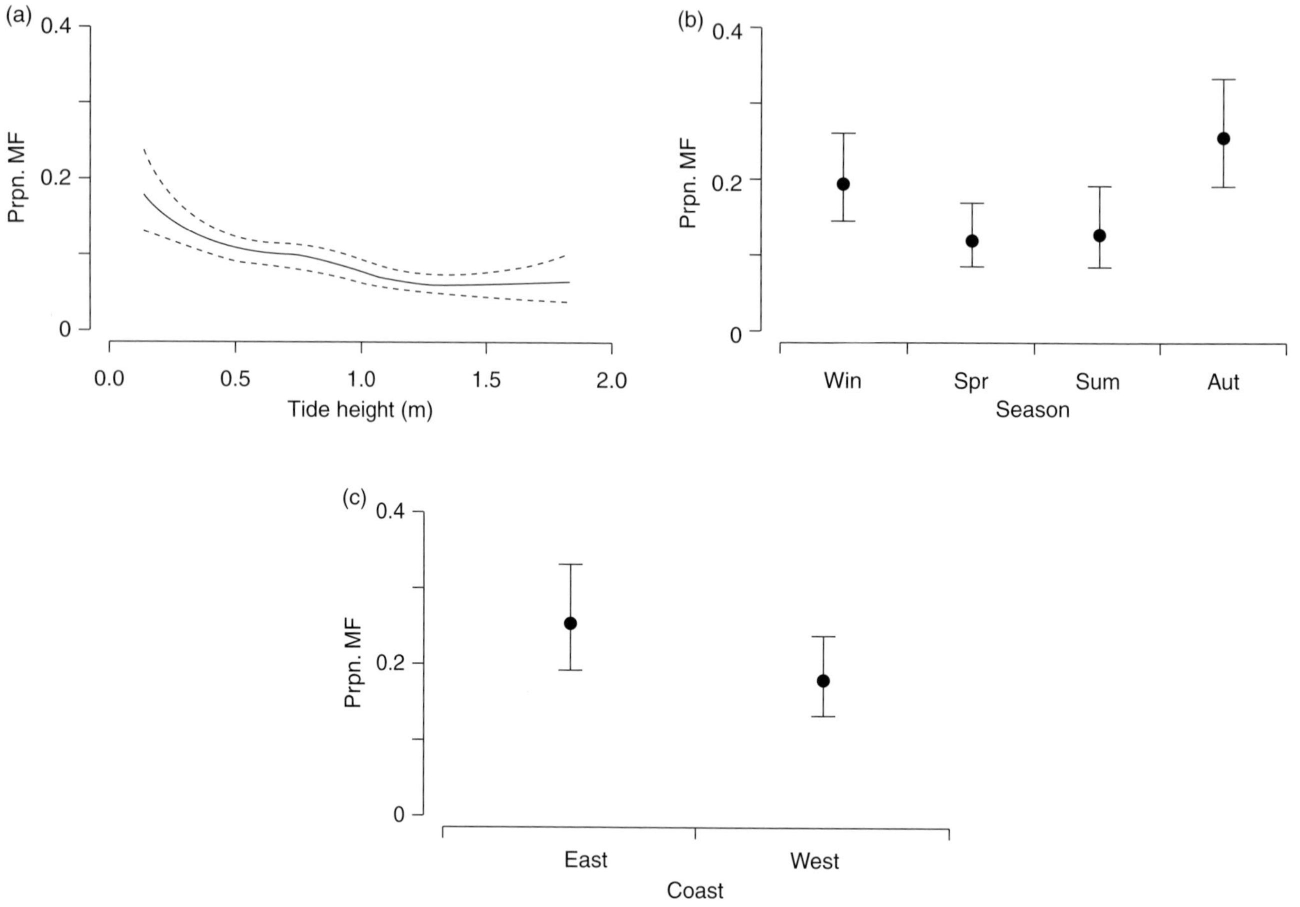

Figure 20.5 Relationships between predicted average proportion of the troop marine foraging during a given hour (on both coasts) and tide height (a), season (b) and coast (c), based on results of a GAM. Dashed lines (in a) and error bars (in b–c) indicate 95% confidence intervals.

Table 20.2 Summary statistics for predictor variables assessed for significance in GAMs fitted to hourly average proportions of the troop marine foraging.[1]

	Both coasts	
Predictor variable	χ^2	***p*-value**
te(Tide height)	**45.85**	< 0.001
Season	**72.50**	< 0.001
Coast	**26.21**	< 0.01

Notes: te = tensor spline smooth; significant values ($p < 0.05$) are in boldface.

effects of wave height and wind speed. It is further possible that other as yet unknown biotic factors including food quality and individual attributes (e.g. skill and strength required to harvest intertidal organisms) may influence both seasonal and individual variation in the frequency and intensity of marine foraging. These remain important areas for future research in further exploring the apparent paradox of why baboons living in a nutrient-poor environment consume such small quantities of an abundant protein-rich resource which they have already learned to exploit.

Chapter 21

Primates and Flooded Forest in the Colombian Llanos

Xyomara Carretero-Pinzon and Thomas R. Defler

Introduction

Tropical wetlands are dynamic and complex systems with a range of species compositions and ecologies, imposing challenges to the primates living in them. These challenges include reduced mobility, reduced botanical diversity (Chapters 2, 4 and 6), and the presence of aquatic predators. Despite these challenges, flooded forests and mangroves are important for primate species including the *igapó*, Pantanal and Chaco fooded forest in three Neotropics (Chapters 7, 8, 22, 27 and 32). Riverine and gallery forests are also important for range-restricted species such as pygmy marmosets (Defler 2010) and the Tana River red colobus (*Piliocolobus rufomitratus*) in Kenya (Wieczkowski 2004; Chapter 31).

The Colombian *llanos* is one such region. Highly diverse, it is composed of natural savannas (seasonally flooded and non-flooded), *gallery forest* (also often seasonally flooded) and never-flooded lowland rain forest. The region's range of geomorphology and geographic and vegetation formations was used by Lasso *et al.* (2010) to define biogeographic regions within the *llanos* (see Figure 21.1a; Table 21.1). The *llanos* differ from other regions of Colombia and within the Orinoco River basin in their flood dynamics and soils, which influence forest productivity, diversity and resource availability (especially fruits, leaves and insects) for primates. Consequently, flood dynamics and soils shape both vegetation form, and the way it is used by primates (Lasso *et al.* 2010, 2011). Some areas may only be flooded for hours or days, others for several months. Based on seasonal patterns, and the duration and the intensity of seasonal floods, flooded forest in the Colombian *llanos* has been classified as riverine, gallery forest, alluvial forest, low river terrace, alluvial terrace forest, foothill forest and *Mauritia flexuosa* swamps (Lasso *et al.* 2013; Mora-Fernandez *et al.* 2011; Pinzon-Perez *et al.* 2011).

Orinoquia, known in Colombia as the *llanos orientales*, comprises all tributary rivers and streams of the Orinoco River in Colombia (388 101 km^2), and continues on as the Venezuelan *llanos* to the north and east in Venezuela (644 423 km^2) as a single ecoregion named the Orinoquian Llanos, an area of some 1 032 524 km^2 (Dominguez 1998). Swampy areas dominated by *Mauritia flexuosa* (Arecaceae) form one of the most typical forest types of the region. Such swamps are characterized by a high level of subsoil water and poor drainage due to impermeability of underlying soils (Lasso *et al.* 2013). These palms often form pure stands at the headwaters of rivers and creeks and may be surrounded by grassy savannas, except where the palms have contact with the multi-species gallery forest. Known in Spanish-speaking South America as *morichales* or *cananguchales* (and *buritizais* in Brazil), they are important for the faunal diversity of the Orioquia region, since many mammals, including primates, eat the pulp surrounding the plum-sized fruits. Local primate species recorded doing this include: red howlers (*Alouatta seniculus*), black-capped capuchins (*Sapajus apella* (= *Cebus apella*)), white-fronted capuchins (*Cebus albifrons*), Colombian squirrel monkeys (*Saimiri cassiquiarensis albigena*), collared titi (*Cheracebus lugens*) and Brumback's night monkey (*Aotus brumbacki*) (Carretero-Pinzon 2008; Defler 2010; Trujillo-Gonzalez *et al.* 2011; Vincelli 1981).

The Orinoquia region is important for a wide variety of vertebrates, including fish (658 species, 56 endemic to Colombia), amphibians and reptiles (266 amphibians and 290 reptiles in Colombia and Venezuela, 86 and 26 endemics, respectively), birds (761 species in Colombia, 82 endemic to both countries) and 318 species of mammals (of which 6% are IUCN Red-Listed) (Acevedo-Charry *et al.* 2014; Defler & Rodríguez 1998; Lasso *et al.* 2010). Although not as diverse as in the Amazonian region, the diversity of primates in the Orinoquia region in Colombia shows interesting adaptations and patterns. The highest species diversities are found towards the transitional Amazon-Orinoquia bioregion (nine species), and the Inirida fluvial border between Colombia and Venezuela (nine species), while around the southwest part of La Macarena bioregion, ten sympatric species are known (Defler 2010; Figure 21.1a).

Because of relatively little mammal survey effort in these areas, there is a lack of clearly defined distribution limits for some of the region's primate species (Defler 2010; Lasso *et al.* 2010). For example, the eastern limits of both the dusky titi (*Plecturocebus ornatus*) and Brumback's owl monkeys (*A. brumbacki*) are poorly known, although defining them is of conservation importance (Defler 2010). For some Llanos primate species, distributional limits appear to be determined by landscape constraints, such as lower forest and higher savanna cover compared to the more continuous rainforest in Amazonia. These more open vegetation forms may present challenges to some primate species due to a reduction of canopy structure and plant diversity and concomitant reduction in resource access and availability (Stevenson & Aldana

Table 21.1 Description of Orinoquian Bioregions (Colombia only), based on their geomorphology, hydrology, soils, landscape and vegetation cover defined by Lasso *et al.* (2011).

Bioregion	Description	Vegetation
Andean	Mountain region with altitudes from 1000 to 4000 m in the east part of the Eastern chain of Colombian Andes. Rivers in this bioregion transport high amounts of sediments. Precipitation reaches up to 5000 mm in some parts.	*Páramo* (high-altitude grassland), cloud forest and sub-andean forest affected by human activities.
Piedmont	Transition from Andes formation towards the Llanos. Altitudes 500–1000 m. High fertility and precipitation.	Gallery forest, lowland forest, urban areas, agro-ecosystems, plantations.
La Macarena	Share features of surrounding bioregions with elements from all of them.	Similar vegetation to Orinoquia – Amazon transition area in the lower lands with savanna areas and floristic elements of tepui (table-top mountains). In the upper parts, vegetation is similar to the Andes region.
Llanos	Alluvial and plains with < 200 m elevation. Rivers with high sediment loads.	Savannas, flooded plains, gallery forest associated with water courses.
Orinoquia – Amazon transition Area	High terraces and lowland forest south of Vichada river with some segments of savannas. Similar to Amazonian forest.	Flooded and unflooded lowland forest.
Guyana	Guyana shield formation from the Precambrian period. Precipitation between 1100 and 4500 mm. Temperature between 28 and 36°C in lowlands and up to 0°C on tepuis. Oligotrophic rivers with low conductivity and low sediment loads.	Isolated hills and tepui formations with plain extensions with savannas and lowland forest.

2008). This is reflected in a reduction in primate diversity in the Llanos subregion of Colombia (Figure 21.1a), in comparison, for example, with the La Macarena subregion (Defler 2010).

Long-term primate studies in the Orinoquian Region of Colombia have been carried out in four main areas, including three bioregions of the Orinoquian region (see Figure 21.1a and Table 21.1 for bioregions descriptions): one in La Macarena (Tinigua National Park), two in the Llanos bioregion (Tuparro National Park and San Martin zone), and one in the Orinoquia-Amazon transition bioregion (San Jose del Guaviare zone; Zarate & Stevenson 2014; Figure 21.1b). This chapter provides an overview of these long-term studies in the Colombian *llanos* (except for the one in the Orinoquia–Amazon transition bioregion) as well as unpublished data of habitat use for the Colombian squirrel monkeys (*Saimiri cassiquiarensis albigena* = *S. s. albigena*) in La Macarena and Llanos bioregions. The studies reviewed here focus on primate use of flooded habitats and include published information on habitat use of flooded habitats by various primate species. Research gaps are identified and future directions for primate studies in regionally flooded habitats are suggested.

Methods

Study Areas

The Orinoco region in Colombia has been a frontier region since the sixteenth century and it continues to be so today (Rausch 1994, 1999, 2013). The main drivers of environmental change of this developing frontier are the continuous migration of people from all parts of the country, plus petrol exploitation, agro-commodities (palm oil plantations that are replacing savanna, pastures and other land forms), livestock (a regional land-use driver since the earliest colonization), illegal crops and infrastructure, especially near to the Andes (piedmont, La Macarena and Orinoquia–Amazon transition area; Figure 21.1a) (Ecopetrol 2015; Etter *et al.* 2006; Fedepalma 2014; Lopez-Hernández *et al.* 2005; Rausch 1994, 1999). This colonization frontier has increased habitat loss and fragmentation which has prompted changes in the flood dynamics of the bioregions reviewed in this chapter (Lasso *et al.* 2013).

La Macarena Bioregion: Tinigua National Park

The Centro de Estudios Ecológicos La Macarena (CIEM) is located at 2°40′N and 74°10′W, on the right bank of the Duda river in Tinigua National Park (Figure 21.2a). There are two seasons: dry season (December–March with maximum rainfall of 100 mm) and wet season (April–November with maximum rainfall of 3000 mm) (Kimura *et al.* 1994). CIEM was founded as a collaborative effort between the Universidad de Los Andes, Bogotá and Miyagi University, Sendai, Japan. Tinigua National Park forest types have been characterized by Hirabuki (1990) and Stevenson *et al.* (1994, 2004), who described three main forest types: (1) mature forest (51%), with trees 20–25 m in height, a continuous canopy with emergent trees reaching 30 m and located on hill ridges; (2) open-degraded forest (40%, including 1% classified as secondary), with few trees of 20–25 m in height, a thick understory or undergrowth and abundant bamboo and vines, located on erosion fronts, small valleys and brooklets, with a discontinuous canopy and (3) flooded forest (8%, including 2% riparian forest), with discontinuous canopy dominated by *Ficus* spp. (Moraceae), *Inga* spp. (Fabaceae) and *Cecropia* spp. (Urticaceae), located on river margin floodplain (Stevenson *et al.* 2004). Primate studies occurred from 1986 to 2002, when ecological and behavioural studies of seven species were undertaken: *A. seniculus, Ateles belzebuth, Lagothrix*

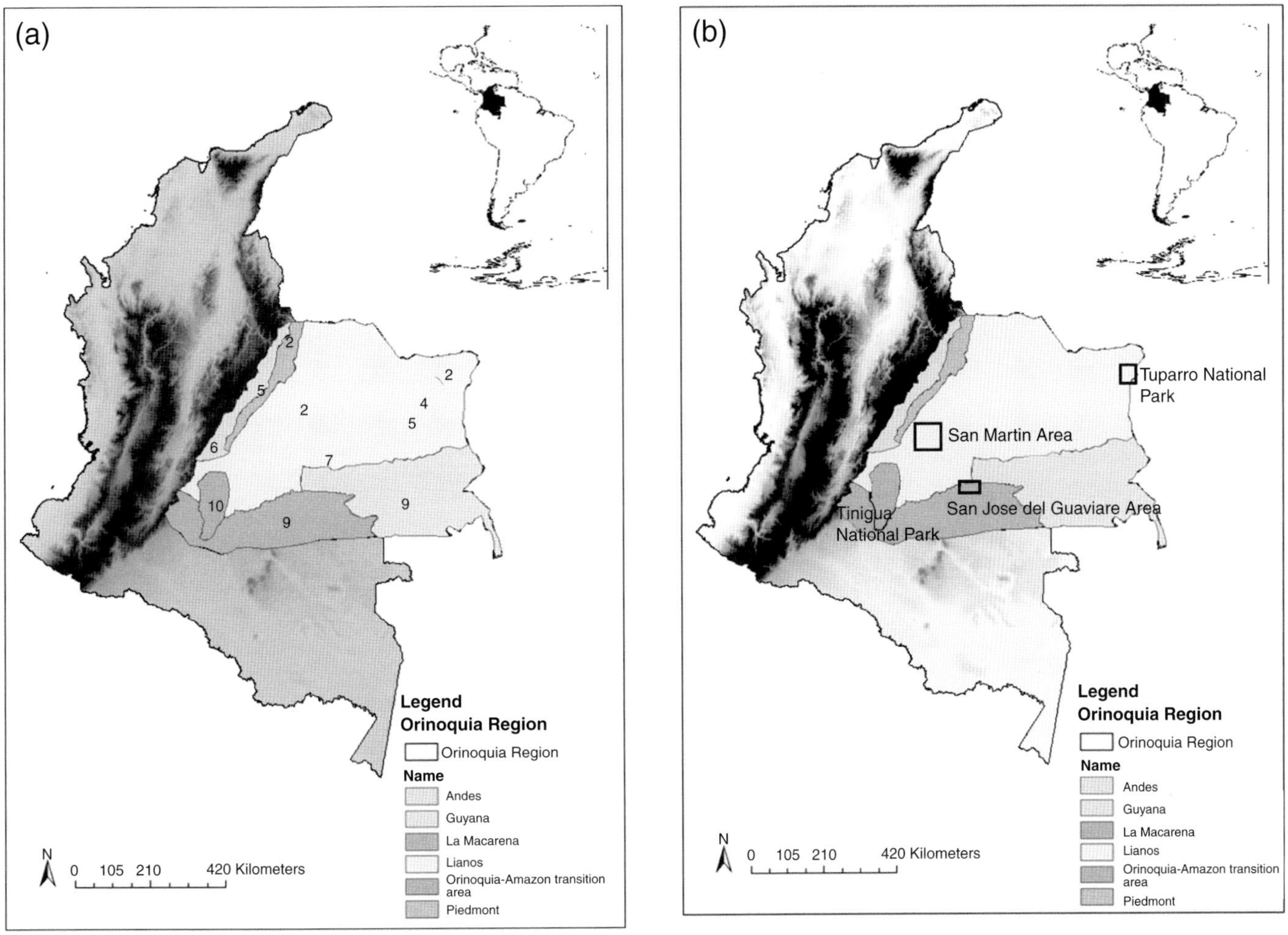

Figure 21.1 (a) Biogeographic regions of the Colombian Orinoquia, as defined by Lasso *et al.* (2010). The numbers represent primate species richness based on Defler (2010). (b) Sites of long-term primate studies in the Orinoquia region of Colombia. (A black and white version of this figure appears in some formats. For the colour version, please refer to the plate section.)

lagothricha, *P. ornatus*, *S. apella*, *S. c. albigena* and *A. brumbacki*. Well-defined trail systems of 6 km^2, which cover all three forest types, have been used in all the published and unpublished data presented here (Izawa & Tokuda 1988).

Information on primate species habitat use at this site come from two sources: (1) published studies in which habitat use was estimated from focal and scan sampling, and *ad libitum* observations (Ahumada *et al.* 1998; Polanco-Ochoa & Cadena 1993; Stevenson *et al.* 1994; Stevenson 2006; Yoneda 1990; Table 21.2) and (2) unpublished data on the Colombian squirrel monkey (*S. c. albigena* (= *S. s. albigena*)) collected by Carretero-Pinzón from January to May 1999 (methodology described below). Published studies were located in Google Scholar using 'Tinigua', 'habitat use' and 'primates' as key words, yielding ten papers. A complementary search was made on available journal volumes produced by Los Andes and Miyagi universities from 1988 to 2004, affording an additional five titles (Table 21.2).

Data (unpublished) on one group of *S.c. albigena* were collected during group follows from January to May of 1999 (total sampling: 369 hours). The group's home range included two study camps with a well-defined trail system and their range overlapped with a well-known group of *S. apella*. Body scars and natural variation in pelage characters allowed individual identification of *S. c. albigena* (see descriptions in Mitchell 1990). Observations were made from 6:00–18:00, using scan sampling (Altmann 1974), with data collected every 5 minutes (1 minute of continuous sampling followed by 4 minutes of no collection). Data collected included food type (arthropods and fruits, the major food types of *Saimiri*), and habitat type (Carretero-Pinzón 2000). The total number of scan samples from each of the three habitat types (mature, degraded and flooded) was compared using a Kruskal–Wallis test, and pairwise differences in monthly habitat use were investigated with a multiple range test (Zar 1996). Statistical analysis was conducted in STATGRAPHICS PLUS 2.0.

Los Llanos Bioregion: Tuparro National Park

Tuparro National Park (548 000 ha) is located in the extreme east of Colombia (Figure 21.1b). Land cover is some 80% *llanos*, natural savannas, and 20% forest (mostly riparian or gallery types) (Defler 1985). Average precipitation varies from 2477 mm in the park's western portion to 2939 mm some 150 km away in its easternmost extent. Information presented here

Table 21.2 Published studies at Tinigua National Park of primate use of flooded habitats.

Species	Method	References
Ateles belzebuth, Alouatta seniculus, Lagothrix lagothricha, Plecturocebus ornatus (=*Callicebus cupreus ornatus*), *Saimiri cassiquiarensis albigena* (=*S. sciureus albigena*), *Sapajus apella* (=*Cebus apella*)	Census observations	Yoneda 1990
Plecturocebus ornatus (= *C. ornatus* = *C. cupreus ornatus*)[a]	Focal animal and *ad libitum* sampling	Polanco-Ochoa & Cadena 1993
Ateles belzebuth	Instantaneous and focal animal sampling	Ahumada *et al.* 1998
Lagothrix lagothricha	Instantaneous and continuous sampling of focal animal	Stevenson *et al.* 1994, 2006

[a] Taxonomy used here follows Byrne *et al.* (2016). Classification of *Saimiri* is also supported by genetic data from Ruiz-Garcia *et al.* (2014) and Lynch Alfaro *et al.* (2014).

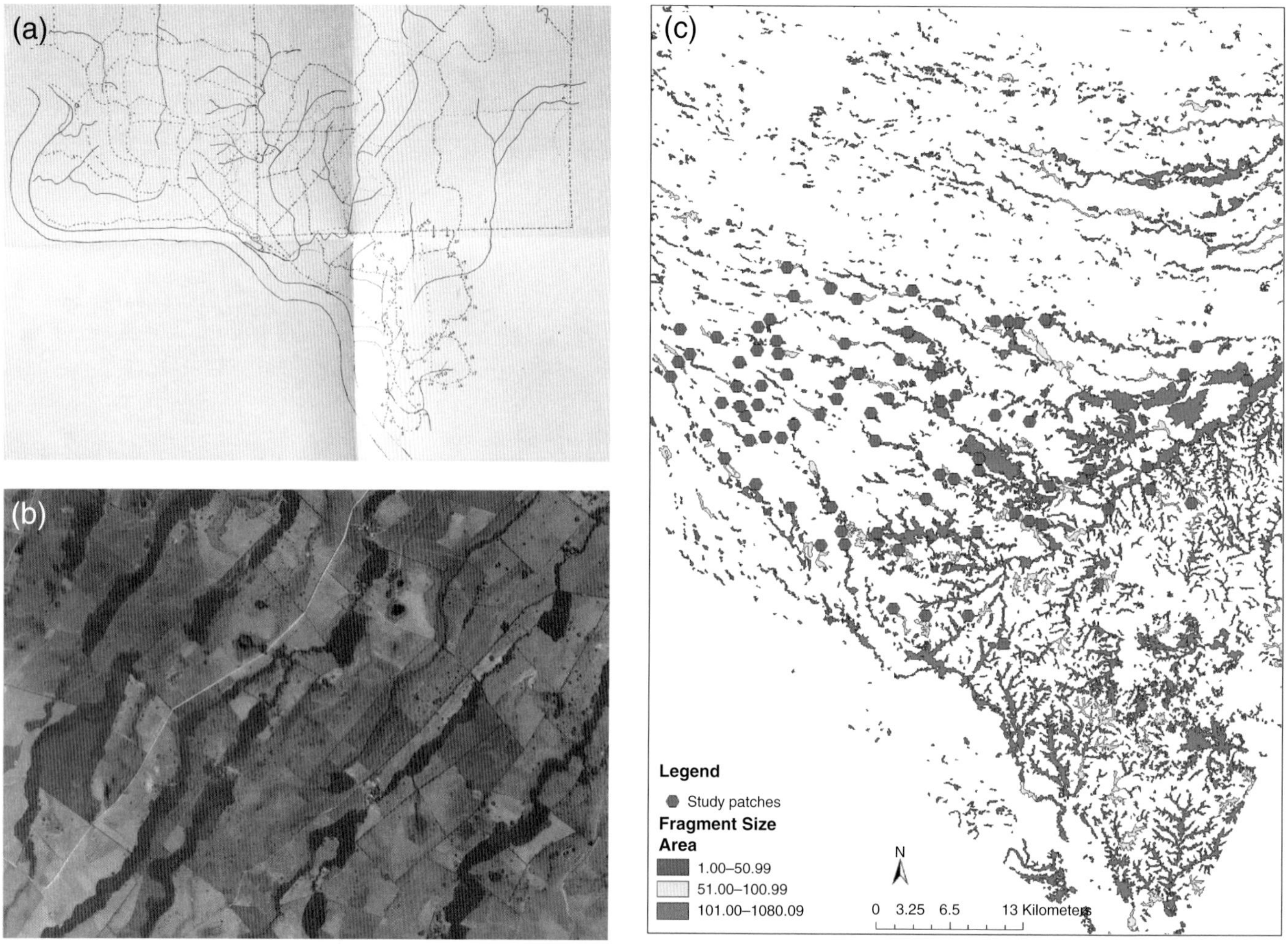

Figure 21.2 (a) Study site at Tinigua National Park (darker line represents the home range of the study group of Colombian squirrel monkeys [*S. c. albigena*], Carretero-Pinzón 2000); Study site at San Martin area: (b) Fragments used by the study group of *S. c. albigena* (red arrows show natural hedgerows (i.e. living fences) used by the study group); (c) Forest fragments surveyed from 2004 to 2014. (A black and white version of this figure appears in some formats. For the colour version, please refer to the plate section.)

is from observations made throughout the park via motorbike (1976–1981) and canoe on all navigable rivers (1981–1983), when all observed groups of *S. apella* and *C. a. albifrons* were mapped (Defler 1985).

In this study, though overall there were intergradations and clinal variation, the habitat along the rivers used by these two capuchin species was classified into five categories: (1) simple early successional communities of a few plant species (notably the regionally common pioneer species *Campsiandra comosa*, Caesalpinaceae). Found on the inside of river bends on loose, sandy substrate, these small stands are generally < 10 m tall; (2) low forests, usually found on a substrate of clay or sand in abandoned river beds parallel to the river course, with several plant species up to 10–15 m tall, especially spiny palms (*Bactris bidentula* and *Bactris brongniartii*; Galeano & Bernal 2010); (3) forests approximately 15 m high with underbrush more developed and plant diversity greater than habitat type 2, and containing greater numbers of many spiny *Bactris* palms (same species, but more dense); (4) well-developed forests 15–20 m high, containing underbrush, sometimes swampy; and (5) forests found on the highest land, on loamy or clay soil, with vegetation reaching heights of 20–25 m, *Attalea regia* and *Oenocarpus bataua* palms often present, and *Bactris* spp. rare or absent. This was the most plant species diverse of the five types of semi-deciduous forests observed in the park equivalent to a typical dense gallery forest.

No attempts were made to characterize forest or habitat types that were not near the rivers and that could not be seen from the canoe. The above subcategorized cline of vegetation types is based primarily on inundation duration. Based on more than five years of residence across all seasons, the first habitat type was the first to become flooded and the last to be drained during the year, whereas habitat type 4 flooded only during high water. Habitat type 5 never flooded (Defler 1985).

In addition to the capuchins, other primate species present in El Tuparro are *C. lugens* (not much attracted to flooded forest), *A. brumbacki* (observed both in flooded forest and unflooded forest) and *A. seniculus* (very evident in flooded forest during early leaf flushing). Altitudes in southern Vichada department range from 80 to 215 m, with the lower altitudes covered with savannas and rivers, while higher altitudes represent the granitic 'Inselbergs' close to the Orinoco River.

San Martín

This area is located near the town of San Martín, Meta Department, in the Colombian Llanos Bioregion. Elevation is around 350 m (Figure 21.2b). The regional wet season is April–November (average rainfall of 1777 mm), with a December–March dry season (average rainfall of 357 mm). The annual average temperature is 26°C (Carretero-Pinzón 2008). The original land cover was a mosaic of lowland forest (including *Mauritia flexuosa* swamps), gallery forest and natural savannas. Now, fragments of gallery forests, lowland forest and *Mauritia flexuosa* swamps, ranging from 0.2–1080 ha in area, and natural savannas are surrounded by pastures planted with exotic grasses and used for livestock, small-scale agriculture and palm oil plantations. Five sympatric primates have been recorded in the area: *S. apella, A. seniculus, S. c. albigena, C. ornatus* and *A. brumbacki* (Carretero-Pinzón 2013). Regional primate data come from two datasets collected by Carretero-Pinzón: (1) ecological and behavioural observations on a *S. c. albigena* group (August 2005–January 2007: methodology described below); and (2) *Ad libitum* observations made from 2004 to 2014 in 100 habitat fragments including *M. flexuosa* swamp fragments (n = 20, size range: 1.7–545.5 ha (SD: 116.6)), lowland rainforest fragments (n = 29, size range: 2.2–1080 ha (SD:229.13)), and gallery forest fragments (n = 51, range size: 0.2–322.1 ha (SD: 146.68)).

A group of 43 individuals of *S. c. albigena* was observed at two farms, Santa Rosa and Arrayanes (3°3′30′N , 73°35′40′W). The combined 3098 ha in which those farms are located, forms- a fragmented landscape of 3098 ha of which < 60% of the original vegetation cover remains (10–60% of remaining forest is classified as a fragmented landscape by McIrtyre & Hobbs 1999). This landscape is composed of small (1–10 ha), medium (10–100 ha), and large (100–1000 ha) fragments, most of which are connected by natural hedgerows and surrounded by pasture (Figures 21.2b, 21.3). For the purpose of this study we distinguished three different habitat types: (1) forest edge; (2) natural hedgerows (lines of unplanted native trees left standing during forest clearance to divide pastures; Carretero-Pinzón *et al.* 2010); and (3) forest fragments. Statistical methods were consistent with the La Macarena study. Species richness for plants DBH > 10 cm in 50 m × 10 m botanical sampling plots in study fragments ranged from 19 to 24 species (Carretero-Pinzón, unpublished data).

Table 21.3 Forest fragment classification based on vegetation height and composition.

Forest type	Vegetation features
Lowland rainforest fragments	Height: 15–25 m. High diversity of plants and understory vegetation (≥ 60 plant species). Presence of emergent trees up to 30 m height. Include patches flooded in high water of small rivers.
Gallery forest fragments	Height: variable but mostly with trees of 10–20 m. Variable diversity with understory vegetation of multiple strata. High presence of pioneer plants. Mostly around water courses.
Mauritia flexuosa swamp fragments	Height: 10–20 m. Little diversity of plants with dominance of *Mauritia flexuosa* palm and little or no understory vegetation (≤ 40 plant species). Surrounding water courses (lentic and lotic waters). High proportion of organic matter in decomposition.

Data presented here are based on 1113 h of follows from August 2005–January 2007 using the same sampling protocol as the La Macarena study. Annually, the two main forest fragments in the group's home range were briefly flooded. Natural hedgerows were used by the study group to move between fragments and to access two smaller patches (0.5 and 2 ha), one of which was frequently flooded during the wet season (Carretero-Pinzón, pers. obs.).

In an area located from 3°45′N, 73°43′W to 3°38′N, 73°14′W and 3°26′N and from 73°26′W to 3°32′N 73°12′W, 100 forest fragments varying from 0.5 to 1080 ha were surveyed for primates (Figure 21.2c) between 2004 and 2014. These were divided into *M. flexuosa* swamp, lowland rainforest and gallery forest (Table 21.3). The majority (88%) of the observations reported here come from census surveys, mainly during the day (90%), but including some nocturnal censuses (10%).

Results

La Macarena Bioregion: Tinigua National Park

Literature Review

Published studies of primate habitat use of flooded habitat in Tinigua National Park have focused on three

Table 21.4 Habitat use of six primate species in Tinigua National Park based on Yoneda (1990).

Species	Habitat preferred (Yoneda 1990)	Habitat preferred (this study)
Ateles belzebuth	Closed forest	Mature forest
Lagothrix lagothricha	Closed forest	Mature forest
Alouatta seniculus	Unclosed forest, followed by riparian forest	Degraded forest, followed by flooded forest
Callicebus ornatus	'Shabby forest'[a]	Degraded forest
Saimiri cassiquiarensis albigena	'Shabby forest'[a]	Degraded forest
Sapajus apella	Closed and unclosed forest	Degraded and mature forest

[a] For more detailed description of this type of degraded forest, see Polanco-Ochoa and Cadena 1993.

species (*P. ornatus*: Polanco-Ochoa & Cadena 1993; *A. belzebuth*: Ahumada *et al.* 1998; and *L. lagothricha*: Stevenson 2006). Previously, Yoneda (1990) documented habitat use of seven primate species based on short-term studies (Table 21.4). Polanco-Ochoa and Cadena (1993) provided more detailed information on *P. ornatus* habitat use, noting a preference for degraded forest (vine thicket and intermediate forests), and the importance of flooded forest for feeding by the focal group especially during the rainy season.

For *A. belzebuth*, Ahumada *et al.* (1998) found that flooded forests are especially important during the low fruit production season. The importance of flooded forest during this season lies in their higher fruit production (compared with non-flooded forests (mature forest in Ahumada *et al.* 1998)), explained by the presence of large fig trees (*Ficus* spp.; Ahumada *et al.* 1998). In contrast, Stevenson *et al.* (1994) and Stevenson (2006) reported a single group of *L. lagothricha* to consistently use mature (non-flooded) forest over three years, while the proportion of time they spent in flooded forest increased during the rainy season as a consequence of a similar or higher average value of fruit production reached in this forest type relative to mature forest (non-flooded forest, Stevenson *et al.* 1994; Stevenson 2006).

Colombian Squirrel Monkey (*S. c. albigena*) Use of Flooded Habitats

Forest types were used differently throughout the study period, with degraded (42%, SD: 216.8) and flooded habitat (40%, SD: 290.8) used more than mature forest (18%, SD: 128; H: 27.32; $p < 0.05$). Flooded forests were used by the study group differently between months with a higher use of flooded forests at the beginning of the wet season (April–May; Figure 21.3), the months of increased fruit production (Stevenson *et al.* 2004). Flooded forest was not used in January and mature forest was not used in May by the study group (Figure 21.3).

Los Llanos Bioregion: Tuparro National Park

Cebus a. albifrons often used flooded forests (habitat types 1 and 2), but these were seldom used by the *S. apella* groups (Defler 1985). In an analysis of 70 contacts with these two species of capuchins, Defler (1985) found that preferences for the five differentiated riverside habitat types differed significantly between the two species ($F = 28.48$, $df = 1.69$, $p < 0.001$; Figure 21.4).

Ad libitum observations also suggested concentrated use of flooded forest by *A. seniculus*, particularly when new leaf flushing was widespread (Defler 1985), with *A. seniculus* densities appearing to increase greatly during the early flooded season (Defler, pers. obs.). Systematic observations inclusive of counts made throughout the year are needed to verify this, but strong support comes from another study site in Amazonian Colombia (Caparu Research Station, Apaporis River, Vaupes), where multi-year censuses carried out along a lake edge showed that *Alouatta* densities and activity greatly increased in flooded forest during the high water (Defler 2013). Low-water censuses often detected no *Alouatta* groups for months, while high-water censuses often detected five to six groups during the 10 km canoe census (Defler 2013).

San Martín

Habitat Use Observations of a *S. c. albigena* Group

Natural hedgerows were used by the study group to access three small forest fragments, one of which flooded briefly during the rainy season. This habitat is not planted or anthropogenic, but composed of lines of native trees, with up to 56 native plant species (in this area), and heights between two and 15 m. The *Saimiri* group used them to access non-flooded forest fragments, even when this involved using a flooded section to do so. The study group used forest fragments (58.7%, SD: 351.2), natural hedgerows (19.9%, SD: 177.4) and the edge of forest fragments (21.4%, SD: 193) differently each month (Figure 21.5), with both natural hedgerows and fragment edges used more intensively in the rainy season (May–August). Over the 13-month study period, the three habitat types were used to different extents ($H = 223, 216$; $p < 0.05$).

Through the *ad libitum* observations, five primate species were found in fragments of lowland forest, gallery forest and *M. flexuosa* swamps (Table 21.5). *Alouatta seniculus* and *S. apella* were the most common primates in the area, followed by *P. ornatus* and *S. c. albigena*. *Aotus brumbacki* was observed in fewer fragments but across all habitat types. Primate species were recorded in *M. flexuosa* swamps in both rainy and dry seasons, when it was flooded and unflooded. *Aotus brumbacki* was recorded as present year round in at least two *M. flexuosa* swamp fragments (2.3 ha and 1.8 ha) (Vargas 2011; Carretero-Pinzón, unpublished data). Trees were in fruit during both seasons but detailed information on fruit production

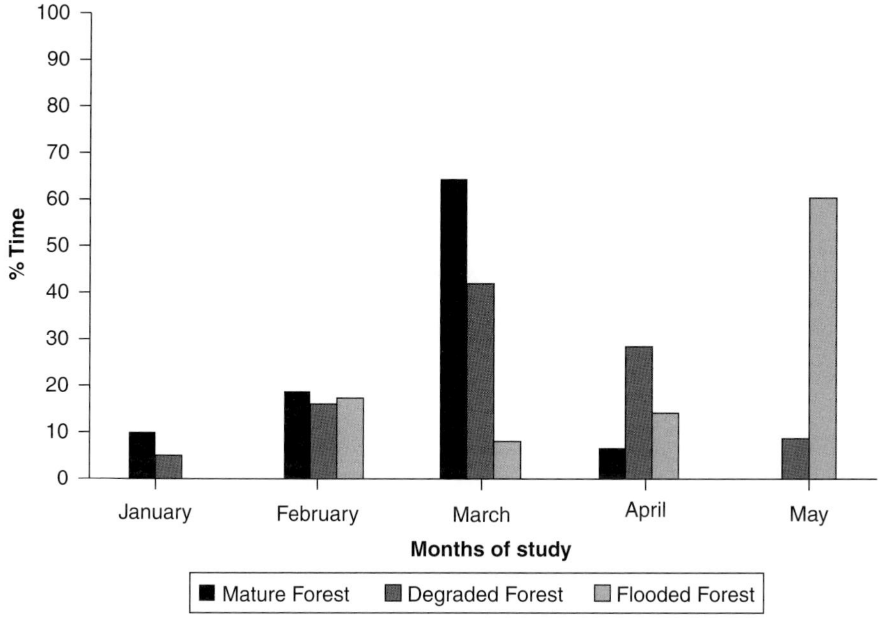

Figure 21.3 Habitat use by a group of *Saimiri cassiquiarensis albigena* from January to May 1999 (369 ho) in Tinigua National Park, Colombia.

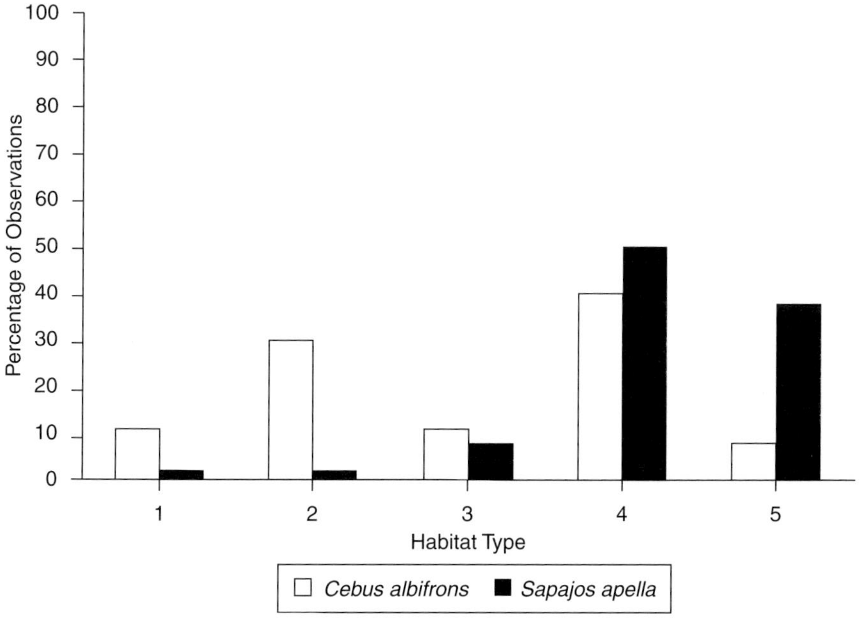

Figure 21.4 Percentage of contacts with *Cebus albifrons* ($n = 32$) and with *Sapajus apella* ($n = 38$) in each habitat type: (1) simple early successional communities; (2) low forests; (3) forests approximately 15m high with underbrush more developed; (4) well-developed forests; and (5) forest found on the highest land (for more detail, see Methods and Defler 1985).

differences between the two seasons is not available (Carretero-Pinzon, pers. obs.).

Discussion

Although all the bioregions presented here are different in their geomorphology, plant diversity, precipitation and flood dynamics (Lasso *et al.* 2010, 2011), it is possible to construct general trends for flooded habitat use by this regional primate assemblage. In Colombia, flooded habitats have been described as having both lower plant density and diversity compared with mature forests and degraded forest in the same area (Ahumada *et al.* 1998; Defler & Defler 1996). This might be expected to translate into low levels of fruit availability and, since frugivorous primate biomass is positively related to fruit availability in general (Hanya *et al.* 2011; Stevenson 2001), to low levels of primate biomass and species diversity in flooded forests. However, some plant species found in flooded forest (e.g. *Laetia corymbulosa* [Salicaceae], *Cecropia membranacea* [Urticaceae], *Ficus* cf. *americana* [Moraceae] and *Inga cylindrical* [Fabaceae]) produce large crops and seem to be of key importance for primates as their fruiting coincides with those periods when non-flooded forests have low fruit abundance (Ahumada *et al.* 1998; Stevenson *et al.* 2004). In addition, Amazonian *igapó* contemporaneous fruiting by several species such as *Moronobea coccinea* (Clusiaceae), *Goupia glabra*

Table 21.5 Number of fragments in which the presence of five primate species was confirmed in the San Martin area.

Primate species	Presence/absence of primates inside fragments	Habitat types		
		No. of lowland forest fragments	No. of gallery forest fragments	No. of *Mauritia flexuosa* swamps
Alouatta seniculus	Observed	27	45	18
	Not observed	2	6	2
	Undetermined*	0	0	0
Sapajus apella	Observed	25	44	14
	Not observed	2	6	6
	Undetermined*	2	1	0
Callicebus ornatus	Observed	23	31	9
	Not observed	6	19	11
	Undetermined*	0	2	0
Saimiri cassiquiarensis albigena	Observed	22	35	14
	Not observed	5	15	6
	Undetermined*	2	1	0
Aotus brumbacki	Observed	3	12	3
	Not observed	0	1	0
	Undetermined*	26	38	17

*Surveys were either not made or effort was insufficient to confirm primate species presence.

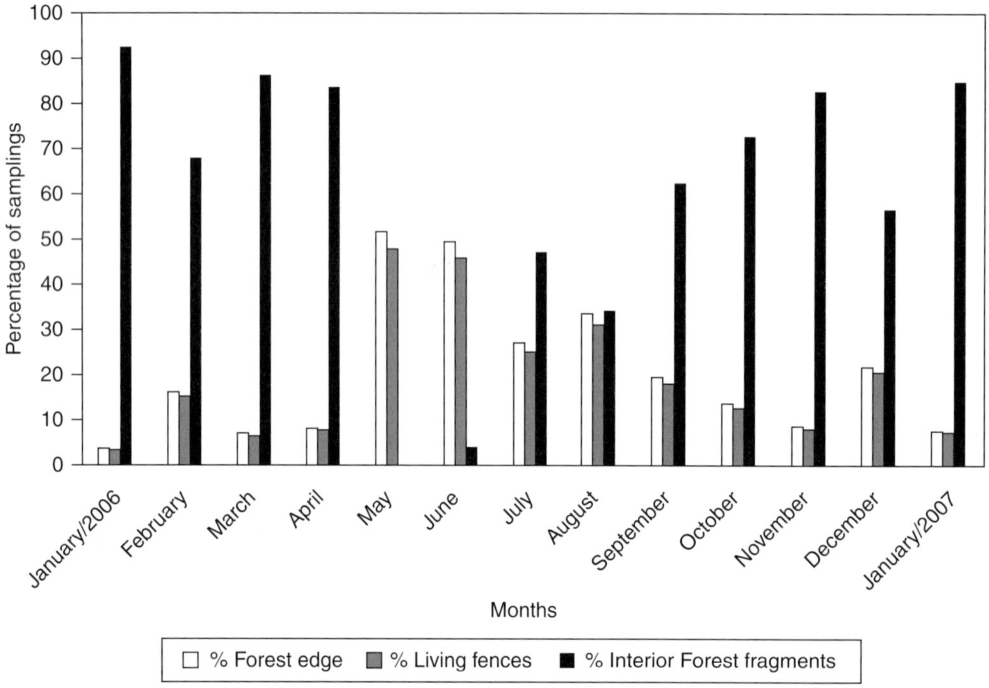

Figure 21.5 Habitat use by a group of *Saimiri cassiquiarensis albigena* from January 2006 to January 2007 in San Martin, Meta Department, Colombia.

(Celastraea), *Pachira aquatica* (Bombacaceae), *Couma macrocarpa* (Apocynaceae), *Duguetia* spp. (Annonacee), *Pouteria* spp. (Sapotaceae), *Manikara* sp. (Sapotaceae), *Mauritia flexuosa* (Arecaceae) produce a spike in fruit production that can attract many primate species to the flooded forest during the early rainy season, especially since, for up to three months, the availability of fruits on these trees far surpass availability inland (Defler & Defler 1996). During March–April at the beginning of the seasonal flooded forest there were 12 fruit crops per hectare available, while low-water and drained forest produced one to four crops from September to February. Unlike flooded forests studied in the Duda river basin of Colombia there were no species of the family Moraceae found in the *igapó* forest. This family of plants is highly important to primates in other (more fertile) regions but apparently not so important in infertile *igapó* habitats (Defler & Defler 1996; Defler unpublished data). Such occurrences underscore the importance of flooded forest for primate species during the rainy season in the Colombian *llanos*.

Indeed, the studies at La Macarena indicate that flooded habitats are important during the early part of the rainy season for at least four of the region's primate species (*P. ornatus, A. belzebuth, L. lagothricha* and *S. c. albigena*). This seems to be related to an increase in fruit production of, among others, the four species mentioned above (Ahumada *et al.* 1998, Stevenson 2006). Such seasonal resource availability is particularly important for frugivorous primate species, including *L. lagothricha, A. belzebuth* and *P. ornatus*. For species such as *S. c. albigena*, which consume a high proportion of arthropods in their diet (78% and up to 82%; Carretero-Pinzon 2000, 2008, respectively), flooded habitats are also important during the rainy season as a fruit source, when fruit production is lower in non-flooded forest.

Yoneda (1990) studied *S. apella, A. seniculus* and *A. brumbacki* at this site. While *S. apella* and *A. brumbacki* do not favour flooded forests, *A. seniculus*, a largely folivorous species, does appear to use flooded habitats for feeding, showing much the same seasonal patterning as the more frugivorous species mentioned above. However, detailed information about when and how often this occurs has not yet been established for *M. flexuosa* swamps and other riverine habitats in the Colombian *llanos*. There are some published comments about *A. seniculus* for an Amazonian site (Caparú station, Apaporis River, Vaupes) by Defler and Defler (1996) and Defler (2013), where leaf flushing and fruit production in the early high-water period seems to attract this species into the habitat.

In the Los Llanos bioregion, primate diversity is reduced and forest fragments are more dispersed and more degraded, and fruit production of some forests are lower than at the La Macarena Bioregion site (Carretero-Pinzon 2008; Stevenson *et al.* 1998). Here, flooded forests and *M. flexuosa* swamp became more important, since those are the main forest types available for primate species, and so represent a key resource that sustains continued primate survival in the region.

Differences in feeding strategies provide one possible explanation of the differential use of flooded and non-flooded forest by the two capuchins species in El Tuparro Park based on their feeding behaviour. Terborgh (1983) describes *C. albifrons* as an 'opportunistic nomad', travelling long distances to look for concentrated patches of abundant food. In contrast, *S. apella* uses resources more evenly, travelling less between resource patches. This explanation was proposed as a strategy to minimize competition between these two sympatric species at Terborgh's Cosha-Cashu study site in Manu National Park, Peruvian Amazon. At El Tuparro, a much less diverse habitat than Cosha-Cashu, these two species are parapatric. Although competition between these species is difficult to measure, it is not the only possible explanation (Defler 1985), as historical factors (such as vast areas of Colombian *llanos* flooded in recent Pleistocene and Holocene times) combined with *C. albifrons* longer distance foraging strategy have shaped the present distribution of these two species of capuchins in flooded habitats (Defler 1985; Terborgh 1983). Certainly, *S. apella* has been observed using flooded forest and *M. flexuosa* swamps in other parts of their distribution (Gomez-Posada 2003). Its relative low use of flooded forests observed at Tuparro may, therefore, be determined by a difference in foraging strategy coupled with a low diversity of resources in these forests (Defler 1985).

In landscapes where more habitat loss and fragmentation occur (for example, the San Martin area in the Los Llanos bioregion), flood dynamics and presence of primate species also change. At this site, fragments of gallery forest seem to support more primate species while *M. flexuosa* swamp fragments support fewer species. This pattern can be explained by the presence of fewer edible plant species available in these swamp fragments as well as by the consequences of habitat loss and fragmentation effects on primate species due to isolation and fragment size (Benchimol & Peres 2013; Harcourt & Doherty 2005). The importance of natural hedgerows and forest edges for species such as *S. c. albigena*, which have large space requirements (home ranges in continuous areas reach 240 ha, Carretero-Pinzon 2000), seem to be underscored by its use in the rainy season. During the rainy season, fruit production inside of forest fragments is lower compared with fruit production available in forest edges and natural hedgerows, increasing its use by the *S. c. albigena* group studied (Carretero-Pinzon 2008). These landscape elements (natural hedgerows) are important since they allow primates access to flooded areas in which fruits and other resources such as small vertebrates and insects, part of *Saimiri*'s diet, can be found (Carretero-Pinzon, unpublished data). This is important in an already reduced home range.

Flooded habitat use by primates is still not well known in the Colombian Orinoquia. The Orinoquia region is one of the most important regions in Colombia for petrol exploitation, infrastructure projects and palm oil plantations. Deforestation and fragmentation rates are increasing due to the implementation of these ongoing development projects. Consequently, there is a need for primate research inside flooded and non-flooded forest in order to prevent local extinctions of primates (Castiblanco *et al.* 2013; Ecopetrol 2015; Etter *et al.* 2006; Fedepalma 2014). Although undergraduate theses have been written based on research at several sites in the whole Orinoquia region (around 23 citations in three of the main Colombian university libraries from 1990 to 2014), only a few have been published ($n = 6$).

The importance of *M. flexuosa* swamp fragments to primates in the Los Llanos bioregion calls for additional research and conservation attention. For example, the studies summarized here found several primate species to use flooded habitat, but it remains unknown exactly how these fragments are exploited and if the primates observed in them are permanent residents or present only during particular seasons. It seems likely that their use varies according to the availability of palm fruits. The rapid transformation of Colombian Orinoquia landscapes makes it urgent that we understand the importance of flooded habitats for especially endemic primate species of this region (*P. ornatus, S. c. albigena* and *A. brumbacki*), as well as understand the processes affecting these flooded habitats under ongoing habitat loss and fragmentation threats.

Acknowledgements

We thank A. Barnett and I. Matsuda for inviting us to contribute this chapter and their valuable help to improve this document. Fieldwork at the different sites described in this chapter would not have been possible without the help and permissions from various institutions, grants and individuals who made it possible. At La Macarena site: J. Ahumada, K. Izawa, C. A. Mejia and A. Nishimura, and field assistants Alvaro, Meyo and Ramiro; at San Martin area: M. Ruiz-Garcia, as well as the Sanchez-Rey, Novoa and Enciso families for their long-term support and permission to XCP to live on their farms. Funding for fieldwork (2004–2014) was provided by Becas IEA (2009, 2011; Conservation International Colombia, Fundación Omacha and Fondo para la Acción Ambiental y la Niñez), RHD Grant from the School of Geography, Planning and Management, The University of Queensland (2013–2014), IdeaWild and private funding (support from farms). All the students who worked at the San Martin area over the past years: S. Escudero, M. Beltran, A. M. Aldana, J. Torres, J. Ospina, J. Ramos, A. Fajardo, C. Gaitan and M. Vargas. Work in El Tuparro from 1977 to 1981 was supported by INDERENA (Ministry of Agriculture) and the American Peace Corps, and in 1982 by World Wildlife Fund-US and INDERENA. All permits were issued from INDERENA. TRD is particularly grateful for the support given by Jorge Hernández-Camacho (RIP) and to the personnel of INDERENA, especially José Vicente Rodríguez Mahecha.

Chapter 22

Primates of the South American Pantanal Wetland

Seasonal Effects on Their Habitats and Habits

Cleber J.R. Alho and Fernando C. Passos

Introduction

Primates of inland wetlands are considered at several points in this book (e.g. Chapters 21, 23–29, 32 and 34–35). Here we consider those in one of the world's largest tropical wetland areas, the Pantanal. The Pantanal is fed by tributaries of the Paraguay River and covers 140 000 km^2 in Brazil (mostly within the state of Mato Grosso do Sul, but also Mato Grosso), with additional areas extending to neighbouring Paraguay and Bolivia. Because of its central position in South America (Figure 22.1), the biome is a meeting point for a number of phytogeographic provinces, including Amazonia (to the north), the *cerrado* savanna (to the east), the Atlantic forest (to the South), and the Bolivian and Paraguayan Chaco (to the west) (Mittermeier *et al.* 1990).

A portion of the Pantanal wetland is under managed protection. The 1350 km^2 Pantanal Matogrossense National Park was designated a Site of International Importance under the Ramsar Convention in 1993, while the Serviço Social do Comércio (SESC) Pantanal Private Natural Heritage (*Reserva Particular do Patrimonio Natural SESC Pantanal)* covers some 879 km^2 and is also Ramsar designated. The region is also recognized under the Brazilian Constitution as an area of national heritage. The area has enormous productivity, is important for migrating birds and has considerable populations of high-profile species such as caiman, jaguar, storks and macaws (Alho 2005; Alho & Sabino 2012). While many aspects of its fauna and flora have been well researched, the primates, somewhat surprisingly, have received much less attention (Brito *et al.* 2009).

Pantanal Vegetation

Some 1900 plant species are known from the region, with most habitats created by fluvial morphogenesis. The complex combination of topography, and variations in the extent and duration of inundation mean that some 12 subregional ecosystems have been defined (Alho 2005), each with distinct ecological, geological and hydrological characteristics. However, savanna habitats (*cerrado*) predominate, covering more than 45% of the region, including patches of forested savanna or *cerradão*, while 24% is composed of forest habitats that are deciduous and have a canopy averaging 15–20 m in height (Alho 2005).

The Pantanal occupies a huge, gently sloping basin, where elevation ranges from 80 to 150 m asl. Annual rainfall varies between 1000 and 1500 mm across the basin, with most rain falling between November and March (Mittermeier *et al.* 1990). During this period, some 80% of floodplains are submerged (Alho 2005). Flood dynamics cause rapid changes in vegetation cover and promote great diversity in pioneer vegetation types, from herbaceous to woody formations (Pott *et al.* 2011). This dynamism promotes vegetational mosaics. In addition, the biome is surrounded by a variety of vegetation types: Humid Chaco to the south, the dry forests of Arid Chaco to the southwest, the Chiquitano dry forests to the west and northwest, while cerrado savannas occupy the north, east and southeast (Alho 2005). Each contributes faunal and floral elements. Consequently, there is a continuum from semi-deciduous forests to dry and forested savanna *(cerradão)* within the Pantanal basin. Primates are restricted to arborescent vegetation; the regional forest types used by the Pantanal's primates are given in Table 22.1.

Pantanal Primates

An established list of 174 Pantanal mammals species (Alho 2005; Alho *et al.* 2011a; Cáceres *et al.* 2008) includes eight primates. The primates of the Pantanal are well adapted to the region's wooded habitats, but are most common in riparian forests and pockets of semi-deciduous forests occupying higher areas (Figure 22.2).

The eight primates are: Aotidae: *Aotus azarae* (Humboldt 1811); Atelidae: *Alouatta caraya* (Humboldt 1812); Cebidae: *Sapajus apella* (Linnaeus 1758), *Sapajus cay* (Illiger 1815); *Callithrix penicillata* (E. Geoffroy 1812); *Mico melanurus* (E. Geoffroy in Humboldt 1812) (=*Callithrix melanura*); Pitheciidae: *Callicebus donacophilus* (d'Orbigny 1836), *Callicebus pallescens* (Thomas 1907). If the entire Paraná-La Plata Basin is included (Paraguay, Paraná and La Plata Rivers), the number of primate species rises to 13, with the inclusion of: Atelidae: *Alouatta clamitans* Cabrera 1940; Cebidae: *Sapajus nigritus* (Goldfuss,1809); *Callithrix aurita* (Geoffroy in Humboldt 1812); *Leontopithecus chrysopygus* (Mikan 1823), and Pitheciidae: *Callicebus personatus* (E. Geoffroy 1812) (see Cáceres *et al.* 2008; Chiarelli 1972; Gregorin 2006; Padua and Padua 2000; Rylands 1993). In addition, the muriqui (*Brachyteles arachnoides;* E. Geoffroy 1806) occurs in this region, but does not occupy flooded habitats (Mendes *et al.* 2008), while *Saimiri boliviensis* (I. Geoffroy and de Blainville 1834) is reported to occur close to the Bolivian Pantanal

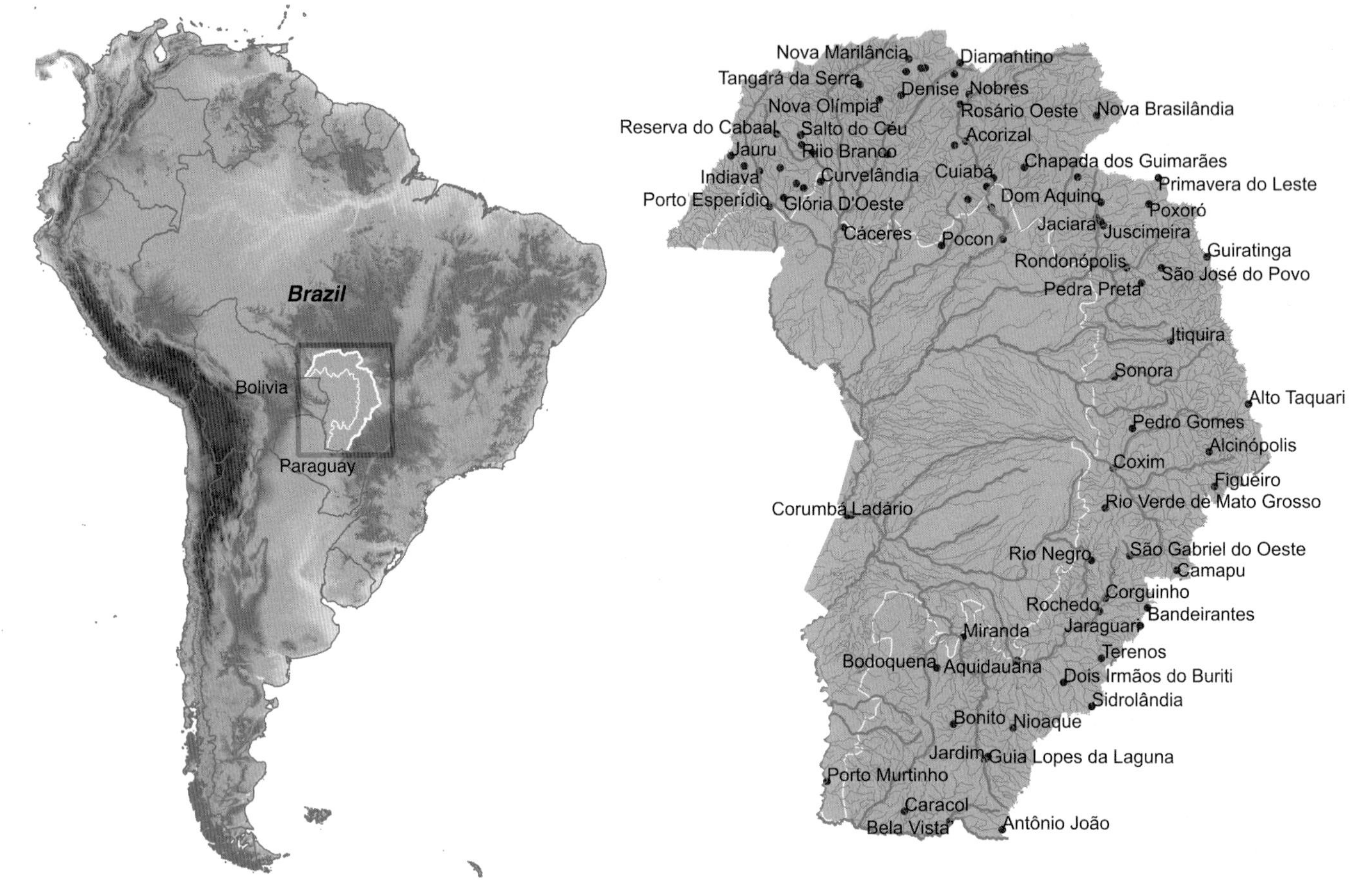

Figure 22.1 The Pantanal wetland located in the centre of South America, surrounded by the upland plateaus, where headwaters of the Paraguay River which feeds the wetland are located. These occur mostly in Brazil (140 000 km^2), whose habitats are biogeographically influenced by the *cerrado* biome (east), Amazonia (north), Atlantic forest (south), as well as the Chaco at the Bolivian and Paraguayan borders (west). (A black and white version of this figure appears in some formats. For the colour version, please refer to the plate section.)

(Martínez *et al.* 2010), and so may also occur there. Records of *Sapajus nigritus* in the Pantanal (e.g. Bastazini 2011) appear to be the result of misidentication, as this species occurs further east in the Mata Atlantica (Lynch Alfaro *et al.* 2012).

There are no primate species endemic to the Pantanal biome. The Paraguay River and its northern tributaries have their headwaters near the transition with the Amazon region and have functioned as corridors for some primate species, which have reached the Pantanal through continuous riparian forests (Alho 2005; Pott & Pott 1994). In addition, the major tributaries of the left bank of the Paraguay River, from the Cuiabá River down to the Apa River, have headwaters in the *cerrado* biome, and so have played a similarly important role in the ingress of savanna-inhabiting primate species into the Pantanal. To the south of the Pantanal, in the Brazilian state of Mato Grosso do Sul, a vegetation transition zone exists connecting plant physiognomies of Amazonia and Chaco to the Pantanal through the uplands of Maciço do Urucum and Serra do Amolar (Cáceres *et al.* 2008). Thus, there is a bridge linking these biomes to the Pantanal, and some Amazonian species such as *Sapajus apella, Saimiri boliviensis, Mico melanurus* and *Callicebus donacophilus* also occur in areas surrounding the outer Pantanal. Other species exhibit similar habitat use and requirements across the broader geographical range. *Alouatta caraya*, for example, occurs in eastern Bolivia, central and southern Brazil and almost all of Paraguay and northern Argentina (Rylands *et al.* 1996). Where this distribution overlaps with the limits of the *cerrado* biome, the species is mostly found in gallery (Henriques & Cavalcante 2004), semi-deciduous forests (Gregorin 2006) and natural wooded forest isolates (de Carvalho de Oliveira *et al.* 2003). It also occurs in flooded areas of the Bolivian Pantanal (Wallace & Rumiz 2010).

Within the Pantanal, flood dynamics cause rapid changes in vegetation cover (a seasonal shrinking and expansion of habitats), and also influence forest phenology (leaf flushes peak in the rainy season). Arboreal and forested habitats of the Pantanal have the highest mammal densities and are the most important environments in terms of relative energy availability and consumption (Desbiez *et al.* 2010a; Rímoli *et al.* 2008). Pantanal mammal assemblages vary between habitats, with frugivorous mammals having a higher energy consumption than browsers and grazers (Desbiez *et al.* 2010a). Most food, notably fruit, is produced in forested habitats, emphasizing the importance of these habitats for primates (Figure 22.3).

While some studies of Pantanal primates have been conducted, and have focused on habitat use and behavioural interactions (Rímoli *et al.* 2008), many gaps in ecological knowledge persist. Consequently, much of the summary

Table 22.1 Vegetation types used by primates in the Pantanal.

Habitat type	Characteristic plant species
Seasonal deciduous forest (Regional name: *Floresta Estacional Decídua* or *Mata Aluvial*)	Majority of the trees (> 50%) drop their leaves during the dry season. Tree species include *Anadenanthera colubrina* (Fabaceae–Mimosoideae, *angico*), *Aspidosperma pyrifolium* (Apocynaceae, as *quebrancho*), *Myracrodruon urundeuva* (Anacardiaceae, *aroeira*) and *Tabebuia impetiginosa* (Bignoniaceae, *piúva*).
Seasonal semi-deciduous forest (Regional name: *Floresta Estacional Semi-decídua* or *Mata Seca*)	Some 20% of tree species drop their leaves during the dry season: *Albizia inundata* (Fabaceae–Mimosoideae, *canafístula*), *Astronium fraxinifolium* (Anacardiaceae, *gonçaleiro*), *Copernicia alba* (Arecaceae, *carandá*), *Protium heptaphyllum* (Burmanniaceae, *almécega*).
Riparian forest (Regional name: *Floresta Ripária* or *Mata de Galeria* or *Mata Ciliar*)	Present along every water course in seasonally flooded areas. Include trees such as *Andira inermis* (Fabaceae–Papilionoidae, *morcegueira*), *Inga vera* (Fabaceae-Mimosoideae, *ingá*), *Albizia inundata* (Fabaceae–Mimosoideae, *canafístula*), *Sapium obovatum* (Euphorbiaceae, *sarã-de-leite*) and *Vochysia divergens* (Vochysiaceae, *cambará*).
Woody savanna (Regional name: *cerrado* or open *cerrado*)	*Cerrado* is the classic savanna biome of central Brazil, and is well represented in the Pantanal, with shrubs and scattered trees that have thick bark and tortuous trunks. Characteristic species include *Annona dioica* (Annonaceae, *articum*), *Buchenavia tomentosa* (Combretaceae, *tarumarana*), *Curatella americana* (Dilleniaceae, *lixeira*), *Dimorphandra mollis* (Fabaceae–Caesalpinoidae, *fava-de-anta*).
Forested *cerrado* woodland (Regional name: *Cerradão* or *Cordilheira*)	A denser savanna woodland, semi-deciduous, with trees reaching 8 to 15 m, growing on sandy soils. Charactertized by *Byrsonima crassifolia* (Malpighiaceae, *canjicão*), *Caryocar brasiliense* (Caryocaraceae, *piqui*), *Dimorphandra mollis* (Leguminosae–Caesalpinioideae, *fava-de-anta*), *Eriotheca gracilipes* (Bombacaceae, *paina*), *Qualea grandiflora* (Vochysiaceae, *pau-terra*).

Figure 22.2 Major vegetation formations of the Pantanal, with open field and patch of forest. Photo: C. Alho.

Figure 22.3 Riparian forest inhabited by black howler monkeys and other primate species in the Pantanal. Photo: S. Mamede.

below is based on these, plus the authors' own, unpublished, observations covering over two decades of study in the biome.

Howler Monkeys

The black howler monkey (*Alouatta caraya*) is the most common arboreal mammal in the Pantanal (Alho *et al.* 2000, 2011a). Indeed, *A. caraya* is one of the most abundant wild mammals in this biome. Field surveys carried out on managed cattle ranches in the central Pantanal found howler densities varied by habitat, from 4.95 individuals/km^2 in forest patches within savanna, to 11.2 in continuous forest habitats (Desbiez *et al.* 2010a). These densities are very similar to those for key terrestrial Pantanal mammal species, such as capybara (*Hydrochoerus hydrochaeris*) and pampas deer (*Ozotoceros bezoarticus*; Alho *et al.* 2011a). Pantanal habitats used by black howlers include semi-deciduous forest on higher ground (locally known as *cordilheira*), riverine and gallery forests, forested patches within savanna (*cerradão*), and patches of arboreal savanna (*capão de cerrado*; Figure 22.3).

There are also local-scale variations in within-habitat suitability. For example, riparian trees such as *Vochysia divergens* (Vochysiaceae, *cambará*) may dominate permanently flooded areas so as to form monodominant areas (locally named *cambarazais*). This modified habitat is not favourable for howler monkeys since the homogeneous vegetation offers reduced food item variety, and howlers require habitats with diverse species of trees to supply their feeding needs (C. Alho, unpublished data).

The diet of the black howler consists of leaves, leaf stalks, buds, fruits, and flowers (Rímoli *et al.* 2008). Sharply defined annual patterns of precipitation mean that many Pantanal tree species are deciduous or semi-deciduous. In such species the flush of young leaves generally begins at the start of the rainy season (November to March). Fruits are usually produced from the end of the rainy season and into the dry season (Pott and Pott 1994). Consequently, food supply for this species can vary seasonally (Bravo & Sallenave 2003; Ludwig *et al.* 2008; Rímoli *et al.* 2008), and while feeding occurs in the forest canopy, howlers occasionally come down to the ground (notably during the dry season) to drink standing water or to supplement their diets with herbaceous plants. Along with many other mammals, they also visit mineral licks in the Pantanal (Coelho 2006).

The phenological rhythms of productivity of new shoots, flowers and fruits depend on the annual seasonal flooding of the Pantanal. The deciduous and semi-deciduous forest habitats, including the forested savanna (*cerradão*), inhabited by abundant populations of black howler monkeys, contain trees that produce flowers fruits, and leaves at slightly different times, prolonging the temporal sequence of diet item availability. Thus, trees like *Handroanthus impetiginosus* (Bignoniaceae, *piúva-da-mata*) produces flowers in the dry season (May–September); *Protium heptaphyllum (almecega)* produces new leaves, and *Enterobium contortisiliquum* (*ximbuva*) new shoots, at the beginning of the wet season (October). Meanwhile, *Inga uruguensis* (*ingá*), common in Pantanal cerrado, produces flowers in the wet season, while its fruits, are produced at the end of the wet season and early dry season (February–April). All of these items are consumed by howler monkeys.

Similarly, in riparian forest trees, *Andira cuyabensis (morcegueiro)* produces perfumed flowers at the beginning of the wet season; *Inga vera* (*ingá*) produces flowers mid-wet season and fruits end of wet season and early dry (February–April); *Sapium obovatum* (*sarã*) produces flowers and fruits during the wet season, as does *Albizia inundata* (*canafístula*). The howler eats these all in temporal sequence of appearance.

Frugivory in black howlers (Figure 22.5) is more commonly observed in the rainy season (Lázaro Jr. & Rímoli 2009). Black howlers in some areas of the Pantanal, while predominantly folivorous (leaves and buds, 77% of diet), also take available fruits in the wet season (8%), and flowers (15%) during the dry season (Nantes & Rímoli 2008). Feeding howlers may be followed by coatis and peccaries, which feed on fallen fruits (Desbiez *et al.* 2010b).

Figure 22.4 Patches of tree–savanna (*cerrado*) vegetation, showing xeromorphic features due to the dry season, which are home to *Sapajus cay*. Photo: S. Mamede.

Figure 22.5 The howlers *Allouatta caraya* consume jenipapo fruit (*Genipa americana*) which appears at the end of the dry season and beginning of the wet season. The howlers also consume fruits of wild figs (*Ficus* spp., ingá (*Inga* spp.), trauma (*Vitex cymosa*), and ximbuva (*Enterolobium contortisiliquum*). Photo: S. Mamede.

Seasonal variation in resources abundance, particularly food and places for reproduction, potentially influences the social systems of this primate species and may determine local abundance. Further research could examine how such seasonal variation in resources affects reproductive strategies, group sizes and other aspects of feeding and reproductive biology in relation to the flooding regime of the Pantanal (a subject which has been studied elsewhere, see Chapter 32).

Outside the Pantanal biome, on the upper Paraná River, *A. caraya* social groups vary from 6 to 18 individuals, with sex ratios of one adult male to two adult females (Aguiar *et al.* 2009). In Argentina, within the La Plata River system, *A. caraya* occurs at high densities in inundated forests, gallery forests and Chaco forests in the provinces of Formosa, Chaco, northwest of Santa Fé, north of Corrientes and extreme west of Missiones (Brown & Zunino 1994, and see Kowalewski *et al.*, Chapter 32, this volume). However, the species is ecologically flexible and occurs also in extremely dry forests in Paraguay (Giordano & Ballard 2010).

Capuchins

The tufted capuchin (*Sapajus apella*) occurs throughout the Amazonian region, including the Brazilian states of Maranhão, Tocantins and Mato Grosso (Silveira *et al.* 2008), entering the Pantanal through a vegetation transition near the Bolivian frontier (Lynch Alfaro *et al.* 2012a). The species also occurs in the municipality of Palmeira in the state of Mato Grosso, just at the southern limit of the Pantanal (Silva Jr. 2001). After *A. caraya*, it has long been considered the second most abundant primate in the Pantanal (e.g. Mittermeier *et al.* 1990).

Food items eaten by this species include fruits, flowers, leaves, buds, nectar, insects and small vertebrates (Freese & Oppenheimer 1981; Terborgh 1983). In line with its broad diet base, *S. apella* also uses a wide range of habitats, such as pristine forests, secondary forests, gallery forests, dry forests, patches of forested savanna (*cerradão*), semi-deciduous forest, patches of arboreal savanna (*cerrado*), as well as disturbed habitats (Silveira *et al.* 2008). Tufted capuchins foraging in gallery forests may knock down fruits (*Ficus, Psidium*) that are subsequently dispersed by fish (including *Brycon hilarii*, Characidae: Reys *et al.* 2009).

The second type of capuchin in the Pantanal, Azaras or hooded capuchin (*Sapajus cay*), is an abundant species that occurs mainly in unflooded habitats. In the Pantanal, *S. cay* has been observed in gallery forest, forested savanna (*cerradão*) semi-deciduous forest, patches of arboreal savanna (*cerrado*), and also in degraded habitats near ranch houses (C. Alho and F.C. Passos, unpublished data). They are able to move through open areas to reach patches of woodland or forested habitats. The species is frugivorous–insectivorous, eating fruits (including acuri palm fruit, *Attalea phalerata*), seeds, arthropods, frogs, nestlings and even small mammals, stems, flowers and leaves (Kinzey 1997). As extractive and manipulative foragers they will even steal from corn and banana plantations, and occasionally prey on young domestic chickens (C. Alho and F.C. Passos, unpublished data). Collared peccaries (*Pecari tajacu*) in the Pantanal are reported to feed on fruits individual *S. cay* knock down when foraging (Tortato *et al.* 2014).

Outside the Pantanal, *S. cay* in Brazil is bounded to the east by the Araguaia River and to the southeast by the Paraná River states of Mato Grosso and Mato Grosso do Sul. Occurrence of this species has also been recorded along the Paraguay River and La Plata River in Argentina (Province of Salta) and Bolivia (Brown & Zunino 1994; Silva Jr. 2001; Silveira *et al.* 2008).

Marmosets

The black-tufted ear marmoset (*Callithrix penicillata*) has a wide distribution in Brazil, occurring from São Paulo state to Maranhão and Piauí states. It occurs mainly in the *cerrado* biome, within which it occupies gallery forests, dry forests, open *cerrado* savanna, denser *cerradão* savanna, and disturbed or secondary growth woodland and forest (Corrêa and Coutinho 2008). The species enters the Pantanal *via* intrusions of *cerrado* vegetation, mainly of savanna patches. It also occurs in the *cerrado* biome of Mato Grosso do Sul, on the outskirts of the Pantanal (Melo *et al.* 2008). The diet of these marmosets is composed of fruits, flowers, nectar, plant exudates (gums, sap, latex) and some small animal prey items (insects, spiders, lizards, frogs, snails and nestling birds). However, exudates from *cerrado* trees form a notable component of their diet (Fonseca & Lacher 1984).

The black-tailed marmoset (*Mico melanurus*) occurs in savanna habitats of the Pantanal National Park, a region with a high degree of seasonal inundation. It is also present at the Taiamã Ecological Station, on the banks of the Paraguay River, north of the Pantanal, as well as occurring in the Bolivian Pantanal, near the Brazilian border (Porcel *et al.* 2010). This species has a wide geographic distribution, being the only species of the genus *Mico* to occur beyond Brazilian boundaries, and occurs in the Amazon region (states of Amazonas, Rondônia, Mato Grosso), reaching the *cerrado* biome in Mato Grosso and Mato Grosso do Sul (Hannibal & Neves-Godoi 2015). It also occurs in the Chaco of Bolivia and Paraguay, though it is very poorly known there (Ferrari 2008; Mercado & Wallace 2010).

In the National Park of Chapada dos Guimarães, and in the Serra das Araras, the upland region surrounding the northern Pantanal, it is reported in savanna-dominant vegetation (Alho 2000; Rylands *et al.* 2008). The intrusions of savanna (*cerrado*) in the Pantanal favours this species in unflooded habitats, where it has a diverse diet, which includes items from *cerrado* plants (exudates, fruits, flowers, nectar, seeds) and some small invertebrate and vertebrate prey (Rylands *et al.* 2008). A study carried out in the *cerrado* along the Manso River, a secondary tributary of the Paraguay River, in Chapada dos Guimarães, found *M. melanurus* to living at a low natural density (Marques *et al.* 2011). Those monitored by radiotelemetry showed home ranges varying from 0.40 to 3.48 km^2.

Night Monkey

Azaras night monkey (*Aotus azarae*) ranges from the north of Argentina, through to the Paraguayan *Chaco*, and Bolivia.

In Brazil, it is known only from the Pantanal wetland, where few occurrences have been recorded on the right bank of the Paraguay River (Brown & Zunino 1994; Cunha 2008). In Bolivia, this species also occurs in wetter habitats of the Pantanal, along the Brazilian border, but its status is uncertain, as its nocturnal habits make it difficult to observe (Romero-Valenzuela & Rumiz 2010). The species is essentially frugivorous, supplementing its diet with leaves, flowers, nectar, insects, invertebrates, and fungi (Fernández-Duque 2007). While it is mainly crepuscular, it is sometimes active diurnally on cloudy or overcast days.

The species is also flexible in its habitat requirements. Outside the Pantanal, *Aotus azarae* occurs in the Argentinian Chaco and tolerates extreme seasonal fluctuations as well as variations in photoperiod throughout the year (Fernandez-Duque *et al.* 2002). In the xerophytic northern Paraguayan Chaco, it lives in low canopy scrub forest and high canopy forest. This area supports the cactus *Cereus peruvianus*, which is also present in the southwestern part of the Pantanal. Given the flexibility of *A. azarae* habitat selection, it is possible that it may also occur in this habitat within the Pantanal – a subject in need of investigation.

Titis

The Bolivian titi monkey (*Callicebus donacophilus*) occurs from west of the Paraguay River, through the Paraguayan and Bolivian *Chaco*, to the southern Pantanal in the Brazilian state of Mato Grosso do Sul (Bordignon *et al.* 2008; Rumiz 2012). It inhabits riparian forest areas, gallery forests, flooded environments, and open areas, moving through the lower levels of woody vegetation and trees. The species is primarily frugivorous, but also consumes leaves, seeds and insects, all food items dependent on habitat seasonality.

The Chacoan or white-coated titi monkey (*Callicebus pallescens*) is the Pantanal's second member of the genus. Its distribution appears restricted to the Chaco and western Pantanal, but there is also a record for the region of Acurizal in Brazil (17°45′S, 57°37′W: Rumiz 2012). In the western Pantanal, this species is frequently observed in non-flooded habitats, and has its highest densities in the highlands at Urucum, Castelo and Santa Teresa, by the Serra do Amolar ridge on the Pantanal wetland's western border (Porfino *et al.* 2014; Tomas *et al.* 2010). The known distribution of this species includes the Bolivian Pantanal all along the Brazilian border (Martínez & Wallace 2010). Reviewing the distribution of *Callicebus pallecens* and *C. donacophilus*, Rumiz (2012) concluded that the geographic limit for both species is poorly known in Bolivia. The identity of the species living along the border regions of Bolivia and Brazil along an area running from Chiquitanas forest, Bolivia, through to the Pantanal in Brazil is still unclear and needs verification.

In the upland region of Corumbá, on the banks of the Paraguay River, densities of *C. pallescens* reach some 11 groups/km^2 (Tomas *et al.* 2010). This is a very high density; *Callicebus nigrifons*, living in remnants of Atlantic forest in Brazil, had upper recorded densities of 0.14 groups/km^2 (Trevelin *et al.* 2007); while *C. personatus* also occurring in Atlantic forest, lived in densities of 3.7 groups/km^2 (Price *et al.* 2002); and *C. aureipalatii* from Bolivia, has been recorded at 6.2 individuals/km^2 (Wallace *et al.* 2006). The reason for such high densities is, as yet, unstudied.

Conclusions

In the Pantanal, while many detailed ecological studies have been done across a variety of taxa (see Alho 2005; Alho & Silva 2012; Heckman 1998; Tomas *et al.* 2011 for reviews), primates lag behind, with most studies in the region dedicated to documenting wild species occurrence. Only a few studies (e.g. Aguiar *et al.* 2009; da Cunha & Byrne 2006, 2013; Larazo 2013; Ludwig *et al.* 2008; Rímoli *et al.* 2012; Trevelin *et al.* 2007) have concentrated on primate habitat use, ecology and behavioural interactions, and these have mostly concentrated on the black howler.

Research Priorities

Given the overall lack of studies of Pantanal primates, specific, focused, investigations are urgently needed to improve our knowledge base on the ecological characteristics of primates living within the Pantanal. These should pay special attention to the community interrelationships among species. In addition, for species the dynamics and structure of their populations, their abundance, behavioural interactions, diets, habitat use and other aut- and synecological relations to the natural habitats should be investigated. This should be done both as a function of the typical regional seasonality, and with a view to estimating the capacity of the various primate species to resist abnormal climatic events and their resilience to anthropic change.

All eight primate species occurring within the Pantanal habitat mosaic also have wide geographical distributions outside this biome. This, and the presence of primates across the mosaic of Pantanal's habitats, offers opportunities for comparative research on adaptation to a heterogeneous complex of habitat types. Such studies could occur within the Pantanal across different habitats or between the Pantanal biome as a whole and habitats elsewhere.

For example, *Aotus azarae* occurs in the Panatanal and further south in the colder drier Argentinian Chaco. Fernandez-Duque *et al.* (2002) found that in the latter habitat a high percentage of births occur between September and October. Such timing means that the young will be well developed by May when periods of low temperatures must be faced. Does *A. azarae*, in the Pantanal similarly adjust its reproductive periodicity to, perhaps, flood-related pulses in food availability? We do not know, but such knowledge could be key to making sure local management practices are coordinated to minimize impacts on seasonally breeding species.

Natural habitats of the Pantanal are highly heterogeneous and accommodate a rich diversity of species of plants and animals. The degree to which each primate species in the Pantanal competes with other species, what resources they use and how the local annual flood cycles change their availability

Figure 22.6 Deforestation and newly established pastures for cattle farming negatively impact the system's functional diversity. Primate species are an intrinsic part of the Pantanal ecosystem and will be adversely affected by such changes. Photo: C. Alho.

in time and space, is poorly understood, although initial studies have been made (Lazari *et al.* 2013; Mamede & Alho 2006), these were not purely primate focused. Similarly, studies are needed to clarify the role of seasonal flooding in population dynamics and seasonal changes in habitat use. How both of these factors influence primate reproductive strategies, social structure, and use of space also remain to be explored. The role of the Pantanal's strong flood pulses as drivers of primate species ecology remain a near mystery, despite their known key role in the lives of other species (Lazari *et al.* 2013; Mamede & Alho 2006). Such topics are of great ecological interest and conservation importance, yet remain to be quantified for Pantanal primates.

Primates are often good proxies for the ecological health of a habitat or region (Caro 2010). But key baseline data are essential to monitoring the effectiveness of management and conservation policies. In the Pantanal this would ideally include information on primate population sizes and distribution. Camera trapping could facilitate such surveys (Coelho 2006; Trolle & Kéry 2005).

Conservation Challenges in the Pantanal

The Pantanal is still considered to be a well-conserved region, but recently, traditional extensive cattle-ranching practices have been changing. In the past, while cattle ranching required large areas, it was low-impact, but practices have become more intensive, requiring conversion of more natural vegetation to pasture. Also in recent decades, large-scale agriculture has seen crops planted intensively, resulting in extensive land conversion, especially in the highlands. The Pantanal floodplain still maintains more than 80% of its natural vegetation, while the surrounding upland plateaus (*planaltos*) have retained only 40% of their original vegetation cover (Figure 22.6). This is cause for concern as the highlands are where the springs and headwaters of rivers that feed the Pantanal wetland are located.

Major environmental threats include:

- Deforestation with conversion of natural vegetation into pasture and agricultural crops, responsible for 18% of vegetation conversion in the Pantanal and more than 45% in the upland plateaus (World Wildlife Fund 2017).

- River flow alteration by hydroelectric plants in the surrounding highlands (small hydropower plants), altering the discharge of nutrients and suspended matter and hence the cycling of nutrients in affected water bodies.
- Introduction of exotic species. Land use and human occupation within the natural habitats of the Pantanal have facilitated the introduction of invasive species of plants and animals, including domestic species. *Callithrix penicillata* is considered an invasive species elsewhere in Brazil, including in some habitats at the meeting points of the *cerrado* and Pantanal biomes.
- Wild fire is a major threat. Ranchers in the Pantanal set fire to vegetation during the dry season as a management technique to clear the vegetation not used by cattle.
- Hunting pressure is a well-reported threat to primate conservation, responsible for reducing wild mammal populations, including primates.
- The resurgence of yellow fever in southern Brazil and Argentina in the years of 2007–2009, and 2017–2018, caused a strong negative impact on the regional primate population (Fialho *et al.* 2012; Holzmann *et al.* 2010). There is currently no evidence of the virus in Pantanal howlers. However, as arbovirosis has been detected in *Sapajus* and *Alouatta* in the nearby Brazilian state of Mato Grosso do Sul, epidemiological monitoring programmes should be established to monitor disease in Pantanal primates (Batista *et al.* 2013).

In addition there is evidence that increasing road traffic is linked to increasing roadkill, including primates (de Souza *et al.* 2015). The Pantanal remains rich in wildlife, including the occurrence of eight primate species, but some 95% of the area remains in private ownership, with livestock ranching as the main economic activity. As a consequence, increasing habitat alteration and loss due to deforestation and other environmental threats are significant and bring wildlife protection and human economic desires into direct conflict. New conservation units are being established (MMA-ICMBio 2017), but there is a need to ensure arboreal and forested habitats are included so as to protect populations of primates and other tree-dependant species. Meanwhile, wildlife tourism has grown in the Pantanal and has been supported species conservation (Tortato *et al.* 2017). Further knowledge of the Pantanal primates, will be key for effective planning of biodiversity conservation and management in this important wetland. Primates' well-established role as flagship species could also apply to the Pantanal.

Acknowledgements

We are grateful to Dr Adrian Barnett (Roehampton University, England), Ikki Matsuda (Primate Research Institute, Japan), and Katarzyna Nowak (University of the Free State, South Africa) for their encouragement and help in preparing this manuscript. José Rímoli provided useful information. Celina Alho helped to draft this chapter. Funds were granted to Fernando Passos (CNPq 303757/2012–14).

Chapter 23

Endangered Range-restricted Flooded Savanna Titi Monkey Endemics *Plecturocebus modestus* and *P. olallae*

Identifying Areas Vulnerable to Excess Flooding, Fire and Deforestation in Southwestern Beni Department, Bolivia

Teddy Marcelo Siles Lazzo, Robert B. Wallace and Jesus Martinez

Introduction

Deforestation is the most important threat for primates around the world (Marsh *et al.* 2013), especially in tropical forests where most primate species occur. Forests cover around 31% of the Earth's surface (FAO 2010, 2013), but are increasingly threatened by deforestation: between 2000 and 2012 a total of 230 million ha of forest were lost (Hansen *et al.* 2013), and global deforestation rates were 0.14% per year between 2005 and 2010 (FAO 2013). In Latin America, 9% of total forest cover has disappeared over the last 20 years (FAO 2013; Marsh *et al.* 2013). Therefore, understanding how primate populations respond to deforestation and associated habitat fragmentation will be crucial to developing appropriate effective conservation actions and policies for neotropical primates (Arroyo-Rodríguez *et al.* 2013; Marsh 2013; Marsh *et al.* 2013; Mercado & Wallace 2010; Wallace *et al.* 2013a).

Two Endangered, range-restricted and endemic Bolivian titi monkeys, *Plecturocebus modestus* and *P. olallae*, occur in a naturally fragmented forest-savanna landscape in southwestern Bolivia. Titi monkeys are small (1.0–1.5 kg), lack a prehensile tail, and typically live in groups of between 2 and 5 individuals (van Roosmalen *et al.* 2002). The titi monkeys comprise one of the most diverse groups of Neotropical primates with 33 species in three genera: *Cheracebus, Plecturocebus* and *Callicebus* (Byrne *et al.* 2016; Dalponte *et al.* 2014; Defler *et al.* 2010; Gualda-Barros *et al.* 2012; van Roosmalen *et al.* 2002; Wallace *et al.* 2006). Most of these species remain poorly studied (van Roosmalen *et al.* 2002; Bicca-Marques & Heymann 2013). Bolivia holds six species of the genus *Plecturocebus* (*P. donacophilus, P. pallescens, P. olallae, P. modestus,P. aureipalatii*), including an as yet-unidentified *Plecturocebus* species in northern Bolivia (Felton *et al.* 2006; Martinez 2010; Martinez & Wallace 2007, 2010, 2013; Wallace *et al.* 2006, 2010).

Two Bolivian species are endemic to the southwestern part of the Llanos de Moxos, Beni Department, *P. modestus* and *P. olallae* (Martinez & Wallace 2013; Wallace *et al.* 2006, 2013a, b). Following their original collection and description (Lönnberg 1939), no further data existed for these two species until they were 'rediscovered' by the Wildlife Conservation Society (WCS) in 2002 (Felton *et al.* 2006). To address the lack of information about the biology and ecology of these primates, WCS began strategic studies to detail their distribution, genetic diversity, taxonomic status, conservation status, abundance, and threats (Barreta 2007; Lopez-Strauss 2007; Lopez-Strauss & Wallace 2015; Martinez & Wallace 2007, 2011, 2013; Wallace *et al.* 2013a, b).

The geographical ranges of both endemic species are restricted (Lopez-Strauss 2007; Martinez 2010; Mercado & Wallace 2010; Martinez & Wallace 2007, 2011; Wallace *et al.* 2013b). Recent studies have found that *Plecturocebus modestus* inhabits only 8966 km^2 of forest in the region, and *P. olallae* just 267 km^2 (Lopez-Strauss 2007; Martinez & Wallace 2007; Wallace *et al.* 2013b). Unsurprisingly, given such highly restricted distributions, estimates of current population sizes are also very small: estimates of 20 027 individuals for *P. modestus* at a density of 4.94 groups/km^2, and 1927 individuals for *P. olallae* at a density of 5.94 groups/km^2 (Lopez-Strauss 2007; Lopez-Strauss & Wallace 2015; Wallace *et al.* 2013a, b).

At the global scale, the conservation status of the known distribution areas of these endemic primates can be considered to be in good condition according to a global map of human influence across the Earth's surface, called The Human Footprint analysis (Fearn 2010; Sanderson *et al.* 2002). Nevertheless, several types of human activities occur in the region (Killeen 2007; Killeen *et al.* 2002; Muller *et al.* 2014), including cattle ranching, colonization and the creation of new human settlements and agricultural land, as well as uncontrolled use of fire in land management.

Forests in southwestern Beni Department, Bolivia, are naturally fragmented and occur as islands within a vast continentally significant grassland matrix, the Llanos de Moxos wetland, much of which has a regular seasonal flooding pattern (Navarro 2001, 2002). However, more severe flooding events occurred in the rainy seasons of 2007, 2008 and 2014 with major impacts on domestic animals reported (*Sistema Nacional de Seguridad Alimentaria Alerta Temprana*, www.geobolivia.org), and anecdotal observations hinting there were also significant wildlife losses (J. Martinez, pers. obs. 2008; G. Ayala, pers. comm. 2014).

This suggests that these forests and the primate species inhabiting them may be significantly influenced by flooding events. Nevertheless, studies are lacking on the ecology of these forests, and the effects on biodiversity of both natural and anthropogenic factors including flooding and uncontrolled fire. This paucity of knowledge represents a huge challenge for the effective conservation of these endemic primates and their habitats, especially since relevant research activities are only very recent.

Though the focus of this volume is flooded habitats, in some areas fire is inundation's confederate, especially under conditions of drought. In others, it is part of a more regular cycle. In the Beni study area, extreme fire events, both natural or anthropogenic, are more likely to occur after a heavy wet season, because of the increase in the amount of fuel remaining in the forest understory following heavy and prolonged flooding (Cochrane 2003; Roman-Cuesta *et al.* 2014). This relationship between heavy flooding and fire is also well documented in tropical forests (Cochrane 2003), and is especially prevalent as a result of non-ENSO (El Niño Southern Oscillation) years of severe drought (Alencar *et al.* 2006). For example, in the 1995 non-ENSO year fire intensity was not as high as in the 1998 ENSO year, and taller forests in Amazonian Brazil were more fire resistant than lower forests (Alencar *et al.* 2006). Differences in rainfall during ENSO and non-ENSO years, combined with deforestation, forest degradation and fragmentation, compromise the capacity of a forest to retain sufficient moisture to mitigate and prevent forest fires (Alencar *et al.* 2006; Cochrane 2001, 2003; Cochrane & Schulze 1998; Tacconi *et al.* 2006).

The areas inhabited by *P. olallae* and *P. modestus* have a high risk of extensive inundation (as occurred in 2007 and 2008), and the impacts of both heavy flooding and fires is thought to have a significant impact on these locally endemic primates (Martinez & Wallace 2007, 2013; Wallace *et al.* 2013b, c). This is supported by evidence of *P. olallae* group displacement when its original territory was lost due to fires, probably associated with extreme flooding in 2007 (Martinez & Wallace 2011). In this case, although fire was in the grassland and did not reach the forest, the smoke and heat apparently were enough for monkeys to search for another territory, but no information is available about whether they succeeded. Additional research would be needed to determine how these primates respond to flooding and fire frequency and intensity; however, even the frequency of natural versus anthropogenic fires in this region is not known.

In the natural savannas of the Beni, cattle ranchers burn grasslands during the winter to generate new spring grass. Sometimes these grassland fires spread to adjacent forest patches, increasing deforestation risk due to the increased likelihood of subsequent conversion of fire-damaged forest to agriculture or pastures (Killeen 2007). Recent studies indicate a direct relationship between climatic anomalies and fire frequency in the broad study area: the interactions between fire response per biome and past climatic conditions show that Atlantic Multi-decadal Oscillation and Niña events have a large influence on fire frequency (Roman-Cuesta *et al.* 2014). While such effects have long been part of natural cycles, anthropic and fire-management policies in the Beni may be changing their intensity and nature, as well as influencing related flooding cycles.

The 'Corredor Norte' road improvement initiative is another emerging major threat for the two titi monkey species (Martinez & Wallace 2007, 2010; Wallace *et al.* 2013b). Bisecting the distributional range of these primates, this road connects all the small towns in the area and may increase the pressure on the forest islands due to increased ease of access, which leads to significant human colonization and subsequent reductions in forest cover (Fleck *et al.* 2006, 2007; Killeen 2007; Vera-Diaz *et al.* 2007). Construction designed to pave the road and raise it to a level above seasonal floodwaters may also drastically alter local hydrology, permanently changing drainage and flooding patterns.

The aim of this study was to identify the most vulnerable forest patches and assess future deforestation risks in the distribution range of each endemic titi monkey species, using spatial representations of the main factors thought to influence forest cover in the region (Forrest *et al.* 2008; Wallace *et al.* 2013b). These factors included the 2007 and 2008 flooding of forest patches, fire frequency over the past decade, as well as the direct influence of human populations, including main and secondary roads. Flooding frequency and intensity and fire risk are inextricably linked and identifying vulnerable areas is an important step towards developing conservation actions for *P. olallae* and *P. modestus*.

Methods

Study Area

The study area was in southwestern Beni Department, Bolivia, in the provinces of General Jose Ballivian and Yacuma. The Llanos de Moxos biogeographic unit is a very large area of inundated savanna interspersed with scattered habitat islands of tropical forest. The Yacuma River and several large lakes form the hydrological backdrop for seasonally inundated forest patches in the area (Navarro *et al.* 2001; Navarro & Maldonado 2002). The study area covers 11 249.7 km^2, with three polygons representing the distribution area of the two focal primate species (Figure 23.1). These polygons were defined by previously reported collection points, observation points, behavioural studies and habitat use of these regional endemics (Barreta 2007; Felton *et al.* 2006; Lopez-Strauss 2007; Lopez-Strauss & Wallace 2015; Martinez & Wallace 2007, 2011, 2013; Mercado & Wallace 2010; Wallace *et al.* 2013a, 2013b).

Plecturocebus modestus populations are found within two polygons located at the northern and southern limits of the study area. The northern *P. modestus* polygon covers 3606.76 km^2, with the area of the polygon analysed in this study being 3148.18 km^2, or around 80% of the total distribution of this species (Felton *et al.* 2006; Lopez-Strauss 2007; Lopez-Strauss & Wallace 2015; Martinez & Wallace 2007, 2013; Wallace *et al.* 2013b). The southern *P. modestus* polygon covers 1365.07 km^2. *Plecturocebus olallae* was only present in one polygon located in the centre of the study area. It covers 267.44 km^2

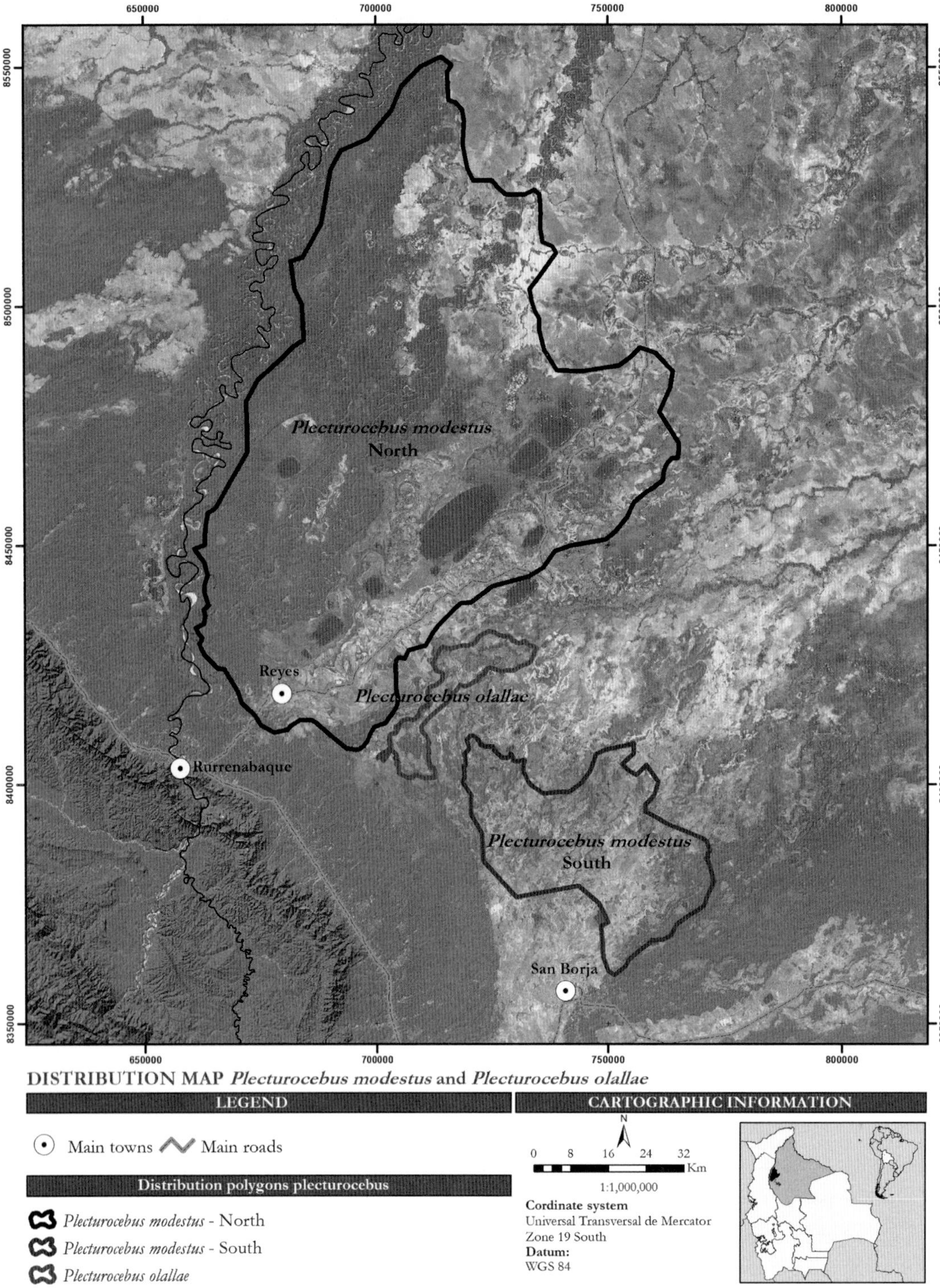

Figure 23.1 Natural fragmentation in the home range of *Plecturocebus olallae* and *Plecturocebus modestus*.

(Felton *et al.* 2006; Lopez-Strauss 2007; Lopez-Strauss & Wallace 2015; Martinez & Wallace 2007, 2013; Wallace *et al.* 2013b). In this polygon, most vegetation is riverine and seasonally flooded gallery forest along the Yacuma River (Navarro & Maldonado 2002). The flora of this region is not well known (Killen 1993; Zenteno 2010), but includes a diversity of Fabaceae species (e.g. *Tabebuia* spp., *Inga* spp.), figs (*Ficus* spp.), palms (e.g. *Attalea, Mauritia*), and several Graminae species in the grassland matrix.

Taking into account the importance of ecotourism in the area, the municipalities recognized the need to preserve the local natural richness against the negative effects of human activities. In 2007, the Santa Rosa del Yacuma municipality created a 616 453 ha municipal protected area, *Reserva Municipal Las Pampas del Yacuma* (OM 15/2007). Similarly, in 2008 the neighbouring Santos Reyes municipality created a similar protected area of 217 129 ha: *Area Natural de Manejo Integral Municipal Santos Reyes* (OM 25/2008) (Lopez-Strauss

Figure 6.1 Getting support from whatever is safe: partially flooded forest, Borneo. Photo: Ikki Matsuda. (A black and white version of this figure appears in some formats.)

Figure 7.1 A pair of thirsty rhesus macaques drinking vigilantly at a sweet water hole in tiger habitat in the Jingakhali Forest in the Sundarban Tiger Project area, India. Photo: Subrata Pal Chowdhury. (A black and white version of this figure appears in some formats.)

Figure 9.3 Northeastern coast of Maranhão state, Brazil. a: Groups of *S. libidinosus*, Rio Preguiças mangrove, Barreirinhas (2°37′21.7′ S; 42°41′18.5′W) use tools. b: Groups of *S. libidinosus*, Rio Novo mangrove, Paulino Neves (2°42′52.5′ S; 42°31′38.5′W) do not use tools. Tidal floodplain forest = *Várzea de maré*; Apicum = high salinity area inside or behind the mangrove forest that does not support mangrove or other tree species. (A black and white version of this figure appears in some formats.)

Figure 12.3 Adult male Zanzibar Sykes's monkey *Cercopithecus mitis albogularis* in mangrove at Vanga, Kenya. Photo: Y.A. de Jong and T.M. Butynski (www.wildsolutions.nl). (A black and white version of this figure appears in some formats.)

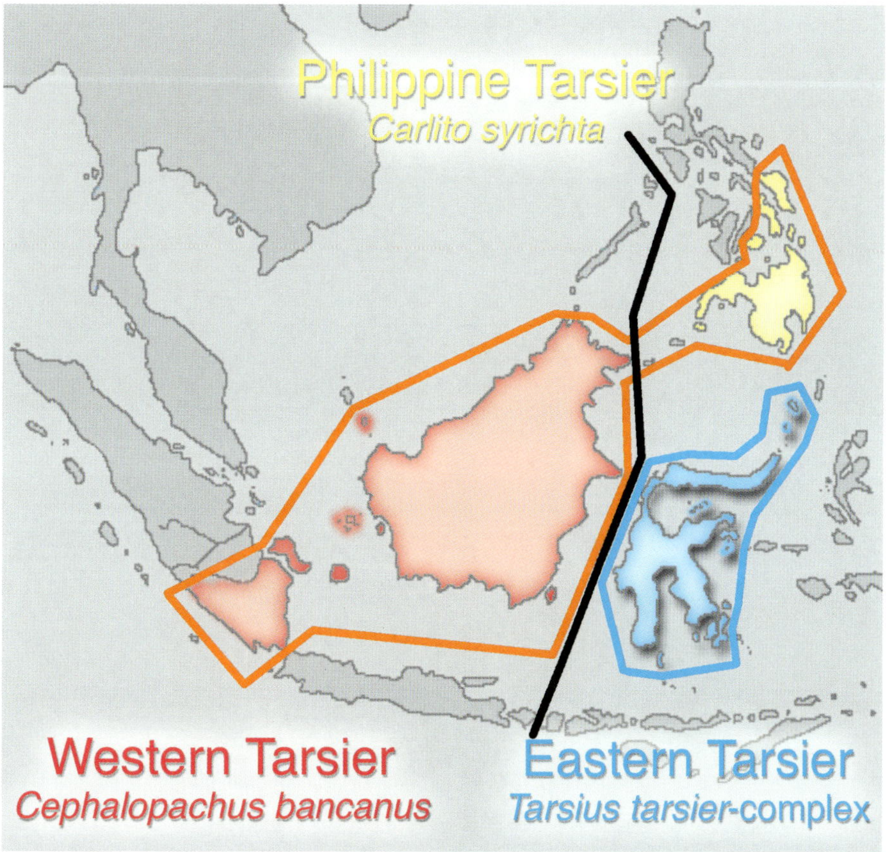

Figure 15.1 Tarsier taxonomy and biogeography. (A black and white version of this figure appears in some formats.)

Figure 16.2 Arboreal perch of a rhesus monkey in mangroves. (A black and white version of this figure appears in some formats.)

Figure 16.3 Rhesus monkey foraging in the mud flats during low tide. (A black and white version of this figure appears in some formats.)

Figure 16.4 A juvenile rhesus monkey searches for crabs. (A black and white version of this figure appears in some formats.)

Figure 17.3 Angola Pluto monkey *Cercopthecus mitis mitis* eating aquatic plant leaf in Kwanza River mangroves, Angola. Photo: Tommy Pedersen. (A black and white version of this figure appears in some formats.)

Figure 17.5 Long-tailed macaque *Macaca fascicularis* swimming in Muara Angke wildlife reserve in North Jakarta, the last remaining mangrove in Jakarta, Indonesia. Photo: Nurul Winarni. (A black and white version of this figure appears in some formats.)

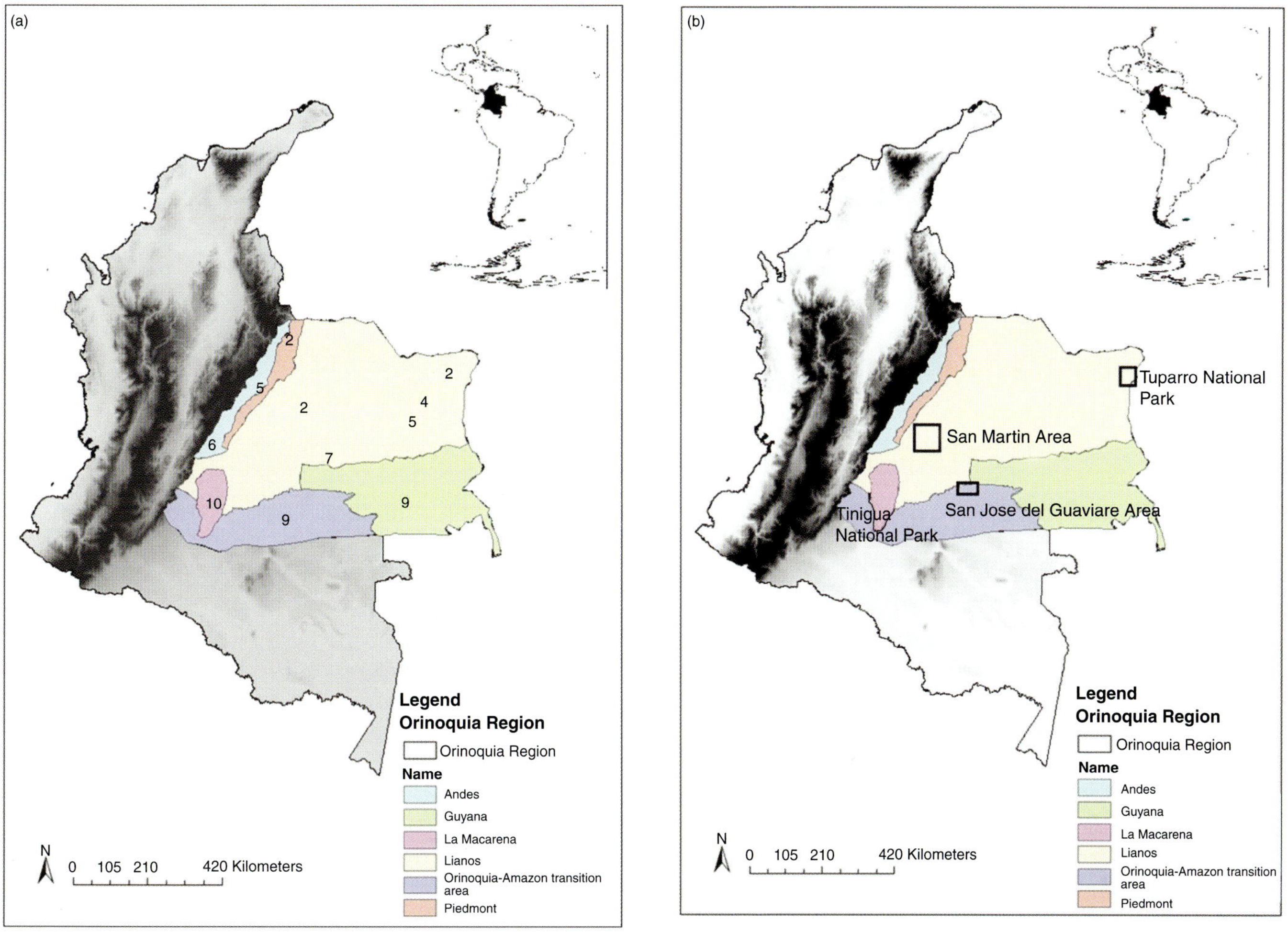

Figure 21.1 (a) Biogeographic regions of the Colombian Orinoquia, as defined by Lasso *et al.* (2010). The numbers represent primate species richness based on Defler (2010). (b) Sites of long-term primate studies in the Orinoquia region of Colombia. (A black and white version of this figure appears in some formats.)

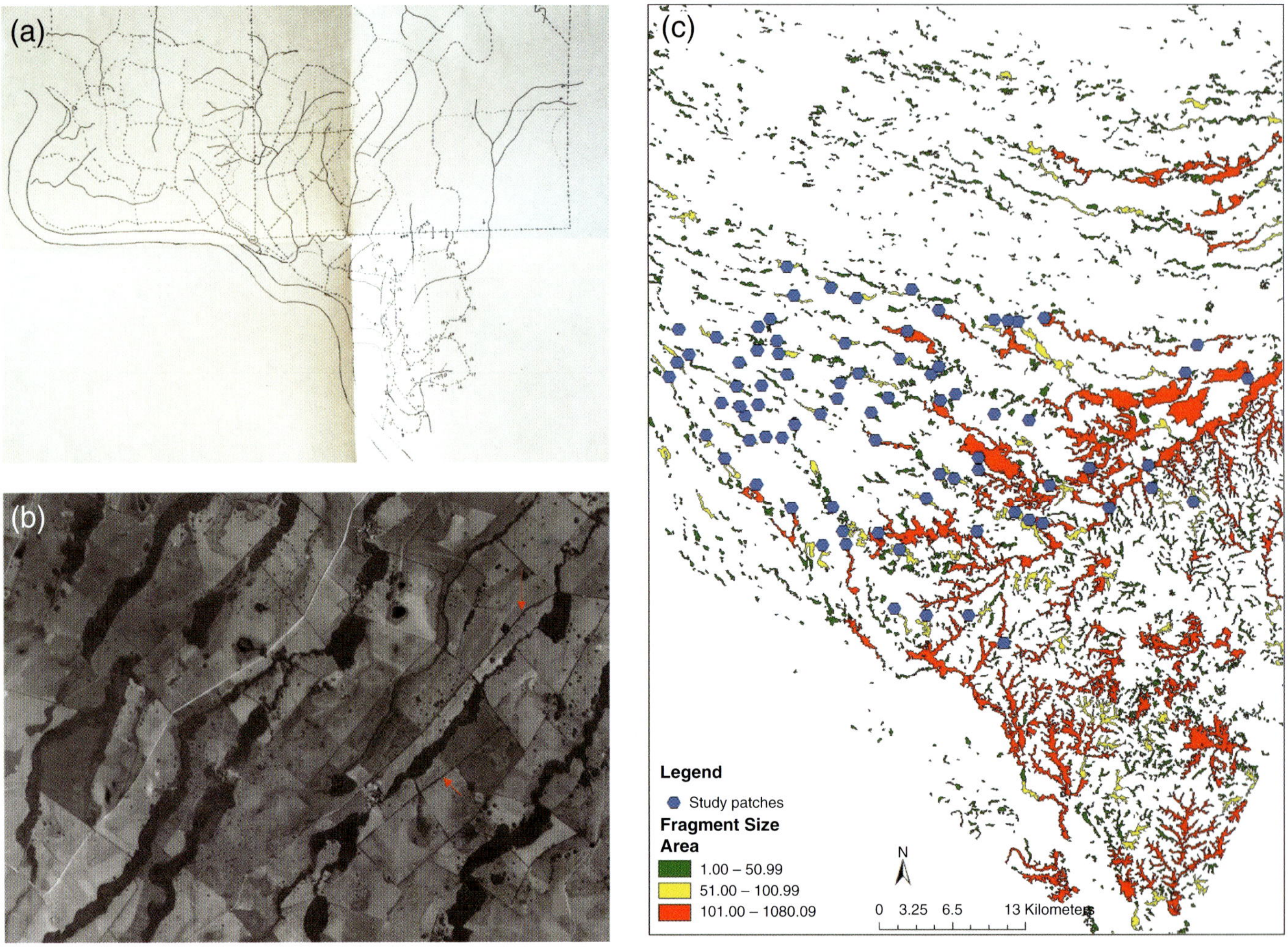

Figure 21.2 (a) Study site at Tinigua National Park (darker line represents the home range of the study group of Colombian squirrel monkeys [*S. c. albigena*], Carretero-Pinzón 2000); Study site at San Martin area: (b) Fragments used by the study group of *S. c. albigena* (red arrows show natural hedgerows (i.e. living fences) used by the study group); (c) Forest fragments surveyed from 2004 to 2014. (A black and white version of this figure appears in some formats.)

Figure 22.1 The Pantanal wetland located in the centre of South America, surrounded by the upland plateaus, where headwaters of the Paraguay River which feeds the wetland are located. These occur mostly in Brazil (140 000 km^2), whose habitats are biogeographically influenced by the *cerrado* biome (east), Amazonia (north), Atlantic forest (south), as well as the Chaco at the Bolivian and Paraguayan borders (west). (A black and white version of this figure appears in some formats.)

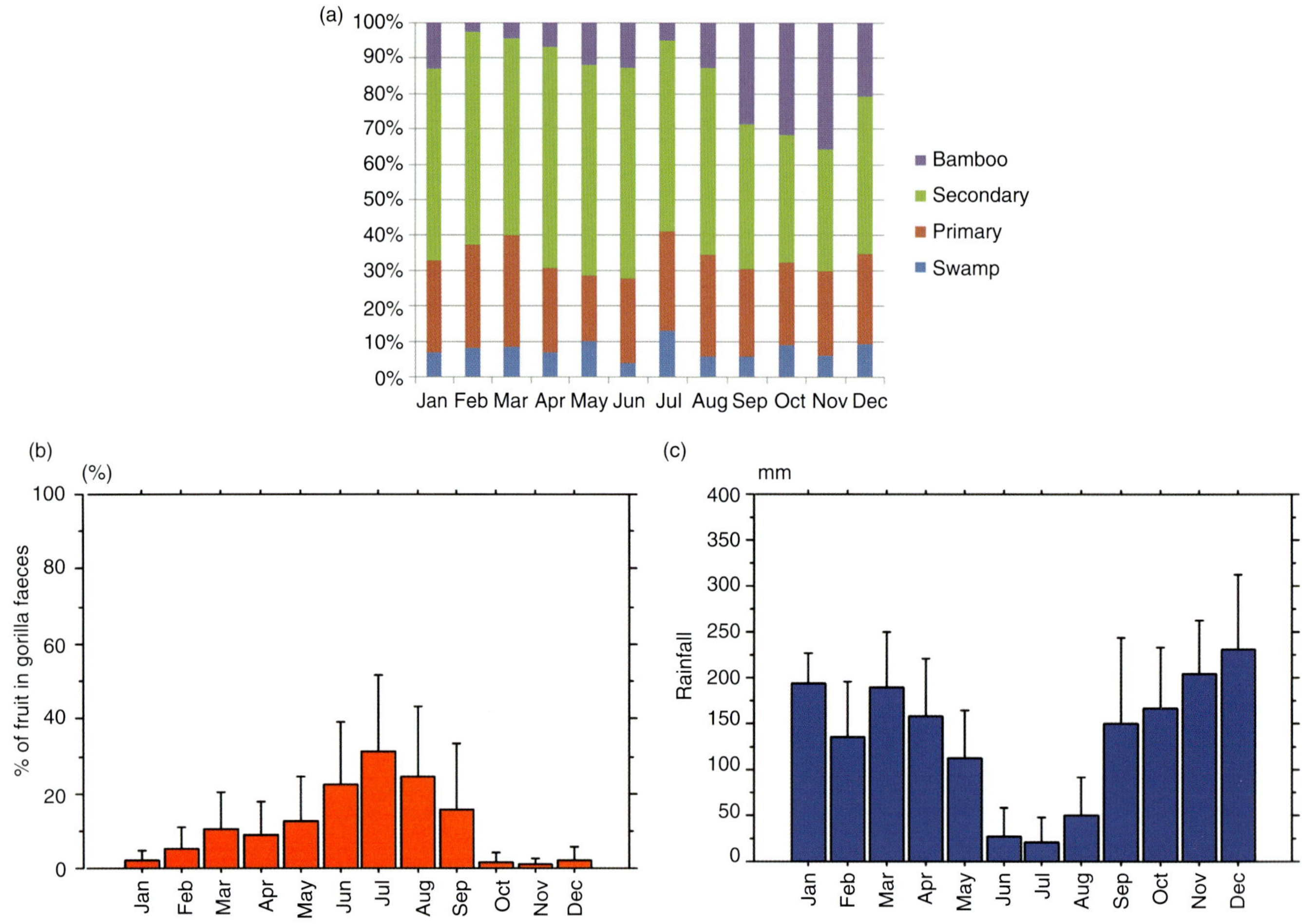

Figure 24.5 The mean monthly proportion of each type of vegetation along daily path routes (a), the mean monthly proportion of fruit remains per faecal sample (b), and the mean monthly rainfall (c) at Kahuzi. (A black and white version of this figure appears in some formats.)

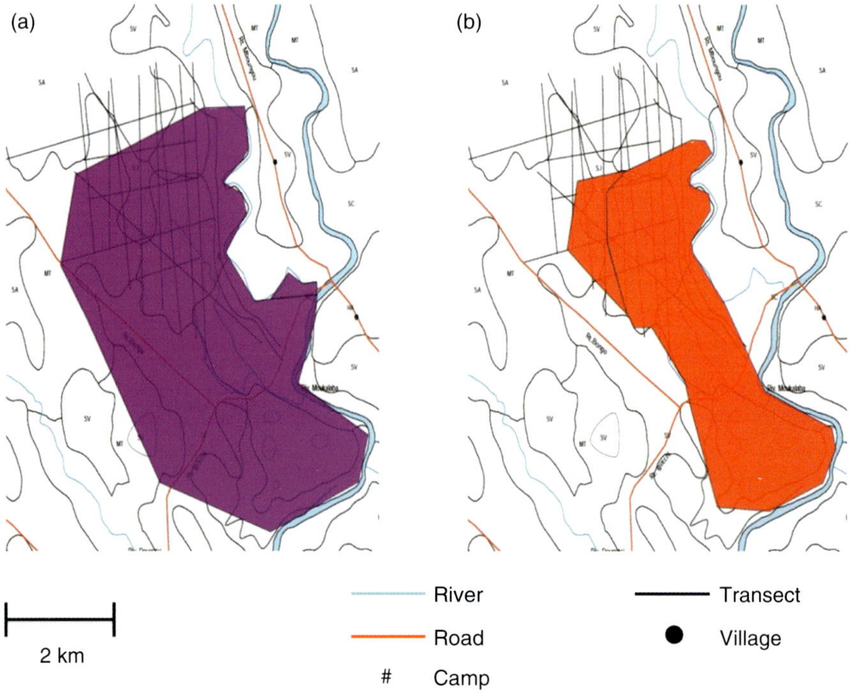

Figure 24.6 Home range of the study group in the fruiting (left) and the non-fruiting season (right) at Moukalaba. (A black and white version of this figure appears in some formats.)

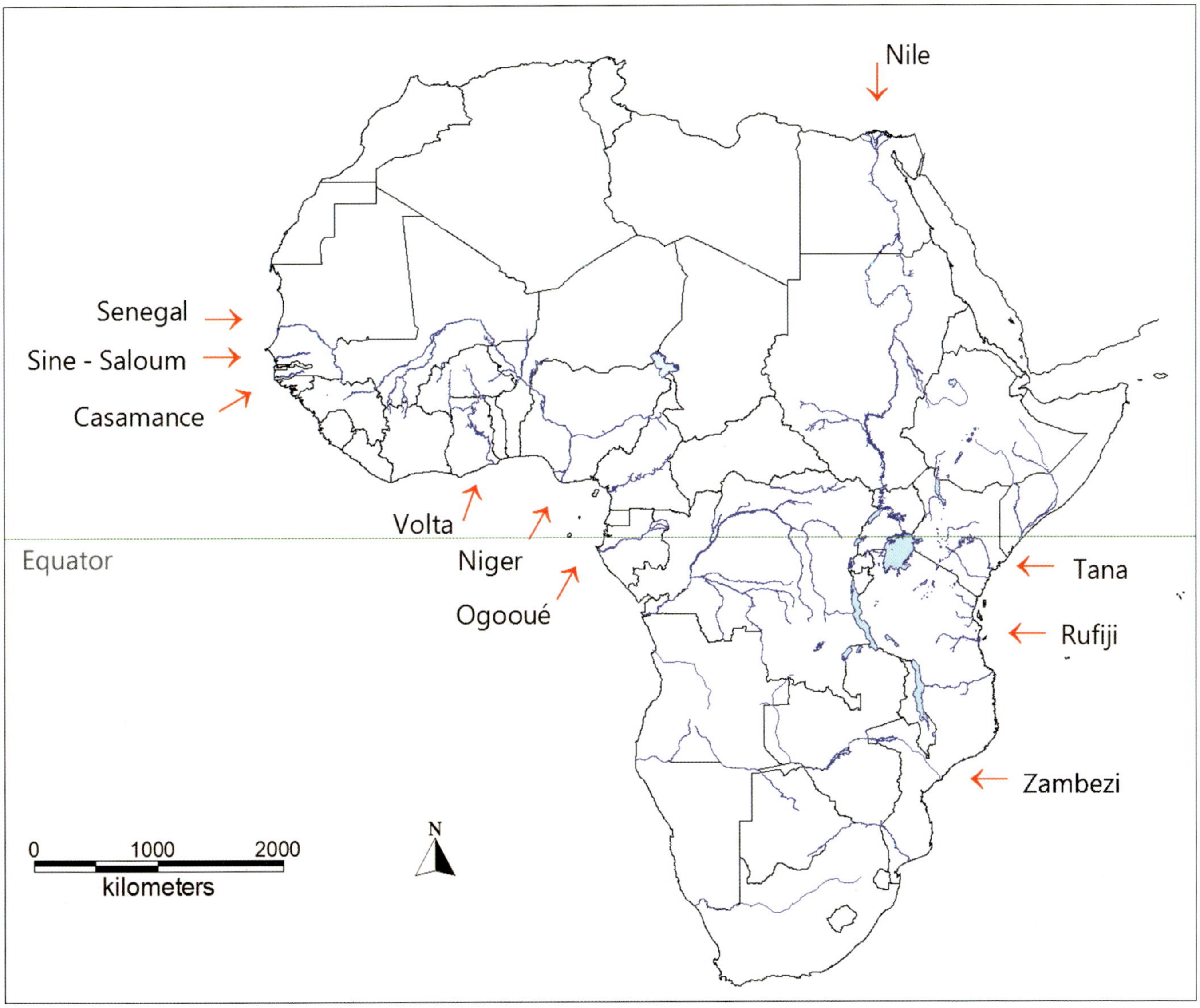

Figure 30.1 Africa's ten largest coastal deltas. Map by Yvonne de Jong & Tom Butynski. (A black and white version of this figure appears in some formats.)

Figure 30.2 Adult male Peters's Angola colobus *Colobus angolensis palliatus*. This eastern Africa endemic subspecies is not known to occur south of the Rufiji River and Rufiji Delta, Tanzania. Photo: Yvonne de Jong & Tom Butynski, wildsolutions.nl. (A black and white version of this figure appears in some formats.)

Figure 30.3 Adult male Tana River red colobus *Piliocolobus rufomitratus rufomitratus*. Fewer than 1000 individuals remain of this 'Endangered' subspecies, which is endemic to the forests of the lower Tana River and Tana Delta, Kenya. Photo: Yvonne de Jong & Tom Butynski, wildsolutions.nl. (A black and white version of this figure appears in some formats.)

Figure 30.5 Adult male Pousargues's monkey *Cercopithecus mitis albotorquatus*. This 'Vulnerable' subspecies is present in the forests of the lower Tana River and Tana Delta, Kenya. Photo: Tom Butynski & Yvonne de Jong, wildsolutions.nl. (A black and white version of this figure appears in some formats.)

Figure 30.9 Juvenile Kenya coast dwarf galago *Galagoides cocos*. This is one of at least three species of galago present in the Tana Delta, Kenya. Photo: Yvonne de Jong & Tom Butynski, wildsolutions.nl. (A black and white version of this figure appears in some formats.)

Figure 30.10 Adult female and juvenile Sclater's monkeys *Cercopithecus sclateri*, in Lagwa, Imo State, Nigeria. This 'Vulnerable' species is endemic to southern Nigeria. This is one of 17 species of non-human primate present in the Niger Delta. Photo: Lynne R. Baker. (A black and white version of this figure appears in some formats.)

Figure 30.11 White-throated monkey *Cercopithecus erythrogaster pococki* in Okomo National Park, Nigeria. This 'Vulnerable' subspecies is endemic to southwest Nigeria. This is one of 17 species of non-human primate present in the Niger Delta. Photo: Noel Rowe. (A black and white version of this figure appears in some formats.)

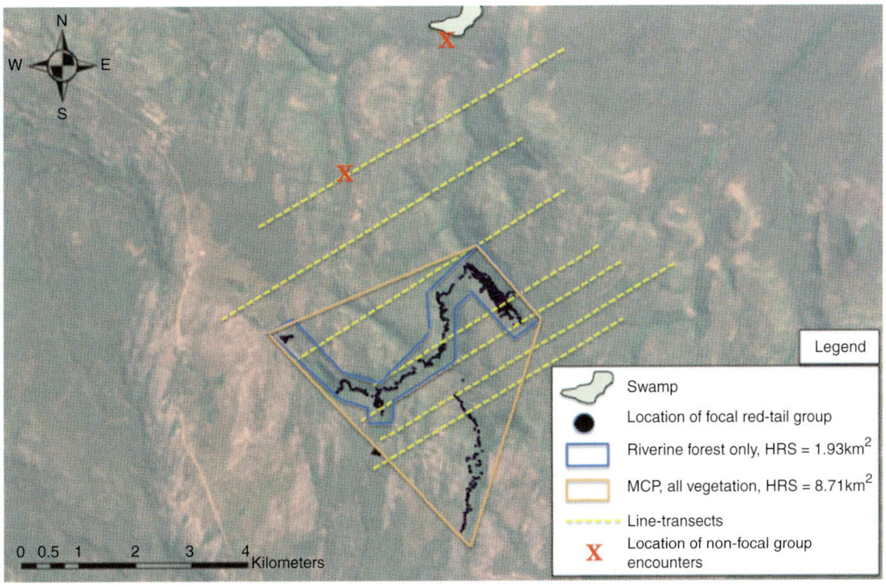

Figure 33.2 Issa Study Site, Ugalla: Red-tail home range for February–April 2012. Polygons surrounding the focal group follows describe the imprecision of methods e.g. orange polygon (8.71 km^2), all vegetation types included, versus blue polygon (1.93 km^2, riverine forest only with ~ 100 m buffer). (A black and white version of this figure appears in some formats.)

Figure 36.2 Balbina Dam Reservoir in Brazil (low topographic relief). (A black and white version of this figure appears in some formats.)

Figure 38.1 The Kinabatangan floodplain. Forest fragments interspersed with oil palm plantations, human settlements and local agriculture. Photo: HUTAN. (A black and white version of this figure appears in some formats.)

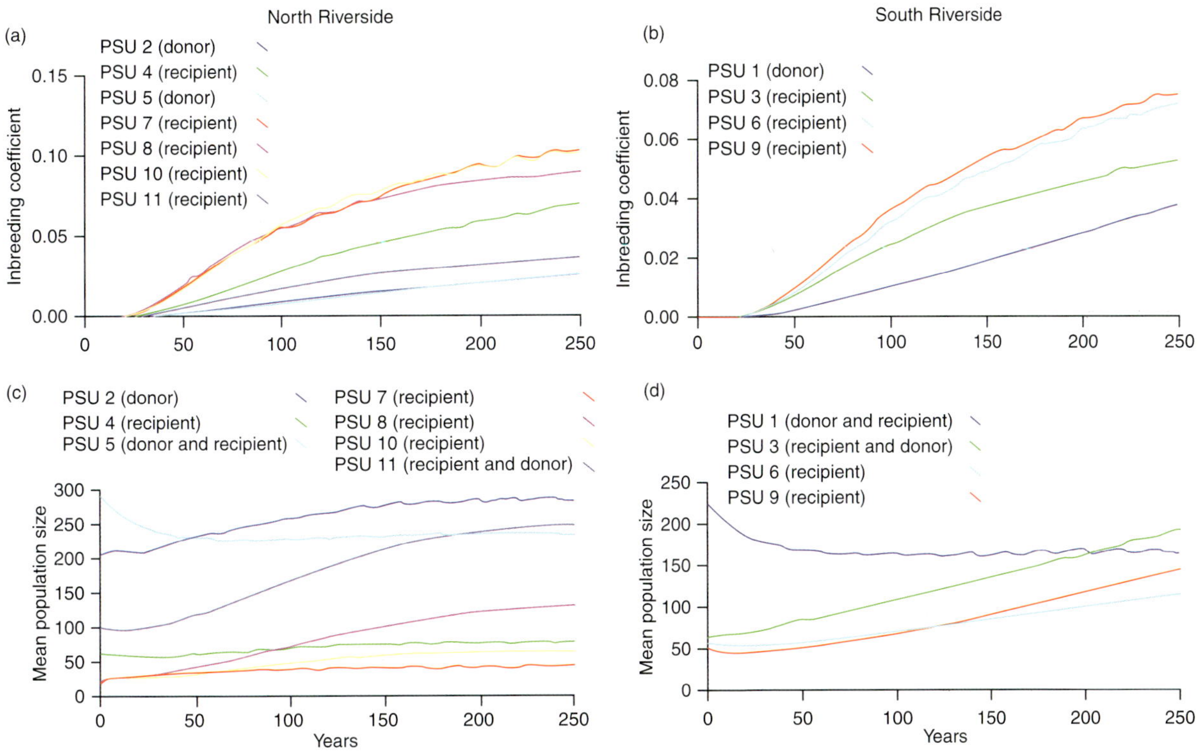

Figure 38.4 Effect of corridor connection and translocations on within-PSU inbreeding coefficients and mean population sizes. (a) Effect of the mixed approach on the inbreeding coefficients of the PSUs on the north riverbank; (b) effect of the mixed approach on the inbreeding coefficients of the PSUs on the south riverbank; (c) effect of the mixed approach on the population sizes of the PSUs on the north riverbank; (d) effect of the mixed approach on the population sizes of the PSUs on the south riverbank. (A black and white version of this figure appears in some formats.)

Figure 39.6 Olive baboon *Papio anubis* crossing the Arli River, jumping to avoid walking or swimming in the water, Burkina Faso. © Galat-Luong A.-IRD. (A black and white version of this figure appears in some formats.)

Figure 39.7 Group of olive baboons, *Papio anubis*, resting on the bank of the Arli River, Arli National Park, Burkina Faso. © Galat-Luong A.-IRD. (A black and white version of this figure appears in some formats.)

Figure 39.10 Olive baboons, *Papio anubis*, eating *Mutelidae* mussels in the Arli River, Burkina Faso. © Galat-Luong A.-IRD (A black and white version of this figure appears in some formats.)

Figure 40.2 Sclater's monkey *Cercopithecus sclateri*, an 'Endangered' primate that occurs only in southeastern Nigeria. Photo: Lynne R. Baker. (A black and white version of this figure appears in some formats.)

Figure 40.3 Invasive water hyacinth *Eichhornia crassipes* in a river in the Niger Delta. Photo: Lynne R. Baker. (A black and white version of this figure appears in some formats.)

Figure 40.6 Cut planks of timber in the Stubbs Creek Forest Reserve in southeastern Nigeria. Photo: David L. Garshelis. (A black and white version of this figure appears in some formats.)

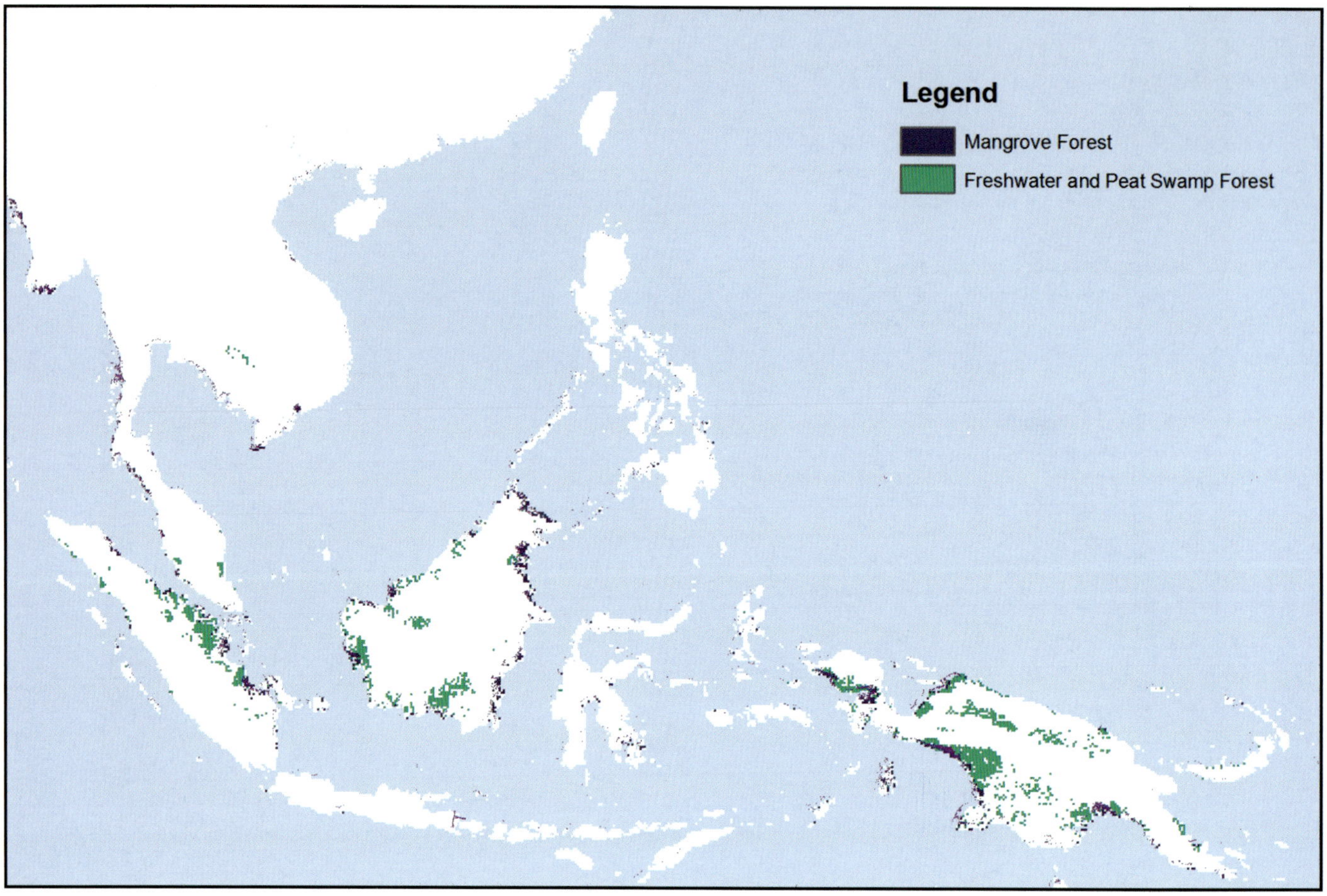

Figure 44.1 Map of Southeast Asia and flooded forests found in this region. From an amalgamation of maps from the following sources: Giri *et al.* (2011). Meittinem *et al.* (2011) and Stigbig *et al.* (2002, 2004). (A black and white version of this figure appears in some formats.)

2007; Martinez & Wallace 2007; Mercado & Wallace 2010; Wallace *et al.* 2013b). In both areas, the presence of the endemic *Plecturocebus* species is considered important, and to date efforts to implement management activities have conceptually embraced the conservation needs of these primates. For example, the zoning proposals of both municipal protected areas explicitly consider the distribution ranges of both endemic titi monkeys. Thus, the municipalities and many of the region's cattle ranchers are willing to actively apply conservation interventions to preserve these two titi monkeys, which are considered emblematic species of the region (Martinez *et al.* 2011).

Second, a recent outreach campaign (2011–2012) communicated information about the titi monkeys and their conservation to local students and authorities, aiming to raise awareness about the singularity of these threatened endemics and their protection needs. Large audiences attended the presentations and local media helped distribute content about the conservation of titi monkeys and the forest they inhabit. The participation of local authorities was remarkable and included organizing public events where children expressed their comprehension and interest in biodiversity conservation. The local population clearly embraced the titi monkeys as symbols of their natural world, and was also aware of current threats (Carvajal *et al.* 2013; WCS, unpublished data).

Previous studies, carried out between 2002 and 2013, identified the known distribution areas of both species through direct observation and local interviews based on confirmed presence and reported absence (Felton *et al.* 2006; Lopez-Strauss 2007; Lopez-Strauss & Wallace 2015; Martinez & Wallace 2007, 2011, 2013; Mercado & Wallace 2010; Wallace *et al.* 2013a, 2013b). Together, over the last decade, these studies increased known distribution points for the two species from one to 53 for *P. modestus*, and one to 21 for *P. olallae* (Martinez & Wallace 2007, 2013; Wallace *et al.* 2013c).

In the analysis herein, we focus on a 486 779.46 ha area where a deforestation study was conducted from 2005 to 2009 (WCS, unpublished data), and which represents 60.7% of the total distribution range (801 500.12 ha) of the two *Plecturocebus* species (Figure 23.2).

Selection of Forest Patches

Using data from a deforestation study comparing Landsat satellite images from 2005 with images from 2009 (WCS, unpublished data), and information about the home range of both species (4 to 9 ha; Martinez 2014), forest patches in the range of each species were selected for analysis. All forest patches ≥ 4 ha were selected for *Plecturocebus modestus*. Given their extreme range restriction (Lopez-Strauss 2007; Martinez & Wallace 2011), all forest patches for *P. olallae* were included, irrespective of their size. The very small distribution of *P. olallae* (50 km^2) makes this species more vulnerable to habitat loss and degradation within its range (Wallace *et al.* 2013a, b), where it inhabits riverine and gallery forests with small trees and high vine densities (Martinez & Wallace 2007, 2010).

Threat Assessment

We considered five factors for assessing future deforestation risk: (i) landscape metrics, (ii) changes in flooding patterns, (iii) fire frequency, (iv) roads and buffered areas of influence and (v) human-populated places and buffered areas of influence. Each factor was measured at different scales using the most recent publically available data (see below).

Landscape Metrics

We described fragmentation levels within each of the three titi distribution polygons with three selected landscape metrics from FRAGSTATS (McGarigal & Marks 1995), a spatial analysis software that quantifies several landscape features. For the fragmentation analysis, we used a deforestation analysis using Landsat images (Painter *et al.* 2013). Forests in the region are naturally fragmented, and there is documented human presence for at least three to five thousand years (Mann 2000), consequently we were not able to distinguish between natural fragmentation and historical anthropogenic deforestation. The metrics we used were:

(a) *Number of Patches* (Rutledge 2003) of forest within each estimated range polygon,
(b) *Largest Patch Index* (LPI): percentage of a polygon occupied by the largest forest patch (McGarigal & Marks 1995; Rutledge 2003),
(c) *Mean Distance Between Patches*: based on Euclidean distance calculation between forest patches using edge to edge distance. This value increases if the patches are isolated and smaller, and therefore affect the movement and dispersal of species (McGarigal & Marks 1995; Midha & Mathur 2010; Rutledge 2003).

Flooded Areas from 2007 to 2008

Forests in the study area are adapted to regular annual flooding, but it is usually ephemeral, lasting only a fortnight (Navarro *et al.* 2001; Navarro & Maldonado 2002). However, recent flooding events (2007, 2008) have been more extensive, with each event lasting one month, as well as more intensive with significantly raised flooding levels; however, details of how precisely these changes have affected titi monkey habitat is unknown. It is important to emphasize that although protracted flooding has occurred before, such extreme events in two successive years have not previously been reported, and the lack of studies in the area impede the assessment of the consequences of flooding on wildlife or human activities. To identify the forest patches affected by flooding during the 2007 and 2008 events we used SINSAAT (Sistema Nacional de Seguridad Alimentaria Alerta Temprana, www.geobolivia.org) maps of these extreme events. This early response system used the MODIS FAS images and is part of the Environmental and Water Management Ministry of the Bolivian government.

Fire Frequency in the Ranges of Titi Monkeys

Heavy flooding events are known to increase forest fire risks because the extensive debris left in their wake functions as potential fuel for fire (Cochrane 2003). Forest fires are clearly

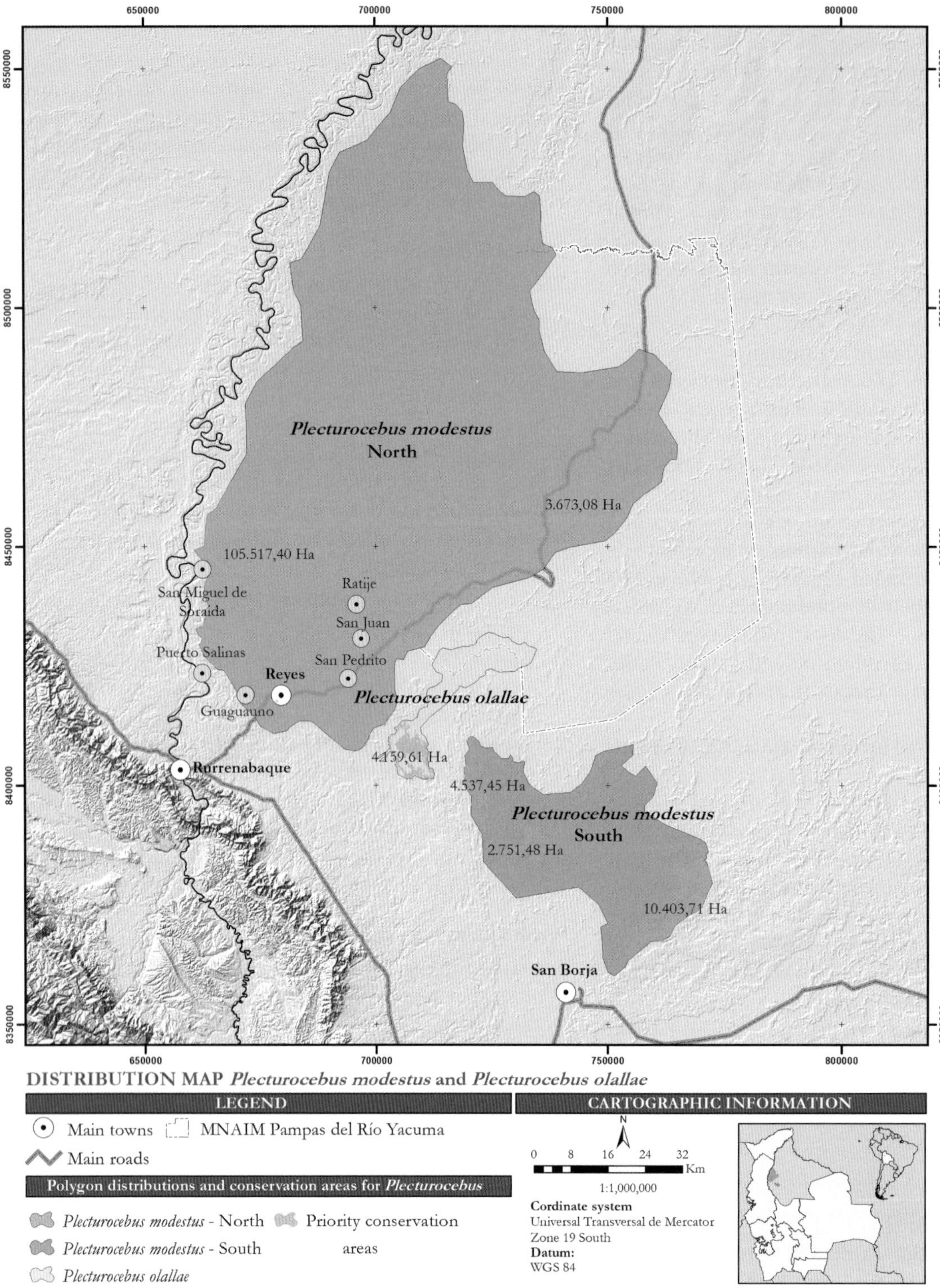

Figure 23.2 *Plecturocebus modestus* and *P. olallae* distribution polygons in southwestern Beni Department, Bolivia.

linked to further future risk of deforestation. To characterize fire frequency within the distribution of each titi monkey species, we used hot spots registered on a monthly basis between 2000 and 2012 by NASA EOSDIS (Earth Observing System Data Information System; https://earthdata.nasa.gov/data/near-real-time-data/firms) with the MODIS Satellite (Moderate Resolution Imaging Spectro-radiometer) (Giglio *et al.* 2009; Roy *et al.* 2008), and MODIS Burned Area (Tsela *et al.* 2010). After combining MODIS hot spots and MODIS Burned Area information using the MAP ALGEBRA of GDAL QGIS® 1.8, the results were converted to raster grids, with the RASTER Calculator of GDAL QGIS® 1.8 used to identify fire frequencies and produce a fire frequency map. For each year, ten monthly images were cumulatively combined, with January and February excluded from analysis due to a lack of hotspots from those months during most years used in the analysis.

Fire frequency in each 500 × 500 m MODIS analysis cell was categorized according to the number of sample years in which fire occurred in that cell, using values of 0 to 12 years (Roy *et al.* 2008; Roy & Boschetti 2009). We classified cells into two broad categories: those having < 4 years with fire detected, and those having 4 to 8 years with fire detected (between 2000 and 2012 for both).

Roads' Area of Influence for Future Deforestation

Roads have a major influence on deforestation rates, both in the broader Amazon region (Killeen 2007), and in the general study area (Forrest *et al.* 2008). Five main roads, including the Corredor Norte, occur within the study area. We assessed the potential influence of roads on the distribution of both titi monkey species via the threat of current and future deforestation. We created a buffer of 2 km on each side of the roads based on previous findings on the influence of roads on deforestation extent in the study area (Forrest *et al.* 2008).

Urban Area of Influence

We used the population census of 2001 (www.ine.gob.bo) to identify each urban zone of influence, and classified urban zones into two classes according to their population size: small towns (between 100 and 1000 people) and villages (fewer than 100 people). After classification, we applied a buffer of 1 km for villages and 2 km for small towns as areas vulnerable to potential future deforestation. A previous analysis of deforestation in the region (Forrest *et al.* 2008) showed that net change in vegetation cover starts decreasing at a distance of 2–3 km from human settlements. In the current analysis we considered that small towns influence up to a distance of 2 km, whereas deforestation around smaller villages is only significant up to a distance of 1 km.

The landscape metrics helped identify the priority distribution polygons for conservation actions for each species. In our qualitative assessment of future deforestation risk we simply assessed whether or not, and to what extent, individual forest patches within each polygon were at risk from each of the following variables: forest fires, flooding events, proximity to roads and proximity to human settlements.

Results

Plecturocebus modestus Northern Polygon

This, the largest polygon, covers 28% of the entire study area. Within it, 44% of the area has some kind of forest cover, with a total of 589 forest patches. In this polygon, we have 83 records of *P. modestus*: 65 observations, one collected specimen, 15 heard, plus two places where absence of *P. modestus* has been confirmed. During the 2007 heavy floods, this polygon was not affected, but 17 forest patches with a total area of 11 495.79 ha (or 8.9% of total forest area) were affected by the 2008 flooding (Figure 23.3). Presence of *P. modestus* was confirmed in two of the 17 forest patches. One 3670 ha forest patch with confirmed *P. modestus* presence (23 records: four heard and 19 observations) was affected by 2008 floods, had no presence of fire, but is surrounded by populated places and roads, making it potentially more vulnerable to deforestation.

Of the 589 forest patches in this polygon, only seven were frequently affected by fire (4–6 years), but, in total, 76.5% of the forest area had been affected by fire at least once in the last decade. The largest forest patch in this polygon (105 580.71 ha) had, in sum, been impacted by fire over four successive years at different locations, and has towns, villages and roads surrounding it, making it potentially vulnerable to future deforestation.

Overall, in terms of future potential deforestation risk, roads influence 83% of the total forest area (172 forest patches), including most of the smallest patches, seven of which have confirmed *P. modestus* presence. Human population centres influence 15 of the largest forest patches, representing 84% of the total forest cover in the study area. Two of those forest patches had confirmed presence of *P. modestus*.

Finally, the largest forest patch accounts for 18.48% of this polygon (Large Patch Index or LPI: 18.48). The mean distance between patches is 123.78 m (Table 23.1).

Plecturocebus modestus Southern Polygon

This smaller polygon represents 12% of the total study area, with 396 forest patches totalling 34 852.97 ha of forest cover, representing 25.5% of the total polygon area; the largest forest patch is 6527.04 ha. This distribution polygon is more vulnerable than the northern one because both the number of forest patches, and the size of the largest patches, indicate a higher overall level of fragmentation. This polygon has eight locations where animals are believed to occur, though only five have presence confirmed. There are three fragments in the centre of the polygon where absence is confirmed. These encompass a landscape dominated by grasslands with only small forest patches without the high abundance of vines characteristic of sites inhabited by titi monkeys. Four of the areas with confirmed presence are located in the northern part of the polygon and one is in the south (Figure 23.4), within the largest forest patches.

Impacts of flooding varied between years. In 2007, flooding affected more forest patches (11), but only 6.3% of the forest coverage, while the 2008 event affected a greater overall area (13.2% of the forest coverage in the polygon), but this was restricted to only two forest patches.

Thirteen forest patches were affected by fire, representing 6.3% of the total forest area or 1.5% of the total area. Eleven of the affected forest patches were small (4–30 ha), and two were larger (300–1000 ha). However, according to previous distributional studies, none of the forest patches affected by fire had confirmed *P. modestus* presence.

For future deforestation risk, around 44.8% of forest area is near some form of human population centre (12 forest patches), and 110 forest patches (37.3% of total forest area) are at risk from roads.

Landscape statistics show that this distribution polygon has a LPI of 2.4 (2.4% of area occupied by the largest forest patch), and a mean distance between patches of 147 m (Table 23.2). The forest in this polygon is relatively more fragmented compared to the other two we assessed.

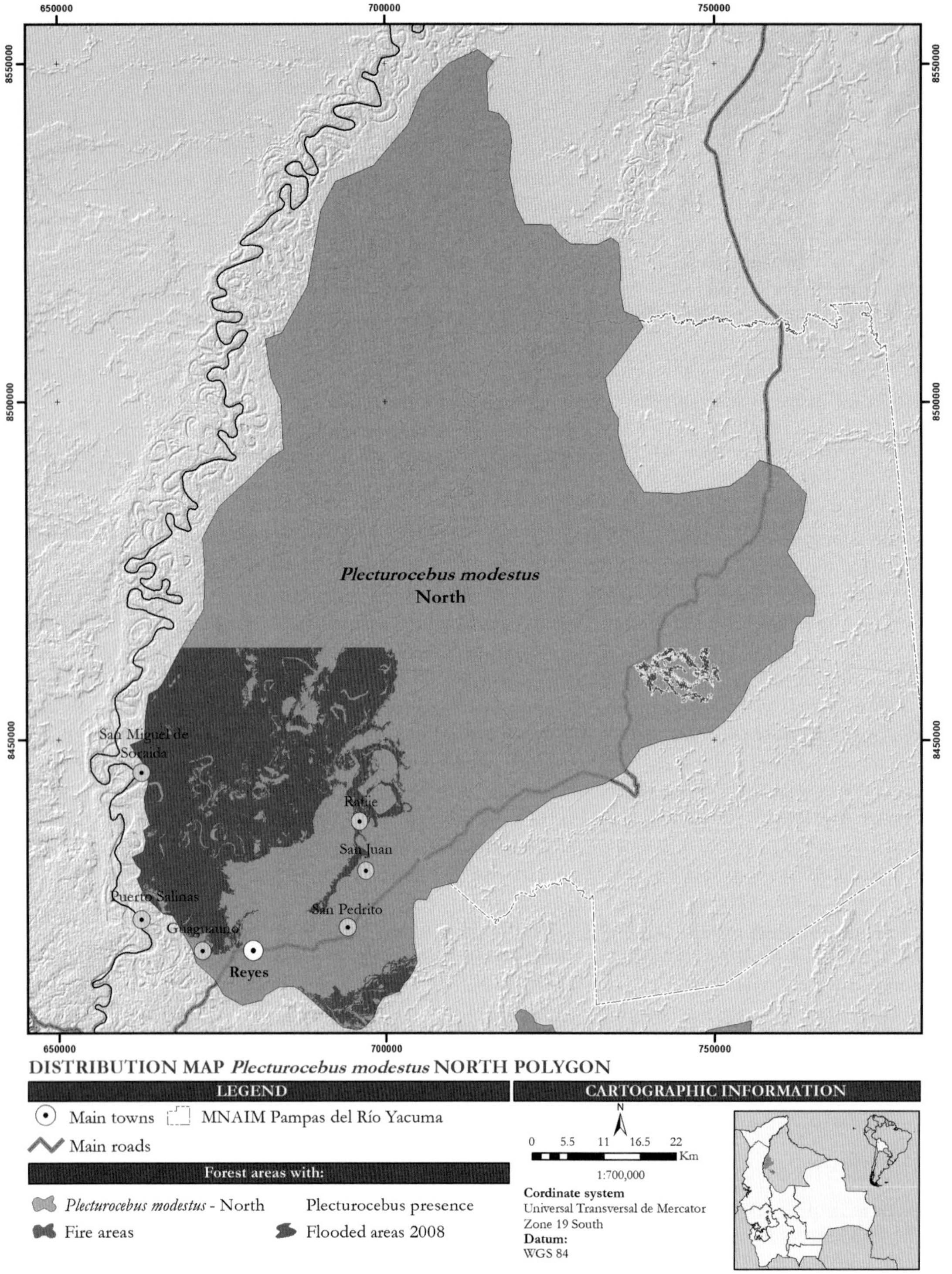

Figure 23.3 *Plecturocebus modestus* north polygon indicating flooding, fire and *Plecturocebus* presence.

Plecturocebus olallae Polygon

This is the smallest polygon with an overall area of just 26 744.2 ha, around 2.3% of the entire study area. It has a total of 622 forest patches, and approximately 23% of the polygon has some kind of forest cover; however, reflecting the high fragmentation level for this polygon, only three forest patches are larger than 100 ha (Table 23.3). Within this area, there are 17 records of *P. olallae*: 16 observations and one collection. In this polygon, 308 forest patches representing 81% of forest cover were affected by the more prolonged 2008 floods (Figure 23.5).

The largest forest patch (4159.61ha) representing 67% of the total forest area is near a populated place, and has confirmed presence of *P. olallae*. There are 129 forest patches influenced by roads, representing around 75% of the total forest area, but presence of *P. olallae* has only been confirmed in nine of these.

Landscape statistics show that 3.94% of the polygon is occupied by the largest forest patch (LPI: 3.94). The mean distance

Table 23.1 Total polygon area, total forest area and areas affected by fire, flooding, human populations and roads in the *Plecturocebus modestus* north polygon.

	Categories	Total area (ha)	Forest patch area (ha)	Average area (ha)	# Patches	% Total area	% Total forest area	Records
P. modestus North	All patches	314 818.71	140 216.58	238.05	589	44.5		83 records 1 collection 65 direct observations 15 heard 2 confirmed absence
	Largest forest patch		105 580.71			33.53	75.2	
	Fire presence		107 288.73	15 326.96	7	34.1	76.5	
	Flooding 2007		–	–	–	–	–	
	Flooding 2008		12 427.92	731.05	17	3.9	8.9	
	Populations 1 km buffer		117 918.95	3 369.11	35	37.5	84.1	
	Populations 2 km buffer		105 656.99	13 207.12	8	33.6	75.4	
	Roads 2 km buffer		116 426.25	672.89	172	37.0	83.0	
	Confirmed *P. modestus* presence		15 876.27	933.89	17	5.0	11.3	

between forest patches is 119.7 m (Table 23.4), again demonstrating that forest in this polygon is extremely fragmented.

Discussion

Recent studies of deforestation analysing data between 2001 and 2009 covering the entirety of the distribution of both species of titi monkeys in Bolivia showed forest loss of 0.17% for *P. modestus* and 0.39% for *P. olallae* (Wallace *et al.* 2013b). Deforestation rates for the three polygons in our study are low compared with the national average of 0.5% for the 2000–2010 period (Killeen *et al.* 2007, 2008; Wallace *et al.* 2013b). Nevertheless, it gives cause for concern since this forest is naturally fragmented, and recent human influence is increasing fragmentation levels.

Our analysis of future deforestation risk is limited by the differing scales of information available for each variable. For example, a 30 m resolution for forest cover compared to a 500 × 500 m resolution for MODIS fire data – as well as uncertainty as to which of the variables is most important in predicting future deforestation. Nevertheless, a qualitative risk assessment was possible by comparing the landscape metrics at the forest polygon scale and then assessing forest patches according to fire and flood risk and proximity to roads and human settlements.

In the northern *P. modestus* polygon, the largest forest patch is located between the Beni and Yacuma rivers within the Pampas del Yacuma Municipal Reserve, which represents a deforestation and fragmentation buffer for *P. modestus*. This distribution polygon is therefore an important opportunity for *P. modestus* conservation, particularly because fragmentation levels in the southern *P. modestus* polygon are very much greater than in the northern one. There are seven human population centres around the northern polygon: Guaguauno, Puerto Salinas, Ratije, San Juan, San Pedro, Soraida and Reyes the biggest regional town, and the local municipal centre. In the southern polygon, the only major conurbation is San Borja, but there are several small villages of less than 300 people, and two large cattle ranches: *Las Palmeras* and *El Triunfo*.

For *P. olallae*, the situation is alarming as the species is so far only recorded in two of the three larger forest patches. According to previous studies, the most important forest patches for conservation action are located in the western portion of the *P. olallae* distribution. Prioritizing conservation actions for this area is of paramount importance given the ongoing road development project *Corredor del Norte* in the area (Wallace *et al.* 2013b).

The distribution polygons of both species are located in naturally flooded forest and flooded savanna, with annual rainfall close to 3000 mm (Navarro & Maldonado 2002). Typical annual flooding events last approximately two weeks, but three events in 2007 and 2008 included in analyses herein, and an even more dramatic event in 2014, lasted for more than a month. These extreme-flooding events probably affect the natural flooding cycle and mid-term water flow of the area, as well as local vegetation that is not adapted to extended inundation. To date, however, there are no studies on the effects of these extreme-flooding events on the ecology of the region, or more specifically their impact on general forest ecology, nor on the ecology of the two *Plecturocebus* species.

Fire is a traditional savanna management practice for cattle ranchers in the area (Killeen 2007; Navarro *et al.* 2001; Navarro & Maldonado 2002; Wallace *et al.* 2013b), but controlled fire management is lacking, and land clearing and burning of old grasslands is affecting the distribution of both titi monkey species. Fire frequency has increased in the past decade because of the development of new roads and the improvement of existing roads that have promoted increases in human settlement and land-management activities which, in turn, have increased fire risks (Cochrane 2001; Killeen 2007).

The effects of fire on titi monkeys need to be assessed beyond simply the loss of forest habitat. Considering the monogamous social organization of titi monkeys, their territoriality

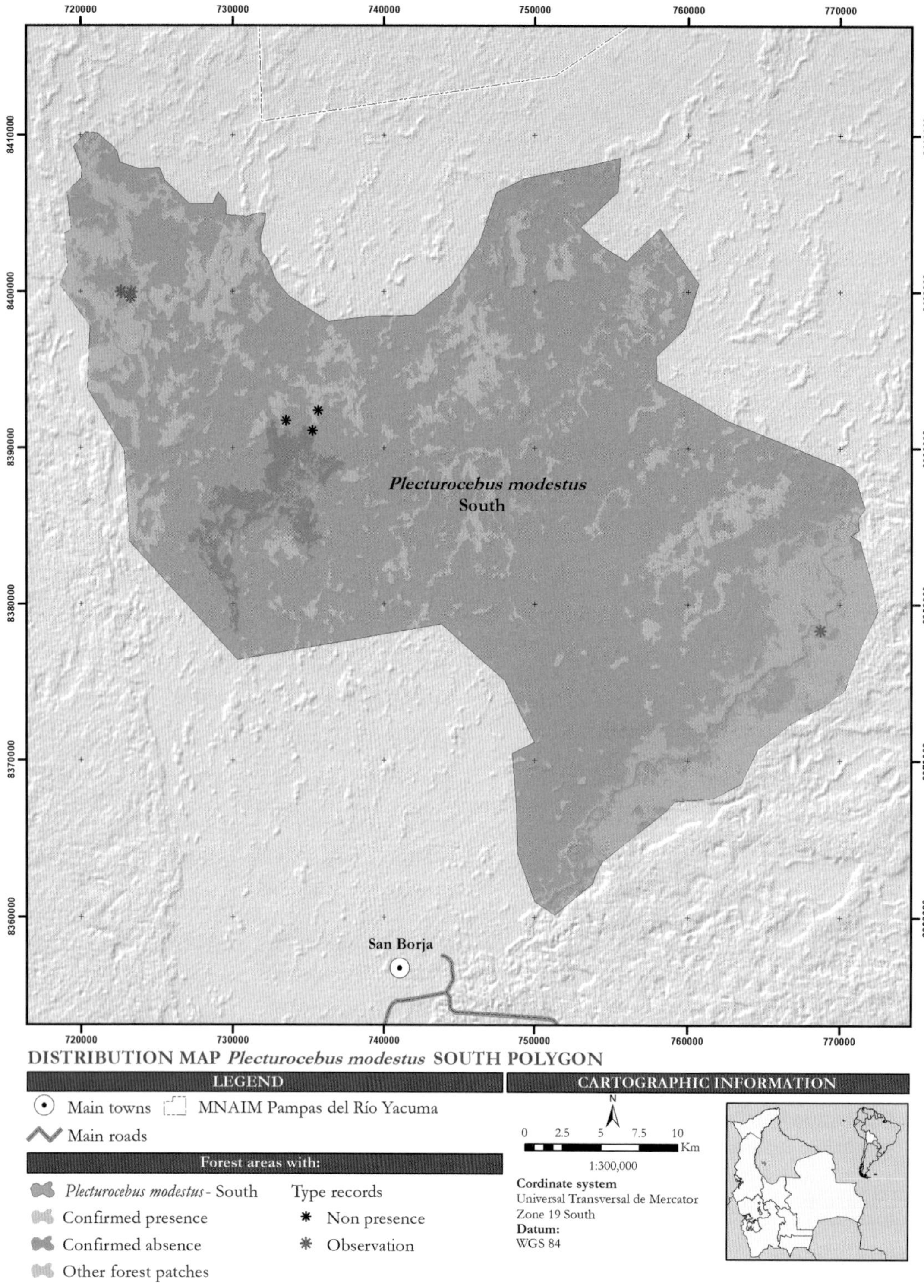

Figure 23.4 *Plecturocebus modestus* south polygon, indicating confirmed presence and absence.

and the high fragmentation levels of forests reported herein, indirect effects of fire, such as habitat degradation, as opposed to destruction, and even the effects of heat and smoke, must be considered.

The plant species within the forests of southwestern Beni are adapted to being underwater for a few weeks during the wet season (Navarro 2001; Navarro & Maldonado 2002). However, in the wet season of 2013–2014 an unprecedented and extreme-flooding event occurred, considerably worse than the ones of 2007 and 2008 (Balcazar 2014), with some areas along the Beni River still flooded in mid-July 2014 (G. Ayala, pers. comm., 2014). This level of flooding may well modify the plant composition and/or food availability for primates such as titi monkeys due to changes in soil nutrients (Haugaasen & Peres 2005a). The effects of variations in food availability are currently unknown, but could affect the persistence of titi

Table 23.2 Total polygon area, total forest area and areas affected by fire, flooding, human populations and roads in the *Plecturocebus modestus* south polygon.

	Categories	Total area (ha)	Forest patch area (ha)	Average area (ha)	# Patches	% Total area	% Total forest area	Records
P. modestus South	All patches	136 507.55	34 852.97	88.01	396	25.5		8 records
	Largest forest patch		6 527.04			4.78	18.7	5 direct observation
	Fire presence		2 203.2	169.47	13	1.6	6.3	3 confirmed absence
	Flooding 2007		2 190.15	199.1	11	1.6	6.3	
	Flooding 2008		4 588.21	2 294.1	2	3.4	13.2	
	Populations 1 km buffer		18 049.87	784.77	23	13.2	51.8	
	Populations 2 km buffer		–	–	–	–	–	
	Roads 2 km buffer		12 992.65	118.11	110	9.5	37.3	
	Confirmed *P. modestus* presence		11 064.49	5 532.24	2	8.1	31.7	

Table 23.3 Total polygon area, total forest area and areas affected by fire, flooding, human populations and roads in the *Plecturocebus olallae* polygon.

	Categories	Total area (ha)	Forest patch area (ha)	Average area (ha)	# Patches	% Total area	% Total forest area	Records
P. olallae	All patches	26 744.23	6175.10	29.27	622	23.1		17 records
	Largest forest patch		4159.61			15.5	67.3	16 direct observations
	Fire presence		435.00	1.10	395	1.6	7.0	1 collection
	Flooding 2007		–	–	–	–	–	
	Flooding 2008		5010.37	16.26	308	18.7	81.1	
	Populations 1 km buffer		4159.61	4159.61	1	15.6	67.4	
	Populations 2 km buffer		–	–	–	–	–	
	Roads 2 km buffer		4638.17	128.79	129	17.3	75.1	
	Confirmed *P. olallae* presence		4182.38	1508.69	9	15.6	67.7	

monkey groups and this should be evaluated as a potential risk for their conservation.

In addition, the unprecedented 2014 flooding may significantly increase fire risk if drought occurs later in 2015 because of increased dead vegetation as potential fuel material during the coming dry seasons. In the study area, the improved road infrastructure may promote colonization providing the 'spark' and increase significantly the risk to titi populations. Given increased fragmentation and degradation in combination with these potential increases in fire risk, the forests in the study area may well be more vulnerable to future fires (Tacconi *et al.* 2006), making the *Plecturocebus* population more susceptible to fire events and their longer-term effects on the ecology of the region. New incomers to the region may not be sensitized to the conservation needs of the endemic titi monkeys, and so continuation of previously successful environmental outreach efforts should also be prioritized.

Conclusion

This study establishes fragmentation levels for the major distribution areas of two endemic species of titi monkey. It also serves as a preliminary evaluation of fire and flooding risks. Our results point to high probability of forest loss which would dramatically reduce available habitat for *P. olallae* and *P. modestus*. Considering research on these species is relatively recent, as is the case for most Pitheciids (Veiga *et al.* 2013), more detailed studies are required to document and monitor how flooding events and associated fire affect the biology and distribution

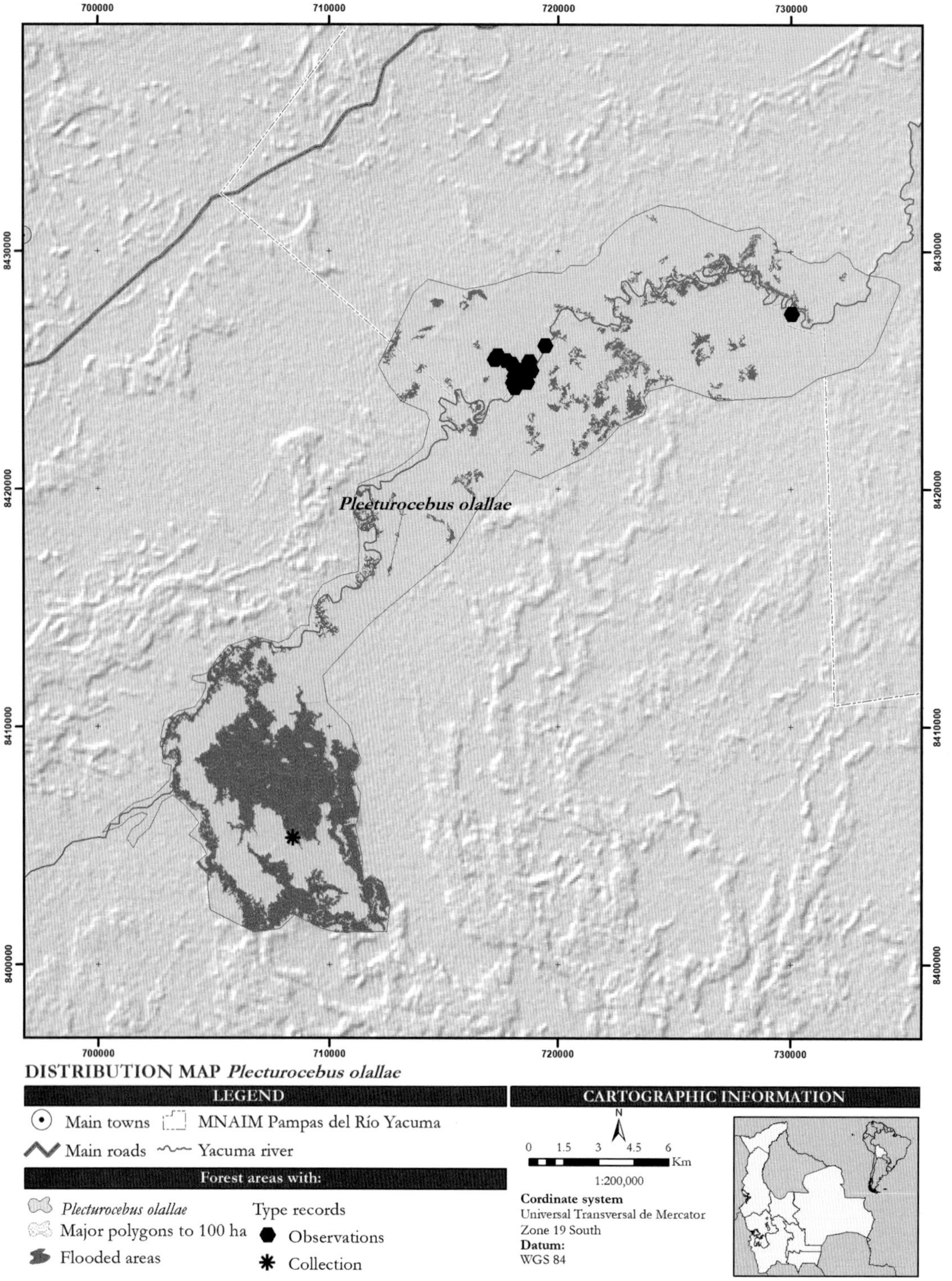

Figure 23.5 *Plecturocebus olallae* polygon indicating largest forest patches, flooded areas and *Plecturocebus* presence.

Table 23.4 Landscape metrics for the three distribution polygons.

Categories	Large patch index	# Patches	Mean distance between patches (m)
P. modestus north	18.48	589	123.78
P. modestus south	2.42	396	147.02
P. olallae	3.9	622	119.7

of these endangered titi monkeys. We identified several forest patches that are particularly important for the conservation of these highly range-restricted primates. Our findings underscore the need for more fieldwork in the southern polygon of *P. modestus*, as to date only a few observations exist. For the moment, two very large municipal protected areas will help aid conservation efforts. Considering their protection levels are not the same as a national level protected area, and that they are still developing their management strategies, substantial coordination and strengthening of efforts together with local people must be prioritized if titi monkeys, and other biodiversity, are to be protected.

Acknowledgements

We would like to thank the Wildlife Conservation Society, the Gordon and Betty Moore Foundation, Primate Conservation Inc., Margot Marsh Biodiversity Foundation, the BP Conservation Leadership Program and the Conservation International Primate Action Fund for their continued financial support. We would like to thank the National Directorate for the Protection of Biodiversity for help in acquiring necessary research permits, as well as the collaboration of the Municipalities of Reyes, San Borja and Santa Rosa del Yacuma, and especially the Nogales cattle ranches for access to study sites.

Chapter 24

Use of Swamp and Riverside Forest by Eastern and Western Gorillas

Juichi Yamagiwa, Yuji Iwata, Chieko Ando and A.K. Basabose

Introduction

The current distributions of African great apes overlap extensively in Central Africa (Caldecott & Miles 2005). Early studies reported distinct niche separation between gorillas and chimpanzees which reduced their direct encounters (Jones & Sabater Pi 1971; Schaller 1963). Folivorous gorillas tended to range in secondary, regenerating forests, feeding on terrestrial herbs, while frugivorous chimpanzees tended to range in primary forests, searching for arboreal fruits in both montane (Uganda) and lowland (Equatorial Guinea) tropical forests. However, more recent studies on sympatric populations of gorillas and chimpanzees have shown an extensive overlap in diet and ranges (Doran *et al.* 2002; Kuroda *et al.* 1996; Tutin & Fernandez 1993a; Yamagiwa & Basabose 2006a, b). In lowland tropical forests, gorillas feed regularly on fruits and insects, as do chimpanzees, but seasonally consume a greater variety of fruits than chimpanzees (Kuroda *et al.* 1996; Rogers *et al.* 2004; Yamagiwa *et al.* 1994). Both western and eastern lowland gorillas range extensively in the primary forests that chimpanzees usually inhabit (Lopé: Tutin & Fernandez 1984; Ndoki: Kuroda *et al.* 1996; Kahuzi: Yamagiwa *et al.* 1996; Mondika: Head *et al.* 2011; Doran-Sheehy *et al.* 2004). Among the types of vegetation within the distributions of these sympatric populations, bamboo (*Arundinaria alpina*) forest constitutes the upper limit of chimpanzee distribution, while it provides gorillas their preferred food (bamboo shoots) seasonally in montane forests (Casimir & Butenandt 1973; Yamagiwa *et al.* 1992). In both montane and lowland tropical forests, swamps are important parts of gorilla's home ranges, while they are rarely used by chimpanzees (Head *et al.* 2012; Nishihara 1995; Williamson *et al.* 1988; Yamagiwa & Basabose 2006b; Yamagiwa *et al.* 1992).

Swamps are frequently used by gorillas. There are many swamps and swamp forests flooded seasonally or throughout the year in all gorilla habitats. For example, the open clearings locally called '*bai*' are observed in dense forests of Republic of Congo, Central African Republic and Cameroon. *Bai* are swamps of several hundred metres (maximum diameter) characterized by low numbers of woody plants and dense herbaceous vegetation with mineral-rich soil and water (Magliocca & Gautier-Hion 2002). Large mammals, such as elephants, buffalos and antelopes frequently appear in *bais*. Western lowland gorillas also frequently visit *bais* and feed on aquatic herbs (Kuroda *et al.* 1996; Magliocca *et al.* 1999; Magliocca & Gautier-Hion 2002; Nishihara 1995; Parnell *et al.* 2002). High availability of several kinds of fruits also attracts them and influences their movements (Doran-Sheehy *et al.* 2004). In the montane forests of Uganda, Rwanda and Democratic Republic of Congo (DRC), swamps with diameters reaching several kilometres are commonly observed. They are composed of a mixture of herbs, vines, shrubs and small trees, and both mountain gorillas (*Gorilla beringei beringei*) and eastern lowland gorillas occasionally visit them to feed on vegetative foods (Casimir 1975; Ganas *et al.* 2008; Nkurunungi *et al.* 2004; Watts 1984; Yamagiwa *et al.* 2005). Swamp forests or gallery forests inundated seasonally are observed widely along tributary streams of large rivers in lowland tropical forests. Both chimpanzees and gorillas frequently range in swamp forests, and gorillas inhabit them at high densities in Gabon and Congo (Fay & Agnagna 1992; Tutin & Fernandez 1984).

What factors attract gorillas into swamps, and what characteristics of swamps are important for helping gorillas survive in both montane and tropical forests? To answer these questions, we analysed data from our long-term research on the ranging patterns of gorillas in Kahuzi-Biega National Park (montane forest in DRC) and Moukalaba-Doudou National Park (tropical forest in Gabon). In Kahuzi, we have habituated a group of gorillas and have recorded their daily ranging since 1994 (Yamagiwa & Basabose 2006a, b; Yamagiwa *et al.* 1996, 2005, 2012). In Moukalaba, we have also habituated a group of gorillas and have recorded their daily ranging since 2004 (Ando *et al.* 2008). Moukalaba lacks any *bai* swamp, but annual flooding of the Moukalaba River during the rainy season creates swamp forest (gallery forest) along the river system, into which gorillas frequently venture. We use data on the ranging of gorillas in these flooded areas to examine their importance for the persistence and conservation of gorillas.

Study Areas and Methods

Kahuzi-Biega (6000 km^2) and Moukalaba-Doudou (5028 km^2) National Parks are situated at the eastern and western ends of the Congo Basin, respectively. Gorillas inhabiting these areas are classified as different species (*Gorilla beringei graueri* at Kahuzi and *Gorilla gorilla gorilla* at Moukalaba). At both sites, a group of gorillas have been habituated for more than ten years (Ando *et al.* 2008; Yamagiwa *et al.* 2005).

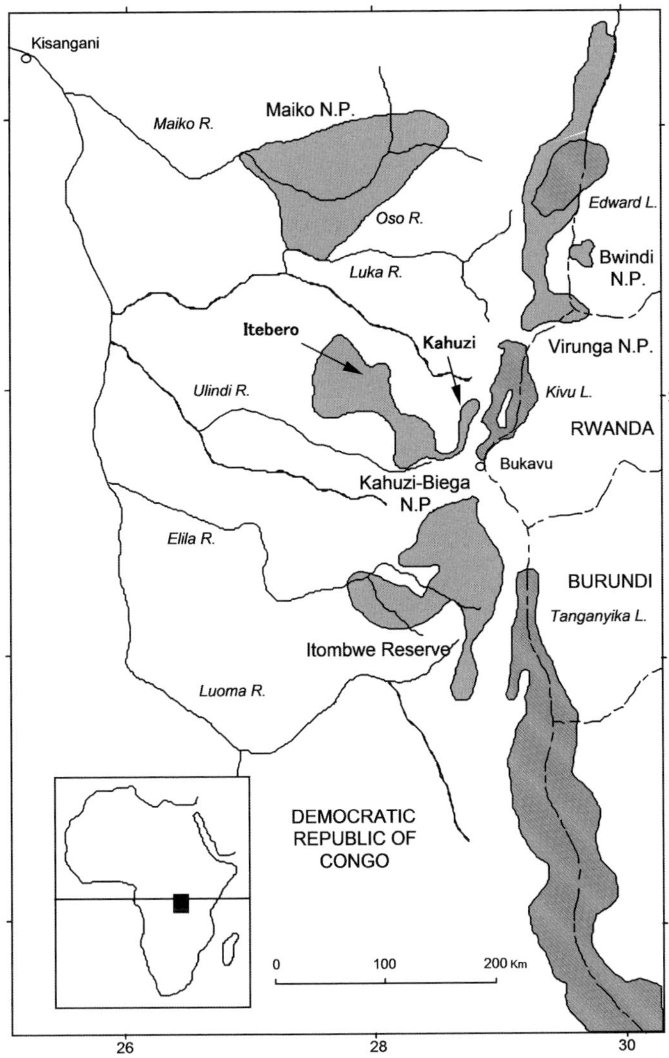

Figure 24.1 Map of Kahuzi-Biega National Park, DRC.

Kahuzi

The study area of Kahuzi covers about 100 km² along the eastern border of the park at an altitude of 2050–2350 m (Figure 24.1), where the study group ranges sympatrically with ten species of primates, including chimpanzees (Mankoto *et al.* 1994). The area consists of bamboo *Arundinaria alpina* forest, primary montane forest, secondary montane forest, *Cyperus latifolius* (Cyperaceae) swamp, and other vegetation areas, as described by Casimir (1975) and Yumoto *et al.* (1994) (Figure 24.2). The mean annual rainfall measured from 1994 to 2002 was 1658 mm (range: 1409–2180 mm) with a distinct dry season (June to August). The mean monthly temperature was 20.2°C (maximum: 26.5°C; minimum: 13.8°C).

Phenological data on fruit were obtained from 28 species of tree (24 species from which fruit were eaten by gorillas or chimpanzees and four species from which fruit were not eaten by apes). For each species, fruits (ripe and unripe) of at least ten reproductively mature trees (range: 10–13) were monitored twice each month. The monthly datum on the presence of fruit was the average of the two records. To estimate fruit abundance (biomass and number) of tree species, we used DBH (Chapman *et al.* 1992; Yamagiwa *et al.* 2005). We calculated a monthly fruit index (F_m) as:

$$F_m = \sum_{k=1}^{S} P_{km} B_k$$

where P_{km} denotes the proportion of the number of trees in fruit for species k in month m, and B_k denotes the total basal area per hectare for species k. The fruit index was calculated in each month, and its seasonal fluctuation was compared between primary and secondary forests.

Due to the outbreak of human warfare, we could not monitor fruiting phenology in August 1996, March 1997, January and August 1998, and February and April 2001. Based on the estimation of fruit abundance, we classified study

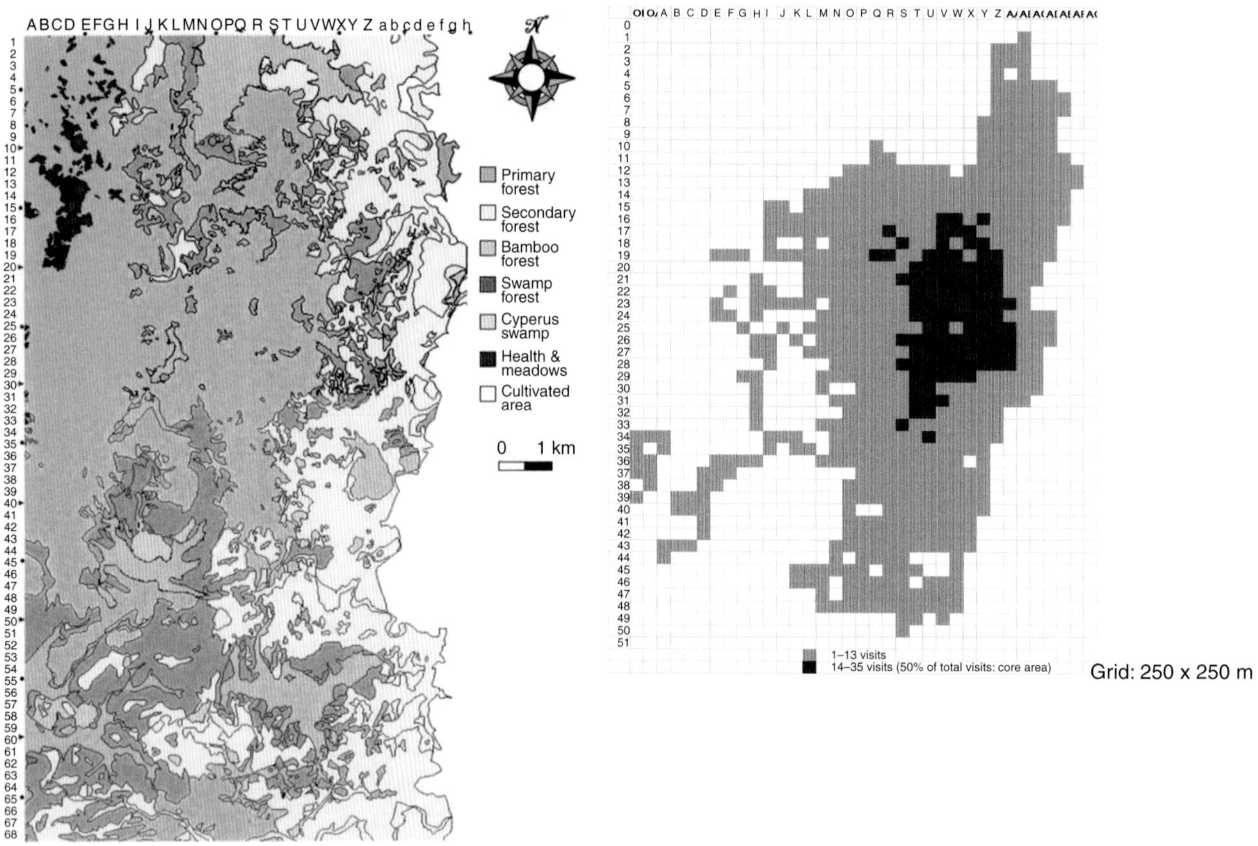

Figure 24.2 Vegetation of the study area and home range of the study group of gorillas between 1994 and 2004 at Kahuzi. Core area is indicated by the dark-coloured grids.

periods into a fruit season (April–August) and a non-fruit season (September–March).

Moukalaba

The study area of Moukalaba covers approximately 120 km^2 in the southeastern part of the park with an altitude of 50–800 m (Figure 24.3). The vegetation of this study area is divided into mixed-species primary forest, secondary forest, *Musanga cecropioides* (Urticaceae) dominated forest, temporarily inundated gallery forest (swamp forest), and savanna. Gorillas live sympatrically at high densities with 12 primate species including chimpanzees (Takenoshita & Yamagiwa 2008). Mean annual rainfall (2002-2006) was 1777 mm (range: 1583-2163 mm), with a mostly distinct dry season (May-September) and rainy season (October-May) (Takenoshita *et al.* 2008).

Fruit Phenology

We plotted nine transects (4-6.7 km in length) in 2003 and since then have monitored fruit phenology monthly using the fallen fruit census method (Furuichi *et al.* 2001), which we modified slightly by using the number of fruit clusters as an index of fruit abundance (Takenoshita *et al.* 2008). A fruit cluster is an aggregation extending 1 m on each side of a transect that includes fallen fruit from a single tree. When there was a large contiguous cluster of fruit that came from several trees of the same species, we divided the cluster by the number of fruiting trees. Thus, the number of clusters matched the number of fruiting plants that dropped fruit within the census belt. We recorded the tree species, the number of fruit in each cluster, and whether the majority of fruits were ripe or unripe. The number of fallen fruit clusters was positively correlated with the number of fallen fruit species (= 88.4, $p < 0.001$). Both fruit abundance and diversity increased in the rainy season (from October to March), and decreased at the end of the rainy season and in the dry season (from April to September).

Group Movements and Ranges

At both Kahuzi and Moukalaba, daily travel of the study groups was tracked by one or two teams of field assistants plus researchers, and recorded using the GPS. As done in previous studies on mountain gorillas (Fossey & Harcourt 1977; Vedder 1984; Watts 1998a), we plotted daily movements of each gorilla study group on a grid of 250 × 250 m quadrats superimposed on a 1:25 000 vegetation map of the study area.

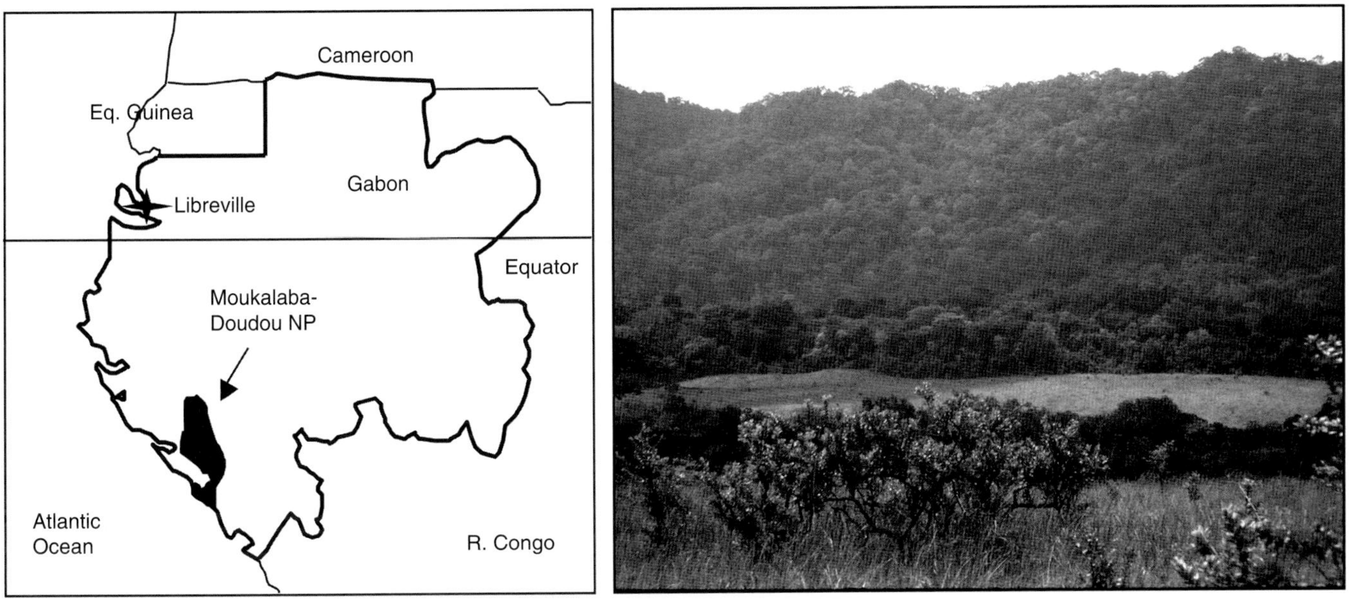

Figure 24.3 Map of Moukalaba-Doudou National Park, Republic of Gabon.

We recorded the study group's fresh travel routes between nest sites on consecutive nights for 1440 days over 92 months (average: 16 days per month, range: 5–29) between 1994 and 2004, with the exception of the period between 1996 and 2001 when we could not record them due to war. Total home range size was the summed area of all quadrats that a group entered during the study period. The frequency of visits was counted as one quadrat per day, with multiple visits to the same quadrat within a day counted as one visit. The core area was defined as the summed area of quadrats that contained the highest frequency of visits until 50% of total visits were recorded (Robbins & McNeilage 2003; Watts 1998a).

Gorilla Diet

The diet of gorillas was evaluated from direct observation, feeding remains along fresh trails and faecal analysis (Tutin & Fernandez 1993b). Fresh (up to 1 day old) faeces were collected mainly at nest sites, washed in 1 mm mesh sieves, dried in sunlight, and stored in plastic bags. The contents of each sample were examined macroscopically and listed as seeds, fruit skins, fibre, leaves, fragments of insects and other items. The portion by volume of each of these contents was estimated at 5% intervals. Fruit seeds and skin were identified at the species level macroscopically. We used feeding trails to analyse which food materials were eaten by gorillas in particular vegetation. While following the trail of a group, we recorded the presence of each food item the first time we encountered it each day. We collected 114 feeding trails more than 400 m from 107 days during 11 months (October 2006–August 2007). When we encountered feeding trails we also recorded vegetation type.

The R 2.6.1 package (version 3.0.2, R Development Core Team 2013) was used for statistical analyses.

Results

Use of *Cyperus latifolius* Swamp by Eastern Lowland Gorillas in Kahuzi

Based on the number of grid squares visited by the study group, the annual home range was 15.5 km^2 (range: 13.2–18.2 km^2), the total home range for 92 months was 42.3 km^2, and the core area occupied 15% of the total home range (Figure 24.2). The number of new grid squares visited by gorillas tended to increase every year.

The proportion of annual travel routes which entered the *C. latifolius* swamp was 7.1% on average (n = 11, range: 1.7–9.3%, SD = 5.5; Figure 24.4). In 1994–1997, when the bamboo flowered and shoots were less available, the study group of gorillas rarely entered the bamboo forest. During those periods of low shoot availability, the study group visited the bamboo forest occasionally, probably to search for bamboo shoots, and returned to the previous ranging area in the primary or the secondary forest after a few days. During those years when bamboo shoots were less available, gorillas increased their use of *C. latifolius* swamps and secondary forests, though they also often entered the *C. latifolius* swamp even in those years when bamboo shoots were available.

The *C. latifolius* swamp was thus used by the study group across the entire year, though at a low frequency compared with the use of other habitats (average = 7.4%, range: 3.7–12.1%, SD = 2.4; Figure 24.5). Their ranging patterns, especially daily path lengths, appeared to be more influenced by the availability

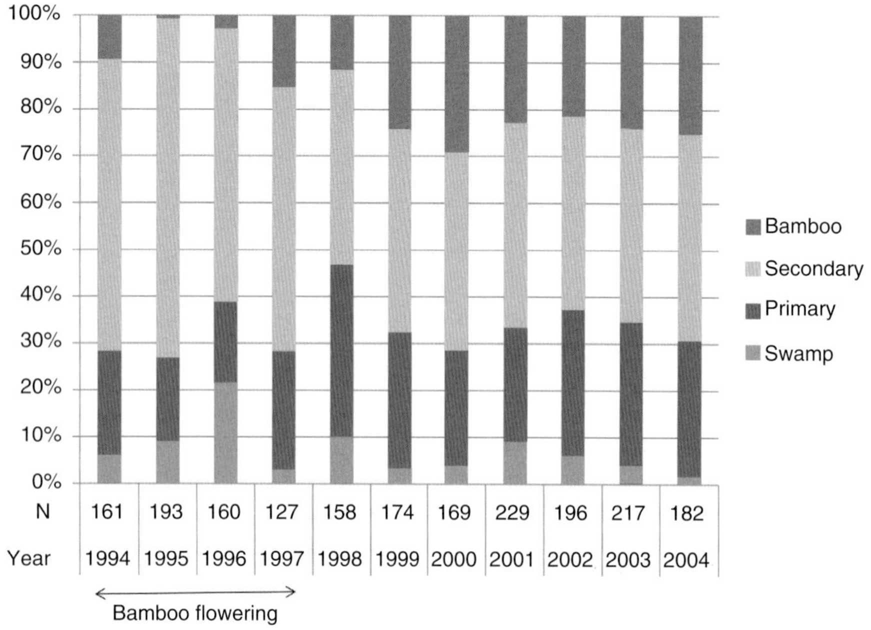

Figure 24.4 The proportion of annual travel routes in swamp by the study group of gorillas at Kahuzi. n = number of 250 × 250 m quadrats entered by the study groups in each year.

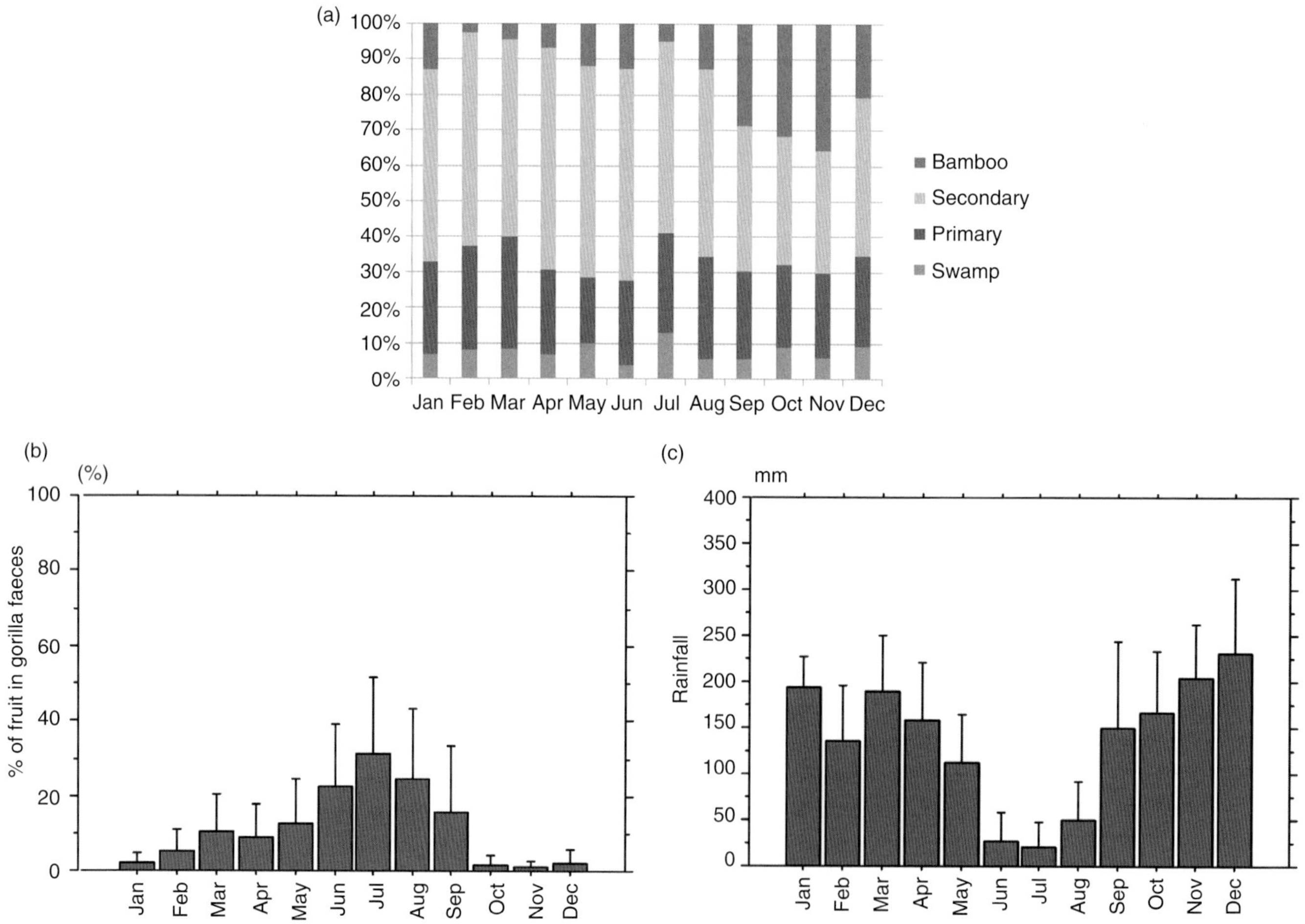

Figure 24.5 The mean monthly proportion of each type of vegetation along daily path routes (a), the mean monthly proportion of fruit remains per faecal sample (b), and the mean monthly rainfall (c) at Kahuzi. (A black and white version of this figure appears in some formats. For the colour version, please refer to the plate section.)

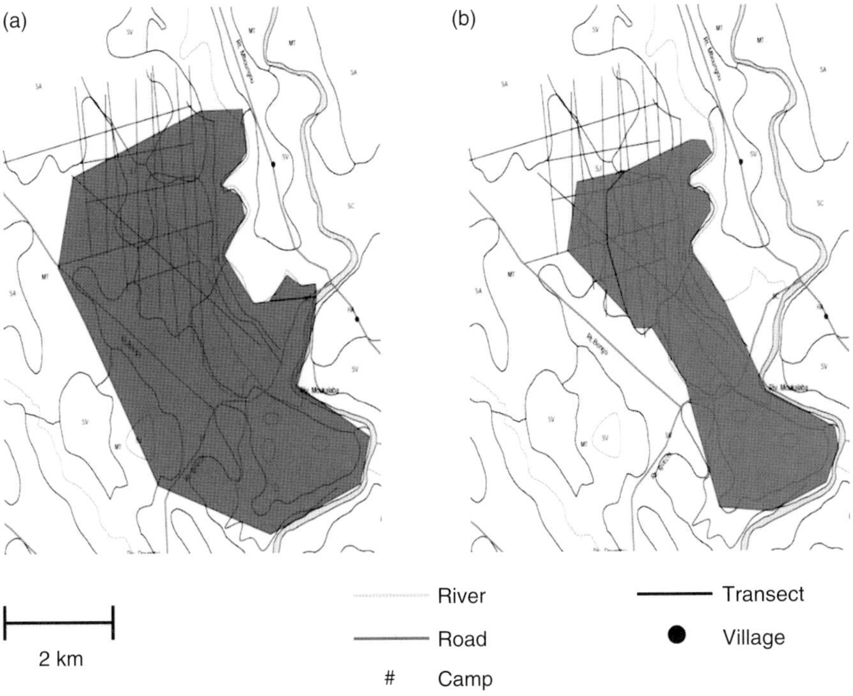

Figure 24.6 Home range of the study group in the fruiting (left) and the non-fruiting season (right) at Moukalaba. (A black and white version of this figure appears in some formats. For the colour version, please refer to the plate section.)

of their preferred foods, i.e. fruits; gorillas tended to expand their daily path length with increasing fruit consumption ($R^2 = 0.096$, $p < 0.01$, by simple regression analysis). They also shifted their monthly range to include the bamboo forest in September–January when bamboo shoots appeared. However, they did not change their use of *C. latifolius* swamps when bamboo shoots were available, suggesting that, even though the frequency of swamp usage was low, *C. latifolius* swamps provide the gorillas with indispensable benefit(s).

In the *C. latifolius* swamp, gorillas always fed on the base of mature leaves of *C. latifolius*. This plant dominates the swamp phytocommunity and is the major food of gorillas in the swamp. The diversity of plants here was extremely low compared to that of other habitat types in the study site; the density of woody plant species above 10 cm DBH in the *C. latifolius* swamp area's transect was less than one-third of those in primary and secondary forests (Yamagiwa *et al.* 2008). Only five species of woody plants were found in the *C. latifolius* swamp. Among these, *Galiniera coffeoides* (Rubiaceae; fruit, leaf, pith and bark) and *Syzygium parvifolium* (Myrtaceae; fruit, leaf and bark) were the preferred foods of gorillas (Yamagiwa *et al.* 2005). In addition, several climbing herbs such as *Urera hypselodendron* (Urticaceae; leaf and bark) and *Basella alba* (Basellaceae; leaf) were available in small amounts in the swamp and eaten by gorillas.

The study group also used the *C. latifolius* swamps for sleeping. When the study group entered a *C. latifolius* swamp, they usually built nests on the ground in and beside the swamp. Leaves of *C. latifolius* were always used as the materials of ground nests in the swamp.

Use of Gallery Forest by Western Lowland Gorillas in Moukalaba

The diet of western gorillas in Moukalaba is relatively frugivorous compared to that in Kahuzi. The mean number of fruit species found in faecal samples was 4.1, and the mean percentage of fruit remains in faecal samples was 99% in Moukalaba, compared with 0.79 and 12% in Kahuzi (Yamagiwa & Basabose 2006a). Nonetheless, the amount of fruit consumed by gorilla in Moukalaba differed across seasons, in that gorillas tended to decrease their consumption of fruits in relation to the general decrease in the availability of fruit (Iwata 2010).

The percentage of gallery forest within the home range of the study group was 11.4%. From August 2006 to July 2008, the ranging area of the study group differed across seasons being larger in the fruiting season (12.1 km^2) than in the non-fruiting season (10 km^2) (Figure 24.6), although the group used the gallery forest in both seasons to different extents (Figure 24.7). They increased ranging in the gallery forest in April, May and June when the percentage of fruit found in faecal samples decreased; the monthly proportion of gallery forest in the ranging area was negatively correlated with the monthly mean proportion of fruit remains per faecal sample ($R^2 = 0.6743$, $p < 0.001$, by simple regression analysis; Figure 24.8).

The study group of gorillas regularly consumed terrestrial herbs in the three types of vegetation (secondary forest, *Musanga* forest and gallery forest). In the gallery forest, they consumed fruits less frequently and leaves more frequently than in other habitats (Figure 24.9). We collected 31 feeding trails in gallery forest for 31 days. In gallery forest, leaves and bark of

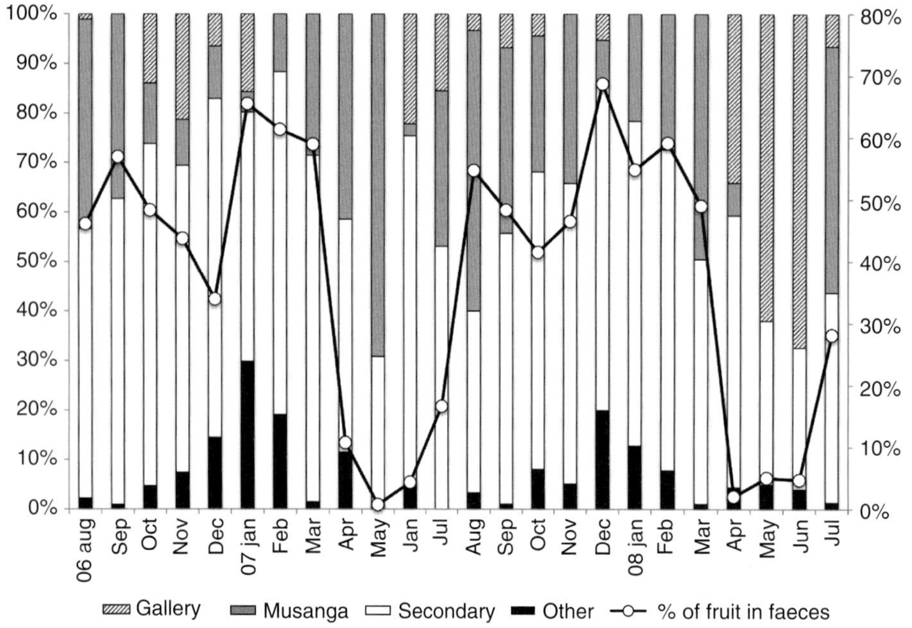

Figure 24.7 Monthly proportion in use of different types of vegetation and frugivorous diet by the study group of gorillas at Moukalaba.

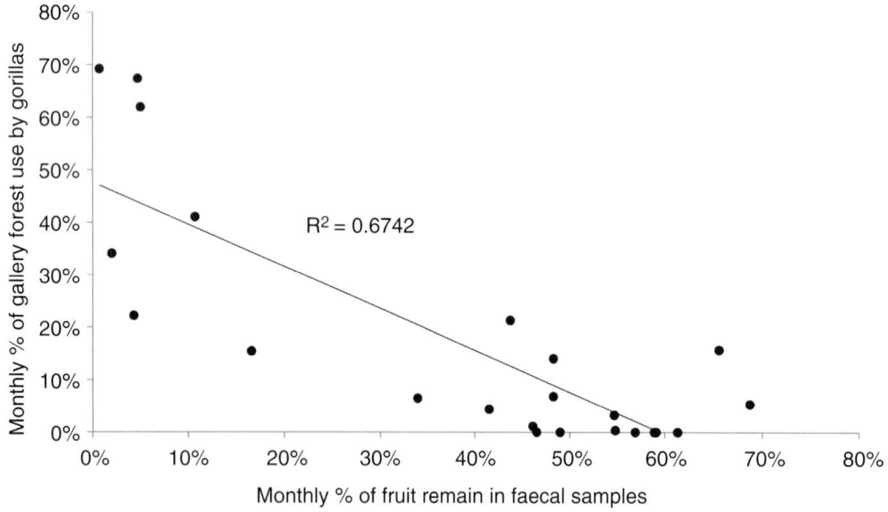

Figure 24.8 Correlation between fruit consumption and gallery forest use by gorillas at Moukalaba.

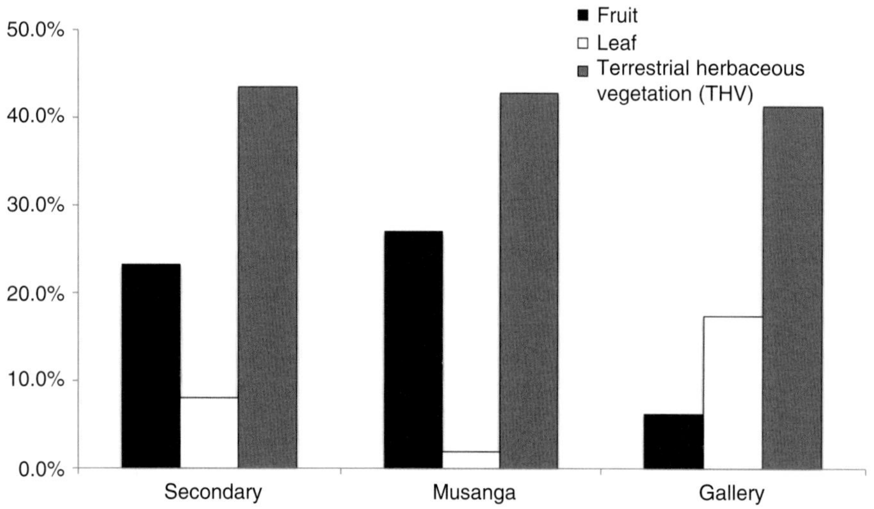

Figure 24.9 The proportion of food type in gorillas' faecal samples collected in each vegetation at Moukalaba.

Table 24.1 Proportion of feeding trails of the study group in which each food item was detected in gallery forest at Moukalaba.

Plant form	Species (food item)	% of days
Tree	*Ficus* spp. (bark)	24
	Ficus spp. (fruit)	8
	Ficus spp. (leaf)	32
	Meiocarpidium lepidotum (fruit)	4
	Milicia excelsa (bark)	80
	Milicia excelsa (leaf)	80
	Myrianthus arboreus (bark)	8
	Myrianthus arboreus (leaf)	4
	Panda oleosa (fruit)	4
	Salacia sp.	8
	unknown species (fruit)	8
	unknown species (leaf)	8
Herb	*Aframomum leptolepis* (pith)	36
	Anchomanes difformis (leaf)	4
	Costus albus (pith)	4
	Haumania liebrechtsina (pith)	80
	Haumania liebrechtsina (shoot)	68
	Marantochloa cordifolia (pith)	16
	Megaphrynium macrostachyum (pith)	36
	Megaphrynium macrostachyum (shoot)	36
	Palisota sp. (pith)	12
Vine	*Alafia caudata* (pith)	8
	Cissus dinkragei (fruit)	8
	Landolphia sp. (fruit)	4
Total		25 days

Millicia excelsa and pith and shoot of *Haumania liebrechtsina* constituted the major portion of the diet of the study group in the gallery forest when fruit was scarce (Table 24.1). The *Millicia excelsa* and *Haumania liebrechtsina* were widely available in the inundated forest. Thus, these species may supply food types to gorillas during periods of low fruit availability.

Gorillas in Moukalaba tended to build arboreal nests (63.2%). In the gallery forest, we found that 85.3% of nests were built in trees by the study group during the rainy season. In the gallery forest, most tree nests were constructed in *Anthonotha* sp. (Caesalpiniaceae; 44.2%). The second most chosen tree for nest construction was *Meiocarpidium lepidotum* (Annonaceae; 5.2%). Interestingly, *M. excelsa*, the species that was the most favoured leaf food material, was never chosen for nest construction. This, therefore indicated a high selectivity of nest material by gorillas in the gallery forest, although we lack data on the relative densities of tree species used in nest construction.

Discussion

This study demonstrates that swamp forests, i.e. *C. latifolius* swamps in Kahuzi and gallery forest in Moukalaba, constitute important food resources and nest sites for both eastern and western lowland gorillas. In addition, our results potentially support the fact that vegetative foods in swamp forests, such as leaves of *C. latifoilius* and *M. excelsa*, are fallback foods of gorillas at Kahuzi and Moukalaba, respectively. At Kahuzi, our study group of gorillas continuously used *C. latifolius* swamps eating its leaves throughout the year and, therefore, its leaves might function as supplementary food during times when their preferred foods are scarce. In particular, for gorillas at Moukalaba, the monthly mean proportion of fruit remains per faecal sample possibly reflected the extent to which gorillas consumed their preferred foods, i.e. fruits. Fruit consumption was negatively correlated with gorillas' usage of gallery forest suggesting that in this habitat they consume the leaves of *M. excelsa*, an important fallback food.

Both chimpanzees and gorillas prefer ripe fruits and overlap extensively in their fruit choices where they coexist sympatrically (Kuroda *et al.* 1996; Stanford & Nkurunungi 2003; Tutin & Fernandez 1993; Yamagiwa & Basabose 2006a; Yamagiwa *et al.* 1996). However, while both chimpanzees and gorillas change their diets in response to annual periods of fruit scarcity, they adapt differently. Although they also increase consumption of leaves, barks and animal matter to supplement the shortage of fruits (Goodall 1986; Nishida 1976; Reynolds *et al.* 1998; Tutin & Fernandez 1993; Yamagiws & Basabose 2009), chimpanzees tend to seek fruit persistently, decreasing party size and dispersing widely (Basabose 2004; Newton-Fisher *et al.* 2000; Wrangham 1986). In contrast, gorillas always consume a certain amount of leaves, barks and pith of herbaceous vegetation throughout the year, but increase its consumption during periods of fruit scarcity (Doran *et al.* 2002; Tutin & Fernandez 1993; Yamagiwa & Basabose 2006a; Yamagiwa *et al.* 1994). In contrast to the fission/fusion-based strategy used by chimpanzees to adapt to periods of low fruit availability, gorillas do not change group size or cohesiveness, but decrease their daily path lengths (Doran-Sheehy *et al.* 2004; Goldsmith 1999; Yamagiwa & Basabose 2006b).

Marshall and Wrangham (2007) classify fallback foods into 'staple' and 'filler'. Staple fallback foods can seasonally constitute the entire diet, and their ecological characteristics (hardness, patch size or dispersion) shape the digestive system and socioecology of primate species eating them. Filler fallback foods never constitute the entire diet and the socioecological adaptations of the consuming species to their characteristics are less complete than those responding to staple fallback foods. Fallback foods with lower nutritional density comprising non-reproductive plant anatomy such as bark and leaves inherently require more processing and promote anatomical adaptation, while fallback foods with higher nutritional density such as fruit and seeds, which are often mechanically protected and difficult to find, require behavioural innovation and tool use (Lambert 2007). Therefore, gorillas are regarded as a good example for the former fallback strategy, i.e. staple fallback foods, and chimpanzees for the latter, i.e. filler fallback foods (Lambert 2007; Marshall & Wrangham 2007; Yamagiwa & Basabose 2009). Differences in the use of swamps by gorillas versus chimpanzees may reflect their different fallback strategies: *C. latifolius* swamp sustains rich herbaceous vegetation as staple fallback foods for gorillas but furnish only rare filler fallback foods for chimpanzees.

Swamps dominated by *C. latifolius* (and the accompanying low diversity community of woody plants) constitutes 7% of the highland sector in Kahuzi-Biega National Park (Goodall 1977). The fact that the mean proportion of the study group's annual travel routes entering the *C. latifolius* swamp was 7.1% in this study indicates that gorillas do not avoid the swamp. Furthermore, this study showed that gorillas in Kahuzi constantly use swamps for feeding and nesting sites throughout the year. Gorillas always fed on *C. latifolius* leaf bases whenever they entered a swamp, suggesting that *C. latifolius* constitutes a staple fallback food by gorillas at Kahuzi, and that the other swamp-derived plants may also be seasonally important (Yamagiwa *et al.* 2005)

A folivorous foraging strategy of primates without specialized gut morphology is characterized by avoidance of repeated use of the same food resources in order to avoid accumulation of antifeedant and deterioration of food patches. The foraging strategy of eastern lowland gorillas at Kahuzi can be regarded as a folivorous one, similar to that observed in mountain gorillas inhabiting high montane forest (2800–3700 m above sea level) in the Virungas. Mountain gorillas at Virungas tend to seek high-quality herbaceous vegetation and to avoid previously used areas that contained trampled vegetation (Vedder 1984; Watts 1987, 1998b). Such movements may result in low site fidelity and extensive overlap among the home ranges of neighbouring groups. Gorillas at Kahuzi also used a wide range evenly to avoid overuse of herbaceous vegetation (Yamagiwa & Basabose 2006b). The small percentage (7.4%) of monthly travel routes in the *C. latifolius* swamp may reflect this folivorous strategy.

Gorillas at Kahuzi are using the swamps, scattered widely throughout their home range, to supplement their dietary changes, via shifts or changes in their ranging patterns. Although eastern lowland gorillas seem also to use a folivorous strategy, their ranging patterns were, in addition, influenced by fruit availability. Their habitat, i.e. the montane forest at lower altitudes (1800–2600 m asl) is composed of diverse woody plants in primary and secondary forests. Seasonally, this provides gorillas with various kinds of fruits, and they extend their daily path lengths with increasing fruit availability (Yamagiwa *et al.* 2003). These characteristics are similar to mountain gorillas in Bwindi Impenetrable National Park at altitudes of 1200–2600 m asl, Uganda (Ganas & Robbins 2005; Nkurunungi *et al.* 2004; Robbins & McNeilage 2003). As with fruits, bamboo shoots may also have an important influence on the ranging patterns of gorillas. Gorillas at Kahuzi seasonally shift their range to the bamboo forest when bamboo shoots are available (Casimir & Butenandt 1973; Goodall 1977) as reported in mountain gorillas in the Virungas (McNeilage 2001; Vedder 1984; Watts 1998b). In addition, when bamboo shoots were available, the gorillas at Kahuzi extended their daily path lengths and monthly ranges just as observed for mountain gorillas at Bwindi (Ganas & Robbins 2005; Robbins & McNeilage 2003) and western lowland gorillas (Cipolletta 2004; Doran-Sheehy *et al.* 2004; Goldsmith 1999).

In addition to their role in providing the gorillas with important food sources, the swamps at Kahuzi may also provide the gorillas with nesting sites that maximize their avoidance of predators such as leopards. Although gorillas at Kahuzi usually build nests on the ground (87.9%; Yamagiwa 2001), immatures tend to build nests in trees, especially if their protector male has recently died. Since predation by leopards has been reported both for mountain and western lowland gorillas (Fay *et al.* 1995; Schaller 1963; Tutin *et al.* 1992), *C. latifolius* swamps, with fewer tree stands but plenty of water, may buffer gorillas against predators, inhibiting their access to the nest builders. In the Kahuzi swamps, all nests built by gorillas were found on the ground. Thus, the presence of a swamp is very important for gorillas in Kahuzi not only as a feeding site for staple fallback foods, but also as a safe sleeping site and refuge from terrestrial predators.

Western lowland gorillas at Moukalaba used the gallery forest as an important feeding site. However, they did not use it constantly but tended to exploit it primarily at times of low fruit availability elsewhere. The gorillas at Moukalaba increased their leaf consumption, especially *M. excelsa*, during periods of fruit scarcity, although they consistently consumed terrestrial herbs, particularly from families Marantaceae and Zingiberaceae, in the three types of vegetation (*Musanga*, secondary and gallery forests: Figure 24.7) over the entire year. These findings suggest that terrestrial herbs may constitute a staple fallback food, and that leaves and bark of *M. excelsa* may play the role of filler fallback food for gorillas in Moukalaba.

In contrast to the gorillas at Kahuzi, the ranging patterns of gorillas in Moukalaba did not reflect a folivorous strategy. They occasionally re-used previous nest sites after a short interval, and the frequency of such use increased with increasing consumption of fruits (Iwata & Ando 2007). Western lowland gorillas at Mondika, Central African Republic, also re-used areas more frequently and consistently compared to folivorous mountain gorillas (Doran-Sheehy *et al.* 2004). Indeed, this type of ranging is not only observed in frugivorous gorillas but also in other frugivorous primates, such as spider monkey and gibbons, as a way to harvest fruits efficiently by adjusting search paths and patch re-visitation (Garber 1989; Waser 1984). The frugivorous chimpanzees at Kahuzi also exhibit this type of ranging pattern (Yamagiwa & Basabose 2006b). It may result in high site fidelity (stable use of the same site as a home range) observed in western chimpanzees at Taï Forest (Lehmann & Biesch 2003) and eastern chimpanzees at Gombe (Williams *et al.* 2004). The study group of gorillas showed high site fidelity. Their annual range in 2007–2008 was 12.4 km^2, and the entire range over several years may not have exceeded twice this area (Y. Iwata, unpublished data). On the other hand, at Kahuzi, the mean annual range of the study group was 15.5 km^2 ($n = 5$, range: 13.2–18.2 km^2), but the whole range over 5 years was 42.3 km^2, due to shifting annual home ranges. In contrast, the home range of the study group at Moukalaba did not distinctly shift between the fruiting and non-fruiting seasons, although the range expanded in the former season. The presence of gallery forest, which provides gorillas not only with diverse fruit species, but also with a large amount of edible leaves as filler fallback foods, may explain their high site fidelity in Moukalaba.

As for gorillas at Kahuzi, the swamp forest at Moukalaba, i.e. the gallery forest, may also provide gorillas with suitable sites for nest building. Leopards are known to attack gorillas in the habitats of western lowland gorillas (Fay *et al.* 1995; Tutin *et al.* 1992) and frequently appeared in the home range of the study group. Thus, the gallery forest may provide the gorillas at Moukalaba with safe sites for nest building. In addition to the benefit of avoiding leopards, the gallery forest may also help gorillas at Moukalaba to avoid elephants. It has been shown that gorillas tend to avoid contact with elephants in Rio Muni (Jones & Sabater Pi 1971), being excluded by the elephants in Loango (Head *et al.* 2012), and the presence of elephants coincided with a high proportion of tree nests at Lope (Williamson 1989). Indeed, elephants frequently appear in the home range of the Moukalaba study group, where gorillas appeared to avoid contact with elephants. Due to the presence of elephants and the shortage of nesting materials on the ground at Moukalaba, gorillas at Moukalaba tended to build arboreal nests more frequently (63.2%) than in other habitats.

In the habitats of western lowland gorillas, the large open swampy clearings known as *bai* are abundant. A *bai* attracts many species of mammals, such as elephants, buffalos, antelopes, monkeys and great apes (Blake 2002; Fishlock *et al.* 2008; Vanleeuwe *et al.* 1998). Gorillas frequently appear in the *bai* to feed on aquatic plants, such as *Enydra fluctuans* (Asteraceae) and *Hydrocharis chevalieri* (Hydrocharitaceae), these being characterized by high mineral and/or protein contents (Magliocca & Gautier-Hion 2002; Nishihara 1995). The availability of ripe fruits within a *bai* also influences the frequency and time of their use by gorillas (Doran-Sheehy *et al.* 2004). More than 30 groups of gorillas used the same *bai* in Maya Nord and Mbeli Bai, Republic of Congo (Magliocca & Gautier-Hion 2002, 2004; Parnell *et al.* 2002).

Despite their highly frugivorous diet, western gorillas do not show territoriality, and the home ranges of neighbouring groups extensively overlap each other (Bermejo 2004; Cipolletta 2004), as observed in eastern gorillas inhabiting montane forests (Robbins & McNeilage 2003; Schaller 1963; Yamagiwa *et al.* 1996). The home range of gorillas in Moukalaba does not include a *bai*, but these individuals live at a higher density than in other habitats (Takenoshita & Yamagiwa 2008). The lack of territoriality, the high diversity of fruit and the presence of inundated (gallery) forest may support the high biomass of gorillas in Moukalaba.

In summary, our findings suggest that flooded habitats are important both for folivorous eastern gorillas and frugivorous western gorillas as feeding and nesting sites. Swamps and gallery forests supported gorillas in fruit-poor periods annually and seasonally. *C. latifolius* was used as a staple fallback food in montane forest and *M. excelsa* was used as a filler fallback food. The gorillas' use of flooded areas and their fallback strategies may enable them to live at high densities in both montane and lowland tropical forests in Central Africa.

In the highland sector of Kahuzi-Biega National Park, gorillas live at a higher density and range over a wider area than do sympatric chimpanzees (Yamagiwa & Basabose 2006b). Frequent use of bamboo forest and swamp by gorillas may be a critical factor in determining these differences (Basabose 2002, 2005; Yamagiwa & Basabose 2006a). At Moukalaba, conventional nest surveys and camera trap surveys conducted concurrently, found that groups of gorillas were relatively homogeneously distributed across vegetation, while groups (parties) of chimpanzees were more abundant in montane forest (Nakashima *et al.* 2013). This difference in distribution may reflect gorillas' usage of diverse vegetation, including swamps.

Finally, inundated forests and swamps have been regarded as relatively worthless for cultivation and human occupation. This may explain why these habitats still support healthy populations of gorillas. However, recent logging, mining and farming activities have rapidly invaded such flooded areas and destroyed the habitats of gorillas (Caldecott & Miles 2005). Our study shows the important roles of flooded areas in the subsistence of gorillas, especially under conditions of fruit scarcity. Recent surveys have confirmed a wide distribution of gorillas living at high densities in swamp forest and *terra firma* mixed forest in northern Congo (Poulsen & Clark 2004; Rainey *et al.* 2010). If flooded areas were to vanish, this could result in the disappearance of gorillas and the appearance of 'vacant forests' as has already happened elsewhere in terrestrial upland areas. Wise management of the parks and forests that support apes and other species should thus consider flooded habitats as vital areas for protection.

Acknowledgements

This chapter was originally prepared for the symposium on '*Swamp Primates: Ecology and Conservation in Monkeys and Apes in Flooded Habitats*' which was held at the 24th Congress of the International Primatological Society at Cancun, Mexico, on 14 August, 2012. We would like to express our hearty thanks to Adrian. A. Barnett and Ikki Matsuda for giving us the opportunity to present our work at the symposium. This study was financed by the Ministry of Education, Culture, Sports, Science and Technology, Japan (No.162550080, No.19107007, No. 24255010 to J. Yamagiwa); by Kyoto University Global COE Program (Formation of a Strategic Base for Biodiversity and Evolutionary Research): and by JST/JICA, SATREPS (Science and Technology Research Partnership for Sustainable Development), and was conducted in cooperation with CRSN and ICCN in DRC, and IRET, CENAREST and ANPN in Gabon. We thank Dr S. Bashwira, Dr B. Baluku, Mr M.O. Mankoto, Mr B. Kasereka, Mr L. Mushenzi, Ms S. Mbake, Mr B.I. Iyomi, Mr C. Schuler, Mr R. Nishuli, Dr A. Ngomanda, Dr A. Mougougou and Mr G.-M. Moussavou for their administrative help and advice. We are also greatly indebted to all of the guides, guards, and field assistants in the Kahuzi-Biega National Park, and field assistants of PROCOBHA and the people in the villages of Doussala, Konzi and Mboungou for their technical help and hospitality throughout our fieldwork.

Chapter 25

Use of Inundated Habitats by Great Apes in the Congo Basin

A Case Study of Swamp Forest Use by Bonobos at Wamba, Democratic Republic of the Congo

Saeko Terada, Janet Nackoney, Tetsuya Sakamaki, Mbangi Norbert Mulavwa, Takakazu Yumoto and Takeshi Furuichi

Introduction

The Congo Basin is a large intracratonic depression that spans Central Africa and contains the second largest tropical forest in the world. Geological research suggests that the Congo Basin originated from a Neo-Proterozoic rift, and evolved through a long and complex process governed by the interaction between tectonic and climatic factors in a variety of depositional environments (Kadima *et al.* 2011). Many rivers flow through the centre of the basin, forming an inundated area where the ground is seasonally or permanently flooded along the rivers. The Congo River is the widest representative river in the basin and flows from the highlands of the southwestern part of the basin to the Atlantic Ocean. As a result, the basin is comprised of a mosaic of non-inundated forests, savannas, seasonally or permanently inundated forests, marshes, and rivers. Like many moist tropical forest areas, the Congo Basin is known for high biodiversity of various taxa such as plants, insects, birds, reptiles and mammals, including endangered mammals such as forest elephants, okapis and great apes (de Wasseige *et al.* 2012).

All of the great apes in Africa – the bonobo (*Pan paniscus*), chimpanzee (*Pan troglodytes*), and eastern and western gorillas (*Gorilla beringei* and *G. gorilla*) – are Endangered and inhabit the Congo Basin (Figure 25.1). Of these, only bonobos inhabit part of the low-lying central region on the south bank of the Congo River in the Democratic Republic of the Congo (DRC), where they are endemic. Although bonobos do not inhabit their range sympatrically with the other species of great apes, some populations of chimpanzees and gorillas have sympatric distributions on the north bank of the Congo River.

Various populations of all great ape species in Africa use inundated areas for ranging, feeding and sleeping (Hashimoto *et al.* 1998; Kempf 2009; Poulsen & Clark 2004; Yamagiwa *et al.* 1992). The importance of inundated areas for great apes in Africa has recently been highlighted by studies revealing higher densities of gorillas and chimpanzees in swamp than non-swamps forests (Poulsen & Clark 2004; Rainey *et al.* 2010). However, general conclusions concerning the degree of use of inundated areas, and the mechanisms of such habitat use, have not been described for bonobos because of the limited number of bonobo study sites. Inundated areas have the potential to provide important habitat for bonobos as well as other apes, and recent studies on bonobos have emphasized the importance of conducting more research in such inundated zones (Hickey *et al.* 2013; Inogwabini *et al.* 2012; Mulavwa *et al.* 2010). Ground surveys in mostly inundated areas are, however, particularly limiting due to the lack of accessibility.

Bonobos mainly use non-inundated tropical forests, but their exact distribution and ecological niche remain unclear. For example, whereas chimpanzees are known to live in forest–savanna mosaics, researchers reported new bonobo populations living in forest-savanna mosaics only after the 1990s (Myers Thompson 1997; Serckx *et al.* 2014). Nishida (1972) also noted that bonobos inhabiting forests could cross through grasslands near Lac Tumba. Based on these reports, it is clear that bonobos can live in drier and more open habitats than previously thought (Thompson 2002). At the same time, bonobos can inhabit areas where inundated habitats constitute most of their home range. Bonobo populations were recently found at the border area of the estimated distribution range in the Lac Tumba landscape, where approximately 60% of the studied area consists of seasonally or permanently inundated forests (Inogwabini *et al.* 2007; Mwanza 2003). Though the northwestern part of the basin, which is mostly inundated, is excluded from the estimated distribution despite the uncertainty regarding their presence within this area (IUCN & ICCN 2012), such inundated areas might be suitable habitats for bonobos.

In this chapter, we focus on the selection of inundated areas as habitats by bonobos. We also present a review of the use of inundated areas by other great apes in the Congo Basin, highlighting differences in use of swamp forests by other bonobo populations, chimpanzees and gorillas. Although our understanding of the habitat types in the bonobos' distribution range has increased over the last three decades, habitat selection and the mechanisms influencing such selection are still unclear. In what situations might bonobos choose to use inundated areas, and what qualities of the inundated areas

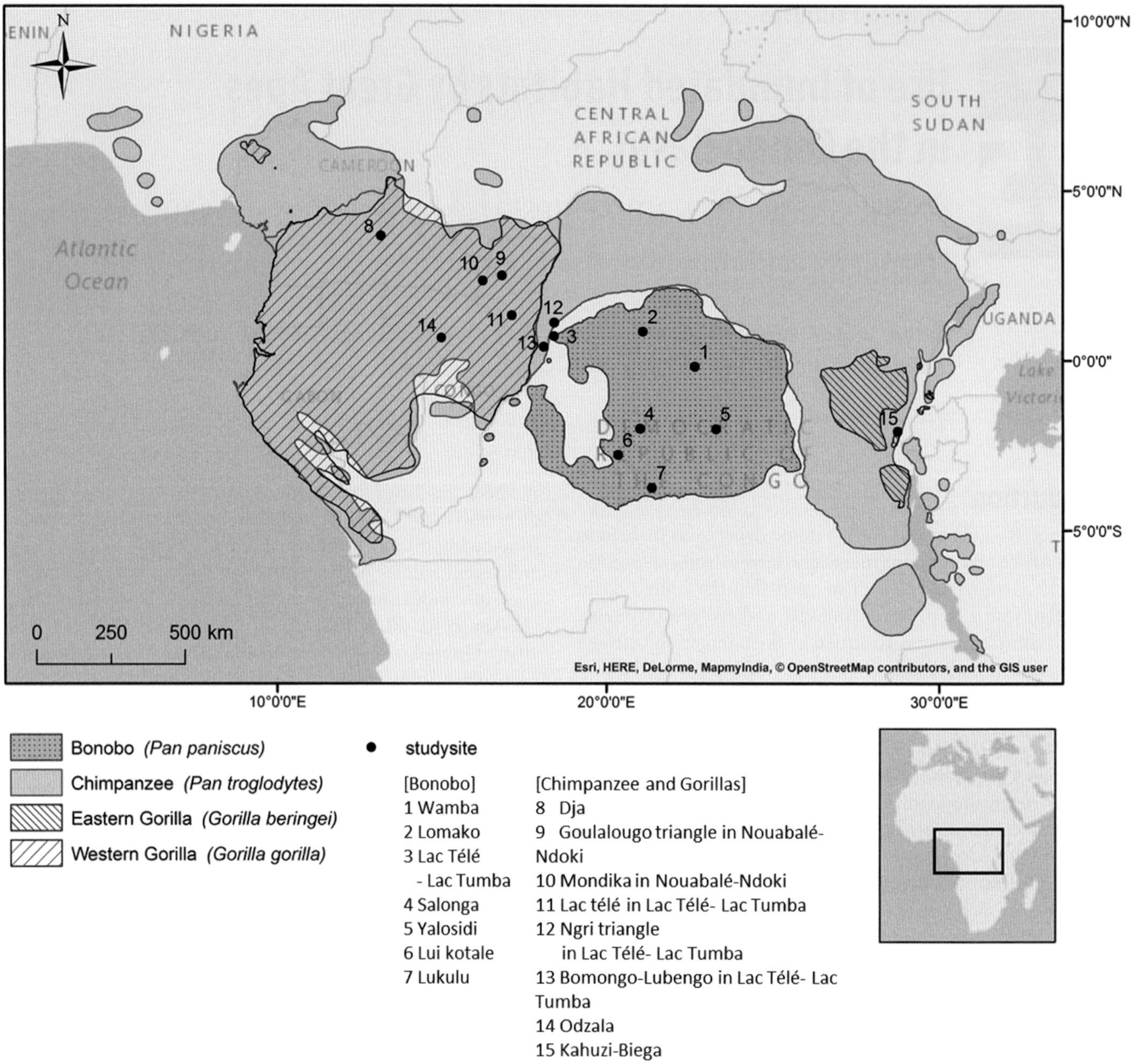

Figure 25.1 Distribution map of great apes in the Congo Basin, Central Africa.

might attract them? Do bonobos differ from other great apes in Africa with regards to their selection of inundated habitats?

To answer these questions, we compared the amount of habitat used by bonobos to the availability of each habitat type found within their home range. Table 25.1 shows a list of habitat types, availability of each habitat type and use frequency by bonobos at several study sites that included inundated areas. As mentioned above, currently, only a few populations living in the marginal zone of the bonobos' distribution are known to use dry grasslands or mosaic forests around savannas. In contrast, inundated areas constituted the home ranges of many populations of bonobos at almost all of the study sites (Table 25.1). Most of the research on habitat use of bonobos is based on nest surveys. Research based on direct observations is scant and usually short term (lasting a few months), and therefore does not always capture the full seasonal spectrum of variation across a given year. Continuous direct observation of great apes is difficult because of low visibility in the rain forest and because unhabituated great apes tend to avoid humans.

Although still fragmentary, the literature does contain some evidence on how bonobos use inundated areas. First, nest surveys have shown that inundated forests can provide sleeping sites for bonobos. Bonobo nests have been found in inundated areas at several study sites, although only in a small proportion, as shown in Table 25.1 (Kano 1983; Mohneke & Fruth 2008; Mulavwa *et al.* 2010; Reinartz *et al.* 2006). Indeed, based on the low proportion of nests found there, bonobos generally seem to avoid sleeping in inundated areas (Mohneke & Fruth 2008). However, Inogwabini *et al.* (2012) reported finding most of bonobo nests in permanently or seasonally inundated forests in the northern part of the Lac Télé–Lac Tumba Landscape (Bolombo–Losombo) (Table 25.1). Therefore, the overall importance of inundated areas as sleeping sites for bonobos is still under dispute and may vary from site to site.

Bonobos may also exploit inundated habitats for food. Several studies have reported that bonobos feed in inundated areas, both at the boundaries of their range and in areas within their typical distribution (Lack Tumba: Horn 1980; Wamba: Kano & Mulavwa 1984; Lukuru: Thompson 2002; Yalosidi: Uehara 1988; Lomako: White 1992). However, while observations of feeding activity as snap shots or short-term records have been reported, only a few studies have more thoroughly documented the relative use by bonobos of different habitats (Table 25.1), except for White (1992) at Lomako using focal observations and Hashimoto *et al.* (1998) at Wamba using group tracking. Therefore, it is difficult to discuss the frequency of use and importance of inundated forests to bonobos for ranging and feeding. Nest counting is a reasonable way to detect distribution extent and home range of bonobos, but habitat selection by bonobos needs to be quantified by continuous activity data as a step towards understanding the suitability of inundated areas for bonobos.

Habitat Use of Bonobos at Wamba

We present our findings on the habitat use of a group of bonobos at Wamba based on continuous direct observations over one full year (Terada *et al.* 2015). Conducting continuous observations of bonobo behaviour for longer time spans is possible in Wamba due to more than 40 years of researcher presence and habituation (Furuichi *et al.* 2012). The site is appropriate for this particular form of study because it includes a range of forest types, including swamp forest, and these are under lower hunting pressure than other areas within the bonobo's range.

We address the following: (1) bonobo selectivity of swamp forest for ranging, feeding, and sleeping; (2) food items consumed by bonobos in swamp forest; and (3) seasonal variation in bonobo use of swamp forest, with reference to fruit abundance phenology in each forest type included in this study.

Methods

Study Site

We studied a group of bonobos, called E1, at Wamba (0°11′N, 22°37′E), in the Luo Scientific Reserve, located in the northern part of the DRC. Annual rainfall was 2526 mm and monthly rainfall ranged between 76–378 mm during the study period. There are two rainy seasons each year, with the heaviest between October and December and the lighter between May and June; our study year was similar to previous years (Hashimoto *et al.* 1998; Mulavwa *et al.* 2008). See Terada *et al.* (2015) for further details on the study site and methodology.

Forest Classification

We made a 60-m spatial resolution vegetation classification map of the study site (Figure 25.2a) by overlaying two satellite-derived maps: a land-cover map from OSFAC (2010) and an inundation probability map from Bwangoy *et al.* (2010). We classified the study area into three forest classes as follows (see Idani *et al.* 1994 for details on vegetation types):

1. Primary forest and old secondary forest (P/OS): Mixed mature forests dominated by Caesalpinioideae in Fabaceae, and a monodominant forest of *Gilbertiodendron dewevrei* (Fabaceae) in non-inundated areas. Most areas were closed forests, but some parts were semi-open with dense Marantaceae and various types of understory.
2. Young secondary forest and agricultural land (YS/Ag): Mixed young secondary forests dominated by Caesalpinioideae in Fabaceae and Euphorbiaceae with a lower tree density, and cultivated or fallow land in non-inundated areas. There was a dense understory dominated by Marantaceae, Zingiberaceae and Commelinaceae.
3. Swamp forest (Sw): Mixed mature primary forests dominated by Caesalpinioideae in Fabaceae and Euphorbiaceae, such as *Uapaca heudelotti*, in almost permanently inundated areas located along rivers, and with a dense understory dominated by Marantaceae, Zingiberaceae, Commelinaceae and Arecaceae.

Bonobo Group Tracking

We followed a group of bonobos called E1 for 226 days in total, averaging 18.9 ± standard deviation of 3.4 days per month, from September 2007 to August 2008. The bonobo group was comprised of 23–24 individuals during our observation period (Furuichi *et al.* 2012). We followed them from morning-sleeping nest sites to night-sleeping nest sites, when possible, tracking by direct observation or via footprints. We only used data from days when it was possible to follow the study group for more than 1 h. While tracking the group, we recorded our locations by using the GPS at 1-minute intervals. We recorded their range locations (n = 106 801) where we visually observed bonobos or fresh footprints, corresponding to 1780 h in total or 7.9 ± 3.1 h/day. We also marked the feeding locations (n = 4390) where we observed at least one individual feeding or if we found food remains. At each feeding location, we recorded food items when possible (both the food eaten by bonobos and food remains) and the type of food source (including trees, herbaceous plants, crops, insects or mushrooms) using Idani *et al.* (1994) to identify the plants to species where possible. We recorded 139 night-sleeping sites (61% of 226 observation days) by visual observation.

Monitoring Fruit Availability

We monitored monthly fruit availability using the fallen fruit census method (see Terada *et al.* (2015) and Furuichi *et al.* (2001)) to examine how fruit availability affects the relative habitat use of bonobos. We used five transects that covered a total of 22.5 km (15.3 km, 4.2 km and 3.0 km in P/OS, YS/Ag,

Table 25.1 List of habitat types, availability, and use by bonobos at bonobo research sites.

Site	Habitat types	Inundated (S: seasonal, P: permanent)	Monodominant	Primary (Pr) /secondary (Se)	Availability (%)		Use (%)					Sources
						Target area/ length	Nests		Definition	Range	Definition	
1. Wamba	Mixed primary and old secondary forests *Including monodominant forest dominated by Gilbertiodendron dewevrei			Pr and Se (old)	65.5	Range area of a group (83.8 km²)a	94.3a	87.0b	Sites of night nest groups (a: $n = 139$, b: $n = 147$) by group tracking	78.9	Group tracking in 1-min intervals (1780 h over 1 year)a	Terada *et al.* 2015, and Mulavwa *et al.* 2010b
	Mixed young secondary forest			Se (young)	15.6		0.7a	0b		5.9		
	Swamp forest	**Inundated (P)**		**Pr**	**18.9**		**5.0a**	**13.0b**		**15.2**		
2. Lomako	Primary forest			Pr	75.2	Range area of all focal animals (7.6 km²)	–			93.4	Focal tracking in 2-min intervals (258.5 h from October to July)	White 1992
	Secondary forest			Se	9.9		–			4.5		
	Monodominant Gilbertiodendron dewevrei forest		Mono	?	2.3		–			1.9		
	Rarely or typically inundated swamp forest	**Inundated (S and P)**			**12.6**		–			**0.2**		
	Swamp grassland	**Inundated (P)**			**0**		–			**0**		
3. Bolombo-Losombo, in the Lac Télé- Lac Tumba Landscape	Mixed mature forest on terra firma				7.8	Reconnaissance & line transects (112 km, every 100 m)	12.5		Nest sites ($n = 23$) by the reconnaissance & transects	–		Inogwabini *et al.* 2012
	Secondary forest on terra firma			Se	3.8		0			–		
	Flooded (seasonally inundated) forest	**Inundated (S)**			**26.9**		**70**			**-**		
	Swamp forest	**Inundated (P)**			**61.5**		**17.5**			**-**		
4. Salonga	Mixed mature/woody forest				23.6	Line transects (65.18 km, every 100 m) *Permanently inundated forest was excluded from the survey	19.1a	18.7b	a: Nest sites ($n = 47$) b: Nests ($n = 257$) by the transects	–		Reinartz *et al.* 2006
	Mixed mature/Marantaceae forest				36.8		68.1a	77.8b		–		
	Mixed mature/liana forest				6.4		0	0		–		
	Mixed mature/understory open forest	**Non-inundated or Inundated (S)**			4.4		0	0		–		
	Old secondary/woody forest			Se (old)	2.7		0	0		–		
	Old secondary/Marantaceae forest			Se (old)	10.5		12.8a	3.5b		–		
	Old secondary/liana forest			Se (old)	17.8		0	0		–		
	Young secondary/ Marantaceae forest			Se (old)	0.1		0	0		–		

	Monodominant *Gilbertiodendron dewevrei* forest/woody forest		Mono	Pr	0.4		0	0		–		
	Monodominant *Gilbertiodendron dewevrei* forest/understory open forest		Mono	Pr	3.7		0	0		–		
	Inundated areas	**Inundated (P)**			**–***		**–**	**–**		**–**		
5. Yalosidi	Primary forest *Including monodominant *Gilbertiodendron dewevrei* forest		Partly mono	Pr	–		52.7		Tree nests (n = 1356) excluding ground nests (n = 16) over a 3-month period (November to January) by observation	–	Encountered locations for 3.5 months	Kano 1983
	Aged Secondary forest			Se (old)	–		21.1			–		
	Young Secondary forest			Se (young)	–		17.5			–		
	Oil Palm forest			Se (young)	–		0			–		
	Swamp forest	**Inundated (S and P)**		**?**	**–**		**8.7**			**–**		
	Marsh grassland	**Inundated (P)**		**?**	**–**		**0**			**–**		
6. Lui-Kotale	Mixed forest on terra firma			Pr	73	Line transects (26.3 km, every 50 m)	97.4		Nests (n = 261) by the transects	–		Mohneke and Fruth 2008
	Monodominant *Monopetalanthus* sp. forest		Mono	Pr	3.3		0			–		
	Monodominant *Gilbertiodendron dewevrei* forest		Mono	Pr	3.1		*0.4*			–		
	Temporarily inundated mixed forest	**Inundated (S)**		**Pr**	**3.6**		**0**			**–**		
	Permanently inundated mixed forest	**Inundated (P)**		**Pr**	**17**		**2.2**			**–**		
7. Lukuru	Climax undisturbed forest			Pr	64.6	Main study area (71.7 km^2)	100		Nests (n = 99) by passive walking survey	38	Encountered locations by visual observations or hearing vocalizations (n = 121)	Myers Thompson 1997
	Secondary forest			Se	11.3		0			56.2		
	Dry grassland			?	20.2		0			4.1		
	Riparian forest on seasonally moist soil	**Inundated (S)**		**?**	**3.9**		**0**			**1.7**		

and Sw, respectively) (Figure 25.2a). We recorded fallen fruit clusters within 1 m of either side of each transect covering 4.5 ha, twice per month. All ripe fruit originating from the same host plant were counted as one 'fallen fruit cluster', so that the number of fallen fruit clusters found per kilometre approximately matched the number of ripe fruiting trees per kilometre. We calculated the number of clusters of ripe fruit per kilometre for each forest type by averaging the two records for each month.

Analyses: Habitat Selection, Food Categories and Effects of Fruit Availability

We examined the bonobos' selection of each forest type for ranging, feeding, and night sleeping respectively, by comparing the proportion of each forest type used by the bonobos relative to its availability within the bonobos' home range. We defined a minimum convex polygon (MCP) for all range locations and used the proportion of the area of each forest type within the MCP as an index of availability. We summed the number of range locations, feeding locations, and night-sleeping locations in each forest type as indexes of habitat use. For the analysis, we used a subset of ranging and feeding location data collected at 1-h intervals (ranging $n = 1812$, feeding $n = 1652$) to lower the spatial autocorrelation. We used all night-sleeping location data because these were collected just once per day. We used a chi-square test and Manly's resource selection ratio (w_i) (Manly *et al.* 2002). The formula for w_i is as follows:

$$w_i = o_i / \pi_i$$

Here, o_i is the proportion of the number of recorded locations in forest type i to the total number of recorded locations, and π_i is the proportion of the area of forest type i to the entire ranging area. Here, w_i indicates the use intensity relative to the availability of forest type i. When w_i is significantly larger or smaller than 1, it indicates that forest type i is positively or negatively selected. We applied a chi-squared test of independence to compare the ratio of food categories consumed across forest types. We also tested whether monthly fruit availability in P/OS, which comprised most of the study area, affected feeding frequency in Sw for each month.

Results

Habitat Selection over a Year

The study group used swamp forest (Sw) for all three focal activities, i.e. ranging, feeding and night sleeping (Figures 25.2b and 25.3), although the group mainly used non-inundated primary forest and old secondary forest (P/OS) during the year. Sw comprised 15.2% of the yearly range area following P/OS (Figure 25.3). With regard to ranging (Figure 25.2b), the group avoided villages centred in the largest area of young secondary forest and agricultural land (YS/Ag), and tended to pass through Sw instead of crossing through villages. The group ranged, fed and slept most frequently in P/OS followed by Sw, and least frequently in YS/Ag (Figure 25.3). Although our study is the first in this area to continuously observe and document the monthly bonobo habitat use over the course of an entire year, previous research in Wamba over the past 20 years has also recorded comparable ranging patterns (Furuichi *et al.* 2008; Hashimoto *et al.* 1998; Kano & Mulavwa 1984).

The bonobos' selection of habitat types is shown in Table 25.2. The bonobos ranged, fed and slept disproportionally in each forest type in the MCP (ranging $\chi^2 = 165.8$, $df = 2$, $p < 0.0001$; feeding $\chi^2 = 157.9$, $df = 2$, $p < 0.0001$; sleeping $\chi^2 = 49.6$, $df = 2$, $p < 0.001$). The selection ratio showed that the group ranged, fed and slept with positive selection in P/OS and negative selection in YS/Ag and Sw. Furthermore, according to the selection intensities shown by the absolute values of the selection ratios (w_i), the relative importance of Sw appeared to differ between the three activities. Both the percentages of recorded locations in each forest type and the corresponding selection ratios showed that P/OS was chosen first, Sw second, and YS/Ag third (Figure 25.3 and Table 25.2). For the three focal activities, the bonobo group showed a strong preference for night sleeping in P/OS; the selection ratio for Sw was the lowest for sleeping, followed by feeding and ranging. The annual frequency of Sw use may vary from year to year; Mulavwa *et al.* (2010) reported that 13% of the group nest locations were in Sw after observing the same group for two years, which was higher than the percentage observed in this study (5.1%). In contrast to Sw, bonobos used YS/Ag almost equally for feeding and ranging, but seldom slept in it (Figure 25.3 and Table 25.2). This was probably because there were almost no high trees for making sleeping nests in this habitat, as well as a higher risk of being detected by humans.

Food Consumed in Swamp Forest

We recorded 5993 feeding records of various food items at 4390 feeding locations (Figure 25.4). The proportion of the eight food categories consumed differed by forest type ($\chi^2 = 622.6$, $df = 14$, $p < 0.001$). Across all three forest types, herbaceous plants were the main food consumed based on the number of records. Bonobos consumed various types of food derived from trees in Sw as well as in P/OS, and in YS/Ag fed more often on herbaceous plants (77% of the 589 records in YS/Ag) compared to the other two forest types (Figure 25.4). Interestingly, the proportion of food types consumed were similar for both P/OS and Sw, with the exception of fungi and vines. Fungi (all truffle mushrooms) accounted for 10% of the 730 food records in Sw and only 0.3% of the 4674 food records in P/OS. Of the 103 fungi records, 75% were consumed in Sw. It is known that bonobos can dig in river beds or inundated ground to feed on various foods, including fungi (such as truffles), dragonfly larvae and earthworms (Kano & Mulavwa 1984). Sw might provide such food that can only be derived from inundated ground, as well as abundant herbaceous plants. In contrast, vines accounted for 9.6% of the food records in P/OS and only 3.7% of the records in Sw. This seems to be due to the abundance of fruit vines, which includes fruits selectively eaten by bonobos such as the fruit of *Landolphia* spp. (Apocynaceae) in P/OS (Kano & Mulavwa 1984).

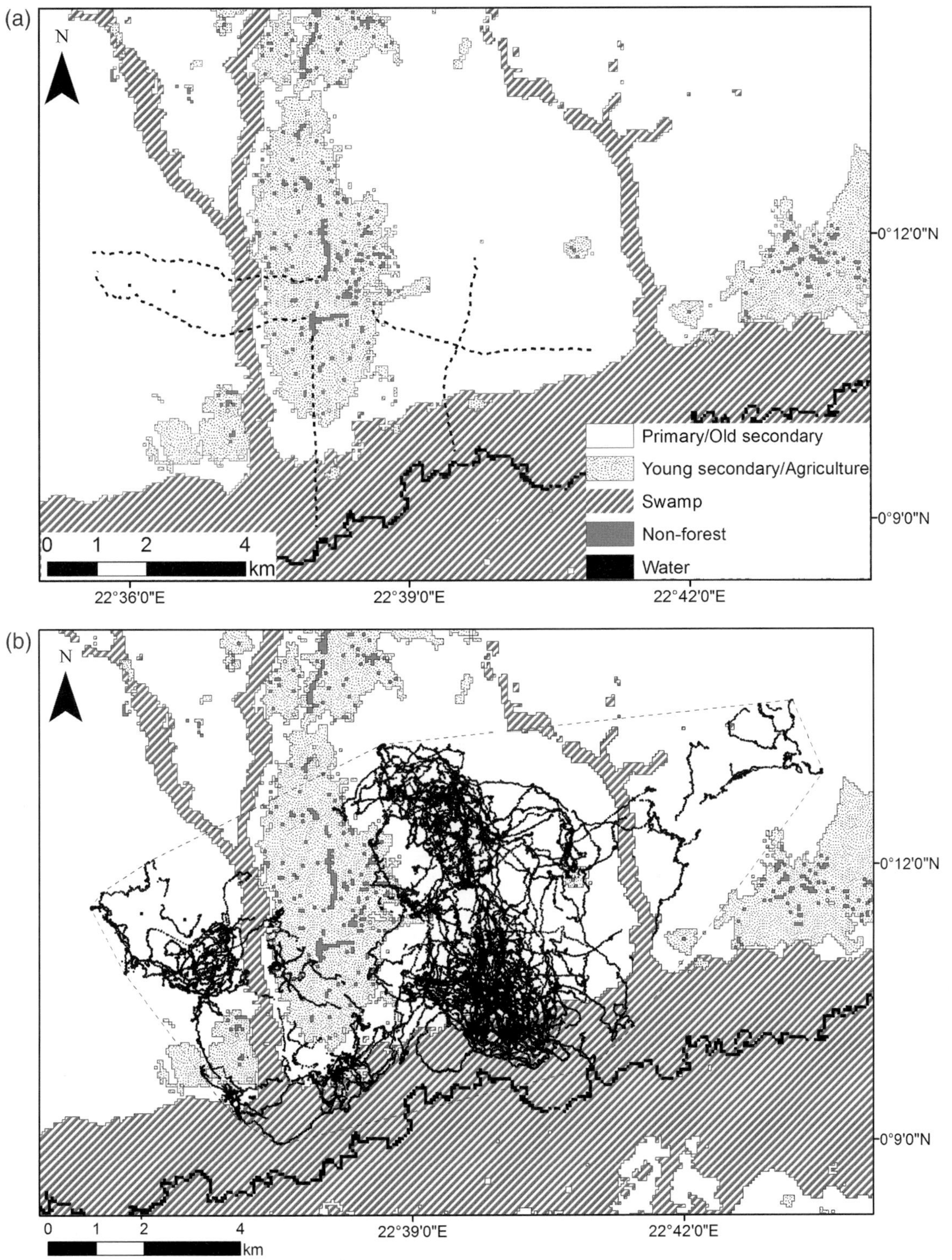

Figure 25.2 (a) Forest classification map of the study site with line transects for fruit monitoring. A wide river is visible across the bottom of the map (Terada *et al.* 2015: a permission has been granted by the publisher, John Wiley and Sons). (b) The range locations of the bonobo group are indicated by dots, and the dotted line outlines a MCP defining the group's ranging area (Terada *et al.* 2015: a permission has been granted by the publisher, John Wiley and Sons).

Seasonality of Swamp Use and Fruit Availability

The study group only used inundated forest at high frequency during specific months. Although the proportion of ranging, feeding and night-sleeping locations used across forest types varied from month to month, the monthly proportion of use among the three forest types (Figure 25.5) was similar to the overall annual proportion (Figure 25.3) except for June. The bonobos used Sw for ranging and feeding almost every month. However, in June, the bonobo group used Sw for both ranging and feeding more frequently than P/OS. The group also slept more frequently in Sw than in P/OS only during June (four of nine nests were found in Sw). The bonobos used Sw for sleeping only during two other months, September ($n = 1$) and December ($n = 2$), although they used Sw for ranging more than 10% in the remaining 5 months (from July to December, except for October). From January to May, the group ranged in Sw less than 10% and

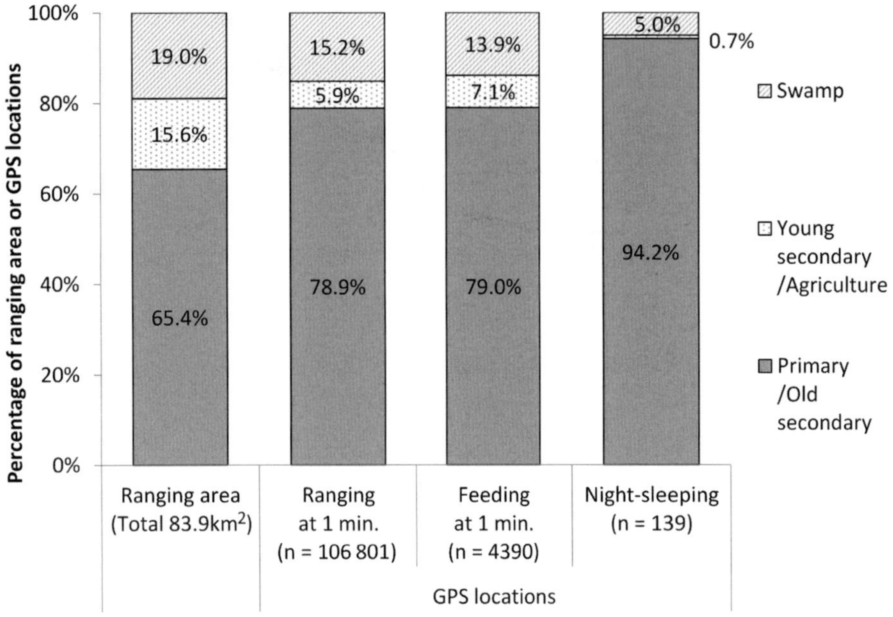

Figure 25.3 Proportion of the range area and the number of ranging, feeding, and night-sleeping locations recorded in each forest type (Terada *et al.* 2015: a permission has been granted by the publisher, John Wiley and Sons).

never used it for sleeping. They used YS/Ag almost every month for ranging and feeding and slept there during only one night in June.

The frequency with which the bonobos used Sw was not related to fruit availability in either Sw or in the bonobos' main habitat (P/OS). Fruit availability was generally higher in the non-inundated forests than in the inundated forests (Figure 25.6). Monthly fruit availability showed a different trend in Sw than in the two other forest types in non-inundated areas. Both fruit availability in P/OS and in YS/Ag peaked in August and was relatively high during the three months (October–December) that corresponded to one of the two annual rainfall peaks (Figure 25.6). In contrast, fruit availability in Sw was the lowest of the three habitat types in all months except January and February, when Sw fruit availability peaked and exceeded that in YS/Ag. However, the bonobos seldom used Sw during these months (Figure 25.5). Furthermore, in June, when the bonobos spent most of their time in Sw, fruit availability in Sw was low compared to that in the other forest types. According to a regression test, fruit availability in P/OS did not affect the feeding frequency in any of the three forest types, though the p values for P/OS and YS/Ag were marginal (P/OS: estimate = 0.001, SE = 0.0008, $p = 0.07$; YS/Ag: estimate = −0.001, SE = 0.0005, $p = 0.07$; Sw: estimate = 0.0003, SE = 0.001, $p = 0.79$). This indicates that the study group did not seem to use Sw for fallback food resources when fruit was scarce in their main habitat.

Discussion

Mechanisms and Implications of Swamp Use by Bonobos at Wamba

The importance of food in inundated areas relative to that in other forest types in the bonobos' home range may be a key factor influencing the seasonal variation of the bonobos' use of swamp forest in Wamba. The foods consumed by the bonobos in swamp forest included two types: (1) foods that are abundant and available all year and (2) foods with seasonal fluctuations. For the first type, the swamp is rich in THV including *Sclerosperma mannii* (Aracaceae), a 2–3 m tall palms whose pith has been found to be one of the bonobos' favourite foods (Idani *et al.* 1994; Kuroda 1979). Mushrooms are another example of a special food type only found in swamp habitat, although the seasonality and distribution of mushrooms there is still unclear. These foods are reliably and consistently available, and this may be one of the reasons that bonobos use inundated forests as part of their range area throughout the year. In contrast, food with seasonal fluctuations seemed to attract bonobos to Sw only during the fruiting season. For example, bonobos exclusively used Sw in June, which corresponded to the fruiting peak of one of their favourite food species, *Uapaca heudelotii* (Euphorbiaceae) (from June to July and from November to January) (Mulavwa *et al.* 2010). This species is found only in Sw, and their fruit is frequently eaten by bonobos (Idani *et al.* 1994). Bonobos at another site (Kokolopori) also relied heavily on one dominant species, *Guibourtia demeusei* (Fabaceae), in Sw during a specific study period (November and December) (Georgiev *et al.* 2011).

The relative phenological timing of important food species in inundated and non-inundated areas may be an important factor in seasonal site-specific habitat use. For example, though the fruits of *U. heudelotii* normally have two peaks in a year at this study site, the study group used Sw with high frequency during only the first peak (June to July). This may be because the fruit availability in non-inundated forests was relatively low in June, but was high during the second peak (November to January) (Figure 25.6). Furthermore, in November and December, other favourite bonobo foods such as *Landolphia* spp. and *Dialium* spp. (Fabaceae) are also abundant in non-inundated forests during the second peak of *U. heudelotii* (Kano & Mulavwa 1984).

Table 25.2 Summary of the bonobos' selection of three forest types by Manly's resource selection ratios. Availability is shown as the size and percentage of each forest type within the MCP. Index of use is the number and percentage of ranging, feeding or night-sleeping locations recorded in each forest type (ranging and feeding locations were taken at 1 h intervals, while night-sleeping locations were taken at 1 day intervals). Manly's selection index is indicated with '*w*'. When *w* is significantly larger or smaller than 1, the forest type is positively or negatively selected. '+/–'show significant selection by 95% confidence intervals.

Forest types	Availability		Use														
	Range area		1. Ranging locations					2. Feeding locations					3. Night-sleeping locations				
	Size (km²)	(%)	Location number	(%)	*w*			Location number	(%)	*w*			Location number	(%)	*w*		
Primary/old secondary	54.9	(65.5%)	1416	(78.1%)	**1.19**	+	*	1306	(79.1%)	**1.20**	+	*	131	(94.3%)	**1.44**	+	*
Young secondary/ agriculture	13.1	(15.6%)	103	(5.7%)	**0.36**	–	*	99	(6.0%)	**0.38**	–	*	1	(0.7%)	**0.04**	–	*
Swamp	15.9	(18.9%)	293	(16.2%)	**0.85**	–	*	247	(15.0%)	**0.79**	–	*	7	(5.0%)	**0.26**	–	*
Total	83.9		1812					1652					139				

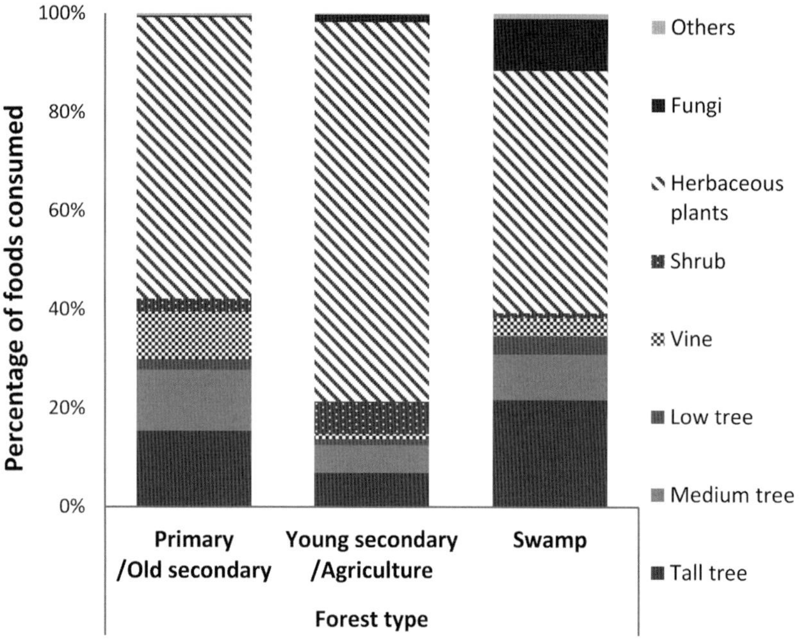

Figure 25.4 Proportion of the number of food items recorded during group tracking (recorded as either food eaten by bonobos or food remains) for each forest type (Terada *et al.* 2015: a permission has been granted by the publisher, John Wiley and Sons).

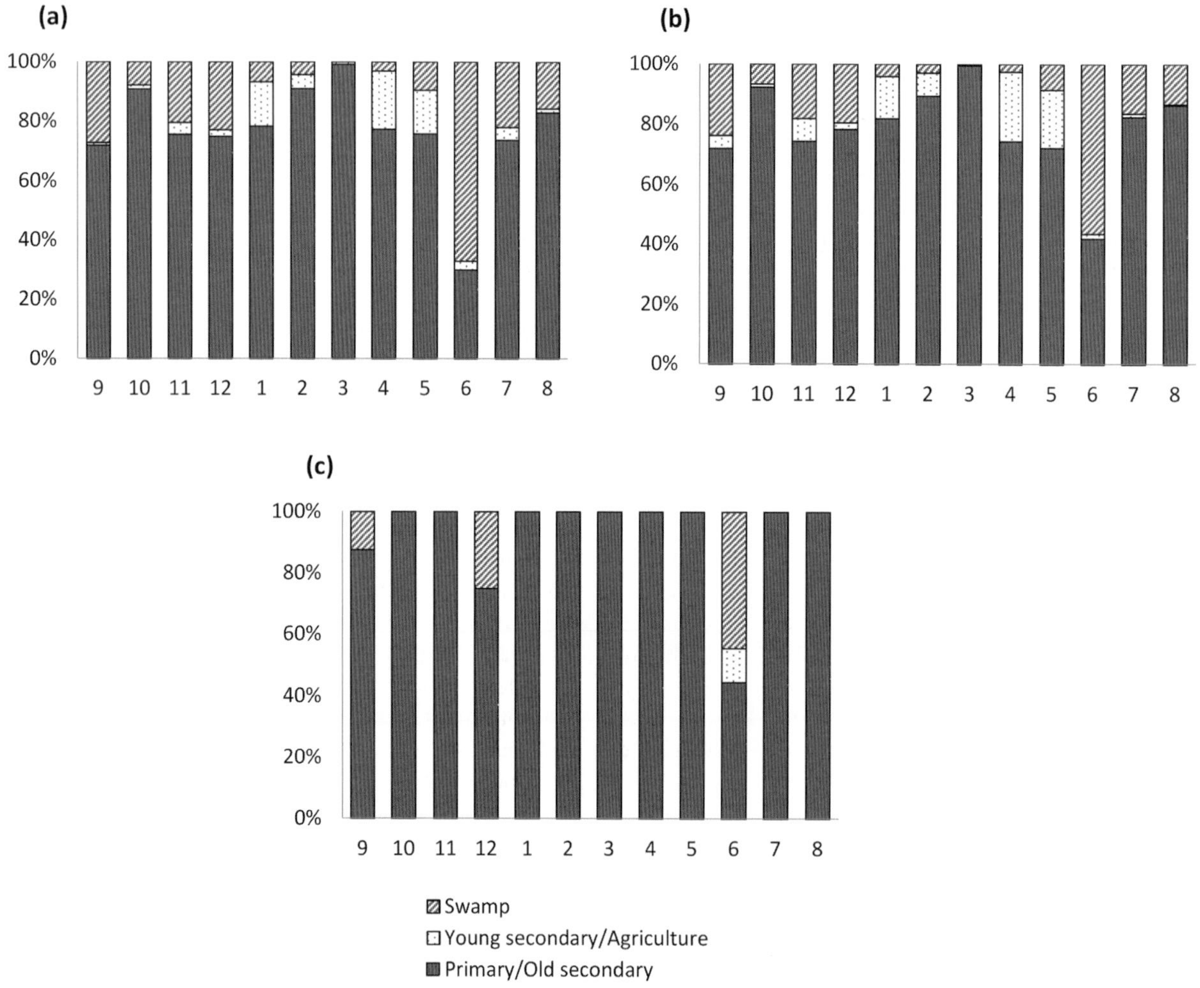

Figure 25.5 For each forest type, the percentage of monthly locations used for (a) ranging, (b) feeding and (c) night sleeping (Terada *et al.* 2015: a permission has been granted by the publisher, John Wiley and Sons).

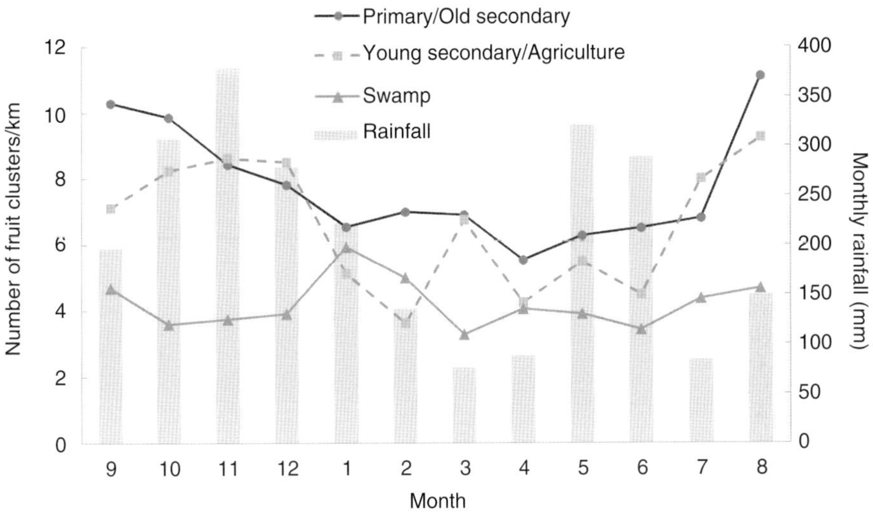

Figure 25.6 Monthly rainfall and the density of fallen fruit clusters for each forest type as observed by line-transect census.

The studied bonobos may sleep in inundated areas during the night only during those periods when they are using inundated areas as their main habitat, which is probably related to seasonal food availability there. They may use Sw in two different ways for night sleeping: (1) to stay short term for feeding, returning to non-inundated areas for night sleeping, or (2) to stay long term for both feeding and eventually sleeping. Kano (1992) showed use of Sw most similar to (1) above, based on observations of the same bonobo group. He reported that the group did not sleep in Sw even on those days when they ranged and fed in Sw. Our study also showed Sw use similar to (1) above in almost every other month throughout the year. This pattern was applicable only during seasons when bonobos likely used Sw as a supplementary area for ranging and feeding on stable foods. On the other hand, the bonobo group showed Sw use similar to (2) above in June only when Sw was used as a main habitat.

Such fine-scaled information based on ground observations can help improve habitat suitability modelling efforts by linking with coarser data derived from remote sensing. For example, a recent modelling effort used nest location data and treated both non-inundated and inundated forests equally for the percentage-forest variable to produce the first range-wide map of bonobo habitat suitability (Hickey *et al.* 2013). In contrast, the current study at Wamba highlighted the importance of inundated areas, and suggested that their use may be underestimated if only using nest count locations and without monitoring use of inundated areas for other activities. If ground research could first distinguish the importance of inundated areas for each of the bonobos' activities, and then help to classify those habitat types using remote sensing techniques, habitat preference considering activities other than nesting might be extrapolated to a wider area.

Comparisons Between Bonobo Populations

In order to understand bonobos' general habitat use, a comparative study across sites using comparable quantitative data on habitat availability and habitat use is needed. Table 25.1 shows that habitat compositions, including a percentage of inundated areas in the bonobos' home ranges, vary across research sites. Mixed primary forests and/or non-inundated forests comprised the main habitats of bonobos at most study sites. However, bonobo habitats also included secondary forests, monodominant *Gilbertiodendron* (Fabaceae) forests, and inundated areas such as mixed flooded forests, swamp forests and grasslands. Inundated areas constitute a supplemental habitat for bonobos at most research sites, where they make up between 3% to slightly less than 20% of land cover. In Wamba, inundated swamp forest constituted a relatively high percentage (18.9%) of available habitat compared to other bonobo study sites, but it was still a 'supplemental' habitat.

Although the number of populations and study sites that can be compared are limited, and the methods used to evaluate habitat availability differs between studies, it is clear from Table 25.1 that bonobos' use of inundated areas may well vary depending on the characteristics of the study area. Bonobos living in closer proximity to habitats dominated by inundated areas might depend on them more often than bonobo populations living in areas dominated by non-inundated areas, according to habitat availability. For example, a population of bonobos inhabiting a mostly inundated area (Bolombo-Losombo), in the Lac Télé–Lac Tumba Landscape slept more frequently in inundated habitats than in non-inundated habitats (Table 25.1). Inogwabini *et al.* (2012) also reported that the encounter rate for nesting sites (0.21/km) in this site was higher than in several forest blocks in Salonga National Park (0.03–0.1/km), where inundated areas were not dominant (see Table 25.1 and Reinartz *et al.* 2006 for details). Furthermore, for this population, although the percentages of habitat availability and nest sites found in all of the inundated areas were similar (availability 88.4%, nest locations 87.5%), the nest sites located mainly in seasonally flooded forests, and the number of nest sites was lower in permanently inundated forests than the expected proportion based on habitat availability. For bonobos, permanently inundated forests may have low suitability for nesting, but seasonally inundated forests

Table 25.3 List of habitat types, availability and use at chimpanzee study sites in the Congo Basin, Central Africa.

Species	Site	Country	Sympatric ape sp.	Habitat types	Inundated (S: seasonal, P: permanent)	Monodominant
Pan troglodytes troglodytes	**8. Dja Faunal Reserve (Ntonga)**	Cameroon	*Gorilla gorilla gorilla*	Primary forest		
				Old secondary forest (little undergrowth)		
				Secondary forest (patches of undergrowth)		
				Young secondary forest (dense undergrowth)		
				Old logging road forest		
				Riverian forest	**Inundated (S)**	
				Swamp forest	**Inundated (P)**	
	9. Goulalougo triangle in Nouabalé-Ndoki NP	Congo	*Gorilla gorilla gorilla*	Mixed forest on terra firma with closed canopy		
				Mixed forest on terra firma with open canopy		
				Monodominant *Gilbertiodendron* dewevrei forest		
				Swamp forest	**Inundated (S and P)**	
	10. Around Mondika in Nouabalé-Ndoki NP	Congo	*Gorilla gorilla gorilla*	Mixed-species forest		
				Gilbertiodendron forest (riverland forest with closed canopy or inland forest with many gaps)		Mono
				Swamp forest (riverside raphia forest, flooded swamp forest, marshy grassland)	**Inundated (P)**	
	11. Lac télé community reserve in Lac Télé-Lac Tumba Landscape	Congo	*Gorilla gorilla gorilla*	Mixed forest on terra firma		
				Seasonally flooded savanna	**Inundated (S)**	
				Seasonally flooded and riperian forest	**Inundated (S)**	
				Swamp forest	**Inundated (P)**	
	12. Ngri triangle in Lac Télé-Lac Tumba Landscape	DRC	No?	Mixed mature forest on terra firma		
				Secondary forest on terra firma		
				Savannahs		
				Flooded (seasonally inundated) forest	**Inundated (S)**	
				Swamp forest	**Inundated (P)**	
	13. Bomongo-Lubengo in Lac Télé-Lac Tumba Landscape	DRC	No?	Mixed mature forest on terra firma		
				Secondary forest on terra firma		
				Flooded (seasonally inundated) forest	**Inundated (S)**	
				Swamp forest	**Inundated (P)**	

Availability			Use (%)				Source
Primary (Pr)/ secondary (Se)		Target area/ length	Nests	Definition	Range	Definition	
Pr	17.6	Line transects (57.7 km, every 50 m)	27.4	Nest sites (*n* = 91) by the transects	–		Dupain *et al.* 2004
Se (old)	10.8		22.0		–		
Se	39.3		34.1		–		
Se (young)	15.6		12.1		–		
Se (old)	2.1		0.0		–		
	5.3		**2.2**		**–**		
	9.3		**2.2**		**–**		
	33.4	Line transects (222 km)	32.6	Nest sites (*n* = 368) by the transects	30–35	Direct observations (*n* = 258) by reconnaissance surveys for about 26 months (estimated % from the fig. of the source article)	Morgan *et al.* 2006
	40.1		9.5		40–45		
	21.3		56.5		25–30		
	5.2		**1.4**		**< 5**		
	59.3	Line transects (19.47 km) - *In total study area (about 20 km²), 15–20% is swamp forest	–		80–100	Observations of individuals (*n* = 115) by the line census	Mitani *et al.* 1993
	39.3		–		0–5	(Estimated % from the fig. of the source article)	
	***1.4**		**–**		**0–15**		
	9.6	Total area of study site (4, 551 km²)	3.6	Nest sites (*n* = 197) by the transects	–		Poulsen & Clark 2004
	15.5		**0**		**–**		
	25.5		**10.1**		**–**		
	49.4		**86.3**		**–**		
	27.6	Reconnaissance & line transects (186 km, every 100 m)	12.5	Nest sites (*n* = 20) by the reconnaissance and transects	–		Inogwabini *et al.* 2012
Se	13.9		–		–		
	2.8		–				
	30.7		**70.0**		**–**		
	24.9		**17.5**		**–**		
	7.0	Reconnaissance & line transects (126 km, every 100 m)	0.0	Nest sites (*n* = 3) by the reconnaissance & transects	–		Inogwabini *et al.* 2012
Se	3.0		0.0		–		
	9.0		**0.0**		**–**		
	81.0		**100.0**		**–**		

(continued)

Table 25.3 *(cont.)*

Species	Site	Country	Sympatric ape sp.	Habitat types	Inundated (S: seasonal, P: permanent)	Monodominant
	14. Odzala NP (Lokoué)	Congo	*Gorilla gorilla gorilla*	Mixed forest with a closed-canopy cover		
				Marantaceae mixed forest with a closed-canopy cover		
				Marantaceae mixed forest with a low canopy		
				Marantaceae mixed forest with an open canopy		
				Monodominant *Gilbertiodendron* dewevrei forest		
				Swamp or gallery forest	**Inundated (S and P)**	
Pan troglodytes schweinfurthii	**15. Kahuzi-Biega NP (eastern part)**	DRC	*Gorilla beringei graueri*	Primary montane forest		
				Secondary montane forest		
				Bamboo forest (*Arundinaria alpina*)		Mono
				***Cyperus latifolius* swamp**	**Inundated (P)**	

may provide suitable nest sites to complement non-inundated forests.

The combination and distribution patterns of habitat types, which determine the relative importance of each habitat type, differ among study sites and even among the home ranges of separate populations at the same study site. Therefore, the use frequency and selection of inundated areas may differ widely between studied populations. Furthermore, as discussed above regarding our study in Wamba, phenology of specific food species in different habitat types appears to influence the relative importance of each habitat for bonobos over the course of a year.

However, it is difficult to draw broad conclusions about the suitability of inundated areas for bonobos across their wider range due to the limited number of reports and the inconsistency of the methods used to determine habitat availability and habitat use. Additionally, there are inconsistencies in habitat definitions and terminology between bonobo study sites (Kortlandt 1995; Reinartz *et al.* 2006). To find a consistent mechanism for habitat choice, we would need comparative data based on uniform or comparable methods focused on the distribution of vegetation types that determine the food and nest resources available for bonobos.

Comparison with Chimpanzee Populations

Table 25.3 shows the habitat types, availability of each habitat, and habitat use by chimpanzees across the study sites shown in Figure 25.1. According to this overview, chimpanzee habitat use shares several characteristics with that of bonobos. First, the home range of some populations of chimpanzees includes habitats with a low percentage of inundated areas, similar to bonobo populations. Second, the nests of some chimpanzee populations that inhabit areas where most of the habitat consists of inundated areas in the Lac Télé–Lac Tumba Landscape (Ngri triangle and Bomongo-Lubengo) were often found to be within seasonally or permanently inundated forests (Inogwabini *et al.* 2012; Poulsen & Clark 2004), similar to the bonobo population studied by Inogwabini *et al.* (2012) in the same landscape (Bolombo-Losombo). It is noteworthy that such chimpanzee populations living in mostly inundated areas used such habitats more frequently than expected based on habitat availability. Chimpanzees are well known for using forest–savanna mosaics while populations of bonobos living in such habitats are rare, but like bonobos, chimpanzees also seem to be able to inhabit areas that are mostly inundated.

The habitat combinations and distributions, which in turn determine resource distribution in home ranges, may

Availability			Use (%)				Source
Primary (Pr)/ secondary (Se)		Target area/ length	Nests	Definition	Range	Definition	
	8.0	Line transects (29.5 km) *Gilbertiodendron forest was not included in line transects	- Nests were found in all forest types - Significantly many in Marantaceae mixed forest	Nests by the transects	–		Devos *et al.* 2008
	23.0				–		
	51.0				–		
	11.0				–		
	-*				–		
	7.0				**–**		
	13.5	Total area of study site (33.1 km²)	49.0	Night nests (*n* = 310) by group tracing for a year	10.55	Observed sightings of individuals by group tracing	Basabose 2005
	64.1		51.0		67.82		Basabose & Yamagiwa 2002
	17.6		0.0		5.93		
	4.8		0.0		**6.6**		

generally explain the pattern of use of inundated areas by the two *Pan* species. Similarly to bonobos, chimpanzees showed different habitat preferences for sleeping and for ranging. In the Goualougo triangle within Nouabalé–Ndoki National Park, Morgan *et al.* (2006) reported the highest abundance of chimpanzee nests in non-inundated monodominant *Gilbertiodendron* forests and the highest number of direct observations of chimpanzees in mixed open forests in non-inundated areas (Table 25.3). They considered that canopy cover may be an important factor for nest site habitat selection because almost all of the monodominant *Gilbertiodendron* forests have a closed canopy (Morgan *et al.* 2006). In contrast, swamp forest comprised only 5% of the area surveyed, but still appeared to be underutilized based on the low percentages of both nests and individuals of chimpanzees observed in the inundated areas. In this area, swamp forest does not appear to be attractive for nesting because of a lower canopy cover compared to that in the monodominant *Gilbertiodendron* forest. Also swamp forest appeared to have less food availability compared to mixed forests in non-inundated areas, therefore chimpanzees may range mainly in mixed open forests in non-inundated areas. Because most of the habitat use data for both chimpanzees and bonobos are collected from nest surveys, further research is needed to compare habitat selection between chimpanzees and bonobos.

It is still unclear whether the preference for inundated areas is the same between bonobos and chimpanzees. Although chimpanzees can occur in drier habitats than bonobos, like bonobos, they also cross streams and feed on aquatic plants and characteristically perform water displays by striking the water or throwing objects (Kempf 2009). However, the frequency with which they enter the water may differ from bonobos. Bonobos use not only inundated forests, but also marsh grasslands to feed on particular aquatic or amphibious herbs and grasses (Kano 1983; Uehara 1990). Additionally, in several study sites, bonobos are known to sometimes enter the water, using bipedal walking, and feed on aquatic algae (Thompson 2002; Uehara 1990). Although, based on our long-term observations in Wamba, bonobos generally avoid entering streams or puddles in swamp forest, bonobos may be able to adapt to water if there are positive reasons to use it. Chimpanzees also have been observed entering a stream and feeding on grasses and algae growing in freshwater at the Mahale Mountains National Park, in Tanzania, a long-term study site on chimpanzees (Nishida 1980; Sakamaki 1998). Although the intraspecific variation seems to be large, both *Pan* species can adapt to inundated habitats and water if there is a need to do so based on availability and distribution of habitat types in their ranging area. Furthermore, the number of populations that have adapted to inundated habitats and water may be larger for bonobos than chimpanzees because most bonobo populations are exposed to inundated habitat conditions in the Congo Basin.

Use of Inundated Habitats by Sympatric Chimpanzees and Gorillas

When considering differences in swamp habitat use between the two *Pan* species, it is important to also consider the influence of sympatric gorillas. Observed patterns in the relative use of swamps by the two *Pan* species may be unrelated to intrinsic species differences, but linked instead to food competition with gorillas or other species.

As shown in Table 25.3, most studied populations of chimpanzees live sympatrically with gorillas, whereas bonobos have no sympatric distributions with other great apes (Figure 25.1). Within the purview of our review (the Congo Basin), we found three types of use of inundated habitats by chimpanzees and gorillas. First, chimpanzees infrequently use inundated forests, like bonobos, at most study sites, but gorillas use inundated forests as a part of their main habitat (Dupain *et al.* 2004; Kuroda *et al.* 1996; Mitani *et al.* 1993). These patterns can be explained by more frugivorous feeding strategies and higher frequencies of arboreal nest construction by chimpanzees than by gorillas (Dupain *et al.* 2004; Yamagiwa *et al.* 1992; Chapter 24), particularly because inundated areas generally seem to have more abundant herbaceous vegetation, less abundant fruit, and fewer trees in which chimpanzees could nest. For example (see Table 25.3), in Dja Faunal Reserve in Cameroon, chimpanzee nests were only rarely found in inundated forests, while gorilla nests were found mainly in a combination of inundated and secondary forests (Dupain *et al.* 2004). In Nouabalé-Ndoki National Park in the Republic of Congo, chimpanzees were observed mainly in mixed forests in non-inundated areas and they rarely entered swamps to forage, while gorillas were observed in both mixed forests and swamp forest, and fed on aquatic plants in marshy grasslands (Kuroda *et al.* 1996; Mitani *et al.* 1993). Chimpanzees and gorillas in the montane forest of Kahuzi-Biega National Park, DRC, also showed similar habitat selection for nesting and ranging (Basabose 2005; Basabose & Yamagiwa 2002; Chapter 24).

Both chimpanzees and gorillas use inundated habitats extensively. In the Lac Tele Landscape, an area dominated by inundated habitats (Table 25.3), both chimpanzee and gorilla densities were high in permanently flooded swamp forest compared to in other habitat types. According to Poulsen and Clark (2004), more information on the fruiting patterns and fruit abundance in each of these habitats is needed to explain the high density of chimpanzees in swamp forest, while the high density of gorillas may be explained by predictably and consistently abundant ground vegetation.

In contrast to reports from other study sites, at both the Goulalougo triangle in Nouabalé-Ndoki National Park and in Lokoué in Odzala National Park, Republic of Congo (Table 25.3), gorilla nests were infrequently observed in swamp forest, although gorillas appeared to prefer more open and Marantaceae-dominated forests than chimpanzees (Devos *et al.* 2008; Morgan *et al.* 2006). Open and Marantaceae-dominated forests seem to provide both nesting sites and feeding places for gorillas; however, this does not explain why gorillas seldom nest in swamp forest.

As described above, there are considerable intersite differences in habitat use by chimpanzees and gorillas, although both species seem to use each habitat based on their respective feeding strategies. Furthermore, it remains unclear whether interspecific interactions affect the relative habitat use of each species. Additional research and monitoring in the forests of the Congo Basin would help shed light on the factors involved in the habitat selection of chimpanzees and gorillas (Morgan *et al.* 2006; Poulsen & Clark 2004) before a valid comparison between the swamp use of the two *Pan* species can be made.

Conclusion

Bonobos appear to use inundated habitats occasionally. Inundated habitats may help sustain bonobos by providing them with specific foods such as fungi and alternative sleeping sites. It is therefore important to consider inundated habitats in conservation planning. Here, we highlighted the use of inundated habitats by bonobos for various activities including ranging, feeding and sleeping. Such field-based information can be useful for informing conservation prioritization efforts at multiple scales, which combine information obtained at a fine scale at small sites with coarser habitat models that examine suitability over broader areas. It is also important to clarify habitat suitability by understanding the ecological factors that drive species habitat selection, and by comparing this selection across several different study sites and groups. Further inter- and intraspecific comparative research may help us understand the niche separation between bonobos and chimpanzees – two *Pan* species that occur non-sympatrically – and their relative use of inundated habitats.

Acknowledgements

We would like to thank the Research Center for Ecology and Forestry (CREF), Ministry of Scientific Research of the DRC for offering continued support for our studies at Wamba, and researchers and local staff who worked with us. We especially thank and remember Kumugo Yangozene, who obtained field data with us; he passed away on 2 January 2014. We would like to dedicate this chapter to him. We also thank Dr Takayoshi Kano, Dr Suehisa Kuroda and other researchers and local communities living in Wamba for their efforts to sustain long-term studies there, as well as Dr Yoshihiro Nakashima, Dr David Sprague, Dr Hiroyuki Takemoto and members of the Primate Research Institute of Kyoto University for providing valuable suggestions. This research was supported by the Environment Research and Technology Development Fund

(D-1007) (to Furuichi); the Global Environment Research Fund of the Ministry of the Environment, Japan (F-061) (to Dr Toshisada Nishida); Japan Society for the Promotion of Science (JSPS) Grants-in-aid for Scientific Research (17255005 and 22255007, to Furuichi and 25257487 to Yumoto); the JSPS Asia-Africa Science Platform Program (2009–2011, 2012–2014 to Furuichi); the JSPS Primate Origins of Human Evolution (HOPE) project (AS-22-027 and ITP-23-006, to Sakamaki and AS-24-031 and ITP-25-002 to Terada); the JSPS Grants-in-aid for JSPS Research fellowship (14J00182 to Terada).

Chapter 26 Differences in Population Density of Orangutan Between Flooded and Non-flooded Forests

Tomoko Kanamori, Noko Kuze and Shiro Kohshima

Introduction

Orangutans (*Pongo* spp.) are the only great ape taxa living in Asia. Until recently, it was thought that two species, *P. abelii* and *P. pygmaeus*, inhabit the islands of Sumatra and Borneo, respectively (Goossens *et al.* 2009). However, in 2017 it was reported that an isolated population inhabiting the area distant from *P. abelii* on the Sumatra Island is a new species described as Tapanuli orangutan (*P. tapanuliensis*) (Nater *et al.* 2017). There are two species of orangutans in Sumatra, while the Bornean species is divided into three subspecies: *P. p. pygmaeus* in west Borneo, *P. p. wurmbii* in west and central Borneo, and *P. p. morio* in east Borneo (Goossens *et al.* 2009; Groves 2001; Nater *et al.* 2017). These three species are classified as a Critically Endangered species; only 14 613 in Sumatra and 54 000 in Borneo are estimated to remain in the wild (IUCN 2017; Wich *et al.* 2008, 2016). At present, 75% of orangutans range outside of national parks (Meijaard & Wich 2007), in areas where the forests have been severely degraded by illegal logging, mining, fires and encroachment by palm oil plantations, and 25% of orangutans are not safe at all because illegal logging occurs inside of most national parks (Nellemann *et al.* 2007). Orangutan populations will decline in the future, so a greater understanding of the relationships between orangutan density and habitat quality is essential for their conservation.

Orangutans mainly inhabit seasonally flooded forests including peat swamp, freshwater swamp and riverine forests, as well as non-flooded forests, dry land forests dominated by tree species of the Dipterocarpaceae family. Husson *et al.* (2009) presented a comprehensive review of orangutan density from 110 different locations, using published and unpublished survey data in the period from 1933 to 2005. They found that compared with dry land forests, orangutan densities in flooded habitats – including riverine and peat swamp forests were often higher. Nonetheless, although Husson *et al.* (2009) compared orangutan densities between flooded and non-flooded forests, they did not distinguish between the different forest types like primary, secondary and logged forests. In addition, they did not consider the influence of supra-annual masting events, which occur approximately every 2–10 years, the majority of trees fruit in synchrony (Ashton *et al.* 1998; Sakai 2002), on reported orangutan densities. Though they occur mostly in lowland forest types in Southeast Asia, they are characteristic of dry land forests, especially lowland dipterocarp forests, and are less prominent in flooded forests (Appanah 1993; Matsuda *et al.* in this volume). Several studies have reported that orangutans flexibly shift their distribution according to changes in fruit availability in non-flooded forest habitats, e.g. *P. abelii* in Ketambe, Sumatra (Buij *et al.* 2002) and *P. p. morio* in Ulu Segama Malua, Borneo (Alfred *et al.* 2010). Furthermore, in non-flooded forests of the Danum Valley, orangutan density after the mast fruiting of September 2005 increased to more than five times that of the average density during non-masting periods (6.8 ind/km^2, Kanamori *et al.* 2017). This underscores the necessity of considering the effects of masting events when comparing orangutan densities reported by different studies.

Thus, in order to identify the ecological factors affecting orangutan density, it is essential to use long-term monitoring to determine the influence of masting events and seasonal fruiting on orangutan movement patterns. Additionally, such studies should cover extensive areas. However, most previous studies of orangutan density have not considered the effects of the migration following masting events or normal seasonal fruiting. Orangutans use wide home ranges, e.g. 850 ha for adult females and greater than 2500 ha for sub-adults and adult males (Singleton & van Schaik 2001). In this study, to further understand differences in population density of orangutans between different forest types, we compared orangutan densities in flooded (riverine and swamp forests) and non-flooded (dry land forests) only using the standardized data from primary forests during non-masting periods, and discussed possible environmental factors influencing orangutan density.

Methods

It is very difficult to estimate orangutan density by direct observation because they are highly arboreal and semi-solitary. Consequently, density estimation by the "Nest counting method" based on the number of observed nests (sleeping platforms) in the area has been used at most study sites (Ghiglieri 1984). In this method, nests are recently counted by aerial survey by helicopter, drones or ground surveys (Ancrenaz *et al.* 2004; Buij *et al.* 2003; Koh & Wich 2012; van Schaik 1995; van Schaik *et al.* 2005). Although, Husson *et al.* (2009) used data collected both by aerial and ground survey, in the current study we only used orangutan densities estimated by

Table 26.1 Population density of orangutans in flooded and non-flooded forests in Sumatra (a) and Borneo (b).

Dominant habitat	Species	Island	Region	Altitude (m)	OU density (individual/km²)	Reference
(a)						
Flooded	*P. abelii*	Sumatora	Suaq Balimbing	10	4.0	Singleton 2000
			Trumon-Singkil	10	4.0	van Schaik *et al.* 2001
		Average			**4.0**	
Non-flooded	*P. abelii*	Sumatora	Ketambe 1	346	1.2	Wich *et al.* 2004
			Ketambe 2	705[a]	2.0	
		Average			**1.6**	
(b)						
	P. p. wurmbii	West Borneo	Gunung Palung[a]	50	4.1	Johnson *et al.* 2005
Flooded		Central Borneo	Tanjung Putting	10	2.7	Husson *et al.* 2009[b]
			Sebangau 1 (TI forest)	10	2.1	Morrogh-Bernard *et al.* 2003
		Average	Sebangau 2 (LP forest)	10	1.0	
					2.5	
Non-flooded	*P. p. wurmbii*	West Borneo	Gunung Palung[a]	50	3.2	Johnson *et al.* 2005
	P. p. morio	East Borneo	Gunug Gajah (Sungai Gie)	300	0.5	Marshall *et al.* 2006
			Ulu Segama	250	1.5	MacKinnon 1972
			Danum Valley	250	1.3	Kanamori *et al.* 2017
		Average			**1.6**	

[a] Forests of Gunung Palung are a mosaic of flooded and non-flooded forests.
[b] The data of Tanjung Puting came from Husson *et al.* (2009) who standardized the original data of Galdikas (1988).

nest count surveys along ground-based line transects. (Thomas *et al.* 2001) and following the formula: $D = n \,/\, (L \times 2w \times p \times r \times t)$, where D = orangutan density (individuals/km²), n = number of nests observed along the transect, L = length of the transect covered (km), w = estimated width of the strip of habitat actually censused (m), p = proportion of nest builders in the population, r = number of nests built per day per individual and t = time during which a nest remains visible (in days).

Of the 110 orangutan density locations reviewed by Husson *et al.* (2009), 43 had logging damage. We only used the data from primary forest without logging damage in our analysis, as many studies have reported orangutan density to be lower in logged and secondary forest than in primary forest (Felton *et al.* 2003; Husson *et al.* 2009; Johnson *et al.* 2005; Morrogh-Bernard *et al.* 2003; Rijksen 1978; van Schaik 1995, but see Ancrenaz *et al.* 2004; Marshall *et al.* 2006). We used only data collected during non-masting periods when comparing between-site orangutan densities.

Results

Figure 26.1 shows the location of study sites described in Table 26.1, with a comparison of orangutan densities between flooded and non-flooded forests in Sumatra (A) and Borneo (B). Overall, combined average orangutan density in Bornean and Sumatran flooded forests (3.0 ± SD0.5 individual/km², range: 1.0–4.1 individual/km², $n = 6$) was significantly higher than in non-flooded forests (1.6 ± SD0.4 individual/km², range: 0.5–3.2 individual/km², $n = 6$, Mann–Whitney U-test: $U = 7$, $p < 0.05$). This tendency was also observed on each island; the average density in flooded forests was higher than in non-flooded forests in Sumatra, i.e. flooded forest: Suaq Balimbing vs. Ketambe1 in non-flooded forest, and in Borneo, i.e. Tanjung Puting vs. Danum Valley, though we could not test this due to the small sample size in Sumatra.

The average orangutan density on Sumatra including flooded and non-flooded forests was higher than that in Borneo, i.e. on Sumatra: 2.8 ± SD0.7 individual/km² (range: 1.2–4.0; $n = 4$), on Borneo: 2.0 ± SD0.4 individual/km² (range: 0.5–4.1, $n = 8$), although the difference in the density was not significant (Mann–Whitney U-test: $U = 20$, $p = 0.49$). On the other hand, the average orangutan density in flooded forests was higher in Sumatra (4.0 individual/km², $n = 2$) than in Borneo (2.5 ± SD0.7 individual/km², $n = 4$), while density in non-flooded forests on both islands was the same (Sumatra: 1.6 individual/km², $n = 2$, Borneo: 1.6 ± SD0.6 individual/km², $n = 4$), though we could not test this due to the small sample size.

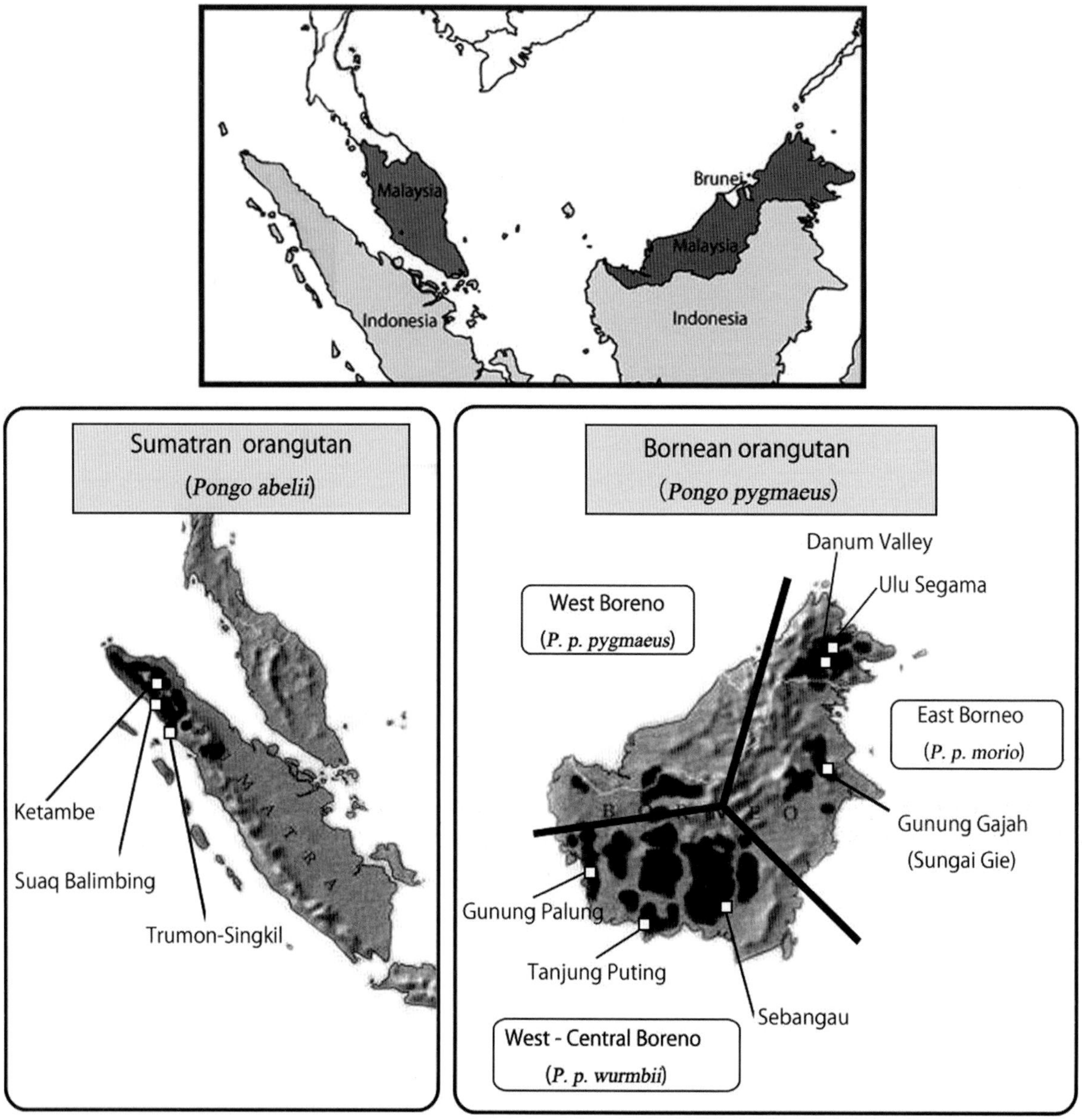

Figure 26.1 Study sites of orangutans in primary forest. The location of each study site is marked with a white box.

The average density of each species and subspecies living in both flooded and non-flooded forests was as follows: *P. abelii* in Sumatra: 2.8 ± SD0.7 individual/km^2 ($n = 4$) and range: 1.2–4.0 individual/km^2; *P. p. wurmbii* in Borneo: 2.6 ± SD0.5 individual/km^2 ($n = 5$), range: 1.0–4.1 individual/km^2; *P. p. morio* in Borneo: 1.1 individual/km^2 ($n = 3$), range: 0.5–1.5 individual/km^2.

Discussion

Difference Between Flooded and Non-flooded Forests

Here we have shown that, in general, orangutan density is significantly higher in flooded forests than in non-flooded forests, and this occurs on both Borneo and Sumatra. The higher density of orangutans in flooded forests may be explained by a higher fruit availability. Indeed, Marshall *et al.* (2009) analysed the relationship between orangutan density and fruiting tree percentage (three sites in Sumatra and six sites in Borneo) using data only from primary forests and found that orangutan density was positively correlated with fruiting tree percentage[1] during the low-fruit periods. This indicates that sites which experience less extreme periods of food shortage can support higher densities. Furthermore, based on the calculation using the supplemental data set provided by Wich *et al.* (2011), in flooded forests – including peat and riverine forests – over half of each of the periods (63.3%) studied were 'high' fruiting months calculated by the same method used by Marshall *et al.* (2009) based on a *z*-score; while, in non-flooded forests (or dry land forests) over the half of each studied periods were 'low' (59.4%). This suggests higher overall fruit availability in flooded forests than that in non-flooded forests. In addition, Matsuda *et al.* (Chapter 4, this volume) compared fruit availability in non-flooded (dry land forest) and flooded forests (riverine and mangrove forests) in Borneo, and showed that fruit availability in flooded forests is higher and more stable than in dry land forests. This lends support to the idea that higher orangutan density in flooded forests is related to higher fruit availability in that habitat.

Difference Between Sumatra and Borneo

Our results showed that, overall, orangutan density did not differ significantly between Sumatra and Borneo. This contrasts with a previous study that reported orangutan density on Sumatra was significantly higher than on Borneo (Husson *et al.* 2009). On the other hand, the density in flooded forests was higher on Sumatra than on Borneo, but in non-flooded forests there was virtually no difference between the islands in this study (noting that a statistical test was not performed due to the small sample size). Similarly, Husson *et al.* (2009) reported highly significant differences between the islands for orangutan densities in flooded forests, but not for non-flooded forests.

In agreement with the pronounced differences between the islands in density of orangutans inhabiting flooded forests, Wich *et al.* (2011) report that the estimated mean percentage of fruiting trees per month on Sumatra was significantly higher than that on Borneo together with the fact that the difference in the fruiting tree percentages between Sumatra and Borneo in flooded forests (23.8%) were larger than that in non-flooded forests (10.9%). These results suggest that greater density of orangutans in Sumatran flooded forests is linked to enhanced fruit availability in this habitat, while the minor differences in the density of orangutans in non-flooded forests on the two islands reflects the smaller difference in fruit availability between these habitats.

The fact that orangutan density in non-flooded forests was not different between Sumatra and Borneo in this study, provides insight into variation in feeding strategies for orangutans inhabiting different islands. Given the apparent links between density in flooded forests and their fruit availability, it remains an open question why a similar density of orangutans has been observed in Bornean and Sumatran non-flooded forests, despite a substantial difference in their fruit availability (Wich *et al.* 2011). A possible explanation is that Bornean orangutans, especially *P. p. morio* populations in east Borneo, have adapted to survive long periods of extremely low fruit availability by feeding on leaves and cambium and so are not dependent on fruits as Sumatran orangutans are. For Bornean orangutans the mean percentage of fruits in their diet has been reported to be somewhat lower (*P. p. wurmbii*: 65.5%; Galdikas 1988; Knott 1998, *P. p. morio*: 56.2%; Mackinnon 1974; Rodman 1988) than in Sumatran orangutans (*P. abelii*: 66.9%; Fox *et al.* 2004; Wich *et al.* 2006). Additionally, the influence of masting must be considered: fruit feeding by Bornean orangutans greatly decreased during periods with extremely low fruit availability that follow masting events. For example, over a 27-month study in the Danum Valley (non-flooded forest), the time percentage dedicated to fruit feeding by *P. p. morio* decreased from 100% during mast fruiting to 11.7% in the periods of extremely low fruit availability that followed the mast fruiting events (Kanamori *et al.* 2010).

Anatomical characteristics support the behavioural traits of *P. p. morio*: the sub-species has the most robust jaw of all orangutan taxa (Taylor 2006), indicating that it is adapted to survive long periods of extremely low fruit availability by consuming harder foods such as bark instead of the softer ripe fruits common in the orangutan diet (Taylor 2009; Taylor & van Schaik 2007).

For future, more finessed, analyses of the factors influencing orangutan densities in flooded and non-flooded forests, it may well be important to consider the effects of leaf quality. This is because amount feeding on other plant parts including leaves exceeded that on fruits when fruit availability is low, though orangutans are generally frugivorous as discussed above. As shown in Matsuda *et al.* (2013), the protein–fibre ratio of leaves was significantly higher in flooded forest than non-flooded forest. This is important since this metric is regarded as a strong indicator of leaf quality for primates (Chapman & Chapman 2002; Wasserman *et al.* 2003), especially colobines. Thus, it is possible that differences in orangutan density between flooded and non-flooded forests are affected not only by fruit availability but also by leaf quality.

Husson *et al.* (2009) reported that in west and central Borneo *P. p. morio* density was significantly lower than *P. p. wurmbii*. This corroborates our results, though our data (Table 26.1(b)) suggest that the difference was caused by the fact that the studied *P. p. wurmbii* population lived in forest mosaics comprised of both non-flooded and flooded forests (with probably more foods available in the latter), whereas all studied *P. p. morio* were exclusively restricted to non-flooded forests (where fewer foods are available). Therefore, the density differences between these orangutan sub-species is most likely due to differences in the food availability within the habitats they inhabit.

Conclusion

We have found strong evidence to indicate that higher orangutan densities in flooded rather than non-flooded forests is probably due to the longer periods of high fruit availability in the former. Differences in orangutan densities between Sumatra and Borneo was observed only in flooded forests probably because of the large difference in fruit availability of flooded forest, a pattern not repeated in non-flooded areas. Thus, this study emphasizes the importance of flooded forests for orangutans and clearly illustrates how the survival of the two is intertwined. Future studies using long-term monitoring to relate orangutan movement patterns to changes in fruit availability and leaf quality over wide areas is a prerequisite for further understanding of the ecological factors affecting orangutan density. In addition, future studies should focus on more detailed comparison of fruit availability including nutrient content of fruits, and their size. Lastly, non-dietary considerations such as sleeping site density and presence of predators or competitors for food should also be considered.

Acknowledgements

We would like to thank the Sabah Biodiversity Council, Danum Valley Management Committee, Sabah University and Sabah Wildlife Department for permission to conduct studies in Sabah. We are grateful to Dr Waidi Sinun of Yayasan Sabah, Mr George Hong of Borneo Nature Tours and Dr Glen Reynolds of Danum Valley Field Centre. We give special thanks to our research assistants for their help in the field. We are also grateful to Associate Professor Ikki Matusda of Chubu University and anonymous reviewers for their kind and valuable comments on our manuscript. This study was financially supported by JSPS Core-to-Core Program, A. Advanced Research Networks, Fujiwara Natural History Foundation, Mitsui & Co., Ltd.

Note

1 Monthly fruit availability was classified from the monthly percentage scores of fruiting trees (or *z*-scores: Sokal & Rohlf 1994), which is the monthly value minus the mean divided by the standard deviation. In this classification, months with a *z*-score < -1 were classified as low fruit periods (low), a *z*-score between -1 and 1 and medium fruit periods (mid), and *z*-scores > 1 as high fruit periods (high).

Chapter 27

Primates in Amazonian Flooded Habitats

Adrian A. Barnett

Introduction

Amazonia is generally thought of as synonymous with lowland tropical rainforest (the 'rainforest' of common parlance), but seasonally inundated habitats cover some 20% of the region's 7 million km^2. They include those flooded by whitewater and blackwater rivers, as well as palm- and reed-dominated swamps. There are species, such as the white bald uacari (*Cacajao c. calvus*), Vanzolini's squirrel monkey (*Saimiri vanzolinii*) and Cazuza's saki (*Pithecia cazuzai*), that are unique to flooded forests. Others, like the golden-backed uacari (*Cacajao ouakary*), use them seasonally, while some, like the pygmy marmoset (*Cebuella pygmaea*), which heretofore preferred to live permanently on the very margins of flooded forests, now appear to be moving out of them and colonizing nearby human-made habitats.

Unlike other large areas of tropical forest, such as those in Borneo, or central or western Africa, the Amazon basin is based around a single hydrological entity – the Amazon river and its tributaries, such as the meandering Purus and Juruá rivers (Goulding *et al.* 2003). In central and eastern Amazonia, especially, there are few highland areas, so that topography varies little over huge swathes of land. In consequence, floodplains are extensive; more than 300 000 km^2 in the Amazon basin is subjected to a seasonal flood pulse. Within this, there are a variety of flooded vegetation types, seasonally inundated for periods varying from a few weeks to several months (see Chapter 2), with the average period of inundation being some 210 days a year (Junk & Furch 1993).

Overall, some 80% of the Amazonian floodplains consist of dense canopy forests, with the other 20% being represented by grasslands, lakes, flooded fields and swamps (Parolin *et al.* 2004). Within this complex are areas dominated by palms (generally *Mauritia flexuosa*, but also *Mauritiella armata* and *Euterpe oleracea* among others), tall herbaceous vegetation (in which the giant aroid *Montrichardia arborescens* is often prominent), areas of floating grass (with *Paspalum repens* and *Echonochloa polystachya* as principle components), and shrubby white sand savannas. There are also sandbanks that may be covered with short-lived herbs during their brief seasonal exposure, though specialized, more persistent shrubs (e.g. *Tessaria integrifolia*, Asteraceae; *Eugenia* and *Myrcia* spp., Myrtaceae) and grasses (notably *Gynerium sagittatum*) have a permanent presence (Kalliola *et al.* 1991). Additionally, extensive archipelagos of islands exist whose vegetation is largely or totally inundated each year (e.g. Anavilhanas on the Rio Negro, Brazil; Latrubesse & Stevaux 2015), and very large riverine islands (e.g. Bananal on the Rio Araguaia, Brazil; de Morais 2006). While not all of these habitats are commonly used by primates (Table 27.1), there are two inundated forest types that play an important role in the ecology of many Amazon primate species: *igapó*, which occupies the narrow ribbon-line floodplains of nutrient- and sediment-poor blackwater rivers, and *várzea*, which occurs on the extensive floodplains of nutrient- and sediment-rich whitewater rivers. Their main differences are listed in Table 27.2 (and see Figure 27.1).

The majority of tropical inland swamps are dominated by herbaceous vegetation (see Chapters 24, 25, 34 and 37). With some 80% of its flooded habitats covered by tree-dominated ecosystems (Junk *et al.* 2014), Amazonia is clearly an exception to this general rule. However, although flooded forests form a prominent part of the Amazonian landscape, their importance to the region's primates has never received a general review. This is an attempt to summarize what is known about the ecology of primate species in flooded habitats in the Amazon basin.

Methods

The available literature was surveyed using Google Scholar, Primate Lit and Web of Science, and using key words: *Amazonia, primate, flooded, aguajal, aningal, beach, buritizal, campina, igapo, Mauritia, Montrichardia, praia, playa, sandbank, várzea, white sand* and the Latin names of all known Amazonian primate genera. Medical and epidemiological studies (e.g. Davies *et al.* 1991; Lourenço-de-Oliveira & Deane 1995), were not included in the analysis. Material was divided into a review of the habitats occupied, followed by the species recorded in these habitats and what has been investigated. The results of the studies are then discussed and some recommendations are made for future research. Taxonomy follows Rylands and Mittermeier (2009), except for the following: *Cacajao* (*melanocephalus* group: Ferrari *et al.* 2014), *Pithecia* (Marsh 2014), *Saimiri* (Lynch Alfaro *et al.* 2015) and capuchin monkeys (Lynch Alfaro *et al.* 2012). The literature used is spread across several decades, hence, when germane, the former taxonomic name by which a species was referred to in an older publication is given in parentheses.

Table 27.1 Amazonian flooded vegetation types and their use by primates.

Vegetaion type	Local names	Characteristic plants	Botanical studies	Used by primates?	Primate studies
Beaches and beach vegetation	*Capoeira*	Short-lived herbs, and specialized shrubs with drought/inundation resistance (e.g. *Tessaria integrifolia*, Asteraceae; and some *Eugenia* and *Myrcia* spp., Myrtaceae), plus *Gynerium sagittatum* (Poaceae)	Some, including Kalliola *et al.* 1991	No regular use reported	No dedicated studies known
Blackwater seasonally flooded forest	*Igapó*	Between 60 and 150 trees > 10 cm DBH per ha. Fabaceae and Euphorbiaceae dominate. Lecythidaceae and Sapotaceae common. *Aldina, Macrolobium, Escheweilera* and *Pouteria* characteristic. Vegetation in diffuse horizontal bands between bank and riverfront, reflecting inundation tolerance of seeds, seedlings and adults	Many, including Ferreira *et al.* 2010c; Parolin *et al.* 2003	Seasonal use by a wide variety of primates. Often the majority of species in an area. No permanently resident endemics reported	Several, including: Barnett 2010; Defler 1999, 2001; Haugaasen 2004
Floating meadows	*Mureru*	Dense mats of floating or rooted vegetation, Dominated by grasses and hydrophytes. Extensively fed on by manatees (Colares & Colares 2002).	Several, including Junk 1970	None reported	No dedicated studies known
Montrichardia swamp	*Aningal*	Part of a complex, but apparently little-researched community that may have up to 24 species of large flowering plant	Some, including Gordon *et al.* 2000, and Pinheiro & Gonçalves Jardim 2015	None reported	No dedicated studies known
Palm swamp	*Aguajal, Buritizal*	Palms, especially *M. flexulosa*, may constitute > 50% of individuals, but up to 25 other lowering plants may occur	Sander 2014, and references therein	Extensively	Aquino 2005; Pontes 1997; Kasecker 2006
White sand scrub forest	*Campina scrub, Campina forest* and *Campinarana*	A graded series of habitats from grassland, sand dotted with shribs to denser woodland. Distinctive and sclerophytic, phenol-rich leaves. Malphigiaceae, Myrtaceae, Rubiaceae, Vochysiaceae characteristic of a specialist species-poor flora	Anderson 1981	Rarely, and then only generalist species on occasion	Cordeiro 2008
Whitewater seasonally flooded forest	*Várzea*	Complex structuring of vegetation by inundation tolerance of adults and seedlings. 60–120 tree species > 10 cm dbh/há. Fabaceae and Euphorbiaceae dominate. Lecythidaceae and Sapotaceae common. *Hevea, Escheweilera* and *Pouteria* characteristic	Many, including Ferreira 1997; Ferreira *et al.* 2010c; Parolin *et al.* 2004	Seasonal use by a wide variety of primates. Often the majority of species in an area. Some permanently resident endemics reported	Many, including: Ayres 1986; Haugaasen 2004; Bowler 2008

Results

Habitats Occupied

Of the six broad habitat types distinguished, only seasonally flooded forests on whitewater and blackwater rivers (*várzea* and *igapó*, respectively) are known to have permanent primate residents, and only in these habitats have primates been studied in any detail (Table 27.3). Other flooded habitats appear to be a transient, opportunistic or seasonal resource, or not used by primates at all. *Mauritia*-dominated palm swamps have fruiting seasons lasting 3–4 months, when very large volumes (Pacheco Santos 2005) of highly nutritious (Darnet *et al.* 2011) fruits are produced in some areas (e.g. the savannas of Roraima state, northern Brazil). Fruit production in *Mauritia*-dominated palm swamps (*buritizal* – Portuguese; *aguajales* – Spanish) often occurs when other regional habitats are producing little or no fruit. This can make them very important for wildlife, including primates (Pontes 1997, 1999). Such palm swamps are also very attractive to primates when they occur interspersed with *várzea*; this is again because of the volume of fruits they produce. In some areas, buriti fruit pulp dominate, for example, the diet of the Peruvian red uacari (*Cacajao calvus ucayalii*) even though the fruit part is soft and this primate is morphologically specialized for eating hard-husked fruits (Bowler & Bodmer 2011). Other monkeys, including *Aotus, Ateles* and *Sapajus* also exploit this seasonal resource (de Carvalho 1961; Peres 1993a; Pontes 1999). In some areas, other palms with preference for swampy areas, such as *Oenocarpus bataua*

Table 27.2 Principle differences between *igapó* and *várzea* flooded forests.

Character	Difference between *igapó* and *várzea*
Inundation duration	Up to 210 days at deepest parts of floodplain (Ferreira 1997).
Depth of inundation	Up to 14 m in both systems (Goulding *et al.* 2003).
Extent of floodplain	Very rarely more than 250 m wide in *igapó*, up to 90 km wide in *várzea* (Goulding *et al.* 2003).
Topographic form	In *igapó*, there is a shallow gently inclined floodplain, small hillocks (*borokotoh*) locally increase species diversity (Barnett *et al.* 2015b); In *várzea* there is complex system of levees (*restingas*) backswamps and sloughs (*chavascais* or *tahuampa*) and oxbow lakes (Nebel *et al.* 2001). The *retsinga* vegetation is further subdivided into high *várzea* (high *restinga*) which is inundated by > 3 m of water for less than 50 days a year, and low *várzea* which is covered by more than 3 m of water for longer than this (Wittman *et al.* 2006).
Sediment load of water	Sediment loads of the Rio Negro (8 × 10^6 tonnes) described by Latrubesse and Franzinelli (2005). as 'insignificant in relation to the huge water discharge'; while the Madeira (a whitewater river with approx. the same discharge rate as the Negro) discharges 400 × 10^6 tonnes of sediment a year into the Amazon (Meade 1996).
pH and nutrient content of water	Around 6.8–7.2 in whitewater (neutral), 4.5–5.7 (very acid) on blackwater rivers. Blackwater rivers are richer in Al, Fe and Mn (often remarkably so), otherwise they are very poor in elements and nutrients – some, like Ca, Mg and K, being present only as traces (Ribeiro & Darwich 1993).
Humic and fulvic acid content of water	Present in both river systems at high levels, and derived from tannins and lignins leached from leaves, bark etc.; 30–40% higher in blackwater rivers and further offset by absorption of tannins etc. by sedimentary particles in whitewater systems (Ertel *et al.* 1986).
Soil chemistry	Sandy, acidic, low nutrient content in *igapó*; muddy, basic soils, high nutrient content in *várzea* (Haugaasen & Peres 2006).
Phenology	Peak flowering often coincides with the falling river phase and dry season, while the peak of fruiting generally coincides with rising river level and the rainy season. Over 70% of flooded forest tree species have flowering periods significantly related with the variation of water level or precipitation, and 85% have fruiting periods significantly related with variation of water level or precipitation (Ferreira & Parolin 2007). There are no variations in these patterns between the two habitats.
Tree species diversity	Both habitats have lower species diversity and higher equitability than terra firme. Species diversity comparisons between *igapó*/*várzea* complicated by existence of large number of forest sub-types (some 17 in all: Ferreira. 1997). Both may have from 40–150 species/ha, but with few rare species. Species richness and diversity often not significantly different; density and basal area often significantly lower in the *igapó* forest than *várzea* forest. Low (10%) number of tree species in common (Ferreira *et al.* 2010c).

(a) (b)

Figure 27.1 The extensive and topographically complex floodplain of Amazonian *várzea* (a), contrasted with the narrow and simple inclination of the *igapó* floodpain (b). Photos: Adrian Barnett

(=*Jessenia bataua*: *pataua, sehe* or *hungurahua*), may serve a similar function (Peres 1994a; Rylands & Keuroghlian 1988; Spironello 2001), as may stands of *Euterpe* species (*açai*) (Zona & Henderson 1979).

Swamps dominated by *Montrichardia arborescens* (Arecaceae) are highly productive (Gordon *et al.* 2000; Villarubia & Cova 1993), and support a rich and varied wildlife (Cintra 2015; Señaris & Ayarzagüena 2004; Zapata-Ríos *et al.* 2005). However, primates appear not to be among them, possibly because the stems are insufficiently robust to support primate movement and Amazonian primates are unlikely to wade extensively as do the African apes (Chapters 25 and 31). This may also explain why primates are not seen in the floating grasslands that border the Amazonian *várzeas* despite the extraordinary productivity of these habitats (50–100 tonnes of biomass/ha, compared to some 30 t/ha for adjacent inundated forest: Morison *et al.* 2000) – although the abundance of caiman and other large ambush predators (Arantes *et al.* 2010; da Silveira *et al.* 2013; Marioni *et al.* 2008) may be an additional factor (terrestrially active primates in the Amazon are known to avoid foraging in high risk areas; Barnett *et al.* 2012a).

Riverine sand banks have a highly specialized fauna and flora, most of which has a weedy ecology (Adis *et al.* 1998; Parolin 2003), adapted to exploit transient periods of non-inundation, and so are of little food value to foraging primates (although such banks may be used to access river water

Table 27.3 Primate species recorded in Amazonian freshwater flooded habitats.

Site	Habitat type	No. spp in flooded forest/ no. in regional assemblage	Primate species	References
Rio Tapiche, Peru	*várzea*	11[a]	*Alouatta seniculus, Aotus nancymae, Cacajao calvus, Callicebus cupreus, Cebuella pygmaea, Cebus albifrons, Lagothrix lagoticha, Pithecia monarchus, Saguinus fuscicollis, Saimiri boliviensis, Sapajus apella.*	Bennett *et al.* 2001
Rio Yavari, Peru				Bowler 2008
Jau National Park, Brazil	*igapó*	5/8	*Alouatta maconnelli, Aotus vociferans, Cacajao ouakary, Cebus albifrons, Saimiri sciureus*	Barnett 2010; Barnett *et al.* 2002, 2005
Mamirauá, Brazil	*várzea*	4/8	*Allouatta jurua, Cacajao calvus, Saimiri vanzolini, Sapajus macrocephalus*	Ayres 1986
Rio Purus, Brazil	*igapó*	7/13	*Allouata seniculus, Callicebus torquatus, Cebus albifrons, Lagothrix lagotricha, Pithecia albicans, Saimiri* cf. *ustus, Sapajus apella*	Haugaasen 2004
Rio Purus, Brazil	*várzea*	9/13	*Allouata seniculus, Ateles chamek, Cebus albifrons, Lagothrix lagotricha Pithecia albicans, Saguinus fuscicollis, Saguinus mystax, Saimiri* cf. *ustus, Sapajus apella*	Haugaasen 2004
Rio Purus, Brazil	*igapó*	6/10	*Alouatta puruensis, Cebus albifrons, Pithecia albicans, Saguinus* sp., *Saimiri* sp., *Sapajus apella*	Kasecker 2006
Rio Purus, Brazil	*várzea*	3/10	*Alouatta puruensis, Saimiri* sp., *Sapajus apella*	Kasecker 2006
Western Amazonia, Brazil (8 sites)	*várzea*	10/17	*Ateles paniscus, Alouatta seniculus, Callicebus cupreus, Cebus albifrons, Cacajao calvus, Pithecia* spp., *Saimiri* spp., *Sapajus apella*	Peres 1997b

[a] *várzea* only.

for drinking: e.g.. *Ateles*, Torbjørn Haugaasen, pers. comm., 2015). Larger, higher, banks are often the traditional sites at which river turtles (mostly *Podocnemis* spp.) lay eggs (Alho & Pádua 1982; Valenzuela 2001). Predation on this abundant resource has been recorded by opportunistic mammals, including capuchins (Alves-Júnior *et al.* 2012; Barnett *et al.* 2002; P. Cook & J. Hawes, unpublished data; Salera Junior *et al.* 2009), but the topic remains essentially unresearched.

The Nature and Extent of Current Studies

As is inevitable, some species and sites are much better known than others. Many primate species have been recorded in flooded forests, but their ecology has not been studied in detail (e.g. Tables 27.1 and 27.3). Others have received more focused study. The pygmy marmoset (*Cebuella pygmaea*) is most common near rivers and in riverine forests (Soini 1982, 1993). Its tiny size (at around 100 g it is the smallest non-strepsirrhine primate), unusual diet (some 80% entirely tree gum) and complex social system have attracted considerable research effort, including diet (Ramirez *et al.* 1977; Yépez *et al.* 2005), habitat preferences (Hernández-Camacho & Defler 1985), chemical (Ziegler *et al.* 1990) and vocal communication (Converse *et al.* 1995; de la Torre & Snowdon 2002; Snowdon & Elowson 1999; Snowdon & de la Torre 2002), positional behaviour (Jackson 2011), and social organization (Heymann & Soini 1999). In addition, de la Torre *et al.* (2000) have studied the impact of ecotourism and human activity on the species, while Bicca-Marques and Calegaro-Marques (1995) and Van Roosmalen and Van Roosmalen (1997) have reported range extensions.

Members of the genus *Cacajao*, the uacaris, have also received considerable attention. Despite the difficulties of working with these animals (Pinto *et al.* 2013), or possibly because of them, uacaris are now some of the best known of Amazonian flooded habitat primates. Of the various subspecies of bald uacari (*Cacajao calvus*), studies have been conducted of their associative behaviour (Leonard & Bennett 1996), diet (*C. c. calvus* – Ayres 1986a, 1989; *C. c. ucayalii*, Aquino & Encarnación 1999; Bowler & Bodmer 2011; Ferrol-Schulte 2008; Swanson Ward & Chism 2003), habitat use (*C. c. calvus* – Ayres 1986a; *C. c. ucayalii* – Aquino 1988; Bowler 2007; Ferrol-Schulte 2008; Heymann & Aquino 2010), parasitology (Conga *et al.* 2014, 2015), positional behaviour (Walker & Ayres 1996), reproductive biology (Bowler *et al.* 2013) and social organization (*C. c. ucayalii*, Bowler & Bodmer 2009; Bowler *et al.* 2012; Gregory & Bowler 2015). Range extensions have been reported by de Alcântara Cardoso *et al.* (2014); Vieira *et al.* (2008) (*C. c. calvus*); Peres 1990, 1997 (*C. c. novaesii*); Silva Júnior and Queiroz (2008); Vieira *et al.* (2008) (*C. c. rubicundus*); Bowler *et al.* (2009); Swanson Ward and Chism (2003); Vermeer *et al.* (2013) (*C. c. ucayalii*); and of a taxon that may, or may not, be *C. c. calvus* (Silva Júnior & Martins 1999).

Of the two species of black-faced uacaris (*sensu* Ferrari *et al.* 2014), the diet and habitat use of *C. melanocephalus* has been studied in detail by Boubli (1999), and Boubli and Tokuda (2008).

Boubli (1993) provides range extension details, while Boubli and de Lima (2009) model the species' potential geographical distribution based on its fundamental niche. Additional ecological data from a short survey is provided by Lehman and Robertson (1994). For *C. ouakary* studies have been conducted on diet and seasonal habitat use (Barnett 2010; Defler 1999, 2001), density (Defler 1999, 2001), insectivory (Barnett *et al.* 2013c), sleeping trees (Barnett *et al.* 2012d), terrestrial foraging in the low-water season (Barnett *et al.* 2012a), and the extent of seed-dispersion made by this erstwhile seed predator (Barnett *et al.* 2012b). *Cacajao ouakary* has been found to preferentially select fruits infested by insects (Barnett *et al.*, unpublished data), and to bite fruits at their weakest points (Barnett *et al.* 2016). For at least one tree (*Macrolobium acaccifolium*, Fabeaceae), they avoid the individuals that are infested with ants (Barnett *et al.* 2015a). The movements of *C. ouakary* groups through the igapó forest canopy canopy increase foraging rates in such sit-and-wait predators as jacamars and nunbirds (*Galbula* and *Monasa*), and reduce predation risks for bark-fossicking ant-birds (Barnett & Shaw 2014). Predation on *C. oukary* by harpy eagles has been reported (Barnett *et al.* 2011), and they are known to generalize their alarm reactions to aerial and ground predators from general models of predator appearance (Mourté & Barnett 2014). As *Cacajao melanocephalus*, Bezerra, Barnett *et al.* (2011) published an ethogram of *C. ouakary*, while Bezerra *et al.* (2010a) and Bezerra *et al.* (2011) studied the vocal repertoire of the species, and Bezerra *et al.* (2012) reported on the bioacoustics properties of the calls (all as *C. melanocephalus*). The species also provided a model experiment for the use of call playback as a survey method (Bezerra *et al.* 2010b).

At Mamirauã on the Rio Solimões, central Amazonia, Queiroz (1995) compared the ecology of sloths and howler monkeys (*Alouatta macconnelli* – as *A. seniculus*) in *várzea* forest. Peres (1997a) included flooded forests in his Neotropics-wide analysis of the biotic and abiotic factors influencing *Alouatta* densities. The other flooded forest inhabiting species to receive detailed attention is *Saimiri vanzolinii*, a species restricted to a small section of the Rio Solimões in central Amazonia, which has been studied for its general ecology (Paim 2008), vocalizations (Paim & Queiroz 2009), and distribution and interactions with the two other species of *Saimiri* in the region (Paim *et al.* 2013).

All other published primate studies in Amazonian flooded forests have, while in some cases including data on individual species, had a community level approach. These have included short (e.g. Ayres & Milton 1981; Branch 1983; Lehman & Robertson 1994), and more extended surveys (e.g. Bennett *et al.* 2001; Peres 1993b), and long-term and detailed ecological studies (Haugaasen *et al.* 2007; Haugaasen & Peres 2005a, b, c; Peres 1993a, 1997a, b).

Discussion

The Influence of Flooded Habitats on the Ecology of Primates

Where the phenology of *terra firme* and flooded forest has been studied quantitatively, it has been shown that the two have asynchronous peaks of leaf flush and fruit production (e.g. Barnett 2010; Barnett *et al.* 2012a, 2013d; Haugaasen & Peres 2005b), and there is a seasonal shift of mammals between the two forest types. This not only occurs with primates (Haugaasen & Peres 2005a), but also with terrestrial mammals, birds (e.g. Haugaasen & Peres 2007), and bats (Bobrowiec *et al.* 2014; Marques *et al.* 2012). This movement generally follows the peaks of fruit production in flooded forests, which themselves follow the timing of seasonal changes in river levels, as many of the trees, palms and lianas have fruits that are either water or fish dispersed (Anderson *et al.* 2009; Correa *et al.* 2007).

However, exactly how primate species move between habitats and for how long appears to vary greatly between locations and vegetation types. For example, at Jaú in central Amazonia, generalist foragers such as *Saimiri cassiquiarensis* (= *S. sciureus* in Barnett *et al.* 2002, 2005) and *Cebus albifrons* remain in the *igapó* throughout the year. During the low-water season, *Cacajao ouakary* forages for germinating seeds on the *igapó* floor as well as foraging in neighbouring *terra firme*, although it has not been recorded moving more than 500 m into this habitat, except to visit buriti palm swamps (Barnett 2010; Barnett *et al.* 2005; Simone Iwanaga, pers. comm., 2008). *Pithecia chrysocephala* and *Alouatta macconnellii* both use *igapó*, but visits appear to be occasional, short, incursions in search of specific resources. This contrasts with, for example, the behaviour of *C. ouakary* at Caparú on the Rio Apaporis, Colombia. Here, when the *igapó* was unflooded, uacari groups would leave the *igapó* and travel for many kilometres into the *terra firme* (Defler 1999). Individuals that remain in flooded forest alter their diets, eating more insects and leaves and fewer fruits, as resource availability changes (e.g. Barnett 2010 for *C. ouakary* where group sizes also alter with the resource availability profile). It would appear they can be classified as *igapó*-dwelling species that seasonally visit *terra firme*. Meanwhile, on the Rio Purus, Haugaasen (2004) found that *Ateles chamek*, *Cebus albifrons*, *Lagothrix lagotricha* and *Pithecia albicans* visit floodplain forests sporadically to exploit pulses of fruit availability. Such primates appear to be primarily *terra firme* species that make occasional use of flooded habitats. On the Rio Tapajós, Brazil, *Ateles chamek*, *A. marginatus* and *Chiropotes albinasus* also display this behaviour (Adrian Barnett, unpublished data), as does *Ateles belzebuth* at Tinugua National Park, Amazonian Colombia. Here the animals spent most of their time in *terra firme*, but would enter into flooded forest, especially when figs (*Ficus*, Moraceae) were in season, eating them in such quantity that they were the fifth. ranked diet species (Ahumada *et al.* 1998). Shaffer *et al.* (Chapter 28) also report extensive use of flooded forests in Guyana by *Chiropotes chiropotes* and *Ateles paniscus* (as well as *Alouatta*, *Cebus* and *Sapajus*), and their apparent avoidance by *Saguinas midas*.

Some species that appear to remain in flooded forest year round tend to have small home ranges and reliable food sources (*Cebuella pygmaea*, gum, Soini 1993; *Aotus* sp., insects, Adrian Barnett, unpublished data), while others (e.g. *Cebus albifrons* at Jaú, *Sapajus apella* on the Purus) are generalist foragers (Fragaszy & Visalberghi 2004). It is currently uncertain

Figure 27.2 A white-fronted capuchin forages from a patch of seeds on the floor of unflooded *igapó*, Jaú National Park, Amazonian Brazil. Photo: Carol Antunes.

whether such differences reflect different resource availability patterns or regional traditions (Barnett *et al.* 2013a), but the situation with *A. belzebuth* (mentioned above) illustrates that it is important to consider regional variations in vegetation composition: the flooded forests at Tinugua had an overstory where *Ficus* dominated, while in some others (at Jaú, for example) *Ficus* is notably absent (Barnett 2010).

One complicating element is the use of terrestrial resources by primates when *igapó* and *várzea* are unflooded. Volumes of fruits and germinating seeds on the floor of such forests can be substantial (Barnett *et al.* 2012a), and these attract terrestrial which move into these areas from the adjacent *terra firme* (Antunes *et al.*, unpub. data; Bodmer 1990; Haugaasen & Peres 2007), including ungulates as well as rodents and the carnivores that prey on them. There are no purely terrestrial neotropical primates, but species do descend to the ground for a variety of reasons (*Ateles*: Campbell *et al.* 2005; *Cacajao, Chiropotes* and *Pithecia*: Barnett *et al.* 2012b; *Cebus albifrons*: Defler 1979a, b). In flooded forests, members of the genus *Cacajao* are known to descend to the ground specifically to forage on fallen fruits (Ayres 1986a; Barnett *et al.* 2012a), as do *Cebus albifrons* (Carol Antunes, pers. comm. 2015: Figure 27.2), that occur in large 'seed mats' that start when floating seeds abut an instruction and persist as water levels fall (Barnett *et al.* 2012a; Ferreira *et al.* 2010a; Piedade *et al.* 2010). While *Sapajus apella* descend specifically to eat fallen *Astrocaryum jauari* palm fruits (Torjbørn Haugaasen, pers. comm., 2015).

Flooded forests are far from homogeneous in composition, even within the same overall forest type (see Worbes *et al.* 1992 for *várzea*, and Montero *et al.* 2014 for *igapó*). Consequently, across geographically distinct populations of the same primate species, seasonal movements between seasonally flooded and never-flooded habitats might be delayed, never occur, or not occur to the same extent in depending on resource availability. There are also cultural factors that determine whether patterns of resource exploitation become part of a local behavioural repetoire (Barnett *et al.* 2013b; Perry *et al.* 2003). Clearly, it will also depend on how viable such movement might be, with variables including the distance involved in moving temporarily to another site as well as the extent and nature of the resources available once arrived. Research on such aspects is sorely needed.

Of all Amazonian flooded forest-inhabiting primates, *Cebuella pygmaea* appears to be the most extreme habitat specialist, preferring forest with no more than two or three metres of standing water for more than three months a year (Soini 1988). Unexpectedly, it appears to have expanded its niche to include forests regenerating from small-scale agriculture (Hernández-Camacho & Defler 1985)).

Because of the higher sediment load of the rivers that flood them (Table 27.2), the fertility of *várzea* soils is much greater than those of *igapó* (Furch 1997; Irion 1978), leading to more productive forests (Cintra *et al.* 2013; Junk *et al.* 2011).This has been observed to have effects on tree species growth rates (da Fonseca *et al.* 2009), plant community structure and composition (Haugaasen & Peres 2006), seed mass (Parolin 2000) and leaf nitrogen levels (Kreibich *et al.* 2002). On the Purus River, close proximity of *igapó* and *várzea* allowed Haugaasen

& Peres (2005a) to directly compare the density and biomass of primate species between the two flooded habitat types, and with never-flooded *terra firme*. They found the biomass of the *várzea* primate assemblage to be more than twice that of *terra firme*, which is similar to the results from the Juruá reported by Peres (1997b). In an analysis of 23 sites in Amazonia, Peres (1997a) found that proximity to a whitewater river was the best single predictor of high *Alouatta* densities. Such results are in line with general observations on the effects of soil (and hence plant) productivity on primate species richness and density (Kay *et al.* 1997; Oates *et al.* 1990). Similar effects have been observed for bats (Pereira *et al.* 2009) and birds (Beja *et al.* 2010; Remsen & Parker 1983).

It should be noted that *várzea* itself is not a single entity. *Várzeas* growing on Pleistocene deposits on the Juruá are narrower than those on Holocene deposits on the Japurá, and have different patterns of tree species richness (Assis *et al.* 2015a). Nevertheless, the former has twice the number of primate species (4–7) than does the latter (3–4), even though *várzea* floodplains on the Japurá may be 90 km wide and consist of a complex slough-and-levee system. However, while the latter is more extensive and topographically complex, it is isolated from *terra firme* and so has only floodplain specialist primates such as *Cacajao c. calvus* (Ayres 1986a; Peres 1997b).

Ubiquity of Use

Whenever studies of regional primate assemblages have included their use of flooded and non-flooded forest types, it appears that use on either a seasonal, near-permanent or permanent basis is a common feature of Amazonian primates. Only a small minority of primate species do not use flooded forests at all. For example, on the Rio Purus, central Brazil, Haugaasen and Peres (2005a) found that only one species (*Callicebus cupreus*) of the 12 locally recorded visited neither *igapó* nor *várzea*, while 11 species were recorded from *igapó* and eight in *várzea*. In a comparison of 23 sites in western Amazonia. Peres (1997b) found that *terra firme* had higher primate species richness than *várzea* (10–14 versus 3–7 species), and that along the Juruá river, all primate species present in *várzea* also occurred in *terra firme*. However, 10 species used *terra firme* only, with species like *Lagothrix lagotricha* appearing to actively avoid flooded forest in some areas (see also Soini 1986; Stevenson *et al.* 1994). At Jaú, Barnett *et al.* (2002) and Barnett (2010) reported that only two of eight primate species did not visit *igapó* (*Saguinus inustus, Sapajus apella*). As might be expected species absent from flooded forests are often understory and mid-story species, notably members of the genus *Saguinus* (Peres 1997b). Their role as insectivores may well be taken by *Saimiri* species which has a slightly broader diet, uses higher canopy and often reach their highest abundance in riverine forests (Peres 1993, 1997b; Peres 1997b).

Within-species differences between sites exist for a number of ecological aspects. For example, while *Sapajus apella* was one of the commonest species in flooded forests on the Rio Purus (Haugaasen & Peres 2005a), and is reported as a frequent visitor in much of Amazonia (see Chapter 28), but was never recorded entering *igapó* at Jaú (Barnett *et al.* 2002; Adrian Barnett, unpublished data). *Cacajao ouakary* was seen in association with other primate species during less than 4% of group encounters at Jáu, while at Caparú, such encounters were very common, and sometimes lasted several days (Barnett, Bezerra *et al.* 2013). Differences also occur in diet: use of exudates from 18 plant species were recorded across four populations of pygmy marmosets in eastern Ecuador by Yépez *et al.* (2005), but populations differed in the total number of species used and in the preferred species and preferences for most-used exudate species among populations did not appear to be related to the availability of these species in each population. Use was non-random, and not simply a reflection of the availability of a species.

Ayres (1986a, 1989) proposed that the genera *Cacajao* and *Chiropotes* were ecological competitors and lived in effective allopatry, by virtue of habitat choice. However, it is now known that, in addition to *Cacajao* and *Chiropotes*, members of the genus *Pithecia* may also be canopy-dwelling specialists in hard-husked immature fruits (Norconk 2011). It is therefore interesting to note that where the genera overlap geographically one species appears to occur in *terra firme* and another use mostly flooded forest (e.g. *P. chrysocephala* and *C. ouakary* at Jaú), whereas on the Purus, where it is the only pitheciid, *P. albicans* uses flooded forest (Haugaasen & Peres 2005a). However, on the Rio Tapajós where *Chiropotes albinasus* occurs with *Pithecia mittermeieri* on the west and *Pithecia* sp. on the eastern bank, again only one species (*C. albinasus*) was ever seen in seasonally flooded forest (Adrian Barnett, unpublished data). This mirrors the observation by Boubli (1999) and Boubli and de Lima (2009) of *Cacajao* in non-flooded forest in areas from which *Chiropotes* were absent. In this context, observations in Guyana by Shaffer *et al.* (Chapter 28) are difficult to interpret because, although *Chiropotes* was commonly seen in flooded forests and *Pithecia* was not, the number of encounters with the latter genus was very low. The situation on the Solimões, where the newly described *P. cazuzae* and *P. vanzolini* appear to share the *várzea* with *Cacajao calvus* (Marsh 2014), awaits ecological investigation.

In addition to ecological factors, hunting can also have a profound effect on primate community composition (Peres 1990; Peres & Dolman 2000) and this may have obscured some aspects of the studies considered here. For example, the failure to record *Ateles* and *Lagothrix* in flooded forests may be artefactual, as with Jaú, where the animals were hunted out some 60–100 years ago (Barnett *et al.* 2002). Peres (1997a) also recorded the local extinction of *Ateles* and *Lagothrix* due to hunting at several sites in western Amazonia. At Capurú, *Cacajao ouakary* competed directly with *Lagothrix lagotricha* for the immature seeds of *Micanda spruceana* (Euphorbiaceae), which, once mature, became too tannin-rich for either species to eat (Barnett *et al.* 2013b; Defler 2004). Such interactions may mean that the post-hunting changes in primate assemblage composition may lead to changes in resource use in the surviving species. Such aspects have not been researched.

Inundation tolerance of the tree species means species composition changes with distance from the bank, both in *igapó* (Ferreira *et al.* 2010a) and *várzea* (Assis *et al.* 2015b). Consequently, species form diffuse ribbons within the floodplains, so spatial movements may also occur as species fruit sequentially (e.g. as occurs with *Pouteria* species: Adrian Barnett, unpublished data for *igapó*; Alencar 1994 for *terra firme*). However, this is in sufficient to maintain populations in isolation and in all well-studied populations, all *igapó*-dwelling species also occur in *terra firme*. This is not the always case with *várzea*. Here the size and extent of *várzea*, plus the fact that it is spatially heterogeneous (*restinga* and *chavascal*; see Table 27.2), means that if the *várzea* is extensive enough then bands of primates may be able to move between sub-habitats, and never leave this seasonally flooded forest type. Such events may account for the fact that, all Amazonian primates known to be restricted purely to flooded forests (that is: *Cacajao c. calvus, Saimiri vanzolinii*, and – probably – *Pithecia cazuzae* and *P. vanzolinii*) are from *várzea*.

What We Know We Do Not Know: Research Foci for the Future

There are, clearly a great many research projects that can be undertaken that could enhance our knowledge of flooded habitat primates, including our ability to plan for their conservation. Above and beyond multi-site ecological studies of flooded habitat primate populations to give the broadest possible information baseline, the following topics appear to be of pressing importance.

1. How important are *igapó* and *várzea* for the regional primate population? Can species that migrate between these and *terra firme* survive when one of these components is removed? Considering the high rate at which flooded habitats are disappearing in the Brazilian Amazon, there is an urgent need to understand the consequences of such landscape simplification on primate species. Yet, this is currently unknown as all existing work has been done in areas which, even if they experience hunting, essentially have their forest cover intact. A series of studies outside protected areas, that assesses resilience to anthropic impacts such as low- and medium-impact timber removal is clearly required.
2. Through the work of Carlos Peres and others, flooded forest primates of southwestern Amazonia are comparatively well known. However, throughout the Amazon, there are several river basins that have yet to be surveyed. Given the widespread plans for the development of hydroelectric dams in the region (Fearnside 2006, 2014; Tundisi *et al.* 2014; Chapter 36), the disproportionate impact that these have on low-lying flood plain habitats (dos Santos Junior *et al.* 2015; Pringle *et al.* 2000), and the often highly restricted nature of Amazon primate distributions (Ayres & Clutton-Brock 1992; Silva & Oren 1996), priority should be given to surveys in the flooded habitats of rivers where the primate fauna is currently unknown.
3. Priority should also be given to studies on species for which flooded forests are considered the likely prime habitat, but about which nothing is known because they are so recently described. Examples are *Pithecia cazuzae* and *P. vanzolini*, both described by Marsh (2014) and both thought to dwell (probably exclusively) in *várzea*.
4. Further studies are needed of the ecological interconnectivity between *terra firme* and flooded habitats, including thorough longitudinal studies of changes in diet, group size and ranging patterns, as groups move seasonally between habitats.

In addition, the other flooded habitat types in Amazonia, including *Montrichardia* swamps, *campina* and *campinerana* should be surveyed: the discovery of very different ecologies or even new species, is tantalizingly possible. Studies should also be undertaken on primate ecology in *Mauritia* swamps. Furthermore, among many other questions of ecological interest, the exploitation by primates of river turtle nesting areas remains unresearched. It is possible that it is not ubiquitous, indicating that there might be variations in local assessability of nesting beaches, or – a common event in primates that are innovative foragers (e.g. Perry *et al.* 2003) – the result of local traditions.

Concluding Remarks

When considering small mammal species diversity and flooded forests, Pereira *et al.* (2013) noted that "the ecological heterogeneity created by seasonal floods is central to maintaining diverse assemblages in this region". This appears true too for primates in Amazonian flooded habitats, where the flooded forests have clearly had an impact on the primates of the Amazon in terms of abundance, behaviour and possibly appearance (the naked red face of *Cacajao calvus* – Barnett 2010), and may even have been responsible for the evolutionary split in the pitheciin lineage that led to the modern genera *Cacajao* and *Chiropotes* (Ayres & Prance 2013). They are clearly a key, but rather overlooked, part of the regional ecology.

As noted by Haugaasen and Peres (2005a), Amazonian flooded forests (especially *várzea*) differ greatly from *terra firme* forest in the abundance, biomass composition and ecology of their primate assemblages. Although few primate species reside within these forests on a permanent basis, the forests are used seasonally in ways that suggest they are extremely important to the long-term survival not only of the region's primates, but of a variety of other wildlife, ranging from bats to birds and fish (Padoch *et al.* 2000). Based on data from several sites across the Amazon, Ahumada *et al.* (1998) showed that the density of primarily *terra firme* dwelling primate species actually increases as the proportion of seasonally inundated habitat rises towards 50%, since it provides resources when foods are lacking in non-flooded habitats. For this reason they concluded that flooded forests are a keystone habitat for Amazonian primates.

It is therefore of extreme importance that the landscape heterogeneity provided by adjacent flooded and non-flooded forest is maintained. Both *igapó* and *várzea* may have lower species diversities than *terra firme* for several key indicator groups (Padoch *et al.* 2000), but this does not mean that their conservation should be ignored. Rather, these ecosystems should be treated as part of a broader vision for the conservation of an integrated whole that includes the key role such forests have during those periods of the year when there are few food resources in *terra firme* (Goulding *et al.* 2003; Peres & Terborgh 1995). Many of the primate species that use flooded forests are well-known seed dispersers (e.g. *Alouatta*, Giraldo *et al.* 2007; Juillot 1996: *Ateles* – Zhang & Wang 1995: *Lagothrix*, Stevenson, 2000. See Andresen (1999) for review); even uacaris, generally considered seed predators, have been shown to act as seed dispersers in some cases (Barnett *et al.* 2012b). If the diverse landscape offered by a mixed flooded–unflooded system, and the buffering effect that the two asynchronously fruiting systems provide, is lost, primate populations may well decline and with them the seed dispersal services they provided (Ahumada *et al.*1998; Peres & Palacios 2007). It is therefore important that conservation planning in Amazonian forests operate at the landscape scale, and that *igapó* and *várzea* forests are viewed as vital complements to *terra firme*, and planned considered to achieve the effective conservation of all three as an integrated whole. Similar conclusions were reached by Pereira *et al.* (2009) for bats, and the importance of flooded forests for fish conservation (and the survival of commercial fisheries in the Amazon region) has long been championed (e.g. Barthem & Goulding 1997; Pusey & Arthington 2003).

The examples of between-population heterogeneity in primate responses given above show that multiple studies will be needed to assess the ecology of the Amazon's flooded forest primate communities. This is not only because variation in assemblage composition, but also in the suite of competitors and predators, abiotic factors such as rainfall soil and fertility (and hence plant phenology and nutritional content, and variation in chemical composition in non-predictable ways between sites – 'chemotypes'; Wallaart *et al.* 2000).

It should also be recognized that primates are conservation flagship species *par excellence* (Russon & Wallis 2014). A flooded forest Amazonian primate, the white bald uacari (*C. c. calvus*), was the original motivation for the establishment of the Mamirauá Sustainable Development Reserve (see Ayres 1986b; Chapter 41), and is one of the main attractions for the reserve's ecotourism sector (Peralta 2013). Under well-managed conditions animals habituate and tourist presence appears to have little impact of their daily lives (Paim *et al.* 2012; Storni *et al.* 2007). Given the comparative ease with which primates in flooded forests can be seen, when compared to their generally poor visability in *terra firme* forest, and the enthusiasm people display for viewing primates (Russon & Wallis 2014), the development of primate-based ecotourism in Amazonian flooded forests may be a viable way of persuading local communities to give value to standing flooded forests. But, of course, for the exercise to appeal to visitors, the ecological information on the species being observed must be available. For this, and many other reasons, primate research in Amazonian flooded habitats has much to do and to aim for.

Acknowledgements

I thank an anonymous reviewer for extremely swift and helpful advice on the earlier versions of this manuscript, and Torbjørn Haugaasen and Joseph Haws for very useful comments and to them and Carol Antunes for sharing unpublished information.

Chapter 28

Primate Community Structure at Three Flooded Forest Sites in Guyana

Christopher A. Shaffer, Barth Wright and Kristin Wright

Introduction

In tropical South America, large areas of forest are subject to seasonal inundation (Junk 1989; Prance 1979). These areas differ considerably from *terra firma* forests in soil fertility, floristic composition, diversity, and biomass (Junk 1989; Terborgh & Petren 1991). Correspondingly, the structure of mammalian communities, including species richness and abundance, varies considerably between flooded and non-flooded forests. While there have been several studies documenting mammalian community structure in the flooded *várzea* and *igapó* forests of Western Amazonia (Haugaasen & Peres 2005a, b; Peres 1997), mammalian communities in the flooded forests of the Guiana Shield, especially the primates, have received relatively little attention. In this chapter, we describe the use of lowland flooded forests by primates in Guyana, South America. The two species of central interest are the black spider monkey (*Ateles paniscus*) and the Guianan bearded saki (*Chiropotes chiropotes*). Both species have been reported to avoid flooded forest, preferring *terra firma* (Ayres 1989; Ayres & Prance 2013; Lehman 2004). For example, Lehman (2004) found that *A. paniscus* was rarely found in flooded areas and suggested that the distribution of this species in Guyana is reflective of an avoidance of inundated forest. Among the pitheciines, *Cacajao calvus*, a species that often inhabits seasonally inundated habitats (Aquino 1998; Boubli & Lima 2009; Bowler 1997; Bowler & Bodmer 2009), has been held up as the ecological contrast to the seemingly *terra firma* – preferring *Chiropotes* (Ayres 1989; Ayres & Prance 2013). Here we use survey, behavioural and botanical data from three sites in central and southern Guyana to assess the extent to which different species of primate utilize seasonally inundated habitats, with particular emphasis on *Ateles* and *Chiropotes*.

Forests of Guiana Shield

The Guiana Shield (spellings abound due to the multinational colonialization of the region), is a 1 800 000 km² land mass in northern South America (Norconk *et al.* 1996) and encompasses the interior regions of southeastern Colombia, northern Brazil (particularly the province of Roraima), Venezuela and the Guianas (Guyana, Suriname and French Guiana) (Hammond 2005). Exposures in the Guiana Shield are Precambrian remnants of the western reaches of Gondwanaland (Hammond 2005; Norconk *et al.* 1996). Uplift and erosion since the Precambrian have eliminated the earliest substrates (Hammond 2005). These geological processes, as well as continuous shifts in the enormous riverine basins, particularly in Guyana (termed locally the 'land of many waters') have led to a patchwork of soil and forest types that vary over small-scale space and result in numerous microhabitats (Hammond 2005; ter Steege 1993; Hammond, pers. comm.).

In Guyana, several types of forests are characterized by seasonal inundation, including mangrove forest, coastal swamp forest, savanna swamp forest and Mora forest (Clarke *et al.* 2001; ter Steege 1990, 1993; ter Steege & Persaud 1991; ter Steege *et al.* 2000). Mangrove forest, characterized by high densities of the species *Avicennia germinans* (Acanthaceae) and *Acrostichum aureum* (Pteridaceae), is found along the coast and the banks adjacent to the mouths of several large rivers. *Rhizophora* spp. (Rhizophoraceae), *Laguncularia racemosa* (Combretaceae), *Pterocarpus officinalis* (Fabaceae) and *Euterpe oleracea* palms are found in more inland, brackish areas. Coastal swamp forests are found in the permanently flooded coastal plain. Common and locally dominant species in these forests include *Symphonia globulifera* (Clusiaceae), *Tabebuia insignis/fluviatilis* (Bignoniaceae), *Pterocarpus officinalis, Pentaclethra macroloba* (Fabaceae), *Vatairea guianensis* (Fabaceae) and *Euterpe oleracea*. In the savannas of southern Guyana, a distinct type of swamp forest is found along rivers adjacent to gallery forest. This forest is seasonally inundated for up to five months of the year and includes tree species such as *Caryocar microcarpum* (Caryocaraceae), *Macrolobium acaciifolium* (Fabaceae), *Senna latifolia* (Fabaceae), *Zygia cataractae* (Fabaceae) and *Genipa spruceana* (Rubiaceae). The swamp forests in the mixed forests of central and southern Guyana are also distinct, and are found in low-lying areas adjacent to rivers and creeks. Mora (*Mora excelsa*, Fabaceae) forest is found throughout the country along rivers and creeks, and is especially abundant along the Essequibo River (Clarke *et al.* 2001; ter Steege 1990). These two forest types are described in more detail below.

Biogeography of Primates in Guyana

Primate communities throughout South America show a broad range of diversity, from sites with as many as 14 sympatric species to as low as 3 or fewer (Palminteri *et al.* 2010; Peres & Dolman 2000; Pontes 1997). Much of this variation may be the consequence of biogeographical or ecological factors. Edaphic factors appear to influence differences in primate diversity and density between northern (Guiana Shield and Roraima Brazil) and central South America. Northern forests are typically oligotrophic, consistent with a closed nutrient cycle and soils that have not been subject to sedimentation since the Precambrian (Peres & Dolman 2000). Central Amazonian forests are often mesotrophic forests adjacent to whitewater floodplains that have received relatively recent nutrient-rich soil deposition, and alluvial forests that are directly subject to frequent sedimentation during flooding (Peres & Dolman 2000). Productivity is higher in the Central Amazonian forests for these reasons and more primate species, in relatively higher densities, occur within them (Palminteri *et al.* 2010).

With 86% forested land, 8–9 primate species (Lehman 2004) and an estimated 13.6 primate groups per 100 km^2, Guyana maintains forests of relatively low primate diversity (Iwokrama International 1998; Norconk *et al.* 1996; Sussman & Phillips-Conroy 1995) compared to Western Amazonia. This low primate species diversity (8 total species) is typical for the Guiana shield (de Thoisy *et al.* 2008; Lehman 2004; Norconk *et al.* 1996; Sussman & Phillips-Conroy 1995). However, Guyana differs from both Suriname and French Guiana in having a more heterogeneous distribution of species diversity (Lehman 2004). Only four species of primate have been reported west of the Essequibo and north of the Rupununi rivers (Lehman 2000; Sauther *et al.* 1998; Sussman & Philips-Conroy 1995). These are the black spider monkey (*Ateles paniscus*), the red howling monkey (*Alouatta seniculus macconnelli*), the wedge-capped capuchin (*Cebus olivaceus*) and the white-faced saki (*Pithecia pithecia*). Eight species have been identified to the east and south of these rivers. These include the four aforementioned species, as well as the tufted capuchin (*Sapajus apella*), the Guianan bearded saki (*Chiropotes chiropotes*), the golden-handed tamarin (*Saguinus midas*) and the squirrel monkey (*Saimiri sciureus*) (Lehman 2000). These patterns of primate diversity in Guyana may be the consequence of habitat preference or historical factors such as regional extinctions during glacial periods (Lehman 2000). Given the seeming propensity of certain primates such as *Ateles* and *Chiropotes* to avoid seasonally flooded forest, present primate distributions in Guyana may also be influenced by the presence and complexity of riparian networks throughout the country (Lehman 2004).

Based on previous faunal and floral observations, we hypothesized that the number of sightings would be limited for the black spider monkey and Guianan bearded saki in seasonally inundated forest (e.g. Mora forest) throughout the year. We based this hypothesis on the assumption that low number of feeding trees in inundated forest would limit the use of these habitats by these frugivorous primates, as opposed to the presence of water on the forest floor. If the latter were true we would expect markedly fewer sightings in inundated forest when flooded.

Methods

Study Sites

Iwokrama

The site of Turtle Mountain (4°75′W, 58°78′E) (named after the approximately 950 m hill to its west and an abundance of giant river turtles, *Podocnemis expansa*) is located within the Iwokrama Reserve, which consists of 360 000 ha of relatively undisturbed rainforest and natural scrub areas to the west of the Essequibo River (Iwokrama International 1998) (Figure 28.1). The site is generally oligotrophic with increased soil nutrition near rivers where the land is frequently subject to sedimentation during periodic flooding (Iwokrama International 1998). Two sites were established at Turtle Mountain, one within the reserve on the western side of the Essequibo River and one on the eastern side of this river (Figure 28.2). This study focuses on the western site, where the great majority of *Ateles* observations were made during a 14-month study from October 1999 to December 2000. Five forest types were identified at the western site, including high and low mixed forest, liana forest, mountain savanna forest and the seasonally inundated Mora swamp forest (Mittermeier & van Roosmalan 1981; Wright, pers. obs.).

At Turtle Mountain, an estimated 2044 mm of rain fell from December 1999 to November 2000, and only three non-consecutive months received < 100 mm of rainfall (Figure 28.3). The peak months of rainfall were from April to August (1260 mm). Forest inundation occurred from May through much of July and was primarily seen in the Mora forest.

Upper Essequibo Conservation Concession

The Upper Essequibo Conservation Concession UECC is an 81 000 ha protected area located between 3°40′–3°20′N and 58° 25′–58° 5′W on the Essequibo River (Figure 28.4). The study concession encompasses a variety of forest types characteristic of lowland rainforests in Guyana, including Mora forest (dominated by *Mora excelsa*: Fabaceae), Greenheart forest (dominated by *Chlorocardium rodiei*: Lauraceae), Wallaba forest (dominated by *Eperua falcata*: Fabaceae), and other types of mixed forest (Fanshawe 1952; Shaffer 2012; ter Steege 1993). Forest throughout the concession is undisturbed. To quantify the extent of flooded forest in the UECC, we created a hydrology model in ArcGIS 10.1 (ESRI) that flooded the Essequibo River and creeks within the concession boundaries. We used the approximate maximum rise in the river during the peak of the wet season during Shaffer's long-term study (Shaffer 2012). This provided a rough estimate of the extent of seasonally inundated habitat in this part of southern Guyana. This analysis showed that approximately 19% of the forest within the concession was inundated seasonally (Figure 28.3).

The two forest types that make up the majority of these inundated areas are Mora forest and marsh forest. Mora forest is found on alluvial silt along the Essequibo and riverine flats throughout the UECC (ter Steege 1990). As its name suggests, this forest type is dominated by *Mora excelsa*, with almost all large canopy trees and emergents being Mora. Mora forest surrounding base camp was very tall (canopy height 35–45 m, emergents reaching 50+ m) and many of the canopy trees were extensively buttressed. The canopy was extremely dense, with a particularly sparse understory. Other common species in this forest type are *Eperua falcata* (Wallaba), *Eschweilera sagotiana* (Black kakarelli), *Carapa guianensis* (Crabwood), *Clathrotropis*

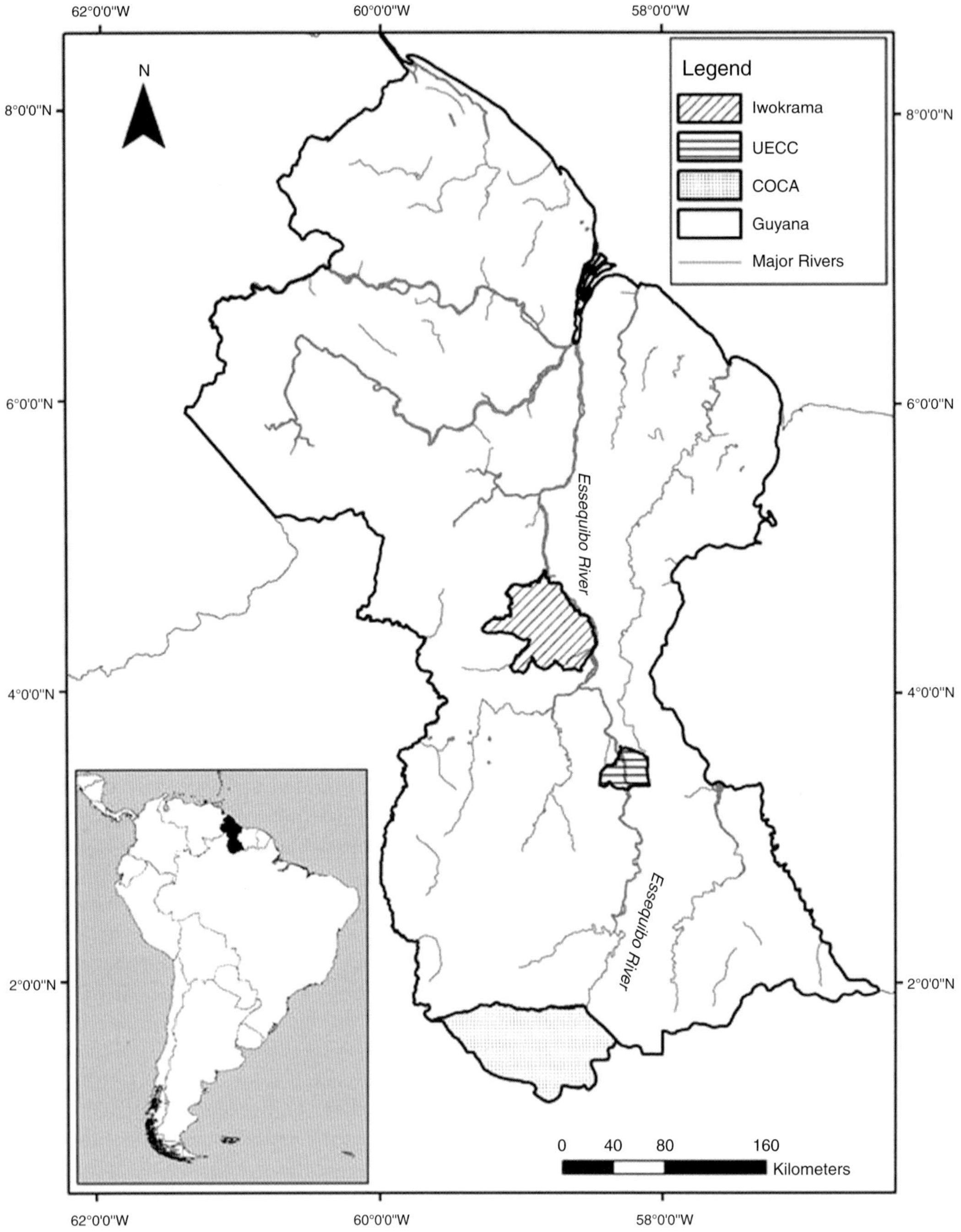

Figure 28.1 Map of Guyana showing locations of the three research sites.

brachypetala (Aromata) and *Couratari gloriosa* (Wadara) (Shaffer 2012). This forest type was seasonally inundated from the beginning of June to the end of August (Shaffer 2012, 2013a).

Marsh or swamp forest is also seasonally inundated, and the marsh forest adjacent to base camp was flooded between May and early September. This forest occurs in low-lying areas adjacent to the Essequibo River and immediately surrounding creeks, as well as areas with poor drainage. Marsh forest is characterized by a low canopy (10–17 m) and a low density of trees with DBH greater than 30 cm. Common tree species included *Abarema jupunba* (Huruasa), *Pradosia schomburgkiana* (Liquorice tree), *Pterocarpus officinalis* (Swamp corkwood), *Eschweilera corrugata* (Wina kakaralli) and *Gustavia augusta* (Kakaralli). In addition, palms, especially *Euterpe edulis* (Manicole), *Attalea regia* (Kokorite) and *Jessenia bataua* (Turu), are frequent (Shaffer 2012; ter Steege 1993).

Konashen Community Owned Conservation Area (KCOCA)

The Konashen Community Owned Conservation Area (KCOCA) is a 625 000 ha conservation concession in southern Guyana adjacent to the border with Brazil (Figure 28.1). The area is owned and managed by indigenous Waiwai

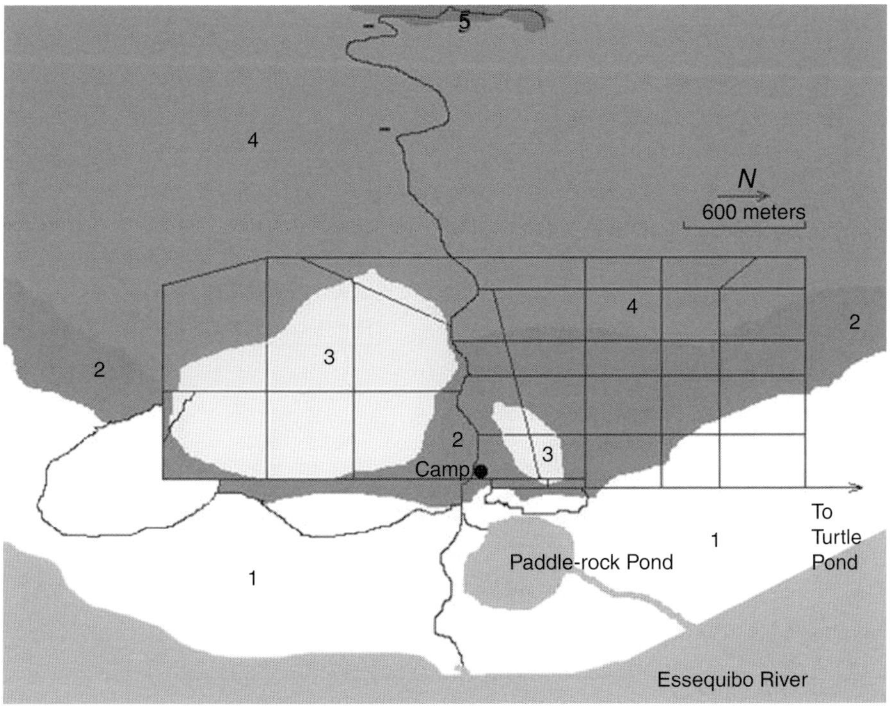

Figure 28.2 Trails and forest types at the Turtle Mountain West Bank site. (1) Mora swamp forest (seasonally flooded); (2) low rainforest; (3) liana forest; (4) high rainforest; (5) mountain savanna forest.

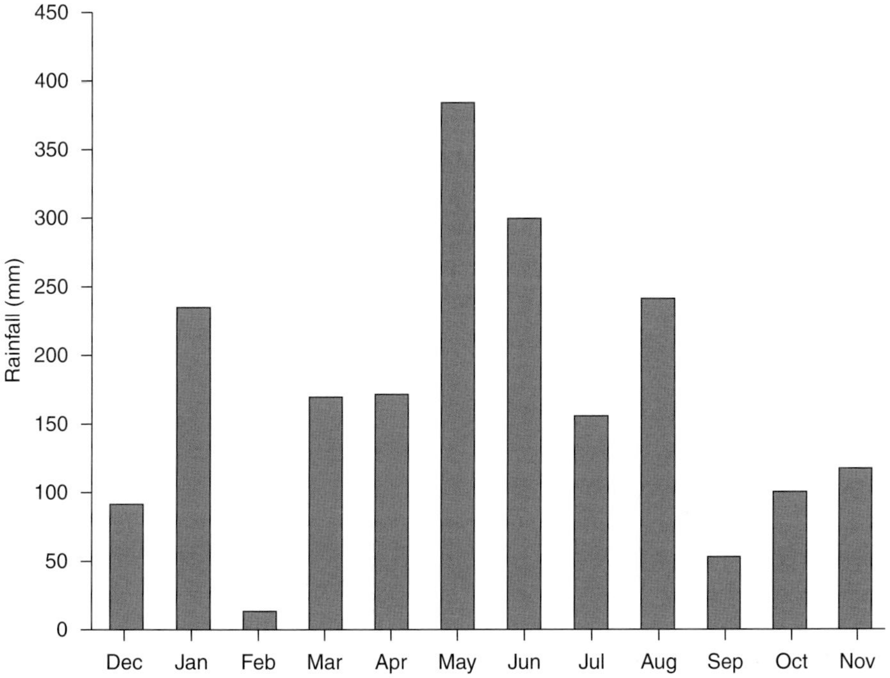

Figure 28.3 Rainfall December 1999 to November 2000 at Turtle Mountain, West Bank, Iworkrama (Total = 2044 mm).

horticulturalists who are concentrated in the 225-person village of Masakenari. Most of the vegetation in the KCOCA consists of mixed lowland rainforest (Alonso *et al.* 2008). The area encompasses the upper watershed of the Essequibo River and contains extensive areas of seasonally inundated swamp forest along the Essequibo, Kassikaityu and Kamoa rivers. Owing to the extremely low human population density in the area (0.36 individuals/km^2), most of the forest is undisturbed. The Waiwai practice subsistence primate hunting, especially of *Ateles paniscus* and *Chiropotes chiropotes*, and densities of these primates are considerably lower within 6 km of Masakenari (Shaffer 2013a). Therefore, we eliminated transect results from within 6 km of the village from our analysis

Survey Methodology

Our primary data set for this chapter consists of survey data taken from the three sites listed above in central and southern Guyana. Surveys were conducted by Wright and Wright in Iwokrama from October 1999 to December 2000. A trail

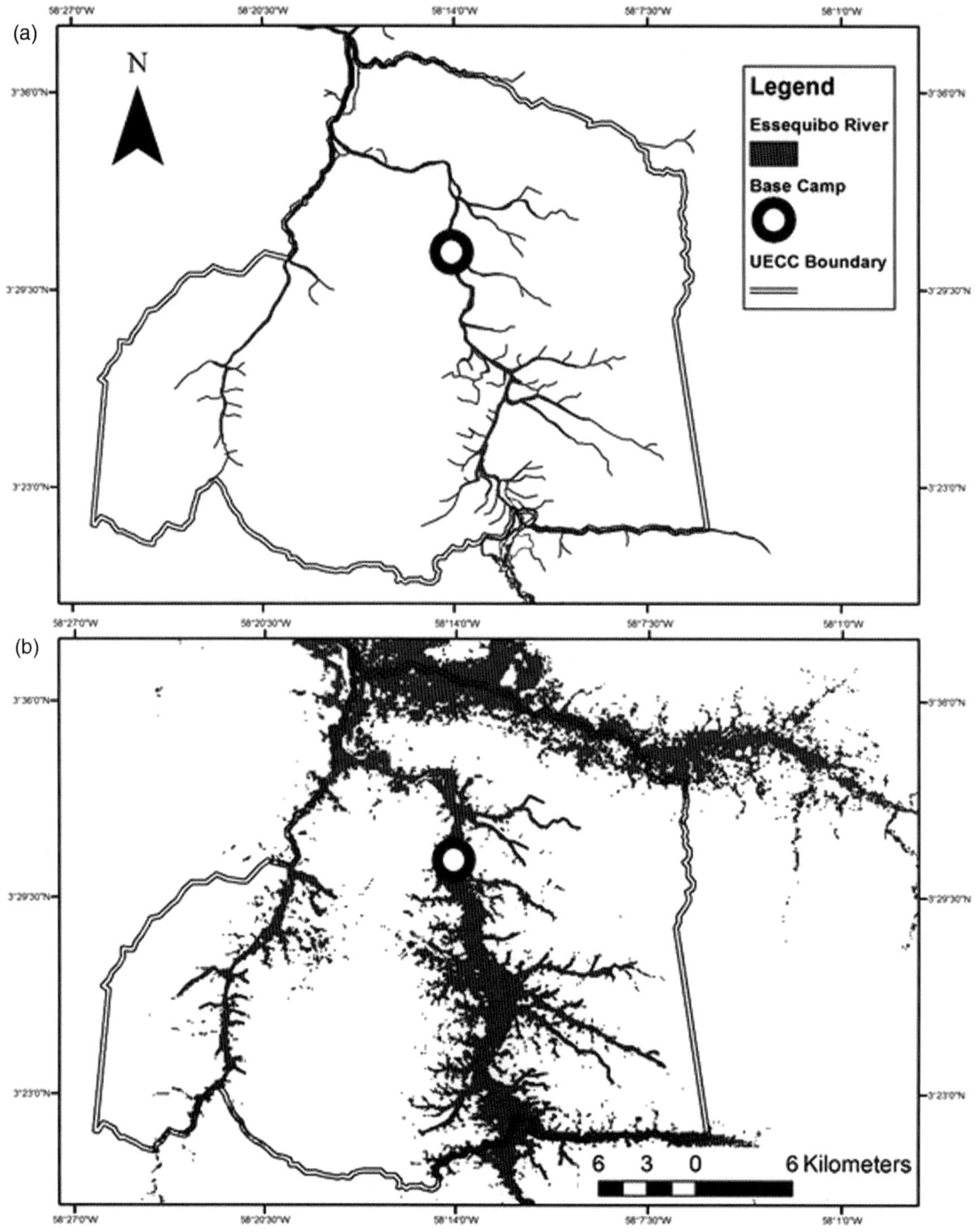

Figure 28.4 Map of the approximate extent of the Essequibo River throughout the Upper Essequibo Conservation Concession during (a) September–October (b) June–August.

system of approximately 2 km^2 was cut with the assistance of forest rangers and local foresters in the reserve below Turtle Mountain (Figure 28.2). These trails were mapped using a Brunton™ transit and GPS points were taken at all trail intersections. During primate follows or on days when walking phenology transects, sightings of all primate species were recorded. Trails were walked at a moderate pace in both the morning (beginning at approximately 06:00 and continuing until approximately 12:00) and afternoon (at approximately 14:00 to 17:30). Only visual sightings were recorded. These data were opportunistic in that the typical survey/census method of standardized times for data collection and consistent use of a single transect were dispensed with due to the need to follow primates once sighted. Upon sighting an individual animal or group, data were collected on the species sighted, the number of individuals encountered, the sex of the sighted individuals, their location along the trail system, the distance of the first-sighted individual from the trail and the type of forest in which they were found. In our results for this study, we focus on total sightings and sightings in flooded versus non-flooded forest.

Shaffer (2012) conducted surveys at two sites in southern and central Guyana. At the Upper Essequibo site, surveys were conducted over three weeks in 2004 and three weeks in 2008. At the KCOCA site, surveys were conducted over a 4-week period in 2013 and 2015. In all cases, a combination of line transect and boat surveys were used. Line transects in the UECC consisted of pre-cut forestry transects and trails cut for a long-term study of the behavioural ecology of *C. chiropotes*. Line transects at the KCOCA site consisted of pre-cut trails. Surveys were conducted in the morning from 06:00 to 10:00 h and in the afternoon from 14:00 to 18:00 h. Line transects were walked at a speed of 1.5 km/h and boat surveys were conducted by paddling at approximately 3 km/h. When primates were detected, the time, species identity, group size, group spread (m) and sighting location along the transect (using GPS) were recorded (Brockelman & Ali 1987; Peres 1999).

To assess primate species preference for flooded or non-flooded forest, we combined sightings data for the UECC and KCOCA sites and compared group sightings to predicted values using a chi-square test. Predicted values were calculated based on the total number of groups sighted and the transect distance surveyed in flooded and non-flooded habitats. Sightings data that were significantly higher or lower than expected by chance suggest either a preference or avoidance of that habitat type. The alpha level for all statistical tests was set at $\alpha \leq 0.05$.

Phenological Monitoring at UECC Site

To assess the availability of food for primates in flooded versus *terra firma* forest at the UECC site, Shaffer (2012) compared monthly fruit abundance in transects that were inundated for at least 1 month ($n = 11$) and those that were not inundated ($n = 20$) during a 12 month period (January to December 2008). Most areas in the inundated category where consistently inundated from June through August and water depth ranged from 0.33 m to greater than 5 m. All trees ≥ 10 cm DBH in these 5 × 50 m transects were marked for identification, and monitored once a month for the presence of ripe fruit, immature fruit or flowers (see Shaffer 2013a). The abundance of plant parts was rated on a relative scale of 0–4, with scores indicating the percentage of total crown volume that contained fruit. A score of (0) indicated no fruit, (1) indicated < 25% of the crown contained fruit, (2) indicated 25–49%, (3) indicated 50–74% and (4) indicated 75–100%. A resource availability index for fruit for each month was calculated by multiplying basal trunk area by phenology score for each tree, adding these scores, and dividing them by total monitored basal area (Shaffer 2013a).

Behavioural Data Collection: *C. chiropotes* at Southern Sites

Here we also report more detailed data concerning the use of flooded habitats by one group of bearded sakis. These data were collected by Shaffer as part of a long-term study of the behavioural ecology of *C. chiropotes* at the Upper Essequibo Concession site (Shaffer 2012). While total contact hours with the group were relatively few during the high-water months (20.1 h), these data provide information on the time bearded sakis spend in flooded parts of their habitat and the types of food they consumed in these areas. To analyse this data, we compared contact time with the group in inundated forest and contact time in non-flooded areas during June, July and August. All trees used for feeding by the study group during this period were marked and identified. Due to the difficulties of maintaining contact with the group in flooded areas, this method almost certainly underestimates the percentage of time bearded sakis actually spent in flooded forest.

Results

Primate Observations at Turtle Mountain

Four of the eight primate species found in Guyana were sighted within the Iwokrama Reserve on the west side of the Essequibo River (Figure 28.5). These include *A. paniscus* ($n = 81$ sightings), *A. seniculus* ($n = 79$ sightings), *C. olivaceus* ($n = 80$ sightings) and *P. pithecia* ($n = 17$ sightings). Sightings of *C. olivaceus, A. paniscus* and *A. seniculus*, accounted for approximately 90% of all observations based on *ad libitum* records over the course of 305 days of observation.

The western site was host to the majority of large-bodied frugivorous primates, and was notable for frequent observations of *Ateles paniscus* ($n = 81$), a primate that was initially reported absent from the area (Lehman 1999). The presence of *A. paniscus* at this site is indicative of the limited survey work that has been conducted in the interior of Guyana, particularly west of Essequibo, and in the southeast of the country (Norconk *et al.* 1996).

Contrary to our expectations, Mora forest was frequently used by all four primate species at the western site at Iwokrama. This flooded forest was the preferred habitat for *A. seniculus* (Figure 28.5). *Ateles paniscus* exhibited the most equal use of all forest types, but Mora forest still proved important for this species. Observations in Mora forest ranked second (28%) after high forest (26%) for *Ateles*. During the period of inundation *A. paniscus* was observed over water in Mora forest in 30% of observations (13 of 43). During this same period, *A. seniculus* was observed over water in 62% of all observations (29 of 47), *C. olivaceus* 22% (8 of 36) and *P. pithecia* 50% (5 of 10).

Primate Observations in Southern Guyana

All eight of the primate species were observed at both sites in southern Guyana (Figure 28.6). *Chiropotes chiropotes* (23 group sightings), *Alouatta seniculus* (15 group sightings) and *Sapajus apella* (11 group sightings) were the three most commonly observed primates in the UECC while *A. seniculus* (15 group sightings), *Saimiri sciureus* (13 group sightings), and *C. chiropotes* (11 group sightings) were most commonly observed in KCOCA. *Pithecia pithecia* and *Cebus olivaceus* were rarely observed at either site.

Of the eight primate species found in Guyana, only two, *Saguinus midas* and *C. olivaceus*, were not observed in

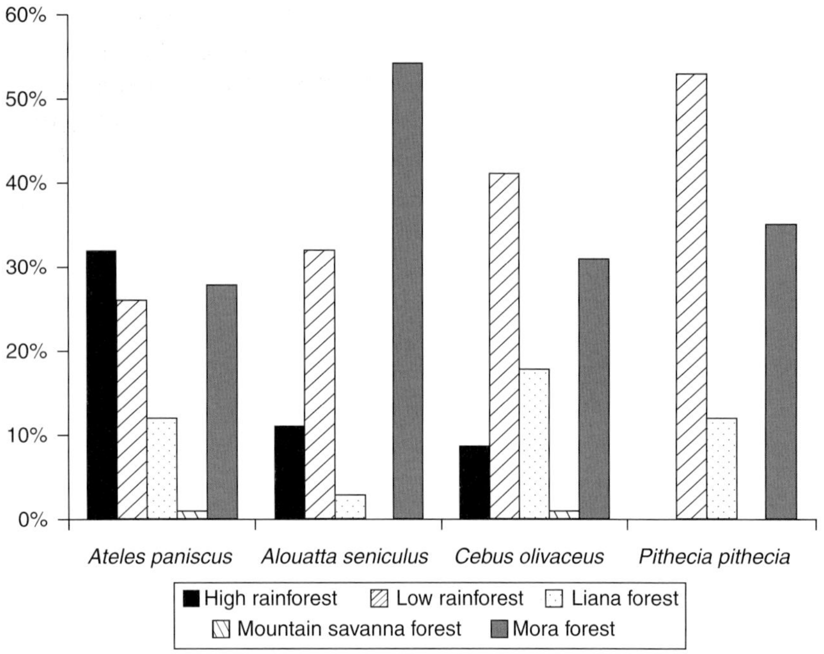

Figure 28.5 Forest use by type for the four primate species at the west bank site of Turtle Mountain, Iwokrama.

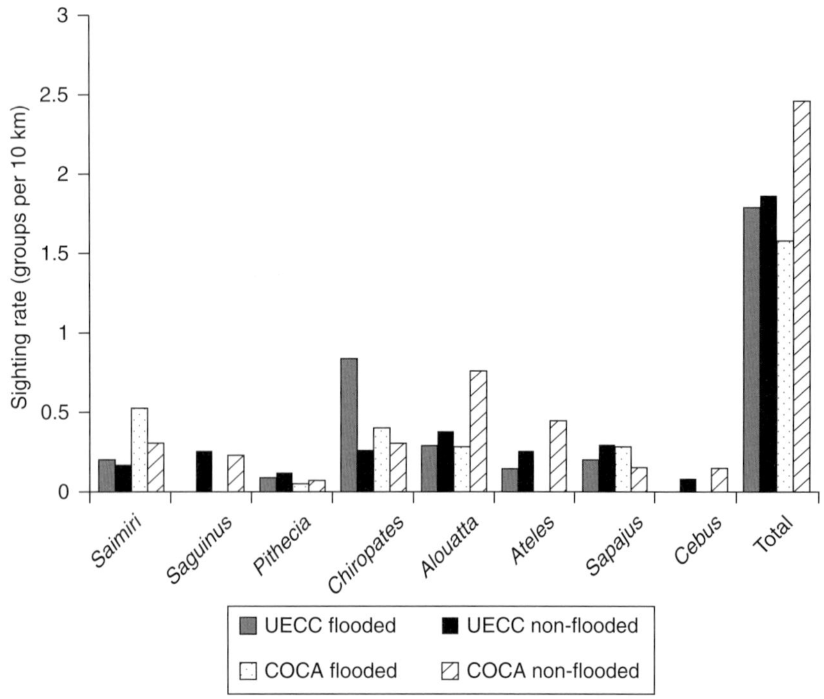

Figure 28.6 Sighting rates of primates in flooded and non-flooded forests in the Upper Essequibo Conservation Concession (410 km surveyed) and the Konashen Community Owned Conservation Concession (300 km surveyed).

flooded forests at either the UECC or KCOCA (Table 28.1). *S. midas, C. olivaceus* and *A. paniscus* were observed significantly less frequently in flooded forest than expected by chance. Several primate species, including *A. seniculus, C. chiropotes, S. sciureus* were commonly observed in flooded forest at both sites, and *C. chiropotes* was observed significantly more often in flooded forest than expected by chance (Table 28.1). Species use of flooded forest was similar at both sites with the exception of *A. paniscus. Ateles paniscus* was observed in flooded forest 33% of the time in the UECC, but never observed in flooded forest in the KCOCA. This may reflect increased hunting pressure in riparian forest at the KCOCA site rather than habitat preference. Two species, *S. sciureus* and *C. chiropotes*, were either more commonly observed in flooded forest or equally observed in flooded and non-flooded forest at both sites.

To more thoroughly assess the use of flooded forest by *C. chiropotes*, we compared contact time in inundated areas and non-inundated areas from June through mid-August in the UECC. During this period, the bearded saki study group spent

Table 28.1 Combined sightings of primate groups for surveys at the Upper Essequibo Conservation Concession and the Konashen Community Owned Conservation Area. Sightings that were significantly higher than expected by chance for one of the habitat types are shown in bold.

Taxa	Sightings (Groups)		Total	χ^2	*p*-value	Preference
	Flooded (370 km)	Non-flooded (360 km)				
Saimiri	13	8	21	1.06	0.304	–
Saguinus	0	9	9	**9.26**	**0.002**	Non-flooded
Pithecia	3	4	7	0.172	0.678	–
Chiropotes	24	10	34	**5.38**	**0.020**	Flooded
Alouatta	11	19	30	2.36	0.124	–
Ateles	3	12	15	**5.67**	**0.017**	Non-flooded
Sapajus	9	9	18	0.01	0.953	–
Cebus	0	4	4	**4.11**	**0.043**	Non-flooded
Total	63	75	138	1.41	0.236	–

Table 28.2 The top ten species eaten (percentage of feeding time) by bearded sakis in inundated forests from 15 June to 15 August 2008 in the Upper Essequibo Conservation Concession.

Plant species	Family	% of *Chiropotes* feeding time[a]	Taxa eating	Dispersal mechanism	Forest type
Mora excelsa	Fabaceae	23.91	*Chiropotes*	Hydrochorous	Mora
Eschweilera sagotiana	Lecythidaceae	18.11	*Chiropotes*	Synzoochorous	Mora
Gustavia augusta	Lecythidaceae	16.43	*Chiropotes*	Synzoochorous	Marsh
Carapa guianensis	Meliaceae	11.36	*Chiropotes*	Hydrochorous, synzoochorous	Mora, Marsh
Moutabea guianensis	Polygalaceae	10.15	*Chiropotes, Ateles*	Endozoochorous	Mora
Protium decandrum	Burseraceae	6.56	*Chiropotes, Ateles*	Endozoochorous	Mora
Terminalia dichotoma	Combretaceae	4.22	*Chiropotes*	Hydrochorous	Marsh
Passiflora spp.	Passifloraceae	3.64	*Ateles, Chiropotes, Cebus*	Endozoochorous	Mora
Hymenaea courbaril	Fabaceae	3.12	*Chiropotes, Cebus*	Synzoochorous	Marsh
Pterocarpus officinalis	Fabaceae	2.50	*Chiropotes*	Hydrochorous	Marsh

[a] Percentage of all feeding records (5:43 h) for bearded sakis in inundated forest from 15 June to 15 August 2008.

approximately half (9.4 of 20.1 h) of their time in inundated habitats. Most of this time was spent in Mora forest (6.8 h). Bearded sakis fed from at least 20 species in inundated areas and the top ten (based on feeding time) are listed in Table 28.2. Many of these species are hydrochorous and their fruiting coincides with the period of maximum inundation (ter Steege 1990; van Roosmalen 1985). During a 15-month study period, the bearded saki group home range was approximately 1000 ha (Shaffer 2013b). About 36% of this home range was subjected to seasonal inundation (23% was Mora forest and 13% was marsh forest).

Phenology of Flooded Forests

At the UECC site, while the phenological patterns in flooded and non-flooded transects were quite similar, the inundated forests contained a higher abundance of fruit during the months when they were flooded (June, July and August) (Figure 28.7). The largest different in abundance between the two forest categories was in July, when inundated forests contained almost twice as much fruit as *terra firma* forests. Much of this variation was driven by the fruiting patterns of *M. excelsa*, the monodominant species in Mora forest. *Mora excelsa* shows two fruiting peaks, one in November and one in June–July (ter Steege 1990). At this site, most of the large *M. excelsa* trees fruited during June and July, and this species made up a high proportion of the total monitored basal area in Mora forest (63%). Other common tree species in Mora forest that fruited during the inundated months were *Eschweilera sagotiana*, and *Carapa guianensis*, both important bearded saki food species (Shaffer 2013a). In swamp forest, common species that fruited

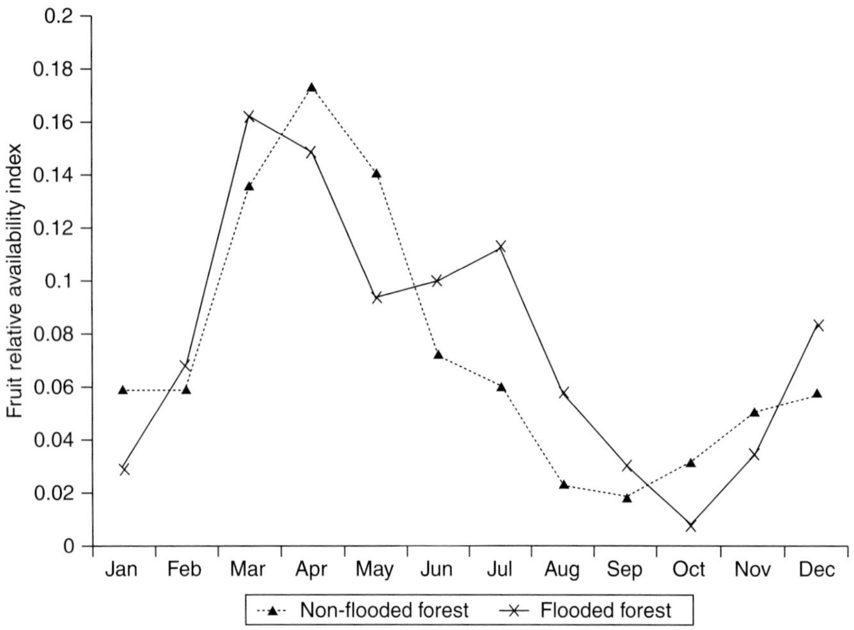

Figure 28.7 Phenology of fruit availability in transects located in flooded and non-flooded forests at the Upper Essequibo Conservation Concession site. Flooded areas were defined as areas that were inundated for at least one month of the year while non-flooded areas were not subjected to seasonal inundation (see Methods).

during the inundated months were *Pterocarpus officinalis* and *Eschweilera corrugata*.

In the forests of Guyana, in general, the months of June through August are those with greatest fruit scarcity (ter Steege 1990). However, the results reported here show that several of the most common tree species in inundated forests, especially *M. excelsa*, have fruiting peaks that coincide with seasonal flooding. Therefore, primates that are able to exploit these species might be expected to frequent flooded areas during the peak wet season.

Discussion

Many researchers have reported the extensive use of flooded habitats by several primate species throughout Amazonia and the Guyana Shield, including *Saimiri* spp., *S. apella* and *Alouatta* spp. (Haugaasen & Peres 2005; Lehman 2004; Marsh 2004; Palminteri *et al.* 2011; Peres 1997; Santos 2010). The riparian preferences of *Saimiri* spp. are particularly well documented (Haugaasen & Peres 2005; Lehman 2004; Marsh 2004; Peres 1997). *Sapajus apella* is also frequently reported in flooded habitats, although not to the extent of *S. sciureus* (Defler 1985; Haugaasen & Peres 2005; Santos 2010; Wallace *et al.* 1998). In his extensive surveys of northern and central Guyana, Lehman (2004) found positive correlations between sighting rates and flooding intensity for both *S. apella* and *S. sciureus*. Similarly, Levi *et al.* (2013) found a positive relationship between density and seasonally flooded forest for *S. apella* and *S. sciureus* at 94 sites throughout southwestern Guyana. Therefore, the frequency of *S. apella* and *S. sciureus* sightings in inundated areas in this study is not surprising.

However, the use of flooded forest by *Ateles* and *Chiropotes* is poorly documented. Lehman (2004) and Lehman *et al.* (2006) found that spider monkeys rarely ranged into flooded habitats and suggested that the lack of fruiting Moraceae trees in flooded forest explained their infrequent usage. They also did not observe bearded sakis in flooded forest. In contrast, our results suggest that *Chiropotes* and *Ateles* are frequently found in seasonally flooded habitats in central and southern Guyana and that, at least in the case of *Chiropotes*, they may actually prefer these habitats to non-flooded forests during the peak wet season months.

While some populations of *Ateles* appear to avoid inundated habitats, other researchers have reported seasonal use of flooded forest (Galvis *et al.* 2012; Haugaasen & Peres 2005; Peres 1997). The discrepancy between our results and those of Lehman (2004) likely reflect a difference in sampling intensity and the fact that Lehman's surveys were concentrated in northern Guyana where flooding is rarer. van Roosmalen (1985) and Haugaasen and Peres (2005) suggested that, while use of flooded environments by spider monkeys is relatively rare compared to other species, seasonal foraging incursions into flooded habitats may be critical during periods of resource scarcity. The extensive use of Mora forest by *Ateles* at Iwokrama may reflect the relative paucity of resources in non-flooded forests at the site.

Bearded sakis have traditionally been characterized as *terra firma* primates, in contrast to closely related *Cacajao* (Ayres 1989; Ayers & Prance 2013; Lehman 2004; Mittermeier & van Roosmalen 1981; Norconk 2011). While the characterization of *Cacajao* as flooded forest specialist is clearly too simplistic (see Heymann & Aquino 2010), many populations of the genus are either entirely restricted to flooded habitats or migrate between *terra firma* and flooded forest depending on the seasonal availability of resources (Ayres 1989; Aquino 1998; Bowler 2007; Bowler & Bodmer 2011; Barnett *et al.* 2013b). This has led some researchers to suggest that the highly granivorous uacaris are able to reach high densities in flooded environments due to the abundance of large-seeded plant species and lack of mammalian competitors (Bowler 2007; Norconk 2011). Our phenological results suggest a similar abundance of large-seeded plant species in inundated forests in southern Guyana, providing a potentially invaluable resource base for the primates able to exploit them. As many of these species fruit during the peak wet season months, when fruit availability is relatively low, they provide a predictable food source for seed predators during an otherwise lean period (Shaffer 2013a). Interestingly, survey results from both the Upper Essequibo site and the KCOCA site show a high abundance of *Chiropotes* compared to other surveys in northern Guyana and central Brazil (Lehman 2004; Shaffer 2014; Sussman & Phillips-Conroy 1995). In addition, group sizes for *Chiropotes* at UECC and KCOCA appear to be higher than at other sites in Suriname and Central Amazonia (Shaffer 2014). Given the intensity of use of inundated forest by *Chiropotes* in this study, and the abundance of large-seeded plant species in flooded habitats, especially from the *Chiropotes* preferred family Lecythidacae, it is possible that extensive areas of flooded habitat in southern Guyana support higher populations of these primates.

Conclusion

The results of our study at three inundated sites in central and southern Guyana indicate that Guianan bearded sakis (*Chiropotes chiropotes*) and, to a lesser extent, black spider monkeys (*Ateles paniscus*) exploit flooded forests during some parts of the year. We suggest that the high availability of certain plant food species in these areas during the flooded periods, including many hydrochorous fruits, provides critically important resources for these primates when fruit availability in other forest types is low. Far from avoiding flooded forest, bearded sakis (and perhaps spider monkeys) may be highly dependent on seasonally inundated forests in Guyana. These results extend our knowledge of the habitat requirements of the primates in Guyana and have important conservation implications. However, more detailed research, particularly behavioural data collection in flooded areas, is necessary to better describe the importance of these habitats for *Chiropotes* and *Ateles* and to characterize the structure of primate communities in other areas of the Guiana Shield.

Chapter 29

Primates of the Peat Swamp in Borneo and Sumatra

Susan M. Cheyne, Marcel Quinten and Keith Hodges

Introduction

Southeast Asia is the most important region in world for tropical peatland, containing the largest share of this resource (247 778 km^2, 56% of the best estimate value), and Indonesia has the largest area (20 950 km^2, 47% of the total best estimate), followed by Malaysia (25 889 km^2; 6% (Page *et al.* 2011)).

Peatlands are those wetland ecosystems characterized by the accumulation of organic matter (peat) derived from dead and decaying plant material under conditions of permanent water saturation fed by prolonged rainfall (e.g. Figure 29.1). Peat swamp habitats are characterized by poor drainage (due to topography and rainfall), permanent waterlogging and substrate acidification (Page *et al.* 1999), and represent important natural ecosystems with high value for biodiversity conservation, climate regulation and human welfare. Peatlands are unique, complex ecosystems of global importance for biodiversity conservation at genetic, species and ecosystem levels. They contain many plant and animal species found only or mainly in peatlands and support a high level of biodiversity of both flora and fauna (Page 2002; Page *et al.* 1997). Peatland-living species are adapted to the special acidic, nutrient poor and waterlogged conditions of their habitat. They are vulnerable to changes resulting from direct human intervention, changes in their water catchment and climate change, that may lead to loss of habitats, species and associated ecosystem services (Harrison *et al.* 2005, 2009; Husson *et al.* 2002; Meijaard 1997; Page 2002). The biodiversity values of peatlands demand special consideration in conservation strategies and land-use planning.

Unfortunately, a substantial proportion of the original peatland habitat in Southeast Asia has already been degraded or lost (Figure 29.2). As of 2006, approximately 45% of Southeast Asia's peatland had been deforested (Hooijer *et al.* 2010), and projections of deforestation under a business-as-usual scenario indicate that just under half of the 68 000 km^2 of peat swamp forest remaining in Kalimantan may be lost by 2020 (Fuller *et al.* 2011). For Borneo and Sumatra, it is estimated that of the *c.* 146 000 km^2 peatland area, only about 35% remains forested (Miettinen *et al.* 2012a), with estimates of annual deforestation rates in the region (1990–2010) of around 3% and 5% for Borneo and Sumatra, respectively (Miettinen *et al.* 2012b). The principle causes of tropical peat swamp forest (TPSF) degradation in all areas are deforestation (legal and illegal logging), conversion to oil palm and other crop plantations and fire (Aldhous 2004; Miettinen *et al.* 2012b).

The clearing for agriculture alone has led to reduction in the C-fixation capacity of 5–9 Mt/yr (Page *et al.* 1999). TPSF is heavily reliant on high water tables. In Central Kalimantan, in 1997, large-scale canals were created for the ill-conceived conversion of TPSF to agricultural land for rice production and by illegal loggers to extract tree trunks, resulting in high levels of drainage and thus oxidization of the peat. This drying combined with El Niño weather events result in increased occurrence of uncontrollable wild fires and in 1997/98 alone, the peat burnt in Central Kalimantan released carbon in the atmosphere equivalent to a third of the world's global annual emissions (Aldhous 2004).

Tropical peatlands are relatively unknown compared to other tropical forest habitats, however, studies in the last 10 years have already indicated a high level of function and worth at both a local and global level. The ecosystem is vital in hydrological processes and nutrient cycling, and economically through the provision of sustainable livelihoods from the plant species that are found there and nurseries for fish species. Desiccation of peat halts the microbiotic conditions vital to the creation of new peat deposits. The remaining peat begins to erode rapidly which causes the peat structure to collapse and compact, the chemical composition alters and invasive species dominate the new vegetation as a lack of suitable germination conditions with high light exposure and lack of water prevent the natural establishment of pioneer species (Graham *et al.* 2013; Harrison *et al.* 2010b). Seed dispersal from the surrounding forest is greatly reduced and active restoration through seedling transplants is perhaps the best option for forest recovery. Pilot studies in the Sabangau have highlighted the importance of primates as primary seed dispersers, especially orangutans and gibbons in the peat swamp forest (Nielsen *et al.* 2011).

Primate Diversity and Population Density in TPSF

Of the large mammal species adapted to peatland habitats, primates are the most specious. Across Borneo, Sumatra and the Mentawai Islands there are 35 primate species (54 subspecies), of which 23 have been recorded in TPSF (Table 29.1),

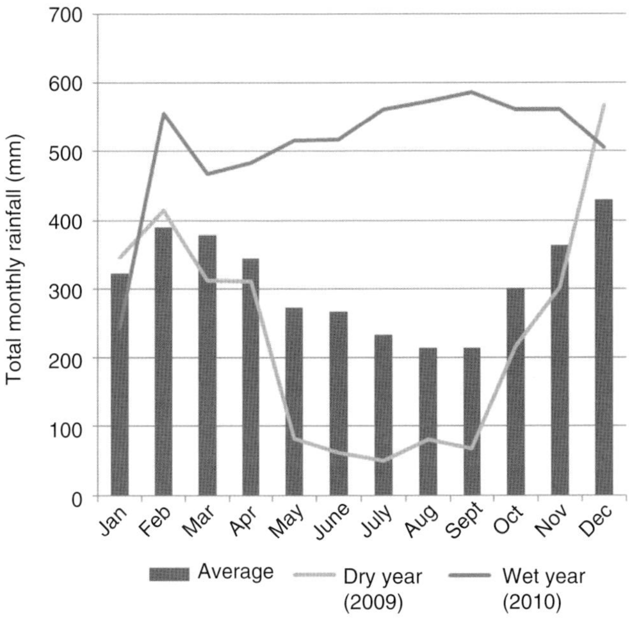

Figure 29.1 Average rainfall in Sabangau, Central Kalimantan, Indonesia (data from 2004–2012) and examples of dry and wet years.

although detailed information is available for only 12. Hence, TPSF provides a habitat for at least 65% of the region's primates, and this may be as high as 77% if four additional species (*Presbytis potenziani, Presbytis siamensis, Macaca pagensis* and *Nycticebus coucang*), known to use 'swamp' forests in general, are also included (Table 29.1). Of the species listed in Table 29.1, 24 are IUCN listed as Vulnerable/Endangered, 4 as Critically Endangered and 1 as Data Deficient. Nowak (2012) writes that the two top primates sites (including TPSF and mangrove habitat) in terms of number of species and endangered status are both in Indonesia (Gunung Palung and Sabangau).

Some species are found in TPSF in densities comparable to those of other habitats e.g. gibbons and orangutans, underlining the potential importance of TPSF habitat for these species. Actual importance, however, depends on how widely used the habitat is and what proportion of total population size is found in TPSF habitat: neither of which is known in sufficient detail for most primate species. This lack of data serves to highlight just how little is known about primates in this habitat and that this should be a priority for future studies.

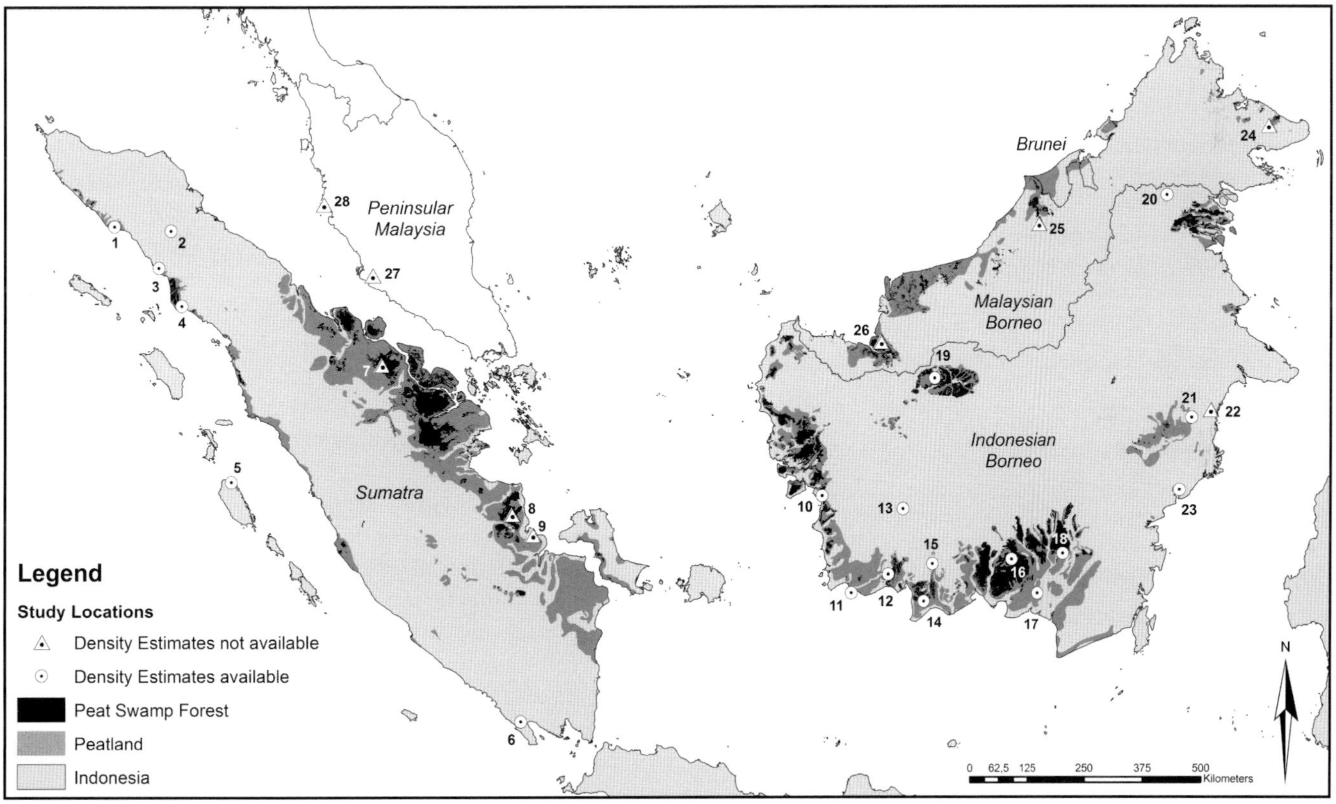

Figure 29.2 Extent of peatland in Borneo and Sumatra and proportion of peatlands still covered by TPSF, based on Miettinen *et al.* (2012) and Association of Southeast Asian Nations (ASEAN) Secretariat and Global Environment Centre (2012). Triangles indicate locations where primates have been found in TPSF; circles are locations yielding detailed primate density estimates. Location details: SUMATRA: 1. Tripa Swamps; 2. Ketambe; 3. Kluet/Suaq Swamps; 4. Singkil Swamps; 5. Peleonan Swamp (Siberut); 6. Bukit Barisan Selatan NP; 7. Siak Area; 8. Berbak NP; 9. Sembilang NP; BORNEO: 10. Gunung Palung NP; 11. Tanjung Keluan; 12. Lamadau Reserve; 13. Belantikan; 14. Tanjung Puting NP; 15. Seruyan; 16. Sabangau Catchment (+ Katingan); 17. Kahayan / Mangkutup; 18. Mawas; 19. Danau Sentarum NP; 20. Sebuku / Sembakung NP; 21. Kutai NP; 22. Sungatta; 23. Sungai Wain/Balikpapan; MALAYSIA: 24. Kinabatangan; 25. Luagan Bunut NP; 26. Maludam NP; 27. Kuala Langat; 28. Perak. (Sources: Ampeng *et al.* 2009; ASEAN Secretariat and Global Environment Centre 2012; Whitten 1982a; Meijaard & Nijman 2003; Norhayati *et al.* 2004; Sangchantr 2004; PT Arara Abadi *et al.* 2005; Wich *et al.* 2008, 2009; Wilson *et al.* 2013).

Table 29.1 Currently recognized primate species of Borneo and Sumatra and their use of Peat Swamp Forest (Classification according to Wilson *et al.* (2013)).

Species and taxonomy in PSF	Present (countries)	Species' range status	IUCN Red List status
Cercopithecidae			
Presbytis bicolor (Black-and-white Langur)	N	S	Data Deficient
Presbytis chrysomelas (Sarawak Surili)	Y	K, M, B	Critically Endangered
Presbytis canicrus (Miller's Langur)	N	K	Endangered
Presbytis femoralis (Banded Surili)	Y	S, M	Near Threatened
Presbytis frontata (White-fronted Langur)	N	K, M	Vulnerable
Presbytis hosei (Hose's Langur)	N	K, M, B	Vulnerable
Presbytis melalophos (Black-crested Sumatran Langur)	N	S	Near Threatened
Presbytis mitrata (Mitered Langur)	N	S	Endangered
Presbytis potenziani (Mentawai Langur)	Sw*	S	Endangered
Presbytis rubicunda (Maroon Leaf Monkey)	Y	K, M, B	Least Concern
Presbytis sabana (Sabah Grizzled Langur)	N	M	Endangered
Presbytis siamensis (White-thighed Surili)	Sw*	S, M	Near Threatened
Presbytis siberu (Siberut Langur)	Y	S	Endangered
Presbytis sumatrana (Black Sumatran Langur)	N	S	Endangered
Presbytis thomasi (Thomas's Langur)	Y	S	Vulnerable
Trachypithecus cristatus (Silvery Lutung)	Y	S, K, M, B	Near Threatened
Nasalis larvatus (Proboscis Monkey)	Y	K, M, B	Endangered
Simias concolor (Pig-tailed Langur)	Y	S	Critically Endangered
Cercopithecinae			
Macaca fascicularis (Long-tailed Macaque)	Y	S, K, M, B	Least Concern
Macaca nemestrina (Sunda Pig-tailed Macaque)	Y	S, K, M, B	Vulnerable
Macaca pagensis (Mentawai Macaque)	Sw*	S	Critically Endangered
Macaca siberu (Siberut Macaque)	Y	S	Vulnerable
Hylobatidae			
Hylobates abottii (Abott's Gray Gibbon)	Y	K, M	Endangered
Hylobates agilis (Agile Gibbon)	Y	S, M	Endangered
Hylobates albibarbis (Bornean White-bearded Gibbon)	Y	K	Endangered
Hylobates funereus (East Bornean Gray Gibbon)	Y	K, M, B	Endangered
Hylobates klossii (Kloss Gibbon)	Y	S	Endangered
Hylobates lar (White-handed Gibbon)	Y	S, M	Endangered
Hylobates muelleri (Müller's Bornean Gibbon)	Y	K	Endangered
Symphalangus syndactylus (Siamang)	Y	S	Endangered
Hominidae			
Pongo abelii (Sumatran Orangutan)	Y	S	Critically Endangered
Pongo pygmaeus (Bornean Orangutan)	Y	K, M	Endangered
Lorisidae			
Nycticebus coucang (Greater Slow Loris)	Sw*	S, M	Vulnerable

Table 29.1 (*cont.*)

Species and taxonomy in PSF	Present (countries)	Species' range status	IUCN Red List status
Nycticebus menagensis (Bornean Slow Loris)	Y	S, M, K, B	Vulnerable
Tarsiidae			
Tarsius bancanus (Horsfield's Tarsier)	Y	S, K, M, B	Vulnerable

* Species reported from 'swamp' forest – no further distinction of swamp type available Abbreviations: Y = yes; N = No; Sw = Swamp Forest; S = Sumatra; K = Kalimantan; M = Malaysia; B = Brunei.

Table 29.2 List of density estimates from TPSF locations for orangutans (all from Wich *et al.* (2008)) listed according to increasing density estimates.

Region	Species	Individuals/km^2
Borneo	*P. p. morio*	0.5 (0.2–0.8)
Borneo	*P. p. wurmbii*	1.56 (0.08–4.09)
Borneo	*P. p. pygmaeus*	2.6 (0.43–4.09)
Sumatra	*P. abellii*	1.52 (0.38–4)

Table 29.3 List of density estimates for gibbons from TPSF locations listed according to increasing density estimates.

Region	*Species*	Gibbon density (groups/km^2)	References
Borneo	*H. albibarbis*	2.35 (1.08–3.92)	Cheyne 2011, 2012
Sumatra	*H. lar*	3.6	Palombit 1997

Orangutans

All known orangutan species and subspecies are found in TPSF. Densities are reasonably high (range: 0.08–4.09 individuals/km^2 in TPSF compared with 0.03–6.36 individuals/km^2 for all orangutan habitats (based on a survey of 55 published studies of which 47 reported density information from TPSF (Table 29.2)). The amount of TPSF remaining in Borneo is roughly 31 700 km^2 so with a conservative estimate of 1.5 individuals/km^2, TPSF in Borneo supports approximately 45% of the total population (Singleton *et al.* 2004). For Sumatra, the remaining TPSF amounts to 18 500 km^2 and using the same estimates of density, TPSF in Sumatra supports about 40% of the population of the Sumatran species, although this is within only 12% of total available orangutan habitat (Singleton *et al.* 2004).

Gibbons

Of eight species found in Borneo and Sumatra (including Mentawai Islands), density data exist for all, although the majority of data are from Borneo. Of 22 populations, the majority (17) can be considered to exist in moderate densities, being below the figure defined as high density (5–6 groups/km^2), but with only 5 being below the threshold (2 groups/km^2) used to indicate low density (Brockelman & Srikosamatara 1993). Overall, densities of gibbons in TPSF range from 1 to 4 groups/km^2 (Table 29.3), well within the values of 0.06–9 groups/km^2 reported for all gibbons (based on a survey of 67 published studies of which only 29 reported density information (Table 29.3)). Thus, TPSF would appear to be potentially important habitat for all the region's gibbons, but particularly for *H. albibarbis*.

Previous studies have found that primate densities are influenced by the quality of their habitat (e.g. Hamard *et al.* 2010; Marshall 2010). Wildfires and logging were found to negatively affect the abundance of orangutans (Felton *et al.* 2003) and to result in lower densities of gibbons, principally due to a decrease in food availability (Johns 1987, 1988; Johns & Skorupa 1987; O'Brien *et al.* 2003, 2004). Gibbons were, however, able to persist in disturbed forests due to their dietary flexibility (Cheyne 2010; Harrison *et al.* 2005), but their reproductive potential was lowered by the shift towards folivory (O'Brien *et al.* 2004). Furthermore, the density of the Southern Bornean gibbon (*Hylobates albibarbis*) was found to be negatively correlated with elevation in the Gunung Palung National Park, West Kalimantan, corresponding to fewer large trees and a lower availability of gibbon food items at higher altitude (Marshall & Leighton 2006; Marshall 2009).

There is a severe lack of data on gibbons in non-protected areas or small forest areas which may also contain viable populations. Current data suggest the population of *H. albibarbis* to be between 75 000 and 130 000 individuals throughout its range of which at least 50% are found in TPSF. Populations of *H. mulleri, H. abbottii* and *H. funerus* combined are estimated at between 270 000–330 000 individuals with less than 25% found in TPSF (S.M. Cheyne, unpublished data), though there are limited estimates for TPSF locations for these species. In the absence of a full population and habitat viability analysis (PHVA), all survey data are vitally important to obtain accurate population estimates of all four species.

Mentawai Species

The Mentawai Islands off the western coast of Sumatra are home to six endemic primate species, of which at least four species inhabit TPSF (Table 29.4). Data presented here come from surveys on Siberut, the northernmost of the archipelago's four islands, where an area in the north was surveyed, representing *c.* 6% (12.5 km^2) of the island's swamp forest. Very little information is available for the species on the three southern islands (Sipora, North and South Pagai): *P. potenziani* and *M. pagensis*

Table 29.4 Showing density estimates for the endemic Mentawai primates (Quinten *et al.* 2010).

Region	Site	Species	Individuals/km²
Mentawai Islands	Peleonan Forest	*Hylobates klossii*	1.0 (95% CI 0.3–2.8)
Mentawai Islands	Peleonan Forest	*Macaca siberu*	35.8 km (95% CI 25.5–50.4)
Mentawai Islands	Peleonan Forest	*Simias concolor* (ssp. *siberu*)	65.5 (95% CI 41.9–102.6)
Mentawai Islands	Peleonan Forest	*Presbytis siberu* (formerly: *potenziani* ssp. *siberu*)	2.7 (95% CI 1.3–5.3)

Table 29.5 List of density estimates from TPSF locations for red langurs listed according to increasing density estimates.

Region	Site	Individuals/km²	Sources
Borneo	Gunung Palung	2.52	Marshall 2010
Borneo	Sabangau	17.29 (± 10.70–28.13)	Ehlers-Smith *et al.* 2013
Borneo	Sungai Wain Protection Forest	27	Nijman & Nekaris 2012

have both been reported to use swamp forests (Fuentes 1996; Sangchantr 2004; Wilson & Wilson 1976), but there are no accounts or density estimates for peat swamp habitat.

The density estimate for *S. concolor* does not differ significantly from that obtained previously for lowland rain forest on mineral soil (Waltert *et al.* 2008), whereas that of *M. siberu* in TPSF was nearly twice as high. Densities of *P. potenziani* and *H. klossii* were approximately one-third and one-tenth, respectively, of those of the same species in the adjacent mineral soil rainforest (Quinten *et al.* 2010). Collectively, these data suggest that TPSF is an important habitat for primates (particularly *M. siberu* and *S. concolor*) in northern Siberut, and possibly more generally, throughout the Mentawai Islands.

Red Langurs

Red langurs (*Presbytis rubicunda*) in TPSF range from 2.25 to 19.74 individuals/km² whereas the range for all red langurs is 1.24 to 21.56 individuals/km² (based on a survey of 12 published studies of which only 3 reported density information from TPSF (Table 29.5).

Previous surveys indicate *P. rubicunda* occurs at a relatively high density in the pristine lowland dipterocarp forests of Lanjak Entimau, Sarawak (Blouch 1997) and at intermediate levels in those at Sepilok, Sabah (Davies 1984), but are lower in Barito Ulu, Central Kalimantan (McConkey & Chivers 2004) and the Gunung Palung ecosystem in West Kalimantan (Marshall 2010). Swamp forests and montane forests above 750 m are shown to support the lowest densities in Gunung Palung (Marshall 2010), suggesting that populations of *P. rubicunda* are not likely to be viable in the large tracts of montane forests remaining (Marshall 2010). Surveys in Sabangau Forest found no red langurs in the low interior forest (LIF) habitat and also found a very low stem density of large, preferred fruit bearing trees in the LIF in comparison to the Mixed Swamp Forest (Ehlers-Smith *et al.* 2013), which supports the hypothesis already established in Gunung Palung that red langur populations are limited by the availability of preferred, not fallback foods (Marshall 2010).

Table 29.6 Density estimates of proboscis monkeys.

Site	Habitat	Individuals/km²	Sources
Balikpapan Bay	Mangrove	2	Scott 2012
Kinabatangan	Riverine	4.1–7.9	Boonratana 2000b, Goossens *et al.* 2002, Sha *et al.* 2008
Sangata River	Riverine	4	Bismark 2010
Tanjung Puting	TPSF	2	Bismark 2010

Proboscis Monkeys

The Bornean endemic proboscis monkey (*Nasalis larvatus*) is highly reliant on the quality of the wetland environments, especially mangrove and riparian forest (Bismark 2010) and, as a result, is relatively intolerant of any form of habitat disruption (Stark *et al.* 2012). The major cause of the recent decline in the number of proboscis monkeys has undoubtedly been the widespread habitat destruction over much of its geographic range (Scott 2012; Stark *et al.* 2012).

Proboscis monkeys are more widely distributed than has been thought previously; they occur throughout Kalimantan, the Indonesian part of Borneo, from the coastal areas to the headwaters of probably all major rivers (e.g. Cheyne *et al.* 2013b; Meijaard & Nijman 2000a). For the continued survival of the species the populations in Sabah, Sarawak and Brunei are still of great significance (Stark *et al.* 2012), as they are considerably larger than those in Kalimantan, though the large areas of Kalimantan remain unsurveyed. The available data (Table 29.6) suggest that the proboscis monkey occurs at similar densities in TPSF and mangrove, whereas densities are at least two times higher in riverine habitats; there is a significant lack of up-to-date data from TPSF.

Nocturnal Species

The cryptic nature of nocturnal primates means that density data are scarce for all habitats. Estimates for *N. coucang* range

Table 29.7 Data from TPSF for the two species of slow loris and tarsiers.

Site	Species	Encounter rate Individuals/km	Source
Perak, Malaysia	*N. coucang*	0.4–1.0	Nekaris *et al.* 2008
Sabangau, Indonesia	*N. menagensis*	0.19–0.33	Nekaris *et al.* 2008
Sabangau, Indonesia	*T. bancanus*	0.003	Blackham 2005

from 0.4 to 1.63 individuals/km^2 based on eight sites and for *N. menagensis* from 0.01–0.36 individuals encountered/km based on three sites (Table 29.7).

Slow lorises have historically been considered Data Deficient or Least Concern (IUCN) and to be common throughout their range, but at present all species are considered globally threatened (Nekaris *et al.* 2008). The lack of data demonstrated above and the limitations of collecting accurate data on loris (and tarsier) density emphasize the urgent need from more information on these species in the wild.

Primate Behavioural Ecology in TPSF

Feeding Ecology and Fallback Foods

Fallback foods (FBFs) are 'foods whose use is negatively correlated with the availability of preferred foods' (Marshall & Wrangham 2007: p. 1220). Due to the detailed nature of feeding ecology, data are generally only available for habituated primates and we present data on orangutans, gibbons, siamangs and red langurs. Bornean orangutans generally eat less fruit and insects, and more bark and leaves, than their Sumatran counterparts (Fox *et al.* 2004; Morrogh-Bernard *et al.* 2010). The data suggest that orangutans consume bark and, to a lesser extent, leaves as filler FBFs in TPSF in Borneo, with greater importance of figs at at least some Sumatran sites (Harrison 2009; Knott 1998; Leighton 1993).

Gibbon diets are similar to those of orangutans, with a high degree of overlap (Cheyne *et al.* 2005), except that gibbons do not feed on bark or pith (Cheyne 2010; Chivers 2001; Leighton 1987; Marshall 2004; Vogel *et al.* 2009). Gibbons specialize on ripe, non-fig fruit and fall back on flowers and young leaves when preferred foods are unavailable (Cheyne 2010; Chivers 2001; Leighton 1987; Marshall 2004; Vogel *et al.* 2009). Leaf consumption in TPSF habitats varies between sites, and has been recorded as 3% in Gunung Palung, Borneo (Marshall 2004), 4% in Ketambe, Sumatra (Palombit 1997) (Palombit 1997), and up to 25% in Sabangau (Cheyne 2010), Figs are generally more commonly eaten (Chivers 2001), comprising, e.g. 6% of the diet in Sabangau, Borneo (Cheyne 2010), 23% in Gunung Palung (Marshall 2004) and 45% in Ketambe (Ungar 1995, Palombit 1997). The wide variation in fig consumption suggests that figs are not a consistent FBF, and the prevalence of figs in the diet is habitat related. In summary, gibbons use both figs and leaves as filler FBFs in dipterocarp forest, but leaves and flowers in TPSF (Cheyne 2010; Cheyne *et al.* 2005).

Red langur feeding ecology in TPSF has only recently been investigated. The first detailed study in Sabangau (Ehlers-Smith *et al.* 2013) concluded that there was a strong preference for seeds, and therefore fruit, as relative fruit consumption consistently exceeded relative availability for 12 months of the year. Leaves and flowers were not preferred as relative consumption was less than relative availability for 12 and 11 months of the year, respectively. There were no significant negative correlations between the availability of the preferred food, i.e. fruit, and consumption of any other food class (i.e. leaves, flowers, pith, invertebrates, fungi), thus the langurs did not rely on any fallback foods. This is most likely due to the relatively constant availability of ripe and unripe fruit as a higher-quality food class year round (Ehlers-Smith *et al.* 2013).

Ranging

An animal's home range is the area that it uses in its normal movements. It usually visits all of its range at least once a year, although in some long-lived animals with longer-term fluctuations in food availability and distribution, all parts of the range may not be covered every year (Jolly 1985). If part of the home range is advertized and defended against intruders and thus used more-or-less exclusively, this part is often referred to as the territory, the definition used here.

Gibbon home ranges in TPSF vary from 0.07 to 0.53 km^2 based on nine available data points. Orangutan female home ranges vary from 0.4 to 8.5 km^2 based on 12 sites with ranges for TPSF between 1.5 and 8.5 km^2 (Singleton *et al.* 2009). Very few studies have managed to obtain reliable estimates of male home range sizes, but Singleton and van Schaik (2001) suggest minimum home range size estimates of ~25 km^2 for non-dominant flanged and most unflanged males in a Sumatran TPSF. The estimates for flanged males range from 0.6 to 25 km^2 for all sites with the TPSF range of 5.6–25 km^2 based on seven sites. The estimates for unflanged males range from 1.75 to 25 km^2 for all sites with the TPSF range of 2.6–25 km^2 based on four sites. Red langur home ranges vary from 0.2 to 1.08 km^2 based on one study site (Ehlers-Smith *et al.* 2013). Proboscis monkeys maintain relatively large home ranges (> 7.7 km^2) and relatively large group sizes in mangroves where > 55% of the diet comprises fruit (Bennett & Sebastian 1988), but ranges are smaller in TPSF (1.3–2.2 km^2) where fruit consumption is less (11–40%; Boonratana 2000b).

Ecological heterogeneity is believed to be a factor determining home range size with food availability influencing the size of territory that can be actively patrolled and defended (Marshall 2004). The availability of mates and competition from conspecifics will also play a role in influencing home range size.

Adaptations to Smoke Caused by Fires in TPSF

Gibbons are characterized by their species-specific vocalizations and both frequency and duration of singing are known to be affected by environmental factors, such as rainfall and smoke from forest fires (Cheyne 2007; Cheyne *et al.* 2007; Hamard *et al.* 2010). For example, periods of severe smoke coverage in

Sabangau, were associated with a significant reduction in the number of days spent singing and duration of singing bouts by *H. albibarbis* when compared to non-smoke periods (Cheyne 2007). Behavioural effects of such smoke-related changes in singing patterns in gibbons are hard to predict without more data, but reduced singing for several months a year (when singing is normally at a peak; Cheyne 2010), could be detrimental for territorial spacing/defence, communication and, ultimately, reproduction.

Predators

Possible predators of primates are present in TPSF and even the large orangutan is susceptible to opportunistic predation. Diurnal predators include raptors, e.g. crested serpent eagles (*Spilornis cheela*, suspected), changeable hawk-eagle (*Spizaetus cirrhatus*, confirmed; Fam & Nijman 2011) and snakes, e.g. reticulated pythons (*Python reticulata*, confirmed; Clarke *et al.* 2006; Uhde & Sommer 2002). Confirmed nocturnal predators include felids, e.g. clouded leopards (*Neofelis diardi*) (Morino 2010; Phoonjampa *et al.* 2010; Uhde & Sommer 2002; Van Schaik 1983) and possibly tigers (*Panthera tigris* ssp. *sumatrae*). Gibbons in Sabangau showed a significant preference for trees without lianas. In Siberut, this has been shown to be to avoid predators, humans and biting ants (Whitten 1982b). Gibbons appear to select sleeping trees with reduced predator access, i.e. tall crowns, less accessible crowns (in terms of surrounding tree canopy) and no lianas (Cheyne *et al.* 2013b).

Summary of Adaptations to TPSF

The case studies presented above highlight the behavioural and dietary flexibility of primates living in TPSF. The habitat characteristics and food availability are unique, as are the human-induced disturbances in the form of fire. The significant differences in population density in TPSF demonstrate that some concessions must be made by primates in this habitat, lower densities, larger home ranges and inferred longer daily path length, and thus increased energy expenditure. The degree to which environment/ecological conditions affect social behaviour/organization is still a subject of further study and no conclusions can be drawn at the present time.

Conservation/Conclusions

Peatlands play a special role in maintaining biodiversity at the species and genetic level as a result of habitat isolation and at the ecosystem level as a result of their ability to self-organize and adapt to different physical conditions. Science is only just beginning to understand the complex and unique differences in behavioural ecology of the wildlife in these habitats.

From Table 29.1, we know about 70% of the regions primates are found in TPSF, so in terms of number of species, TPSF are important. For orangutans, we can estimate that 40%, for *H. albibarbis* (50%) and the *H. muelleri*, *H. funerus* and *H. abbotti* gibbons (20–25%) (Chivers *et al.* 2013) of remaining populations are found in TPSF. So for these ape species at least, TPSFs are already home to very substantial proportions of the total populations. For other species, we do not know, but given the rate of loss of other tropical rainforest habitat, we can reasonably assume that TPSF already contain significant proportions of the remaining populations of most, if not all, of the 23 species listed. Further, given that TPSF is still generally less accessible that other rainforest, especially areas far from rivers, they may become increasingly important as refuges for the more threatened species as their overall ranges decline and populations become more fragmented.

Of particular concern is the impact of habitat fragmentation on primates, especially the aboreal species, since it generally impedes dispersal and can limit gene flow in the same way as natural barriers (Goossens *et al.* 2005), although some fragments are still capable of supporting substantial populations of primates (Cattau *et al.* 2015). Recent surveys indicate that there is a minimum patch size required for supporting a viable population of orangutans (> 3.5 km^2; Cattau *et al.* 2015) but no such estimates exist for other primates in TPSF. More information is needed regarding the response of primates to other patch characteristics, such as patch shape, core–area ratio or distance to another fragment/forest. Additionally, little is known about how primates respond to a non-forest matrix. If individuals can traverse grassland, shrubland or burned area, then small forest patches might supplement the habitat area provided by larger patches. Arboreal primates such as gibbons, orangutans and red languers have been observed to travel on the ground for over 100 m (S.M. Cheyne, pers. obs.) Additionally, the small patches could promote the movement of individuals between patches by serving as stepping stones through a harsh matrix.

Logging (legal) is often not adequately policed in terms of which species and sizes of trees are extracted. Uncontrolled illegal logging in protected (and unprotected areas) is extremely hard to monitor. Both types of logging often target tree species of importance in the diets of the primates including both keystone and fallback species (Cheyne 2008, 2010; Cheyne *et al.* 2005; Harrison *et al.* 2010b; Marshall 2010).

A series of recommendations has been developed from research activities and workshops which will have direct and indirect impacts on the long-term conservation of primates in TPSF. Specifically the issue of increasing fragmentation needs to be addressed through stricter control of the issuing of permits for plantations (oil palm and acacia in particular), which should not be given for areas where there is standing forest, rather only for areas which are already deforested. Drainage is also an important issue, since drainage of the peat through the creation of canals results in a drop in the water table, oxidization of the peat and loss of cohesiveness. This exacerbates the fires caused by drainage which destroy TPSF habitat and create palls of smoke that can last for several months and are detrimental to primate health (and to humans). Restoration of the hydrological balance is crucial through damming of canals and the prevention of new canal digging. The issue of land-use planning needs further attention to determine clear boundaries between protected areas and districts agreed between local government and the forestry management. Development and implementation of a control system by the Department of Forestry and local government for logging concession and plantation companies is crucial to halt increasing fragmentation. Additional recommendations on activities to ensure the continued survival of TPSF and the primates which inhabit it are presented in Campbell *et al.* (2008), Cheyne *et al.* (2012) and Harrison *et al.* (2012).

Only with sufficient information on feeding behaviour (energy intake and food selection), ecology and the effects of habitat degradation on primate density and population numbers, combined with reforestation efforts and increased protection can the TPSF habitat be managed to ensure the long-term survival of the primates and this multifaceted ecosystem.

Chapter 30

Primates of Africa's Coastal Deltas and Their Conservation

Thomas M. Butynski and Yvonne A. de Jong

Introduction

There is little information on the non-human primates (hereafter 'primates') of Africa's deltas, or on the importance of these deltas to the conservation of primate diversity on the continent. This chapter is concerned with the conservation of Africa's ten largest coastal deltas and their importance to the maintenance of primate diversity. This chapter also draws attention to (1) the need for much more research on the distribution, abundance and conservation status of the primates that inhabit Africa's large coastal deltas, and (2) the fact that the biological values of most of these large coastal deltas are being rapidly degraded and are in dire need of targeted conservation actions.

Deltas

Definition

A 'delta' is a tract of alluvial land, often more-or-less triangular in shape, enclosed or traversed by the diverging mouth of a river (Shorter Oxford English Dictionary 2007). Big rivers that flow over large expanses of ground where there is little altitudinal gradient form complex divergent drainage systems at their mouth. This causes the water flow to slow and sediments to be deposited into expanses of wetlands and shallow water, creating a delta. Deltas may be 'inland', where the river flows into a swamp and/or lake, or 'coastal', where the river meets the sea.

Africa's Inland Deltas

Large, inland bodies of water in Africa, such as the Okavango (Botswana; *c.* 16 000 km^2), Sudd (South Sudan; *c.* 30 000 km^2), and Inner Niger (Mali; *c.* 38 000 km^2) all display large triangular (or fan-shaped) geological features that are the result of deposits from their primary rivers and, therefore, can be referred to as 'deltas'. In the case of the Sudd and Inner Niger, their deltas are not immediately apparent as each occurs within a vast area of swamp, marsh and floodplain. The Okavango, Sudd and Inner Niger are Africa's three largest inland deltas. All, however, rank low in terms of their importance to the conservation of Africa's primate diversity as none supports more than four primate species, not one of which is globally threatened (i.e. all are 'Least Concern'; IUCN 2017). The reasons for the relatively low primate species richness in these inland deltas are, no doubt, related to the fact that all lie well outside Africa's Rainforest Biotic Zone (Happold & Lock 2013). Although inland deltas are important for the regional and national primate diversity they hold (for example, the Okavango is a stronghold for all three of Botswana's species of primate), no inland delta is vital to the maintenance of the continent's primate diversity. As such, this review is confined to coastal deltas.

Africa's Coastal Deltas

Coastal deltas are comprised of a complex mixture of marine, freshwater and terrestrial environments, as well as gradients and transition zones among these environments. This complexity is enhanced by daily tides, storm tides, seasonal influxes of freshwater, nutrients from rivers, and other events. Coastal deltas are, therefore, extremely variable and dynamic, both in space and time, and in terms of their structures, habitats and biotas. Coastal deltas support habitat mosaics that may include lagoons, estuaries, lakes, rivers, marshes, swamps, swamp forests, mangroves, floodplains, *terra firma* forests, woodlands, savannas, bushlands, dunes, beaches and other habitats. These yield a dense array of ecotones that further add to the complexity of coastal deltas. Not surprisingly, coastal deltas often support a relatively high biological richness and typically have considerable conservation value (Hughes & Hughes 1992; Ramsar 2015).

Approach and Methods

The primate taxonomy followed in this chapter is that presented in Butynski *et al.* (2013), with the following exceptions: (1) three species of potto *Perodicticus* are recognized (Oates 2011; Stump 2005); (2) the genus *Piliocolobus* is recognized (Groves 2001, 2007); and (3) the Niger Delta Red colobus *Piliocolobus epieni* is elevated from a subspecies (Groves 2007b; Oates 2011; Oates & Werre 2009). The information presented derives from a detailed review of the literature, extensive correspondence with colleagues, and our own work in the Tana Delta, Kenya, and along the Lower Rufiji River, near the inland apex of the Rufiji Delta, Tanzania.

Surprisingly, no authoritative list or database of Africa's coastal deltas exists. The several sources found (e.g. Hughes & Hughes 1992; ProtectedPlanet 2015; Ramsar 2015; World Delta Database 2015) were either incomplete and/or held errors. This study, therefore, reverted to a review of all of Africa's larger rivers to assess whether they support one of the continent's

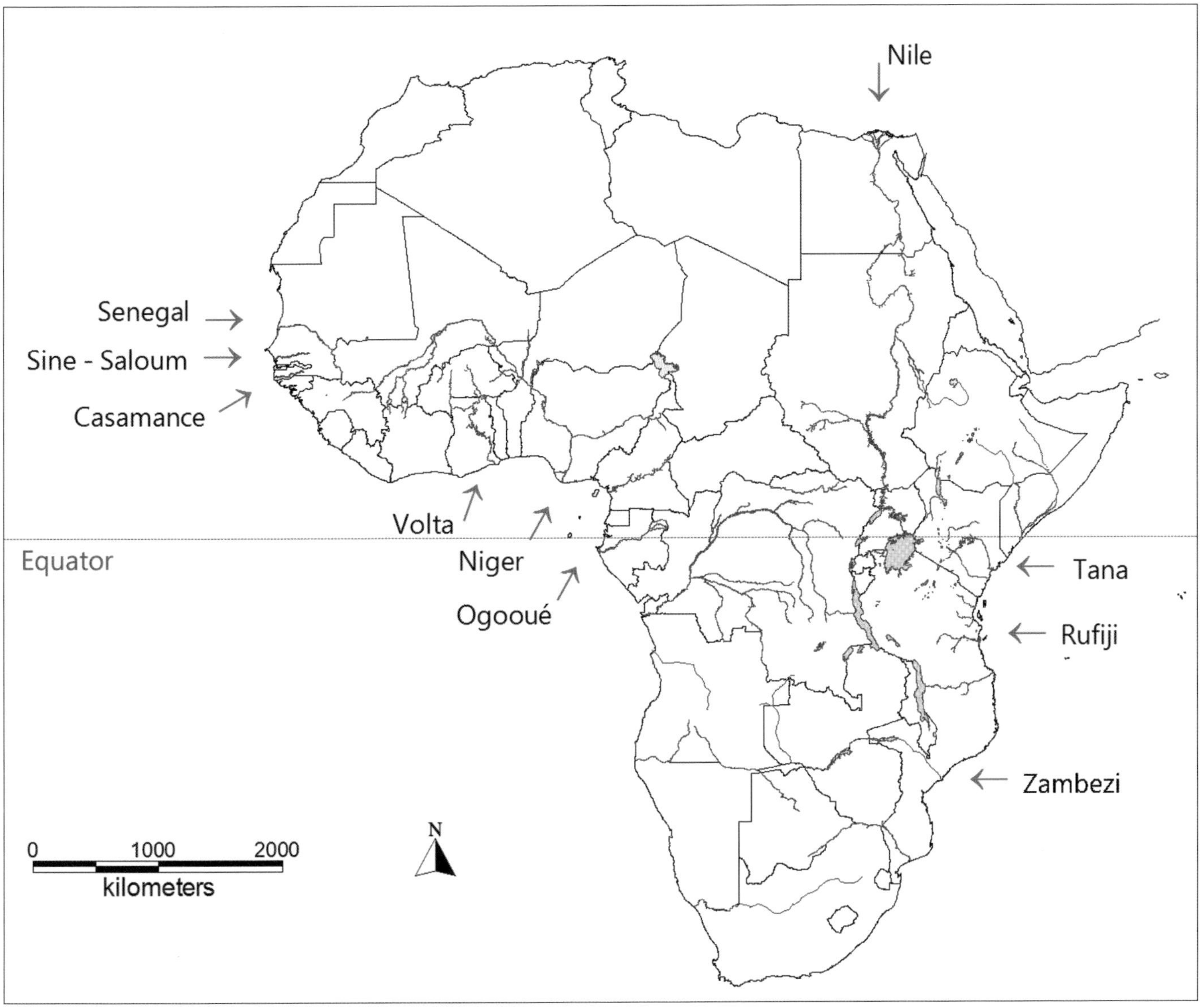

Figure 30.1 Africa's ten largest coastal deltas. Map by Yvonne de Jong & Tom Butynski. (A black and white version of this figure appears in some formats. For the colour version, please refer to the plate section.)

ten largest deltas (i.e. deltas > 700 km^2). From an initial set of purported coastal deltas for Africa, the number was reduced to the ten largest (Figure 30.1; Table 30.1). The following 24 large rivers (listed clockwise around the coast of Africa, beginning with Somalia) either do not terminate in a delta or, if they do, their delta is < 700 km^2: Shabeelle-Jubba, Sabaki, Pangani, Wami, Ruvu, Rukwa, Limpopo, Maputo, Pungwe (Pungue), Mkhuze, Orange, Cunene, Cubango, Congo (= Zaire), Campo, Mbini (= San Benito), Muni, Sanaga, Cross, Ouémé, Sassandra, Gambia, Bandama, and Comoe.

With the exception of the Tana Delta and the Sine-Saloum Delta, authoritative lists for the primate taxa present in each of the largest deltas were not found. When lists were available, they had mistakes and/or omissions. In particular, lists for the strepsirrhine fauna were either not available or incomplete/incorrect. The lists presented in Table 30.2 are, therefore, derived through a detailed review of the literature and extensive consultation with colleagues. Despite these efforts, a number of questions remain as to which taxa occur in several of the deltas. These questions cannot be resolved without additional field work.

The ten largest coastal deltas were ranked according to their importance for the conservation of Africa's primate diversity (Table 30.3). The conservation rating system used followed the approach of Oates (1996), which is based on 'degree of threat' and 'taxonomic distinctiveness'. Unlike Oates (1996), however, the rating system applied here uses the degree of threat category presented in the current *IUCN Red List of Threatened Species* (IUCN 2017). Species were given a numerical point value based on their current degree of threat status as follows: 'Least Concern' or 'Near Threatened' = 1 point; 'Vulnerable' = 2; 'Endangered' = 3; 'Critically Endangered' = 4. Taxonomic distinctiveness was assessed on a 2-point scale. A species received 2 points if it has no more than one close relative (i.e. a member of the same species-group, subgenus, or genus). All other species received 1 point.

Table 30.1 Overview of Africa's 10 largest coastal deltas. Main sources: Hughes & Hughes (1992), ProtectedPlanet (2015), Ramsar (2015), and World Delta Database (2015).

Delta (km^2)	Main supporting river(s)	Country	Non-Ramsar protected areas that include at least part of the delta or are adjacent to the delta (km^2)	Ramsar Sites that include at least part of the delta, year of listing, and size (km^2); Montreux Record listing also noted
Nile (22 000)	Nile	Egypt	None	Lake Burullus: 1988 (462). Placed on Montreux Record in 1990.
Tana (1636)	Tana	Kenya	None. Tana River Primate National Reserve *c.* 40 km up-river of delta	Tana River Delta: 2012 (1636)
Rufiji (720)	Rufiji	Tanzania	Mangrove-Rufiji Forest Reserves	Rufiji-Mafia-Kilwa: 2004 (5970)
Zambezi (12 000)	Zambezi	Mozambique	Marromeu Complex Game Management Area	Marromeu Complex: 2004 (6880)
Ogooué (5500)	Ogooué	Gabon	None Proposed: Ogooué National Park (centred on lakes Onangué, Oguémoué, Evaro, Ezanga)	Bas Ogooué: 2009 (8627). Does not include the marine delta or mangroves. Includes a long stretch of the Ogooué R. and part of Abanga R. (H. Rainey, pers. comm.; J.-P. Vande Weghe, pers. comm.).
Niger (46 420)	Niger	Nigeria	Upper Orashi Forest Reserve Lower Orashi Forest Reserve Edumanom Forest Reserve Apoi Creek Forest Reserve Ikebiri Creek Forest Reserve Egbedi Creek Forest Reserve Taylor Creek Forest Reserve Nun River Forest Reserve Osomari Forest Reserve Olague Forest Reserve Kwale Forest Reserve	Upper Orashi Forests: 2008 (252) Oguta Lake: 2008 (6) Apoi Creek: 2008 (292)
Volta (1925)	Volta	Ghana	Songor UNESCO-MAB Biosphere Reserve	Anlo-Keta Lagoon Complex: 1992 (1278) Songor Lagoon: 1992 (287)
Casamance (2500)	Casamance	Senegal	National Park de Basse Casamance Forêt de Boukitingo Forest Reserve Forêt d'Oukout Forest Reserve Forêt de Diantene Forest Reserve Forêt Classée des Narangs Forêt Classée des Kalounayes Forêt Classée des Bayottes	None
Sine-Saloum (1800)	Sine, Saloum	Senegal, Gambia	Saloum Delta National Park and Biosphere Reserve (Senegal) Niumi National Park (Gambia)	Delta du Saloum, Senegal: 1984 (730)
Senegal (4254)	Senegal	Senegal, Mauritania	Réserve spéciale de faune de Guembeul (Senegal) Forêt de Leybar Forest Reserve (Senegal) Forêt Classée de Maka Diama (Senegal) Langue de Barbarie National Park (Senegal) Diawling National Park (Mauritania) Delta du Fleuve Sénégal UNESCO-MAB Biosphere Reserve (Senegal and Mauritania)	Gueumbeul, Senegal: 1986 (7) Bassin du Ndiaël, Senegal: 1977 (100). Placed on Montreux Record in 1990. Djoudj, Senegal: 1977 (160). Place on Montreux Record in 1993 and removed in 2009. Chat Tboul, Mauritania: 2000 (155) Parc National du Diawling, Mauritania: 1994 (156) Placed on Montreux Record in 2002 and removed in 2009.

Results

In Africa, the diversity of primates, as with many other taxonomic groups, is inversely related to latitude (Cowlishaw & Hacker 1997; Harcourt 2006), with the Guineo-Congolian Lowland Rain Forest Zone (White 1983) supporting the highest diversity. This is followed by the tropical montane forests and then by the moist forests of coastal East Africa (Barnes 1992). As expected, primate diversity in Africa's coastal deltas varies greatly (from 0 to 17 taxa), generally declining with increasing

Table 30.2 Primate taxa present in or near Africa's 10 largest coastal deltas. Main sources: Oates (2011) and Butynski *et al.* (2013).

Delta	Primate taxa in and/or near the delta and category of threat (IUCN 2017)	Ecoregion (Olson *et al.* 2001)	Vegetation
Nile	No non-human primates	Nile Delta Flooded Savanna	Grassland, swamp, marsh (Baha El Din 1999).
Tana	Tana River red colobus *Piliocolobus rufomitratus rufomitratus* (Endangered) Tana River mangabey *Cercocebus galeritus* (Endangered) Northern yellow baboon *Papio cynocephalus ibeanus* (Least Concern) Hilgert's vervet *Chlorocebus pygerythrus hilgerti* (Least Concern) Pousargues's monkey *Cercopithecus mitis albotorquatus* (Vulnerable) White-tailed small-eared galago *Otolemur garnettii lasiotis* (Least Concern) Kenya lesser galago *Galago senegalensis braccatus* (Least Concern) Kenya coast dwarf galago *Galagoides cocos* (Least Concern)	Northern Zanzibar-Inhambane Coastal Forest Mosaic	Riverine forest, grassland, woodland, acacia bush, coastal forest, beach, dune, swamp, mangrove (De Jong & Butynski 2009; Hamerlynck *et al.* 2012; Robertson & Luke 1993; Tana River Delta 2015).
Rufiji	Peters's Angola colobus *Colobus angolensis palliatus* (Least Concern) Southern yellow baboon *Papio cynocephalus cynocephalus* (Least Concern) Reddish-green vervet *Chlorocebus pygerythrus rufoviridis* (Least Concern) Tanzania Sykes's monkey *Cercopithecus mitis monoides* (Least Concern) Miombo silver galago *Otolemur crassicaudatus monteiri* (Least Concern) Pangani small-eared galago *Otolemur garnettii panganiensis* (Least Concern) Matundu dwarf galago *Galagoides zanzibaricus udzungwensis* (Least Concern) Mozambique dwarf galago *Galagoides granti* (Least Concern)	East African Mangroves	Swamp, sand ridge, grassland, shrub, woodland, forest, mangrove (Ochieng 2002).
Zambezi	Southern vervet *Chlorocebus pygerythrus pygerythrus* (Least Concern) Samango monkey *Cercopithecus mitis erythrarchus* (Least Concern) Miombo silver galago *Otolemur crassicaudatus monteiri* (Least Concern) Mozambique dwarf galago *Galagoides granti* (Least Concern)	East African Mangroves Zambezian Coastal Flooded Savanna	Dry forest, *Acacia-Borassus-Combretum* woodland, riverine forest mosaic, acacia thicket and savanna, doum palm savanna, wetland pan, grassland, mangrove, swamp forest, papyrus swamp, cultivation, dune (Beilfuss *et al.* 2001).
Ogooué	Western lowland gorilla *Gorilla gorilla gorilla* (Critically Endangered) Central chimpanzee *Pan troglodytes troglodytes* (Endangered) Gabon black colobus *Colobus satanas anthracinus* (Vulnerable) Red-capped mangabey *Cercocebus torquatus* (Vulnerable) Mandrill *Mandrillus sphinx* (Vulnerable) Grey-cheeked mangabey *Lophocebus albigena* (Least Concern) Northern talapoin monkey *Miopithecus ogouensis* (Least Concern) Black-footed crowned monkey *Cercopithecus pogonias nigripes* (Least Concern)	Central African Mangroves Atlantic Equatorial Coastal Forests	Sand-dune, sandbank, alluvial plain, mangrove, mudflat, grassland, swamp, papyrus, flooded forest, forest, savanna, coastal *Dalbergia* thicket (Birdlife International 2015; Latour 2005; Ramsar 2015; J.-P. Vande Weghe, pers. comm.).

(continued)

Table 30.2 *(cont.)*

Delta	Primate taxa in and/or near the delta and category of threat (IUCN 2017)	Ecoregion (Olson *et al.* 2001)	Vegetation
	Eastern putty-nosed monkey *Cercopithecus nictitans nictitans* (Least Concern) Red-tailed moustached monkey *Cercopithecus cephus cephus* (Least Concern) Grey-tailed moustached monkey *Cercopithecus cephus cephodes* (Least Concern) Milne-Edwards's potto *Perodicticus edwardsi* (Least Concern) Golden angwantibo *Arctocebus aureus* (Least Concern) Gabon squirrel galago *Sciurocheirus gabonensis* (Least Concern) Southern needle-clawed galago *Euoticus elegantulus* (Least Concern) Demidoff's dwarf galago *Galagoides demidovii* (Least Concern)[a] Thomas's dwarf galago *Galagoides thomasi* (Least Concern)[a]		
Niger	Nigeria-Cameroon chimpanzee *Pan troglodytes ellioti* (Endangered) Olive colobus *Procolobus verus* (Near Threatened) Niger Delta red colobus *Piliocolobus epieni* (Critically Endangered) Red-capped mangabey *Cercocebus torquatus* (Vulnerable) Olive baboon *Papio anubis* (Least Concern) Common tantalus monkey *Chlorocebus tantalus tantalus* (Least Concern) Mona monkey *Cercopithecus mona* (Least Concern) Martin's putty-nosed monkey *Cercopithecus nictitans martini* (Least Concern) Sclater's monkey *Cercopithecus sclateri* (Vulnerable) White-throated monkey *Cercopithecus erythrogaster pococki* (Vulnerable) Milne-Edwards's potto *Perodicticus edwardsi* (Least Concern) Benin potto *Perodicticus potto juju* (Least Concern) Calabar angwantibo *Arctocebus calabarensis* (Least Concern) Cross River squirrel galago *Sciurocheirus alleni cameronensis* (Least Concern) Nigeria needle-clawed galago *Euoticus pallidus talboti* (Least Concern) Demidoff's dwarf galago *Galagoides demidovii* (Least Concern)[a] Thomas's dwarf galago *Galagoides thomasi* (Least Concern)[a]	Niger Delta Swamp Forest Central African Mangroves	Mangrove, freshwater swamp, swamp forest, freshwater forest, lowland forest, beach (Baker 2005; Bocian 1999; Ikemeh 2014a,b, 2015; Ikemeh & Oates 2017; Luiselli *et al.* 2015; Petrozzi *et al.* 2015; Powell 1993; UNDP 2011; Werre 2000, 2001a,b; Werre & Powell 1997; Chapter 40).
Volta	Olive baboon *Papio anubis* (Least Concern) Common tantalus monkey *Chlorocebus tantalus tantalus* (Least Concern) Mona monkey *Cercopithecus mona* (Least Concern) Thomas's dwarf galago *Galagoides thomasi* (Least Concern)[a]	Central African Mangroves	Savanna, grassland, mudflat, mangrove, riverine forest (Olson *et al.* 2001; Ramsar 2015).
Casamance	King colobus *Colobus polykomos* (Vulnerable) (extirpated) Temminck's red colobus *Piliocolobus badius temminckii* (Endangered) Sooty mangabey *Cercocebus atys* (Near Threatened) (extirpated) Guinea baboon *Papio papio* (Near Threatened) Western patas monkey *Erythrocebus patas patas* (Least Concern)	Guinean Mangroves Guinean Forest-savanna Mosaic	Mangrove, gallery forest, swamp forest, savanna woodland, palm-pandan swamp, reed swamp, grassland, forest (Birdlife International 2015; ProtectedPlanet 2015; Ramsar 2015).

Table 30.2 *(cont.)*

Delta	Primate taxa in and/or near the delta and category of threat (IUCN 2017)	Ecoregion (Olson *et al.* 2001)	Vegetation
	Green monkey *Chlorocebus sabaeus* (Least Concern) Campbell's monkey *Cercopithecus campbelli* (Least Concern) Senegal lesser galago *Galago senegalensis senegalensis* (Least Concern) Thomas's dwarf galago *Galagoides thomasi* (Least Concern)[a]		
Sine- Saloum	Temminck's red colobus *Piliocolobus badius temminckii* (Endangered) Guinea baboon *Papio papio* (Near Threatened) (introduced species) Western patas monkey *Erythrocebus patas patas* (Least Concern) Green monkey *Chlorocebus sabaeus* (Least Concern) Senegal lesser galago *Galago senegalensis senegalensis* (Least Concern)	Guinean Mangroves Guinean Forest-savanna Mosaic	Mangrove, dune forest, dry forest, dry woodland, gallery forest, savanna (Galat-Luong & Galat 2005; Ramsar 2015).
Senegal	Western patas monkey *Erythrocebus patas patas* (Least Concern) Senegal lesser galago *Galago senegalensis senegalensis* (Least Concern)	Sahelian Acacia Savanna	Savanna, acacia shrub, beach, salt flat, dune (Ramsar 2015; World Delta Database 2015).

[a] *Galagoides demidovii* and *Galagoides thomasi* are similar species that are broadly sympatric. As such, they have often been confused, both in the field and in the literature. While the best information available has been used to designate these two species in this table, their presence needs to be validated for all four deltas for which they are listed.

Table 30.3 Summary of the significance for primate conservation of Africa's ten largest coastal deltas.

Delta	Primate conservation rating and (rank)	Number of primate taxa	Country endemic primate taxa	Threatened primate taxa (IUCN 2017)*	Main threats to the delta
Nile	0 (10)	0	None	None	Rapid human population growth and unsustainable use of natural resources. Dams; water diversion; habitat loss to agriculture (Baha El Din 1999; Hughes & Hughes 1992).
Tana	22 (3)	8	*Piliocolobus rufomitratus rufomitratus* *Cercocebus galeritus*	*Piliocolobus rufomitratus rufomitratus* (Endangered) *Cercocebus galeritus* (Endangered) *Cercopithecus mitis albotorquatus* (Vulnerable)	Rapid human population growth and unsustainable use of natural resources. Forest degradation and loss due to extraction of forest products, invasive species (especially *Prosopis juliflora* [Leguminosae]), river-edge agriculture and expansion of irrigation; poaching; reduction in flooded area due to five hydroelectric power dams and a sixth planned dam (Hamerlynck *et al.* 2012; Mbora & Butynski 2009; Wieczkowski & Butynski 2013).
Rufiji	18 (4)	8	*Cercopithecus mitis monoides*	None	Rapid human population growth and unsustainable use of natural resources. Harvest of mangrove and other wood products; clearance of mangrove for fish and prawn farming and agriculture; industrial and agricultural pollution; increased siltation; changes in flood hydrographs through hydropower dam development; water extraction (Birdlife International 2015; Burgess *et al.* 2004; Duvail *et al.* 2014; Ochieng 2002).

(continued)

Table 30.3 *(cont.)*

Delta	Primate conservation rating and (rank)	Number of primate taxa	Country endemic primate taxa	Threatened primate taxa (IUCN 2017)*	Main threats to the delta
Zambezi	9 (7)	4	None	None	Rapid human population growth and unsustainable use of natural resources. Disturbed annual flood cycles due to dams (e.g. Kariba Dam, Cahora Bassa Dam); construction of roads, railways, factories and dykes; loss of habitat to agriculture (particularly sugarcane plantation) and settlement; hunting; fishing (Beilfuss *et al.* 2001; Ramsar 2015; World Delta Database 2015; WWF 2015; K. Tinley, pers. comm.).
Ogooué	46 (1)	17	*Cercopithecus cephus cephodes*	*Gorilla gorilla gorilla* (Critically Endangered) *Pan troglodytes troglodytes* (Endangered) *Colobus satanas anthracinus* (Vulnerable) *Cercocebus torquatus* (Vulnerable) *Mandrillus sphinx* (Vulnerable)	Very low human population density (probably < 0.02 people/km^2). Relatively few threats. Some over-hunting (e.g. *C. satanas*, *L. albigena*) (Latour 2005; Ramsar 2015; J.-P. Vande Weghe, pers. comm.).
Niger	45(2)	17	*Piliocolobus epieni* *Cercopithecus sclateri* *Cercopithecus erythrogaster pococki*	*Pan troglodytes ellioti* (Endangered) *Piliocolobus epieni* (Critically Endangered) *Cercocebus torquatus* (Vulnerable) *Cercopithecus sclateri* (Vulnerable) *Cercopithecus erythrogaster pococki* (Vulnerable)	Rapid human population growth and unsustainable use of natural resources. Dams and their impact on the hydrological balance; hunting; deforestation due to agriculture, logging and settlement; invasive species; oil exploration, extraction and pollution (Baker 2005; Blench 2007; Bocian 1999; Ikemeh 2014a,b, 2015; Ikemeh & Oates 2017; James *et al.* 2007; Kadafa 2012; Luiselli *et al.* 2015; Moffat & Lindén 1995; Oates & Werre 2009; Oates *et al.* 2004; Petrozzi *et al.* 2015; Phil-Eze & Okoro 2009; Powell 1997; UNDP 2011; Werre 2001a,b; Werre & Powell 1997; Chapter 40).
Volta	6 (8)	4	None	None	Rapid human population growth and unsustainable use of natural resources. Salt mining; harvesting of mangroves; habitat loss due to extraction of forest products and agriculture; fishing; hunting (Ramsar 2015).
Casamance	Historic: 22 Present: 16 (5)	7 (was 9 but 2 species extirpated)	None	*Colobus polykomos* (Vulnerable) (extirpated) *Piliocolobus badius temminckii* (Endangered)	Rapid human population growth and unsustainable use of natural resources. Harvesting of mangrove; habitat loss to agriculture; dams; greatly increased salinity; hunting (Birdlife International 2015; Blesgraaf *et al.* 2006; Ramsar 2015; A. Galat-Luong & G. Galat, pers. comm.).
Sine-Saloum	13 (6)	5 (1 species introduced)	None	*Piliocolobus badius temminckii* (Endangered)	Rapid human population growth and unsustainable use of natural resources. Habitat loss to agriculture; exploitation of wood products; hunting; fire (Galat-Luong & Galat 2005; Oates 2011; Ramsar 2015).
Senegal	5 (9)	2	None	None	Rapid human population growth and unsustainable use of natural resources. Diama Dam (at river mouth) and Manantali Dam; invasive plants; increasing salinity; habitat loss to agriculture; fishing (Birdlife International 2015; Hamerlynck & Duvail 2003).

distance from the equator (Table 30.3): e.g. Ogooué Delta (*c.* 100 km south of equator; 17 primate taxa), Niger (*c.* 500 km north of equator; 17 taxa), Rufiji (*c.* 900 km south of equator; 8 taxa), Zambezi (*c.* 2100 km south of equator; 4 taxa), and Nile (*c.* 3500 km north of equator; 0 taxa).

Africa's coastal deltas provide habitat for a wide range of primates, from tiny (< 70 g), arboreal, nocturnal and omnivorous primates (e.g. Demidoff's dwarf galago *Galagoides demidovii*), to one of the world's largest primates (> 190 kg), the semi-terrestrial, diurnal and herbivorous western gorilla *Gorilla gorilla*. No fewer than 57 primate taxa (47 species) occur (or did occur) in Africa's ten largest coastal deltas (Table 30.2). Of Africa's 95 primate species, 49% are (or were) represented in the ten largest coastal deltas. Of Africa's 25 primate genera, 20 (80%) are present in at least one of these ten deltas. None of these deltas supports the same array of primate taxa. Four deltas (Tana, Rufiji, Ogooué and Niger) support one to three (seven total) 'nationally endemic' primate taxa. Some of these seven taxa are endemic to the delta, or to the delta and the river from which the delta derives (Table 30.2). Five deltas support at least one threatened primate species (Table 30.3). The Ogooué Delta holds the largest number of threatened primate species at five. A total of 12 threatened primate taxa occur in the ten largest deltas. The number was 13, but king colobus *Colobus polykomos* was extirpated from the Casamance Delta, Senegal, prior to 1988 (A. Galat-Luong & G. Galat, pers. comm.).

Coastal deltas and their supporting rivers may serve as absolute or partial barriers to the movement of some primate taxa (Booth 1958; Butynski *et al.* 2013; Grubb 1982; Harcourt & Wood 2012; Oates 2011; Oates *et al.* 2004). For example:

- The Rufiji River and Delta (Tanzania) appear to be the southern limit for Peters's Angola colobus *Colobus angolensis palliatus* (Figure 30.2; Rodgers 1981; W. Jubber, A. Perkin & O. Hamerlynck, pers. comm.), Matundu dwarf galago *Galagoides zanzibaricus udzungwensis*, and Uganda lesser galago *Galago senegalensis sotikae*, and the northern limit for Mozambique dwarf galago *Galagoides granti* (Butynski *et al.* 2006, 2013; De Jong 2012).
- The Zambezi River and Delta (Mozambique) represent the northern limit of grey-footed chacma baboon *Papio ursinus griseipes* (Cowlishaw 2013), and southern limit of southern yellow baboon *Papio cynocephalus cynocephalus* (Altmann *et al.* 2013).
- The Niger River and Delta (Nigeria) serve as the western limit of Sclater's monkey *Cercopithecus sclateri* (Baker & Olubode 2008; Oates & Baker 2013), Calabar angwantibo *Arctocebus calabarensis* (Oates & Ambrose 2013), and Milne-Edward's potto *Perodicticus edwardsi*, and as the eastern limit of Benin potto *Perodicticus potto juju* (Oates 2011). The western limit of Nigeria needle-clawed galago *Euoticus pallidus talboti* is the eastern edge of the delta (Ambrose & Oates 2013). Although white-throated monkey *Cercopithecus erythrogaster pococki* occurs on both sides of the Niger River and in the Niger Delta, it is absent from the east sector of the delta (Baker 2005; Oates 2011, 2013; Oates *et al.* 2004). The Niger Delta is also at, or very near, the eastern limit of olive colobus *Procolobus verus* in southern Nigeria, although this species occurs slightly to the east of the Niger River in central Nigeria (Anadu & Oates 1988; Oates 2013).

Figure 30.2 Adult male Peters's Angola colobus *Colobus angolensis palliatus*. This eastern Africa endemic subspecies is not known to occur south of the Rufiji River and Rufiji Delta, Tanzania. Photo: Yvonne de Jong & Tom Butynski, wildsolutions.nl. (A black and white version of this figure appears in some formats. For the colour version, please refer to the plate section.)

Primate Conservation in Africa's Coastal Deltas

As highly productive ecosystems, coastal deltas are important for both the biodiversity they maintain and the ecosystem services that they provide. Nonetheless, people often unsustainably exploit deltas for water, agriculture, fish, bushmeat, wood products, oil and other resources (Hughes & Hughes 1992; Ramsar 2015). Coastal deltas are also negatively affected by invasive species, pollution, climatic change, and upstream dams that affect both the overall flow of water and the annual flood regime. The situation is often made worse as a result of corruption and inadequate governance, planning, management, security and law enforcement. The main threats to each of Africa's ten largest coastal deltas are summarized in Table 30.3.

The relative importance (i.e. rank) of each delta for African primate conservation is also presented in Table 30.3. Of the ten coastal deltas included in this review, the Niger, Ogooué, Tana, Rufiji, and Casamance Deltas are, in that order, of greatest importance for the conservation of Africa's primate diversity. Although the Niger and the Ogooué Deltas attained the same rating (i.e. 46), the Niger Delta is ranked higher because it supports three country-endemic primate taxa, whereas the Ogooué Delta supports only one (Table 30.3). The Niger Delta also supports the only African primate that is endemic to a delta, *P. epieni*. The Casamance Delta would have ranked third (tied with the Tana Delta), if not for the extirpation of two species (*C. polykomos* and sooty mangabey *Cercocebus atys*) as a consequence of hunting and habitat loss (A. Galat-Luong & G. Galat, pers. comm.). Of the ten coastal deltas considered in this chapter, the Ogooué Delta is, by far, the least degraded and least threatened by human activities.

The Tana and Niger Deltas are presented here as case studies to illustrate the importance of large coastal deltas for conserving primate diversity, and as examples of the threats that typically affect Africa's large coastal deltas and their biodiversity. Of the ten largest coastal deltas in Africa, these two deltas are the best studied in terms of their primate faunas and threats.

Kenya's Tana Delta

The Tana Delta covers an area of *c.* 1636 km^2 (Figure 30.1) and is part of the Coastal Forests of Eastern Africa Biodiversity Hotspot. The inland apex of this delta lies 3 km south of Garsen at Idsowe Bridge. From here to the Indian Ocean is *c.* 41 km. This delta extends *c.* 50 km along the coast from Kipini in the northeast to Mto Kilifi in the southwest (UNESCO 2013). See maps in Duvail *et al.* (2012) and Tana River Delta (2015).

The Tana Delta supports a fine mosaic of riverine forest, woodland, bushland, grassland, mangrove, estuary, dune and beach. Characteristic trees include *Ficus* spp. (Moraceae), *Phoenix reclinata* (Palmae), *Acacia robusta* (Mimosaceae), *Populus ilicifolia* (Salicaceae), *Blighia unijugata* (Sapindaceae), *Sorindeia madagascariensis* (Anacardiaceae), *Diospyros mespiliformis* (Ebenaceae), and *Mimusops obtusifolia* (Sapotaceae) (Robertson & Luke 1993; UNESCO 2013).

The Tana Delta and delta fringe support eight primate taxa (Butynski & Mwangi 1994; De Jong & Butynski 2009, 2012), of which three (38%) are threatened and two are endemic to the region (Table 30.3).

The 'Endangered' Tana River red colobus *Piliocolobus rufomitratus rufomitratus* (Figure 30.3) is endemic to a stretch of *c.* 60 km of highly fragmented evergreen forests along the lower Tana River and into the northern Tana Delta. Fewer than 1000 individuals remain (Mbora & Butynski 2009), occupying an area of < 13 km^2 (Butynski & Hamerlynck 2016; Butynski & Mwangi 1994; Hamerlynck *et al.* 2012).

The 'Endangered' Tana River mangabey *Cercocebus galeritus* (Figure 30.4) is also endemic to the forests of the lower Tana River and northern Tana Delta. About 2000 individuals occupy *c.* 26 km^2 (Butynski & Hamerlynck 2016; Butynski & Mwangi 1994; Hamerlynck *et al.* 2012; Wieczkowski & Butynski 2013).

Figure 30.3 Adult male Tana River red colobus *Piliocolobus rufomitratus rufomitratus*. Fewer than 1000 individuals remain of this 'Endangered' subspecies, which is endemic to the forests of the lower Tana River and Tana Delta, Kenya. Photo: Yvonne de Jong & Tom Butynski, wildsolutions.nl. (A black and white version of this figure appears in some formats. For the colour version, please refer to the plate section.)

The 'Vulnerable' Pousargues's monkey *Cercopithecus mitis albotorquatus* (Figure 30.5) is endemic to forest along the north coast of Kenya (probably extending up the south coast of Somalia), the Tana Delta, and along the Tana River upstream to Kora National Park and Meru National Park off the northeast side of Mt. Kenya. The Extent of Occurrence of *C. m. albotorquatus* is roughly 10 250 km^2. This subspecies is common at many sites (De Jong 2012; De Jong & Butynski 2011). The northern yellow baboon *Papio cynocephalus ibeanus* is one of the eight species of non-human primate in the Tana Delta (Figure 30.6).

Three, perhaps four, species of strepsirrhine (Figures 30.7, 30.8 and 30.9) are present, making this site as rich as any in Kenya for strepsirrhines (De Jong & Butynski 2012). None of these species is threatened.

Due to the small area of forest, serious threats, high primate diversity, and presence of two endemic 'Endangered' primate taxa, the forests of the Tana Delta and lower Tana River represent the most important site in East Africa for primate conservation actions (De Jong & Butynski 2012). These forests face serious threats from the five upstream hydroelectric dams, a rapidly growing human population, unsustainable exploitation of wood products, clearance for agriculture and settlements, oil exploration, corruption, inter-ethnic violence, and insecurity (De Jong & Butynski 2012; Duvail *et al.* 2012; Hamerlynck *et al.* 2012; Mbora & Butynski 2009; Wieczkowski & Butynski 2013). The threats continue to mount, particularly

Figure 30.4 Adult female Tana River mangabey *Cercocebus galeritus*. Fewer than 2000 individuals remain of this 'Endangered' species, which is endemic to the forests of the lower Tana River and Tana Delta, Kenya. Photo: Julie Wieczkowski.

Figure 30.5 Adult male Pousargues's monkey *Cercopithecus mitis albotorquatus*. This 'Vulnerable' subspecies is present in the forests of the lower Tana River and Tana Delta, Kenya. Photo: Tom Butynski & Yvonne de Jong, wildsolutions.nl. (A black and white version of this figure appears in some formats. For the colour version, please refer to the plate section.)

from poorly conceived, unsustainable, major 'development' projects. None of these projects has received adequate review and inputs from the local communities, the conservation community or the Government of Kenya. Here are some examples of development projects approved or proposed that will affect the Tana Delta: construction of a sixth hydroelectric dam on the Tana River (the High Grand Falls Dam will be the biggest on the Tana River); a 300 km^2 sugarcane plantation by Mat International; a 400 km^2 rice, maize and sugarcane plantation by the Tana River Development Authority; and a 400 km^2 fruit and vegetable plantation by the State of Qatar (Duvail *et al.* 2012; Tana River Delta 2015).

There is, however, some recent good news. Beford Biofuels, a Canadian company seeking to establish a 640 km^2 biofuels plantation (mostly *Jatropha curcas* [Euphorbiaceae]) in the delta has, after great opposition, aborted its plan (Tana River Delta 2015). Similarly, G4 Industries is pulling back from its plan to place 280 km^2 of the delta under oil seeds production.

In 2012, the Tana Delta became a Ramsar Site (Convention on Wetlands of International Importance), 'an intergovernmental treaty that provides the framework for national action and international cooperation for the conservation and wise use of wetlands and their resources' (Ramsar 2015). Otherwise, the Tana Delta holds no legally protected area or protection status.

The widely recognized importance of this region has led to various conservation initiatives that aim to enable local people to conserve the biodiversity, improve livelihoods, rehabilitate degraded areas, and establish sustainable income-generating projects, including ecotourism. Several conservancies, all founded and managed by local communities, have been established in the region. The Kenya Wildlife Service, Kenya Forests Working Group, Nature Kenya, East Africa Wildlife Society, A Rocha Kenya, Kenya Wetlands Forum, Royal Society for the Preservation of Birds, World Wildlife Fund, Northern Rangelands Trust, and others, have been active in promoting the conservation of the Tana Delta, facilitating local involvement in conservation initiatives, and in limiting unsustainable, environmentally and socially damaging 'development' projects.

Nigeria's Niger Delta

The Niger River drains much of West Africa's water into the Niger Delta, by far Africa's largest delta (46 420 km^2; Figure 30.1)

Figure 30.6 Adult female northern yellow baboon *Papio cynocephalus ibeanus* with infant. This is one of the eight species of non-human primate in the Tana Delta, Kenya. Photo: Yvonne de Jong & Tom Butynski, wildsolutions.nl.

Figure 30.7 Adult white-tailed small-eared greater galago *Otolemur garnettii lasiotis*. This is one of at least three species of galago present in the Tana Delta, Kenya. Photo: Yvonne de Jong & Tom Butynski, wildsolutions.nl.

and the third largest contiguous area of mangrove in the world (6700 km^2) (Hughes & Hughes 1992; Spaulding *et al.* 2010; Ramsar 2015). This delta, extending *c.* 500 km along the coast, from the Benin River in the north to the Imo River in the east, incorporates *c.* 65% of Nigeria's coastline. Here, the Niger Delta is taken to include the deltaic plain and floodplain of the lower Niger River (see maps in Baker 2005; Ikemeh 2014a, b; Luiselli *et al.* 2015; Chapter 40).

The Niger Delta is part of the Guinean Forests of West Africa Biodiversity Hotspot (Oates *et al.* 2004). This delta is comprised of four major ecological zones: coastal barrier islands, mangrove forest, freshwater swamp forest, and lowland forest (Ikemeh 2015; Luiselli *et al.* 2015; Moffat & Lindén 1995; Petrozzi *et al.* 2015; Powell 1993, 1997; Werre 2001a). The freshwater swamp forest zone covers *c.* 11 700 km^2 and is the second largest swamp forest in Africa. More than 80% of the delta is seasonally flooded (Moffat & Lindén 1995). Historically, the entire lower delta plain was covered by mangrove forest, while the seasonally inundated portion was covered by freshwater forest. Common trees in the freshwater forest include *Ctenolophon englerianus* (Linaceae), *Uapaca staudtii* and *Uapaca heudelotii* (Euphorbiaceae), *Hallea*

Figure 30.8 Subadult Kenya lesser galago *Galago senegalensis braccatus*. This is one of at least three species of galago present in the Tana Delta, Kenya. Photo: Yvonne de Jong & Tom Butynski, wildsolutions.nl.

Figure 30.9 Juvenile Kenya coast dwarf galago *Galagoides cocos*. This is one of at least three species of galago present in the Tana Delta, Kenya. Photo: Yvonne de Jong & Tom Butynski, wildsolutions.nl. (A black and white version of this figure appears in some formats. For the colour version, please refer to the plate section.)

ledermannii (Rubiaceae), *Alstonia boonei* (Apocynaceae), *Symphonia globulifera* (Clusiaceae), *Pycnanthus marchalianus* (Myristicaceae), *Xylopia staudtii* and *Hexalobus crispiflorus* (Annonaceae), and *Klaineanthus gaboniae* (Euphorbiaceae). Palms (Palmae) are common (e.g. *Raphia vinifera, Raphia*

Figure 30.10 Adult female and juvenile Sclater's monkeys *Cercopithecus sclateri* in Lagwa, Imo State, Nigeria. This 'Vulnerable' species is endemic to southern Nigeria. This is one of 17 species of non-human primate present in the Niger Delta. Photo: Lynne R. Baker. (A black and white version of this figure appears in some formats. For the colour version, please refer to the plate section.)

hookeri and *Elaeis guineensis*) (Bocian 1999; Werre 2000, 2001b; J. Oates, pers. comm.).

The Upper Guinea (West Africa) fauna and the Lower Guinea (Central Africa) fauna converge in the Niger Delta. One result is an exceptionally rich primate community of 17 species (Table 30.2). Of these, two species (*P. epieni* and *C. sclateri*) and one subspecies (*C. e. pococki*) are endemic to Nigeria (Table 30.3). Furthermore, *P. epieni*, is endemic to the delta (Grubb & Powell 1999; Oates 2011). Five (29%) of the 17 primate species are threatened (Table 30.3).

The 'Critically Endangered' *P. epieni* is restricted to a small area of freshwater swamp forest in Bayelsa State, central Niger Delta (Ikemeh 2014a, 2015; Oates & Werre 2009; Werre 2000, 2001a). Habitat degradation, loss, and fragmentation as a result of farming and logging, together with hunting, led to an estimated population decline of > 80% between 1979 and 2009 (Oates & Werre 2009; Werre 2001a). This decline has continued, if not accelerated (Ikemeh 2014a). Estimated at > 10 000 individuals in 1996, with a geographic range of 1500 km^2 (Oates & Werre 2009; Werre 2000; Werre & Powell 1997), surveys conducted in 2013 suggest < 1000 *P. epieni* survive and that the range has declined to *c.* 78 km^2 (Ikemeh 2015; Ikemeh & Oates 2017).

Cercopithecus sclateri (Figure 30.10) is a 'Vulnerable' species endemic to southern Nigeria, from the eastern Niger Delta to the Cross River (Baker & Olubode 2008; Oates *et al.* 2004). This species persists in secondary lowland, gallery, riverine, and swamp forests, and at several sites with a mix of agriculture, degraded forest, and fragmented forest. Their small size, cryptic behaviour, and non-preferred status among hunters (relative to other monkeys) have benefited this species (Baker & Olubode 2008; Oates & Baker 2013).

Figure 30.11 White-throated monkey *Cercopithecus erythrogaster pococki* in Okomo National Park, Nigeria. This 'Vulnerable' subspecies is endemic to southwest Nigeria. This is one of 17 species of non-human primate present in the Niger Delta. Photo: Noel Rowe. (A black and white version of this figure appears in some formats. For the colour version, please refer to the plate section.)

Cercopithecus e. pococki (Figure 30.11) is a 'Vulnerable' subspecies that is endemic to southwest Nigeria (Oates *et al.* 2004). While *C. e. pococki* is sometimes common in lowland forest, including human-modified habitats, it is now absent or rare in many areas due to habitat loss and intense hunting (Oates 2011, 2013).

The 'Vulnerable' red-capped mangabey *Cercocebus torquatus* is patchily distributed in the delta in freshwater forest and mangrove. This species has been extirpated from many sites and is now rare in the eastern sector (Baker 2005; Oates *et al.* 2004; Petrozzi *et al.* 2015; Powell 1995).

The 'Endangered' Nigeria-Cameroon chimpanzee *Pan troglodytes ellioti* is rare in the Niger Delta, with recent, confirmed records only for the eastern floodplain of the lower Niger River and within and around Edumanom Forest Reserve in the southeast sector (Baker 2005; Bocian 1999; Ikemeh 2014b; Luiselli *et al.* 2015; Oates *et al.* 2004; Petrozzi *et al.* 2015; Powell 1993, 1995; Chapters 40 and 43). Habitat loss and hunting are the major threats for *P. t. ellioti* (Ikemeh 2014b; Oates 2011).

Seven species of strepsirrhine, belonging to five genera, occur in the Niger Delta (Jewell & Oates 1969; Oates 2011; Pimley 2009; J. Oates, pers. comm.; Table 30.2). For its size, the Niger Delta may hold the highest diversity of strepsirrhines in Africa. All seven species are able to survive in degraded, secondary and fragmented forest habitats in farmland mosaics. In addition, all are small (< 2 kg), cryptic, nocturnal and of little interest to hunters. As such, none of these strepsirrhines is threatened.

The Niger Delta is comprised of all of Bayelsa State, much of Rivers State, and a small part of Delta State (Chapter 40). The delta supports a high human population density; in 2006, there were 181 people/km^2 in Bayelsa State and 498 people/km^2 in Rivers State (National Population Commission 2010). The central part of the delta, however, has a relatively low human population density as it is remote, without roads, and once received pronounced and erratic floods (now reduced by dams, particularly the Kainji Dam in western Nigeria).

Despite the abundance of forest, fish, and oil in the Niger Delta, the region is one of extreme poverty (Moffat & Lindén 1995; UNDP 2011). The main socioeconomic activities in the Niger Delta are fishing, farming (including oil palm plantations), trading, removal of wood products, and oil extraction (Blench 2007; Ikemeh 2014a,b, 2015; Moffat & Lindén 1995; Werre 2000, 2001a,b; Werre & Powell 1997; Chapter 40). The Niger Delta is the centre of the Nigerian oil industry (the tenth largest in the world) and is criss-crossed by oil and gas pipelines. In 2006, 11 oil companies operated 159 oil fields and 1486 oil wells in the delta (Kadafa 2012). There were many more oil fields offshore and four oil tanker ports (Hughes & Hughes 1992; UNDP 2011). Oil spills have had a major negative impact on the natural vegetation, crops, aquaculture, and people of the delta (Egberongbe *et al.* 2006; Ikemeh 2014b, 2015; Kadafa 2012; Maiangwa & Agbiboa 2013). The Niger Delta is one of the world's five most severely oil-affected ecosystems. Nine to 13 million barrels (or 1.5 million tons) of oil are estimated to have spilled into this delta during 1956–2006. This is the equivalent of one *Exxon Valdez* oil spill each year. One result is tens of billions of dollars of environmental damage. Additional environmental damage from oil operations has resulted from road construction, forest clearance, dredging and filling, canalization, gas flaring, and increased population pressure from immigration to the region.

Another major threat to the Niger Delta is the construction of dams along the course of the Niger River, particularly the Kainji Dam. These have, over the past 50 years, greatly disrupted water flow regimes and sediment deposition. This has led to increased riverbank and coastal erosion, increased damage to crop-lands and infrastructure, and declines in fishery productivity and human health (Moffat & Lindén 1995). Three additional major dams are under construction at this time: Fomi in Guinea, Taoussa in Mali, and Kandjadji in Niger.

Some of the other threats to the delta include over-fishing, logging, hunting, conversion of forest to agriculture, inadequate farming practices, poorly planned infrastructure development, urbanization, invasive species (e.g. water hyacinth *Eichhornia crassipes* [Liliidae] and Nypa palm *Nypa fruticans* [Palmae]), and global warming (Baker 2005; Blench 2007; Giosan *et al.* 2014; Ikemeh 2014b, 2015; James *et al.* 2007; Moffat & Lindén 1995; Phil-Eze & Okoro 2009; UNDP 2011). There are no effectively protected conservation areas in the delta (Oates *et al.* 2004; UNDP 2011; WWF 2015; Chapters 40 and 43), although three Ramsar Sites and at least 11 forest reserves have been designated (Table 30.1).

The legacy of environmental degradation, together with the long-standing poverty, insecurity, corruption, ethnic conflict, and political instability (Kadafa 2012; Maiangwa & Agbiboa

2013), have combined to have a heavy negative impact on the delta's primates. All primate taxa face some form of threat, and all diurnal primates are in decline (Ikemeh & Oates 2017; Werre 2001a; Werre & Powell 1997). As the Niger Delta is the delta with the highest priority for primate conservation in Africa, it is essential that (1) the remaining primate habitats are effectively protected; (2) corridors between key primate sites are established; (3) public awareness is raised for the environment and biodiversity; (4) logging and hunting are sustainably managed; and (5) the oil and gas industry implements best practices for biodiversity conservation and provides significant financial support for environmental protection and ecosystem restoration.

Some improvement to the present situation and trend might be achieved by more government and community participation in conservation projects (Ikemeh 2014b; Phil-Eze & Okoro 2009; UNDP 2011), particularly those that integrate the improved livelihoods of local people with ecosystem restoration. This might be funded by the big oil companies. One such initiative is the 'Sustainable Livelihoods and Biodiversity Project' (implemented by Wetlands Africa, Living Earth Nigeria Foundation, Nigerian Conservation Foundation, and Shell). In addition, the United Nations Development Program (UNDP) and the Global Environmental Facility (GEF) have established the 'Niger Delta Biodiversity Project' (UNDP 2011). This 5-year (2013–2017), US$14 million, project established the 'Niger Delta Biodiversity Trust'. Some of the funds required for this ambitious project are coming from the oil and gas industry, which is expected to contribute substantially to the Trust.

Some Observations

Some drivers that may affect efforts to conserve coastal deltas and their biodiversity, including their primates, are mentioned above. Three of these seem to be of particular relevance:

- The high human population growth rate over much of Africa, particularly in the coastal deltas, is seldom mentioned in the literature. This study never found reference to the 'need' to curb this growth, let alone suggestions or recommendations for 'how' to curb this growth. Given that people in and around most of Africa's largest coastal deltas are already living in extreme poverty, and that these sites are often considered the 'poor house' of their respective nations, it is difficult to see how the situation for the people, or for the biodiversity, of these sites will be better in 20 years when the human population has more than doubled. In short, the root-cause for the biodiversity crisis in Africa's largest coastal deltas has been, and is being, ignored.
- Although most deltas have at least one Ramsar Site (Table 30.1), it is surprising how little mention is given to the Ramsar Sites in the literature related to the conservation of the ten deltas reviewed in this chapter. This suggests that at least some Ramsar Sites are not having the positive impact on the conservation of deltas that one might assume or hope for. The nation state is the only authority for the implementation of the Ramsar Convention at Ramsar Sites within its territory. Every 3 years, during the Conference of Contracting Parties, there is the opportunity to express concern and to list threatened or destroyed Ramsar Sites on the Montreux Record (Ramsar 2017). 'The Montreux Record is a register of wetland sites on the List of Wetlands of International Importance where changes in ecological character have occurred, are occurring, or are likely to occur as a result of technological developments, pollution or other human interference. It is maintained as part of the Ramsar List' (Ramsar 2017). By placing a Ramsar Site on the Montreux Record, the Conference requests the country to 'take swift and effective action to prevent or remedy such changes'. Some deltaic Ramsar Sites in Africa (e.g. Bassin du Ndiaël in the Senegal Delta, and Lake Burullus in the Nile Delta) have been on the Montreux Record for over two decades (Table 30.1). In many cases, the major change in ecological character is linked to changes in the water supply and flooding pattern due to the construction of dams or increased water abstraction upstream – far from the Ramsar Site.
- It appears that most of the effective initiatives for improving the livelihoods of people living in and around coastal deltas – for sustainable development, environmental protection, and conservation of biodiversity – are coming from local individuals or communities, often with support from national and international conservation non-governmental organizations (NGOs). Governments, with all their power to take the lead and to do good, appear to be primarily involved with facilitating the unsustainable use of natural resources and with approving, if not promoting, 'development' and growth activities that further impoverish the people and the environment. Similarly, big companies, particularly those representing the oil, gas, logging, and agricultural industries, have yet to act responsibly towards either the people or the environments of Africa's coastal deltas.

Conclusions

Africa's coastal deltas are of major importance as sites for the maintenance of the continent's primate diversity. Africa's ten largest coastal deltas hold (or did hold) at least 57 primate taxa, 49% of Africa's 95 primate species and 80% of Africa's 25 primate genera. Seven of the taxa are endemic to a delta and its up-river forests, and 12 are threatened. Those coastal deltas nearest the equator hold the highest number of primate taxa, endemic primate taxa, and threatened primate taxa. Most significant among these are the Niger, Ogooué, Tana, Rufiji, and Casamance Deltas. The Ogooué Delta is the least threatened and most undisturbed large delta in Africa. This is attributed to the relatively low human population density in the region. Primate populations in all ten deltas (including the Ogooué Delta) have been reduced and fragmented. Two species of primate have already been lost from the Casamance Delta.

With the exception of the Ogooué Delta, the conservation values of the ten largest coastal deltas in Africa are in rapid decline as they all lack an effective protected area system, remain under extreme threat, and can expect their human populations to double before 2038. In the wake of the World Commission on Dams (see WCD 2000), there is some reluctance by the 'traditional' major multilateral and bilateral donor agencies to fund large dams. With the advent of new donors/builders (e.g. China, India, Brazil, and oil-rich countries), however, the standards and recommendations for maintaining downstream ecosystems, particularly in deltas, have not been incorporated into national policy, nor into private sector or donor guidelines (Duvail *et al.* 2014; Fujikura & Nakayama 2009). At this time, new dams are planned for almost all major river systems in Africa, including the Tana and Niger. This entails a major risk of further losses of deltaic habitats and the biodiversity they support (Chapter 36). As such, the long-term survival of the primate faunas of nine of Africa's ten largest coastal deltas seems bleak unless much more attention is given to their conservation by local people, national governments, the international community, and those big companies that are exploiting their resources.

If there is hope, it appears to lie in grassroots initiatives led by those who recognize that environmental conservation and sustainable use of natural resources is a prerequisite for development, if not for survival. There is some indication that the awareness necessary for a significant surge in the number of local conservation actions is afoot for at least some of the ten coastal deltas covered in this chapter, and that those institutions, agencies, and companies with the power, expertise and monies necessary to support these actions may now be willing to do so.

Acknowledgements

We thank Jean-Pierre Vande Weghe, Ken Tinley, John Oates, Lynne Baker, James Culverwell, Anh Galat-Luong, Gérard Galat, Andrew Perkin, Juliet King, Greg Botha, Michael Lawes, Ara Monadjem, Walter Jubber, Hugo Rainey, Rachel Ikemeh, and Oliver Hamerlynck for information related to one or more of the deltas covered by this review, particularly on which primate taxa are present and the current threats. We are grateful to Lorna Depew, John Oates, Lynne Baker, Carly Butynski, Anh Galat-Luong, Gérard Galat, Fiona Maisels, Kate Nowak, and Olivier Hamerlynck for their valuable comments on the manuscript. Lynne Baker, Julie Wieczkowski, and Noel Rowe kindly provided photographs.

Chapter 31

Primates of Riverine and Gallery Forests

A Worldwide Overview

Shawn Lehman, Kerriann McCoogan and Adrian A. Barnett

Introduction

What are Gallery Forests?

The tropical forest biome is composed of a complex mosaic of different habitat types, ranging from tropical cloud forests to seasonally dry tropical forests (Dirzo *et al.* 2011). Gallery forests are narrow strips of tropical forests that commonly follow watercourses through savanna regions in dry tropical habitats (Beard 1955; Felfilli 1995). The borders are typically abrupt though the widths are variable (Beard 1955). Gallery forests contain a variety of sub-types, but are composed of two broad types: swamp or seasonal formations (Beard 1955). Gallery forests are usually rich in woody plant species typical of continuous forests, while riverine forests often contain habitat-specific species (Kellman *et al.* 1998).

From a conservation perspective, gallery forests are highly important because, as one of the few examples of naturally isolated tropical forest fragments, they can lend insights into the impacts of prolonged fragmentation (Kellman *et al.* 1998). In addition, gallery forests can play an important role for species movement. For example, the tropical gallery forests of the Brazilian *Cerrado* grassland form an ecological link between the Amazonian forests to the west and north, and the Atlantic Rain Forest to the east and south; they have provided key biogeographical bridges in the past and do so too for current regional fauna (Mares & Ernest 1995).

What are Riverine Forests?

Riverine forests (also known as riparian forests) are a type of gallery forest, occurring along the banks of rivers (Gautier-Hion & Brugière 2005). River dynamics play a major role in shaping riverine forest formation due to variations in hydrogeomorphology. including duration and intensity of flooding, sedimention/erosion processes and floodplain width, angle and complexity (Salo *et al.* 1986). This results in a diverse and varied network of forest communities in these areas (Salo *et al.* 1986).

Forests that run adjacent to rivers can provide a high-quality habitat for primate species. For example, riparian forests are more likely to retain their leaves throughout the dry season, and thus can provide both food and shade for primates and other animals (Fedigan & Jack 2001). In some tropical areas, floodplain forest soils receive an annual influx of alluvial nutrients (Peres 1997). Although primate species diversity is typically higher in non-flooded forests than in flooded forests, primate community biomass can be much higher in flooded forests (Haugaasen & Peres 2005). Nonetheless, the linear composition of forests that run along bodies of water present unique challenges to primates that forage in these habitats, typically resulting in extensive day ranges as animals move up and down the narrow forest in their search for food. In this chapter, we review the biology of riverine and gallery forests and the consequences for primates of living in such highly seasonal, very linear and restricted habitats.

Gallery and Riverine Forests in the Neotropics

In Central and South America, the most common type of gallery forests occur as narrow strips of tropical forest that occur alongside rivers, on mountain sides, and in deep gorges near the tops of sandstone cliffs (Felfilli 1995). Some of the better known examples of gallery forests in South America are those found in the *Cerrado*, a savanna-scrub ecoregion found in Brazil (Felfilli 1995). Within it, gallery forests occur as narrow strips on either side of the rivers and streams, and are characterized by dense forests with trees reaching heights of 20 m (Mares & Ernest 1995). These forests are typically backed by non-forest habitats. The soils within *Cerrado* gallery forests often have a strong seasonal variation in their soil-water regime, although because they often run alongside rivers, these gallery forests are less impacted by fire than other *Cerrado* areas (Felfilli 1995).

The gallery forests of Central Brazil are among the richest neotropical habitats for small- to medium-sized mammals (Mares & Ernest 1995). For both fauna and flora, such forests contain many species that are also found in the rainforests of the Amazon and the Atlantic coast of Brazil (Pennington *et al.* 2000). Gallery forests form corridors that can act as refuge for mammals living in the savanna habitat and movement along them can lead to temporal changes in species richness (Mares & Ernest 1995).

Longitudinal studies of primate community structure have provided important insights into spatial variations in species assemblages in flooded (i.e. riparian) and non-flooded forests in Amazonia (Haugaasen & Peres 2005; Peres 1997). Although flooded forests have lower primate species diversity than non-flooded forests, primate density and biomass are typically much higher in flooded than non-flooded forests. Specifically, small-bodied insectivores, such as tamarins (*Saguinus* spp.) are rarely found in flooded forests, whereas other comparably sized insectivores, such as squirrel monkeys (*Saimiri* spp.), exist at very high densities in flooded forests. In some areas, large-bodied folivores, such as *Alouatta* spp., may reach very high densities in such seasonally flooded forests.

A survey of the primate community in a gallery forest in humid southwestern Brazil (Henriques & Cavalcante 2004), recorded black-and-gold howler (*Alouatta caraya*), black-striped capuchin (*Cebus libidinosus*), and black-tufted marmoset (*Callithrix penicillata*). While species richness was comparable to that in other South American gallery forests, it was low compared to that in continuous Neotropical forests, probably due to natural habitat fragmentation and highly seasonal fruit production (Henriques & Cavalcante 2004). At the species level, South American primates differ in their ability to survive in gallery forests. For example, in Venezuela, howler monkeys are better able to survive in inhospitable gallery forests in the Llanos grasslands than spider monkeys living in this same region (Ford 2006: Ch. 21). In Santa Rosa National Park, Costa Rica mantled howler monkeys (*Alouatta palliata*) are known to prefer riparian forests, ranging along narrow riverside forest strips (Fedigan & Jack 2001).

Primates in the genus *Cacajao* (uacaris) are among the best-studied riparian specialists (Ayres 1989; Barnett, Castilho *et al.* 2005; Bowler & Bodmer 2009: Chapters 27 and 41). Most populations of the three uacari species prefer riparian forests, although they will sometimes range into other, non-flooded habitats (e.g. never-flooded rainforest). All uacaris are highly frugivorous, preferring to eat the hard, fat-rich seeds of tough, young fruit (Barnett *et al.* 2005). When fruit is ripe, these monkeys avoid competing with other frugivores by supplementing their diet with insects, flowers, and even seed germinating on the floor of seasonally unflooded riverine forests (Barnett *et al.* 2012c). Due to the long, narrow topography of flooded forests, uacaris have highly extended day ranges, which vary from 2.3 to 7.3 km/day (Barnett & Brandon-Jones 1997). Thus, their preference for unripe fruit in riparian forests places considerable energetic demands on the animals.

In Central America, gallery forests typically run alongside rivers. One example of such a forest is the Community Baboon Sanctuary in Belize, which consists of a 47 km^2 forest alongside the Belize River (Ostro *et al.* 1999). Running adjacent to mixed broad-leaved secondary forest, this tract of forest consists of many tall trees and occurs in a narrow band and is seasonally flooded (Ostro *et al.* 1999).

One of the occupants of Central American gallery forests, like the Community Baboon Sanctuary, are howler monkeys (Horwich & Johnson 1986), who are often found at elevations below 400 m (Estrada *et al.* 2004). Howler monkeys are behaviourally and ecologically flexible which makes them better-equipped than most primates to survive in the sometimes challenging environment of the riverine forest. When hunting pressure is negligible, howler monkeys can survive in disturbed forests, forest fragments, as well as within close proximity to human populations (Crockett 1998). However, especially narrow strips of forest can be a challenge even for howler monkeys due to overcrowding. For example, the black howler monkey (*Alouatta pigra*) in Belize is found at densities of up to 178 individuals/km^2, which has been suggested to result in overcrowding in the fragmented riparian strips of forest (Estrada *et al.* 2004; Ostro *et al.* 1999).

Gallery and Riverine Forests in Madagascar

Western Madagascar is characterized by dry forest, with a cline of increased aridity from north to south (Goodman *et al.* 2005). Gallery forests grow mainly along the margins of rivers within these dry regions. For example, in Berenty Reserve in southern Madagascar, there are two large areas of gallery forest: Bealoka and the main Berenty reserve, which have been described as 'natural islands' of rich habitat that were initially formed by ancient oxbow lakes or an entire river arm (Jolly *et al.* 2006). These gallery forests are of utmost conservation importance. By flying over the Mandrare Valley in 2004, researchers discovered that they are the only true blocks of gallery forests remaining below the headwaters (Jolly *et al.* 2006).

Research has shown that the gallery forest density and distribution can be crucial to the survival of lemur populations (Sussman & Rakotozafy 1994: Ch. 5). At Berenty, ring-tailed lemur population density is higher near the river (175 individuals/km^2) than for the whole reserve (115 individuals/km^2) (Sussman & Rakotozafy 1994). In addition, groups of ring-tailed lemurs living nearer to the river have smaller home ranges (average of 17 ha) than those further away from the river (average of 32 ha). These home range sizes were found to be correlated with the average number of large trees – there were larger trees closer to the river, and thus possibly more food available (Sussman & Rakotozafy 1994). Lemur species respond in different ways to gallery forests, and this includes some lemurs that may not do as well in gallery forests. Verreaux's sifaka has larger groups, smaller home ranges and a higher density of individuals in the areas further away from the river at other dry forest site in Madagascar, which researchers attribute to different dietary preferences in contrast to ring-tailed lemurs (Verreaux's sifaka being more folivorous).

The types of tree species and other vegetation that grow alongside rivers may also influence how and why certain Malagasy primate species can survive within riverine forests. Berenty gallery forests are characterized by a dominance of tamarind trees (*Tamarindus indica*, Fabaceae), which typically grow on moist soils (Sussman & Rakotozafy 1994). There is some debate surrounding whether or not the tamarind species was introduced to Madagascar or if it is, in fact, native. However, research has shown that the tamarind trees are a stable component of the forest, and thus may be native (Sussman & Rakotozafy 1994). It is therefore considered that tamarind trees may be a keystone species for the lemurs of Berenty, with the density of the lemur population being strongly linked to the density of tamarind trees.

Gallery and Riverine Forests in Africa

The continent and nearby islands of Africa cover approximately 30.2 million km^2. Within this, gallery and riparian forests are found along stretches of many major rivers, as well as along rivers and streams that meander across savannas and through valleys in mountainous regions of sub-Saharan Africa (Naiman & Décamps 1997). For example, there are long stretches of gallery forest along the banks of rivers and seasonal lakes in regions surrounding the lowland rain forests of Central Africa. These gallery and riparian forests function as important buffers protecting the water in rivers and lakes from excessive sedimentation and runoff of contaminants produced by adjacent human settlements (Lowrance *et al.* 1997).

The structure and composition of gallery and riparian forests are influenced by the flooding produced by seasonal rains throughout Africa (Eeley *et al.* 2002). Only a few evergreen trees achieve maximum height and diameter in well-drained areas compared to the more abundant, smaller trees that dominate areas subject to more frequent and heavy flooding. The distribution and abundance of trees within riparian and gallery forests are also influenced by soil properties, grazing by wild and domesticated animals, and cultivation practices of local people (Hoegberg 1986). As a result of global climate change, rainfall patterns are changing throughout Africa, resulting in increased stress on plants that must cope with new and unpredictable changes in the timing and intensity of flooding patterns (Hughes 2000).

In West and Central Africa, riparian and gallery forests contain a remarkable variety of birds, reptiles and amphibians, and mammals, including up to 278 bird species just in some regions of West Africa (Mittermeier *et al.* 1998). Many relatively small-bodied mammals, such as red-flanked duikers (*Cephalophus rufilatus*, Bovidae; 12 kg) and African clawless otters (*Aonyx capensis*, Mustelidae; 14–30 kg), share these habitats with the much heavier hippopotamus (*Hippopotamus amphibious*, Hippopotamidae; 1300–1600 kg) and African forest elephant (*Loxodonta cyclotis*, Elephantidae; 2700–5900 kg). A variety of primates range seasonally into riparian and gallery forests in this region, and some are only found in these habitats. Gautier-Hion and Brugière (1995) reviewed primate census data and found that six species are restricted to riparian forests: De Brazza's monkey (*Cercopithecus neglectus*), Gabon talapoin (*Miopithecus ogouensis*), agile mangabey (*Cercocebus agilis*), black-and-white colobus (*Colobus guereza*), sooty mangabey (*Cercocebus atys*), and Pennant's red colobus (*Piliocolobus pennantii*). It was suggested that these primates survive because they are comfortable on the ground, and so they have access to a wider variety of foods compared to other primates in adjacent forests (Gautier-Hion & Brugière 2005).

In East Africa, the Tana River mangabey (*Cercocebus galeritus*) and the Tana River red colobus (*Piliocolobus rufomitratus*) are two of the best-studied primates to use riparian and gallery forests (Medley 2002). These monkeys are only found along or near the Tana River in Kenya. The Tana River red colobus, in particular, is considered one of the world's most endangered primates because it is found only in gallery forests along a 57–60 km strip bordering the river (Groves *et al.* 1974). Both the Tana River mangabey and the Tana River red colobus are considered to be under threat due to the current and trending loss resulting from human encroachment (Decker & Kinnaird 1992).

Gallery and Riverine Forests in Asia

Tropical regions of Asia cover vast areas of land and include over 17 000 islands in the Sula, Java and Banda Seas. This huge area is characterized a monsoonal rainfall pattern, where shorter periods of extremely intense rainfall are interspersed with pronounced dry seasons. Monsoonal systems are also associated with the development of hurricanes, which can strongly impact local forests and animals, including primates (Lugo 2008). Monsoon rains also promote extensive networks of rivers and streams that course down mountains and through many valleys, producing a variety of mangrove and swamp forests near river mouths (Dudgeon 2000a). Consequently, there are long stretches of riparian forests but relatively few gallery forests in this region compared to the other major regions covered in this chapter (e.g. Africa, South America and Madagascar). Some of the main rivers relevant to primates in riparian forests are the Mekong and Irrawaddy rivers in Southeast Asia, Kapuas River in Borneo and the Solo River in Indonesia.

Asian riparian forests are dominated by a variety of tree species in the plant family Dipterocarpaceae, as are many other forest habitats in this region (Janzen 1974). Dipterocarps are among the tallest of all forest trees, with some individual trees as tall as 70 m. Latitude strongly impacts the distribution and abundance of dipterocarps in riparian forests by creating varying patterns of temperature seasonality. The distribution and abundance of riparian trees in the Dipterocarpaceae also result from spatial variations in soil properties, flooding patterns, and cultivation practices (Paoli *et al.* 2005). Dipterocarps are intensively exploited in the timber trade, which has resulted in the loss of entire forests across vast regions of Southeast Asia. These forested regions have also been replaced by enormous coffee, palm oil and coconut plantations, all with negative impact on native primate populations (Chapters 26, 35, 38, 44).

Asian riparian forests contain high levels of species richness for birds, reptiles, fish and mammals. Indonesia alone is home to more species of plants and birds than are found in all of Africa, and over 35% of this has been recorded to use riparian forests (Dudgeon 2000b). Riparian forests are exploited seasonally by the largest mammals in Asia, such as Asian elephants (*Elephas maximus*, Elephantidae), Javan rhinos (*Rhinoceros sondaicus*, Rhinocerotidae), and Asian black bears (*Ursus thibetanus*, Ursidae), as well as by many primate species, such as long-tailed macaques (*Macaca fascicularis*) and silvered langurs (*Trachypithecus cristatus*) (Sodhi *et al.* 2004). Smaller-bodied mammals, such as oriental small-clawed otters (*Aonyx cinerea*, Mustelidae) and fishing cats (*Prionailurus viverrinus*, Felidae), are endemic to riparian forests. Orangutans (*Pongo pygmaeus* and *Pongo abelii*) and proboscis monkeys (*Nasalis larvatus*) represent two of the best-studied primates in Asian riparian forests (Chapters 13, 38, 44).

Orangutans are found in riparian and swamp forests on the islands of Borneo and Sumatra (Chapters 4, 26, 29). The riparian habitat has been found to have significant consequences for this species. Research on orangutan genetics in forest fragments along and across rivers indicates a low likelihood of individuals crossing a river, meaning that orangutans on opposite riverbanks represent separate genetic populations (Goossens *et al.* 2004). However, the same research project found strong evidence for orangutan migration along riverbanks, at least until the fragments became isolated from each other due to deforestation. Recent genetic studies indicate that anthropogenic habitat changes over the last 200 years, such as deforestation and forest fragmentation, have resulted in the loss of 95% of orangutans in northern Borneo (Goossens *et al.* 2006). Consequently, all orangutans are considered to be Endangered (Chapter 29).

Proboscis monkeys are named for their long, pendulous noses, which are much larger in males than females. These monkeys are often found in all-male groups or in one-male, multi-female groups in flooded forests (e.g. riparian forests and swamps) across Borneo (Meijaard & Nijman 2000b: Chapters 7 and 13). Although proboscis monkeys are predominantly arboreal, they are known to be good swimmers and will wade across shallow river beds to reach food trees (Chapter 13). Within these flooded forests, proboscis monkeys feed predominantly on leaves and seeds from June to December, but will switch their diet to fruit and flowers during January through May (Yeager 1989). Although their preferred foods come from only four tree species, they will exploit 51 other tree species across the year. In 2000, proboscis monkeys were elevated to the endangered conservation category due to the deleterious effects of hunting and habitat destruction in riparian and coastal forests in Borneo (Chapter 44).

Primate Conservation in Gallery and Riverine Forests

Many of the primates in gallery and riparian forests face a precarious future. This conservation issue is due to the combined effects of dams on rivers, exploitation of forests by local people and commercial hunting of primates. Dams built along rivers to produce electricity and reduce catastrophic floods have created large reservoirs that flood out forest habitats for primates (Thomas 1996: Ch. 36). Forests down river from the dams suffer from reduced flooding, which prevents the deposition of mud and other organic nutrients that serve as fertilizer and nutrients for plants in gallery and riparian forests. Local people clear forests for agriculture and increase their exploitation of trees for building materials and secondary forest products in shrinking forest fragments (Kramer *et al.* 1997). Long-term studies have revealed that some primates have yet to recover from logging that occurred 28 years ago, despite decades of regrowth in forests (Chapman *et al.* 2000). The other, more immediate threat to primates in riverine and gallery forests results from illegal hunting for food for local consumption and for the bushmeat trade, where wild-hunted meat is sold in markets locally, nationally, and even internationally (Chapman *et al.* 2006). Sadly, the bushmeat trade is escalating, resulting in the catastrophic loss of entire primate populations in many regions of Africa and elsewhere.

Summary

Tropical gallery forests are narrow strips of forests that commonly follow watercourses through savanna regions in dry tropical habitats. Riverine or riparian forests also occur along riverbanks, but are backed by continuous forest, rather than an open habitat. Gallery and riverine forests are found in all parts of the world where primates live, including the Neotropics, Madagascar, Africa and Asia. In each of these geographic regions, the gallery and riverine forests have locally specific habitat structures, including forest size, soil quality and species structure. Such properties can influence local primate densities, ranging patterns and even the genetic structure of primate populations. Primate responses to gallery and riverine forests can vary based on their individual behavioural and ecological strategies. For example, howler monkeys, which exhibit ecological flexibility, are able to survive in circumstances where food may be scarce. Orangutans, on the other hand, are considered less flexible and their inability to move between riverine habitats has resulted in genetic shifts in populations. Ultimately, continuing and expanding the study of the primate species that inhabit the world's gallery and riverine forests is of key importance, not only because many species and their habitats are threatened, but because studies can identify extinction resilience features that may aid their effective conservation and, as flagship taxa, the habitats themselves.

Chapter 32

Life-history Traits and Group Dynamic in Black and Gold Howler Monkeys in Flooded Forests of Northern Argentina

Martin M. Kowalewski, Romina Pavé, Vanina A. Fernández, Mariana Raño and Gabriel E. Zunino

Introduction

Mammal populations living on islands frequently differ from mainland conspecifics (or congenerics) in behaviour and/or morphology. This is often attributed to the ecological effects (changes in resource base) and genetic effects (limited genetic interchange and/or founder effects). There are a number of mainland-based non-human primate species that also have island-inhabiting populations (*Papio ursinus griseipes*: Cheney *et al.* 2004; *Alouatta caraya*: Kowalewski & Zunino 2004; *Alouatta palliata*: Duarte Dias & Rodríguez Luna 2006; Chapters 27 and 31). The current chapter deals with one of these, the black and gold howler (*A. caraya*). In northern Argentina and southern Brazil, this species has populations that live on islands, characterized by flooded forests, along the Parana and Uruguay rivers, while mainland populations inhabit natural fragmented forests and gallery forests along small rivers (Brown & Zunino 1994; Kowalewski & Zunino 2004; Ludwig *et al.* 2008; Miranda 2009; Zunino *et al.* 2001).

Howlers comprise some 12 species and range from southern Mexico to northern Argentina (Cortés-Ortiz *et al.* 2015). Populations of several different howler species are known to live in flooded forests, including *A. caraya* (Aguiar 2006; Aguiar *et al.* 2007a,b; Ludwig *et al.* 2008; Kowalewski & Zunino 2004), *A. guariba* (Aguiar *et al.* 2007a), and *A. palliata* (Froehlich *et al.* 1981). In Argentina, *A. caraya* reaches the southernmost limit of the species and the genus (Cortés-Ortiz *et al.* 2015), with the most southerly of all living at 29°30′S, in flooded forest on islands in the Parana River (Brown & Zunino 1994). Black and gold howler monkeys (*A. caraya*) are a folivore–frugivore arboreal and diurnal primate characterized by bisexual dispersal (Di Fiore *et al.* 2011). In the particular case of *A. caraya* the adults are sexually dimorphic (males are larger than females) and sexually dichromatic (adult males are black and females are golden, all infants are gold and males turn black during sexual maturation).

On these islands, *A. caraya* are found at higher densities compared to mainland populations (Aguiar *et al.* 2009; Kowalewski & Zunino 2004; Rumiz 1990). For example, on Isla Brasilera, an island with continuous forest in the Middle Parana River, the ecological density of *A. caraya* is 3.25 individuals/ha, while in mainland fragmented forest at the EBCo site, some 27 km to the southeast, the ecological density is 1.04 individuals/ha (Kowalewski & Zunino 2004; Zunino *et al.* 2007). EBCo and Isla Brasilera are long-term sites and studies have been carried out on demography, ecology, behaviour and disease dynamics of black and gold howlers (i.e. Kowalewski 2007; Kowalewski & Zunino 2004; Fernández 2014; Oklander *et al.* 2014; Pavé 2013; Pavé *et al.* 2012; Rumiz 1990; Zunino *et al.* 2007). The EBCo site is characterized by semi-deciduous natural forest fragments (up to 20 ha) in a grassland matrix, while Isla Brasilera is a 290 ha island on the Parana River characterized by a continuous forest that suffers periodical floods. Studies at EBCo began during the 1980s, and on Isla Brasilera at the end of the 1990s.

Here, we present life-history traits from a long-term study on Isla Brasilera, and compare this with previously published or available data from EBCo. It is suggested that some life-history traits of non-human primate species follow variation in ecological variables such as food availability and forest/vegetation structure and composition (Altmann & Alberts 2005; Borries *et al.* 2001; Carnegie *et al.* 2011; Hauser & Fairbanks 1988; Lycett *et al.* 1998). For example, Hauser & Fairbanks (1988) found differences in the length of the interbirth interval between two populations of *Cercopithecus aethiops*, inhabiting dry woodland and swamp, with swamp-dwelling females (n = 14) exhibiting shorter interbirth intervals than females (n = 10) from the woodland (1 versus 2 yrs). The habitats varied possibly in food availability; while the swamp had consistent availability of high-quality food throughout the year (characterized by a high availability of a particular *Acacia* sp. whose resources comprised 50% of the diet), the woodland was the less favourable environment because it was dominated by another *Acacia* sp. whose resources were poorer in quality.

Our two study sites differ in forest continuity (Kowalewski & Zunino 2004) and in the diversity and availability of food throughout the year (Kowalewski & Zunino 2004; Pavé 2013); for example, during the winter season (June–August), food resources at EBCo (mainland) decrease twice as quickly as the food on the island (Kowalewski & Zunino 2004; Pavé, 2013; Figure 32.1). However, differences in food availability between sites do not appear to be an important factor influencing interbirth interval lengths, infant mortality rates,

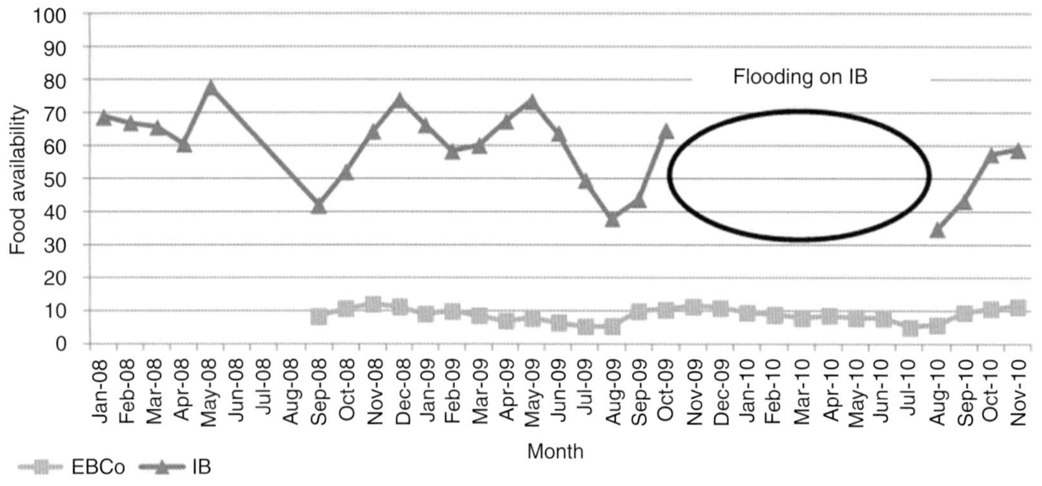

Figure 32.1 Mean monthly food availability on Isla Brasilera (IB; between January 2008 and November 2010) and EBCo (between September 2008 and November 2010). Each point represents the sum of the monthly abundance index of each plant structure (mature leaves, new leaves, shoots, immature fruits and mature fruits) for all plant species studied (17 spp. from EBCo and 11 spp. from IB). This figure was obtained from Pavé (2013).

mother–infant relationship (Pavé *et al.* 2012, 2015), or activity budget (Bravo & Sallenave 2003; Fernández 2014; Kowalewski 2007). In the flooded forest, forest continuity allows overlap of howler home ranges, and the existence of a higher proportion of multi-male groups (Kowalewski & Garber 2010; Kowalewski & Zunino 2004; Pavé *et al.* 2012). This high overlapping of home ranges (up to 80%), results in a different group dynamic in comparison to the mainland, where the relative costs of moving between groups and fragments is much higher (Kowalewski *et al.* 2011; Oklander *et al.* 2010).

Since access to mainland population sites is easier than to the island populations, behavioural and demographic data from the islands come at a slower pace. However, it is important to know the life-history traits from long-term studies on island populations to develop an understanding of the set of adaptations that howlers are capable of in the extreme portion of their southern distribution (Kowalewski *et al.* 2015). Data from long-term studies are critical for answering questions on primate life history, adjustments to ecological changes, and population dynamics (Glander 1981).

Consequently, the goal of this chapter is to provide the first compilation of life-history traits coming from a long-term study on black and gold howlers living in a continuous forest characterized by periodical floods in northern Argentina. Although certain aspects of *A. caraya*'s behaviour, ecology, and life history are highly conservative, and vary little from site to site (i.e. body weight, activity budget, mother–infant relationship, gestation length, interbirth interval), patterns of group structure and composition, diet, birth seasonality, sexual dispersal and frequency of infanticide are found to vary between mainland and island habitat populations (Bravo & Sallenave 2003; Kowalewski & Zunino 2004; Pavé *et al.* 2012, 2015; Rumiz 1990; Rumiz *et al.* 1986; Zunino *et al.* 1986). Consequently, we expected that life-history traits in the continuous forest on the islands would not be different from those of mainland forest populations but, instead, be a continuum within the broad set of phenotypically flexible strategies characteristic of howlers (Agrawal 2001; Jones 2005, but see Garber *et al.* 2015).

Methods

Study Sites

Populations of *Alouatta caraya* were studied at two geographically proximate sites in northern Argentina (Figure 32.2). One, EBCo, is a mainland site. It is located in Corrientes Province (27°30′S, 58°41′W), and is characterized by semi-deciduous forest fragments (up to 20 ha). Normally, each fragment is inhabited by a single group, although up to four howler groups have been found in some fragments (Zunino *et al.* 2007). The other site is the 292 ha Isla Brasilera in Chaco Province (27°18′S, 58°38′W). Isla Brasilera is a typical island of the Middle Parana River, which experiences at least one flooding event annually, though these vary in intensity and duration (Bonetto 1986). A previous study showed that an extraordinary flood, that lasted 6 months between 1997 and 1998, did not alter plant species richness or the floristic composition of the forest (Kowalewski & Zunino 2004), indicating a high level of flood tolerance. The constant accumulation of nutrients from the river sediments during flooding results in these environments being highly productive (Neiff 1978). The islands of the Parana River have a high density of fruit-producing tree species (for example *Cecropia pachystachya*, Cecropiaceae; *Banara arguta*, Salicaceae; *Ocotea diospyrifolia*, Lauraceae; *Inga uraguensis*, Fabaceae; and *Eugenia burkartiana*, Myrtaceae), food resources that are important to the howlers (Fernández 2014; Kowalewski 2007; Zunino *et al.* 2001). Isla Brasilera does not have permanent human settlements, and attacks from potential predators, such as the crab-eating fox (*Cerdocyon thous*), pampas fox (*Pseudalopex gymnocercus*), jaguarundi (*Herpailurus yaguarondi*), and yellow anaconda (*Eunectes notaeus*), on howlers have never been observed in the almost 20 years research which has been conducted there. Although, agonistic interactions, such as alarm vocalizations and chases with these potential predators, have been observed (0.29 ± 0.4 interactions with mammals every 100 h, n = 24, and 0.6 ± 0.8; Fernández, unpub. data). Isla Brasilera is not a closed system because howlers migrate

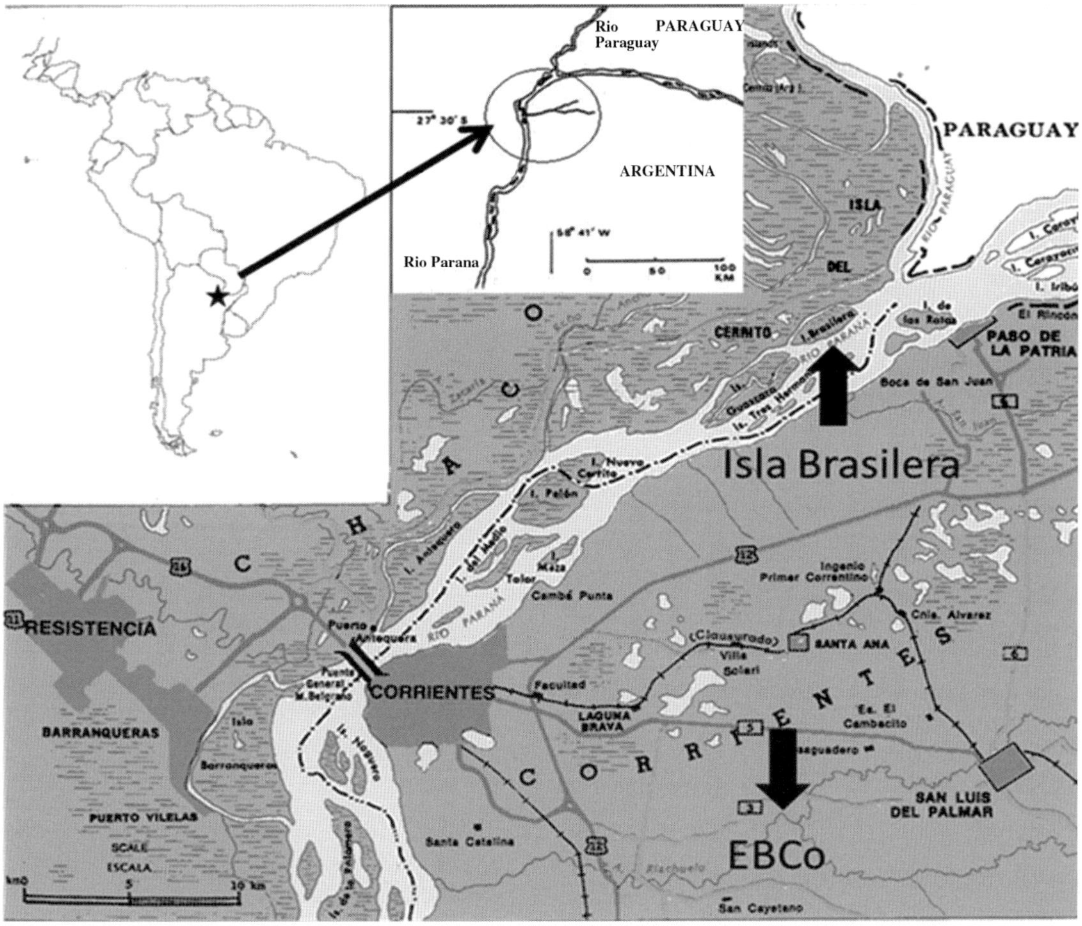

Figure 32.2 Location of both study sites in northern Argentina.

by crossing the river between the island and the closest and largest island of the system (Isla del Cerrito 12 000 ha, a distance of 64 m) by swimming or by floating on vegetation rafts formed during flooding events. Previous studies have also shown that food availability varies between the sites (Table 32.1; Figure 32.1), generally being more seasonal in the fragmented forest with peaks of low food availability during the winter.

Temperature, precipitation and photoperiod are similar at both study sites (Rumiz *et al.* 1986). The climate of both sites is subtropical with a mean annual temperature of 21.6°C (varying from a maximum absolute temperature of 40.5°C in summer and a minimum of 0.4°C in winter), and a mean annual rainfall of 1200 mm (Servicio Meteorológico Nacional de Argentina).

Life-History Traits at the Flooded Forest

Life-history data presented in this chapter are original data, but they are contextualized by the results of several previous studies on demography and behaviour, carried out at the study sites by the authors and collaborators (Fernández 2014; Kowalewski 2007; Kowalewski & Garber 2010; Kowalewski & Zunino 2004; Pavé 2013; Pavé *et al.* 2012). At EBCo, we collected data monthly on two groups from September 2005 to August 2006, September 2007 to March 2008, and August 2011 to July 2012; six groups from September 2008 to December 2010, and 20 groups from June 2013 to July 2014 with total of 259 group-months. At Isla Brasilera, we collected data monthly on four groups from April 2003 to December 2004, September 2005 to September 2006, and September 2008 to November 2010; 29 groups from October 2006 to July 2008, and 22 groups from June 2013 to July 2014, with a total of 690 group-months. Groups were recognized by their location and composition, and we identified individual howlers by age, sex and natural and/or artificial markings (coloured ankle bands or ear tags, and natural ear cuts), using the age–sex categories of Rumiz (1990). A particular characteristic of this species is that males are black and females are golden. Males and females are born golden and males turn black as they mature (Bicca-Marques & Calegaro-Marques 1998; Rumiz 1990). During each encounter with a group, we recorded the age and sex of adults, subadults, juveniles and infants as well as changes in group membership (births, deaths or disappearances, emigrations and immigrations). In each group we have identified a central or dominant male based on priority access to females (i.e. only the central male engaged in mate guarding), visibly larger body size and the initiation of howling vocalizations followed by travel to confront neighbouring groups during intergroup encounters (Kowalewski & Garber 2010). Similar measures have been used to identify a central (or

Table 32.1 Ecological, demographical and social characteristics of the study sites in northern Argentina.

Forest type	Flooded forest (Isla Brasilera)	Semi-deciduous forest (EBCo)	Sources
Forest continuity	Continuous	Fragmented (up to 20 ha)	1–2
Vegetation (Richness and plants diversity)	53 spp; H'index range 1–2.5	61 spp; H'index range 1–13	1–8
Food availability (monthly average)	High and constant throughout the year (59%)	Low throughout the year and seasonal (fall in winter) (9%)	2, 6, 9
Ecological density	3.25 individuals/ha	1.04 individual/ha	2–3, this study
Year of census Groups with 1 male / > 1 male, total number of groups	1997–1999: 37%/67% (*n* = 27) 2006–2008: 10%/90% (*n* = 29) 2013–2014: 32%/68% (*n* = 22)	1984: 73%/27% (*n* = 11) 1999: 57%/43% (*n* = 23) 2003: 80%/20% (*n* = 34) 2013–2014: 65%/35% (*n* = 20)	1–5, this study
Rate of intergroup encounters	2 encounters/day (*n* = 2 groups, 12 mo)	2007: 0.75 encounter/day (*n* = 2 groups, 6 mo) 2010: 0.75 encounter/day (*n* = 2 groups, 9 mo) 2011: ≈1.3 encounter/day (*n* = 2 groups, 12 mo)	6–8, this study
Home range size (ha)	2003–2004: 5.92 ± 0.04 (*n* = 2 groups, 12 mo) 2008–2010: 4.1 ± 1.01 (*n* = 5 groups, 16 mo)	2004: 9.25 ± 2.45 (*n* = 2 groups; 11 mo) 2007: 4.5 ± 1.8 (*n* = 2 groups, 6 mo) 2008–2009: 8.14 ± 4.38 (*n* = 4 groups; 26 mo) 2010: 9.2 ± 0.8 (*n* = 2 groups, 9 mo) 2011: 10 ± 2 (*n* = 2 groups, 12 mo)	6–10, this study
Overlapping of home range (%)	70.5 ± 6.5 (*n* = 2 groups, 12 mo)	2007: 53 ± 32 (*n* = 2 groups; 6 mo) 2010: 56.5 ± 18.5 (*n* = 2 groups; 9 mo) 2011: 55 ± 25 (*n* = 2 groups; 12 mo)	6–8
Distribution of births (mean, n)	1997–1999: throughout the year (*n* = 140 births, 21 mo) 2006–2008: August (*n* = 113 births, 27 mo)	May-June (*n* = 101 births, 26 mo)	2, 4, 9
Female life-history traits			
Age at first reproduction (years)	5.4 ± 0.73 (*n* = 4 females)	4.5–5 (*n* = 2 females)	6, this study
Gestation (days)	187 ± 7 (*n* = 5 births)	160 ± 9 (*n* = 3 births)	11, this study (data on EBCo were calculated constructing hormonal profiles with daily urine collection)
Weaning age (months)	12.14 ± 1.46 (*n* = 7 infants)	9 ± 1.52 (*n* = 15 infants)	9, 12
Interbirth interval (IBI, months)	14.07 ± 1.87 (*n* = 14 IBIs)	15 ± 3.56 (*n* = 43 IBIs)	1, 4, 9
Male life-history traits			
% of infant mortality associated to infanticide/Infanticide rate	18.6% (*n* = 9 infants) RATE: 0.87 ± 0.22 (*n* = 8 infants)[a]	25% (*n* = 8 infants)	4, 13
Dispersal age (years)	4.33 ± 1.16 (*n* = 8 males)	4.9 ± 0.7 (*n* = 6 males)	4, 6, this study
Central male tenure (years)	4 ± 2.02 (*n* = 4 males)	5.5 ± 2.7 (*n* = 8 males)	4, 6, 9, this study
Age of first reproduction (range years)	5–6	4.5–6	this study
Full black colour acquirement (range years)	4–6	4–6	this study

[a] Infanticide rate: calculated as the number of infants that disappeared due to infanticide during the first month of life divided by the number of infants present the previous month in the population.

References: 1: Rumiz 1990; 2: Kowalewski & Zunino 2004; 3: Zunino *et al.* 2007; 4: Pavé *et al.* 2012; 5: Agoramoorthy & Lohmann 1999; 6: Kowalewski 2007; 7: Raño 2010: 8: Fernández 2014; 9: Pavé 2013; 10: Zunino *et al.* in press; 11: Kowalewski & Garber 2010; 12: Pavé *et al.* 2010; 13: Zunino *et al.* 1986.

dominant) male in other howler species (*A. palliata*, Wang & Milton 2003; *A. pigra*, Van Belle & Estrada 2008).

Results

During the last complete study on Isla Brasilera (October 2006 to July 2008), we censused monthly 29 *A. caraya* groups (273 individuals including adults, subadults, juveniles and infants) with an average group size of 9.41 ± 3.33 (range: 3–17) individuals. Of these, 12 groups were unimale/multi-female (1 adult male, 1–3 adult females), 4 were age-graded male/multi-female groups (1 adult male, 1–2 subadult males and 3–5 adult females), 10 were multi-male/multi-female (2–4 adult males, 0–3 subadult males and 2–4 adult females), and three changed from multi-male/multi-female to unimale/multi-female (Pavé *et al.* 2012; this study).

The characteristics found in this population are similar to those previously reported by Kowalewski and Zunino (2004) for the same population between 1997 and 1999. This last study recorded 263 individuals belonging to 27 groups (10 unimale, 5 age-graded male groups and 12 multi-male/multi-female) with an average group size of 9.7 (range: 3–20) individuals. Over the years of study, we have observed changes in the social structure of the groups. For example, of the four most-studied groups on Isla Brasilera (Empanada, Gritones, Marley and Xeneizes), between 2003 and 2010, just one (Marley) remained as multi-male, with the same two adult males throughout the years, while the other three changed either from multi-male/multi-female to unimale/multi-female (Gritones), or from multi-male/multi-female to age-graded male/multi-female and then to unimale/multi-female (Empanada and Xeneizes). Also, in the latter three groups there were changes in the central male position throughout time (Table 32.1). This dynamic group composition, a result of the high level of group overlap, is relatively infrequent among the mostly unimale groups of EBCo. However, after the clear-cutting of some forest fragments during 2004, four groups occupied a 15 ha fragment and coexisted with overlapping home ranges, where previously there were one or two groups with almost no home range overlap. Data compiled from literature and our current studies on life-history traits at EBCo are presented in Table 32.1.

Discussion

Our results show that *Alouatta caraya* densities in the flooded forest were higher than those on the mainland. Due to the continuity of the forest, all flooded forest groups had overlapping home ranges (with at least two neighbouring groups), and their home ranges are smaller than those at mainland sites. Although at EBCo some groups have overlapping home ranges, this is quite rare as the occurrence of more than one group (up to four) within a fragment of 15–20 ha during the last 35 years is a relatively unusual and new event. This group crowding at EBCo usually follows a process of forest fragment clear-cutting nearby. In the flooded forest, the long history of overlapping home ranges probably results in a set of particular strategies that are absent or occur at a lower frequency at the mainland, including, for example, a high rate of mostly non-aggressive intergroup encounters (Kowalewski 2007). In the flooded forest, the high rate of encounters in which males engage in collective howling, as the main defence display (Garber & Kowalewski 2013), may result in the higher proportion of groups with more than one male (Kowalewski & Garber 2015). Moreover, the relationship among males in these multi-male groups are cooperative and nonhierarchical, with all or most resident males having access to receptive females; and their acting collectively to exclude extragroup males (Kowalewski & Garber 2015; see discussion for other species in Nunn 2000; Pereira *et al.* 2000; Strier 2000). At EBCo, we have found that unimale groups tend to become multi-male with the passing of years, notably when a fragment receives new groups as a consequence of deforestation, and the frequency of intergroup encounters increase. Under such circumstances, when group home ranges overlap on the mainland, intergroup encounters are initially more aggressive (Pavé 2013; M. Raño, unpublished data), possibly due to the novelty of the events. Therefore, we suggest that frequent intergroup encounters at Isla Brasilera allow exchange of information on group composition, availability of dispersal partners, and reproductive status of females. Moreover, Kowalewski and Garber (2010) have shown that females may use these encounters to engage in extragroup copulations, and suggest that one prominent female sexual strategy is mating with several males, a strategy that is less common at EBCo (M. Raño, pers. obs.).

With respect to life-history traits focusing on infants, there are no differences on the length of the interbirth interval between the sites, but we have found differences in the weaning age; infants at Isla Brasilera wean 2 months later than those at EBCo (Pavé 2013). However, considering the small sample size for weaning ages at the study sites (5 at Isla Brasilera, 15 at EBCo), we still consider that these differences may be in response to multiple factors, including the age of the mother, food availability during the month when infant becomes 8–9 months old (the age when infants begin to rely more on solid food items than on milk), and the arrival of newborns (Pavé *et al.* 2010, 2015). Age at first reproduction between sites appears to be similar (between 4 and 6 years at both sites), which may simply suggest that this is a more conserved trait.

Since the mainland population has more unimale groups than the island population, a higher infanticide rate should be expected to occur there than at Isla Brasilera (Leland *et al.* 1984). However, during a 27-month period, infant mortality and infanticide appeared to be similar at both sites (Pavé *et al.* 2012; Zunino *et al.* 1986). Also, our study shows that central male tenure is similar, possibly slightly higher on the mainland. Moreover, at both sites, infanticide is associated mainly with male replacement in unimale groups and with intergroup encounters between neighbouring groups (Pavé 2013; Pavé *et al.* 2012; Zunino *et al.* 1986). Age at first reproduction of males and full black colour acquisition seems not to differ at either of the sites. However, males in multi-male groups appear to maintain remnants of their young pelage longer and remain in their natal groups longer than males maturing in unimale groups (Blomquist *et al.* 2009). When males turn black, they are either quickly evicted by resident males or incorporated in multi-male groups (Garber & Kowalewski 2011). A more continuous habitat (as the one offered by the flooded forest in northern Argentina) could increase the frequency of multi-male groups, promoting selection for delayed maturation for males, with the colour change possibly following some hormonal mechanism as a response to a social cue (Blomquist *et al.* 2009). In a study of the reproductive physiology of *A. caraya* in captivity, Moreland *et al.* (2001) reported that ejaculates from subadult males ($n = 3$), contained more morphologically abnormal spermatozoa than adult ejaculates ($n = 3$). In this regard, remaining as a subadult individual may have associated costs of not being a normal reproductive individual.

Many of the strategies presented in this chapter are probably environment-dependent (as opposed to microevolutionary differences among populations), and can be thought of as a norm of reaction as suggested by Blomquist *et al.* (2008). In this context, it has been suggested that differences in food availability between the island and mainland (Figure 32.1, Table 32.1) may drive differences in life-history traits between different populations (Kowalewski & Zunino 2004; Rumiz 1990; Rumiz *et al.* 1986). However, recent studies on spider monkeys (*Ateles chamek*, Felton *et al.* 2009a,b), gorillas (*Gorilla beringei*, Rothman *et al.* 2008), howler monkeys (*Alouatta pigra*, Righini 2014; *Alouatta caraya*, Fernández 2014), and humans (Gosby *et al.* 2011; Simpson *et al.* 2003; Simpson & Raubenheimer 2005) suggest that primates regulate nutrient intake regardless of food availability (i.e. primates maintain a daily constant intake of protein and/or carbohydrates and lipids). In the presence of an unbalanced diet, where certain nutrients are consumed in excess/deficit in order to maintain a nutritional target, the different patterns of responses that arise may depend on the current nutritional profile in the environment which different species or populations currently inhabit (Fernández *et al.* 2014). For example, *Alouatta pigra* present a constant ratio of daily protein and non-protein energy intake, regardless seasonal variation in the diet (Righini 2014), while *A. caraya* have an analogous pattern but only in winter months, whereas in spring months, when the food items consumed varied more in protein content, a protein regulation pattern arose (Fernández 2014). These patterns of macronutrient intake are independent from food availability across sites (Fernández 2014; Righini *et al.* 2014).

Some of the life-history traits we have presented, and which are considered to vary between populations, are possibly responses to differences in structure of the forests rather than to food availability (including the physical structure in terms of height and density of canopy), in populations living in continuous forest as opposed to the fragmented forest that characterized mainland sites in northern Argentina (see Gonzalez *et al.* 2002; Rumiz *et al.* 1986). We suggest that the combination of the continuity of the forest canopy and high density of howlers in the islands, allows the high degree of home range overlap among groups and the set of behaviours that characterized these groups (i.e. high density of non-aggressive intergroup encounters and high degree of female extragroup copulations; Kowalewski 2007; Kowalewski & Garber 2010).

The different social and reproductive strategies that *A. caraya* show in the flooded forest cannot be considered to be isolated or rare events: bearing in mind that the studied populations are at the extreme southern end of the species range, the exhibited behaviours may be thought of as forming part of the complete set of ethological phenotypes that black and gold howlers are capable of, a subsection of the norm of reaction that allows howlers to survive in such non-predictable habitats. In this regard flooded forests in northern Argentina suffer periodical floods of variable intensity and duration (i.e. a flood may last 2 months or more than a year). Pave *et al.* (2012) reported that a flood that lasted 3 months and covered the entire island (except the tree crowns) affected the production of food items such as mature leaves during the winter season. They suggested an association between this food reduction (mature leaves appear to play an important role in the ability of mothers to nurse their offspring) and an increase in infant mortality (including deaths from drowning following falls). In this context, we recommend concentrating conservation studies on understanding the total life-history strategies across species, and the potentially differing responses of populations within species.

Flooded forests inhabited by primates in northern Argentina are very limited in extent and none are included within any existing protected area. In this regard studies on these habitats are important to understand the phenotypical variability of primates and other mammals that extend their distribution towards the south only on islands characterized by flooded forests. Mainland forest have been profoundly changed due to commercial agriculture (i.e. soy plantations) and commercial timber exploitation (Zunino & Kowalewski 2008). In this regard, flooded forests in Argentina are in many cases the only refuge for many animal species (i.e. *Alouatta caraya*) at the southern extreme of their distribution.

Acknowledgements

We are very grateful to all field assistants who helped us during the data collection, especially to Silvana Peker and Nelson Novo, who carried out independent projects in the area. We also want to thank Angel Ramon Martinez, Miguel Blanco and Ramon Romero who helped out during several trips to the field. We also thank Charlotte Bender who helped us with the English version of the manuscript. This study was supported by grants and fellowships of the Leakey Foundation (MK), The Wenner Gren Foundation (#7034 and #7622 MK), American Society of Primatologists (MK), Scott Neotropical Fund of The

Cleveland Zoological Society and Cleveland Metroparks Zoo (MK), American Society of Mammalogist (RP, VF), Idea Wild (MK, RP), Barcelona Zoo (RP), The Scientific Research Society (VF), International Primatological Society (VF, MR), PIP IU 0355 CONICET (MK), Consejo Nacional de Investigaciones Científicas y Técnicas de Argentina (MK, RP, VF, MR). MK thanks BK for teaching him about life history patterns in human primates. The study complied with the current laws and permission of the United States and Argentina (IACUC protocol #01071).

Chapter 33

Riverine Red-tails

Preliminary Data on Forest Guenons in a Savanna Woodland Habitat in the Issa Valley, Ugalla, Western Tanzania

Simon Tapper, Caspian Johnson, Anna Lenoël, Alexander Vining, Fiona A. Stewart and Alex K. Piel

Introduction

Most guenon species studied to date are forest dwelling and arboreal, despite both inter- and intraspecific variation in habitat and behaviour (Cords & Sarmiento 2013). The red-tailed monkey, *Cercopithecus ascanius*, has been studied predominantly in tropical rainforest habitats (Chapman *et al.* 2002; Cords 1987a,b; Lambert *et al.* 2004; Mammides & Cords 2008; Struhsaker & Leland 1979, 1988; Treves 1999). Data from long-term field sites, such as Kibale National Park in Uganda and the Kakamega Forest Reserve in Kenya, have provided the foundation of red-tail population parameters, diet and ranging behaviour. From these studies we know, for example, that red-tail density is lower in disturbed forests (Chapman *et al.* 2010; Fashing *et al.* 2011), red-tails exhibit considerable inter-annual variation in consumed plant species (Chapman *et al.* 2002), and have relatively long day journey lengths relative to home range size (HRS; Cords & Sarmiento 2013; Table 33.1). However, while much is known about the behaviour of forest-dwelling red-tails, less is known about groups that inhabit mosaic habitats.

In the current chapter, we contrast what is known about red-tail monkey behaviour in such forested habitats with very preliminary data from a single focal group living in the thin strips of riverine forest, surrounded by miombo savanna woodland in the Issa Valley, Ugalla, western Tanzania (Figure 33.1). Sympatric chimpanzees rely on woodland plant species for feeding throughout the year in Issa, have low population densities and are hypothesized to have correspondingly large home ranges (Moore 1992). Red-tails exploit a range of food sources (including leaf, flower, fruit and insects), and are usually restricted to forests, given the increased predation pressure they are likely to face in open habitats (Isbell 1994). Given the low availability and linearity of riverine forests on which Issa red-tails depend, we hypothesized that, compared to closed-forest red-tail groups, Issa red-tails would: (1) occur at a lower population density, (2) exploit the miombo woodlands – either for food or for travel and (3) have a larger home range size.

Methods

Study Area

The Ugalla region, approximately 3300 km^2, lies in the north-eastern part of the Greater Mahale Ecosystem, equidistant between Gombe and Mahale Mountains National Park, and inland ~80 km from Lake Tanganyika. The Issa Valley study area, where we collected the current data, lies at approximately 5.50°S, 30.56°E. The study area is ~70 km^2, and includes several vegetation types: miombo woodland, swamp, grassland, wooded grassland and evergreen open- and closed-canopy forest (Hernandez-Aguilar 2009; Stewart *et al.* 2011). For this study, we classified all evergreen forest (primary and secondary with > 80% canopy) as 'closed', and all other deciduous vegetation (swamp, grassland, wooded grassland, woodland) as 'open' (Stewart *et al.* 2011). Open vegetation accounts for approximately ~97% of the region, while closed (riverine forest) represents only ~3% (unpublished data). Miombo woodland consists mainly of two genera of deciduous tree species, *Brachystegia* and *Julbernardia*, which lose their leaves by the end of the dry season. Evergreen forests are almost exclusively located adjacent to rivers, or at points of water-runoff, and remain green year round.

There are two distinct seasons at Issa: the dry season (May to September) and wet season (October to April), with temperatures reaching as low as 14°C in the wet season and as high as 34°C in the dry season (Stewart 2011). Annual rainfall averages ~1200 mm (range 955–1400, n = 5 years, 2002–2003; 2009–2013). In the dry season, from May to October, mean monthly rainfall ranges from 0 to 54 mm (unpublished data). We have monitored phenological patterns of 1017 trees (200 in the forest and 817 in the woodland) marked on line transects every month since 2009 and recent (unpublished) results support those of previous work that there is more fleshy fruit available in woodland vegetation, and in the wet than in the dry season (Hernandez-Aguilar 2006).

The area is home to a diversity of wildlife, including several species of primates (e.g. *Cercopithecus ascanius*, *Chlorocebus pygerythrus*, *Pan troglodytes*, *Papio papio cynocephalus*,

Figure 33.1 An adult male red-tail monkey in the miombo woodland (*Brachystegia*) of Issa (Catie Profaci).

Piliocolobus tephrosceles, *Otolemur crassicaudatus*), large mammals (e.g. Felidae: *Panthera leo*, *P. pardus*, Bovidae: *Alcelaphus lichtensteinii*, *Hippottragus equinus*) (Hernandez-Aguilar 2009), and birds of prey (e.g. *Milvus migrans*, *Polyboroides typus*, and *Terathopius ecaudatus*), among others. We have maintained a permanent research presence at Issa since 2008, focused initially on chimpanzee behavioural ecology (Hernandez-Aguilar *et al.* 2013; Rudicell *et al.* 2011; Stewart & Piel 2013; Stewart & Pruetz 2013; Stewart *et al.* 2011), and more recently on sympatric red-tails and yellow baboons (Johnson *et al.* 2013).

Data Collection

We used three types of data from several different study periods: (1) long-term line-transect data were collected from January 2009 to December 2012; (2) dietary data were recorded opportunistically from May 2011 to April 2012; and (3) focal follows of a single study group of red-tails were conducted from February to April 2012.

We used data collected from line transects to assess population density. We established seven line transects in 2009 to monitor large mammal (including red-tail monkeys) presence and density across time. Over this same period we also recorded all opportunistic encounters with red-tail monkeys across the study area on recconnaissance (recce) walks. We separated the transects between 400 and 800 m (using a randomly selected distance in that range) and all crossed the study area on the same bearing. AP and one of two field assistants walked all transects in 2009 and through March 2010, after which field assistants walked them on their own. Transect teams always consisted of two people, and we tried to walk all transects on consecutive days in the middle of the month, although this was not always possible, for 48 consecutive months. Following Marshall *et al.* (2008), we recorded the perpendicular distance between the nearest monkey and the transect line. Additionally, we recorded all large mammal observations, noting GPS coordinates, vegetation type, species and number of animals. From these 38.6 km of transects, 92.1% represented 'open' habitat, whereas only 8.9% represented 'closed' habitat.

To investigate whether red-tails exploit woodland resources, our initial aim was to establish a comprehensive feeding species list that would allow us to quantify the proportion of food species found in closed versus open habitats. We collected dietary data directly (feeding observations) and indirectly (by sieving fresh faecal samples collected under monkeys). We opportunistically recorded direct feeding observations when a monkey was clearly seen eating any food type (fruit, leaf, flower, insect, etc.). We also collected all fallen feeding remains after the monkeys had departed from the area. All plant parts (e.g. fruits, leaves) were later identified by Professor Henri N'dangalsi of the Department of Botany, University of Dar es Salaam. For faecal dietary data, we collected all fresh red-tail faeces (< 24 h old) both during focal follows and *ad libitum*. Upon return from the field, we washed and sieved faeces using 2 mm metal sieves, subsequently weighing the samples, identifying and counting all seeds, and estimating proportions of fibre, following methods described in McGrew *et al.* (2009).

We collected data on ranging behaviour from February to April 2012. Although we initiated habituation efforts of the focal red-tail group in July 2011, they were not habituated until February 2012. The group was identified by the presence of several distinctive individuals (described in results below). From then, focal follows were conducted 23 times between February and April. To track group ranging patterns, we used the 'track' function of Garmin GPS units to automatically record waypoints every five minutes from as close to the group as possible without disturbing individuals. When we did not see or hear the monkeys for more than 30 minutes, tracking was discontinued and data collection interrupted until we encountered another individual. We also recorded group counts and demographics opportunistically when the monkeys crossed an open gap in the canopy.

Data Analysis

Red-tail monkey population density was calculated from line transects using the program DISTANCE (V.6, see Thomas *et al.* 2010). We tested multiple models, in accordance with recommendations of Buckland *et al.* (2010) and Thomas *et al.* (2010), with a quantitative method for model selection that uses AIC values to evaluate the fit of the model. We selected the model that gave the lowest AIC value, and here report the result of a single model that fit the data ($\chi^2 = > 0.05$). All assumptions of DISTANCE were met. Following methodology of other studies (see Table 33.1), we also calculated population density by dividing our group size (35) by the size of its home range, as estimated using several different methods (see below).

For dietary analysis, we counted the total number of feeding tree species ($n = 23$) that were either savanna woodland, both savanna woodland and forest, or exclusively forest species. We also provide the type of food (e.g. liana, fruits, leaves) consumed from each woodland tree.

Table 33.1 Characteristics of red-tail monkey density and ranging parameters across Africa. In studies with multiple groups, the mean (of all groups) value is given.

	Study Site (# of study groups) [length of study]	Day range (m/day)	Home range size km²	Methodology	Pop. density (individuals/ km²)	Group density (groups/ km²)	Group size (# of individuals)	References[1]
Tanzania	**Ugalla** Issa Valley (1) [3 months]		8.71	MCP100% (forest + woodland)	4.01	0.11	35	**Current study**
			1.93	MCP100% (forest)	18.13	0.52		
			1.51	GCM 100 m × 100 m	23.18	0.66		
			0.78	GCM 50 m × 50 m	44.90	1.28		
			–	Line transects	0.68+0.30	-		
Uganda	**Kibale** Ngogo (3) [12 months]	1390 ± 33	0.35 ± 0.44	MCP100%	-	6.35	17.81	1
	Kibale Ngogo (1)	1000	0.63	GCM 25 m × 25 m	-	-	50	2
	Kibale Kanyawara (3) [4 months]	–	0.21	MCP100%	-	6.3	-	3
	Kibale Kanyawara (4)	1447±253	0.24	GCM	120–158	4–4.5	30–35	4
	Kibale Dura River, Sebatoli & Mainaro HL – K15 (3) UL – K30 (3) [12 months]	640.9 622.9	0.37 ± 0.12 0.26 ± 0.04	Line transects	38.12 135.05	1.04 4.83	15.0 ± 0.5 27 ± 0.9	5, 6
	Budongo Forest Reserve N3 (Logged Group) N15 (Unlogged Group) [4 months]	1297 956	0.199 0.204	GCM 100 m × 100 m	60 19.2	13.3 4.2	15.6 13.8	7
Kenya	**Kakamega** (4) [11 months]	1543 ± 296	37.3 ± 12.7	GCM 50 m × 50m	72	5.17	22.8 ± 9.0	8, 9
Central African Republic		–	0.15	NA	–	–	17–23	10

[1] References: 1, Brown 2011; 2, Windfelder & Lwanga 2002; 3, Wrangham *et al.* 2007; 4, Struhsaker & Leland 1979; 5, Rode *et al.* 2006; 6, Chapman & Lambert 2000; 7, Sheppard 2000; 8, Cords 1987b; 9, Cords 1990; 10, Galat-Luong 1975.

We calculated HRS over a three-month period (February, $n = 3$ days; March, $n = 7$ days; April, $n = 13$ days) using the grid cell method (GCM: Grueter *et al.* 2009). We included all days – full (12 h, $n = 3$) and partial (1 h < 12 h, $n = 20$, mean = 8.15 h) in our calculations; this was to maximize sample size and improve accuracy of HRS estimates. We entered all data into ArcGIS versus 9.3 (ESRI, Redlands, California, USA), and subsequently superimposed a 50 m × 50 m grid. All cells containing data points were summed, and the HRS was calculated by multiplying the total number of cells by the area of one grid cell (*sensu* Grueter *et al.* 2009). We repeated this analysis using a 100 m × 100 m grid in order to compare to previous studies.

Researchers studying home range size in non-human primates and other animals have demonstrated that the type of analytical technique chosen influences HRS estimates (Boyle *et al.* 2009; Grueter *et al.* 2009; Pebsworth *et al.* 2012). We used GCM for our analysis because it has been previously demonstrated that other techniques, such as MCP and kernel density estimates, may over-exaggerate HRS when sample size is small and there are large amounts of unused area (Borger

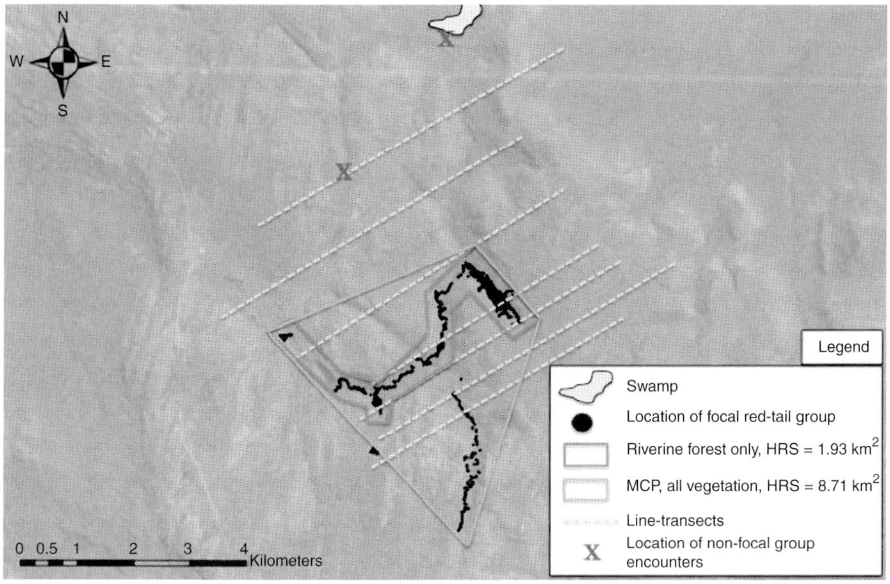

Figure 33.2 Issa Study Site, Ugalla: Red-tail home range for February–April 2012. Polygons surrounding the focal group follows describe the imprecision of methods e.g. orange polygon (8.71 km^2), all vegetation types included, versus blue polygon (1.93 km^2, riverine forest only with ~ 100 m buffer). (A black and white version of this figure appears in some formats. For the colour version, please refer to the plate section.)

et al. 2006; Boyle *et al.* 2009). We also measured the red-tail HRS using MCP to compare estimates across these methods and to other studies.

Results and Discussion

The data reported here offer a snapshot of red-tail behavioural ecology in a mosaic environment, where these guenons live predominantly in riverine forest strips surrounded by miombo savanna woodland. Our results, while preliminary and based on only 3 months of data collection, support our hypotheses that red-tail monkeys in the Issa Valley occur at a lower population density and also exhibit larger home ranges than those resulting from previous (albeit significantly longer) studies in tropical forest. We also found support for our hypothesis that Issa red-tails take advantage of the dominant miombo woodlands that surround the forest strips, using this vegetation type for either foraging or travel. Of course, there are inherent dangers of comparing multi-year studies with that of our 3-month study. The current data may not represent 'normal' behaviour and instead be an anomaly, for example, if there was unusually low food availability during the study period. While we attempt to biologically interpret our results here, small sample size, the brevity of the data collection period, and especially method all may contribute to these differences.

Group Size and Population Density

Between February and April 2012, focal group counts fluctuated between 30 and 35 individuals, with variation likely attributable to observer error (e.g. missed individuals). We are not confident that we reliably distinguished adult females from large juveniles, but we are confident that we regularly observed two to three adult males in or near the group, unusual for this species (but see Galat-Luong 1975). Not only did these individuals exhibit sex-specific characteristics (e.g. dropped testes), they were also much larger than all other group members. We identified each male by specific markings or features (e.g. kinked tail, noticeable scars, etc.) and documented their presence and location opportunistically. Over the course of the study period, we did not observe any intergroup encounter; however, we did observe one other group on multiple occasions (during reccee walks and line transects), in different riverine forest patches outside of the range of the study group (Figure 33.2).

Like chimpanzee population densities at Issa, red-tail monkey population densities (0.46–44.90 individuals/km^2; Table 33.1) are lower than those of red-tail monkeys in tropical forests (Table 33.1). Calculating population density is complicated by the heterogeneity of the Issa habitat. Table 33.1 describes the degree to which population density estimates vary with method and inclusion of different vegetation types. Over the four years of walked transects, a total of 158 individuals were counted from transects, across 19 observations ($\bar{X}$ = 8.31 individuals, ± 5.26). Using DISTANCE, this yielded the lowest density estimate of all methods, 0.68 individuals/km^2 across both open and closed habitats. Line transects may underestimate red-tail density estimates at Issa because they undersample the core area used by the focal group (Figure 33.2). Thus, our line-transect density estimate may be biased by over-sampling miombo woodland, and undersampling the riverine forests, the dominant habitat in which the red-tails live.

Population densities calculated from the home range size of our single study group vary widely depending on the method used to calculate HRS (Table 33.1). Using the MCP method, which includes both 'open' and 'closed' habitats, population density is calculated to be as low as 4.01 individuals/km^2. However if the polygon excludes large tracts of woodland where red-tails were not recorded to range, this estimate increases to 18.13 individuals/km^2. GCM methods give the highest estimates of population density, which are greater still, if smaller grid cells of 50 m × 50 m are used (Table 33.1).

Issa red-tails are primarily forest dwelling, and their home range almost exclusively follows the course of the riverine forests (Figure 33.2). These riverine forest strips can be as thin as one tree in width, and if red-tails remain close to the forest edge, even when foraging or ranging in woodland, then GCM using small grid cells therefore likely provides the most accurate estimate of our focal group's home range by excluding large areas of miombo woodland, which would otherwise be included in MCP calculations (see Figure 33.2). Therefore while MCP may be appropriate for analysing guenon home range size in relatively large homogeneous forests, we suggest that GCM (44.90 individuals/km^2) provides the most conservative estimate of red-tail HRS in this mosaic environment.

Resource Exploitation

Despite the paucity of data described here, even a snapshot of Issa red-tail use of open vegetation – namely for travelling and feeding – reveals a type of behavioural flexibility not yet well described for an otherwise 'forest' guenon. We identified 23 plant feeding species, of which 6 are woodland species, 3 occur in both woodland and forest, 10 are exclusively forest, and 4 have not yet been identified. Of the 6 woodland species, we observed monkeys eating fruit from 3 and flowers from 1; for 2 of the feeding trees, the item being consumed could not be identified. In addition to observing individuals feed in woodland habitats, we also observed individuals to travel in them, sometimes as far as 150 m from the riverine forest edge. Some of these excursions into woodlands were shortcuts that monkeys used when travelling near the junction of two forest strips. As the monkeys became more habituated, we also observed longer feeding bouts (compared to when travelling) in the woodland, which sometimes lasted over 15 minutes in length. In both forests and woodlands, we recorded individuals consuming a variety of plant parts, including fruits, leaves, and flowers. Although we often observed the monkeys chewing on items plucked from tree branches, their active behaviour in plucking the items suggested these were likely insect prey. However, we could not confirm exactly the type of food (insect, exudate or small lizards etc.) consumed during these feeding bouts.

The dietary data that we present here provide a fraction of the species that red-tails consume. However, even this brief list demonstrates that red-tails are able to exploit resources outside of the forest, a behaviour which is common among animals living in fragmented habitats (Asensio *et al.* 2009). There are, however, multiple reasons to treat these data with caution. First, they are sparse, and are likely influenced by those individuals most comfortable feeding in the presence of human observers. Second, we observed monkeys to feed on multiple plant species for which we were not able to obtain identifications, local or scientific. Lastly, red-tails have been described as 'frugivorous insectivores' (Chapman *et al.* 2002), and we could not confirm any observations of insectivory. Future work will look more into to what extent Issa red-tails also consume invertebrates in their diet.

Using GCM 50 m × 50 m, 158 of 310 total cells (describing red-tail home range at Issa) contained at least one data point in the miombo woodland, suggesting that red-tail monkeys are able to exploit open environments by exhibiting behavioural and ecological flexibility (e.g. Donati *et al.* 2011). However, as we cannot yet provide data on total time spent in each vegetation type, our observations of woodland use may be a result of either (1) prolonged woodland foraging bouts or (2) travelling excursions into the woodland. Further research will systematically assess and quantify the reliance on extra-forest resources, and explore whether seasonal changes in fruit availability, leaf cover or water availability influence the intensity and patterns of red-tail use of woodland habitat.

Home Range Size

Regardless of which method is used to calculate home range size, Issa red-tails exhibit a home range that is 2–14 times larger than those of closed-forest habitats of Kibale and Kakamega (Table 33.1), with the exception of the extremely large red-tail group (51 individuals) at Kibale, which fissioned soon after study (Windfelder & Lwanga 2002). As noted above, GCM using smaller (50 m × 50 m) grid cells, provides the most conservative estimate of a home range size of 0.78 km^2 (Table 33.1). It is worth noting, however, that the estimated home range size of the Issa red-tails is likely to increase with further research, given that the current range was calculated from a limited, preliminary study of only 23 full day follows over 3 months. It remains unclear to us why the Issa red-tails exhibit such a large range, as while group spread may be directly affected by the linearity of food sources (e.g. interindividual spacing may increase to reduce intragroup feeding competition; Lambert 2002), absolute HRS must not necessarily increase. It could be that the riverine forests of Issa are less productive than those of other tropical forests, resulting in the monkeys having to forage over a larger area to meet daily nutritional requirements, or that the extreme seasonality of Issa means that large parts of the red-tail range are inadequate for parts of the year. Future analysis of forest productivity as well as intra-annual ranging patterns may inform on the reasons for increased ranging and the role of seasonality in habitat use, respectively.

Conclusion

In this chapter, we have described preliminary results of a long-term study of red-tail monkeys in a mosaic environment. The implications of this work are twofold: First, we have shown that the heterogeneity of the Issa habitat means that the selected method of HRS calculation can result in a twofold order of magnitude difference in calculated home range size or density. Thus, ensuring that cross-population comparisons use similar methods is critical. Second, although studies from tropical forested sites such as Kibale and Kakamega have provided extensive knowledge of red-tail monkey behavioural ecology (Cords & Sarmiento 2013), our long-term study at Issa expands our understanding of the behavioural flexibility of these guenons, specifically the impact of living in thin riverine forest strips on home range size and density. Subsequent work will address to what extent diet, intergroup dynamics, demography, and activity budgets also respond to such conditions.

Acknowledgements

We thank the editors for inviting our brief report on Issa's riverine red-tails, as well as Shedrack and Joffrey Lucas, Busoti Juma, Ndai Sammwely, Msigwa Rashid and Mlela Juma for their invaluable help and enthusiasm in the field. Many thanks also to Jim Moore and especially Marina Cords for their comments on an earlier version of this chapter. We are enormously grateful to the UCSD/Salk Center for Academic Research and Training in Anthropogeny (CARTA) for support of ongoing research at Issa. Finally, we thank the Tanzania Commission for Science and Technology and Tanzania Wildlife Research Institute for permission to work in Tanzania.

Chapter 34

Consequences of Lakeside Living for the Diet and Social Ecology of the Lake Alaotran Gentle Lemur

Patrick O. Waeber, Fidimalala B. Ralainasolo, Jonah H. Ratsimbazafy and Caroline M. Nievergelt

Madagascar is without a doubt the world's primate conservation priority, with very high species and unmatched endemism at the species, genus and family levels. It is the second country on the world list for primate species diversity (in spite of being less than 7% the size of Brazil, the world leader in primate species diversity, and roughly one-third the size of Indonesia, third on the list). All its 107 taxa (including subspecies) are endemic to Madagascar (Mittermeier *et al.* 2010; Tattersall 2013).

Among the lemur species, members of the genus *Hapalemur* are best known for a diet dominated by bamboo (Rand 1935); an unusual ecological specialization among primates. *Hapalemur* can be readily distinguished from *Lemur*, *Eulemur* and *Varecia* by features of the head, including its round shape, large face, short muzzle and ears that are largely hidden by fur. All *Hapalemur* are small to medium sized greyish animals, with moderately long hindlimbs. They prefer vertical resting postures, and leap readily between closely spaced vertical supports. They tend to be crepuscular in their habits. As in *Lemur*, brachial and antibrachial glands are present (Macedonia & Stanger 1994).

The Alaotran gentle lemur (*Hapalemur alaotrensis*) belongs to a genus, which now includes two other extant species: the lesser bamboo lemur (*H. griseus*), the golden bamboo lemur (*H. aureus*); the greater bamboo lemur (formerly known as *H. simus*) having been reclassified to its own genus *Prolemur simus*, after resurrection of the name *Prolemur* by Groves in 2001. The phylogenetic relationships within the species *H. griseus* remain controversial (Fausser *et al.* 2002; Pastorini *et al.* 2002). Four taxa have been recognized: the dark-coated southern subspecies (*H. g. meridionalis*), and the three more closely related taxa: *H. g. occidentalis* from the deciduous forests of western Madagascar, *H. g. griseus* from the humid eastern forests (since 2001 elevated to its own species *H. griseus*: Groves 2001), and the larger *H. g. alaotrensis* endemic to the Lake Alaotra in the east, also its own species since 2001 (*H. alaotrensis*: Groves 2001). However, although (to some extent) distinguishable in ecology and physical features, based on karyotypes and mitochondrial DNA sequences, no clear separation between *H. griseus* and *H. alaotrensis* have been found in studies specifically focusing on these taxa (Nievergelt *et al.* 2002a; Pastorini *et al.* 2002). Remarkably, all bamboo lemurs are forest dwellers, and occupy a variety of forest habitats across Madagascar. The grand exception is *H. alaotrensis*, which is confined to the marshlands of the Lake Alaotra.

Lake Alaotra is located in the Alaotra–Mangoro region in the central-east of Madagascar covering an area of 20 000 ha in the central highlands of the island. It is Madagascar's largest lake and was designated as a wetland of international importance under the Ramsar Convention in 2003. The Government of Madagascar recognized the conservation value of this area by classifying it as a new protected area on 17 January 2007 (Chapter 37). The surrounding 20 000 ha of marshes is dominated by *Cyperus madagascariensis* (Cyperaceae) and *Phragmites communis* (Poaceae). Adjoining this area are 120 000 ha of rice fields within a watershed encompassing a total of 722 500 ha. The endemic *Hapalemur alaotrensis* is the only primate taxon in the world that lives exclusively in a marsh habitat. The species is classified as 'Critically Endangered' according to IUCN Red List criteria due to its extremely small natural geographic range.

Studies on *H. alaotrensis* in its natural habitat are complicated by the restricted access to the marshlands. Most observations are made during the rainy seasons when increased water levels allow for the passage of canoes through the dense reed and sedge beds bordering the lake. With these limitations in mind, the known social groups of *H. alaotrensis* typically include 2–9 individuals (Nievergelt *et al.* 2002b), with only a few exceptions of groups bigger than 13 individuals (Waeber & Hemelrijk 2003). The forest-dwelling *H. griseus* have slightly bigger groups of 3–6 individuals (Overdorff *et al.* 1997; Petter & Peyrieras 1970; Pollock 1986; Wright 1986; Wright & Randriamanantena 1989). As with *H. griseus*, groups of *H. alaotrensis* can contain a variety of combinations of adult individuals; one male and female (with 42% the majority of cases), one male and two females, one female and two males or several males and two females (Nievergelt *et al.* 2002b). The same variability in social structure is observed in *H. griseus* (Grassi 2006; Tan 2007). The home range of *H. griseus* was estimated to be between 6 and 20 ha (Overdorff *et al.* 1997; Wright 1986; Wright & Randriamanantena 1989). For the other species, like *H. aureus*, the family group is estimated to be about 2–4 individuals maintaining a territory of up to 80 ha (Meier *et al.* 1987; Wright & Randriamanantena 1989). In comparison, home ranges of *H. alaotrensis* are much smaller, varying only from 2 to 5 ha depending on group size (Mutschler *et al.* 1999; Nievergelt *et al.* 1998).

Evidence for the social system, mating system and territorial behaviour of *H. alaotrensis* has accumulated over the last 15 years through studies in captivity, the field and by means of genetically determined relatedness of animals from longitudinally followed groups (Nievergelt *et al.* 2002b). The birth season for *H. alaotrensis* typically starts in September and ends in February (Mutschler 1999), making it longer than any other lemur species; also, births are not synchronized within groups (Nievergelt *et al.* 2002b). In the forest dwelling *H. griseus* and *H. aureus*, birth seasons are shorter and take place between October and November and between November and December, respectively (Tan 2000). For all *Hapalemur* species the gestation period is similar, around 137–149 days. In *H. alaotrensis*, females are dominant, with most travelling groups led by females, and with females accessing new food sources first. Interestingly, males do lead channel crossings more often than females (Waeber & Hamelrijk 2003). In *H. alaotrensis*, both males and females are involved in territorial interactions with neighbouring groups, but with males playing more active roles (Nievergelt *et al.* 1998). In *H. griseus*, males actively defend resources, while the females rarely take part in territorial interactions; in this case, males are the primary responsible for defence of the home range (Grassi 2002). Both males and females disperse, but females most likely emigrate one year earlier than males. In *H. griseus*, unlike other lemur species, female dominance may be established before sexual maturity. It seems that for this species females are more active in defining social relationships than males; females are more often the aggressors and males the recipients of aggression (Grassi 2002). As is the case with *H. griseus*, only one male is reproductively active in each *H. alaotrensis* group. Both, monogamous and polygamous mating systems occur, with one (40%) or two related breeding females (60%), respectively (Nievergelt *et al.* 2002b). This high social flexibility has also been shown for *H. griseus* and *Prolemur simus* (Tan 1999). *Hapalemur alaotrensis* shows a high reproductive output; females in both mating systems have been observed to breed across several consecutive years. This is also the case with *H. griseus* in multi-female groups.

A wide array of activity patterns has been reported for *Hapalemur*. While *H. alaotrensis* is cathemeral (irregularly active at any time of night or day), but with two distinct activity peaks, one in the early morning and one in the late afternoon (Mutschler 1999; Mutschler *et al.* 1998), *H. griseus* is diurnal (Overdorff *et al.* 1997; Pollock 1986; Wright 1986) with a tendency towards being crepuscular with peak activities around dawn and dusk (Petter 1962; Petter & Peyrieras 1970). Diurnal activity also appears to be the pattern in *H. aureus* (Meier *et al.* 1987; Wright & Randriamanantena 1989).

Hapalemur alaotrensis lives in a very open habitat with high risk of predation. Potential predators include two species of small carnivores (ring-tailed mongoose, *Galidia elegans*, and Durrell's vontsira, *Salanoia durrelli*); the yellow-billed kite (*Milvus aegyptius*) could be a danger for smaller individuals, and the Madagascan tree boa (*Sanzinia madagascariensis*) is often found resting during daylight in tall grasses at the borders of the marshland, thus representing another potential threat. As the smallest of the diurnal lemurs in the forest, *H. griseus* is at greatest risk of predation. Individuals are vulnerable to attacks from diurnal raptors, scavenging diurnal (*G. elegans*), nocturnal (striped civet, *Fossa fossana*) and cathemeral carnivores (fossa, *Cryptoprocta ferox*), as well as snakes (*S. madagascariensis*) (Csermely 1996; Rakotondravony *et al.* 1998; Wright 1995, 1999). *Hapalemur griseus* appears to use several strategies to avoid predation, the most important of which is the choice of height used for different activities and corresponding predation threat (Grassi 2001). Although the wetlands have weakly developed vegetation structure (i.e. heights of 3–4 m versus up to 15–20 m for forests), similar predator-associated behaviour is shown by *H. alaotrensis*; when a potential aerial predator is approaching, there is a specific call (see Waeber & Hemelrijk 2003) from the vigilant member, all group members then rush to the lowest portions of the marsh vegetation. Here they immediately hide and freeze until the potential threat is over. When there is a ground attack the reverse happens and the group ascends, pushing the most vulnerable younger group members (lighter in body weight) up the vegetation.

Compared to its congeners, *H. alaotrensis* can be considered as one of the most specialized herbivorous lemurs. It displays relatively low dietary diversity, targeting 11 plant species and 16 different plant parts (including pith, stem, leaves, buds, seeds, shoots and flowers; Mutschler 1999). The level of specialization becomes even more apparent when time spent feeding on each plant species is considered; more than 95% of time is spent on *Cyperus madagascariensis* (Cyperaceae), and three species from the family Poaceae: *Phragmites communis*, *Echinochloa crusgalli* and *Leersia hexandra* (in order of feeding preference). For *H. alaotrensis*, staple plant diet species are available year round. This is in contrast to its forest-dwelling close relatives who must adapt to seasonal changes in food plant availability, especially *Prolemur simus* (Tan 1999). At any one locality, the diversity in terms of food plant species used by *H. griseus* is small when compared to other lemurs. However, across its geographic distribution, the dietary flexibility of *H. griseus* is greater than either *H. aureus* or *P. simus* when the number of species eaten is considered. This can be attributed to a wide geographic distribution and range of altitudinal distributions at which *H. griseus* is found (Grassi 2006; Tan 1999).

With its cathemeral behaviour, *H. alaotrensis* has one of the highest activity budgets over the course of 24 h of any folivorous primate, and seems not to reduce energy expenditure at the behavioural level (e.g. increased resting time or reduction of other activities; Mutschler 1999; Waeber & Hemelrijk 2003) as is suggested in other similar sized primates with low-energy diets (e.g. diurnal or nocturnal Indriidae; Ganzhorn 1995). Although food choice is highly selective with specific parts of different food plant species being targeted in comparison to other primates, *H. alaotrensis* has a low-energy diet, with low protein and moderate fibre content (Mutschler 1999). Mutschler (1999) suggests that *H. alaotrensis* has increased digestive capacities due to hindgut fermentation, as is the case with *H. griseus* and *H. aureus* where gut passage time is long (18–36 h) (it is with less than 10 h in *P. simus*; Tan 1998, 1999).

Hapalemur was only able to occupy one wetland in Madagascar (i.e. Lake Alaotra). There are only five fossil deposits with material belonging to *Hapalemur* and *Prolemur*.

Figure 34.1 Madagascar fossil deposits of *Hapalemur* and *Prolemur.*

Fossil material from *Prolemur simus* has been recorded in Ampasambazimba, Ankarana, Anjohibe, Bemaraha and Montagne des Français, while material attributed to *Hapalemur* cf. *griseus* has been recorded from Ankarana and Ampasambazimba in the highlands (Figure 34.1) (Burney *et al.* 2004; Crowley 2010; Godfrey *et al.* 2004, 1996; Tattersall 1973). The question remains, why was *Hapalemur* not able to occupy a greater area? Or, alternatively, why did *Hapalemur* occupy the Alaotra wetlands in first place? Grassi's (2006) study describes the ability of *H. griseus* to adapt to a variety of microhabitats within its current distribution by altering population density, group size, and mating system, and by adjusting its diet according to local resource abundance and its spatial and temporal distribution. Given the genetic, behavioural and ecological similarities of *H. griseus* and *H. alaotrensis*, we hypothesize that the common ancestor to the two taxa was originally forest-living, and used occasionally the adjacent wetlands at Alaotra, and then gradually became more specialized. As is a general trend in the highlands of Madagascar where many wetlands and forests are disappearing due to anthropogenic land transformation (Kull 2012), forests no longer surround the wetlands at Alaotra. The closest primary forest is Zahamena some 30 km eastwards, and here *H. griseus* still occurs. The current situation could, therefore, represent an evolutionary snapshot, where, as described in this chapter, *H. alaotrensis* and *H. griseus* still share a high degree of similarity, indicating that time enough has not yet passed to allow the development of fully distinct wetland-adapted ecological and behavioural features in the former. Given this is the only year-round marsh-living primate, it is unclear what such features would look like. It seems that the marshlands for *H. alaotrensis* mean a 'cul de sac'; spatially this lemur will likely not be able to go back to living in forests (Chapter 37).

Acknowledgements

We would like to thank Lucienne Wilmé for providing the map.

Chapter 35

Non-leaping Slow Lorises

Ecological Constraints of Living in Flooded Habitats

Anna Nekaris, Denise Spaan and Vincent Nijman

Introduction

Slow lorises (*Nycticebus*) are distributed across parts of South and Southeast Asia. They are a unique primate genus in terms of their diet (highly insectivorous and exudativorous), locomotion (inability to leap or jump), strong precision grip with specialized morphological adaptations to hold on to branches, and their venom (Nekaris 2013; Nekaris *et al.* 2013a). Arguably, of all primates, the non-saltatory slow loris is most affected by flooding, since this may entirely restrict its available pathways. Gaps measuring over a metre in arboreal pathways, easily crossed by most primates, may be insurmountable barriers for slow lorises. Several slow loris species are known to cross small gaps by bridging (Nekaris 2001), or positioning their body to ensure that the branch is lowered by their weight, but there are clear limits on where and how such methods can be used. When no arboreal pathways are available, slow lorises come down to the ground to cross gaps (Nekaris 2001; Wells *et al.* 2004).

For an animal that is unable to leap from one branch to another, connectivity of forest strata is of paramount importance. The strong grip and unique locomotion of slow lorises requires that in order to move across gaps in their habitat they must rely on slow and steady climbing, bridging and regularly make use of all levels of forest structure, including the ground (Nekaris 2001). Slow lorises also are adapted to holding onto trunks to feed on plant exudates for extended periods (Starr & Nekaris 2013). In undisturbed forests, slow lorises reside primarily in the understory, where dense vegetation is at a premium, and where they also remain less exposed to forest eagles, one of their main predators (Nekaris & Bearder 2007). Although slow lorises can move in the canopy, predation risk poses a danger. Thus, in areas that are either seasonally or permanently inundated, flooding would pose an ecological barrier.

Despite this, it has been suggested that slow lorises inhabit flooded habitats, including swamps and mangroves (Rowe 1996; Wolfheim 1983); however, to date, no systematic surveys comparing study sites have been done, and little is known about slow lorises in these types of habitats across their range. Here, we provide an overview of slow lorises in flooded habitats. Furthermore, based on a study conducted in a heavily disturbed and non-flooded habitat, we present data on ground use. We predict that slow lorises in a heavily disturbed forest will use the ground more than the 1–3% of ground use observed in less disturbed forests. We then discuss the amount of time spent on the ground in order to evaluate the implications of how slow lorises might survive in flooded habitats.

Methods

Slow Lorises in Flooded Habitats

We reviewed the literature for data on slow lorises in flooded habitats using the Web of Science, Google Scholar (search terms loris; *Nycticebus* and mangrove; swamp; riverine as well as their Malay/Indonesian equivalents), and consulted an extensive literature database on slow lorises maintained by K.A.I. Nekaris. We restricted ourselves to the Sundaic region, i.e. the Thai-Malay Peninsula south of the Isthmus of Kra, Singapore, Sumatra, Java and Borneo. Six species of slow loris are found in this region: the greater slow loris (*N. coucang*) on the Thai-Malay Peninsula, Singapore and Sumatra, the Javan slow loris (*N. javanicus*) on Java, and the Philippine slow loris (*N. menagensis*), Bornean slow loris (*N. borneanus*) and Kayan slow loris (*N. kayan*) in parts of Borneo and off-lying islands. In addition, the Bangka slow loris (*N. bancanus*) is found on the islands of Bangka and Belitung, but no information on this species is currently available. Few in-depth studies have been conducted on slow lorises in this region, and the ones that have were almost invariably conducted in non-flooded habitat types (Barret 1981; Wiens & Zitzmann 2003). Notable exceptions are the studies by Munds *et al.* (2013) and Nekaris *et al.* (2008), which were conducted in semi-inundated riverine forest in the Malaysian part of Sabah, and in peat swamp forest in Indonesian Kalimantan, respectively.

We included only data directly linking slow lorises to flooded forests (swamp, riverine and mangrove). We included riverine forest only if it was clear that this comprised regularly flooded forest occurring on floodplains, and not just dry forest situated along rivers. Likewise, we only included coastal forest when there was evidence of it being influenced by flooding. Beach forest, a distinct coastal forest form, was excluded, as these are almost never inundated (van Steenis 1958). For many Sundaic national parks, information on habitat types is available alongside species lists. The presence of slow lorises in these parks, as well as the presence of flooded habitats in itself, was not sufficient to be included in our overview. Likewise, references to particular flooded forest types as habitats for slow lorises in general faunal or regional compendiums or encyclopaedias without providing additional information were discounted (but see Discussion).

Use of the Ground by Javan Slow Lorises

We studied ground use of Javan slow lorises in and around the village of Cipaganti, West Java, Indonesia (7°6′6–7°7′0S and 107°46′0–107°46′5E). Cipaganti is situated on the eastern slopes of Mt Papandayan, with the study conducted between elevations of 1200–1600 m asl (the area does not flood). The study area is characterized as a mosaic habitat in which crop fields are interspersed with introduced tree species such as *Acacia decurrens* and *Eucalyptus* sp., as well as patches of bamboo (mainly *Gigantochloa apus* and *G. atter*). Commonly grown crops include carrots, cabbage, tea and tomatoes. We collected behavioural data on the Javan slow loris between August 2013 and March 2014 on 18 radio-collared individuals (6 males, 12 females). We collected data at 5-minute intervals in two shifts during all-night follows (roughly from 17:00 to 05:00 h). We followed females on average over 11 nights during which we collected 185 data points. For males, our sample averaged 9 nights and 146 data points. We analysed the proportion of time slow lorises spent on the ground or within 50 cm from the ground – this being the level that even a moderate flood would inundate – and the proportion of time slow lorises hold on to 1, 2 and up to 5 different branches or substrates – reflecting the need for continuity in vegetation structure.

Results and Discussion

Slow Lorises in Flooded Habitat

Mangroves

We found little direct detailed accounts of slow lorises in mangroves (see Nowak 2012). An exception was a study by Nor (1996) who reported on Philippine slow loris on the island of Banggi, off the north coast of North Borneo. He recorded slow lorises at a number of sites on the island, including Sabor and Karakit. The habitat type of Sabor was described as including logged forest, scattered open grassland and mangrove forest on the coast whereas for Sabor, Nor (1996: p. 13) noted 'several specimens were collected from the nearby mangrove forest'. This tentatively suggests that slow lorises are possibly found in, or make use of, mangroves on Banggi. Significant areas of mangroves occur naturally along the east coast of Sumatra, the north coast of Java, throughout the coastal fringes of Borneo, as well as on the numerous smaller islands dotted around the Sundaic region. However, mangroves are severely threatened, and many have been converted to fishponds, agriculture or been logged (Giesen *et al.* 2006; MacKinnon 1996; Valiela *et al.* 2001; Whitten *et al.* 1996, 2000) and, crucially, little work has been done on the nocturnal fauna of mangroves in the region.

As indicated in our Introduction, mangroves have been listed as a suitable habitat for slow lorises in a number of compendiums and encyclopaedias, including ones authored by Nekaris (Nekaris 2013; Nekaris & Bearder 2007), but contemporary data on slow lorises in mangroves are scant. Mangroves in Southeast Asia comprise several zones, including fringe mangroves, central mangroves, back mangroves and brackish stream mangroves (Giesen *et al.* 2006 and see Chapter 2). The former types are inundated daily by the incoming sea and terrestrial movement in these habitats by slow lorises would be difficult. We anticipate that if slow lorises are indeed found in mangroves, they will inhabit mainly the back and brackish stream mangroves as these abut the other habitat types and are, at most, inundated during high tide only. In addition, they may be found in mangroves there where they abut other habitat types habitually inhabited by slow lorises.

Freshwater Swamp Forest

Slow lorises have been recorded in a number of freshwater swamp forests throughout the region. The Javan slow loris has been recorded in the freshwater swamp forest of Ujung Kulon National Park and nearby Rawa Danau Nature Reserve (Table 35.1), but it is unclear how much time the animals spend in these forests, as drier forests sit adjacent to the flooded forests. The extent of freshwater swamp forest in Java has been greatly diminished to such a degree that Ujung Kulon and Rawa Danau are among the few remaining (Whitten *et al.* 1996). In Singapore, the greater slow loris has been recorded in the one remaining swamp forest on the island, Nee Soon Swamp Forest (Fam *et al.* 2014). On Borneo, extensive areas of freshwater swamp forest can be found in, for instance, Danau Sentarum and the Mahakam Lakes areas. Slow lorises appear on many of the regional species lists, but it is unclear to what extent they indeed use the inundated freshwater swamps.

Peat Swamp

Peat swamp forests occur where waterlogged soils prevent dead leaves and wood from fully decomposing. Over time this creates layers of acidic peat; where the soils are better drained, peat swamp forest is replaced by lowland rain forest. Many of the peat swamp forests in Sumatra and Borneo occur along the coastal plain, but there are also river-fed peat swamp forests at higher elevations in the interior. Most of our observations of slow lorises in flooded habitats originate from this forest type (Table 35.1). It is clear that slow lorises are a regular faunal component of peat swamps, but it is unclear to what extent flooding or permanent inundation impedes their movements.

Peat swamps are characterized by their dense vegetation structure (high stem number) and low numbers of rattan stems or other lianas that connect trees (Felton *et al.* 2003; Page *et al.* 1999). A prominent characteristic of periodically waterlogged tropical forests is the undulating microtopography with numerous small hummocks comprising of large woody debris, fallen logs and knee roots. Tall hummocks undergo less frequent flooding during the rainy season compared to low hummocks and peat hollows (Nishimua *et al.* 2007). Hummocks can be used by slow lorises to cross gaps, especially during the drier periods of the year. Felton *et al.* (2003) quantified gap size in an undisturbed peat swamp forest and found on average 1.5 gaps of more than 5 m wide for every 100 m of transect walked, with the largest gaps being over 25 m wide. Gaps smaller than 5 m were not recorded, but it is clear that these forests are far from continuous from the perspective of a slow loris.

Table 35.1 Slow lorises in flooded habitats in Southeast Asia.

Species	Site	Flooded habitat type	References
N. coucang	Kluet, Sumatra	Peat swamp	Van Schaik *et al.* 2009
N. coucang	Bukit Batu, Sumatra	Peat swamp	Fujita *et al.* 2012
N. coucang	Labuan Bilek, Peninsular Malaysia	Coastal forest	Wiens & Zitzman 2003
N. coucang	Nee Soon Swamp Forest, Singapore	Freshwater swamp forest	Fam *et al.* 2014
N. kayan	Muara Kaman, Borneo	Peat swamp and riverine	V. Nijman, unpub. data
N. menagensis	Kinabatangan, Borneo	Semi-inundated lowland forest, seasonal freshwater swamp	Boonratana 1997
N. menagensis	Kinabatangan, Borneo	Semi-inundated lowland forest	Munds *et al.* 2013
N. menagensis	Banggi Island, Borneo	Mangrove	Nor 1996
N. menagensis	Tanjung Putting, Borneo	Peat swamp, freshwater swamp	Nekaris & Nijman 2013
N. menagensis	Tuanan, Borneo	Peat swamp	S. Wich, pers. comm. 2013
N. menagensis	Sebangau, Borneo	Peat swamp	Nekaris *et al.* 2008
N. javanicus	Ujung Kulon, Java	Freshwater swamp	Gurmaya *et al.* 1992; Hoogerwerf 1970
N. javanicus	Rawa Danau, Java	Freshwater swamp	Melish & Dirgayusa 1996

Riverine Forest

As with many of the other flooded habitat types, riverine forest is highly threatened in the region. Especially in Borneo where road infrastructure is poorly developed, rivers are used as the main transport arteries (Meijaard & Nijman 2000b). The forest on the river banks are often the first to be converted to other land uses. The only area from which we have information on slow lorises inhabiting the flooded riverine forest comes from the Kinabatangan River in northern Borneo. Here, Philippine slow lorises occur at low numbers. Munds *et al.* (2013) explored habitat use of slow lorises in these forests and found them to use higher levels of the forest than other slow lorises, possibly in response to the presence of sympatric tarsiers (and thus partially avoiding competition).

Terrestrial Movement by Javan Slow Lorises

We collected 2111 data points on substrate use (i.e. the number of substrates slow lorises held onto at any given point in time) and 2772 data points on substrate height (i.e. the type of substrate it was in contact with and its height) during 178 nightly follows. Slow lorises were observed to use a range of different substrates, most frequently tree branches (61.5% of the time) and tree trunks (20.1%), as well as bamboo (17.9%), the latter being their preferred substrate for sleeping during the day. Slow lorises have a strong precision grip enabling them to hold on to several substrates simultaneously. Males held on to two or more branches or other substrates at the same time for 46% of the time and females for 24%. Clearly, slow lorises need to have a good grip on their immediate environment (Figure 35.1). Slow lorises are indeed largely arboreal: the ground was used for a mere 0.4% of the time by males and 0.9% of the time by females. This translates to an average of less than 3 minutes for males and over 6 minutes for females spent on the ground per night. The time individuals spent on the ground in each individual observation bout was mostly short, ranging from less than 1 minute to over 20 minutes.

Not only did females spend more time on the ground, they also made more use of the lower levels of the forest than males (Figure 35.2). The distance covered while walking on the ground differed greatly: mostly this was a couple of metres, up to 20 m, with one record of a male crossing a 300 m freshly ploughed field.

Conclusion and Avenues for Further Research

We have provided an overview of slow lorises in flooded habitats and found them to be primarily present in peat and freshwater swamp forest, and to a lesser extent in riverine forest, with limited data available on their presence in mangrove forests. Our study from Java shows that slow lorises do come down to the ground to cross gaps in areas where the canopy is not continuous. While this only represents a small proportion of their time, typically less than a couple of minutes per night, the importance of being able to come down to the ground in order to cross over to other parts of their home range is clear.

Other non-salatory arboreal mammals that live in flooded habitats have adapted ways to cross gaps in the canopy, such as sloths that are able to swim between mangrove trees (Esser *et al.* 2010). On the other hand, some species have not adapted to the barriers posed by an inundated habitat. Population densities of the pygmy three-toed sloths, for instance, are assumed to be in decline as a result of mangrove cutting and their inability to come to the ground (Kaviar *et al.* 2012). Slow lorises' terrestrial behaviour shows that they have the ability to adapt to a mosaic environment provided the area is not inundated. In flooded habitats, either seasonally or permanently flooded, where the ground is covered by water, there would be no option for the

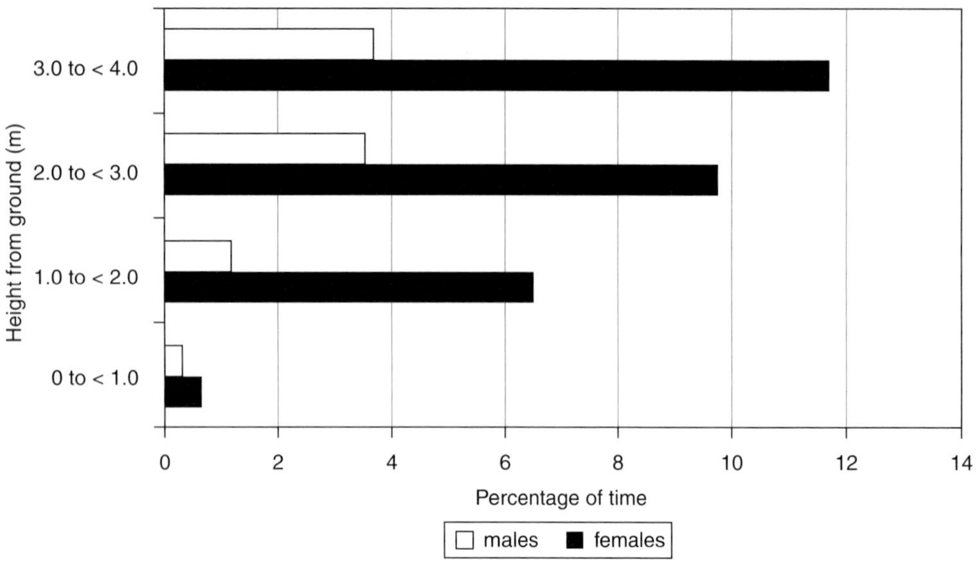

Figure 35.1 Use of the ground and lower levels by male and female Javan slow loris (*Nycticebus javanicus*) in Cipaganti, West Java, Indonesia. Note that the majority of time (males 91%, females 72%) is spent in trees and bamboo more than 4 m above the ground.

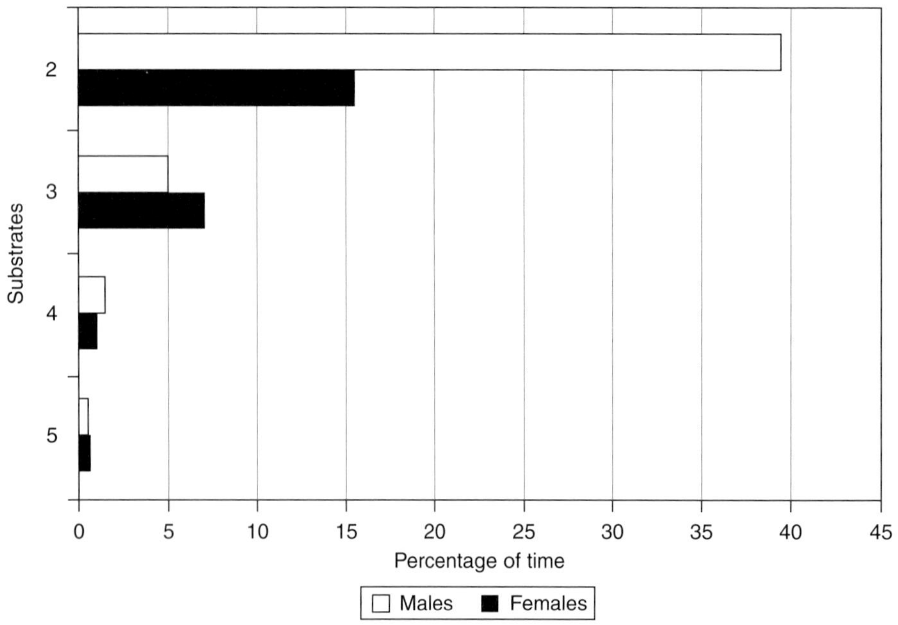

Figure 35.2 Number of substrates held on to by male and female Javan slow loris (*Nycticebus javanicus*) in Cipaganti, West Java, Indonesia. For 46% (males) and 24% (females) of the time the individuals hold on to two or more braches or other substrates.

slow loris to use the ground. In such environments, low levels of forest connectivity could lead to the formation of ecological barriers for this non-leaping primate.

On the basis of our study of the literature and the use of the ground by slow lorises in Java, it is clear that much remains to be learned about slow lorises in flooded habitats. Of the studies listed in Table 35.1 that record the presence of slow lorises in flooded habitats, few, if any, have information on the water levels at the time of recording the presence of the species. Thus, it remains unquantified to what extent slow lorises use flooded habitats at times these are indeed flooded.

Furthermore, it is unclear whether flooding affects slow lorises at the individual level, impeding movement and access to particular resources including mates, or whether flooding has a wider population level effect, with, for instance, lower densities in more frequently flooded habitats. Finally, with all species of slow lorises being considered globally threatened (Nekaris & Streicher 2008a,b; Nekaris *et al.* 2013b; Streicher *et al.* 2008a,b) it is unclear to what extent degradation of flooded forest (logging, drainage, burning, etc.) impedes their ability to make use of these habitats – not being able to come down to the ground because of high water levels and not being able to move from one tree to the next because of the increase of gaps may leave them trapped in an area that is too small for effective gene flow and ultimately, survival.

Acknowledgements

We thank the editors for the opportunity to contribute to this exciting volume and we thank them even more for their patience and understanding. Dendi Rustendi, Aconk Zeleni and Adin Nunur helped with radio-tracking the slow lorises. We also thank the Indonesian Authorities (RISTEK, LIPI) for permission to conduct our studies, and the International Primate Protection League, Leverhulme Trust (RPG-083), People's Trust for Endangered Species, Augsburg Zoo, Phoenix Zoo, Brevard Zoo, Henry Doorly Zoo and the Cleveland Zoological Park and Cleveland Zoo Society for support. We thank the reviewers and the editors for constructive comments and suggestions for improvement.

Chapter 36

Dams

Implications of Widespread Anthropic Flooding for Primate Populations

Amy Harrison-Levine, Herbert H. Covert, Marilyn A. Norconk,
Ricardo Rodrigues dos Santos, Adrian A. Barnett and Philip Fearnside

Introduction

As part of a volume dedicated to non-human primates that inhabit flooded ecosystems, it is important to acknowledge that some species live in habitats that were only recently inundated at the hands of humans. While the majority of non-human primates (hereafter primates) discussed in this book have had time to adapt in various ways to flooded environments, the primates discussed in this chapter have been forced to make significant changes virtually overnight as a result of the construction of a dam and inundation of an associated reservoir.

Over 45 000 large dams (≥ 15 m tall) have been built worldwide (Nilsson *et al.* 2005; World Commission on Dams 2000). These dams and other anthropogenic diversions affect the flow of approximately 60% of the world's 227 largest rivers (World Wildlife Fund 2004). All of the dams reviewed in this chapter (Table 36.1) are considered large dams and two are among the largest in the world, Brazil's Tucuruí Dam and Venezuela's Guri Dam (WCD 2000).

People build dams for many reasons including irrigation, production of hydroelectricity, flood control, ensuring water supplies and improving river navigation ability (Liao *et al.* 1988; World Wildlife Fund 2004). Most large dams are built for the purpose of generating electricity. Indeed, in 2004, WWF reported that almost 20% of the world's electricity was being provided by dams.

Hydroelectric dams have long been touted as a clean alternative for energy production, when compared to fossil fuels (Moore *et al.* 2010). However, recent evidence suggests that, for many years, hydroelectric dams can produce nearly as much, just as much, and sometimes even more greenhouse gas emissions than fossil fuel methods of energy production (Abril *et al.* 2005; Fearnside 2002, 2009; Fearnside & Pueyo 2012; Kemenes *et al.* 2007, 2011). Today, experts debate whether dams are indeed clean energy producers and, therefore, whether the benefits of damming truly outweigh the costs (Fearnside 2011; Poff *et al.* 2003).

Though dam construction peaked in the 1950s through the 1980s, it slowed in the 1990s as studies of their impact became available. Even with this information, 1600 new large dams were under construction in 2004 (World Wildlife Fund 2004) with many more constructed since. A dramatic spike in dam construction is anticipated over the next 10 to 20 years (Tundisi *et al.* 2014), and a high proportion of these are in primate range countries. Finer and Jenkins (2012), for example, report on plans for 151 new dams in the Amazon basin – 60% of which would affect river connectivity and more than 80% of which would result in deforestation due to new roads, transmission lines or inundation. The World Wildlife Fund (2004) estimates that of the world's remaining 64 large free-flowing rivers, at least 17 are in danger of being dammed by 2020, including several within primate habitat countries in South America and Southeast Asia.

Regardless of where the dam is built, economic, social and environmental assessments are typically conducted prior to construction. Environmental Impact Assessments (EIAs) are perhaps the most common tool used to evaluate environmental effects of dam construction and reservoir flooding (Robinson 1992). While their use is not required in all countries (Pack 1994), and the timing of incorporating EIA studies has often come too late to influence dam construction and design decisions (McAllister *et al.* 2001; Rodrigues 2006), efforts to incorporate EIA findings into dam design and mitigation plans are improving (Robinson 1992; Tullos 2008). Still, many countries that conduct pre-construction EIA studies fail to fully implement plans to minimize biodiversity losses (Schneider 2001; Alho 2011).

This chapter will highlight results of a literature review investigating how primates are affected by human-built dams. Particular attention will be given to a few well-documented case studies in South America and Southeast Asia. Our review of the impacts observed at these dam sites will form the basis for a concluding suite of recommended future actions that could help minimize and mitigate adverse effects of dam construction and reservoir inundation on primates.

Overview of Impacts

The initial impact of damming on primates occurs well before the reservoir itself is flooded. Sites are cleared, roads are built, construction crews move in, animal translocations are sometimes undertaken and the loud, dusty business of dam building begins. Collectively, these anthropogenic actions cause noise, air pollution, habitat loss and fragmentation, result in human population increases and relocations, and may either deplete wildlife (via hunting) or increase wildlife population densities in areas into which animals have been displaced (due to animals

Table 36.1 Dams, reservoirs and rescues.

Name of dam	Year reservoir filled	Dam location	Estimated reservoir area (km²)	Mean reservoir depth (m)	Reservoir volume (km³)	Number of Islands	Number of rescued primates	Rescue timing
SOUTH AMERICA								
Afobaka	1964	Suriname	1683[a]	–	–	–	528[a]	During flooding[a]
Balbina*	1987	Brazil	2360[b]; 4437[c] 2996[e]	7.4[b] 4.8[d]	17.5[c]	1500[b]; 3299[c]	–	During flooding[b]
Guri	1986	Venezuela	4240[f]	[< 50][g]	135[f]	>100[f]	–	After first phase of flooding[h]
Petit Saut	1995	French Guiana	365[i]	35[i]	–	>200[j]	225[i]	During flooding[i]
Samuel	1988	Brazil	540[d]	8.4[d]	–	–	1352[k]	During flooding[d]
Tucuruí	1984	Brazil	2430[l]	20.2[d]	45.5[m]	>1600[n]	27 007[l]	Before[n] and during[o] flooding
SOUTHEAST ASIA								
Chiew Larn	1986	Thailand	165[p]	–	–	241[p]	152[p]	During flooding[p]
Na Hang	2002	Vietnam	57[q]	–	2.2[r]	–	–	–

* Balbina Dam reservoir area estimates and estimated number of islands are highly debated.
[a] Price (2011); [b] Fearnside (1989); [c] Cabral *et al.* (2008); [d] Fearnside (2005); [e] Feitosa *et al.* (2007); [f] Alvarez *et al.* (1986); [g] Terborgh *et al.* (1997); [h] Konstant & Mittermeier (1982); [i] de Thoisy *et al.* (2001); [j] Cosson *et al.* (1999); [k] Gribel (1993); [l] Fearnside (2001, 2006); [m] Fearnside (2002); [n] Bastos *et al.* (2010); [o] Ferrari *et al.* (2004); [p] Nakhasathien (1989); [q] Lang (2002); [r] Mahabir (2008).

fleeing from the flood zone or being translocated: Woodford & Rossiter 1993; Schneider 2001; Rodrigues 2006).

During flooding, more terrestrial and riverside habitats are lost. Once the dam is completed and the reservoir begins to fill, animals in the flood zone escape to higher ground, are rescued, or perish. In most cases, it is not known how many animals die as a result of reservoir flooding events (Rodrigues 2006). Animals that survive and move into or are translocated to new habitats face challenges associated with unfamiliarity of surroundings, increased population densities, and competition for limited resources (Rodrigues 2006). Rescue operations that sometimes take place are controversial for several reasons, but can also form the basis of important empirical studies (Schneider 2001).

After flooding, when the total amount of immediate habitat loss is realized and populations that may have temporarily faced high densities decline and return to pre-dam sizes, further impacts continue to occur (Benchimol & Peres 2015a,b; Gibson *et al.* 2013). Most of the lasting effects on primate populations are due to habitat fragmentation. Whether caused by roads, new human settlements, or islands created within the flood zones themselves, habitat fragmentation influences primate behaviour, ranging patterns, diet, population densities and ultimately their ability to successfully reproduce.

The following sections focus on some of the most significant impacts of damming on primate populations including drowning, habitat loss, influx of human populations, rescue operations, movements into adjacent habitats, and habitat fragmentation (Figure 36.1).

Specific Impacts

Drowning

Once a dam is built and the reservoir begins to fill, animals in the flood zone meet one of three fates: flee to higher ground, be rescued or drown. Their likelihood of survival is often directly affected by the depth of the reservoir, the speed at which it is filled and the amount of prior vegetation clearing, especially for primates and other animals that can move across shallow bodies of water from tree to emergent tree. Whereas the location of some reservoirs, such as the one at Na Hang, Vietnam, are typified by significant topographic relief (Lang 2002), others like the Balbina (Cabral *et al.* 2008) and Samuel (de Sá 2004; Fearnside 2005) dams in Brazil are very shallow. Because the trees at the latter two sites did not fully submerge, the chance of surviving inundation were likely higher because primates had the ability to disperse into adjacent habitat (Benchimol & Venticinque 2014).

In most cases, the extent of wildlife lost to entrapment in a reservoir flood zone is unknown. The eerie sight of trees poking out of flooded reservoirs is common (Terborgh *et al.* 1997), as are haunting stories of mammalian skeletons found clinging to treetops during the first dry season after flooding (Luis Balbás, pers. comm. to Harrison-Levine, 1997). Kingston (1986) suspected that a significant portion of the estimated 100 000 primates that were stranded in the Tucuruí flood zone in Brazil either drowned or starved to death on emergent trees. During rescue operations after flooding at Chiew Larn Dam in Thailand, 40 primates were found dead (152 were rescued alive:

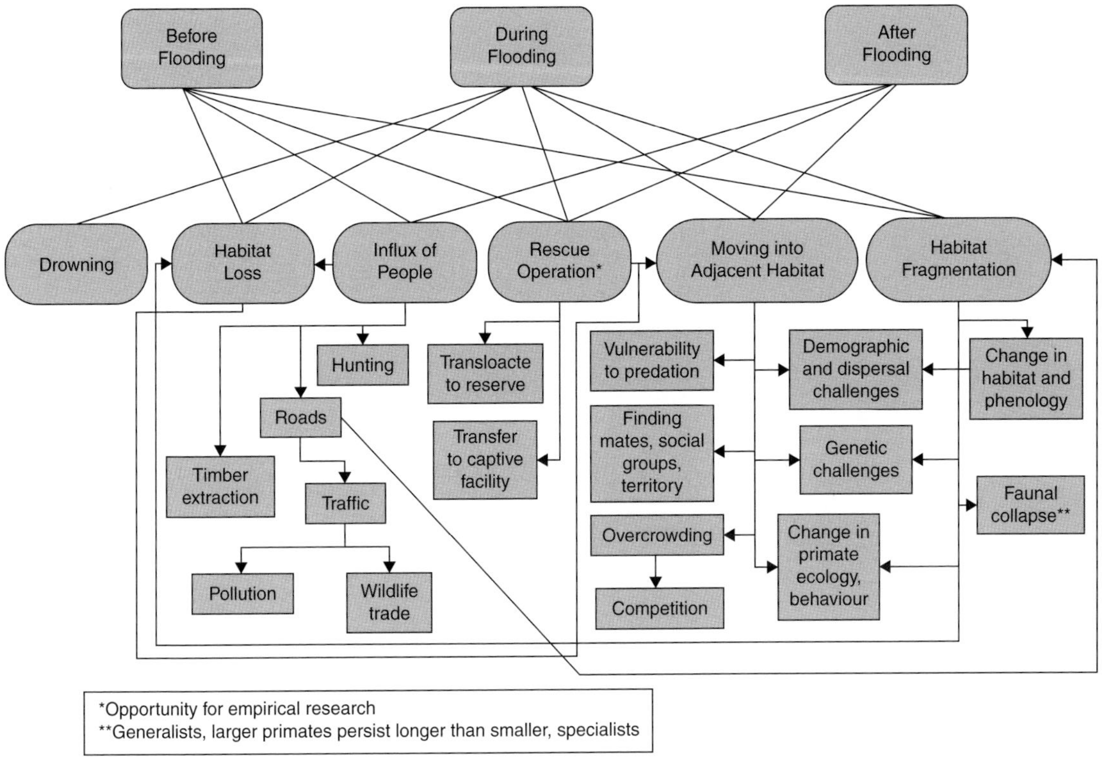

Figure 36.1 Impacts of damming on primate populations.

Nakhasathien 1989). Most primate deaths at Cheiw Larn were the result of starvation and drowning. Although drowning is a heart-wrenching negative impact of damming, it may not be the most significant impact in the long term.

Habitat Loss

Authors agree that, in most cases, habitat loss is the most significant negative impact of dam construction and reservoir inundation on primates (Alho 2011; de Sá 2004; Enari & Sakamaki-Enari 2014; Gribel 1993; Liao *et al.* 1988; Vié 1999). The amount of habitat remaining after flooding depends largely on the topography of the region (Figure 36.2). Areas with significant topographic relief end up with small, deep lakes (Na Hang: Lang 2002). Flood zones with intermediate topography result in hundreds or thousands of land-bridge islands dotted throughout the reservoir (Tucuruí: Bastos *et al.* 2010; Guri: Terborgh *et al.* 1997), and the outcome of inundation in the flattest regions is a flooded contiguous forest (Balbina: Fearnside 1989; Samuel: de Sá 2004 and Fearnside 2005). Even the partially submerged forests at Samuel, with protruding green treetops, however, died within a few years after flooding (de Sá 2004). While many riparian forest trees annually tolerate up to 10 months inundation, and can even withstand two to three years of consecutive flooding, periods longer than this will kill them (dos Santos Junior *et al.* 2013; Ferreira *et al.* 2013).

While the area inundated is an important contributor to total habitat lost, it is not the only factor. Habitats near the dam site are often cleared to build lodgings for construction workers. Land is also cleared to meet the housing and agricultural needs of local people displaced by the reservoir, and the additional habitat loss from these resettlement activities is rarely estimated or included in EIAs (Moore *et al.* 2010; Tan & Yao 2006). At Balbina Dam, for example, approximately 311 km^2 of an indigenous peoples' reserve was in the flood zone, forcing one-third of the surviving members of the tribe to relocate (Fearnside 1989).

Influx of Human Populations

For some primate populations, impacts related to habitat loss and drowning are less important than those resulting from the large increase in human populations. During their construction, the Santo Antônio and Jirau dams (built simultaneously on Brazil's Madeira River) employed 20 000 people and attracted as many as 100 000 others to this area in southwestern Amazonia (Fearnside 2014; Figure 36.3). In addition to resettled human populations and the influx of construction workers, roads are built to provide access to the dam site. These roads provide humans with an entrée into regions that may have previously been quite remote (Boyle 2008; WCD 2000).

One major consequence of access roads is increased traffic. An average of 360 trucks passed through Na Hang each day during the peak of dam construction in early 2003 (Martin 2004). Inside those trucks are people who are likely to cause increases in both legal and illegal trade in wildlife and other forest products (Martin 2004; Thach Mai Hoang 2010; Wolters 2004). Construction and large truck traffic also increase dust in the atmosphere and affect the amount of silica layered on leaves which, in turn, may contribute to dental abrasion and primate mortality (Covert *et al.* 2008).

Figure 36.2 Balbina Dam Reservoir in Brazil (low topographic relief). (A black and white version of this figure appears in some formats. For the colour version, please refer to the plate section.)

The number of workers required for dam construction varies widely from site to site, but labourers typically number in the thousands at peak construction times. An estimated 7000 labourers immigrated into Na Hang, Vietnam, during dam construction – more than doubling the size of the local human population (Martin 2004). Years after a dam is completed, many construction villages become permanent towns with increasing resident human populations.

Around such new pockets of human populations it is not unusual to see increased hunting as well as extraction of timber and non-timber forest products. Both result in substantial impacts on wildlife, including primates (Boyle 2008; Kingston 1986; WCD 2000). At Na Hang Dam, for example, hunting is thought to have been a significant contributing factor in the abrupt decline of the already Critically Endangered Tonkin snub-nosed monkey (*Rhinopithecus avunculus*) (Covert *et al.* 2008; Martin 2004). There were 130 monkeys living at Na Hang before construction, and this relatively large population was considered to be the best hope for conserving the species (Mittermeier *et al.* 2009). However, surveys conducted just 13 years after dam construction estimated that fewer than 40 monkeys remain at Na Hang (Thac Mai Hoang 2010). Because the global population of *R. avunculus* was approximately 300 individuals in 2006 (Mittermeier *et al.* 2009), a loss of over 90 individuals at Na Hang is devastating. An EIA conducted in Vietnam prior to the construction of the Na Hang Dam predicted this severe population decline (Scott Wilson Asia Pacific Ltd. (SWAPL) 2000); but, the dam was still built, and few of the recommended mitigation measures were implemented.

While there is no direct evidence for dam-related hunting at Na Hang, such evidence does exist elsewhere. Forty primates were found dead during rescues at the Chiew Larn Dam in Thailand. Several had bullet wounds that were thought to have been the cause of death (Nakhasathien 1989) – a clear demonstration that hunting was occurring during inundation.

Figure 36.3 Dam-related construction at the Santo Antonio Dam on the Madeira River. Photo: Philip Fearnside

Rescue Operations

At some dam sites, wildlife rescue operations are undertaken before, during and/or after flooding. The number of primates involved varies from hundreds to tens of thousands but for the most part, rescues are poorly documented (Fournier-Chambrillon *et al.* 2000; Schneider 2001). Five operations in the Southeast Asian and South American tropics kept careful records of the number of primates rescued: 152 were rescued alive at Chiew Larn in Thailand (Nakhasathien 1989); 225 at Petit Saut in French Guiana (de Thoisy *et al.* 2001); 528 at Afobaka in Suriname (Walsh 1967); 1352 at Samuel in Brazil (Fearnside 2005); and 27 007 at Tucuruí in Brazil (Mascarenhas & Puorto 1988; Peres & Johns 1991). Most often, primates and other animals were captured, and within hours released into adjacent habitat. At Afobaka (Suriname), for example, the goal was to maintain animals in captivity long enough to restore activity levels before releasing them into nearby forest (Walsh 1967). However, at Samuel in Brazil, primates were sent to a nearby reserve (de Sá 2004), and some rescued from Brazil's Balbina reservoir were transferred to captive colonies (Fearnside 1989).

It is tempting to think of these rescues as success stories, but because little was known about population densities in areas that were flooded or about pre-existing primate densities at most release sites, what proportion of these individuals actually survived in the long term cannot be reliably estimated (de Sá 2004; Peres & Johns 1991). Based on estimates of primate population densities before inundation, what is known – at least at Tucuruí – is that larger-bodied primates (such as *Alouatta*, *Cebus* and *Sapajus*) were more likely to be found than smaller ones (such as *Mico*, *Saguinus* and *Saimiri*), while cryptic species (such as *Aotus*) were almost never rescued (Peres & Johns 1991). At Afobaka, 94% of monkeys rescued were howlers and, although eight species inhabit the surrounding forests, only four (howlers, squirrel monkeys, tamarins and white-faced capuchins) were among those rescued. It is unclear why no sakis, brown capuchins or spider monkeys were rescued (Walsh 1967).

Still, rescue operations allow researchers to collect and test biological samples, conduct health evaluations and place tracking devices on primates (de Thoisy *et al.* 2001; Peres & Johns 1991). They have therefore formed the basis of some interesting scientific studies. Perhaps one of the best-studied dam sites with respect to its impact on primates is the Petit Saut Dam in French Guiana. A total of 124 howlers (*Alouatta*), 6 sakis (*Pithecia*), and 95 tamarins (*Saguinus*) were rescued and each received health evaluations. All primates captured at Petit

Figure 36.4 Bot-fly infested, deceased howler monkey (*Alouatta belzebul*) at Tucuruí Dam. Photo: Simone Martins

Saut during flooding were found to be in good condition. The rescue operation continued post-flooding and, one year after flooding began, 11 rescued howlers showed signs of severe nutritional stress (de Thoisy *et al.* 2001).

Some prior rescue operations, such as the one undertaken at Tucuruí, are thought to have been linked more to public relations than to wildlife conservation (Alho 2011; Fearnside 2001). While it is pleasing to hear that a rescue operation is saving hundreds or thousands of animals, if these are not sent to a captive rescue centre (Kingston 1986), or are subject to a follow-up field study, then the fate of translocated animals remains unknown (Rodrigues 2006). It is probable that most re-released primates ultimately perish. Some animals likely die due to unfamiliarity of the habitat or competition with conspecifics already inhabiting release sites (Alho 2011; de Sá 2004; Griffith *et al.* 1989; Kingston 1986). At Tucuruí, reserves created as release sites were immediately invaded by loggers and hunters (Fearnside 2001). In addition, increases in local human population densities may increase disease transmission (IUCN 1987; Magnusson 1995; Woodford & Rossiter 1993). However, if rescues and releases are carefully planned, they could contribute significantly to our understanding of the impact of damming on primate populations and help avert detrimental impacts (de Sá 2004; de Thoisy *et al.* 2001; Peres & Johns 1991).

Moving into Adjacent Habitat

Regardless of whether primates move into new habitat as a result of land clearing, construction noise (Martin 2004), increased hunting pressure due to the presence of new access roads (Alho 2011; Martin 2004), flooding (Alvarez 1986; de Sá 2004), or are rescued and translocated, primates face myriad challenges post-damming. They may encounter overcrowded habitats and competition with conspecifics or they may be at a disadvantage due to habitat unfamiliarity (Fischer & Lindenmayer 2000; Schneider 2001; Rodrigues 2006).

When a primate is not familiar with its surroundings, it may have difficulty finding food, water, shelter, a social group and a mate, and it may be at a higher risk of stress, disease and predation. Research from an island in Venezuela's Guri Reservoir suggests that primates rely heavily on memory to find resources (Cunningham & Janson 2007). At Chiew Larn (Nakhasathien 1989), Petit Saut (de Thoisy *et al.* 2001) and Balbina (Figure 36.4) primates were found malnourished, injured, infected with parasites and otherwise stressed during post-inundation rescue operations, much of which is hypothesized to have been a consequence of habitat unfamiliarity.

Primate mortality was more often attributed to habitat loss, overcrowding and habitat unfamiliarity than drowning at the Samuel Dam in Brazil (de Sá 2004). In 2004, de Sá found that *Callicebus*, *Pithecia* and *Samiri* species were among the most frequently captured during rescue operations at this site. A temporary increase in population density of these three primates was observed in adjacent habitat just after release, but populations declined in subsequent years, probably due to dispersion or mortality. However, at Santo Antonio Dam, Brazil, telemetry tracking indicated that only 7% of translocated pygmy marmosets (*Cebuella pygmaea*) survived 3 months post-release (Dias *et al.* 2015).

Tracking rescued, released primates has become more common in recent years. While most published post-release studies are short term (and considering many dam-linked projects are consultancies; therefore, biologists may not be given permission to disseminate outcomes when translocations fail) some promising results have been published. Although a 1997 study linked to Brazil's Novo Ponte Dam attempted to radio-track 15 translocated *Callicebus personatus*, authors were unable to follow the animals due to technical difficulties (Neri *et al.* 1997). Since that time, however, several teams have successfully followed primates post translocation. For example, in Belize, a group of translocated Central American black howlers (*Alouatta pigra)* survived at least 1 year and established territory (Ostro *et al.* 2000). Marques *et al.* (2011) used radiotelemetry to follow black-tailed marmosets (*Mico melanura*) translocated as part of a study of wildlife affected by flooding the Manso River reservoir in western Brazil. Of the five animals monitored, two pairs survived at least 8 months, successfully established territories and appeared healthy. In northern Brazil, during the formation of the reservoir on the Madeira River created by the Santo Antônio Hydroelectric Dam, two groups of pygmy marmosets (*Cebuella pygmaea*) were translocated into nearby protected, open tropical rainforest. A 3-month post-release monitoring study using radiotelemetry at this site found that group members remained together and settled within stable home ranges near their release sites (Dias *et al.* 2015).

Similarly, some of the primates rescued at Petit Saut were fitted with telemetry devices. Six sakis and 14 howlers were tracked for an average of about 1 year after translocation (Richard-Hansen *et al.* 2000; Vié *et al.* 2001). Sakis and howlers established home ranges within 1 year and both were also observed integrating with resident groups. Mortality rates were difficult to discern because telemetry collars contributed to individual deaths (mainly due to screw worm larvae infections). However, Vié and colleagues (2001) and Richard-Hansen *et al.* (2000) indicated that sakis and howlers may both benefit from future, well-planned translocation efforts. So although forced movement into new habitat may ultimately result in primate deaths, this evidence supports the idea that translocations can serve to rescue a large number of primates, as long as translocations are well planned (Konstant & Mittermeier 1982; Fischer & Lindenmayer 2000).

None of the studies listed above were of sufficient duration to establish whether the translocations had long-term success. However, some translocated animals do appear to survive and reproduce. Oklander *et al.* (2017) found that populations of southern black howlers (*Alouatta caraya*) showed strong regional genetic structuring across the Argentinian and Paraguayan part of their range. An exception to the pattern was the population in the Chaco National Park, which contained genetic elements from a population 380 km away from the Yaciretá Dam. During dam construction here, many primates were removed from areas to be inundated and transported to other sites, including the park. Similar evidence for regional differences in genetic variability around the Tucuruí Dam will be discussed in the following section.

Habitat Fragmentation

Perhaps the most significant long-term impact of dam construction and reservoir inundation on primates is habitat fragmentation. In the case of damming, habitat fragmentation occurs primarily in two ways: (1) the construction of access roads that bisect habitats and (2) flooding, which often results in a reservoir strewn with numerous islands. Roads are widespread anthropogenic contributors to habitat fragmentation around the world. Many of the primates studied in the investigations reviewed here appeared to treat roads as barriers and avoided crossing them (Richard-Hansen *et al.* 2000). And the reservoirs themselves – especially those with intermediate topographic relief – may contain hundreds or thousands of islands once flooding is completed. The islands that remain are true fragments, with the unforgiving surrounding matrix of water (Anderson *et al.* 2007). Consequently, fragmentation can lead to several impacts on primate populations including the challenges of swimming from one fragment to another, changes in food availability, demographics, travel patterns and rates of social interaction, as well as genetic diversity, and ultimately, faunal collapse.

Many monkeys can swim quite well (e.g. Berman 1977; Anderson, Peignot & Adelbrecht 1992; Chaves & Stoner 2010; Gonzalez-Socoloske & Snarr 2010; Peck *et al.* 2014; Chapter 7) but swimming also comes with some caveats. Predation on swimming monkeys may be high: during the filling of the Balbina reservoir, Barnett (unpublished data) saw caiman and jaguars take swimming howlers, and eagles swoop at animals as they climbed out onto trees. Harrison-Levine *et al.* (2003) reported predator-sensitive behaviours, such as increased vigilance, nervousness and guarding, in sakis drinking at Guri Reservoir's edge when other water sources had gone dry. In addition, dam-promoted inundation generally disrupts phenology even when areas are not directly flooded (as on hilltop islands: Ferreira *et al.* 2013; Kozlowski 2002; Maingi & Marsh 2002; Stave *et al.* 2005). So although most monkeys can swim, the uncertainty of resource availability on the destination island may be just as psychologically and physiologically stressful as the risk of predation.

When a previously contiguous forest is transformed into islands surrounded by water, the sun, wind and water erosion can impact the islands' edges. This phenomenon, referred to as the edge effect, can in turn influence forest structure and lead to further reduction in utilizable primate habitat (Ferreira *et al.* 2012; Laurance *et al.* 1998). The edge effect has more significant effects on water-surrounded islands than on mainland habitats surrounded by a matrix such as pasture, which can act as a better buffer than water (Anderson *et al.* 2007; Norconk & Grafton 2003). This said, edge-related fire damage was an important factor in small islands at Balbina (Benchimol & Peres 2015a). Edge effect outcomes for primates may include changes in their travel patterns, diet and food availability (Cosson *et al.* 1999; Ferreira *et al.* 2012; Norconk 2007; Norconk & Grafton 2003).

Fragmented habitats are often completely isolated from one another and this can have an effect on primate behaviour and ranging patterns, as well as cause devastating long-term impacts on the genetic and reproductive viability of primate populations. Years after flooding at Tucuruí, Silva and Ferrari (2009) returned to the dam to compare the behaviour of island and mainland groups of bearded sakis (*Chiropotes*). They found that island individuals rested more and travelled less than those living on the mainland, and island populations also had lower levels of social interaction. These behavioural patterns can be attributed to comparatively small home range and group sizes on islands, thought to be the result of isolation.

When primate populations are forced to escape upwards – to mountaintops, if these are available – as a reservoir floods, groups remaining on each island may exhibit unusual demographic patterns, including single-sex groups that are unable to breed. At the Guri Dam, post-flooding surveys indicated that at least a few islands contained non-breeding populations of white-faced sakis, *Pithecia pithecia* (Norconk 1997). Other islands had populations with juveniles that may not be able to disperse from their natal group, as is typical for the species. A population that lacks the ability to disperse will face genetic viability ramifications. Indeed, Norconk and Grafton (2003) reported that, over a 10-year period, the sakis on one Guri Reservoir island had not had a single surviving infant.

In 2008, Gonçalves and colleagues reported on a study comparing genetic data collected during the rescue operation at Tucuruí with that collected 15 years later. These authors found that while the genetic diversity of mainland howlers had increased, possibly due to the influx of genes from translocated or fleeing howlers, the genetic diversity of howlers on at least one Tucuruí reservoir island was lower than that of mainland howlers. This could be additional evidence to support post-flooding genetic viability concerns for island populations.

Land-bridge islands, or landscape fragments that have been isolated by rising water levels with an associated reduction in habitat area (Diamond 1972; Terborgh *et al.* 1997), are thought to experience a kind of ecological meltdown (Terborgh *et al.* 2001). Terborgh and other authors (Benchimol & Venticinque 2014; Cosson *et al.* 1999; Wu *et al.* 2004) found that while the population density of some primate species decreases, or primates disappear entirely on smaller islands, the density of other species increases. This in turn decreases primate diversity on islands compared with the mainland (Cosson *et al.* 1999; Terborgh *et al.* 1997). Similarly, at Balbina Dam, Benchimol and Venticinque (2014) found that 60% of primate biodiversity was retained on islands larger than 100 ha. In addition to size, structural complexity of land-bridge islands is also closely related to primate biodiversity at Balbina. Interestingly, these authors noted that larger primates were more likely to be found on islands than small ones, possibly due to their ability to more easily swim from island to island. It has also been observed by many of these researchers that – especially in the absence of top predators – species that exhibit more behavioural and dietary flexibility (such as howler monkeys, capuchin monkeys and some marmosets) will persist longer on island fragments. While habitat fragmentation resulting from anthropogenic damming may have an overall negative impact on less flexible, smaller primate species, more flexible, larger primates do not appear to experience the same types of impact.

Conclusion

Dam construction and reservoir inundation can have devastating effects on habitats and primate populations. Drowning, habitat loss, increased human populations and associated increases in primate hunting, primates' retreat into already occupied habitats and habitat fragmentation are among the most significant impacts. The degree to which each of these affects primate populations varies by site and species. Ultimately, habitat loss, the influx of people and habitat fragmentation represent the most significant dam-related outcomes for primates.

In addition, while rescue operations may seem to be a humane alternative to primates drowning or starving, and could also offer tremendous opportunities for research on effects of flooding and success of translocation, they must only be undertaken under specific circumstances. Those conducting rescues should use and identify well-researched and appropriate release methods and sites. Release sites must have sufficient resources and minimal anthropogenic threats, such as hunting, and be characterized by relatively low population densities of the species to be relocated. The rescue operations themselves should be studied, monitored long term, and successes and failures reported (akin to exemplary work by Neri *et al.* 1997).

Primary aims should include engagement of governments in not only permitting but requiring empirical studies as essential components of the dam-planning process, and working with funding institutions to hold back loans unless scientifically based plans to mitigate implications of dam construction are developed and implemented. The WCD (2000) considers there to be five main limitations to the effective implementation of EIAs: (1) resistant attitudes, (2) insufficient structural integration of EIA recommendations into policy and decision-making, (3) insufficient scope of EIAs, (4) inadequate procedural assessments and (5) poor technical quality of EIAs. Therefore, it is also important for researchers to advocate for the incorporation of scientific findings early on in dam-planning processes; such information and advocacy could be critical to decisions about dam building, as well as during dam site selection and dam design. As a high-profile taxonomic group that is comparatively easily studied, primates can play a key role in dam-based EIAs. Scientists should gather baseline data on primates pre-construction so that post-construction and post-inundation comparisons are possible (McCartney *et al.* 2001). Similarly, comparative studies on island and mainland populations are also needed. All primate populations affected by dam construction and reservoir inundation should be part of long-term monitoring programmes (McAllister *et al.* 2001) that investigate the longitudinal impacts on primate behaviour, ranging patterns, diet, health status and genetics.

Chapter 37

Hapalemur alaotrensis

A Conservation Case Study from the Swamps of Alaotra, Madagascar

Patrick O. Waeber, Jonah H. Ratsimbazafy, Herizo Andrianandrasana, Fidimalala B. Ralainasolo and Caroline M. Nievergelt

Introduction

Madagascar is one of the world's least developed countries, ranked 155 out of 166 countries on the human development index (HDI) (2014 estimates). It supports a fast growing population (13th highest birth rate), where 85% of people live in rural areas and survive on less than US$2 per day (CIA 2013; Gaffikin *et al.* 2007). The rural poor in Madagascar rely heavily upon the natural environment and its ecosystem services for their livelihoods. A lack of alternatives and weak environmental governance (Horning 2008, 2012; Waeber *et al.* 2016) is resulting in continued clearance and degradation of forests and wetlands, especially in Madagascar's highlands (Kaufmann 2006; Kull 2012). While the population has grown from less than 7 million in the 1960s to over 20 million in the 2000s, Madagascar has lost about 50% of its forest cover (Harper *et al.* 2007). Equally, all wetlands have been severely degraded especially in the highlands with only a few falling under some form of protection (Bakoarininiaina *et al.* 2006; Rabearivony *et al.* 2008). Wetland resources are repeatedly identified as the most threatened habitats in Madagascar (Rabearivony *et al.* 2010; Thieme *et al.* 2005) which often do not garner the same conservation, research and policy attention as the well-publicized sites within Madagascar's eastern rainforests (Kull 2012). Yet wetlands provide invaluable ecosystem services for local human populations who rely on their water, animal and plant products for survival (e.g. Rabearivony *et al.* 2008). As part of a national strategy declared in the Durban Vision by former President Ravalomanana to protect 10% of the country's terrestrial surface (Rasoavahiny *et al.* 2008), 93 new protected areas were identified in 2003 to support and protect key areas for biodiversity.

The Alaotra Wetlands

Lake Alaotra (751 m asl) is located in the Alaotra-Mangoro of Madagascar in the central highlands (Figure 37.1). At 20 000 ha, it is Madagascar's largest lake. Alaotra was classified as Madagascar's third Ramsar site in 2003 (Ramsar 2013) for its unique natural environment, with the goal of encouraging the sustainable management of the entire wetland which is also of high national agro-economic importance. On 17 January 2007 the government of Madagascar officially recognized the conservation value of this area by classifying Lake Alaotra and its marshes as a new protected area under national law N°381-2007/ MINENVEF/MAEP. In June 2015 it was granted permanent status as a community managed protected area (IUCN Category V).

Lake Alaotra's surrounding 23 000 hectares of freshwater marshes are dominated by cyperus, *Cyperus madagascariensis* (Cyperaceae), and reed, *Phragmites communis* (Poaceae), and adjoining these are 120 000 ha of rice fields within a watershed encompassing 722 500 ha. A recent mechanistic model that identified centres of endemism included areas lying between large watersheds with sources at high elevation (Wilmé *et al.* 2006). The Maningory watershed with its lake and marshes is part of one of Madagascar's 15 centres of endemism, Analanjirofo (Wilmé *et al.* 2012), supporting Critically Endangered and locally endemic mammals such as the Alaotra gentle lemur *Hapalemur alaotrensis* (Andrianandrasana *et al.* 2005) and the recently described *Salanoia durrelli*, a small carnivore (Durbin *et al.* 2010). *Hapalemur alaotrensis* is the only primate taxon in the world that lives exclusively in a wetland habitat (Chapter 34). The species is classified as Critically Endangered (IUCN 2012) due to its extremely reduced geographic range. The only other lemur found in marshes is the brown mouse lemur (*Microcebus* aff. *rufus*).

The human population of Amparafaravola and Ambatondrazaka, the two lake districts of the Alaotra-Mangoro region, has increased from some 110 000 people in the 1960s to over 710 000 in the 2000s (Monographie Régionale Alaotra-Mangoro 2007). The primary economic drivers of growth in the Alaotra region are fisheries and rice production providing one-third of the country's total rice output (Andrianandrasana *et al.* 2005; Ferry *et al.* 2009). The Alaotra fishery is organized and guided by a Fisher Federation made up of local fishermen associations from four zones corresponding to the Lake Subdivision. The Federation is responsible for issuing annual fishing licences (which costs approximately 3 euros). In 2004, over 8000 fishermen were registered, while in 2009, only 616 fishermen were (A. Randriamanarivo, Secrétaire Générale de la Fédération des Pêcheurs, pers. comm., 2012), which is likely due to political instabilities that emerged in 2009 and their impact on governance and outreach nationwide (see Randrianja 2012).

In 2012, it was estimated that over 12 000 active fishermen were depending on the lake's resources (B.J. Rasolonjatovo, pers. comm., 2012). Alaotra provides for the majority of fish production for the urban population of Antananarivo and Toamasina. Annual fish catches at Lake Alaotra were rich and amounted to

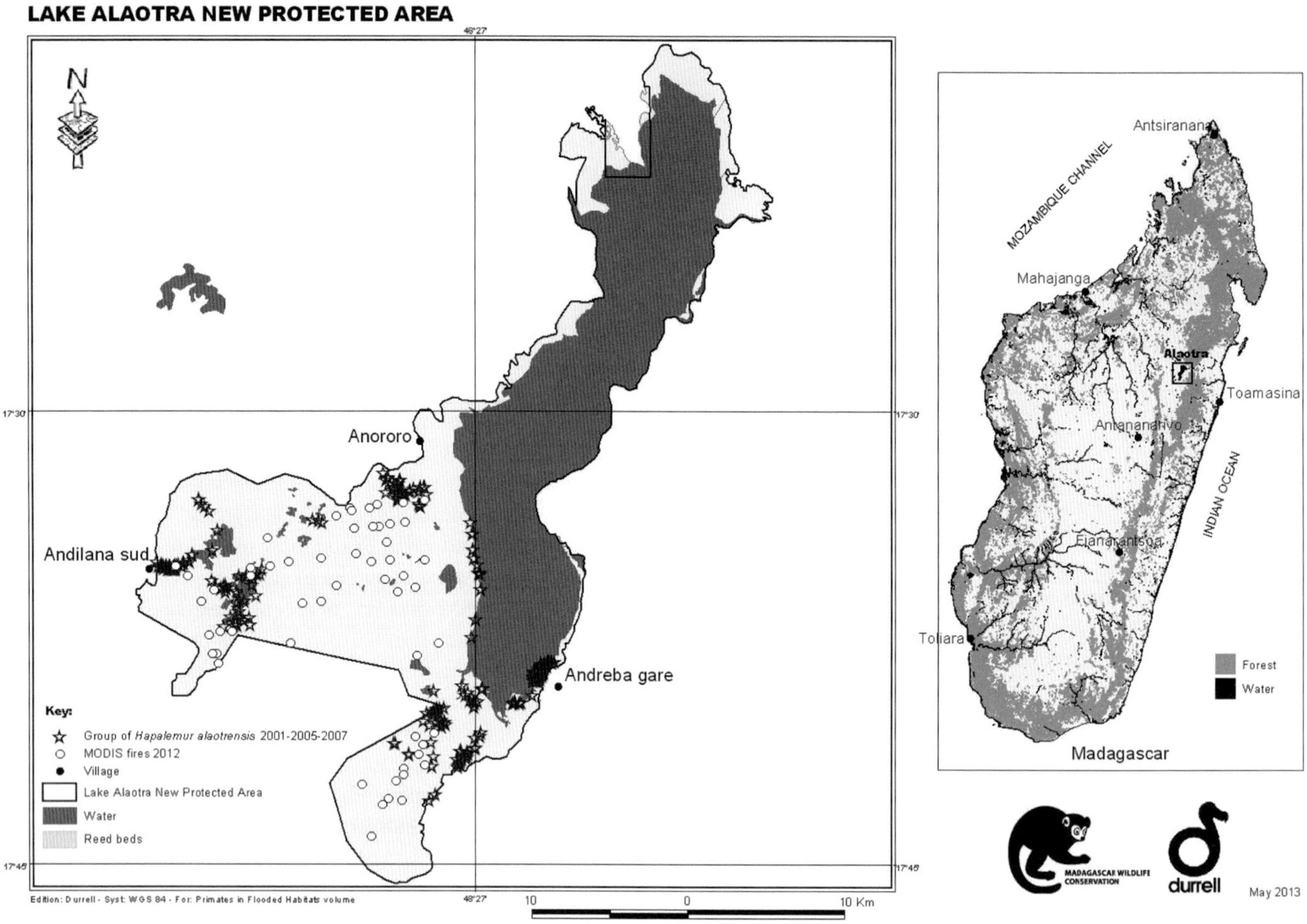

Figure 37.1 The Alaotra protected area hosting the Critically Endangered and endemic *Hapalemur alaotrensis*.

4000 tonnes in the 1960s (Pidgeon 1996). However, fish catches have declined and recent data suggest a low of 600–800 tonnes in recent years (Ratsimbazafy *et al.* 2013). This is probably a result of overfishing, acidification of the lake, introduced fish species and siltation (Andrianandrasana *et al.* 2005; Lammers *et al.* 2015; Razanadrakoto 2004). Almost all marshland bordering the lake has now been converted to rice production or undergone a form of anthropogenic disturbance (Ranarijaona 2007), with some 120 000 ha outputting about 300 000 tonnes of rice per year.

Drivers of Change

The land surrounding the lake and marshland has already been widely deforested for agricultural production and livestock, leading to a continuous increase of soil erosion. This is aggravated especially during the rainy season; consequently, every year a considerable amount of sedimentation reaches the marshes and lake, which has led to a reduction in open water surface of more than 5 km^2 in the past 30 years (Bakoarininiaina *et al.* 2006; Wright & Rakotoarisoa 2003). Another factor bearing negatively on agricultural production and the wetland ecosystem is the sinking hydric level in the entire area (Ferry *et al.* 2009) which is further aggravated by extended drought periods observed in recent years.

The aquatic environment has been seriously altered by sedimentation and by invasive allochtonous plants (mainly *Eichhornia crassipes*) and fish (*Tilapia* spp., *Channa maculata*) (Pidgeon 1996; Lammers *et al.* 2015). Two endemic bird species, the Madagascar pochard *Aythya innotata* and the Alaotra little grebe, *Tachybaptus rufolavatus*, have disappeared from the lake (Wilmé 1994); the former was rediscovered at a lake in northern Madagascar in 2007 (Rene de Roland *et al.* 2007; Rabearivony *et al.* 2010), while the latter is now considered extinct. There has been a general reduction in species lakewide; while Pidgeon (1996) in his study in 1993–1994 found over 70 bird species, Kaufmann (2012), though focusing her survey only on Andreba Gare during the peak dry season, found less than half this number of species.

The conservation situation of Alaotra's flagship species *Hapalemur alaotrensis* (Durbin 1999) is of great concern as population numbers continue to decline. In the 1990s, the total population was estimated at over 11 000 individuals (Mutschler & Feistner 1995, 2001); in 2005, it had shrunk to an estimated 2500 individuals (Ralainasolo *et al.* 2006). Given growing threats, it is now thought to be much lower than the last census estimated. The main causes for the decline of *Hapalemur alaotrensis* have been destruction of its habitat, including conversion of land into rice fields, widespread and repeated burning of remaining areas of marshland and hunting for local consumption (Ralainasolo 2004; Ralainasolo *et al.* 2006; Reibelt *et al.* 2017; Waeber *et al.* 2017). No population estimates for other primate species in the marshes, such as *Microcebus* aff. *rufus*, are available.

Disturbance of habitat especially through fire is posing a threat to the survival of this lakeside lemur. So far, the years 2000 and 2004 have been the most severe fire years on record for the marshes of Alaotra (Ratsimbazafy *et al.* 2013), with 2012 likely representing another extreme fire year. Fires continued to burn into February 2013, which is rather unusual and has to do with a prolonged period of drought. It is apparent that a vast extent of *Hapalemur alaotrensis* habitat has been affected (Figure 37.1). The *Hapalemur alaotrensis* sightings (Figure 37.1) are located mainly around borders with open water, which has to do with the census methodology used (canoeing along water channels and counting lemurs; see Mutschler & Feistner 1995 or Ralainasolo *et al.* 2006 for methodological details). More individuals and groups can be found towards the centres of the marshes, which suggests that *H. alaotrensis* is directly affected by fire.

Burning marshland happens for several reasons (Copsey *et al.* 2009a, b; Guillera-Arroita *et al.* 2010; Ralainasolo *et al.* 2006), but the main ones are: land conversion for rice cultivation and improved access to fishing ponds (e.g. fishing of *Channa maculate*; Copsey 2009a, b). Informal discussions carried out during November 2012 with villagers from the main areas of Anororo, Andilana Sud and Andreba Gare, revealed that the main drivers for this year's fires are likely due to some key people who do not live locally or depend on the lake. They hire the locals to set fires in exchange for little money and afterwards, buy the destroyed marshlands to convert them into profitable land (Waeber & Wilmé 2013).

Habitat fragmentation is certainly a problem for *H. alaotrensis*. This lemur is able to swim short distances if needed; however, it is a very slow swimmer, making it vulnerable to predation. The extent of habitat fragmentation it has endured is still unclear to date, although the monitoring approach suggested by Guillera-Arroita *et al.* (2010) allows for the development of quantified estimations. An example of the consequences of fragmentation is given for Park Bandro (17°37.20′S, 48°30.00′E) located at Andreba Gare on the eastern side of the lake. The park is an 85 ha protected area which is now completely cut off from any other marshes and becoming an island. The park hosts the biggest population of *H. alaotrensis* of about 150 individuals. We have observed many adult individuals showing signs of fight injuries on the face; this possibly represents an indication of a crowding effect of this highly territorial species (Nievergelt *et al.* 1998). Another way to estimate a crowding effect would be to consider group sizes which tend to get bigger when the population is compressed. We have also observed this at the park, however, we have not yet pursued a systematic assessment.

Management and Conservation of the Alaotra Wetlands

There are several management challenges associated with fires in marshlands. In marshes, frequent fires often contribute to weed invasion, inducing loss of native plants and the slowing down or impeding of the recovery of the pre-disturbance state depending on fire frequency and intensity further reducing suitable habitat for *Hapalemur alaotrensis*. In contrast to forests, control and management of fires are more challenging in marshes since the vegetation is floating. Fire breaks cannot be installed *a priori* or during a fire event. Ratsimbazafy *et al.* (2013) suggest that the only effective management approach consists of a fire brigade; however, the costs of it would be exorbitant and not sustainable. Hence, this likely remains an unrealistic mitigation approach.

With increasing climatic variation and change over the past several years, the region risks extended drought periods as well as more abundant dramatic events like heavy cyclones, affecting the entire hydric balance even further. A vicious cycle begins: the search for irrigated and cultivable land pushes the growing human population further into marshland and *Hapalemur alaotrensis* habitat. Increased areas of so called 'riz de contre-saison' (rice paddies established in the dry season during the months of August to December) within the boundaries of the lakefront (i.e. the transition zone from lake to vegetated area) is increasing from year to year. An example is given from Andreba Gare: In 2005, there were only a few such fields visible. In 2012, there were over 30 fields transforming the natural marshes into floating rice fields. A couple of such fields were found in November 2012 within the 2004 established and officially protected Park Bandro boundaries. Ratsimbazafy *et al.* (2013) also report ongoing encroachment of *Hapalemur* habitat for conversion of marshland into rice fields.

The Lake Alaotra region hosts several endemic species, and as such, has certainly served as a refuge (sensu Ganzhorn *et al.* 2014; Mercier and Wilmé 2013; Waeber *et al.* 2015) during drier periods of the Plio-Quaternary. Unfortunately, in less than a century, agricultural and other anthropogenic activities have pushed the remaining impoverished wildlife into ever-smaller pockets of habitat and to the brink of extinction. The emerging question is how to tackle the Alaotra conservation and development challenges? In poverty-stricken areas like Alaotra, few opportunities exist for people to generate income and local employment remains extremely low. Those who have a major impact on threatened resources, e.g. the main resource users or rural poor, are most often influenced by a small number of wealthy elites, who often have little attachment to the resources being exploited (as seen in the case of Madagascar's eastern rainforests, see Innes 2010; Randriamalala & Liu 2010). A growing challenge is how to reconcile a continuously increasing demand for agricultural products (e.g. rice) with a growing number of values and interests including environmental values for conserving biodiversity (e.g. the lakeside habitat of *H. alaotrensis*), while maintaining ecological functions and providing critical ecosystem services that uphold rural livelihoods. Current conservation and research efforts use a socioecological landscape approach to deliver important data and information about the drivers and barriers to livelihood opportunities and threats. These social and development demands are juxtaposed with biodiversity conservation values to ultimately shape and inform policies and decisions about sustainable use and landscape management. This is important at a time of rapid change including in climate (e.g. extended drought periods), economies (e.g. increased staple food prices) and culture (migration into Alaotra). And increased understanding of the linkages and dynamics of livelihood needs and ecosystem services and functions will help shape development within a socially accepted management and policy framework.

Meanwhile, ongoing conservation efforts are a melange of research and *in situ* actions focused on invoking change in human behaviour at several levels of the Alaotra resource user society. Conservation campaigns run by the Durrell Wildlife Conservation Trust, Madagascar Wildlife Conservation, Imperial College and others are using diverse and innovative communication channels including festivals (Andrianandrasana *et al.* 2005; Ratsimbazafy *et al.* 2013), films, radio broadcasts (sensu Waeber & Orengo 2008) as well as educational posters and comic books (Maminirina *et al.* 2006). These mediums will provide the main means for engaging with the public and stakeholders therein to help ensure long-term sustainability.

Beyond ongoing development support to local communities, Durrell engaged 96 local monitors in 2011 to carry out weekly patrols in the Lake Alaotra protected area (see Figure 37.1) to record biodiversity data and illegal activities that are regularly reported to the regional government. The long-term conservation goal for Alaotra is to achieve a community co-managed protected area that is governed through financially sustainable management bodies and led by local communities. Sustainability in resource management and planning of the Alaotra socioecological landscape can only be achieved if the main resource users see ownership in the conservation framework that reflects their needs. Conservation achieved through concrete and participatory schemes will ensure that *H. alaotrensis* (Figure 37.2) and other species at risk of extinction can persist and coexist alongside people.

Figure 37.2 Juvenile Alaotran gentle lemur on reed. Photo: Arnaud De Grave, used with permission.

Acknowledgments

We would like to thank Arnaud De Grave, Le Pictorium Agency (photographer) for providing us with the picture of the Alaotra gentle lemur.

Chapter 38

Landscape Genetics Applied to the Conservation of Primates in Flooded Forests

A Case Study of Orangutans in the Lower Kinabatangan Wildlife Sanctuary

Milena Salgado-Lynn, Mohammad Fairus B. Jalil, Lounès Chikhi, Marc Ancrenaz, Laurentius N. Ambu, Michael W. Bruford and Benoît Goossens

Introduction

Two major threats to biodiversity worldwide are habitat loss and fragmentation caused by human exploitation of environmental resources. One example is Singapore where over 70% of biodiversity has been lost because of large-scale deforestation and habitat modification (Brook *et al.* 2003; Brooks *et al.* 2002, 2006; Chapman & Peres 2001; Fahrig 2003; Fischer & Lindenmayer 2007; Myers *et al.* 2000; Pimm & Raven 2000; Segelbacher *et al.* 2010). In the matrix newly created by land-use changes and transformation, populations of wild animal species often become reduced and isolated. As a consequence, natural processes can be jeopardized and can induce a significant reduction in gene flow, increased genetic drift (and hence differentiation between populations that were once connected). Moreover, the fitness of the surviving and isolated populations can be reduced by inbreeding depression when closely related individuals mate. Combined, these processes reduce genetic diversity and can eventually lead to extinction (Amos & Balmford 2001; Caballero Rodríguez-Ramilo *et al.* 2009; Fernández *et al.* 2008; Frankham 2006; Hager 2003; Holderegger & Wagner 2008; O'Grady *et al.* 2006; Segelbacher *et al.* 2010; Toro & Caballero 2005). In addition to these genetic changes, demographic processes also play a major role in extinction rates. In particular, local extinction of small fragmented populations is increasing and becoming relatively common (Fahrig 2002; Kattan *et al.* 1994; Matthies *et al.* 2004; Michalski & Peres 2005; Tscharntke 1992). Thus, species survival will increasingly rely on the ability of individuals to disperse and move across heterogeneous landscapes between surviving populations or re-colonize empty fragments, and on our ability to maintain connectivity between remaining populations (Fahrig & Merriam 1994). Therefore, habitat connectivity (Lindenmayer & Fischer 2006) is a key issue for the management of endangered species located in multiple-use landscapes (Lindenmayer *et al.* 2008; Taylor *et al.* 2006).

Primates have been greatly threatened by habitat conversion and fragmentation (Arroyo-Rodríguez & Mandujano 2009; Marsh & Chapman 2013). According to Chapman and Peres (2001), primate habitat countries are losing annually *c.* 125 000 km^2 of forest resulting in remnant primate populations being increasingly isolated in highly fragmented and low-quality habitats; this has resulted in the extinction of several populations and species and this extinction process is anticipated to worsen within this decade and the next (Cowlishaw 1999; Cowlishaw & Dunbar 2000). Extensive literature reviews fail to detect clear patterns of the effects of habitat fragmentation and disturbance on primates, probably because the responses to habitat modification depend, among other factors, on the biological characteristics of each taxon and also on the highly variable ways of conceptualizing and measuring fragmentation effects (Arroyo-Rodríguez & Dias 2010; Arroyo-Rodríguez *et al.* 2013). Primates live in fragments throughout the globe (Marsh & Chapman 2013), including gibbons in Java (Nijman 2013), red howlers in Brazil (Boyle *et al.* 2013), howler and spider monkeys in Mexico (Cristóbal-Azkarate & Dunn 2013), red and black-and-white colobus in Uganda (Chapman *et al.* 2013), macaques in Thailand (Aggimarangsee 2013), capuchins in Venezuela (Ceballos-Mago & Chivers 2013) and orangutans throughout their range (Wich *et al.* 2009), to name just a few.

The Lower Kinabatangan Wildlife Sanctuary (LKWS) in Sabah, Malaysia, is an important site for primatology and primate conservation as ten sympatric non-human primate species (including the proboscis monkey (*Nasalis larvatus*), the Bornean gibbon (*Hylobates muelleri*) and the Bornean orangutan (*Pongo pygmaeus*) all endemic to Borneo) can be found at relatively high densities (Ancrenaz 2007). This protected area is located within the Kinabatangan River catchment, an important wetland in Malaysia, consisting of a variety of habitats including seasonally flooded, riverine and swamp forests, dry dipterocarp and mangrove (including nipah palm) forests (Azmi 1998). Most of the area has been extensively logged in the past (McMorrow & Talip 2001) and today only about 65 000 ha of highly degraded forests remain along the Kinabatangan River. The remaining matrix is primarily industrial oil palm monoculture surrounding different sized patches of forest of different quality which are poorly connected or

Figure 38.1 The Kinabatangan floodplain. Forest fragments interspersed with oil palm plantations, human settlements and local agriculture. Photo: HUTAN. (A black and white version of this figure appears in some formats. For the colour version, please refer to the plate section.)

completely isolated (Figure 38.1). In 2005, the LKWS was officially gazetted by the Sabah's State Government comprising ten forest lots (about 26 000 ha) attempting to create a riparian corridor to link seven Virgin Jungle Forest Reserves (about 15 000 ha) and 10 000 ha of forested state and private lands (Ancrenaz *et al.* 2004; Figure 38.2). Due to its characteristics, the LKWS provides an ideal study site in which to assess the impact of forest fragmentation on primate population structure. Additionally, the Kinabatangan River bisects the sanctuary lengthwise (approximately 200 m in width), potentially acting as a natural barrier to primate dispersal.

Orangutans (*P. pygmaeus*) are one of ten primate species that can be found in the LKWS. They present extreme sexual dimorphism in body size and appearance, and also a pronounced bimaturism among sexually mature males (Delgado & Van Schaik 2000; Utami *et al.* 2002). The lifespan of this species in the wild is estimated to be at least 50 years for both sexes (Wich *et al.* 2004) with slow growth and development rates contributing to this trait. This ape is considered to have a 'semi-solitary' social system but van Schaik (1999) described an individual fission–fusion system for the Sumatran orangutan (*P. abelii*) in the swamp forest of Suaq Balimbing, Indonesia. In studies elsewhere, consortships, travel bands and temporary foraging parties at fruiting trees have also been described (Utami Atmoko *et al.* 2009). The mating system seems to be a combination of female choice and male harassment and coercion, with both morphs (flanged and unflanged males) being reproductively successful in the populations (Delgado & van Schaik 2000; Goossens *et al.* 2006a; Utami Atmoko *et al.* 2009; Utami *et al.* 2002). Maturing females tend to remain near the natal area (philopatry), while males disperse (Galdikas 1995; Singleton *et al.* 2009; Singleton & van Schaik 2001; van Noordwijk *et al.* 2009). However, there is evidence from one of the forest fragments in the LKWS's orangutan population for male and female philopatry (or dispersal) (Goossens *et al.* 2006, see section below on population structure, gene flow and dispersal). Additionally, orangutans occur at low densities with natural fluctuations in population parameters depending on forest type (Husson *et al.* 2009; Marshall *et al.* 2009), however, historically they might have occurred at higher densities (Meijaard *et al.* 2010). As the world's largest arboreal mammal, it has been suggested that terrestrial locomotion is part of the Bornean orangutan's natural behavioural repertoire to a much greater extent than previously thought, and is only modified by habitat disturbance (Ancrenaz *et al.* 2014a, b). Currently, most orangutans are increasingly restricted to small forest fragments (Wich *et al.* 2008, 2009).

The orangutan population of the LKWS is one of the best documented in the world, mainly due to efforts of the

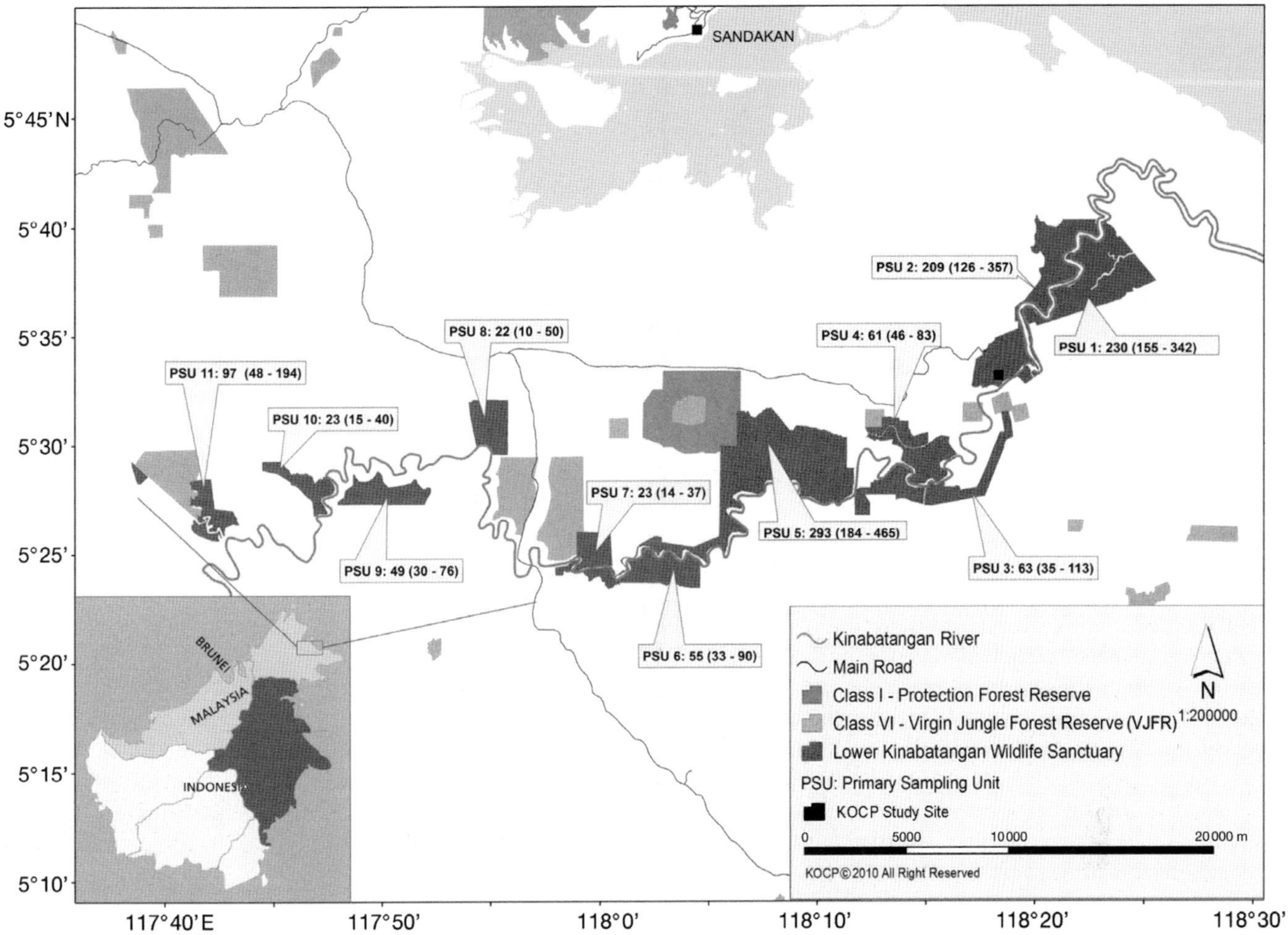

Figure 38.2 Kinabatangan region and the 11 PSUs. Values indicate mean number of orangutans estimated to occur in the respective PSUs in 2004 (range in parentheses). Modified from Ancrenaz *et al.* (2004), with permission from the author.

Kinabatangan Orangutan Conservation Programme (KOCP) by the French NGO, HUTAN. Among other activities, KOCP studies the impact of habitat alteration on orangutan socioecology, and aims to find ways to achieve long-term survival of orangutan populations in exploited areas, especially within and around the Kinabatangan Wildlife Sanctuary. Thus, studies on this population range from systematic estimates of population size and density (Ancrenaz 2007, 2008; Ancrenaz *et al.* 2004, 2005) to endocrine stress responses to habituation and tourism (Muehlenbein *et al.* 2012) and adaptation to newly built environments (Ancrenaz *et al.* 2014a, b). This chapter reviews several genetic studies that were conducted in the area on this population and how they have contributed to the local conservation of the orangutan.

Orangutan Population Structure, Genetic Diversity and Dispersal

Genetic studies on wild orangutans have bloomed in this and the previous decade (Arora *et al.* 2010; Bruford *et al.* 2010; Goossens *et al.* 2004, 2006; Greminger *et al.* 2014; Jalil *et al.* 2008; Morrogh-Bernard *et al.* 2010; Nater *et al.* 2011, 2012; Nietlisbach *et al.* 2012; Sharma *et al.* 2012; Utami *et al.* 2002). This chapter describes, to our knowledge, the first comprehensive population genetic study using samples collected in the wild at a small spatial scale.

In 2004, Ancrenaz *et al.* (2004) divided the Lower Kinabatangan area into 11 'primary sampling units' (PSUs, equivalent to each lot of the LKWS, with lot 10 divided into two distinct parts; Figure 38.2) to estimate orangutan distribution, density and population size. These surveys were conducted by counting orangutan nests along the ground and via aerial transects (Ancrenaz *et al.* 2004). The census estimated a total of 1125 (95% CI 691–1807) individuals, across the 11 PSUs with variable figures ranging from as few as 22 individuals (PSU 8) to as many as 293 (PSU 5) (Figure 38.2). Goossens *et al.* (2005) grouped the 11 PSU into nine sampling units (S1–S9, PSUs 5 and 7 were fused, as well as PSUs 10 and 11) from where faecal samples of 200 wild individuals were collected, representing the largest ever genetic sample from a wild orangutan population. The patterns of genetic diversity and structure within the Lower Kinabatangan orangutan population were then investigated by analysing the genotypes of 14 microsatellite loci. Despite the fragmentation of their habitat, the orangutan population exhibited a high level of genetic variability (e.g. an expected heterozygosity – H_E of 0.74 (Goossens *et al.* 2005). However, this genetic diversity seemed to be the remnant of

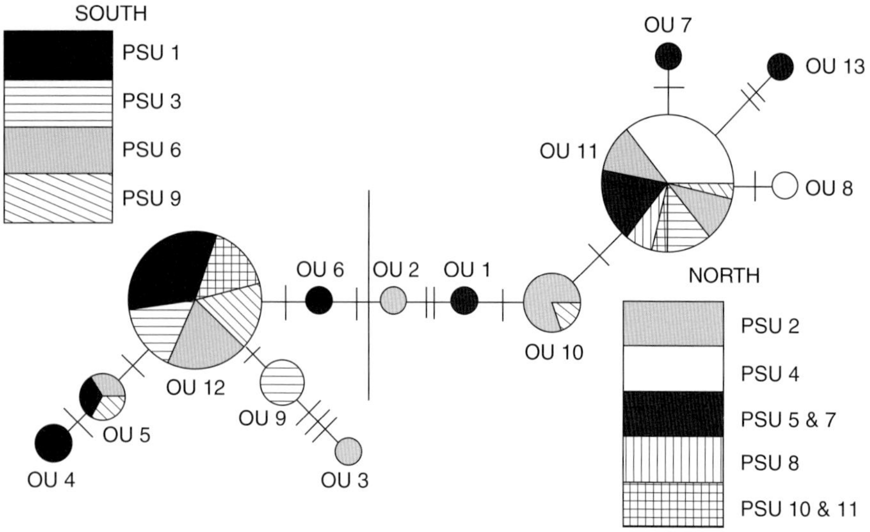

Figure 38.3 Minimum spanning network (MSN) of the Kinabatangan orangutan sequences. Each circle represents a haplotype and the diameter scales to haplotype frequency. The smallest circles represent singletons. Mutational steps are represented by solid black bars on lines connecting haplotypes. Dotted line represents the apparent partition between the populations on both banks of the Kinabatangan River.

an ancient significantly larger population that inhabited the whole region (Goossens *et al.* 2006). Significant genetic differentiation was found between most sampling units but the absolute level of genetic differentiation was limited (average $F_{ST} = 0.04$, $p < 0.001$). This difference was higher between samples separated by the Kinabatangan River than between samples from the same river side ($F_{ST} = 0.06$ versus $F_{ST} = 0.02$, $p < 0.01$). These results indicated the role played by the Kinabatangan River as a natural barrier for orangutan dispersal. To explore the effect of the river on gene flow, Bayesian migration estimation and assignment tests were performed by Goossens *et al.* (2005). The authors found that there was a high frequency of individuals moving between PSUs on the same side of the river. They also found that migration across the river was close to zero. Due to the necessarily limited number of genetic markers and the low level of genetic differentiation, it could not be completely ruled out. Thus, the results of Goossens *et al.* (2005) indicated that orangutans used to move relatively freely between neighbouring PSUs of the LKWS until recently, and that there was a need to maintain migration between isolated forest fragments on the same side of the Kinabatangan River in order to facilitate gene flow (Goossens *et al.* 2005).

The influence of the Kinabatangan River on the population genetic structure of the LKWS orangutans was further confirmed in a study by Jalil *et al.* (2008). In that study, sequences of the mitochondrial DNA (mtDNA) control region were examined for genetic variability and structuring. About 7% (73 individuals from Goossens *et al.* 2005) of the total Kinabatangan population was analysed and 13 haplotypes were identified. A population bottleneck followed by rapid growth and accumulation of mutations was suggested based on the overall high haplotype (0.734 ± 0.035) and low nucleotide diversity (0.008 ± 0.005) found for the whole sanctuary. In addition, the samples on either side of the river were strongly differentiated ($\Phi_{ST} = 0.404$, $p < 0.001$), and a minimum spanning tree analysis on gene genealogies indicated a separation of the haplotypes into two groups, one on the north and one on the south riverbank (Minimum Spanning Network, Figure 38.3), reinforcing the previous inference that the Kinabatangan is (and has been for long periods) a major barrier to gene flow, and this is congruent with evidence elsewhere in Borneo on rivers acting as barriers to the dispersal of this species (Arora *et al.* 2010; Jalil *et al.* 2008).

In addition to the genetic diversity analyses, population structure was further investigated by examining patterns of relatedness and parentage (Goossens *et al.* 2006). Thirteen microsatellite loci (from the 14 used by Goossens *et al.* 2005) were used to genotype 32 identified individuals residing in the KOCP intensive study area located in PSU 2 (north riverbank) of the LKWS (Ancrenaz *et al.* 2004; Goossens *et al.* 2006). The genotypes of 95 individuals from the 200 identified in Goossens *et al.* (2005), but which resided elsewhere in the north bank of the Kinabatangan River, were added to guarantee an unbiased relatedness analysis. The results from the study indicate philopatric behaviour of both male and female orangutans, contrasting with the often reported male-biased dispersal behaviour of this primate (Galdikas 1985, 2008; Houston 2000; MacKinnon 1974; Mitani 1989; Nater *et al.* 2011; Nietlisbach *et al.* 2012; Rijksen 1978; Rodman 1973; Singleton & van Schaik 2002; van Schaik & van Hooff 1996). To be precise, all individuals resident in the KOCP site were on average more related to one another than individuals outside the core area but still within PSU 2 (Goossens *et al.* 2006). This unexpected dispersal pattern, and the high density of males observed in the area, suggest that more work should be carried out to determine whether this is due to the fragmented state of the Kinabatangan forest and its recent reduction in size (Goossens *et al.* 2006b).

Population Decline and Viability

The results presented in the studies by Goossens *et al.* (2005, 2006) and Jalil *et al.* (2008) demonstrated the role played by the Kinabatangan River as a barrier to orangutan movement and gene flow. Additional results of these same studies also suggested that fragments had not yet drifted significantly from each other and were still little differentiated. At the same time, the high densities and the lack of clear differences between males and females in terms of philopatry suggested that some effects on the genetic structure and gene flow were potentially starting to appear. In fact, given the long generation time of orangutans, it is not necessarily surprising that there is a lag in genetic effects brought on by fragmentation. To investigate whether the genetic patterns were showing long-term past or recent events, three different but complementary approaches were used to detect, quantify, and date a putative decline in orangutan populations (Goossens *et al.* 2006b). These methods are described in detail in this study and were named the EWCL (for Ewens-Watterson-Cornuet-Luikart), the Beaumont (Beaumont 1999), and the Storz and Beaumont (Storz & Beaumont 2002) methods.

Regardless of the mutation and demographic models used, the molecular analysis of the microsatellite genotypes from the 200 individuals sampled by Goossens *et al.* (2005) showed strong evidence for a recent and dramatic population decline. Precisely, a particular signature of a population collapse of more than 95% was detected and dated to recent times, and excluding times older than a couple millenia. Thus, the dating strongly suggested that the cause of that decline was unlikely explained by prehistoric hunting and Pleistocene climatic events, nor could it be explained by the arrival of the first farmers in the area. The recent anthropogenic fragmentation of the habitat, namely the exploitation of Sabah's forests which started in 1890, was found to be the only major event that might have significantly influenced orangutan populations in the last decades or centuries. Moreover, the migration patterns reported by Goossens *et al.* (2005) were consistent, as we noted above, with the recent history of logging (1950s) and subsequent oil palm agriculture since the 1970s and 1980s. However, the role of recent forest exploitation in generating bottlenecks in orangutan populations should not be interpolated to every region in Borneo and more theoretical work to understand the multiple demographic events impacting the genome of this species must be encouraged (Arora *et al.* 2010; Meijaard *et al.* 2010; Sharma *et al.* 2012). In addition to the evidence of recent population collapse, Goossens *et al.* (2006b) found extremely low current population size estimates, which were in close agreement with the census estimates of Ancrenaz *et al.* (2004). These two lines of evidence implied the need for immediate conservation efforts to halt genetic drift from quickly eliminating the remaining genetic diversity in the fragmented forests of the Kinabatangan floodplain.

To further assess orangutan population viability in the LKWS, genetic data were incorporated into a stochastic population modelling program under different management strategies to predict the evolution of genetic diversity and demography at different times in the future (Bruford *et al.* 2010). The parameters of the model were based on previous PHVA (Singleton *et al.* 2004), and research and observation in the LKWS (Ancrenaz *et al.* 2004). Different models were designed to test the genetic and demographic consequences of: (1) no intervention, (2) translocations, (3) establishment of forest corridors and (4) a mixed approach combining translocations and corridors. The possible outcomes for the LKWS population are not optimistic under a non-intervention policy where high extinction probabilities (≥5%) are expected for six of the PSUs either including (PSU 4, 6, 7, 8, 9, 10) or excluding (PSU 7, 8, 9, 10) inbreeding depression in the model. These extinctions are predicted to occur within the next 250 years, and even within the next 100 years in the case of PSU 10 (< 5 generations). Furthermore, mean final population size was predicted to decrease in seven of the 11 PSUs (i.e. 64%) when inbreeding depression was not included and in nine of 11 PSUs (i.e. 82%) when inbreeding was included (see Table 1 in Bruford *et al.* 2010). When the model incorporated the translocation of a single adult female every 50 years from PSU 2 to both PSUs 4 and 7 and from PSU 5 to PSUs 8, 10 and 11 (north of the river) and from the south side of the river, from PSU 1 to PSUs 3, 6 and 9, the accumulation of significant amounts of inbreeding within the PSUs having the smallest carrying capacity was not prevented. However, scenarios with more frequent translocations (10 and 20 years) were more successful at controlling inbreeding coefficients in these populations.

As there exist only one or two large source populations donating to several or many small sink populations, the donor populations could become demographically unstable, as was found for PSU 1 if the translocation was conducted every 10 years. In a conservative approach, the third model simulated the establishment of corridors over 100 years (PSUs 4 to 5 and 5 to 7 and PSUs 1 to 3) or 250 years (the remainder). Under this model, corridor reconnection seemed unlikely to occur rapidly enough for the most isolated PSUs, therefore this measure alone might not be able to prevent large-scale genetic and demographic losses nor to prevent extinction in these areas. In contrast to the sole use of translocations or corridors, demographic stability and an inbreeding threshold below 10% were achieved by the mixed approach model where the translocation of one adult female every 20 years was simulated along with the corridor establishment (Figure 38.4). This mixed management approach seemed to be a pragmatic and realistic solution to the current orangutan demographic problem.

Conservation Measures

The Orangutan Action Plan (Sabah Wildlife Department 2011) for the state of Sabah, Malaysia, was the outcome of an extensive consultation process and embodies a consensus of recommendations from relevant stakeholders involved in the management of orangutan populations in Sabah. The conclusions from the studies reviewed in this chapter were incorporated into the actions for the forests of the Lower Kinabatangan floodplain, hence emphasizing the urgent need for habitat restoration and connectivity. However, the creation of habitat corridors for sustainable conservation management can be extremely difficult to achieve due to many factors that can influence the timescale and demographic gains associated with forest corridor establishment, such as financial constraints (land purchase), forest reestablishment rates (which are slow for high canopy dipterocarp forest but which can be much quicker for riparian and seasonally inundated forest) and habitat occupancy and corridor usage dynamics by the faunal community, which could in principle be almost instantaneous or could be a protracted process (Bruford *et al.* 2010). Therefore, as a

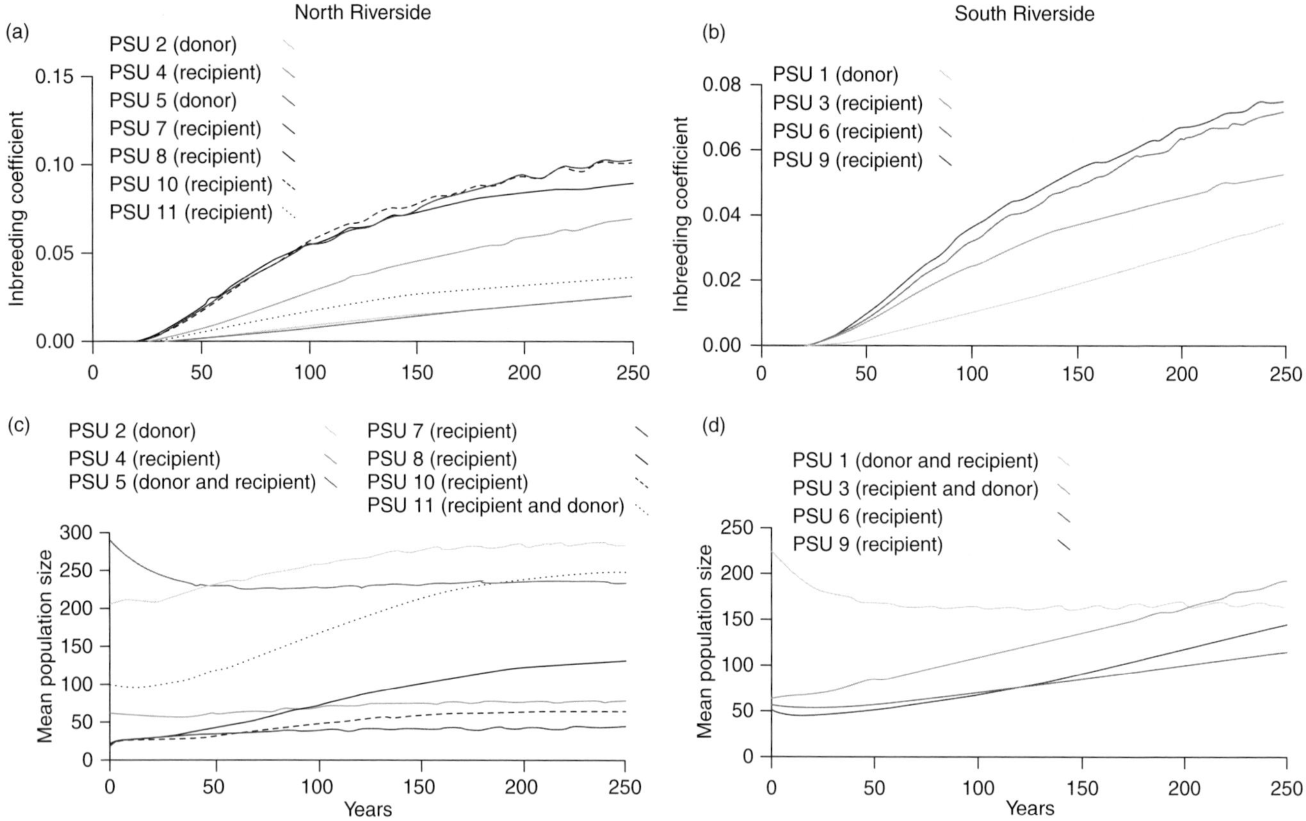

Figure 38.4 Effect of corridor connection and translocations on within-PSU inbreeding coefficients and mean population sizes. (a) Effect of the mixed approach on the inbreeding coefficients of the PSUs on the north riverbank; (b) effect of the mixed approach on the inbreeding coefficients of the PSUs on the south riverbank; (c) effect of the mixed approach on the population sizes of the PSUs on the north riverbank; (d) effect of the mixed approach on the population sizes of the PSUs on the south riverbank. (A black and white version of this figure appears in some formats. For the colour version, please refer to the plate section.)

potentially quicker alternative, and in addition to the restoration and size augmentation of the LKWS, the establishment of orangutan bridges wherever necessary was also recommended (Sabah Wildlife Department 2011). These bridges would temporarily substitute the natural canopy coverage, which is currently lacking, for the orangutans to move freely above the small tributaries of the Kinabatangan River and the drains created by the oil palm plantations. To date, eight orangutan bridges have been set up over four small tributaries of the Kinabatangan River (Figure 38.5; Ancrenaz 2010; DGFC and Sabah Wildlife Department 2010; Lackman & Ancrenaz 2009), with documented success in all of them (Ancrenaz 2010).

The management of small populations of endangered species is complex by nature: in addition to the scientific aspects, we need to incorporate a welfare dimension that we cannot underestimate for iconic species such as the orangutan. As an example, a rescue operation was approved by the Sabah Wildlife Department on August 2012. A small forest on private land nearby PSU 2 (north riverbank) was cleared for oil palm plantation and a male and a female orangutan were urgently transferred to PSU 6 (south riverbank).

A conservative conservation management plan would normally seek to maintain the natural demographic isolation between populations to the north and south of the Kinabatangan River (Goossens *et al.* 2005; Jalil *et al.* 2008). Indeed, a recommendation by Bruford *et al.* (2010) was to maintain these populations as separate Management Units. However, under time and logistical constraints, measures were taken for the sake of the individuals' welfare as well as for the species conservation. It was thus decided to translocate the animals to a PSU under high extinction risk and where continuous monitoring activities are undertaken. After that incident, effort was made to follow the recommendations from Bruford *et al.* (2010) and the subsequent case (September 2013) was that of a flanged male which was rescued in the area of Gomantong (on the north riverbank) and was then translocated to PSU5 (same riverbank); this individual has been observed frequently since, on camera traps (DGFC, camera traps database).

The importance of genetic diversity in population viability and management has been debated in the past (Asquith 2001) and it is not always clear whether demographic, environmental or genetic factors will be the first to impact threatened species. However, it is increasingly recognized that genetic data provide unique information. While the application of genetics (and genomics; Sharma *et al.* 2012) in the management of threatened species is increasing (DeSalle & Amato 2004), there seems to be a general failure to incorporate these types of data into concrete conservation actions. This failure may be due to two main factors: first, the difficulty of interpreting the results of genetic data for non-geneticists; and, second, the difficulty of becoming involved in policy and practical conservation decisions for conservation geneticists. There is also a need

Figure 38.5 Orangutan using the 'orangutan bridge' over the Rasang River, a tributary of the Kinabatangan River. Photo: Ajirun Osman/HUTAN-KOCP.

to develop tools that will help conservation biologists to use and master population genetics concepts (i.e. Conservation Genetic Resources for Effective Species Survival (CONGress), http://www.congressgenetics.eu). The incorporation of genetic data into species action plans has recently been advocated, but will require the above-mentioned difficulties to be overcome (Frankham 2009; Laikre 2010). Genetic data must be integrated with an understanding of landscape dynamics and area-based conservation actions to achieve successful decisions concerning areas, landscapes and species.

Conclusions

Nearly 15 years have passed since genetic samples were collected by Goossens *et al.* (2005) in 2001, and it has been more than half a decade since the publication of the Orangutan Action Plan (Sabah Wildlife Department 2011). During this time, much was achieved to advance the knowledge of the LKWS orangutan's population genetics. It certainly stands out as a major achievement to have the data incorporated into an official management plan. Nonetheless, the orangutan is a slow breeder and it will take time before deleterious genetic effects are detected in this population. Time is pressing, land conversion in the area is still ongoing and the population size is still declining (Alfred *et al.* 2010; Ancrenaz 2008). For some species on the brink of extinction (i.e. the Sumatran rhinoceros, Goossens *et al.* 2013) a 10-year wait might be an unaffordable luxury. Therefore, faster mechanisms for the incorporation of genetic data into management plans should be devised and scientists and policy makers should also make a compromise regarding the extent of genetic information really needed before expediting the successful and urgent protection of a species like the orangutan through restoration of habitat connectivity and other means already known to us.

Acknowledgements

We are grateful to the Sabah Wildlife Department and the Economic Planning Unit for permission to work in Sabah. We would also like to thank the following sponsors who aided in the funding of this work, from orangutan surveys to bridge construction: zoos of Pittsburgh, Cleveland, Columbus, Chester, Houston, Brookfield, Lincoln Park, Apenheul, Beauval and La Palmyre; Bush Gardens, Dierenpark, USFWS GACF, NGS, WWF (US, UK, Netherlands, Malaysia), AOF, BOS USA, Disney's Animal Kingdom Conservation Fund, Borneo Conservation Trust Japan, the Darwin Initiative for the Survival of Species (Grant no. 09/016, DEFRA, UK), Cardiff University, Universiti Malaysia Sabah (short-term grant no. B-0804–11-PR/U009; SLAB UMS), the Laboratoire d'Excellence (LABEX) entitled TULIP (ANR-10-LABX-41), the Instituto Gulbenkian de Ciência, HUTAN, DGFC and several other private sources. We extend our thanks to the KOCP research assistants for all their hard work at the study site and their dedication to the conservation of the orangutan and its habitat in the Lower Kinabatangan Wildlife Sanctuary. Orangutan hair and faeces were exported to the UK under export permit from the Federation of Malaysia (CITES Certificate no. 0467, security stamp No. MY 9123707) and under import permit from the UK (CITES Certificate nos. 236719/01 for shed hair samples and 236719/02 for faecal samples). This chapter is dedicated to the memory of Mohd. Fairus Bin Jalil.

Chapter 39

African Flooded Areas as Refuge Habitats

Anh Galat-Luong, Gerard Galat, Rebecca Coles and Jan Nizinski

Introduction

On the basis of our own ecological and behavioural studies on primates in flooded habitats in West and Central Africa – mangroves, flooded and swamp forests, seasonal grassland floodplains and poor quality soils – we aim to answer the following questions: Does the abundance of monkeys or apes differ in flooded versus non-flooded habitats? If so, what kinds of species adapt to life in flooded habitats (e.g. swamp specialists, opportunistic generalists, other specialists) and how do they adapt? What are the implications of regarding mangroves or other flooded areas with poor soils as primate refuge habitats?

Mangroves

In Senegal, from 1974 to 1976, Galat and Galat-Luong (1976, 1978) and Gatinot (1976) studied the ecology and behaviour of monkeys in what is now the Delta du Saloum National Park a UNESCO/MAB reserve and Ramsar site. Although patas monkeys (*Erythrocebus patas*) and Temminck's red colobus (*Piliocolobus badius temmincki*) were never observed to range in the 23 000 ha of flooded areas and mangrove, green monkeys (*Chlorocebus sabaeus*) successfully adapted to this habitat (Galat & Galat-Luong 1976; Galat-Luong & Galat 2005; Figure 39.1).

Research continued from 1988 to 2002, partly in the framework of an IUCN-ORSTOM pluridisciplinary conservation project (Galat *et al.* 1998b, 2002, 2009a; Galat-Luong & Galat 2005; ISE *et al.* 2000a). Galat-Luong and Galat (2005) showed that over 30 years, with a 300 mm deficit in rainfall, drastic habitat changes occurred in the National Park; *terra firma* forest reduced in size by more than 50% (75% in the Fathala gallery forests used by the red colobus) and woody species diversity decreased by 30%. Despite these significant changes, the red colobus population only decreased from approximately 600 to 500 individuals in the Fathala forest. Galat-Luong and Galat (2005) suggested that their survival was due to five adaptations which emerged in their population during the 30 years: (1) frugivory; (2) terrestriality (Figure 39.2); (3) tendency to form polyspecific associations with green monkeys (Figure 39.2); (4) increased utilization of open habitats; and (5) use of mangrove swamps for refuge and forage (using knowledge socially learned from green monkeys). By exhibiting these adaptations in a very short time, the red colobus, although a leaf-eating specialist, demonstrated its ability to adapt and persist in a degraded habitat.

Figure 39.1 Green monkey, *Chlorocebus sabaeus*, on mangrove roots, *Rhyzophora racemosa*, Rhizophoraceae, Saloum mangrove, Delta du Saloum National Park, UNESCO/MAB Reserve and Ramsar site, Senegal. © Galat-Luong A.-IRD.

Since the 1990s, patas monkeys have also entered mangrove (Galat-Luong & Galat 2005; Labouze *et al.* 1996; Figure 39.3),

Figure 39.2 Red colobus, *Piliocolobus badius temmincki*, on the ground with green monkeys, *Chlorocebus sabaeus*, and a bushbuck, *Tragelaphus scriptus*, Bovidae, Fathala forest, Delta du Saloum National Park, UNESCO/MAB reserve and Ramsar site, Senegal. © Galat-Luong A.-IRD.

and so have introduced baboons (*Papio hamadryas papio*), which, as crop raiders, are hunted intensively (Galat *et al.* 2000; Galat-Luong & Galat 2011, 2013; Galat-Luong *et al.* 2006). Other large vertebrates also use the mangrove as a night (guinea fowl, *Numida meleagris*, Numididae) or day refuge for resting (bushbuck, *Tragelaphus scriptus*, Bovidae; sitatunga, *Tragelaphus spekii*, Bovidae; civet, *Civettictis civetta*, Viverridae; hyena, *Crocuta crocuta*, Hyaenidae). Consequently, mammal diversity has increased in the mangrove over time (Galat-Luong & Galat 2007).

In the Ivory Coast, from 1978 to 1988, Galat-Luong and Galat (1979a) surveyed the very small mangrove of Lagune Ebrié, 1 km from Yopougon-Abidjan. Lesser spot-nosed guenons (*Cercopithecus petaurista*) and olive colobus (*Procolobus verus*) lived so discreetly in this mangrove that they had not been observed previously. Only Lowe's guenons had been studied (Bourliere *et al.* 1969, 1970; Hunkeler *et al.* 1972; Galat-Luong & Galat 1979a) as they also used the lagoon fringe forest where they were hunted regularly, but individuals coming from the mangrove regularly recolonized the riverine forest (Galat-Luong & Galat, pers. obs.). Green monkeys have also been observed at other sites in this same lagoon (Galat 1983; Galat-Luong *et al.* 2011).

Flooded and Swamp Forests

In Senegal, from 1975 to 1976, Galat (1983) and Galat & Galat-Luong (1977) studied the home ranges and densities of green monkey groups in the seasonally flooded Senegal River valley (Figure 39.4). The presence of green monkeys, living in large groups (up to 176 individuals) in the *Acacia nilotica* forest and along the oxbows remains, had been known historically (Buffon 1789). As climatic conditions were harsh (up to 57°C, 20–200 mm rainfall/year), only the seasonal flooding of their home range was able to provide the required food.

In the Central African Republic, in the Ngotto Forest, now a Ramsar site, Galat (1977) compared the abundance of seven primate species in the riparian seasonally flooded area of the forest (palm swamp) versus their abundance in the *terra firma* area (Table 39.1). All seven species were observed in the flooded swamp, showing that not only swamp specialists use this habitat. Interestingly, the most abundant species in this habitat was

Figure 39.3 Patas monkey, *Erythrocebus patas*, in a black mangrove, *Avicenia nitida*, Verbenaceae, Saloum mangrove, Delta du Saloum National Park, UNESCO/MAB reserve and Ramsar site, Senegal. © Galat-Luong A.-IRD.

Figure 39.4 Green monkey, *Chlorocebus sabaeus*, moving in the seasonally flooded gum arabic tree forest, *Acacia nilotica*, Fabaceae, Senegal Valley River, Senegal. © Galat-Luong A.-IRD.

Table 39.1 Number of group encounters per day for seven primate species during surveys of flooded palm swamp and *terra firma* habitats.

Forest habitat	Type	Flooded swamp		Primary *terra firma*		Secondary *terra firma*
		C	E	C	E	E
Primate species						
Cercopithecus pogonias	AF	4.1	1.1	0	0	1.1
Piliocolobus rufomitratus oustaleti	HL	2.2	0	0	0	0
Cercopithecus nictitans	AF	0.8	0	0.9	0	0
Cercopithecus neglectus	AF	0.8	1.5	0	0	0
Cercopithecus ascanius	LF	0.6	0	1.3	0	1.1
Cercocebus galeritus	GF	0.3	0	0	0	0
Colobus guereza	AL	0.3	0.4	0	0	0
Unidentified species		0.6	0	0.4	0	2.2
Total number of encounters (groups)		**9.7**	**3**	**2.7**	**0**	**4.4**
Number of different species observed		**7**	**3**	**2**	**0**	**2**

C: centre of Ngotto Forest, near Mbaere River (> 5 km Badane); E: edge of Ngotto Forest, far from the river (> 5 km Mandoukou). Specialist types: A: All strata; H: high strata; L: low strata; G: ground strata; F: frugivorous omnivore; L: leaf-eating.

the crowned guenon (*Cercopithecus pogonias*), an opportunistic generalist, and not the De Brazza's monkey (*Cercopithecus neglectus*), a swamp specialist in the area (Struhsaker *et al.* 2008). The Oubangui red colobus (*Piliocolobus rufomitratus oustaleti*), ranked second, and was only recorded in the flooded swamp forest (see discussion).

In Burkina Faso, from 2003 to 2004, green monkeys and baboons (*Papio hamadryas anubis*) were surveyed in the Mare aux Hippopotames de Bala Mab Reserve (Galat-Luong & Galat, pers. obs.). The day and night-sleeping trees of both species were located exclusively in the swamp areas. Patas monkeys sometimes came to drink at a water source, remaining in the swamp area for their daytime sleep (see discussion).

In the Ivory Coast, in the N'zo Forest swamps, olive colobus (*Procolobus verus*; Figure 39.5) lived in larger sized and more often monospecific troops than those troops observed in the adjacent Taï National Park (see discussion) (Galat-Luong 1983; Galat & Galat-Luong 1985; McGraw & Galat-Luong 2011).

Seasonal Grassland Floodplain

In Senegal, during the transition period between the dry and wet seasons (May-July), Galat *et al.* (2007) studied the large mammals, including primates, of the Simenti grassland floodplain, part of the Niokolo-Koba National Park and Mab Reserve. This floodplain covers 108 ha in the Sudano-Guinean zone and is seasonally (July–October) watered by the Gambia River. From dawn to dusk, the number of individuals for each species inside the grassland was counted every 15 minutes. In addition, the line-transect method was employed to estimate mammal densities peripheral to the floodplain, using *Distance* software (Laake *et al.* 1996). Primate densities within and peripheral to the floodplain were compared to the density in the National Park as a whole (Galat *et al.* 1998a, b, 2007, 2009b). For green monkeys, densities of 8.93 (n = 2241) and 21.15 (n = 182) individuals/km^2 were observed in the floodplain and the surrounding area, while a lower density of 1.6–1.7 individuals/km^2 (n = 2095) was observed in the National Park as a whole. Baboon densities were 6.71 (n = 2485) and 18.58 (n = 365) individuals/km^2 in the floodplain and its surrounding area, versus a lower density of 6.3–7.3 (n = 14159) individuals/km^2 in the National Park as a whole. Observations of patas monkeys were too infrequent for analyses. In agreement with studies on other mammals (Galat *et al.* 2007), the figures show the high attractiveness of the seasonally flooded grassland, whenever it is not flooded.

Biodiversity and Soil Quality Analysis

In Senegal, Galat-Luong and Galat (2003) conducted a large mammal survey on 53 species, including six primate species, in the regions of Tambacounda and Kolda (about 80 000 km^2) as part of a large pluridisciplinary DDR-IRD study (Direction of Rural Development – Institut de Recherche pour le Development). Although this study was not focused specifically on flooded areas, these regions do contain hydromorphic soils characterized by an excess of water due to a temporary obstruction or fluctuation in the groundwater table, usually in floodplain areas bordering main rivers (such areas are favorable to the installation of rice fields after drainage). The comparison of soil quality maps (appropriateness for human farming, Louhoungou *et al.* 2001) and the distribution of mammal species, including primates (Galat-Luong & Galat 2003), showed that soils of areas occupied by wild fauna were of poorer quality than soils in farming areas; these uncultivable or not cultivated areas provided natural refuges for species avoiding human pressures.

Figure 39.5 Olive colobus, *Procolobus verus*, Taï National Park, Ivory Coast. © Galat-Luong A.-IRD.

Behavioural Data

In this context, it is necessary to examine reports of primate individuals interacting with water. Some water adaptive or related behaviours have been reported above, but additional observations give further insights into five behaviours described below.

Avoiding or Playing with Water

Monkeys often jump to avoid small watercourses (Figure 39.6), while others do not. In eastern Senegal, chimpanzees (*Pan troglodytes verus*) live and build their nests in the valleys during the dry season, but move to the highlands in the wet season when the valleys are flooded (Ndiaye *et al.* 2013). In the Ramsar Azagny National Park, Ivory Coast, introduced chimpanzees are restricted to islands because they fear water. However, some individuals were observed walking in water up to their knees, one using a stick as observed in bonobos (*Pan paniscus*, Williamson *et al.* 2013), and one individual washed his feet while brachiating over a lagoon (Galat-Luong & Galat pers. obs.). Green monkeys drink from branches, avoiding the mud of the river bank. When no branch is available, they clean their hands with the dry soil. Three moustached guenons (*Cercopithecus cephus*), a grey-cheeked mangabey (*Lophocebus albigena*), a patas (*Erythrocebus patas*) and a tantalus (*Chlorocebus tantalus*) in semi-captivity were presented with a small water tank (2 m × 1 m × 10 cm) and hosepipe. Only one guenon played in the water, wading along the base of the tank, or placing its hand over the end of the hosepipe to create a spray to shower itself (Galat-Luong & Galat, pers. obs.). Whether an individual plays with water appears to be an individual preference rather than a habit characteristic of a group, population or species.

Filtering Drinking Water

In Senegal in the dry season, water becomes rare and putrid. Since chimpanzees and baboons prefer clean drinking water, they filter water by digging wells in the sand rather than drinking close-by stagnant water (Galat *et al.* 2008; Galat-Luong & Galat 2000; Galat-Luong *et al.* 2009). This represents another interaction between primates and water, in this case, the water they consume.

Swimming

In Senegal, in the Saloum mangrove, young and sometimes adult green monkeys jump in the sea playing 'king of the mountain'. Some individuals avoid mud flats at high tides, while

Figure 39.6 Olive baboon, *Papio anubis*, crossing the Arli River, jumping to avoid walking or swimming in the water, Burkina Faso. © Galat-Luong A.-IRD. (A black and white version of this figure appears in some formats. For the colour version, please refer to the plate section.)

others cross the flats by jumping bipedally. Others, including a mother with her infant carried ventrally (thus submerged), swam a distance of over 100 m (Galat 1989; Galat-Luong *et al.* 2011; Galat & Galat-Luong 2013). Similar behaviours have also been observed in baboons crossing the Gambia River (*Papio papio*, Senegal), the Arli River, and the Pendjari River (*P. anubis*), Burkina Faso, Benin: some individuals jumped as far as possible into the water from branches, while others swam along the length of the river. A mother invited and waited for her offspring to climb onto her back before crossing (Galat-Luong & Galat, pers. obs.). Whether an individual swims or avoids swimming appears to be an individual preference and may indicate an adaptive advantage.

Staying Near Water

We observed green monkeys spending up to three consecutive days taking advantage of the cool, insect-free mangrove for social activities and sleep (Galat & Galat-Luong 1976). Late evening resting and socializing on river banks, without drinking or foraging, and sleeping in sites overhanging water are frequent: baboons along the Arli River in Burkina Faso (Figure 39.7), along the Gambia River in Senegal (Adie *et al.* 1997; Galat-Luong & Galat 2013); green monkeys along the Senegal River (Galat & Galat-Luong 2013); baboons and green monkeys sharing the swamp sleeping sites in the Mare aux Hippopotames de Bala, Burkina Faso; white-napped mangabeys (*Cercocebus lunulatus*) resting along the Comoe River (Figure 39.8); and green monkeys sharing their riverine sleeping sites along the Comoe River, Burkina Faso (Galat & Galat-Luong 2006).

Foraging on Aquatic Food

We frequently observed green monkeys feeding on the various parts of the mangrove *Rhizophora* sp. (Rhizophoraceae) (see Galat & Galat-Luong 1976, for photos of green monkeys eating the pith of mangrove roots). Green monkeys were recorded eating water lilies (Galat & Galat-Luong 1977; Figure 39.9) and fiddler crabs (*Uca tangeri*, Ocypodidae) using highly adaptive techniques (Galat & Galat-Luong 1976), and also West African marbled lungfish (*Protopterus annectens*, Protopteridae) (Galat *et al.* 2011; Galat-Luong *et al.* 2011). Guinea baboons also eat fiddler crabs in the Saloum mangrove, water plants while

Figure 39.7 Group of olive baboons, *Papio Anubis*, resting on the bank of the Arli River, Arli National Park, Burkina Faso. © Galat-Luong A.-IRD. (A black and white version of this figure appears in some formats. For the colour version, please refer to the plate section.)

wading in the Gambia River, Senegal, and also forage for oysters (*Etheria* sp., Etheriidae) (Galat-Luong & Galat 2013). Olive baboons collect and eat Mutelidae mussels in the Arli River, Burkina Faso (Galat-Luong & Galat pers. obs.; Figure 39.10). Patas monkeys hunt fiddler crabs in the Saloum mangrove, and catch and eat mudfish (*Clarias anguillaris*, Clariidae) in east Senegal (Galat-Luong 1991a; de Jong *et al.* 2011). Feeding on *Rhizophora* mangroves by Temminck's red colobus has been observed in a Gambian mangrove (Ellenberg *et al.* 1988) and by Zanzibar red colobus in Tanzania (Nowak 2008). Oubangui red colobus, although a higher strata specialist, displayed adaptive behaviours, such as wading in water and feeding on African water onion bulbs (*Crinum natans*, Amaranthaceae) and Nymphaeaceae (Galat-Luong & Galat 1979b; Galat-Luong *et al.* 2012), implying that the swamp was a refuge habitat.

Discussion

Mangroves as Refuges

Terra firma primates living in Guinean mangroves in conditions similar to those observed in Saloum include *Chlorocebus sabaeus*, *Cercopithecus campbelli*, *Colobus polykomos*, *Cercocebus atys*, *Piliocolobus badius temmincki* and *Pan troglodytes verus* (Bah *et al.* 1999; Leciak *et al.* 2005; OGM 2006; SDAM 1990). Large vertebrate diversity may increase in mangroves due to colonization by new species (Galat-Luong & Galat 2005, 2007). This increase, however, is deceptive; mangroves are only a last refuge for endangered species in the short term, as species remain dependent on the quality and access to inland coastal habitats.

Since the conservation of these areas is of high priority, the IUCN-ORSTOM 1998–2002 Saloum Mab Reserve pluridisciplinary research and management programme wrote an action plan with the following goals: (1) the rehabilitation of saline soils through the establishment of tree plantations and the management of grazing animals; (2) research integrating species monitoring, education and public awareness of mangroves, birds and mammals (including cetaceans and monkeys); (3) mapping the biosphere reserve zones, including the red colobus sanctuary; (4) developing sustainable tourism management; and (5) in order to ensure the conservation of the red colobus population, preservation of the forest and mangrove (Galat 1991; Galat & Galat-Luong 1995,

Figure 39.8 White-napped mangabey, *Cercocebus lunulatus*, resting along the Comoe River, Burkina Faso. © Galat-Luong A.-IRD.

1999; Galat *et al.* 1998a, b, 2002, 2009a; Galat-Luong 1991b, 1999, 2000, 2001; Galat-Luong & Galat 1999, 2001a, b, 2002, 2005, 2007; Galat-Luong *et al.* 2008; Galat-Luong & Luong 2001, http://www.unesco.org; ISE *et al.* 2000a, b; Tumbarello *et al.* 1995a, b).

Swamp Forests as Refuges

In the Ngotto Forest, the Oubangui red colobus, a canopy and emergent tree specialist, was only observed in the centre of the flooded swamp forest, although the canopy was lower and emergents rarer; density was high and individuals were observed wading in the river feeding on aquatic plants (Galat-Luong & Galat 1979b; Galat-Luong *et al.* 2012). Differences in primate densities across habitat types may be due to environmental differences: forest degradation, with increasing importance of secondary forest near the edge of the forest, and also increased hunting pressure; at the centre of the forest, around the small pygmy camp of Badane, hunting pressure is lower than at the edge of the forest around the larger village of Mandoukou, where guns are used in the *terre firma* forest. Large colobus monkeys are the easiest to hunt and have the best return price for a cartridge.

In the Taï National Park, olive colobus exhibit a hiding strategy against their predators: troops split into very small groups, each group associating in large polyspecific troops (up to seven species; Galat & Galat-Luong 1985). This strategy constrains their abundance, keeping it low, and other primate species are more abundant (olive colobus ranks 7/7; Galat & Galat-Luong 1985). The larger groups of olive colobus living in the N'zo Forest swamps may benefit from the natural protection offered by this less accessible habitat from their main predators, the leopard (*Panthera pardus*) and humans (Galat-Luong 1983). Additionally, the relative abundance of other primate species is lower (Galat-Luong 1983). As a result, these populations may not need this fissioning strategy, and thus, can exhibit stronger group cohesion and live in larger and monospecific groups (Galat-Luong 1983).

Gautier-Hion and Brugiere (2005) compared the structure of 12 Central African primate communities (including that of the

Figure 39.9 Green monkey, *Chlorocebus sabaeus*, eating water lilies in a flood pool at the end of the rainy season, Senegal River valley, Senegal. © Galat-Luong A.-IRD

Ngotto Forest), six in riparian forests and six in adjacent *terra firma* forests. On average, communities in riparian forests had 1.5 times more primate species than communities in *terra firma* forests, since riparian forests sheltered four specialist and six to seven generalist species. In Africa, most riparian-specialist primates are terrestrially adapted and can therefore access a larger food niche than more strictly arboreal species. Thus, protecting riparian forests and adjacent *terra firma* forests, so that the majority of lowland forest diversity is present in the protected area, is highly recommended. The linear shape of riparian forests, promoting gene flow over large distances, and their persistence in anthropic landscapes, since they represent areas of lower value for agriculture and logging than mainland forested areas, predisposes them as biodiversity sanctuaries. The Tanoe swamp forest, Ehi Lagoon in the Ivory Coast, is a case study. Since 1978, Miss Waldron's red colobus (*Piliocolobus badius waldroni*) was thought to be globally extinct (McGraw 2005; Oates *et al.* 2000), but Bi *et al. (*2008), Kone & Akpatou (2004), Kone *et al. (*2007) and Zamblé (2009), discovered a relict population of this critically endangered subspecies (Oates *et al.* 2008b) in this swamp forest, as well as a population of the Endangered roloway monkey *Cercopithecus diana roloway* (Oates *et al.* 2008a).

In the Mare aux Hippopotames de Bala, monkeys took refuge in the swamp areas because they were the only areas, although a Mab Reserve, without disturbing pressures (e.g. human use of natural resources, including poaching).

Seasonal Grassland Floodplains

When resources of a residential habitat are depleting, monkeys and other large mammals take refuge in this grassland floodplains; species congregate at high densities reflecting the attractiveness of this area and the major role it has for biodiversity conservation (Galat-Luong 2000; GEPIS 2000).

Behaviour in Flooded Habitats

The propensity to like or dislike water and swimming appear to vary across primate species and also individuals. Gautier-Hion and Gautier (1971) recorded observations for the Gabonese guenons. Struhsaker (pers. comm.) observed the mona monkey (*Cercopithecus mona*) jumping and swimming in Cameroon mangroves (see Gartlan & Struhsaker 1972), and the Diana monkey (*Cercopithecus diana diana*) swimming 27 m across the Meno River in the Taï Forest, often with head and body completely submerged. In contrast, several authors have concluded that apes under-utilize swamp forests (Blake *et al.* 1995; Fay & Agnagna 1992; Fay *et al.* 1989, 1990; Marchesi *et al.* 1995; Morgan *et al.* 2006; Poulsen & Clark 2004; Reinartz *et al.*

Figure 39.10 Olive baboons, *Papio anubis*, eating *Mutelidae* mussels in the Arli River, Burkina Faso. © Galat-Luong A.-IRD. (A black and white version of this figure appears in some formats. For the colour version, please refer to the plate section.)

2008; Williamson & Usongo 1995). When swamps are used, it is hypothesized that they act as a refuge, offering protection against human disturbances (Blake 1993). The individuals which do not avoid water may well have an adaptive advantage.

Poor Quality Soils

The DDR-IRD pluridisciplinary study in east Senegal showed that poor soil quality is an advantage for biodiversity conservation; human pressures are lower in such areas compared to those areas of higher soil quality. Swamps and other flooded areas are considered to have low-quality soils. At present, however, palm oil production is a major threat to species preservation. The challenge to conserve the Tanoe forest swamp, which holds the last population of the roloway monkey and, potentially, Miss Waldron's red colobus (*Piliocolobus badius waldroni*) is a pertinent example in West Africa, as the area is coveted for palm oil production.

Monkeys adapt to flooded areas. Some species like Allen's swamp monkey (*Allenopithecus nigroviridis*) are swamp specialists (Gautier 1985; Gautier-Hion 1988; Chapter 3), while some generalists, like green monkeys, adapt opportunistically. In addition, other specialists like the 'large tree' specialists red colobus, or the 'desert' specialists patas monkeys, do so when facing pressures such as hunting or habitat loss. Thus, mangroves and flooded areas are increasingly becoming refuge habitats for primate species.

In some cases, high density of a species in a flooded habitat is due to a species-specific adaptation, enabling the species to successfully live in and show habitat preferences for flooded areas. This is the case for Allen's swamp monkey (Gautier, 1985; Gautier-Hion 1988, 2013; Gautier-Hion *et al.* 1999) and the golden-bellied mangabey (*Cercocebus chrysogaster*) which has 58% of its distribution area in swamp forests (Inogwabini & Myers Thomps 2013) and spends 95% of its time in swampy forest, even when flooded (Gautier-Hion *et al.* 1999). As discussed, higher species densities may be a consequence of the habitat acting as a last refuge for such species. We can expand on this point: faunal density, including primate density, is higher in the wild compared to anthropized habitats. When analysing the quality of primate habitats and primate distributions it may appear that primates inhabit the 'poorest quality habitats'. This does not imply that primates prefer such habitats, but it reflects that 'better habitats' are occupied by humans; primates have no other options but to adapt to ensure survival in poor quality habitats.

Conclusions

The examples discussed suggest that there is a continuum between flooded areas as habitats for populations seeking refuge due to their forced displacement by exterior pressures, including hunting by humans or habitat destruction, and as habitats for populations preferentially exploiting these areas seasonally. Flooded areas have a higher productivity due to their humidity compared to the decreasing productivity of their usual range. For each species, some individuals are observed to have a better relationship with water, and presumably act as pioneers when various constraints lead groups to venture into flooded or swampy areas.

Chapter 40

Diversity and Conservation of Primates in the Flooded Forests of Southern Nigeria

Lynne R. Baker and John F. Oates

Introduction

Nigeria covers an area of 923 768 km^2 and thus ranks 14th in size among African nations. However, it has the densest human population of any large country on the continent and is easily Africa's most populous country. According to United Nations estimates, by 2050 Nigeria will be the third largest nation globally, with more than 410 million people (UN 2017). The environmental consequences associated with such a dense and rapidly growing population place immense pressure on the country's natural habitats, primates and other wildlife.

Nigeria has one of the world's richest non-human primate faunas, including 28 species and 18–19 genera. Most of Nigeria's primates occur in low-lying forests in the southern part of the country, in a region that reaches as far as 200–300 km inland from the coast and has a mean annual rainfall of at least 1500 mm (Happold 1987). The southern forest region comprises mangrove, freshwater swamp and lowland moist broadleaf forests (Areola 1982; Poorter *et al.* 2004). However, outside of the relatively intact forests of Cross River State (in southeastern Nigeria along the Cameroonian border) and isolated patches elsewhere (typically in government forest reserves), much of Nigeria's lowland moist forest has been converted to farms, plantations and derived savanna as a result of many years of human activity (Figure 40.1).

Between 1976 and 1995, the area of land in Nigeria devoted to agriculture and plantations (crop and grazing lands, floodplain agriculture, irrigation and livestock projects, and tree and forest plantations) increased by nearly 84 000 km^2. During the same period, more than half of the country's undisturbed lowland moist forest was lost (from 25 951 to 12 114 km^2) (Geomatics International *et al.* 1998). More recent data show that Nigeria converted more land to annual crops than did any other tropical nation for the decade ending in 2008 (Phalan *et al.* 2013).

Although southern Nigeria's forests have suffered widespread destruction and, consequently, the loss and decline of many primate populations, comparatively large areas of seasonal and permanent swamp forest still occur, primarily because such forests are relatively difficult to access and farm. These flooded forests are among the largest tracts of remaining habitat for primates in the country and provide at least a partial refuge for many species. In 1995, freshwater swamp forests were estimated to cover 16 499 km^2, mangrove forests 9977 km^2, and riparian forests 5254 km^2, together comprising 3.5% of Nigeria's surface area (Geomatics International *et al.* 1998).

Threats to these inundated habitats are numerous and accelerating. Within the Niger River Delta, for example, an estimated 213 km^2 of mangrove forest were lost between 1986 and 2003 due to activities related to oil and gas development, dredging, wood cutting, urban development and the spread of an invasive palm (James *et al.* 2007). Many of Nigeria's flooded forests have already been heavily modified by the harvesting of timber trees, and none has effective protection. Some are designated as government forest reserves, but these are managed primarily for timber production, and only limited controls are imposed on hunting and the collection of non-timber forest products, such as edible fruits, nuts, honey, bamboo and rattan cane.

Here we describe the diversity of primates in the flooded forests of southern Nigeria and discuss the value of these forests to primate conservation. Relatively little research has been conducted on primates in these forests, so our report should be regarded as a preliminary assessment. In this account, we follow the primate taxonomy presented in Oates (2011), except that we have adopted the now-widespread usage of *Piliocolobus* as the generic name for red colobus monkeys.

Classification and Distribution of Flooded Forests in Southern Nigeria

This chapter focuses on flooded forests that lie between the coast and the inland lowland forest zone, as well as strips of riparian forest that extend into regions with drier lowland forests or woodlands. The eastern boundary of this region is the Cross River, while the western boundary is at the border with the Republic of Benin (Areola 1982). The Niger River and its delta occur near the centre of this region (Figure 40.1).

The Niger, the longest river in West Africa and the third longest river in Africa, has profoundly influenced the geomorphology, biogeography and political economy of southern Nigeria. It is an important distributional boundary for many mammals, including a number of primates (Booth 1958; Happold 1985; Oates 1988; Chapter 30). The river flows into the Gulf of Guinea through Africa's largest delta. Along with the Ogooué Delta in Gabon, the Niger Delta harbours more primate taxa (17) than any other coastal delta in Africa (Chapter 30). The Delta comprises a core area within three states: Bayelsa, Delta and Rivers. The northern part of Delta State and eastern part of Rivers State are geographically not

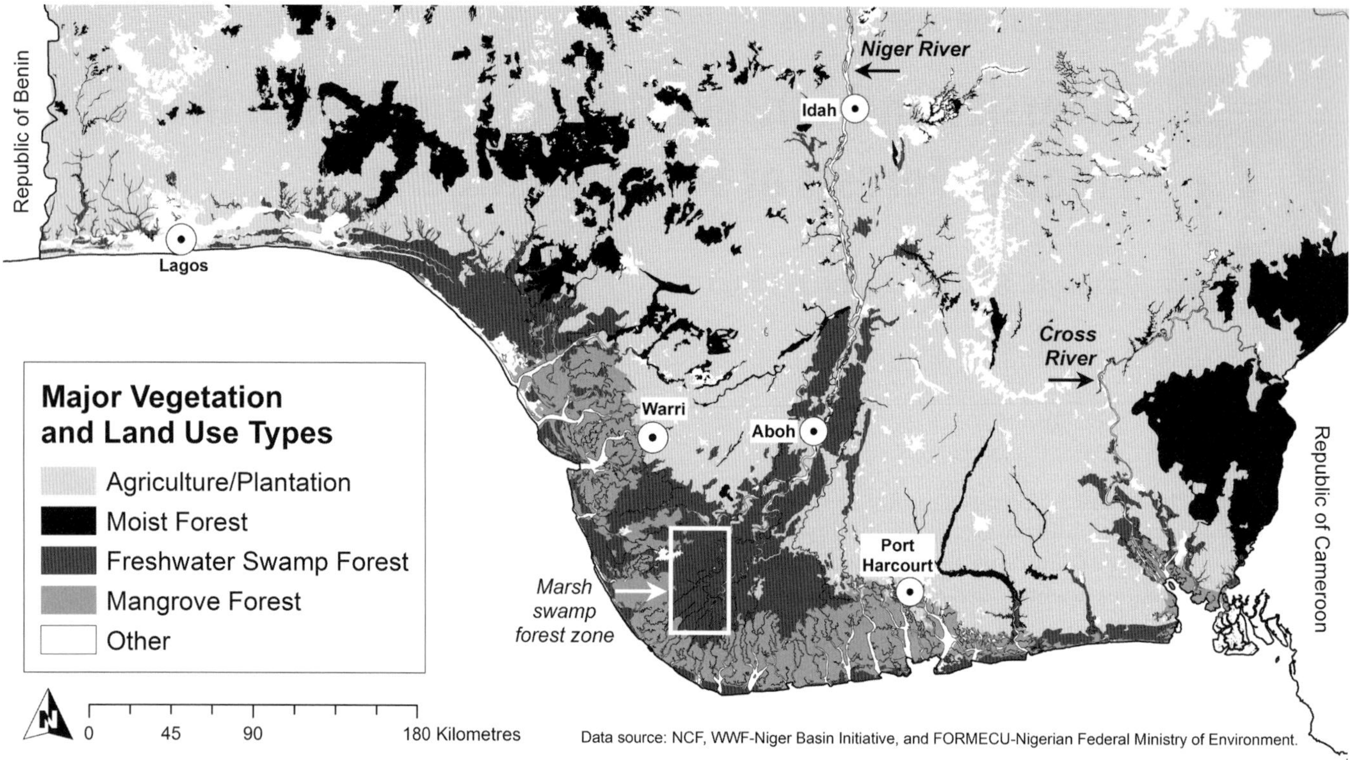

Figure 40.1 Major vegetation and land-use types across southern Nigeria, as assessed in 1993–1995 (Geomatics International *et al.* 1998): 'agriculture/plantation' includes crop and grazing land, floodplain agriculture, irrigation and livestock projects, and forest and tree plantations (most of which contain exotic species); predominantly low-lying 'moist forest,' which includes both disturbed and undisturbed forests and riparian forests; 'freshwater swamp forest,' which is the most extensive freshwater wetland type in Nigeria; and 'mangrove forest.' 'Other' includes all other vegetation and land uses, such as woodlands, grasslands, tidal flats, rocky outcrops, mining sites and urban areas. The general area where marsh swamp forest occurs in the Niger Delta is highlighted.

within the Niger Delta. To the west and east of the Delta lie extensive areas of low-lying forest subject to flooding.

We consider four main types of forest: freshwater swamp, Niger River floodplain, Niger Delta marsh and mangrove. This classification is based on that developed by Powell (1995) in his description of the wildlife of the Niger Delta; this classification, in whole or part, was subsequently followed by other investigators in their accounts of primates and other mammals in southern Nigeria (e.g. Baker 2005; Ikemeh 2015; Luiselli *et al.* 2015; Tooze 1996; Werre 2000, 2001a).

Freshwater Swamp Forest

Extensive areas of freshwater swamp forest occur on the flanks of the Niger Delta, as well as to the east and, especially, west of the Delta inland of the mangrove zone. Smaller patches are found within *terra firma* forests on some coastal barrier islands, in low-lying areas and along rivers in the lowland moist forest zone. These swamp forests are often dominated by *Raphia* palms, such as *R. vinifera* and *R. hookeri.*

Diurnal primates reported to occur in the freshwater swamp forests of southern Nigeria include the red-capped mangabey (*Cercocebus torquatus* – IUCN Red List category[1]: 'Endangered'), mona monkey (*Cercopithecus mona* – 'Near Threatened'), white-throated monkey (*C. erythrogaster* – 'Endangered'), Sclater's monkey (*C. sclateri* – 'Endangered'), putty-nosed monkey (*C. nictitans* – 'Vulnerable'), olive colobus (*Procolobus verus* – 'Vulnerable') and chimpanzee (*Pan troglodytes* – 'Endangered') (Akani *et al.* 2014; Anadu & Oates 1982, 1988; Angelici *et al.* 1999; Baker 2005; Baker & Olubode 2008; Bocian 1998, 1999; Ikemeh 2014b; Luiselli *et al.* 2015; Oates 1989; Petrozzi *et al.* 2015; Powell 1995; Tooze 1996, 1997; Werre 2000).

Sclater's monkey (Figure 40.2) is endemic to Nigeria, as is the Nigerian subspecies of white-throated monkey (*C. e. pococki*). The former occurs only in southeastern Nigeria (Baker & Olubode 2008), while the latter is restricted to a few forests in the southwest of the country, the Niger Delta and a small region of the eastern floodplain along the lower Niger River (Oates 2011). Two subspecies of putty-nosed monkey occur in Nigeria: *C. n. ludio* in the far southeast and *C. n. insolitus* elsewhere. Although the IUCN Red List classifies the species *C. nictitans* as 'Vulnerable', both subspecies face numerous threats, particularly *C. n. insolitus*, which is regarded as 'Endangered'. Chimpanzees in southern Nigeria are generally assumed to belong to the subspecies *P. t. ellioti*, but the taxonomic position of populations west of the Cross River is not fully resolved (e.g. preliminary analysis of a small number of chimpanzee DNA samples from the Niger Delta and from west of the Niger River suggests that while this population is related to *P. t. ellioti*, it is evolutionarily distinct; further research is in progress [C. Hvilsom, pers. comm., 2018]).

At present, a small population of chimpanzees survives within the Edumanom Forest Reserve, a freshwater swamp forest on the eastern flank of the Niger Delta (Akani *et al.* 2014; Baker 2005; Bocian 1998, 1999; Ikemeh 2014b; Petrozzi *et al.* 2015; Powell 1995). Chimpanzees also occur in swamp forest in parts of the Gilli-Gilli and Ologbo Forest Reserves in Edo State

Figure 40.2 Sclater's monkey (*Cercopithecus sclateri*), an 'Endangered' primate that occurs only in southeastern Nigeria. Photo: Lynne R. Baker. (A black and white version of this figure appears in some formats. For the colour version, please refer to the plate section.)

in southwestern Nigeria, immediately north of the Niger Delta (A. Adeleke, pers. comm., 2014; Greengrass 2009). There are unconfirmed reports of their presence in the Agge-Letugbene area in the southwestern corner of the Delta, near the coast (Ikemeh 2014b).

The southeastern Niger Delta is near the eastern edge of the range of olive colobus (Oates 2011). Although there are reliable records of this species from the forests of the Niger River floodplain (see next section), its presence in the freshwater swamp forests on the eastern flank of the Niger Delta (reported in Powell 1995) has not been confirmed, and its status in the region is uncertain. During more recent surveys, for instance, Petrozzi *et al.* (2015) did not record olive colobus in any of five forest reserves located in the floodplain and eastern flank zones.

Nocturnal primates that occur in the freshwater swamp forests of southern Nigeria include Thomas's dwarf galago (*Galagoides thomasi* – 'Least Concern'), northern needle-clawed galago (*Euoticus pallidus* – 'Near Threatened'), Allen's galago (*Sciurocheirus alleni* – 'Near Threatened'), angwantibo (*Arctocebus calabarensis* – 'Near Threatened'), and two species of potto (*Perodicticus potto* – 'Vulnerable' and *P. edwardsi* – 'Least Concern') (J.F. Oates, pers. obs.; Luiselli *et al.* 2015; Pimley 2009; Powell 1995). *E. pallidus*, *S. alleni* and *A. calabarensis* are known only from the Niger Delta eastward (Jewell & Oates 1969; Oates & Jewell 1967; Pimley 2009), and the Niger River separates *P. potto* to the west from *P. edwardsi* to the east (Oates 2011). The subspecies of potto found west of the Niger, *P. potto juju*, is regarded as 'Endangered' on the IUCN Red List.

Demidoff's dwarf galago (*Galagoides demidovii* – 'Least Concern') also probably occurs in these forests. Only quite recently did primatologists realize that two species of dwarf galago (*G. demidovii* and *G. thomasi*) occur widely and sympatrically in the African forest zone (Bearder *et al.* 1995); previously, all sightings of small galagos in Nigeria were recorded as *G. demidovii*. Subsequent to this finding, the few surveys of nocturnal primates undertaken in Nigerian swamp forests confirmed the presence of only *G. thomasi*.

Niger River Floodplain

From below the region of Idah, at about 7°N, extensive floodplains border the Niger. Parts of these are inundated by the annual Niger flood, which begins in August at the end of the rainy season and eases in December. Flooding is most prevalent south of Aboh, west of the Niger at 5°33′N, near the apex of the Niger Delta (Figure 40.1). In the dry season, soil in these flooded areas dries out, except where there are permanent creeks, small lakes and seasonal flood channels (Werre 2001b).

Luiselli *et al.* (2015) compared forest zones within the Niger Delta and found that floodplain forests harbour the most mammal species and the highest percentage of threatened mammals. In fact, this region not only retains primate diversity comparable to that of the protected forests in the biodiverse Cross River State, but also supports a greater diversity of primate species than found across the more degraded lowland forest zone to the east and west of the Niger River.

In the seasonally flooded forests on the floodplain west of the Niger, reliable local reports from the early 1980s suggested the presence of several primates, including dwarf galagos (*Galagoides* spp.), potto (*P. potto*), red-capped mangabey, mona monkey, white-throated monkey, olive colobus and chimpanzee (Anadu & Oates 1982, 1988); the chimpanzee is likely now extirpated from these forests (Oates 1989). The following diurnal primates have been recorded in floodplain forests east of the Niger: tantalus monkey (*Chlorocebus aethiops tantalus* – 'Least Concern'), mona monkey, white-throated monkey, Sclater's monkey, putty-nosed monkey and olive colobus. Nocturnal primates reported from this region by Luiselli *et al.* (2015) include potto, angwantibo and a dwarf galago; however, given that no surveys dedicated to nocturnal primates have been conducted in Niger floodplain forests, other species may occur here.

The putty-nosed monkey was visually confirmed in the southerly portions of the floodplain (e.g. Taylor Creek Forest Reserve) by Tooze (1997), but the species was considered very rare; it was more recently reported as present in this reserve and again noted as rare (Petrozzi *et al.* 2015). Putty-nosed monkey was not confirmed in other surveys in the central and northern reaches of the floodplain (Baker 2005; Ikemeh 2014b; Oates 1989). Similarly, the white-throated monkey is found only in the southern part of the eastern floodplain, being replaced by Sclater's monkey farther north. Very small populations of olive baboon (*Papio anubis* – 'Least Concern') may also persist within the eastern floodplain. Although local informants noted the continued presence of baboons as recently as 2013 (Ikemeh 2014b), we are unaware of any sightings by experts in this region, other than a captive individual seen in 2004 and reportedly captured in the vicinity nearly 20 years earlier (Baker 2005).

Reports of chimpanzees on the eastern floodplain were recorded by Oates (1989), Powell (1995), Baker (2005), Ikemeh (2014b) and Petrozzi *et al.* (2015). Ikemeh detected chimpanzee signs in the northern portion of the floodplain, in the Osomari Forest Reserve, which also harbours Sclater's monkey (Oates 1989). With the exception of the Edumanom freshwater swamp forest, chimpanzees are probably extinct in other (even relatively large) forest tracts between the Niger and Cross Rivers in southern Nigeria (Baker 2005; Gadsby 1989; Luiselli *et al.* 2015; Powell 1995; Table 40.1).

Marsh Forest

Marsh swamp forest occurs in the central Niger Delta, inland of coastal mangroves (Figure 40.1). This freshwater forest is influenced by tides, which dampen the amplitude of the annual Niger flood. As a result, the flood affects only parts of the marsh zone (Powell 1995; Werre 2000). The area of marsh forest unaffected by the annual flood is notable as the only known habitat of the locally endemic and highly threatened Niger Delta red colobus (*Piliocolobus epieni* – 'Critically Endangered'). This habitat is a mosaic of low-lying swamp forest and drier ridge forest. *Raphia hookeri* and *R. vinifera* palms dominate the middle canopy; *Uapaca staudtii* and *U. heudelotii* are the most abundant large trees; and the most common emergent is *Ctenolophon englerianus* (Werre 2000, 2001b), which is restricted to freshwater swamp forests in southern Nigeria, Gabon, Democratic Republic of the Congo (DRC) and Angola. *C. englerianus* is important in the diet of red colobus: It comprised nearly 50% of all feeding observations in a year-long study in marsh forest at Gbanraun, Bayelsa State (Werre 2000).

Only recently discovered, the Niger Delta red colobus was first described by Grubb and Powell (1999) as *Piliocolobus pennantii epieni*. The taxon was elevated to the species level, as *Piliocolobus epieni*, by Groves (2007b). Other monkeys recorded in marsh forest include red-capped mangabey, putty-nosed monkey and mona monkey (Ikemeh 2015; Isoun 2009; Luiselli *et al.* 2015; Werre 2001a); white-throated monkey and olive colobus have been recorded at only a few marsh sites (Ikemeh 2015; Werre 2001a) and may have patchy distributions in this zone. Chimpanzees have never been recorded here. At Gbanraun, Pimley (2009) reported having observed Allen's galago, needle-clawed galago, a dwarf galago (*Galagoides* sp.) and a potto (*Perodicticus* sp.). Pimley (2009) and Luiselli *et al.* (2015) did not record angwantibos in marsh forest; Powell (1995) also noted their absence from this zone.

Mangrove Forest

Nigeria's mangrove forest belt, estimated to cover 7360 km^2, is the most extensive in Africa and accounts for about 5% of the global extent of mangroves (Spalding *et al.* 2010). The mangrove zone runs roughly parallel to the coastline; on the flanks of the Delta, mangroves may reach as far as 50 km inland (Figure 40.1). Small, isolated pockets of freshwater forest occur within the brackish-water mangroves (Powell 1995). Although Nigeria's mangrove forests have received scant attention from primatologists, evidence suggests that these forests may have some importance for primates.

Maps in Rosevear (1953b) show mona monkey as the only primate present in the mangrove forests of the Niger Delta. Other investigators have recorded, in addition to mona monkey, red-capped mangabey and Sclater's monkey using mangrove forests, although these reports were mostly from the edge of the mangrove zone (Gadsby 1989; Powell 1995; Tooze 1996). According to Tooze (1996), mangroves are a potential refuge for primates, which may shift to this habitat to avoid more intense human pressure in adjacent forests (see also Chapters 7 and 39). Tooze also suggested that mona monkeys forage on *Rhizophora racemosa* shoots and root pith, as well as small crustaceans at low tide; Baker (2003) observed the latter foraging behaviour by this species.

In the coastal mangroves of southwestern Cameroon, immediately to the east of Nigeria, mona monkey and red-capped mangabey were the only monkeys observed by Gartlan and Struhsaker (1972), who described monas as 'extremely abundant.' In western Gabon, Cooke (2012) observed red-capped mangabeys most often in mangrove forest, compared

Table 40.1 Key areas for primate conservation among the flooded forests of southern Nigeria. For each site, diurnal primates confirmed as present are noted. Where specified in data sources, primates assessed as extirpated/likely extirpated are also noted.

Name	Major forest type	State	Estimated size (km²)	Diurnal primates			Conservation status
				Confirmed by sight/sound/sign or reliable report	Nearly extirpated or extirpated	Possibly present	
Apoi Creek	Marsh	Bayelsa	65	Red-capped mangabey Mona monkey Putty-nosed monkey Niger Delta red colobus		White-throated monkey Olive colobus	Forest reserve; no effective protection; part of a larger area (292 km²) designated a Ramsar site (2008)
Edumanom	Freshwater swamp	Bayelsa	87	Mona monkey Sclater's monkey Putty-nosed monkey Chimpanzee[a]	Red-capped mangabey	Olive colobus	Forest reserve; no effective protection
Gilli-Gilli	Freshwater swamp and lowland dry forest	Edo	363	Red-capped mangabey Mona monkey White-throated monkey Putty-nosed monkey		Chimpanzee	Forest reserve; some protection (since 2007) due to presence of conservation project managed by the Nigerian Conservation Foundation
Niger Delta National Park (proposed)	Marsh	Bayelsa	140	Mona monkey White-throated monkey Putty-nosed monkey Olive colobus Niger Delta red colobus		Red-capped mangabey	Currently unprotected
Osomari	Eastern Niger floodplain	Anambra	115	Mona monkey Sclater's monkey Tantalus monkey Chimpanzee[a]	Red-capped mangabey	Olive baboon	Forest reserve; no effective protection
Stubbs Creek	Freshwater swamp	Akwa Ibom	310	Red-capped mangabey Mona monkey Sclater's monkey Putty-nosed monkey	Chimpanzee		Forest reserve; no effective protection
Taylor Creek	Eastern Niger floodplain	Bayelsa and Rivers	219	Mona monkey White-throated monkey Putty-nosed monkey Tantalus monkey Olive colobus	Red-capped mangabey Chimpanzee		Forest reserve; no effective protection
Upper Orashi River	Freshwater swamp	Rivers	90	Mona monkey Sclater's monkey Putty-nosed monkey	Red-capped mangabey Chimpanzee	Olive colobus	Forest reserve; no effective protection; part of a larger area (252 km²) designated a Ramsar site (2008)

[a] Recently confirmed as present, but assessed as very rare (Ikemeh 2014b; for Edumanom only, Akani *et al.* 2014 and Petrozzi *et al.* 2015).

with other available habitats. These observations suggest that mangrove forest is more than just a marginal habitat for mona monkey and red-capped mangabey.

Although no distinct surveys of nocturnal primates have been conducted in Nigerian mangroves, the angwantibo is listed as present here by Luiselli *et al.* (2015).

Conservation Status of Flooded Forests in Southern Nigeria

As we have described, many of the seasonally and permanently flooded forests in southern Nigeria are important habitats for a number of threatened and declining primate populations. The relative inaccessibility of these forests has conferred on them a degree of protection. Even so, these forests have been long degraded by unregulated logging and hunting, as well as conversion to cropland (Anadu & Oates 1982; Baker 2005; Ikemeh 2014a, b, 2015; Werre 2001a, b, c). Flooded habitats in the oil-producing region of southern Nigeria experience the additional hazards of pollution and degradation associated with oil and gas development (Moffat & Linden 1995). Further negative impacts are due to alien invasive plant species, notably water hyacinth (*Eichhornia crassipes*) in freshwater swamp forests (Figure 40.3) and Nypa

Figure 40.3 Invasive water hyacinth *Eichhornia crassipes* in a river in the Niger Delta. Photo: Lynne R. Baker. (A black and white version of this figure appears in some formats. For the colour version, please refer to the plate section.)

palm (*Nypa fruticans*) in mangroves. Longer-term impacts due to climate change may particularly affect the Niger Delta, which receives too little sediment to offset rising sea levels (Giosan *et al.* 2014); this is exacerbated by the construction of dams along the Niger River in Guinea, Mali and Niger (Chapters 30 and 36).

Aside from Cross River National Park and the Afi Mountain Wildlife Sanctuary in the forests along the Nigeria–Cameroon border in Cross River State, there are no official protected areas designated for biodiversity conservation in either southeastern Nigeria or the Niger Delta, and only one national park (Okomu) currently exists in southwestern Nigeria. Southern Nigeria does contain several government forest reserves; however, these reserves are managed by individual states for timber production, and hunting and the harvesting of forest products are little restricted or wholly unrestricted. Here we describe several forest reserves and other forest tracts that: (1) contain significant areas of flooded forest, and (2) we consider particularly important for primate conservation (Figure 40.4). Our set of forests overlaps with five of the seven sites selected by Phil-Eze and Okoro (2009) for consideration for conservation attention based on their estimated overall biodiversity richness and threatened status.

The features of the forests we describe here, including their diurnal primate faunas, are summarized in Table 40.1. The exact legal status of each government forest reserve is unclear; although they are described as forest reserves in several documents (e.g. Olaleye & Ameh 1999), not all of them appear to have been formally gazetted by state governments.

Apoi Creek Forest Reserve (65 km^2) and Regional Marsh Forests

Apoi Creek Forest Reserve is located in the marsh forest zone in the central Niger Delta in Bayelsa State. This reserve and the surrounding marsh forest are key areas for the conservation of the locally endemic Niger Delta red colobus. Given the location of the marsh forest and its year-round inundation, the major socioeconomic activity here is fishing; in comparison, hunting has traditionally played a minor role in people's livelihoods. Some local hunters even avoid red colobus, citing

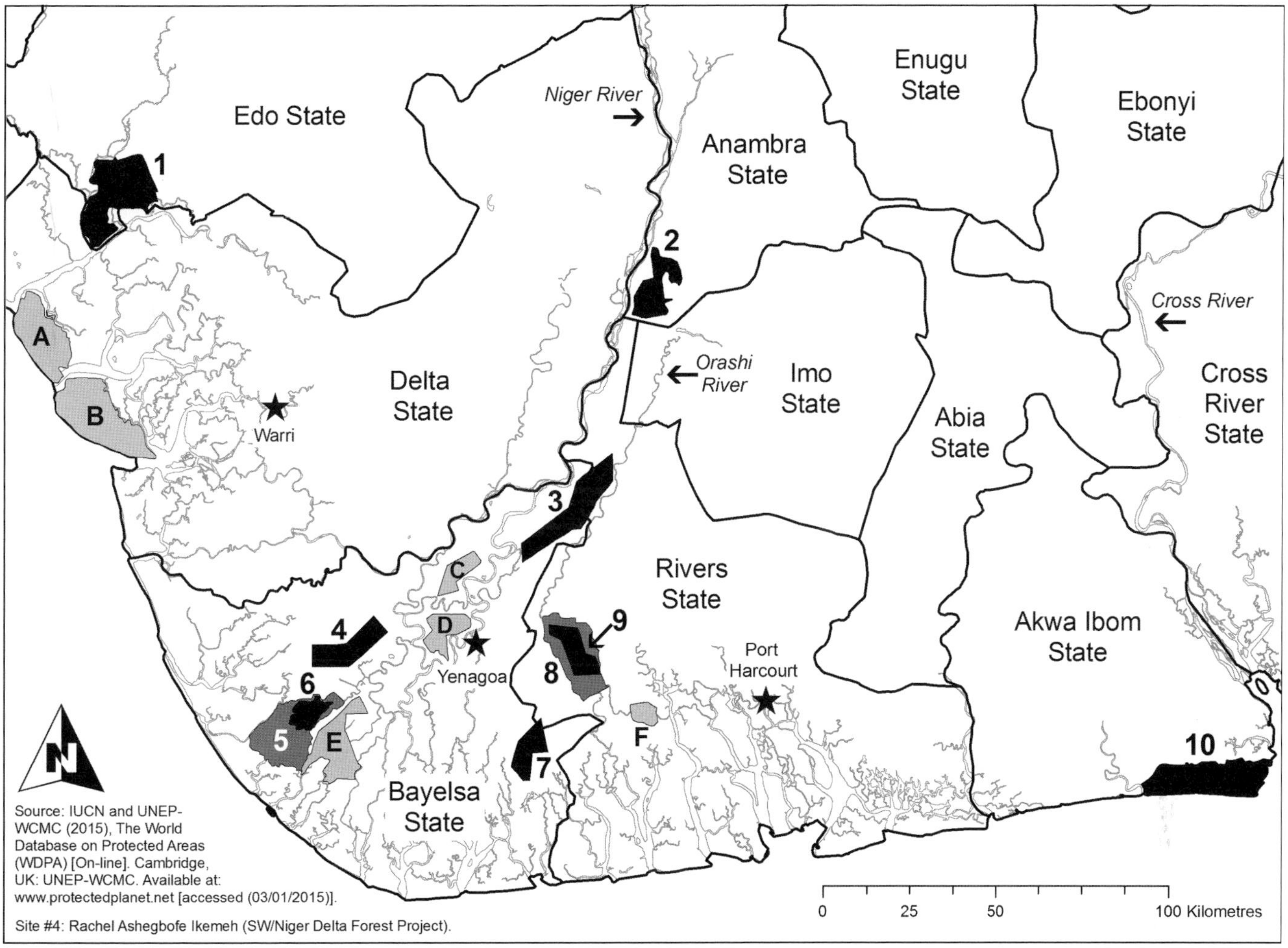

Figure 40.4 Key flooded forests for primate conservation in southern Nigeria: (1) Gilli-Gilli Forest Reserve, (2) Osomari Forest Reserve, (3) Taylor Creek Forest Reserve, (4) proposed Niger Delta National Park, (5) Apoi Creek Forests Ramsar Site (#1751), (6) Apoi Creek Forest Reserve, (7) Edumanom Forest Reserve, (8) Upper Orashi Forests Ramsar Site (#1759), (9) Upper Orashi River Forest Reserve and (10) Stubbs Creek Forest Reserve. The diversity and status of primates in these forest reserves deserve further investigation in terms of their importance for primate conservation in the region: (A) Olague, (B) Uremure-Yokri, (C) Egbedi Creek, (D) Nun River, (E) Ikebiri Creek and (F) Lower Orashi.

its tough meat and foul odour (Werre 2001a) (though hunters have expressed a preference for its meat in other areas; Ikemeh 2014a). Relatively low hunting pressure may contribute to the continued presence and relative abundance of the larger-bodied red-capped mangabey and putty-nosed monkey in marsh forest (Werre 2001a).

Although hunting pressure may be lower in the Apoi Creek area than elsewhere in southern Nigeria, logging is rampant (Figure 40.5). Werre (2001a) noted intensive artisanal logging in the mid-1990s, driven largely by the demand for abura (*Fleroya ledermannii*). During a survey near Gbanraun in 2009, investigators learned that multiple chainsaws were heard daily in the forest (Isoun 2009). More recently, Ikemeh (2015) recorded intensive logging, as well as oil spill pollution related to 'bunkering' (oil theft resulting from pipeline vandalism). Ikemeh also suggested a severe decline in red colobus abundance, as well as significant range contraction.

Although conservation interventions are both underway and proposed in this region, prior efforts have had very limited success. During Werre's study of Niger Delta red colobus in 1996–1997, a proposal to protect 500 ha of forest near Gbanraun (Werre 2000) did not come to fruition. In 2002, the Niger Delta Wetlands Centre (NDWC) attempted to renew local support for a community-based conservation programme for the species, but it did not evolve into an effective conservation programme (Isoun 2009). NDWC later contributed to a proposal to recognize the Apoi Creek Forest Reserve and surrounding forests as a globally important wetland. This region (292 km^2) was designated a Ramsar Site (#1751) in 2008 (Ramsar 2018). Out of the seven areas analysed by Phil-Eze and Okoro (2009), Apoi Creek ranked fourth overall as a priority site for biodiversity conservation. This forest is also under consideration as a national park by the Federal Ministry of the Environment (R.A. Ikemeh, pers. comm., 2014).

Edumanom Forest Reserve (87 km^2) and Upper Orashi River Forest Reserve (90 km^2)

The Upper Orashi River and Edumanom Forest Reserves are freshwater swamp forests located on the eastern flank of the

Figure 40.5 Near Gbanraun and the Apoi Creek Forest Reserve, timber rafts waiting to be pulled by boats to coastal trading and urban centres for processing and sale. Photo: John F. Oates.

Niger Delta in Rivers State and Bayelsa State, respectively. Although both have been notably affected by logging, farming, hunting and oil- and gas-related activities, they appear to be particularly important habitats for the regionally endemic Sclater's monkey. This species was recorded as the most abundant diurnal primate in both sites (Baker & Olubode 2008; Baker *et al.* 2011). While chimpanzees are likely extirpated from the Upper Orashi (Baker 2005; Petrozzi *et al.* 2015; Powell 1995), they persist in Edumanom; this forest is thus of special conservation concern. Not surprisingly, the Edumanom forest ranked second (out of seven) in terms of conservation importance in the Niger Delta (Upper Orashi forest ranked fifth) (Phil-Eze & Okoro 2009).

Conservation interventions are either underway or proposed in both sites. The Upper Orashi River forest falls within a larger area (252 km^2) that was designated a Ramsar Site (#1759) in 2008 (Ramsar 2018), and Edumanom is under consideration as a national park by the Federal Ministry of the Environment (R.A. Ikemeh, pers. comm., 2014).

Gilli-Gilli Forest Reserve (363 km^2)

Gilli-Gilli (also known as Gili-Gili and Gele-Gele) is located in the far south of Edo State, on the north bank of the Benin River, on the western flank of the Niger Delta. It has been long considered a site worthy of conservation for its wildlife. In colonial times, Gilli-Gilli was designated as a game reserve (probably because of the presence of forest elephants), but this designation appears to have lapsed. It was later recommended for national park status by Petrides (1965) as a way of conserving examples of Nigerian freshwater swamp and mangrove ecosystems. At present, it is classified as a forest reserve.

A little over half of Gilli-Gilli was originally covered by lowland high forest. Much of the remaining area was covered by freshwater swamp forest, with a tiny area of mangrove in the southwestern corner (Anadu & Oates 1982; Federal Department of Forestry 1978). Dry-land sections of the reserve have been heavily logged and cultivated since the early 1980s (Mmom & Mbee 2013), but much of the swamp forest is probably largely intact. One of us (J.F. Oates) observed white-throated and mona monkeys in the reserve in 1982, and chimpanzees were reported as present, but rare (Anadu & Oates 1982). More recently, hunters reported that chimpanzees still occur in swampy parts of the reserve (A. Adeleke, pers. comm., 2014). Adjoining Gilli-Gilli to the east is the smaller Ekenwan Forest Reserve, and east of Ekenwan is the Ologbo Forest Reserve; these two reserves, although highly degraded, also include small areas of swamp forest.

In 2007, the Nigerian Conservation Foundation (NCF) began a project in Gilli-Gilli designed to develop a biodiversity action plan and build the capacity of local communities to help manage the forest reserve. NCF established a 36 km^2 central core area where logging and farming are controlled; however, oil exploration and high levels of hunting persist as challenges (A. Adeleke, pers. comm., 2014).

Osomari Forest Reserve (115 km^2) and Taylor Creek Forest Reserve (219 km^2)

Along the eastern Niger floodplain are two key forests: Osomari in Anambra State in the northern reaches of the floodplain and Taylor Creek, which straddles Bayelsa and River States in the lower portion of the floodplain. Little is known about the condition of the Osomari forest. In the late 1980s, Oates (1989) reported many human settlements and little natural forest in parts of the reserve, and a 2013 survey in the southwestern portion of Osomari noted a high rate of logging (Ikemeh 2014b).

Phil-Eze and Okoro (2009) ranked Taylor Creek as the number 1 priority for biodiversity conservation in the Niger Delta. Indeed, this forest may be the last remaining refuge for elephants within the Niger–Cross interfluvial region. It is also important habitat for the white-throated monkey, which may be the most abundant primate in the reserve (Tooze 1997). The Government of Rivers State announced its intention to constitute Taylor Creek as a forest reserve in 1975, but we are not certain that the reserve was later formally gazetted. Management of the forest was complicated by the creation of Bayelsa State from the western portion of Rivers State in 1996, with a resultant division of administrative responsibility. However, there is a proposal under consideration by the Federal Ministry of the Environment to upgrade Taylor Creek to a national park (R.A. Ikemeh, pers. comm., 2014).

Stubbs Creek Forest Reserve (310 km^2)

Located along the coast in the southeastern corner of the densely populated Akwa Ibom State, Stubbs Creek is predominantly a freshwater swamp forest that seasonally floods in some areas. It also contains brackish-water swamp forest, mangroves, secondary forest, farmland, palm bush and abandoned farms. The invasive Nypa palm has significantly affected the reserve's mangroves. Given that fishing is the major socioeconomic activity among local people, hunting is less severe than in other regional forests (Baker 2003). This may explain the continued presence of the red-capped mangabey, reported by locals as the second most common primate in Stubbs Creek (after mona monkey) (Baker 2003, 2005). Mangabeys are otherwise notably rare or absent across the Niger–Cross interfluvium (Baker 2005; Petrozzi *et al.* 2015; Powell 1995; Table 40.1).

In the southwest section of the reserve at the beach line is an oil terminal operated by Mobil Producing Nigeria (MPN). Oil operations and associated activities, which represent the major industry in the region, have attracted new settlers, thus adding to population growth and pressure on natural resources. Although most oil operations are offshore, a major road built by MPN has facilitated exploitation of the reserve. Coastal communities have also complained of oil slicks along the beach, as well as oil in their nets while ocean fishing (Baker 2003). The major threat to the Stubbs Creek forest, however, is the high rate at which trees are being cut for lumber and land cleared for farming (Baker 2005) (Figure 40.6). Conservation measures are urgently needed here. Previous attempts (in the 1990s) to initiate a conservation programme in Stubbs Creek were unsuccessful (Tooze *et al.* 1998). Later efforts (2003–2004), in conjunction with MPN, also did not lead to any effective conservation action.

Riparian Forests

In other parts of southern Nigeria, where the landscape is largely deforested and consists mainly of human settlements, agriculture and farm-bush, strips of swampy riverine forest appear important in maintaining populations of red-capped mangabey and Sclater's, mona and putty-nosed monkeys (Baker 2005; Baker & Olubode 2008; Tooze 1995, 1997). Sclater's and mona monkeys even occur in narrow gallery forests in derived-savanna and woodland regions where more typical savanna species (e.g. tantalus monkey and patas monkey [*Erythrocebus patas* – 'Near Threatened']) are common (Baker 2005).

Proposed Conservation Sites and Forests for Further Investigation

To the northeast of Apoi Creek in Bayelsa State, Ikemeh (2015, and pers. comm.) identified a relatively inaccessible area of marsh forest south and southwest of Kunu, north and west of Azama and east of Azagbene (centred at about 4°54′06N and 5°57′37E). The site may contain one of the largest remaining populations of Niger Delta red colobus. The forest has not been designated a protected area, but given its high conservation potential, Ikemeh (2015) proposed that 140 km^2 be considered for national park status ('Niger Delta National Park') (site 4, Figure 40.4).

A much larger region overlapping this proposed park was previously recommended as a 'Niger Delta National Community Park' (4000 km^2) (ERM 2002). This site forms a triangle starting at the confluence of the Niger and Nun Rivers, extending all the way to the coast in Bayelsa State. This proposed protected area is bounded by the Forcados River and Bomadi Creek to the north and the Nun River and Apoi Creek to the south. The entire region encompasses several government reserves, including Nun River (97 km^2), Apoi Creek (65 km^2), Egbedi Creek (66 km^2), and Pennington Mangrove Forest Reserves. The feasibility of this proposal is questionable, given the complex politics of the Niger Delta and lack of precedents for such a large conservation area of this kind in Nigeria.

The status of primates and their habitats in Ikebiri Creek (192 km^2), Nun River and Egbedi Creek Forest Reserves in Bayelsa State and the Lower Orashi River Forest Reserve (48 km^2) in Rivers State (Figure 40.4) deserves further investigation, in terms of the significance of these sites for primate conservation (though see Petrozzi *et al.* 2015 for limited data on primates in Nun River and Egbedi Creek). Likewise, along the coast of Delta State, two relatively large, adjacent forests should receive a closer look: Olague Forest Reserve, which ranked high (third) among the forests assessed as conservation priorities by Phil-Eze and Okoro (2009), and Uremure-Yokri Forest Reserve (Figure 40.4). The latter reserve, which lies between the Forcados and Escravos Rivers, was included

Figure 40.6 Cut planks of timber in the Stubbs Creek Forest Reserve in southeastern Nigeria. Photo: David L. Garshelis. (A black and white version of this figure appears in some formats. For the colour version, please refer to the plate section.)

in a previous proposal for a wildlife sanctuary called the 'Ogidigben/Escravos Island Community Wildlife Sanctuary' (ERM 2002). At that time, Uremure-Yokri was considered the only remaining intact barrier-island forest in Delta State (ERM 2002).

Just to the south of Uremure-Yokri are the coastal flooded forests between the Agge and Letugbene communities in Bayelsa State. Here, Ikemeh (2014b) received local reports of chimpanzee, as well as red-capped mangabey, mona monkey and putty-nosed monkey, and recommended the region for further investigation.

Conclusion

Threats to the rich primate fauna of southern Nigeria are similar to those faced by primates in many parts of the world: overhunting, habitat destruction and fragmentation, and environmental pollution and degradation. However, these threats, which have long affected the wildlife of this region, are intensified by a rapidly growing human population, the largest in Africa and one of the largest globally. Nigeria's pace of development, fuelled by oil exports and, more recently, growth in non-oil sectors such as agriculture and information technology, resulted in the country surpassing South Africa in 2014 as the largest economy on the continent. A downside of such economic and population growth is ongoing deterioration of natural habitats and negative impacts on populations of non-human primates and other wildlife.

Except in Cross River State in the east of the country, dry-land forests in southern Nigeria have been reduced to fragments, many of which are highly degraded. This makes the freshwater and marsh swamp forests along the coast and within the Niger Delta particularly important for primate conservation. Although these forests are also under severe pressure, they are less affected by large-scale logging and agriculture than most forests farther inland, and many have lower hunting pressure because of their relative inaccessibility. Hunting may also be less intense because fish are usually more important than meat in the diets of coastal and riverine communities.

Of Nigeria's two endemic primate species, Niger Delta red colobus and Sclater's monkey, the former occurs only in the marsh forests of the central Niger Delta, while several of the most important, and possibly the largest, populations of the latter are restricted to swamp forest. In one study in the Edumanom Forest Reserve, for example, Sclater's monkey was by far the most abundant diurnal primate encountered, occurring in an estimated 3.1 groups/km^2 (Baker *et al.* 2011).

Despite the great importance of Nigeria's flooded-forest region for primate conservation, no area within this region has been officially designated a wildlife conservation site, although there have been proposals to create national parks at Apoi Creek, Edumanom, Taylor Creek and elsewhere. We recommend adding the Stubbs Creek and Upper Orashi River Forest Reserves to the list of sites being considered for greater protection. Nevertheless, previous proposals to upgrade forests to national parks, such as that for Gilli-Gilli in Edo State, never materialized. Serious protection of several swamp forest ecosystems and their primates is urgently needed before the habitats are irreversibly degraded and their primates extirpated. In several important flooded forests, some primate species are very rare or already lost (Table 40.1).

We are well aware that the creation on paper of new national parks or other protected areas will not be sufficient to save these Nigerian wetland ecosystems. New protected areas will require effective management, and for monkeys and apes, this must include the careful regulation or prohibition of hunting and other forms of destructive exploitation. For example, Cross River National Park (CRNP) in eastern Nigeria was established in 1991, but from the beginning, it faced high levels of hunting (Effiom *et al.* 2013; Oates 2002; Oates *et al.* 2004). It is now threatened by new concessions for oil palm and rubber plantations (Schoneveld 2014). Many of the people living around and within the park are weakly or not at all supportive of conservation activities because financial support for development activities either ceased after a few years or never materialized (Oates *et al.* 2004). In contrast, in the community-controlled Mbe Mountains (which border the Okwangwo Division of CRNP), a conservation programme managed by a local community organization in conjunction with the Wildlife Conservation Society has had greater success in reducing hunting activity (Dunn 2010). In addition to many other examples, this highlights the need to make genuine community participation a key component in the management of any new protected areas in Nigeria's swamp forests. Indeed, for some sites where government interest or investment is minimal or nil, community-based efforts may be the only solution to conserving important flooded forests and their primate fauna in southern Nigeria.

Acknowledgements

We are grateful to Rachel Ikemeh for information related to the forests and species covered in this chapter. We also thank Thomas Butynski, Mary Glenn and Zena Tooze for their helpful comments on the manuscript.

Note

1 IUCN Red List categories used in this chapter were established at an IUCN Primate Specialist Group meeting in Rome in April 2016. At the time of writing, these categories had not been published on www.iucnredlist.org.

Chapter 41

Mamirauá Reserve

Primate-based Flooded Forest Conservation in the Amazon

Nelissa Peralta, Hani R. El Bizri, Fernanda P. Paim and João Valsecchi

Introduction

Tourism is a fast growing industry worldwide, and often associated with social, economic and environmental impacts, especially in rural areas (Mowforth & Munt 1998). Despite its negative effects, a segment of the industry – ecotourism – has been endorsed as one strategy to promote the conservation of natural areas, linking it to economic returns for local people (Stronza & Pegas 2008). In the Amazon, there are hundreds of ecotourism enterprises that claim to conserve natural areas and their biodiversity, while also producing economic and social benefits to rural populations. In Brazil, ecotourism has been incorporated into national protected area management plans as a means of attributing economic and social value to biological diversity (MMA 2004). As ecotourism relies on natural areas and their 'watchable' species as its main asset base (Wunder, 2000), it provides an incentive for the conservation of those species that are considered to be tourist attractions.

Large charismatic mammals and birds are taxonomic groups that attract ecotourists to particular sites (Cousins 2007; Okello 2008). Primates are among the major ecotourism attractions, especially in tropical forests (Ross & Wall 1992; Weber 1993). In the Amazon, although there is great potential for primate tourism (with 122 primate species in the biome), only a few ecotourism sites rely on primate sightings. One of these sites is the Mamirauá Sustainable Development Reserve in Brazil. Its establishment is closely linked to primate conservation, since the main justification for creating the protected area was the need to protect two primate species (*Cacajao calvus calvus* and *Saimiri vanzolinii*) from local threats. In this chapter, we assess the role of primate conservation and ecotourism in the formation and protection of the Mamirauá Reserve.

Primate-based Formation of the Mamirauá Sustainable Development Reserve

In 1983, the primatologist José Márcio Ayres arrived in the region of what is today the Mamirauá Reserve to study the primate *Cacajao calvus calvus* (Figure 41.1). There, he found one of the most specialized primates in the Neotropical region, whose distribution range is entirely within the white-water river floodplains or *várzeas* (Ayres 1986a). Right from the start, his fieldwork revealed the vulnerability of *C. calvus calvus* from threats to its habitat, mainly from logging since, in the west Amazon, most timber was extracted from *várzea* (Ayres & Johns 1987) and, secondarily, commercial fishing. Another of the region's primate species also vulnerable to habitat loss is *Saimiri vanzolinii*, described by Ayres as having the smallest distribution among South American primates (Ayres 1985). Ayres's research also showed the importance of floodplains to the speciation processes in Amazonian primates (Ayres 1986a; Ayres & Clutton-Brock 1992), recognizing thereby that the maintenance of gene flow and ecological and evolutionary processes should be as important as species diversity when selecting priority areas for conservation (Pinedo-Vasquez & Dávalos 2011).

His studies on primates were important to the development of river-refuge hypotheses (Ayres & Clutton-Brock 1992; Pinedo-Vasquez & Dávalos 2011) and to the conception of conservation based on ecological corridors (Ayres 1986b; Ayres *et al.* 2005). Furthermore, his work showed the importance of *várzea* and the presence of endemic primates in this habitat and their vulnerability to habitat destruction. These were the foundations for a proposal to create a protected area in the region. The protected area was established by the Amazonas State Government and comprised the whole of the then known distribution range of *C. calvus calvus* and *S. vanzolinii* – an area of 11 240 km^2 located within the confluence of the Solimões, Japurá and Auatí-Paraná rivers (Figure 41.2). In addition to these two species, nine more primate species were then protected by the reserve, representing four families and eight genera (Ayres 1985, 1986a, 1986c, 1995; Ayres & Johns 1987; Hershkovitz 1984, 1987; Paim & Queiroz 2009; Valsecchi 2005; Vieira *et al.* 2008; Table 41.1).

The reserve was created as an Ecological Station, a very restrictive category of protected area that does not allow human presence or natural resource extraction (MMA 2004). But anthropological studies carried out by Deborah Lima showed that the area was inhabited by local populations that depended on and had historically used locally available natural resources, and were themselves engaged in a social movement for preservation of the area's lakes (Lima-Ayres 1992). This reinforced Ayres's support of the presence of local people in protected areas, whose livelihoods depended on biodiversity (Ayres & Best 1979). Their main concern was to actually reconcile conservation and development goals: 'protected areas should be seen as an element in a regional human development strategy,

Figure 41.1 The white bald-headed uakari, a primate species whose distribution range falls entirely within the white-water floodplains (*várzea*) in the Mamirauá Reserve. Photo: Felipe Ennes.

considering and incorporating within costs of protection, the needs of local communities and relating social and economic benefits to the sustainable use of local natural resources' (Ayres *et al.* 2005, p. 23).

To recognize that by involving and providing economic incentives for local people would conservation be successfully achieved, he worked towards the creation and implementation of a new category of protected area. The Sustainable Development Reserve[1] was created in 1996, and defined as a natural area that allows the residence of traditional populations, whose livelihoods are based on systems of sustainable use of natural resources, developed along generations and adapted to local ecological conditions, and which perform a fundamental role in the protection of nature and maintenance of biological diversity (MMA 2004). Mamirauá Ecological Station was then changed to this new category of protected area, a Sustainable Development Reserve.

The challenge afterwards was to create a strategy that would incentivize local people to continue to protect and use resources sustainably. During the early nineties, researchers and local leaders set out to elaborate and agree upon a zoning system and set of norms for the use of natural resources. In 1996, a management plan was published. The zoning system destined a core area as a totally protected zone, where human settlements and use of natural resources were prohibited. Surrounding this core area was a sustainable-use zone, where most of the settlements were located and economic productive activities could be carried out. A set of alternative income activities was also proposed as incentives for conservation; among these activities were fisheries management, forest management and ecotourism (SCM 1996). These economic activities would offset pressure from locally threatened natural resources and raise household incomes (SCM 1996). An ecotourism enterprise was planned in Mamirauá, within the totally protected zone near the Mamirauá Lake (Figure 41.2), an area subject to pressure from large fishing vessels that were extracting tons of fish at a time.

Community-based Ecotourism and Conservation

Ecotourism is a segment of the tourism industry advocated as a market tool for conservation (Gossling 1999; Stronza 2007). It is an important source of funding for protected areas, and it has grown especially in developing countries (Davenport *et al.* 2002; Hartshorn 1995). Ecotourism can be seen as a non-extractive strategy to promote the conservation of forests and primates (Care for the Wild International and Pro Wildlife 2007). Ecotourism enterprises usually entail natural areas as destinations and the promotion of biodiversity conservation in those areas, with socioeconomic benefits going to local people (Boo 1992; Honey 1999; IES 1994; Kiss 2004). For practitioners, ecotourism should generate associations between socioeconomic benefits to local populations, and the endorsement of conservation strategies in those natural areas where the projects are developed (Peralta 2013).

In Mamirauá, ecotourism has been in existence since 1998. Seven communities are directly involved in the ecotourism project. The lodge that hosts guests was named after the white bald-headed uacari (*C. calvus calvus*). The Uakari

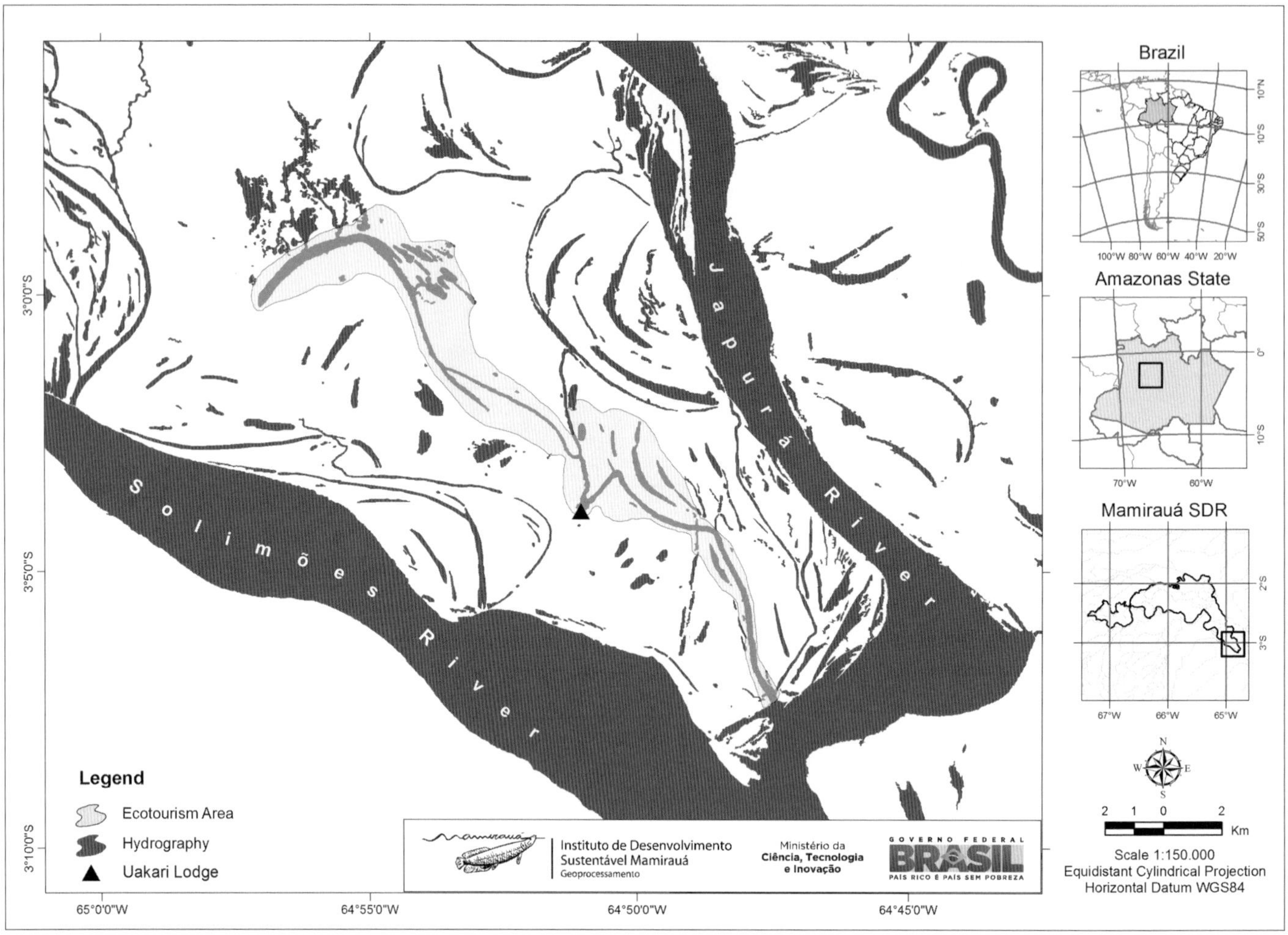

Figure 41.2 Map showing the area designated for ecotourism and the location of the Uakari Lodge in the Mamirauá Sustainable Development Reserve, Amazonas, Brazil.

Floating Lodge has been hosting visitors to Mamirauá since 1999. Seventy per cent of its guests are international visitors and 30% are Brazilian nationals. Travellers visit the natural area due mainly to its natural and cultural attractions – especially good primate-viewing opportunities. Tourists engage in a combination of activities, including excursions to trails and lakes in order to observe wildlife, visit local communities to get to know and understand the local way of life, and visit research stations, where they interact with researchers.

Local inhabitants participate in ecotourism activities working and managing the lodge, planning and evaluating activities, selling agricultural produce and fish to the lodge, hosting visitors and selling handicrafts to visitors to their communities. Since 2002, in the years when the enterprise generated profits, these were shared among the seven communities involved. Communities invested half of the profits to local development projects and the other half in the protection of the area.

A true ecotourism project must minimize its negative impacts, especially in a protected area. To do so in Mamirauá, a visitor management system was established integrating a set of norms, recommendations and good practices. Only 1000 people per year (and 20 people at a time) are allowed at the Uakari Lodge. Trails can only be visited in small groups of up to six people. Large boats are not allowed into the area, and the means of transport must be restricted to motorized canoes using low horsepower engines at low speed. Taking into account that impacts are related not only to the total number of tourists but also to their behaviour towards the environment and local population, information on the norms of the protected area prior to arrival is part of the Uakari Lodge visitor management system. Outings can only occur with the presence of a local guide, and guides are trained extensively to inform visitors of the appropriate conduct during the various activities carried out in the protected area.

Some measures have also been taken to minimize negative social impacts of the activity. Ecotourism was not implemented to substitute traditional economic activities, but to be an extra source of income. In order to not affect other economic activities, a rotation system of work was planned, and other more traditional economic activities, like fishing, agriculture and handicrafts production, are encouraged in synergy with tourism. Specific rules and recommendations were elaborated by locals to guide the conduct of tourists in their communities.

Table 41.1 Primate species whose occurrence was confirmed in Mamirauá Reserve.

Family	Species	Common name
Aotidae	*Aotus cf. vociferans*	Night monkey
Atelidae	*Alouatta juara*	Red howling monkey
	Ateles chamek	Black-faced black spider monkey
Callitrichidae	*Cebuella pygmaea*	Pygmy marmoset
Cebidae	*Saimiri vanzolinii*	Black-headed squirrel monkey
	Saimiri macrodon	Ecuadorian squirrel monkey
	Saimiri cassiquiarensis	Humboldt's squirrel monkey
	Sapajus macrocephalus	Large-headed capuchin
Pithecidae	*Cacajao calvus calvus*	White bald-headed uacari
	Cacajao calvus rubicundus	Red bald-headed uacari
	Pithecia cazuzai	Cazuza's saki

Other good practices are related to the employment of alternative technologies for minimum environmental impacts, such as a system of solid waste management, rainwater collection and the use of renewable energy, such as solar power.

In order to investigate impacts of ecotourism activities on local fauna, a trail monitoring system was implemented. Between 2001 and 2004, data were collected by local guides on 765 tourist trail excursions. On these outings, all primate sightings were registered, as well as records of two species of cracids *Mitu tuberosum* and *Crax globulosa* (CRACIDAE) and the black-fronted nunbird *Monasa nigrifrons* (BUCCONIDAE). About 24 000 sightings were registered in total, with more than half (55%) representing five species of primates (*A. juara*, *C. calvus calvus*, *S. vanzolinii*, *S. cassiquiarensis e S. macrocephalus*). Average sighting rates did not significantly differ over the years (ANOVA statistical testing). From 2005 onwards, the monitoring system changed so as to exclude tourist outings. Two groups of trails were monitored. One group included trails that allowed for intensive tourist use and the other group were of trails where tourist had their access restricted (Paim *et al.* 2012; Storni *et al.* 2007). The study showed that densities of the primates *A. juara*, *S. macrocephalus*, *S. vanzolinii* and *C. calvus calvus*, have been stable between 2007 and 2010, and that their use of areas with tourist trails was adequate (Paim *et al.* 2012). Primates were chosen as the focal taxon for this monitoring programme, because their presence is an indicator of habitat quality, and because the ecosystem services they provide are essential for humans (e.g. Chapman & Onderdonk 1998; Chapman & Peres 2001; Lambert & Garber 1998).

In Mamirauá, four primate species accounted for the majority of hunted mammal biomass between 2002 and 2005 (Valsecchi 2005). Lopes *et al.* (2012) showed the red howler monkeys (*A. juara*) were the preferred game meat in three out of four communities of their study area in Mamirauá. Between 2002 and 2010, 286 individuals of *A. juara* were hunted, representing 3% of the total biomass hunted in the reserve (Valsecchi 2012). Despite this pressure, *A. juara* hunting levels are considered sustainable (Queiroz & Valsecchi 2005).

Tourism has generated around US$800 000 for about 100 families over the years (1999–2011). Although economic benefits have not been high, the income that tourism did provide was important, since locals' annual average family income was US$3319 in 2005 (Peralta *et al.* 2009). The value of tourism was demonstrated by a study that found that average buying power grew by 148% in one of the communities directly involved with tourism (Peralta 2005). Another study found a 34% difference in average income between communities that were involved with ecotourism and those who were not (Peralta 2013). A great amount of income was generated to women and young people, who had previously more restricted access to income-generating activities.

One problem faced by the project has been the instability in the flow of guests to the lodge. Demand is very elastic and is affected by external factors (such as foreign currency exchange rates and the transport infrastructure available). Since the beginning of its operations, the lodge visitor numbers had shown an average annual growth rate of 25%. But in the years 2006 and 2007 the local airport closed down, and this impacted visitation in subsequent years. Another challenge is the lack of formal education of employees and temporary staff, who live locally in rural areas with poor schooling. Although there was success in developing local skills in guiding tours, hotel housekeeping, and other services, in general, locals still lack experience in marketing, product development and financial management, which have been provided by the institution that offers long-term technical assistance to the lodge (Peralta 2013).

But a linkage between tourism and the preservation of the area did occur (Peralta 2013). Locals related the protection of Mamirauá Lake system to tourism. And tourism provided an economic incentive for locals to protect the area and thwart external threats, such as large fishing vessels from urban areas. It has been a motivation for those who benefited from tourism to try and maintain the protection status of a lake that they saw as important for ecotourism activity. One problem is that, despite attempts to diffuse economic benefits as much as possible, due to logistical problems and the instability of tourist demand, only four out of seven communities enjoyed the bulk of the benefits. Those who did not see the activity as profitable to themselves justify destining the lake to fishing management and not to total protection (Peralta 2013).

Conclusions

The Mamirauá Reserve has won the Best Ecotourism Destination Award from Conde Nast Traveler Magazine. It also received a prize for Sustainable Tourism from the Smithsonian Magazine and United States Tour Operators Association (USTOA) Conservation Traveler's Foundation. Community participation and the scientific basis of the project were fundamental to its achievements. The Uakari Lodge is a clear example that ecotourism, if well planned and carefully monitored, may be an effective tool for conservation. It is a direct source of income that is related to the area's preservation and serves as an incentive for the protection of the natural area by local people. Research on primate population densities, habitat use and extraction by people have remained stable since ecotourism was implemented suggesting that it is a sustainable non-extractive activity. Evidence also suggests that given adequate levels of protection, and additional economic opportunities, traditional forms of natural resource use by people in protected areas can be maintained.

Acknowledgements

We are grateful to the Ministry of Science, Technology and Innovation in Brazil for their financial support and to the Community-Based Tourism Programme of the Mamirauá Institute for partnership in collecting data. We are especially thankful to local people, research assistants and guides, all of them responsible for biodiversity conservation in the area.

Note

1 Corresponding to the IUCN category VI.

Chapter 42

Primates in Flooded Forests of Borneo

Opportunities and Challenges for Ecotourism as a Conservation Strategy

Stanislav Lhota, Katherine S.S. Scott and John Chih Mun Sha

Introduction

Ecotourism is a form of nature tourism. It is defined by The International Ecotourism Society as 'responsible travel to natural areas that conserves the environment and improves the well-being of local people'. The IUCN defines it as 'environmentally responsible travel and visitation to relatively undisturbed natural areas in order to enjoy and appreciate nature (and any accompanying cultural features – both past and present) that promotes conservation, has low negative visitor impact, and provides for beneficially active socioeconomic involvement of local populations' (Wood 2002, p. 9). This form of tourism is one of the fastest evolving conservation strategies worldwide, with primate species among the most attractive ecotourism draws (Cowlishaw & Dunbar 2000). The prospects for development of ecotourism and the tangible benefits to conservation that can be derived from such programmes however vary widely and depend on specific local conditions.

The flooded forest is a unique habitat that provides outstanding opportunities for ecotourism but also poses several specific challenges to its development as a primate conservation strategy. Flooded forests in Borneo include mangroves, peat swamps, riversides and floodplain forests; commonly viewed as impenetrable, filthy, full of mud and debris, infested with mosquitoes and sometimes dangerous due to the presence of people-eating crocodiles. Such perceptions are often not too far removed from reality. So what makes swamps attractive destinations for ecotourism? One key attraction for tourists is the challenges involved in exploring these inhospitable habitats. Since it is usually difficult to walk in swampy forests on regular forest trails, boats and boardwalks are the most typical ways of experiencing these forests, and such unconventional approaches to exploration may hold exotic appeal to potential visitors (like canopy walks do elsewhere). Flooded forests can be visited by day or night; with nocturnal boat trips allowing views of vast aggregations of fireflies blinking in synchrony in tree crowns along the riverbanks. Visiting swampy forests on foot may also be attractive for tourists who are slightly more adventurous. For example, in Balikpapan Bay, the visitors (not only foreigners but also Indonesians) enjoy walking short distances across the tall primary mangrove forest during low tide and clambering on the top of the aerial mangrove roots while getting covered in mud. In other destinations (where no crocodiles are present), visitors can wade in flooded mangroves during high tide. Tourists view the flooded forests, particularly the mangrove swamps, as a unique alien environment, which is tempting to experience. Such ecotourism opportunities have been explored in tropical countries around the globe, for example, in India (Thomas & Fernandez 1994).

In Borneo, the most attractive flooded forests are those where large charismatic animals can also be seen, and there are numerous such locations (see Figure 42.1 for location of sites mentioned throughout the text). For example, in Balikpapan Bay (East Kalimantan, Indonesia), two major flagship species which attract visitors to the mangroves are the proboscis monkey (*Nasalis larvatus*) and the snubfin dolphin (*Orcaela brevirostris*); the latter also represents the major tourist attraction in Mahakam Lakes (East Kalimantan, Indonesia), where it occurs in freshwater (Olsder & van der Donk 2006). The Lower Kinabatangan Wildlife Sanctuary (Sabah, Malaysia) is now one of the most popular ecotourism destinations in Borneo; apart from the proboscis monkey, it also offers the chance to view species such as the elephant (*Elephas maximus*), orangutan (*Pongo pygmaeus*) and large estuarine crocodile (*Crocodilus porosus*). Evening boat trips in the Lower Kinabatangan offer a small chance to see some very rare animals, including wild cats such as the Sunda clouded leopard (*Neofelis diardii*) or flat-headed cat (*Prionailurus planiceps*). Tanjung Puting (Central Kalimantan, Indonesia), although initially set up to ferry tourists to see the orangutans living around Camp Leakey and other camps within the National Park, has also seen a rise in tourists wanting to see proboscis monkeys and other wildlife species other than its well-known orangutans (K.S.S. Scott, pers. obs.). Bird-watching opportunities also attract many visitors with specific interest in birds to places like the Lower Kinabatangan or Mahakam Lakes. Mahakam Lakes represent the most extensive swampy area and a prime water bird-watching site in Borneo, with the added opportunities to experience other species mentioned above.

Of the various Bornean primates living in flooded forests (Meijaard & Nijiman 2003), proboscis monkeys play a key role for ecotourism in Borneo. For foreign and local visitors, it is one of the most popular animals, apart from the orangutan. People are amazed by the grotesque visage of this large-bodied and

Figure 42.1 Major political divisions of Borneo and locations of ecotourism sites mentioned in the text.

big-nosed primate. Adult male proboscis monkeys are among the most popular targets of nature photographers. Most importantly, however, compared to orangutans and many other large animals in Borneo, viewing proboscis monkeys is relatively easy due to their habit of congregating along the riverbanks to sleep almost every night in most places (Matsuda *et al.* 2010b; Matsuda *et al.* 2011); therefore sightings of proboscis monkeys are almost guaranteed by the tour guides even during short visits. Furthermore, in some places such as Bako National Park (Sarawak, Malaysia) or Tarakan (North Kalimantan, Indonesia), the habituated proboscis monkeys can be seen easily from boardwalks as well. The popularity of proboscis monkeys is evident from looking at social media, such as Facebook or Twitter, and travel articles, where proboscis monkeys are increasingly featured and marketed as one of the main attractions in Borneo. The proboscis monkey is an endemic species of Borneo, and because of limited captive populations in zoos and sanctuaries (Manansang 2005; Sha *et al.* 2011), visiting Borneo still remains one of the best ways to see the proboscis monkey.

Literature describing ecotourism in Borneo is rich but does not comprehensively evaluate ecotourism as a conservation strategy, including the numerous challenges faced in its implementation. Current information is provided predominantly in tourist guidebooks such as *Lonely Planet* and *Rough Guide*, photographic books and field guidebooks. The potential for further development of ecotourism in Bornean flooded forests have also been described in several publications (Husson & Morrogh-Bernard 2007; Sha *et al.* 2011; Tisdell & Swarna Nantha 2007), but most other information is hidden in 'grey literature', for example, in technical reports, which are difficult to access. Specific experiences by managers of ecotourism projects, especially experiences with project failures and negative impacts, are also often not shared in order to maintain positive images of such projects that are often of national importance, as wildlife protection and ecotourism involving charismatic species are often overseen by governmental agencies. Such information biases can retard the development of ecotourism efforts based on sound conservation strategies.

The objective of this chapter is to outline issues related to the challenges of ecotourism in flooded forests of Borneo, with a focus on primates. We consider pertinent issues to encourage biologists and social scientists to further investigate these topics in order to develop ecotourism as an effective and holistic tool for conservation. We base our discussion on relevant literature for Bornean sites, where available; otherwise we base our suggestions on our personal experience and information from colleagues involved in similar work. We are aware that ecotourism in Borneo is a dynamic and continuously changing topic. This chapter is based in the situation as we have observed it in 2016.

The Economics of Ecotourism

Economics is increasingly dominating the logic behind conservation planning (Turner *et al.* 1993). Under this paradigm, long-term conservation efforts have little opportunity to be sustained unless they also provide direct or indirect economic incentives, which should outweigh the benefits of alternative uses of natural resources. Economic viability of conserving primate habitats for ecotourism depends on use and non-use value of the natural habitat at one end of the spectrum and the costs of conserving it on the other (Tisdell & Swarna Nantha 2007).

Ecotourism provides direct and easily accountable monetary benefits, which can be shared among various stakeholders like entrepreneurs, landowners, local communities and governments (Olsder & van der Donk 2006). However, regardless of how obvious these benefits are, the question remains whether such derived economic benefits are sufficient to cover the costs and to offset other more lucrative options contrary to the conservation cause.

When conservationists speak of economic costs of ecotourism, they tend to consider the cost of infrastructure and management of developing and running ecotourism programmes, but often fail to properly consider opportunity costs involved in preventing other uses of the habitat. In contrast, for the decision-makers, it is usually the opportunity cost of developing a natural area as an ecotourism destination (and therefore forbidding other profitable economic activities within the same area) that really matters. The opportunity cost of using an area for ecotourism (and conservation) is defined as the highest economic return that can be alternatively obtained by using this area for something else (Pearce & Moran 1994). In its simpler form, it means that wilderness has not only to pay for its existence (by providing ecotourism and other services), it also has to compete with numerous profitable anthropogenic activities.

Riversides, lowlands and coastal areas are among the most accessible and most human-populated habitats in Borneo, and as a result the commercial economic value of such lands tends to be very high compared to remote inland areas. Many investors are interested in developing these areas for highly profitable enterprises, which are incompatible with ecotourism and conservation. They include plantations, urban and industrial development. As a consequence, the opportunity cost of conserving the riverside forest for ecotourism, instead of converting it for other uses, is often very high and may pose a major obstacle to ecotourism development.

This, however, does not necessarily apply to swamp forest habitats, which are less suitable for agricultural and infrastructural development. In most mangrove swamp habitats, shrimp and milkfish farming has been the main economic reason for converting and using such forests (Valiela *et al.* 2001). The opportunity cost of habitat conservation for ecotourism (instead of conversion) can therefore be estimated as the potential income from such aquaculture activities, which proved to be highly profitable in some areas (Mahakam Delta, Adang Bay) but not necessarily elsewhere. For example, most of the shrimp and milkfish farms established in Balikpapan Bay went bankrupt. Extensive freshwater swamps have even lower potential for exploitative use; they have been used for selective logging (Bennett 1988), but once the main timber resources are exhausted, there is little potential for further exploitation. As long as the potential for other economic use of the swamps remains limited, the opportunity cost of developing ecotourism is low and these habitats provide opportunities for conservation and tourism.

With the advancement of technology, which makes land reclamation in mangroves and freshwater swamps increasingly feasible, the interests of other businesses (plantations, industry and housing estates development) need to be increasingly considered in the economic equation. The opportunity costs of conserving these habitats for ecotourism is likely to rocket once such new technologies allow further exploitation of swamps at increasingly lower costs. It is therefore increasingly difficult for ecotourism programmes to generate enough funds to cover not only their own infrastructure and management costs but also the opportunity costs, and therefore influence the decision-making processes in favour of conservation. This is most feasible in areas with very little value for other exploitative use (low opportunity costs) and very little threats (low conservation costs); such favourable situations would however be more an exception than norm in most lowland primate habitats in Borneo.

Case Studies

Balikpapan Bay

Balikpapan Bay provides a good example of how the increasing opportunity cost can hamper the conservation of primate swamp habitats for ecotourism. Ecotourism in Balikpapan Bay is still in its infancy but the potential is exceptionally high: the attractions include not only nature, i.e. primary mangroves and rainforest, proboscis monkeys and snubfin dolphins, but also culture, i.e. traditional fishing practices and ways of living. All these attractions can be found within 1–2 hours from the biggest international airport in Kalimantan, where the number of passengers has been increasing over the past decade with approximately 6 million passengers using the airport each year (Citrinot 2014). However, most tourists pass through Balikpapan, finding no reason to stay in the city, as it lacks any major tourist attractions.

In addition to tourists, there is a rich expatriate community in Balikpapan, most of which lacks the opportunity for leisure and recreation. However, not only foreigners but also the local people of this relatively wealthy town, and of nearby Samarinda, represent a highly promising target group for ecotourism; and, yet, despite this potential target audience, awareness about the easily accessible wealth of natural heritage remains low. This suggests that with proper promotion and basic infrastructure, ecotourism in Balikpapan Bay has the potential to provide substantial income for the local people and the government, and can help project a positive image of Balikpapan and Penajam cities and the East Kalimantan province.

Against this backdrop of immense opportunity, why then does the local government not leverage the development of ecotourism in Balikpapan Bay? The apparent answer is that the immediate income derivable from legal and illegal forest conversion greatly exceeds the potential income from ecotourism. It is important to note that sustainability issues do not play a significant role in this decision-making process, and the economic cost of the environmental damage are not taken into account (they become externalities). Extensive areas of riversides and coast of Balikpapan Bay are therefore being converted into palm oil plantations, palm oil bulking stations and refineries, coal terminals, ports, wood chip mills, other industry and housing estates. This is despite the apparent legality issues related to forest conversion particularly along the coast and rivers; where several national and local regulations define the protected green belt along the seashore and riverbanks (National Law UU No. 27/2007, Presidential Decree No. 32/1990, Ministerial Decree No. KB.550/264/Kpts/

4/1984, Circular No. 507/IV-BPHH/1990). These activities are believed to be the main contributing factor for proboscis monkey local extinction risk (Stark *et al.* 2012), and similarly threaten the snubfin dolphins and other species, as well as the aesthetic value of the natural landscape.

Proposals for developing ecotourism in Balikpapan Bay are consistently postponed by the local government, possibly to the point where ecotourism potential could be completely lost. In the new spatial plan (Perda No. 12/2012 Kota Balikpapan), there is practically no area left for long-term viable conservation and ecotourism programmes along the coastal habitats of Balikpapan Bay within the administrative area of Balikpapan city, although huge potential still remains in the more remote areas of Balikpapan Bay, which are under the administration of a different district (Penajam Paser Utara).

Fortunately, things began to change recently. In 2016, the regional office of the Ministry of Environment and Forestry in Balikpapan initiated a project on economic evaluation of recreation and ecotourism environmental services of Balikpapan Bay. This was demanded by a national plan, yet it only became a reality in response to pressure from local NGOs and the public (including a demonstration in front of the Ministry building in Balikpapan). Once long-term sustainability and costs of the negative impacts of industrial development become incorporated into the new economic evaluation, there is a chance that the government reevaluates its policy and spatial planning – if it is not already too late.

Tanjung Puting

Tanjung Puting shows how difficult it is even for a high-profile and well-developed ecotourism programme to compete with other economic priorities like the ever expanding palm oil industry. In Tanjung Puting, proboscis monkeys inhabit two sides of the Sekonyer River where one side is protected and one side is not. The monkeys, however, use both sides of the river and to cross the rivers on a regular basis (Yeager 1991a). The unprotected side of the Sekonyer River is set to be converted to palm oil despite fierce opposition by conservationists, local people and the general public. Despite this outcry, the popularity of the river for ecotourism and the economic benefits generated for the local economy, local authorities are looking at the larger and more immediate financial benefits from palm oil industry; and long-term perspectives are casually overlooked. The orangutans will be protected for the tourists as they are not known to cross the river within the park, but for the proboscis monkeys, it could spell imminent disaster and cause problems of population compression (due to habitat shrinkage) and eventual starvation and decline. Palm oil monocultures will not be able to sustain any proboscis monkey populations and proboscis monkeys will be forced to compete in an area already saturated with orangutans, macaques and other mammals.

Shifting the Posts on Stakeholder Dynamics

The economic viability of ecotourism does not depend only on how large the benefits and the costs are, but also on who shares the benefits and who bears the costs. The most viable solution is where the cost bearers share the benefits (Scheyvens 1999). This situation tends to minimize conflicts. If the beneficiaries and cost bearers differ, the conflict over land use persists and the outcome largely depends on the political power of the competing stakeholders. If the major cost of conserving an area for ecotourism involves opportunity costs (i.e. preventing other uses of the land), then who are the cost bearers? Which stakeholders are prevented from the alternative uses of an area?

A simplified view tends to consider the local communities as the major cost bearers; therefore, involving the local community in the ecotourism programme and sharing the benefits with them should reduce the conflict. This, however, is not always the case. In the Bornean swamps, there may be little use by local communities other than the local fishery, but the local fishery (with the significant exception of shrimp and milkfish farming in mangroves) tends to be highly compatible and can even mutually benefit from ecotourism. There is always some impact of logging by local communities, which does cause conflicts with conservation, but this is currently not a major threat. The larger problem is that vast areas of coastal and riverside forests are being privatized in recent times, for example, in Indonesia (Forest Peoples Programme 2007). The lands are initially divided among members of the community but it usually does not remain in their hands for long. Most people immediately look for the opportunity to sell the newly acquired lands (for which they do not find any other use). Due to these land acquisition practices, vast areas of coastal and riverside forests in Indonesia are now owned by major commercial conglomerates. The major opportunity cost bearers of conservation proposals now switches from the local communities to the land speculators, corporations, investors and local authorities, all of which would benefit highly from forest conversion, rather than conservation or ecotourism.

The case of Balikpapan Bay illustrates how complex the situation has become due to such changes in land ownership. Following large-scale land-grabbing, the major stakeholders are now the corporations and the local and provincial governments, which gain tax benefits and other perks from industrial development. These land areas are slated to fall into the hands of commercial corporations through land speculators where loopholes in the land-use system often allows them to use even the legally protected land for eventual forest conversion. In such a situation, representatives of local communities may get employment and other benefits from the development but this may not be equitable to the sustainable benefits that could have been derived if the land-grabbing did not occur. Even if community-based ecotourism development is allowed by the government and supported by conservationists, there is very little synergy between the beneficiaries and the cost bearers of ecotourism and other development activities, local communities are effectively excluded from important decision-making processes, corporate interests are given utmost priority, and the conflict among various stakeholders is likely to persist.

Ecotourism as Part of a Holistic Economic Approach

It is becoming apparent that in most cases, it cannot be ecotourism alone that safeguards conservation of natural resources. To be an economically viable alternative to conversion of large forest areas, ecotourism needs to be part of a holistic programme that also generates, directly or indirectly, additional benefits for diverse stakeholders (Tisdell & Swarna Nantha, 2007). Other economic benefits of conserving primate habitat are highly diverse, and are often not fully appreciated. These include various indirect societal benefits, which are commonly termed 'ecosystem services' (Nasi *et al.* 2002). Swamp habitats provide extremely important ecosystem services, which can have implications for local and national economies. Flooded forests represent important fishery resources, play a considerable role in water management, prevent sedimentation and natural disasters, and benefit carbon sequestration and biodiversity conservation. These services are, however, difficult to quantify (Daily *et al.* 2009), taking into account factors like national legislation, government policies and distribution of government revenues. More recently, efforts have emerged to trade these societal benefits directly (Bishop *et al.* 2009). These endeavours include attempts to trade carbon, water and even biodiversity.

Ecotourism may still have a key role to play, even in a more holistic economic approach. It cannot be denied that the ecosystem services provided by rivers, swamps and mangroves impact positively on national economies; ecotourism, together with fisheries, are among the few economic benefits that are relatively easy to quantify and appreciated by decision-makers. Incorporating ecotourism may therefore be a good starting point for the holistic evaluation, which would provide sound economic arguments for conserving wildlife swamp habitats in the face of apparently high opportunity costs. This is consistent with the Millennium Ecosystem Assessment (2006), which regards recreation, tourism and education among the ecosystem services of swamp habitats, besides other functions like hydrological regulation, sediment retention, water purification, recharge/discharge of groundwater, climate change mitigation, harvesting fish and other products and cultural values.

Enabling factors like political will are critical to the implementation of such frameworks into feasible programmes. Tisdell and Xue (2013) outline how private interests of developers and politicians hamper the evaluation of ecosystem services and implementation of ecosystem management practices. In many real-world situations, direct monetary benefits often tilt the balance towards considerations of high opportunity costs of preventing habitat conversion to plantations, industry and other businesses, while the extremely valuable but less directly evident economic benefits of maintaining healthy ecosystem services remain largely ignored by decision-makers.

Undesired Impacts of Ecotourism

Ecotourism, in contrast to hunting or logging, represents a non-extractive exploitation of natural resources. In theory, wildlife and other attractions are being viewed and then left where they are. This, however, does not mean that wildlife viewing and associated commercial activities do not have any negative impact on the natural ecosystem.

Direct Habitat Impacts

Ecotourism is usually associated with some infrastructure development, which inevitably causes disturbance to wildlife and habitats. Tourist resorts, restaurants, boat docks, visitor centres and even offices are often developed very close to wild habitats or even inside them. This can be observed at, for example, Lower Kinabatangan, Sabah, where numerous riverside tourist lodges are built within the habitat ranges of primate species (Matsuda, pers. comm., 2015). The siting of such basic infrastructure allows tourists to have closer contact with nature, thus enhancing visitor experiences. Many government- and corporate-sponsored ecotourism development programmes may, however, be over-zealous with infrastructure development, which often leads to building more facilities than are actually needed. Building unnecessary tourist facilities is often viewed by the reserve managers as the best way to obtain project money. Because reserve management is frequently not allocated enough land outside the monkey's habitat for such developments, these infrastructures are often built inside the habitats. Some mangrove conservation areas in Kalimantan are being gradually transformed into grounds full of boardwalks, gazebos, lodges, picnic and resting places, and even captive animal exhibits, leaving little exclusive space for conservation. Tourism departments often manage such developments, rather than environment or forestry departments, for example in Tarakan (Lhota, pers. obs.). The priorities for tourism infrastructure development may thus contradict conservation goals.

Impacts of Visitor Interaction

The effects of visitors on primate behaviour have been widely shown in both wild and captive primates (e.g. Hosey 2005; Mallupur *et al.* 2005; Maréchal *et al.* 2011; Treves & Brandon 2005). Negative impacts of ecotourism range from disturbance to wildlife and habitats to reverse zoonotic disease transmission (Cowlishaw & Dunbar, 2000; Kinnaird & O'Brien 1996).

At flooded forest ecotourism sites, proboscis monkeys are often viewed from boats at their sleeping sites, usually late in the afternoon. Tourists prefer close-up views for taking pictures or even attempting interaction (for example, to make subjects move to a clearer vantage point). The tourist guides or boatmen tend to do their best to satisfy the demands of tourists. Such actions can disturb the monkeys causing them to be displaced from their sleeping sites. The tourists are often not aware of regulations (if there are any) and their justification. In Tanjung Puting, boatmen have been known to literally ram the boats into trees in order for people to take close-up pictures on their low-zoom cameras. While there are rules in the park itself (and guards) in order to protect the orangutans and tourists, waterways do not appear to be governed in the same way. There are no guidelines (e.g. minimum distance maintained between people and monkeys) with respect to viewing the proboscis.

During rainy seasons, monkeys get more respite from the public as the river is higher and the proboscis monkeys sleep further inland (Yeager 1993). Although direct empirical effects of such disturbances on, for example, wild proboscis monkeys have not been shown (no studies have yet been conducted), anecdotal examples of the effects of disturbance on primate species within flooded forests are highly evident. Proboscis monkeys have been observed to move permanently away from heavily visited tourist areas in Sabah (J.C.M Sha, pers. obs.). Serious agonistic interactions among group members can also occur due to human disturbances (Scott, pers. obs.).

Primate provisioning represents another problem associated with primate ecotourism. Feeding wildlife is strictly prohibited or discouraged in most areas but it is being used as a management tool in some places (e.g. Labuk Bay in Sabah; Tarakan in North Kalimantan). Proboscis monkeys are fed unripe bananas (Tarakan: Lhota, pers. obs.) or special baked cakes composed of wheat and rice flour (Labuk Bay: Agaromoorthy & Hsu 2005). The primary reason is to maintain a high density of proboscis monkeys in small forest fragments with limited natural food resources or to habituate the monkeys within visitor areas where they can be easily viewed and photographed by tourists. In areas where the long-tailed macaques are the primary attraction (e.g. Pulau Kwangan in Banjarmasin, South Kalimantan), food (mainly peanuts) is sold to tourists to feed the monkeys for amusement.

Tourists feeding monkeys for their amusement, with no regulation of the amount and kind of food provided, should be strictly prohibited in any genuine conservation programme. Provisioning can result in overpopulation, behavioural changes, increase in conflicts with humans and possible health problems for wildlife (Fa & Southwick 1988). As such, even the controlled feeding by park or tourism management, for the purpose of maintaining a high population density, habituating the monkeys and restricting their ranging patterns is a highly questionable practice. This strategy may provide appealing tourist engagement (such as viewing the monkeys at their feeding site from the comfort of a restaurant) but its conservation value, if any, is highly limited. The mentality behind such efforts needs to change. Instead of maintaining high numbers of monkeys in small tourist-visible areas, where they vitally depend on artificial provisioning, and where their natural behaviour and social structure gets increasingly disrupted, conservation-oriented ecotourism programmes should focus on maintaining sufficient extents of natural habitats, where wild species can meet their dietary requirements and live long term in viable populations under natural conditions where they can be appreciated and respected by visitors.

Increasing evidence is emerging on zoonotic and reverse risks (Chapman *et al.* 2005, Cowlishaw & Dunbar 2000). Such risks increase with the magnitude of interface between humans and wildlife, particularly for human-commensal species like long-tailed macaques (Engel & Jones-Engel 2011). These monkeys have a high propensity to interact with humans and human environments. They scavenge in rubbish bins and around picnic sites, beg or steal food from people, touch people and may sometimes attack and bite them (for a similar situation in Singapore and Peninsular Malaysia, see Sha *et al.* 2009, and Hambali *et al.* 2012, respectively). This provides several possible routes of disease transmission – alimentary, droplet infection, skin contact or saliva or blood – from humans to monkeys or vice versa (for a similar situation in Gibraltar and Bali, see Fuentes 2006). It is likely that even the proboscis monkeys, which rarely interact with humans, face higher risk of disease transmission in the tourism areas, especially if they come into contact with human waste or if direct provisioning is involved. With increasingly close contact between proboscis monkeys and humans (not only due to tourism), but also along waterways contaminated by human pollution, reverse zoonotic transmission that threatens the survival of entire populations may also become a real risk.

The negative impact of disturbance on wildlife by tourists tends to be reduced through habituation, a process during which monkeys gradually lose their fear and interest in boats and tourists. Besides reducing disturbance to primates, habituation also significantly enhances the experience enjoyed by tourists. Habituation is often a by-product of ongoing ecotourism programmes, wherein wildlife gradually gets used to non-threatening human presence. Yet it is never perfect, even 'completely' habituated monkeys are still being disturbed by the presence of tourists to some extent. Habituating primates has its own ethical considerations, which has been detailed by Williamson and Feistner (2003). One pertinent issue is the increased ease with which poachers can access habituated groups that no longer fear humans. Hunting pressure is traditionally higher in Borneo's interior, where the Dayak and other non-Muslim tribes dominate. However, even along Muslim or multi-ethnic coastal areas, where the majority Muslim population does not consume primates, there is some threat of occasional hunting by Dayaks and immigrants from nearby islands, or local people on bird hunting trips who may sometimes shoot at monkeys just for fun. The natural habit that proboscis and other monkeys have of sleeping along riverbanks already renders them very easy targets for hunters (Meijaard & Nijman, 2000b; Nijman 2001; Sha *et al.* 2008) and increased habituation may exacerbate this problem.

Developing Responsible Tourism

Encouraging responsible tourism that benefits local communities and the natural environment is essential for sustainable ecotourism development (Spenceley 2010). Specifically, regulations need to be introduced to minimize disturbances posed by ecotourism. Sha *et al.* (2011) listed a set of 13 rules, which should be followed by tourists and tourist guides on boat trips in Lower Kinabatangan. For example, tourist guides should observe quotas for their tour frequency, a distance of at least 10 m should be maintained between the boat and the monkeys, and not more than three boats should congregate at one point when observing the monkeys. With the increasing numbers of tour operators and tourist resorts, these regulations are not easily enforced. Due to the lack of enforcement by some operators, attempting to guarantee a 'successful' trip for tourists, many permit the feeding of some habituated primates, including orangutans. Such behaviour has also been observed at other tourist sites in Sumatra, Indonesia (e.g. Bukit Lawang: J.C.M. Sha pers. obs.; Singleton, pers. comm.) There are exceptions to such negative trends, with local guides who innately care about their local animal species, are knowledgeable of their behaviour and diet and actively share their observations with tourists to enhance visitor experience. Incentives are needed to encourage such practices for guides to play a leading role in positively influencing tourist behaviour.

Informing and educating tourists is an important means of enforcing regulations that aim to reduce the negative impact of ecotourism. Guidebooks such as *Lonely Planet* and *Rough Guide* are under no obligation to publish information on best practices for tourists, although 'responsible tourism' chapters are provided in recent editions of these guidebooks. Due to the wide readership of these publications, they should arguably play a more important role in promoting ethical tourism as part of corporate social responsibility. In addition, existing guidebooks do not rate programmes according to their conservation contribution. This could be important as a way of influencing visitor choice of responsible programmes. Specialized ecotourism guidebooks (modelled, for example, on a recent ecotourist's guide to the Everglades, Florida; Silk 2016) would allow better opportunities to educate travellers about responsible ecotourism, but no such guidebooks are currently available for Borneo (S.M. Cheyne, pers. comm.).

In reality, not all tourists and tour operators are likely to follow regulations out of pure good will. The only way this could be remedied would be by policing the waterways and tourist trails and imposing fines on the tour operators and tourists who flout the rules. To minimize unnecessary conflict, awareness programmes need to be in place for visitors and tour operators to understand the rationale behind guidelines and the repercussions of undesirable behaviour.

Towards More Conservation-Oriented 'Ecotourism'

Current frameworks and practices for ecotourism do not sufficiently support the ultimate aim of conservation. The issue of long-term population viability of wildlife populations in their natural habitats should be a central goal in any well-designed ecotourism programme. In practice, this consideration tends to be neglected in many commercially oriented 'ecotourism' programmes. It is financially beneficial to conduct tourism in a small area, often on islands, where the monkey density is kept high by provisioning or by 'rescuing' and translocating animals from elsewhere. This is a profitable model – attracting high volumes of visitors and keeping conservation and opportunity costs low. Therefore, for profit-oriented entrepreneurs and governments, maintaining small and intensively managed reserves as tourist attractions are gaining high priority. Such forms of 'ecotourism' should technically not fall under the label ecotourism according to mainstream definitions.

Given that such parks have little value for the conservation of wildlife, particularly for larger species that require extensive habitats, the amount of positive propaganda they increasingly receive may actually be detrimental to conservation. Furthermore, this model imprints a misleading view on many local people who are not exposed to alternative forms of ecotourism efforts more pertinent to conservation.

The importance of sound ecotourism initiatives that contributes to habitat protection and long-term survival of species has to be better emphasized. Although factors like minimum habitat sizes needed to maintain most wildlife species is open to interpretation due to lack of quantitative studies, informed estimates can be made from available data. In the population and habitat viability analysis for Indonesia's proboscis monkeys (Manansang 2005), populations of more than 100 individuals were considered viable. Stark (2008), who employed population viability analysis for Borneo, also considered populations of over 100 individuals to be viable. Applying published estimates of population densities in natural habitats (Boonratana 2000b), such a population would require 1.6 to 3.0 km^2 of continuous riverside habitat, with reference to population density estimates in Sukau and Tanjung Puting (but note that the population in Tanjung Puting may already be overcrowded). Suppose that a reserve extends at least 500 m on both sides of the river and includes good habitat for proboscis monkeys, then 1.6 to 3.0 km of the river should be effectively protected. Much longer segments of the river, however, need protection in reality, where the forest extends less than 500 m from the riverbank, is limited to only one side of the river and/or is degraded and not fully used by the monkeys. In mangroves, the population densities of proboscis monkeys are lower, and more extensive areas are therefore needed; for example, at least 10 km^2 of mangrove habitat would be required for Abai, Sabah. However, many ecotourism conservation areas (tourist parks) are much smaller, just a few hectares in size, and most of them are effectively isolated from other larger continuous primate habitats.

Protecting sufficient food resources for proboscis monkeys is particularly important in mangrove reserves because this habitat is generally species poor and does not provide primates with much food plant choice. Preliminary observations from Balikpapan Bay (S. Lhota, unpublished data) strongly suggest that proboscis monkeys regularly leave mangrove forest to feed on a number of non-mangrove species in surrounding forests. Since it is often cheaper to purchase mangrove forest rather than the higher-prized inland forests, small tourist parks tend to contain only mangroves and no other forest types. It may be impossible to maintain a viable population of proboscis monkeys in such areas, without food provisioning.

Tourist Typology: Maximizing Ecotourism Benefits

Identifying tourist targets is essential to the planning, management, and marketing of tourism to ensure long-term sustainability of ecotourism programmes that match tourism types to resource capabilities (Smith & Smale 1980; Taylor 1986; Wall 1993). Ecotourism benefits can be maximized for both tourists and wildlife with properly managed ecotourism activities (Ballantyne *et al.* 2009), whereby profit driven goals may not necessarily be mutually exclusive from conservation and education objectives (Wright 1993).

High-End Foreign Tourists

Exclusive ecotourism programmes which primarily target high-budget foreign tourists can generate substantial conservation dollars. Wealthy foreign ecotourists are often ready to spend large sums of money for once-in-a-lifetime experiences with wildlife, and may continue their support for associated conservation programmes, if their experiences are positive. Although such programmes generally require much better service standards and infrastructure compared to more generic programmes catered to casual tourists, high-end tourists arrive in limited numbers (causing fewer negative impacts on the ecosystem) but the profit margin per visitor can be reasonably high. For example, at the Lower Kinabatangan, orangutans are among the most attractive animals but they are rather difficult to observe. High-budget tourists, who can pay several hundreds of dollars to be shown a wild orangutan by a highly skilled professional guide, are therefore the most suitable target group. Red Ape Encounters packages are an example of this model. In general, however, such great ape ecotourism is still highly underdeveloped in Indonesia (and also in Malaysia), compared to similar initiatives in African countries such as Uganda, Rwanda and Democratic Republic of the Congo, where tourists spend several millions of US dollars each year for such packages (Husson & Morrogh-Bernard 2007). In Borneo, the main focus of orangutan ecotourism is still related to rehabilitation centres rather than actual wildlife experiences. Alternative ideas need to be explored, for example, a potential option is to sell orangutan trekking permits that could potentially bring significant additional income.

Volunteers

Another example of intensive, low-frequency but high-impact ecotourism can be found in 'eco-volunteerism'. Volunteer ecotourist programmes have multiple objectives – from income generation via volunteer contributions, to taking advantage of volunteers in fieldwork or animal care and rescue, to professional development of volunteers (Wearing & Neil 2001). Volunteers are often students and conservationists embarking on their careers, who may continue to support the programme as their career progresses. There is a strong educational focus in most of these programmes but the capacity for participants can be limited due to the intensive focus on volunteer experience. There are several eco-volunteering programmes in Lower Kinabatangan, Sepilok Orangutan Rehabilitation Center and other locations in Sabah; the Orangutan Tropical Peatland Project's (OuTrop) programme in Sabangau (Central Kalimantan) is one example from Indonesia (Harrison, pers. comm., 2013). Such programmes are currently focused mainly on orangutans.

Lower-Budget Foreign Travellers

Proboscis monkeys, which share the same habitat with orangutans in Lower Kinabatangan and some other flooded habitats, represent a typical tourist attraction for middle-to-low budget foreign travellers. These monkeys may be less attractive compared to orangutans, but they can be shown relatively easily and reliably at their sleeping sites to a large number of people at affordable prices. In Sungai Hitam (Kuala Samboja, East Kalimantan), a motor boat can be rented from the local community for approximately US$20 per trip for 3 to 4 people (Tri Atmoko, pers. comm., 2013). In Balikpapan Bay, proboscis monkeys can be seen from larger boats for US$30 for up to seven people (M.A. Asrani, pers. comm., 2013). Although these prices do not include fees paid to tour agencies, they are still quite affordable for most travellers.

Many tourist agencies that organize proboscis monkey ecotourism therefore focus on foreign travellers (often young backpackers), who usually belong to middle social classes who visit Borneo on limited budgets, albeit in larger numbers. This brings a significant work opportunity for many local people, who usually appreciate it, but the experience by either side is not always positive. Due to their limited budget, travellers often tend to demand 'bang-for-the-buck' services (Balikpapan Bay: S. Lhota, pers. obs.). Programmes focused on lower-budget travellers are less luxurious (and usually less professional) compared to programmes focused on high-budget tourists, which often means bargaining for price. At Tanjung Puting, operators often set their own prices, excluding park fees, and costs may vary widely. However, under this mode of operation, there is no way of distinguishing who the best guides are, unless by word of mouth. Some travellers arrive with unrealistic expectations, or may be given unrealistic hopes by the guides arranging their tours. This is a problem, particularly for wild orangutans, where encounters are difficult. For example, at Tanjung Puting, many visitors expect to see famous orangutans such as 'Tom' or 'Siswe' but sightings are not guaranteed. These travellers often complete their trip dissatisfied (no animal sightings and overpriced) and their perception of conservation may deteriorate.

However, most travellers are often positively influenced and some of them do change their behaviour after visiting such ecotourism sites. Some travellers, for example, began boycotting palm oil products and starting campaigns against its use in Europe, after they have experienced the environmental damage caused by the continuing expansion of palm oil plantations.

Local Visitors

In developing countries the word 'tourists' often synonymous with 'foreign tourists'; this however ignores the fact that the majority of visitors in many tourism sites in Indonesia, Malaysia and Brunei, are Indonesians, Malaysians and Bruneians. Despite their large numbers, local ecotourists in Borneo remain an underrepresented target group in many high-profile ecotourism programmes because targeting the local tourists tends to be a less profitable business model. These visitors often travel at discounted prices. At Balikpapan Bay, boat trips are offered at substantially cheaper prices to allow local people, especially students and organized 'nature lovers' (*Pecinta Alam*) groups to visit. Many less-known mangrove tourist parks allow free entry and/or opportunities to explore by boardwalks without guides, which generates less money per person compared to the guided boat trips.

On the other hand, the prevalence of local tourists in many ecotourism sites offers enormous potential to influence the national awareness of natural heritage value and the need for its protection. Unfortunately, environmental education tends to be poorly managed in most of these lesser-known sites where local people form the main visitor group. Furthermore, it is not an easy task: compared to foreign tourists, who tend to be highly motivated to learn about nature, the local tourists usually visit the nature reserves only for the purpose of recreation and social interaction with their friends and families, and may be less open to learning (Cochrane 2006).

Environmental Education as a Core Tenet for Ecotourism

Even if the main objective of ecotourism is to generate funds that help compensate opportunity and conservation costs of natural attractions, environmental education is always an embedded part of any ecotourism programme. Ecotourism inevitably influences people and their behaviour. Interactions with tourists have the potential to improve local people's attitudes towards conservation and raise their awareness about the values of nature and environmental issues (Ballantyne *et al.* 2007, 2009). Simultaneously, contact with nature is important for spreading conservation commitment among tourists, particularly for the uninitiated. Even in purely economic terms, while ecotourism as an enterprise can create direct economic benefits, its educational impact also increases the 'existence value' of nature, or the value people confer to the intrinsic value of nature (Tisdell *et al.* 2007), which in turn influences the amount of effort, support, and modesty in their needs, people are willing to contribute to conservation. Under certain circumstances, properly managed environmental education may have an even larger conservation impact than the monetary value of funds generated from these programmes (Balikpapan Bay: S. Lhota, pers. obs.).

There are ample examples, worth emulating, of how small-scale ecotourism can contribute to environmental education, especially among local visitors. There are mangrove reserves outside Borneo where environmental education is of high priority, for example, the Sungei Buloh Wetland Reserve in Singapore. In Borneo, the Kota Kinabalu Wetland Center in Sabah has no primate population but provides good opportunities for learning about the mangrove ecosystem. There are also several exemplary non-mangrove tourist areas in Borneo where environmental education is of high priority. They include the Rainforest Discovery Center in Sepilok (Sandakan, Sabah), which also offers an option of a one day hike across Sepilok Virgin Forest Reserve to the coastal mangroves of Sandakan Bay, or Kawasan Wisata Pendidikan Linkungan Hidup (KWPLH, Balikpapan, East Kalimantan), which also provides information on boat trips and proboscis monkey observation in Balikpapan Bay. In the case of KWPLH, six ex-captive confiscated sun bears (*Helarctos malayanus*) that cannot be released back into the forest, are being kept in a large naturalistic enclosure as the main tourist attraction. The enclosure is surrounded by a boardwalk, where guides explain the biology and conservation of sun bears. Next to the enclosures are gazebos with extensive interactive educational panels and replicas of Dayak longhouses that offer venues for meetings, competitions, presentations or ceremonies. Up to 50 000 people, mostly Indonesians, visit the education centre every year (Fredriksson, pers. comm., 2012). The conservation messages conveyed to visitors emphasize protection of wild sun bears in the nearby Sungai Wain Forest Reserve and conservation in general. Admission is currently free and only a small admission fee is expected to be charged in the future. KWPLH provides an example of a tourism attraction where environmental education, rather than fund generation, forms the basis of operations, and clearly represents wider benefits for conservation. Importantly, most of these initiatives do not rely on tourist contributions for sustenance as they are usually government-funded establishments.

Freely accessible and attractive visitors' centres focusing on environmental education, with information available also in local languages, should ideally be developed in every major ecotourism area, including those focused on foreign tourists, to raise environmental awareness of both foreign and local visitors.

Future Directions

The potential for ecotourism is still highly underutilized in mangroves, peat swamps and floodplain forests of Borneo, especially in the Indonesian Kalimantan provinces. It can be illustrated by two sites with exceptionally high ecotourism potential due to their short distance from Balikpapan city and its international airport. Currently, tourists only arrive five times a month to Sungai Hitam in Kuala Samboja (Tri Atmoko, pers. comm., 2013) and four to eight times to Balikpapan Bay (M.A. Asrani, pers. comm., 2013). Experience shows that with logistical improvements and proper promotion, the frequency of tourist visits (both by foreigners and locals) can be increased by approximately ten times, particularly during vacation periods or during periods of high promotion (environmental campaigns, in case of Balikpapan Bay: S. Lhota, pers. obs.).

Given the negative impacts associated with many small ecotourism projects, especially those sponsored by commercial entities or local governments, there is a serious risk of losing public trust in the conservation impact of ecotourism (Self *et al.* 2010). The term 'ecotourism' may gradually be diminished in the foreseeable future if steps are not taken to retain its value as a conservation strategy. Some degree of independent evaluation and certification of ecotourism programmes may be needed (Medina 2005). Such evaluation would however be inefficient and susceptible to corruption if carried out purely on a local scale; alternatively, global scale certification, made by credible international organizations, would be expensive. The latter approach would probably also marginalize small ecotourism entrepreneurs, which may in turn hamper further expansion of ecotourism. In 2002, there were more than 60 ecotourism certification programmes, which adopted the internationally recognized sustainability criteria set by the Global Sustainable Tourism Council. Most of these programmes were local or national, although a few operated internationally (Bien 2008).

These programmes still cover only a very small fraction of ecotourism enterprises.

Another important issue is the involvement of various stakeholders in ecotourism programmes. Local people do not usually have the education level, skills or resources to manage ecotourism programmes. Purely community-based ecotourism (CBET) has also seen limited success (Kiss 2004). Furthermore, after having sold their land to corporations, many local communities already lost their influence on conservation politics in Borneo. Larger business entities and governments, on the other hand, tend to focus on immediate-profit-oriented or propaganda goals that divert attention and resources away from real long-term conservation goals. Ecotourism in their hands may be more about exploitation and greenwashing than conservation. Conservationists, in turn, may provide genuine contributions to such programmes but their influence is very limited and their views are often ignored by other stakeholders. Therefore, there does not seem to be any 'best candidate' formula for managing ecotourism programmes. A multi-stakeholder approach involving local communities, corporations and governments as well as conservationists is needed to develop properly managed ecotourism.

The contributions of conservation scientists need to be expanded. Baseline information is needed that can inform practical management measures, such as determining viable primate population sizes in small tourist parks, improving methods for habitat restoration, identifying factors and consequences for increased tourist–primate interface and ways of minimizing impacts while maximizing visitor satisfaction. Considerations must go beyond the realms of ecological research and take sociopolitical and economic factors into account when justifying indirect long-term environmental services as an alternative to short-term profits (a situation whereby ecotourism alone is likely to lose out against alternative land-use plans). Only with good knowledge of social, political and economic factors, can some of the problems outlined in this chapter be solved: existing laws and regulations to prevent the negative impact of ecotourism can be enforced, various types of tourists can be optimally targeted, and education potential maximized.

These issues apply to flooded forest primate habitats in Borneo but can be also generalized to many other settings. Ecotourism provides unique challenges – and opportunities – for conservationists to improve cooperation among different disciplines and stakeholders and ultimately arrive at holistic solutions.

Acknowledgements

We are thankful to Ikki Matsuda, Arie Soetjiadi, Susan Cheyne, Clement Tisdell, Heather Leasor, Tony Gilding, Alfredo Quarto, Jim Enright, Mark Harrison, Janelle M. Baker, Katarzyna Nowak and Tri Atmoko for sharing their experience and for valuable comments on an earlier draft of the manuscript. We also thank Marc Myers for drawing the map of selected flooded forest ecotourism destinations in Borneo.

Chapter 43

Conservation Value of Africa's Flooded Habitats to Non-human Primates

Katarzyna Nowak, Fiona Maisels, Lynne R. Baker and Hugo Rainey

Introduction

Non-human primates occur in, and use, a variety of permanently and seasonally flooded habitats in sub-Saharan Africa. Such habitats include swamp, mangrove, marsh, littoral, riparian and gallery forests, as well as *bais* (open swampy clearings) and seasonally flooded grasslands (Bennett *et al.*; Chapter 2, this volume). To date, the study and conservation of most African primate species have focused on populations in *terra firma* (never-flooded) habitats, while flooded habitats have been under-emphasized in the primate ecology and conservation literature and under-represented in Africa's protected area network (Gautier-Hion & Brugiere 2005). In Central Africa (from the Nigeria–Cameroon border eastward to the Albertine Rift), only 6.6% of the region's vast swamp forests lie within protected areas, compared to 17.0% of *terra firma* forests; mangroves are better represented, but occur within just 13.5% of protected areas in the region (Table 43.1).

In this chapter, we discuss how African primates use flooded habitats, describe species for which these habitats are critical for conservation, and name the major threats to these environments. Several chapters in this volume contain related data. We do not summarize all of their findings here, but point to relevant chapters. Our intent is to expand on what has been discussed elsewhere in this volume, to further illustrate the role of flooded habitats in the conservation of many of Africa's primate communities. Herein, we follow the primate taxonomy presented in Butynski *et al.* (2013), with these exceptions: (1) for red colobus other than *Procolobus verus*, we use the genus *Piliocolobus* and not *Procolobus* per the forthcoming IUCN Red List; (2) we elevate the Niger Delta red colobus to species level as *Piliocolobus epieni* following Oates (2011), and (3) we follow the latest assessment for lemurs as presented by the IUCN (2017). For threatened taxa only, we note their current conservation status as assessed by the IUCN Red List.

An Awakening: Apes in the Congo Basin

Although the preference of several African primate species for flooded habitats has been known for decades, it was only relatively recently that the conservation significance of inundated forests for primate communities was brought to the fore. In particular, several studies prompted global attention to the central Congo Basin, home to the greatest expanses of flooded forest in Africa. Many of these studies focused on the Lake Télé–Lake Tumba Landscape, Africa's largest wetland and the world's largest swamp forest at 126 000 km^2 (Twagirashyaka & Inogwabini 2009). Partly in the Democratic Republic of the Congo (DRC) and partly in the Republic of Congo (Congo), this region also contains patches of *terra firma* forest and (often seasonally flooded) savanna.

Three ape taxa have long been known to use the swamp forests of Lake Télé–Lake Tumba: the Critically Endangered western lowland gorilla (*Gorilla gorilla gorilla*), Endangered central chimpanzee (*Pan troglodytes troglodytes*) and Endangered bonobo (*Pan paniscus*) south of the Congo River. However, the importance of these forests for primates was not recognized until surveys, beginning in the late 1980s, showed that the area supports high densities of apes, especially western lowland gorillas (Blake 1993; Blake *et al.* 1995; Fay *et al.* 1989; Fay & Agnagna 1992). The area is now considered to be a stronghold for gorillas; the swamp forest they inhabit is impenetrable and possesses poor-quality timber, making it unfavourable to humans and thus more impervious to anthropogenic change (Poulsen & Clark 2004). Subsequent surveys estimated more than 18 000 gorillas in the swamp forests of the Lake Télé–Likouala Landscape in Congo (Iyenguet *et al.* 2012; Maisels *et al.* 2012; Malanda *et al.* 2010; Rainey *et al.* 2010; Stokes *et al.* 2010). Across the Congo River in the Lake Tumba Landscape, bonobos and chimpanzees are both strongly associated with swamp and flooded forests (Inogwabini *et al.* 2012).

Studies elsewhere in the Congo Basin further support the significance of flooded habitats for primate conservation in this region. Western lowland gorillas in southern Cameroon prefer swamp habitats (but avoid riparian forest) when constructing nests (Willie *et al.* 2013). In the Central African Republic, Brugiere *et al.* (2005) found high densities of primates and higher primate species richness in flooded forest compared with *terra firma* forest. A follow-on assessment of 12 primate communities in Central Africa showed that riparian forests support 1.5-times more primate species than *terra firma* forests (Gautier-Hion & Brugiere 2005). Outside the Congo Basin, other flooded habitats are important for primates, including freshwater and mangrove swamp forests of Nigeria, particularly in the Niger Delta (Chapter 40), as well as several other coastal deltas, including Ogooué in Gabon, Tana in Kenya, and Rufiji in Tanzania (Chapter 30).

Flooded Habitats' Built-in Defence

Africa's flooded forests may hold particular conservation value because they are generally less affected by anthropogenic

Table 43.1 Protected areas within inland swamp or mangrove forests in Central Africa (from the Nigeria–Cameroon border eastward to the Albertine Rift). Total forest area: mangroves: 9013 km²; inland swamp forests: 191 463 km².

Forest type	Country	Protected area	Area of PA within flooded habitat type (km²)	% of total forest type
Mangrove	Cameroon	Campo-Ma'an National Park	13	0.14
	Cameroon	Rio Campo Natural Reserve	8	0.09
	Democratic Republic of Congo	Mangroves Natural Reserve	138	1.53
	Equatorial Guinea	Estuario del Muni Natural Reserve	146	1.62
	Gabon	Akanda National Park	58	0.64
	Gabon	Loango National Park	90	1.00
	Gabon	Moukalaba-Doudou National Park	225	2.50
	Gabon	Pongara National Park	534	5.92
	Gabon	Wonga-Wongue Presidential Reserve	3.4	0.04
Subtotal mangrove				13.48
Inland swamp forest	Cameroon	Boumba Bek National Park	22	0.01
	Cameroon	Kom National Park	117	0.06
	Cameroon	Lobeke National Park	9.7	0.01
	Cameroon	Santchou Faunal Reserve	36	0.02
	Central African Republic	Bangassou Forest Reserve	240	0.13
	Central African Republic	Botambi Classified Forest	43	0.02
	Central African Republic	Dzanga-Ndoki National Park	85	0.04
	Central African Republic	Dzanga-Ndoki Special Reserve	113	0.06
	Central African Republic	Mbaere-Bodingue Faunal Reserve	12	0.01
	Democratic Republic of Congo	Lake Tumba-Ledima Reserve	3045	1.59
	Democratic Republic of Congo	Lomako Natural Reserve	2	0.00
	Democratic Republic of Congo	Salonga National Park	1684	0.88
	Democratic Republic of Congo	Yangambi Biosphere Reserve	601	0.31
	Gabon	Minkebe National Park	204	0.11
	Gabon	Mwagna National Park	79	0.04
	Gabon	Wonga-Wongue Presidential Reserve	0.5	0.00
	Republic of Congo	Lake Télé Community Reserve	3226	1.68
	Republic of Congo	Nouabale Ndoki National Park	5.5	0.00
	Republic of Congo	Ntokou-Pikounda National Park	3126	1.63
	Republic of Congo	Odzala National Park	0.8	0.00
Subtotal inland swamp forest				6.69

Note: We acknowledge other sites of importance for which there are no habitat/vegetation data, including two coastal sites in Gabon: Mayumba NP and the Akaka swamps of Loango NP, and one site in the Republic of Congo: Conkouati-Douli NP.

threats. These habitats are often marginal lands for agriculture given the challenges of converting inundated habitat to farmland (Gautier-Hion & Brugiere 2005; Nowak 2012). Hunting pressure may also be lower. Local people living in or near flooded habitats may rely on fishing for their livelihoods and therefore fish as a protein source, reducing hunting pressure on terrestrial wild species (Baker 2003; Werre 2001a). In addition, inundated environments are generally less accessible to hunters. Travelling on foot is hindered by thick, muddy terrain and by the stilt roots or aerial roots of trees typical of these forests, such as species of *Uapaca* (Phyllanthaceae) and *Pandanus* (Pandanaceae) in swamp forests, and *Rhizophora* (Rhizophoraceae) in coastal mangroves. *Pandanus* and *Raphia* spp. (Arecaceae) and rattans (lianescent palms), also common in swamp forest, are generously laced with spines along their leaves and/or stems. In Lake Télé in Congo, Rainey *et al.* (2010) describe *Raphia* swamp forest as providing natural protection against hunting for gorillas. Although colobus monkeys are commonly hunted across Central Africa, populations of Oustalet's red colobus (*Piliocolobus oustaleti* in the forthcoming IUCN Red List) and guereza colobus (*Colobus guereza*) resident in swamp forests of Lake Télé are so well protected by the impenetrability of this habitat that they behave naively towards humans (H. Rainey, pers. obs.).

For similar reasons, mangroves afford protection to some primate populations. Due to the relatively lower hunting pressure and degradation in mangroves versus neighbouring *terra firma* forests, researchers have suggested that primate populations seek refuge in mangroves, including two Endangered red colobus species: the Zanzibar red colobus (*Piliocolobus kirkii*) in Zanzibar and Temminck's red colobus (*Piliocolobus temminckii*) in Senegal (Galat-Luong & Galat 2005; Chapters 7 and 39). In southeastern Ivory Coast, mangroves are important for two Endangered species: roloway monkey (*Cercopithecus roloway*) and white-naped mangabey (*Cercocebus lunulatus*). Researchers suggest that these species have adapted to permanently flooded mangroves to not only avoid severe hunting pressure in adjacent flooded forest (deemed more accessible than mangroves), but also take advantage of the food resources found in mangroves (Bi *et al.* 2013).

The relative inaccessibility of flooded habitats likely contributes to the survival of a population of the Critically Endangered greater bamboo lemur (*Prolemur simus*) in the Torotorofotsy wetlands (Ramsar Site #1453), a large marshland in Madagascar (Dolch *et al.* 2008), as well to the survival of the Critically Endangered Niger Delta red colobus (*Piliocolobus epieni*), found only in marsh forest in the central Niger Delta (Werre 2000, 2001a), and the recently rediscovered Bouvier's red colobus (*Piliocolobus bouvieri*) in the swamps of northern Congo west of the Sangha River (Devreese 2015).

Elsewhere in southern Nigeria, high human population density and associated pressures on natural habitats have led to widespread forest loss and degradation. Most of the largest remaining primate habitats in the region are seasonally and permanently flooded forests, which confer a degree of protection on populations of some species, including the Endangered red-capped mangabey (*Cercocebus torquatus*) and two Endangered endemics, the Sclater's monkey (*Cercopithecus sclateri*) and white-throated monkey (*Cercopithecus erythrogaster pococki*) (Baker 2005; Baker & Olubode 2008; Oates 1989; Oates & Anadu 1988; Chapters 30 and 40). In the Niger Delta, many swamp and riparian forests persist as they are more difficult to access and farm, although some productive flooded habitats, such as forests along the Niger River floodplain, have nevertheless been converted to cropland because of human 'ingenuity' (Ikemeh 2014; Oates 1989).

Diversity of Flooded Habitat Use

Our knowledge of primates in flooded habitats has progressed quickly in recent decades (Table 43.2). Researchers have discovered that primates use flooded habitats for a variety of reasons and to varying extents. Some species vary geographically in their use of inundated habitats. For instance, guereza colobus is a *terra firma* specialist in East Africa, but is associated with water where it occurs in Central Africa (Gautier-Hion & Brugiere 2005). Similarly, Oustalet's red colobus, the most widespread of the *P. rufomitratus* subspecies, uses a variety of tropical forest types (Oates *et al.* 2008), but in the Ngotto Forest in Central African Republic, it inhabits only flooded forests on alluvial river banks (Gautier-Hion & Brugiere 2005). In flooded and swampy areas, this monkey spends more than one-quarter of its time at below 10 m above ground and regularly enters water to collect bulbs of aquatic plants (Galat-Luong & Galat 1979; A. Galat-Luong & G. Galat, pers. comm., 2014).

Other species rely heavily, if not exclusively, on flooded habitats. In Madagascar, the Critically Endangered Lac Alaotra bamboo lemur (*Hapalemur alaotrensis*) feeds almost exclusively on papyrus and reeds in marshland, making it the only primate uniquely adapted to marsh habitat (Chapter 34). Allen's swamp monkey (*Allenopithecus nigroviridis*) occurs only in swamp and riparian forests, where it is sympatric with De Brazza's monkey (*Cercopithecus neglectus*), Tshuapa (or Thollon's) red colobus (*P. tholloni*), guereza colobus monkeys, and either red-tailed (*Cercopithecus ascanius*) or moustached (*C. cephus*) monkeys depending on the location (Gautier 1985; Maisels *et al.* 2006b). De Brazza's monkey is always associated with riverine or swamp forest (Chapter 39; Gautier-Hion & Gautier 1978; Gautier-Hion 2013b). Like De Brazza's, the northern talapoin monkey (*Miopithecus ogouensis*) is considered a flooded forest specialist, preferring riparian habitats (Gautier-Hion 2013c; Maisels *et al.* 2006a); the southern talapoin monkey (*Miopithecus talapoin*) likely has similar preferences (Gautier-Hion 2013d). In addition, both red-capped mangabeys and golden-bellied mangabeys (*Cercocebus chrysogaster*) prefer wet habitats, including seasonally flooded and swamp forests (Ehardt 2013; Ehardt & Butynski 2013). The presence of some primates in parts of their ranges, as well as the extent to which they rely on flooded habitats, may have been overlooked by researchers due to the difficulties associated with surveying wildlife in these habitats (e.g. Maisels *et al.* 2006a, 2007b).

Gorillas were thought to avoid water until the 1970s, when investigators began to learn otherwise. Casimir (1975) found that Critically Endangered eastern lowland gorillas (*Gorilla beringei graueri*) wade through streams up to 60 cm deep in the eastern DRC. Western lowland gorillas in Equatorial Guinea frequently 'bathe' in streams, according to local reports (Groves & Sabater Pi 1985), and they regularly cross streams and nest in sites surrounded by water in Lopé National Park, Gabon (Williamson *et al.* 1988). In northern Congo, gorillas use swamp forests almost exclusively for at least part of the year and feed on the abundant herbaceous monocotyledons (Fay *et al.* 1989). The rich herb layer in *bais* also provides a highly digestible, sodium-rich food source for gorillas (Blake

Table 43.2 Studies of African primates in flooded habitats (non-exhaustive list).

Habitat type	Country	Common name	Genus	Species	Source(s)
Freshwater swamp, including *Raphia* swamp forests	Cameroon	Western lowland gorilla	*Gorilla*	*gorilla gorilla*	Willie *et al.* 2013
	Democratic Republic of Congo	Bonobo	*Pan*	*paniscus*	Inogwabini & Matungila 2009
	Nigeria	Red-capped mangabey	*Cercocebus*	*torquatus*	Baker 2005; Baker & Olubode 2008
	Nigeria	Sclater's monkey	*Cercopithecus*	*sclateri*	Baker 2005; Baker & Olubode 2008; Baker *et al.* 2011
	Republic of Congo	Allen's swamp monkey	*Allenopithecus*	*nigroviridis*	Maisels *et al.* 2006
	Republic of Congo	Chimpanzee	*Pan*	*troglodytes*	Poulsen & Clark 2004; Rainey *et al.* 2010; Stokes *et al.* 2010
	Republic of Congo	Western lowland gorilla	*Gorilla*	*gorilla gorilla*	Blake 1993, Blake *et al.* 1995; Iyenguet *et al.* 2008 & 2012; Kalan *et al.* 2010; Malanda *et al.* 2010; Poulsen & Clark, 2004; Rainey *et al.* 2010; Stokes *et al.* 2010
Mangrove	Gabon	Western gorilla	*Gorilla*	*gorilla*	Chapter 12
	Guinea	Chimpanzee	*Pan*	*troglodytes*	Hockings, pers. comm., 2013
	Madagascar	Pygmy mouse lemur	*Microcebus*	*myoxinus*	Ganzhorn pers. comm., 2013
	Senegal	Green monkey	*Cercopithecus*	*sabaeus*	Galat & Galat-Luong 1976
	Senegal	Senegal red colobus	*Piliocolobus*	*temmickii*	Galat-Luong & Galat 2005
	Zanzibar	Zanzibar red colobus	*Piliocolobus*	*kirkii*	Nowak & Lee 2010
Littoral	Madagascar	Collared brown lemur	*Eulemur*	*collaris*	Ganzhorn *et al.* 2007
	Madagascar	Mouse lemur	*Microcebus*	*spp.*	Ganzhorn *et al.* 2007
	Madagascar	Southern woolly lemur	*Avahi*	*meridionalis*	Ganzhorn *et al.* 2007
Riverine	DRC & Gabon	Western gorilla	*Gorilla*	*gorilla*	Chapter 24
	Kenya	Tana River mangabey	*Cercocebus*	*galeritus*	Wieczkowski & Butynski 2013
	Kenya	Tana River red colobus	*Piliocolobus*	*rufomitratus*	Butynski & Mwangi 1994; Mbora & Meikle 2004
	Tanzania	Red-tail monkey	*Cercopithecus*	*ascanius*	Chapter 33
Marsh swamp	Nigeria	Niger Delta red colobus	*Piliocolobus*	*epieni*	Ikemeh 2015; Werre 2000, 2001a
Seasonally flooded inland delta	Botswana	Chacma baboon	*Papio*	*ursinus*	Cheney *et al.* 2004; Hamilton *et al.* 1976; Johnson 2003
Intertidal zone	South Africa	Chacma baboon	*Papio*	*ursinus*	Chapter 20

Calculated from the World Resources Institute's Atlases of Central Africa: www.wri.org/our-work/project/congo-basin-forest-atlases#project-tabs.

1994; Olejniczak 1994; Vanleeuwe *et al.* 1998). Similarly, chimpanzees and guereza colobus monkeys feed on algae in *bais* in Congo (Devos *et al.* 2002).

Flooded habitats clearly provide some primate populations with unique opportunities for refuge and feeding. Such habitats may have a higher relative abundance of invertebrates; high protein foods, such as shellfish (in mangrove areas); and a high density of monocotyledonous foods. In Salonga National Park, DRC, sympatric Tshuapa red colobus and Angola colobus (*Colobus angolensis angolensis)* are common in flooded forests where they eat a diet full of seeds (Maisels *et al.* 1994). A population of red-capped mangabeys in Gabon occurs in mangrove forest significantly more often than in *terra firma* forest (Cooke 2012); in mangroves, the species feeds on crabs (Cooke 2015). In the Tsinjoriake region of southwestern Madagascar, Endangered ring-tailed lemurs (*Lemur catta*) retreat to flooded habitats during the dry season because of cooler temperatures (J. Youssouf, pers. comm., 2014; Chapter 7).

Using flooded habitats can be risky for primates. Chacma baboons (*Papio ursinus*) can occur in high densities in the Okavango Delta, an inland delta in Botswana (Johnson 2003); the baboons use *terra firma* islands, edged with woodland, but also feed on the floodplain (Hamilton *et al.* 1976). More than any other food item, aquatic plant underground storage organs are the primary fallback foods for one population (Wrangham *et al.* 2009). There are costs associated with exploiting the floodplain, however: baboon mortality is highest during the 3-month peak flood period (June–October) when baboons navigate the flooded plains and move among islands (Cheney *et al.* 2004; R. Seyfarth, pers. comm., 2014). Crocodiles make water crossings dangerous, and the only documented lion attacks on baboons occurred when baboons emerged onto dry islands after crossing water.

These examples illustrate the value – but also risk – of flooded habitats for several African primates; however, for many of these species, additional habitats are required for survival. The Zanzibar red colobus population on Uzi Island, in the Zanzibar archipelago, uses mangroves to a significant degree, yet the species must have access to upland coral rag forest (Nowak & Lee 2011). Exclusive use of mangroves would likely impede this population's health given the high level of tannins and salt in mangrove-derived foods (Nowak 2008). Additionally, consumption of mangrove leaves by this population has resulted in rates of water foraging higher than those known for any other red colobus species or population; an energy-intensive activity, water foraging by the Uzi Island red colobus population occurs daily and throughout the day.

Flooded Habitats: Threats and Conservation

Given that flooded habitats are relatively inaccessible and difficult to convert to agricultural land, they provide a degree of natural protection for many primates in Africa. They also harbour key resources for some populations. Although rivers and flooded zones can act as barriers to dispersal and gene flow (e.g. Chapman *et al.* 1999; Eriksson *et al.* 2004; Quéméré *et al.* 2010; Chapter 38), riparian forests, due to their linear nature, may function as dispersal corridors for certain populations and species (e.g. lemurs, Ganzhorn *et al.* 2006; chimpanzees, McLennan 2008; several guenons, Chapter 39). Protection of riparian forests and adjacent *terra firma* forests may be the best way to safeguard much of Africa's primate diversity (Gautier-Hion & Brugiere 2005). Inundated habitats, consequently, deserve special conservation attention (see Foreword).

Increasing human populations across sub-Saharan Africa are placing greater pressure on already stressed flooded habitats. Threats to these environments are manifold, and many sites lack effective protection. Africa's ten largest river deltas, some of which support at least several threatened primate taxa, are largely unprotected and unrecognized formally as biodiversity sanctuaries (Chapter 30). Only half (10 of 20) of the protected areas within inundated and flooded forests in Central Africa are national parks (IUCN Category II), although one is a Presidential Reserve (which carries a high level of protection). The rest are Faunal Reserves, Forest Reserves, or other less strictly and formally protected areas (Table 43.1).

The very important Lake Télé–Lake Tumba Landscape in the Congo Basin is under threat as a result of commercial hunting, road building, and impacts associated with human refugees. Two protected areas within the landscape cover less than 10% of the region, although several new protected areas have been proposed (Twagirashyaka & Inogwabini 2009). Some African habitats under protection are not secure and require careful monitoring. For example, the Okavango Delta, which comprises one protected area (Moremi Game Reserve) and became a UNESCO World Heritage Site in 2014, faces several risks, including diversion of water upstream for a hydropower plant (Boyes 2014), as well as fencing, overgrazing, overfishing, and climate change (Darkoh & Mbaiwa 2014). Dam construction would force the Delta's primate populations to adapt to anthropic flooding (Chapter 36). Similarly, severe vegetation changes due to dam construction, irrigation projects, and water diversion have negatively affected populations of the Endangered Tana River red colobus (*P. rufomitratus*) (Oates *et al.* 2008; Chapter 30).

Other areas are remarkably underprotected. The Niger Delta region is important for many of Nigeria's primates, including the country's only endemic species (*C. sclateri* and *P. epieni*), but the Delta has no effectively protected areas (Chapters 30 and 40). The region has long experienced degradation due to a number of threats, notably impacts associated with oil-related development and a growing human population. Although flooded forests may limit logging activities where timber cannot be readily cut and transported, loggers in some inundated forests in the Niger Delta float valuable trees in 'timber gutters' (narrow canals dug in the forest) and then assemble these into timber rafts; the rafts are often pulled by boats to coastal trading centres (Baker 2005; Ikemeh 2015; Werre 2000). One economically valuable species, *Abura* (*Fleroya ledermannii* = *Hallea ledermannii* or *Mitragyna ledermannii*, Rubiaceae), a preferred food of *P. epieni*, has been heavily exploited in the Niger Delta, notably within the range of *P. epieni* (Werre 2000). Intensive logging may have contributed to severe population declines of this highly threatened primate (Ikemeh 2015).

Globally, mangroves face threats from shrimp and timber industries and are declining at a rate of 1% each year (Polidoro *et al.* 2010). Although mangroves have been receiving greater attention for the ecosystem services and economic benefits they provide (Polidoro *et al.* 2010; UNEP-WCMC 2006), they generally do not receive the same level of official recognition as *terra firma* forests and may simply be designated as 'Mangrove Forest Reserves' (e.g. in Tanzania; K. Nowak, pers. obs.). This is in contrast to terrestrial protected areas, which have distinct names, identities, biodiversity inventories and internet websites, and can often be immediately recognized by name and reputation (e.g. 'the Serengeti', 'the Virungas'). Such ubiquity can greatly aid conservation efforts (Chapters 7 and 12). Only five of the nine protected areas within the Central African mangrove region are national parks (IUCN Category II); one is a Presidential Reserve; and the other four are Natural Reserves (Table 43.1).

Both research and conservation actions are urgently needed in flooded zones across sub-Saharan Africa. New research is likely to uncover additional evidence of the use of, and reliance on, flooded habitats by primates. Ultimately, effective protection of these habitats will be essential to conserving Africa's primates as pressure on *terra firma* forests escalates. For example, conservation of gorillas and other primates in the Congo Basin requires the integration of riverine, swamp and coastal forests outside of current protected areas into surveys, monitoring, and protection plans (Maisels 2006b, 2007a; Rainey *et al.* 2010). In early 2015, negotiations were underway to expand the boundaries of the Lake Télé Community Reserve; such an extension would benefit large numbers of great apes (Rainey *et al.* 2010). This vast swamp also benefited from particular attention in the most recent IUCN great ape action plan for the region (IUCN 2014). Finally, key sites that lack any protection, such as the Niger Delta, should be prioritized by the conservation community to ensure that the degradation of these areas does not result in unrecoverable loss of Africa's primate diversity.

Acknowledgements

We are grateful to Kate Abernethy and Tom Butynski for their helpful comments which improved our manuscript.

Chapter 44

Southeast Asian Primates in Flooded Forests

John Chih Mun Sha, Shun Deng Fam and Andie Hui Fang Ang

Introduction

Forest habitat loss, fragmentation and degradation have been widely acknowledged as the main threats facing many primate species throughout their range (Chapman & Peres 2001; Cowlishaw 1999; Johns & Skorupa 1987; Mittermeier & Cheney 1987). Southeast Asia is a major biodiversity hotspot encompassing several important ecoregions (Myers *et al.* 2000; Navjot *et al.* 2004; Page *et al.* 1999), and harbouring a large number of highly threatened primate species (at least 15 'Critically Endangered' species as classified by the IUCN Red List of Threatened Species). Flooded forests (which include peat swamp, freshwater swamp, mangrove and riparian habitats) are particularly highly threatened and being lost at an alarming rate that is higher than deforestation of other forest types (Giri *et al.* 2010; Miettinen *et al.* 2011).

General understanding of primate occurrence in flooded forests and the impacts of forest loss on the long-term survival of primate species remains patchy. This is due largely to the lack of research attention on the use of flooded forests by primates, likely as a result of the difficulties in accessing these habitats. In this chapter, we evaluate what is known about flooded forest use by primates in Southeast Asia, highlight possible limitations to current knowledge and discuss conservation implications for primates and flooded forests.

Importance of Flooded Forests for Primates in Southeast Asia

In Southeast Asia, primate species that are widely known to occur in flooded forests include the proboscis monkey (*Nasalis larvatus*) of Borneo (Sha *et al.* 2011); and to a certain extent, the banded leaf monkeys (*Presbytis femoralis*) and Sarawak surilis (*P. chrysomelas*) found in Borneo, Sumatra, Peninsula Malaysia and Singapore (Ang *et al.* 2011; Crockett & Wilson 1980; Lucas *et al.* 1988). These species are examples of flooded forest specialists that persist largely in these forests. With rapid decline and fragmentation of their habitats, many of the populations are disconnected and forced into degraded forests and plantations, leading to local extinctions and low genetic variability (Ang *et al.* 2012; Sha *et al.* 2008).

Flooded forests may also be important for other Southeast Asian primate species that are not known to be flooded forest specialists. For example, the Bornean agile gibbon (*Hylobates albibarbis*) showed high population persistence in peat flooded forests (Buckley *et al.* 2006; Geissmann, 2007). This contrasts with the ranging behaviour of most other gibbon species, which occur predominantly in dry primary, secondary and evergreen forest (Galdikas & Shapiro 1994; Wolfheim 1983). Similarly, the densities and distributions of orangutans (*Pongo* spp.) in flooded forests have been found to be comparable to other forest types, such as lowland dipterocarp forest, with which they are more commonly associated (Morrogh-Bernard *et al.* 2003; Rijksen & Meijaard 1999; Russon *et al.* 2001; Chapter 26). This could be due to less consistent and reliable fruit production in dipterocarp forests, which exhibit supra-annual 'mast' fruiting. For example, the lower variability in food availability in peat swamp forests allows the maroon leaf monkey (*Presbytis rubicunda*) to rely less on fallback foods (Ehlers-Smith *et al.* 2013b).

There is emerging evidence of the importance of flooded forest habitats to many Southeast Asian primate species, of which only a small proportion are known to be specialists for these habitats (Chapters 4 and 13). Much less is known about the possible importance of flooded forests to other primate species that are not specially adapted to use these habitats, but which may use them seasonally or occasionally (Chapters 27 and 43 for examples from the Neotropics and Africa, respectively).

Status of Flooded Forests in Southeast Asia

We define Southeast Asia according to country membership in the ASEAN and Timor Leste, which excludes China and New Guinea. Insular Southeast Asia is thus defined to contain Borneo (Brunei and East Malaysia), Peninsular Malaysia, Indonesia, the Philippines and Timor Leste. Flooded forest habitats that are found in this region include mainly freshwater flooded, peat swamp and mangrove forests (Figure 44.1). This excludes riparian forests which are not covered in existing habitat classification schemes.

Peat swamp forests are waterlogged forests growing on areas where rainfall and topography are conducive to poor drainage, with peat accumulating up to 20 m thick (Page *et al.* 1999; UNDP 2006). An estimated 70% of global tropical peat lands (23–25 Mha) are found within the Southeast Asian region (Joosten 2009; Page *et al.* 2011). The estimated areas of peat swamp coverage (in thousand hectares or t ha) are: Indonesia

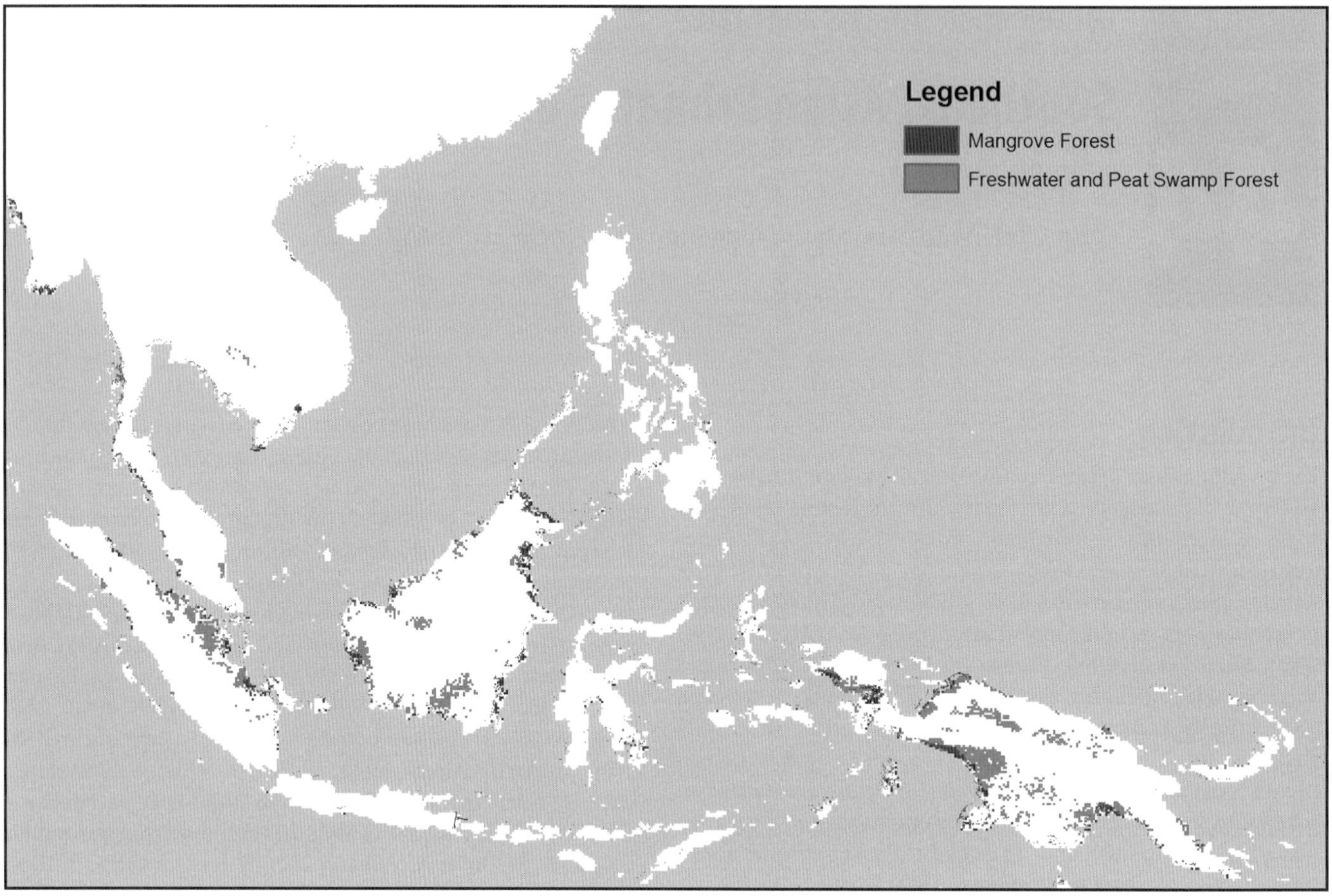

Figure 44.1 Map of Southeast Asia and flooded forests found in this region. From an amalgamation of maps from the following sources: Giri *et al.* (2011). Meittinem *et al.* (2011) and Stigbig *et al.* (2002, 2004). (A black and white version of this figure appears in some formats. For the colour version, please refer to the plate section.)

(20 695 t ha), Malaysia (2589 t ha), Myanmar (123 t ha), Brunei (91 t ha), the Philippines (65 t ha), Thailand (64 t ha), Vietnam (53 t ha), Laos (19 t ha), Cambodia (4.6 t ha) (APFP 2013).

Mangroves are woody vegetation types occurring in marine and brackish environments, generally restricted to the tidal zone characterized by depositional coastal environments (Giesen *et al.* 2006; Giri *et al.* 2011; Chapter 2). Mangroves occur throughout the Southeast Asian region; from the Irrawaddy delta in Myanmar in the northwest, through scattered islands of the Indonesian and Philippine archipelagos to the east (Giesen *et al.* 2006; Giri *et al.* 2011; Spalding *et al.* 1997). Mangrove forests in Southeast Asia represent about one-third of the world's total mangrove cover (FAO 2007). Estimates of current distribution and area coverage of mangrove differs between studies. We report here, the estimates given by FAO (2007) and Giri *et al.* (2011). The estimated areas of mangrove coverage are: Indonesia (3113 t ha), Myanmar (518 t ha), Malaysia (505 t ha), the Philippines (263 t ha), Thailand (240 t ha), Vietnam (157 t ha), Cambodia (69 t ha) and Brunei (18 t ha). Approximately 0.66 t ha of mangrove forests is found in Singapore (Yee *et al.* 2010).

Freshwater swamp forests are temporarily or permanently inundated by freshwater. If peat is present, it is often no more than a few centimetres thick compared to several metres thick in peat swamps (Yamada 1997). These forests can also be inundated by freshwater from river systems (Wyatt-Smith 1961). Freshwater swamps in Southeast Asia are mainly found in the Indochina peninsula, Thailand and Myanmar along the Mekong, Chaophrya and Irrawaddy Rivers (Yamada 1997). Compared to mangrove and peat swamp forest, little research has been conducted on freshwater swamp forests. There is no existing detailed information on the extent and status of freshwater swamp forests in Southeast Asia, apart from some areas where they have been specifically studied, for example, Singapore, where freshwater swamp forest covers approximately 0.087 t ha (Turner *et al.* 1996).

Riparian forests are defined as narrow strips of vegetation growing next to any shore, stream, river, pond, lake or wetland (Naimen *et al.* 2000; Vellidis & Lowrance 2004: Chapters 2 and 31). The spatial extent of the riparian zone is difficult to delineate, but generally encompasses the portion of the terrestrial landscape between the low and high water marks (Naimen & Décamps 1997; Naimen *et al.* 1993). Effects of past and ongoing riparian deforestation and pollution is typical of many Asian rivers, especially in Southeast Asia (Dudgeon 2000b; Iwata *et al.* 2003), and there is a lack of integrated knowledge on riparian forests, in part, due to the lack of primary research and scattered, highly fragmented, literature, some of which are not readily accessible. Riparian forests are not categorically classified in any existing forest classification schemes for Asia.

An Estimation of Flooded Forest Use by Southeast Asian Primates

Current knowledge on primate flooded forest use is generally sporadic in both published and grey literature, and largely based on opportunistic observations from species-specific population surveys or site based surveys of a wider range of animal taxa, rather than focused surveys of primates in flooded habitats. Some compiled information on primate flooded forest use is, however, available. For example, Gupta and Chivers (1999) reviewed the use of five types of forest habitat use by South and Southeast Asian primate species, including flooded forests. More comprehensive reviews of primate flooded forest use include Nowak (2013), which evaluated flooded forest use by primates and felids and how these taxa are threatened with habitat loss. This list was largely based on data contributed to the 'All the World's Primates' database (www.alltheworldsprimates.org), which includes information on the types of habitats primate species use, including peat swamp, mangrove and riverine forests. Another valuable resource is the Southeast Asian Mammal Databank (SAMD) (Catullo *et al.* 2008) which provides basic information on the suitability of different forest types for Southeast Asian mammals, with suitability classified as 'suitable', 'moderately suitable', 'unsuitable' or 'undefined' and assessed using available and published data on species–habitat relationships, elevation range and extent of occurrence. This databank provides information on primate species' use of freshwater swamp, peat swamp and mangrove forests.

Using these available data sources, we evaluated flooded forest use by 71 primate species found in Southeast Asian (Table 44.1). As the data sources included analytical methods for estimating flooded forest suitability for primate species, as well as direct sighting records of species occurrences in flooded forests, we considered a species to use flooded forest if there was at least one data source that included a record of flooded forest use or suitability. In the case of the SAMD databank, species were included when they were indicated to be 'suitable' or 'moderately suitable' for flooded forests. The results from our analysis showed that 49 species (69.0%) of Southeast Asian primates were classified in at least one category indicating the use of flooded forest habitats. Another four species listed in the data sources were suspected to use flooded habitats, but were unconfirmed, potentially bringing the total number of primate species to 53 (74.6%).

Twenty-three species (32.4%) were indicated by at least one source to use freshwater swamp, with another eight species suspected to use this forest type, potentially bringing the total number of primate species to 31 (43.7%). Twenty-four species (33.8%) use mangrove according to at least one source, with another five species suspected to use this forest type, potentially bringing the total number of mangrove-using primate species to 29 (40.9%). Twenty-nine species (40.8%) were indicated by at least one source to use peat swamp with another eight species suspected to use this forest type, potentially bringing the total number of primate species to 37 (51.1%). Twenty-six species (36.6%) were indicated by at least one source to utilize riverine forest.

At the family level, flooded forest use was highest for species belonging to the family Hominidae and Tarsiidae, with at least one flooded forest category indicated for both orangutan species and for all eight tarsier species. Between four to five (80–100%) species from family Loridae were indicated to use at least one flooded forest; between 29 and 32 (65.9% to 72.7%) species from family Cercopithecidae; and six species (50%) from family Hylobatidae. The main primate genera using peat swamp and mangrove forest were from the family Hominidae (1 genera, 2 species), Tarsiidae (1 genera, 7 species) and Loridae (1 genera, 4 species); for freshwater swamp, the dominant families were Hominidae and Loridae; and for riverine forest, Hominidae and Cercopithecidae. There were no reports of mangrove-using species from the family Hylobatidae.

Further analysis was conducted using Nexus data from Perelman *et al.* (2011) which analysed 34 927 bp of sequences from 54 nuclear genes and generated a maximum likelihood tree containing 186 primate taxa and outgroup taxa from Scandentia, Dermoptera and rooted by Lagomorpha. The Nexus file was analysed in MacClade (ver.4); the resultant tree contained 36 primate taxa which had information on habitat use; those taxa with no habitat use data were removed. Flooded habitat use was then mapped as a character onto the tree (Figure 44.2). From this preliminary analysis, it appeared that there was little phylogenetic link between flooded habitat use and major primate taxonomic groups.

We further analysed use of flooded forest by primate species was classified at the country level (Figure 44.3a). The proportion of primates that use flooded forests in each country was compiled from Table 44.1. The proportion of available flooded forest cover (derived from estimated cover of peat swamp and mangrove habitats divided by total land area, using information from the World Bank accessible at http://data.worldbank.org/indicator/AG.LND.TOTL.K2) was also compiled (Figure 44.3b). The proportion of primate species indicated for flooded forest was highest for Indonesia, Malaysia and Brunei, excluding sample sizes of less than five species (i.e. Singapore and the Philippines). There was a significant positive correlation between the proportion of primate species (adjusted values) that use flooded forests and the proportion of available flooded forest habitats (Spearman rank correlation test: $r_s = 0.804$; $n = 10$; $p = 0.005$). These results suggest that the availability of flooded forest habitat is likely to be a more important determinant of flooded forest use by primates rather than inherent phylogenetic characteristics of primate species.

Species Range as Indication of Potential Flooded Forest Use

Since the availability of flooded forest habitat could be an important determinant for use of flooded forests, the percentage of range overlap with flooded forest habitats for each species was examined in more detail. We overlaid IUCN range maps for each species with the flooded forest map in Figure 44.1 and calculated the percentage of overlap using Geographic Information System (ESRI ArcGIS™ 9.3, Environmental Systems Research Institute, USA 2005). We specifically

Table 44.1 List of Southeast Asian primate species and indicated flooded forest use.

Family/Species		Native range, SE Asia	IUCN Red List Status	Habitat				
Common name	**Scientific name**			**FWS**	**M**	**PSF**	**R**	**S**
Tarsiidae								
Western Tarsier	*Tarsius bancanus*	Brunei; Indonesia; Malaysia (Sabah, Sarawak), Indonesia (Sumatra and offshore islands	Vulnerable	SAMD	ATWP	SAMD	–	ATWP, SAMD
Dian's Tarsier	*Tarsius dentatus*	Indonesia (Sulawesi)	Vulnerable	–	SAMD	SAMD	ATWP	ATWP, SAMD
Lariang Tarsier	*Tarsius lariang*	Indonesia (Sulawesi)	Data Deficient	–	SAMD	SAMD	–	SAMD
Peleng Tarsier	*Tarsius pelengensis*	Indonesia (Sulawesi)	Endangered	–	SAMD	SAMD	–	SAMD
Pygmy Tarsier	*Tarsius pumilus*	Indonesia (Sulawesi – Possibly Extinct)	Data Deficient	–	SAMD	SAMD	–	SAMD
Sangihe Tarsier	*Tarsius sangirensis*	Indonesia (Sangihe island)	Endangered	–	SAMD	SAMD	ATWP	ATWP, SAMD
Philippine Tarsier	*Tarsius syrichta*	The Philippines	Near Threatened	SAMD	SAMD?	–	–	SAMD
Spectral Tarsier	*Tarsius tarsier*	Indonesia (Sulawesi)	Vulnerable	–	SAMD	SAMD	–	SAMD
Loridae								
Bengal Slow Loris	*Nycticebus bengalensis*	Cambodia; Laos; Myanmar; Thailand; Vietnam	Vulnerable	SAMD	SAMD	ATWP	–	ATWP, Nowak, SAMD
Greater Slow Loris	*Nycticebus coucang*	Indonesia (Sumatra); Peninsular Malaysia; Singapore; Thailand	Vulnerable	SAMD	ATWP	ATWP	–	ATWP, Nowak, SAMD
Javan Slow Loris	*Nycticebus javanicus*	Indonesia (Java)	Endangered	SAMD?	SAMD?	SAMD?	–	SAMD?
Bornean Slow Loris	*Nycticebus menagensis*	Brunei; Indonesia; Malaysia (Sabah, Sarawak); some islands off the Philippines and Sumatra	Vulnerable	SAMD?	ATWP, SAMD?	ATWP, SAMD?	–	ATWP, Nowak, SAMD?
Pygmy Slow Loris	*Nycticebus pygmaeus*	Cambodia; Laos; Vietnam	Vulnerable	SAMD	SAMD	SAMD	–	SAMD
Cercopithecidae								
Stump-tailed Macaque	*Macaca arctoides*	Cambodia; Laos; Peninsular Malaysia; Myanmar; Thailand; Vietnam	Vulnerable	–	–	–	–	–
Assamese Macaque	*Macaca assamensis*	Laos; Myanmar; Thailand; Vietnam	Near Threatened	–	–	–	ATWP	–
Long-tailed Macaque	*Macaca fascicularis*	Brunei; Cambodia; Laos; Malaysia; Myanmar; The Philippines; Singapore; Thailand; Vietnam; Indonesia	Least Concern	ATWP	SAMD, ATWP	ATWP	SAMD, ATWP	ATWP, G&C, Nowak, SAMD
Heck's Macaque	*Macaca hecki*	Indonesia (Sulawesi)	Vulnerable	–	–	–	ATWP	ATWP
Northern Pig–tailed Macaque	*Macaca leonina*	Cambodia; Laos; Myanmar; Thailand; Vietnam	Vulnerable	–	–	SAMD	ATWP	ATWP
Moor Macaque	*Macaca maura*	Indonesia (Sulawesi)	Endangered	–	ATWP, SAMD	–	–	ATWP,SAMD
Rhesus Macaque	*Macaca mulatta*	Laos; Myanmar; Thailand; Vietnam	Least Concern	SAMD	SAMD, ATWP	SAMD	–	ATWP, G&C, Nowak, SAMD

Southern Pig-tailed Macaque	*Macaca nemestrina*	Brunei; Indonesia (Kalimantan); Malaysia (Peninsular Malaysia, Sabah); Thailand	Vulnerable	SAMD	–	SAMD		G&C, Nowak, SAMD
Celebes Crested Macaque	*Macaca nigra*	Indonesia (Sulawesi)	Critically Endangered	–	–	–	–	–
Gorontalo Macaque	*Macaca nigrescens*	Indonesia (Sulawesi)	Vulnerable	–	–	–	–	–
Booted Macaque	*Macaca ochreata*	Indonesia (Sulawesi)	Vulnerable	–	SAMD	–	ATWP	ATWP, SAMD
Pagai Island Macaque	*Macaca pagensis*	Indonesia (Mentawai Islands)	Critically Endangered	SAMD?	SAMD?	SAMD?	ATWP	ATWP, Nowak, SAMD?
Siberut Macaque	*Macaca siberu*	Indonesia (Siberut)	Vulnerable	SAMD?	ATWP	SAMD?	ATWP	ATWP, Nowak, SAMD?
Tonkean Macaque	*Macaca tonkeana*	Indonesia (Sulawesi)	Vulnerable	SAMD?	SAMD?	SAMD?	ATWP	ATWP, SAMD?
Proboscis Monkey	*Nasalis larvatus*	Brunei; Indonesia (Kalimantan); Malaysia (Sabah, Sarawak)	Endangered	SAMD	ATWP, SAMD	ATWP, SAMD	ATWP	ATWP, G&C, Nowak, SAMD
Bornean Banded Leaf Monkey	*Presbytis chrysomelas*	Brunei; Indonesia (Kalimantan); Malaysia (Sarawak)	Critically Endangered	SAMD	ATWP	SAMD, ATWP	ATWP	ATWP, Nowak, SAMD
Javan Leaf Monkey	*Presbytis comata*	Indonesia (Java)	Endangered	–	–	–	ATWP	ATWP
Banded Leaf Monkey	*Presbytis femoralis*	Indonesia (Sumatra); Peninsular Malaysia; Myanmar; Singapore; Thailand	Near Threatened	SAMD	SAMD	SAMD	–	ATWP, G&C, Nowak, SAMD
White-fronted Leaf Monkey	*Presbytis frontata*	Indonesia (Kalimantan); Malaysia (Sarawak)	Vulnerable	SAMD?	–	SAMD?	ATWP	ATWP, SAMD?
Hose's Leaf Monkey	*Presbytis hosei*	Brunei; Indonesia (Kalimantan); Malaysia (Sabah)	Vulnerable	SAMD?	–	–	ATWP	ATWP, Nowak, SAMD?
Mitred Leaf Monkey	*Presbytis melalophos*	Indonesia (Sumatra)	Endangered	SAMD	–	–	ATWP	ATWP, Nowak, SAMD
Natuna Leaf Monkey	*Presbytis natunae*	Indonesia (Bunguran Island)	Vulnerable	–	–	–	ATWP	ATWP
Mentawai Leaf Monkey	*Presbytis potenziani*	Indonesia (Mentawai Islands)	Endangered	SAMD	–	SAMD	–	ATWP, G&C, Nowak, SAMD
Maroon Leaf Monkey	*Presbytis rubicunda*	Indonesia (Kalimantan); Malaysia (Sabah, Sarawak)	Least Concern	SAMD	–	ATWP, SAMD	ATWP	ATWP, Nowak, SAMD
Thomas's Leaf Monkey	*Presbytis thomasi*	Indonesia (Sumatra)	Vulnerable	SAMD	–	ATWP, SAMD	ATWP	ATWP,Nowak, SAMD
White-thighed Leaf Monkey	*Presbytis siamensis*	Sumatra, Peninsular Malaysia, Thailand	Near Threatened	–	–	–	–	ATWP
Grey-shanked Douc	*Pygathrix cinerea*	Vietnam	Critically Endangered	–	–	–	–	–
Red-shanked Douc	*Pygathrix nemaeus*	Cambodia; Laos; Vietnam	Endangered	–	–	–	ATWP	ATWP
Black-shanked Douc	*Pygathrix nigripes*	Cambodia; Vietnam	Endangered	–	–	–	–	–
Tonkin Snub-nosed Monkey	*Rhinopithecus avunculus*	Vietnam	Critically Endangered	–	–	–	–	–
Pig-tailed Snub-nosed Monkey	*Simias concolor*	Indonesia (Mentawai Islands)	Critically Endangered	SAMD	–	SAMD	–	ATWP, G&C, Nowak, SAMD

(*continued*)

Table 44.1 *(cont.)*

Family/Species		Native range, SE Asia	IUCN Red List Status	Habitat				
Common name	**Scientific name**			**FWS**	**M**	**PSF**	**R**	**S**
Ebony Leaf Monkey	*Trachypithecus auratus*	Indonesia (Java, Bali, Lombok)	Vulnerable	–	–	–	–	Nowak
Barbe's Langur	*Trachypithecus barbei*	Myanmar; Thailand	Data Deficient	–	–	SAMD?	–	SAMD?
Silvered Leaf Monkey	*Trachypithecus cristatus*	Brunei; Indonesia; P. Malaysia	Near Threatened	SAMD	ATWP, SAMD	SAMD	ATWP	ATWP, G&C, Nowak, SAMD
Delacour's Langur	*Trachypithecus delacouri*	Vietnam	Critically Endangered	–	–	–	–	–
Francois's Langur	*Trachypithecus francoisi*	Vietnam	Endangered	–	–	–	–	–
Indochinese Silvered Langur	*Trachypithecus germaini*	Cambodia; Laos; Myanmar; Thailand; Vietnam	Endangered	SAMD	SAMD	SAMD	–	SAMD
Hatinh Langur	*Trachypithecus hatinhensis*	Laos; Vietnam	Endangered	–	–	–	–	–
Laotian Black Langur	*Trachypithecus laotum*	Laos	Vulnerable	SAMD?	–	–	–	SAMD?
Spectacled Leaf Monkey	*Trachypithecus obscurus*	Peninsular Malaysia; Myanmar; Thailand	Near Threatened	SAMD	SAMD	SAMD	–	ATWP, G&C, SAMD
Phayre's Leaf Monkey	*Trachypithecus phayrei*	Laos; Myanmar; Thailand; Vietnam	Endangered	–	SAMD?	SAMD?	–	SAMD?
Capped Leaf Monkey	*Trachypithecus pileatus*	Myanmar	Vulnerable	–	–	–	–	–
Cat Ba Langur	*Trachypithecus poliocephalus*	Vietnam	Critically Endangered	–	–	–	–	–
Shortridge's Langur	*Trachypithecus shortridgei*	Myanmar	Endangered	SAMD?	SAMD?	SAMD?	ATWP	ATWP, SAMD?
Hylobatidae								
Western Hoolock Gibbon	*Hoolock hoolock*	Myanmar	Endangered	–	–	–	–	–
Bornean Agile Gibbon	*Hylobates albibarbis*	Indonesia (Kalimantan)	Endangered	SAMD	–	SAMD	–	ATWP, Nowak, SAMD
Kloss's Gibbon	*Hylobates klossii*	Indonesia (Sumatra)	Endangered	–	–	–	–	Nowak
Lar Gibbon	*Hylobates lar*	Indonesia (Sumatra); Laos; Peninsular Malaysia; Myanmar; Thailand	Endangered	–	–	–	–	–
Moloch Gibbon	*Hylobates moloch*	Indonesia (Java)	Endangered	–	–	–	ATWP	ATWP
Muller's Bornean Gibbon	*Hylobates muelleri*	Brunei; Indonesia (Kalimantan); Malaysia (Sabah)	Endangered	SAMD	–	ATWP, SAMD	–	ATWP, Nowak, SAMD
Pileated Gibbon	*Hylobates pileatus*	Cambodia; Lao; Thailand	Endangered	–	–	–	–	–
Black Crested Gibbon	*Nomascus concolor*	Laos; Vietnam	Critically Endangered	–	–	–	–	–

Red-cheeked Gibbon	*Nomascus gabriellae*	Cambodia; Laos; Vietnam	Endangered	–	–	–	ATWP	ATWP
Northern white-cheeked Gibbon	*Nomascus leucogenys*	Laos; Vietnam	Critically Endangered	–	–	–	–	–
Southern white-cheeked Gibbon	*Nomascus siki*	Laos; Vietnam	Endangered	–	–	–	–	–
Agile Gibbon	*Hylobates agilis*	Indonesia (Sumatra); Peninsular Malaysia, Thailand	Endangered	–	–	–	–	ATWP, G&C, Nowak
Hominidae								
Sumatran Orangutan	*Pongo abelii*	Indonesia (Sumatra)	Critically Endangered	SAMD	SAMD	ATWP, SAMD	ATWP	ATWP, G&C, Nowak, SAMD
Bornean Orangutan	*Pongo pygmaeus*	Indonesia (Kalimantan); Malaysia (Sarawak)	Endangered	SAMD	ATWP	ATWP, SAMD	ATWP	ATWP, G&C, Nowak, SAMD

Habitat – FWS = freshwater swamp forest; M = mangrove; PSF = peat swamp forest; R = riparian forest; S = any one swamp forest type.
ATWP = All the World's Primates http://www.alltheworldsprimates.org/Home.aspx; G&C – Gupta & Chivers (1999); Nowak – Nowak (2013); SAMD = Southeast Asian Mammal Databank www.ieaitaly.org/samd.
A "?" affixed to the SAMD source indicates that forest type suitability is unverified.

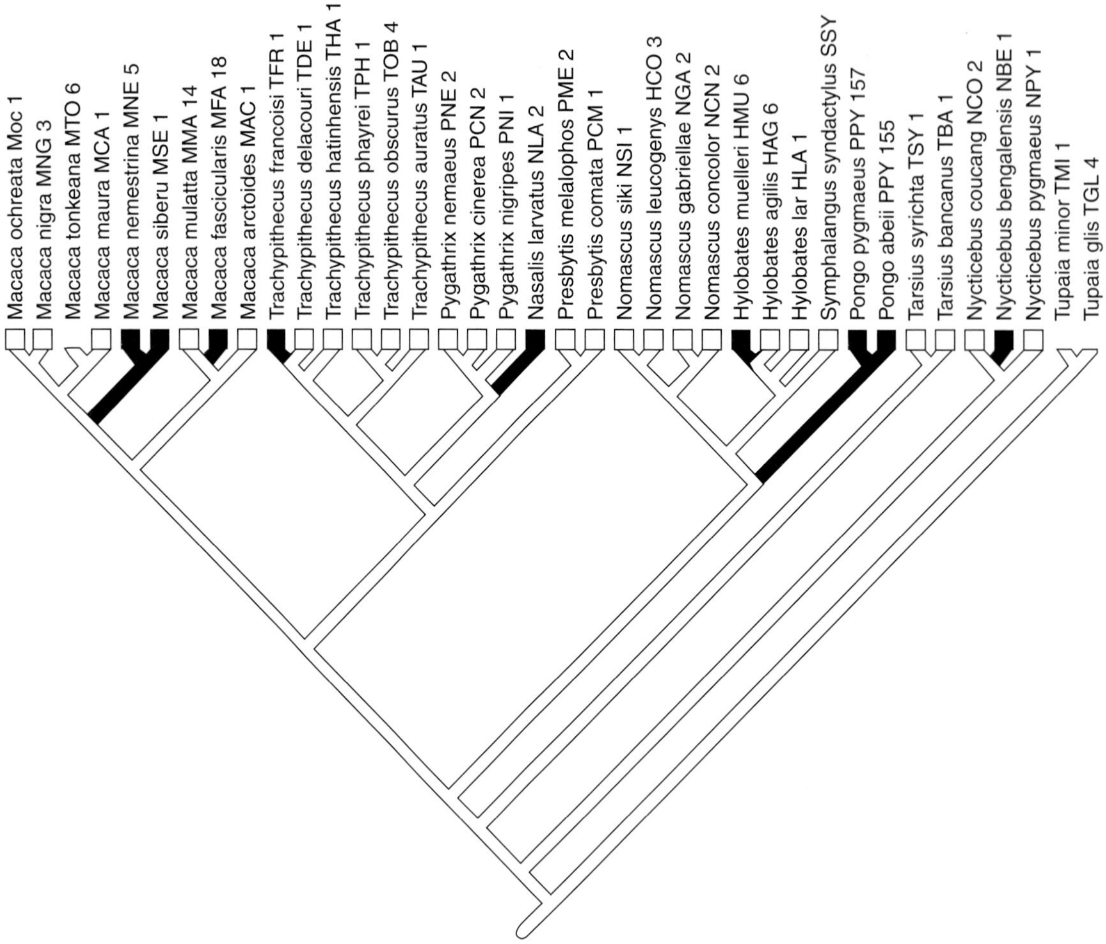

Figure 44.2 Mapping flooded forest use onto a primate phylogenetic tree. Black boxes indicate species which are found in flooded habitat. Nexus data was obtained from the supplementary information from Perelman *et al.* (2011).

identified: (1) species not indicated to use flooded forests, but with existing overlaps with these habitats (Table 44.2a); and (2) species indicated to use flooded forest but with no existing overlaps with these habitats (Table 44.2b). From this analysis, 24% (17 species) that were either not indicated to use flooded forests or not confirmed to do so in published or grey literature, showed overlapping ranges with these forest types within their distributional ranges (between 0.2% and 20.8%). The potential use of flooded forests by these primate species should be further investigated, especially for species that could potentially shift their habitat niches within geographical range limitations, due to habitat loss or other human disturbances (see Chapter 39, plus Chapter 8 for Neotropical examples). Judging from available flooded forest habitats, 10% (7 species) that are known to utilize flooded forest may have been extirpated due to more recent flooded forest habitat loss.

Progressive Knowledge on Primate Flooded Forest Use

In 1999, Gupta and Chivers indicated 12 species of Southeast Asian species were known to utilize flooded forests. Catullo *et al.* (2008) increased this estimate almost three-fold, to 32 species. However, flooded forests use was still indicated as unverified for another 12 species. Some of these species were subsequently identified to use flooded forests by Nowak (2013) and other sources. For example, *Macaca siberu* were found in higher densities in peat swamp forest than in lowland forest (Quinten *et al.* 2009). The same study also verified the presence of *Hylobates klossi* in flooded peat forests in Sumatra, a habitat in which it had neither been previously recorded nor was considered suitable. Blackham (2005) and Nekaris *et al. (*2008) also verified the presence of *N. menagensis* in peat swamps in Sebangau National Park in Kalimantan, albeit at extremely low densities (0.21–0.38 individuals/km). The suitability for and use of flooded forests by this species was previously indicated as unknown. The 'All the World's Primates' database also included some species that may potentially use flooded forest: *Macaca pagensis*, *M. tonkeana*, *Presbytis frontata*, *P. hosei* and *Trachypithecus shortridgei*, all of which were previously indicated by Catullo *et al.* (2008) as to be of unknown suitability for flooded forests. Despite this progressive increase in knowledge of primates' use of flooded forest, the potential use of flooded forest use by several species still requires confirmation, including *T. laotum*, *T. barbei*, *T. phayrei* and *N. javanicus*, some of which have overlapping distribution ranges with flooded forests but are currently recorded as unknown for use of or suitability for these habitats.

Further clarification is also required for species that were indicated to use flooded forests but due to more recent habitat

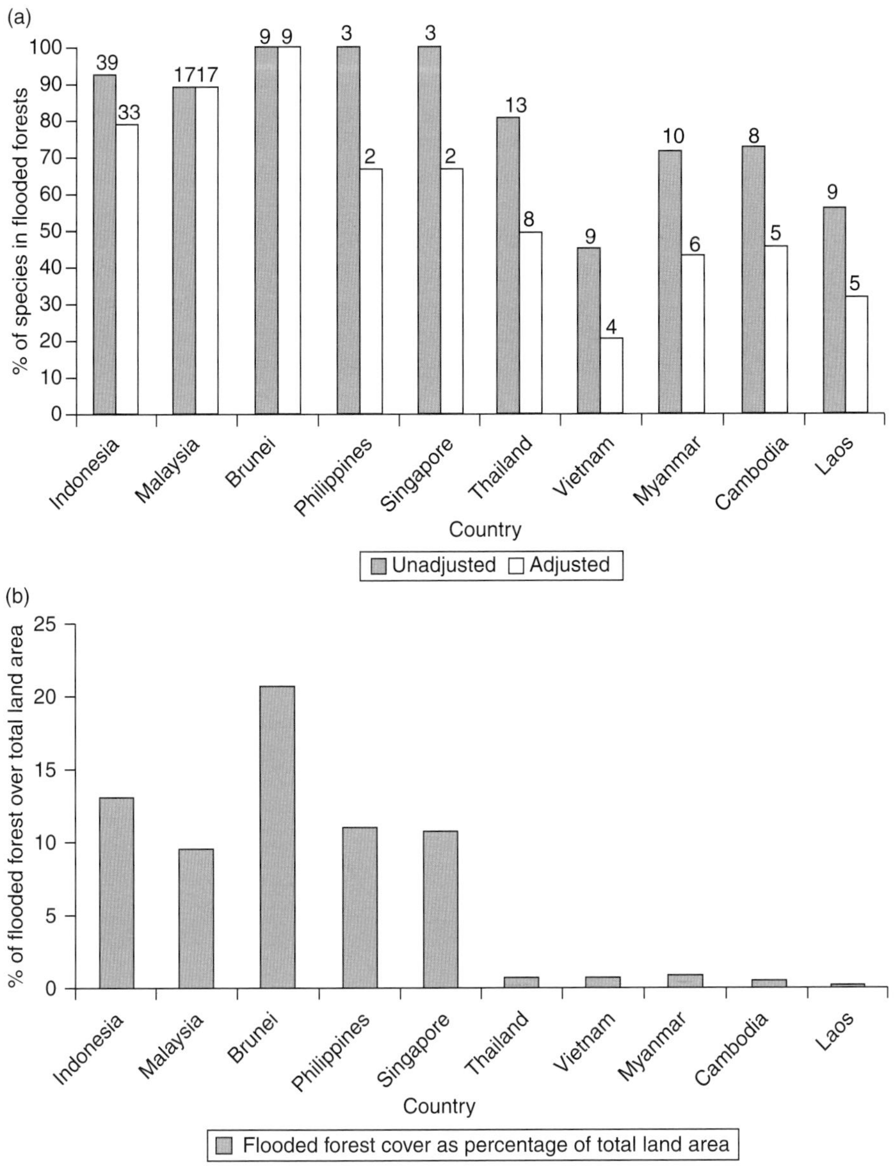

Figure 44.3 (a) Percentage of species in flooded forest by country. For unadjusted figures, flooded forest use for species was assumed across countries; for adjusted figures, species were only included if there were actual range overlaps with flooded forests within the range countries. The number of species is given at the top of each column. (b) Percentage of flooded forests over total land area, classified by country.

loss, may be extirpated from such habitats within their distributional ranges (see Table 44.2b). From these results, it appears that flooded forest use by Southeast Asian primate species is relatively widespread. Considering the progressive knowledge gained on this topic over the past 15 years, there is also realistic potential for flooded forest use by other species of primates currently not indicated to use flooded forests.

Limitations to Current Knowledge

Difficulties in Estimating Flooded Forest Occupancy

Apart from species that specialize in using flooded habitat, such as the proboscis monkey, it is difficult to determine whether flooded habitat use by different primate species is due to historical occupancy or seasonal and/or temporary adaptations to habitat changes and food resource availability. Nowak (2013) attempted to make a distinction between such differences in habitat occupancy by classifying primate species according to historical occupancy (10 species), potential recent niche shifts in response to habitat disturbance (3 species), temporary or seasonal extension of range for feeding or shelter (3 species). However, 10 species were unclassified, and the above occupancy types were 'best guesses' based on available data.

Difficulties in determining the nature of flooded occupancy highlights problems associated with determining habitat use and suitability for many species, particularly for those that naturally occur at low densities. An example comes from the various *Nycticebus* spp., for which low densities and lack of historical range data, means that it is not likely, even with significant survey effort, to determine with certainty whether certain habitats were historically occupied and preferred over other habitat types.

Underestimation of Flooded Forests Extent

An analysis of available information on primate use of flooded forest is limited, partly due to the unavailability of

Table 44.2a Species with potential flooded forest use and their location

Species with potential flooded forest use	Location
Macaca arctoides	Indo-Burma
Macaca assamensis	Indo-Burma
Macaca hecki	Sulawesi
Macaca nigra	Sulawesi
Macaca nigrescens	Sulawesi
Macaca tonkeana	Sulawesi
Presbytis frontata	Borneo
Presbytis natunae	Bunguran Island
Trachypithecus barbei	Indo-Burma
Presbytis siamensis	P. Malaysia/Sumatra
Trachypithecus francoisi	Indo-Burma
Trachypithecus phayrei	Indo-Burma
Trachypithecus poliocephalus	Indo-Burma
Hoolock hoolock	Indo-Burma
Hylobates lar	Indo-Burma/P. Malaysia/Sumatra
Hylobates moloch	Java
Hylobates pileatus	Indo-Burma

Table 44.2b Species potentially extirpated from flooded forests and their location.

Species potentially extirpated from flooded forests	Location
Macaca maura	Sulawesi
Presbytis comata	Java
Trachypithecus laotum	Laos
Trachypithecus shortridgei	Indo-Burma
Nycticebus javanicus	Java
Tarsius lariang	Sulawesi
Tarsius pumilus	Sulawesi

comprehensive habitat maps for flooded forests and because, even where maps are available, discrepancies in mapping are evident. There are currently no maps available that detail all flooded habitat types within Southeast Asia. Stigbig *et al.* (2002) provided a SPOT-Vegetation satellite image of coarse spatial resolution (1 km) generated from satellite images acquired for the period 1998 to 2000 for insular Southeast Asia. The relevant flooded forest categories available from this map were mangrove forests and flooded forests (combining both freshwater and peat flooded forests). Stigbig *et al.* (2004) provided a similarly-derived vegetation map for continental Southeast Asia, from which the same flooded forest categories were derived. Meittinem *et al.* (2011) derived forest cover from Moderate Resolution Imaging Spectroradiometer images in 2010. This map included flooded classifications for mangrove and peat flooded forests, but not for freshwater flooded forests. Giri *et al.* (2011) provided an updated map on mangrove forests, but this map used the GLC 2000 map classifications and updated with Landsat imagery from the USGS and other secondary sources. The lack of a comprehensive and consistent habitat maps for flooded forests in Southeast Asia is thus a strong limiting factor for a more comprehensive analysis of primate flooded forest overlap and utilization. Due to the general inaccessibility of these habitats, it is also difficult to ground-truth estimates from satellite imagery analyses.

This problem is exacerbated by the exclusion of riverine habitats in current mappings of flooded habitats. Riverine forests are dipterocarp forests along rivers which, in terms of species composition, are distinct from those in the hinterland (Whitmore 1984). Densities of proboscis monkeys have been found to be comparatively higher along riverine forests compared to mangrove forests, until recently regarded as the main habitat of proboscis monkeys (Sha *et al.* 2008). Primate censuses conducted along riverine systems also yielded many sightings of other primate species, some at relatively high densities, e.g. the extensive distribution of species like the long-tailed macaque (*Macaca fascicularis*), pig-tailed macaque (*Macaca nemestrina*), silver-leaf monkey (*Presbytis cristata*) and orangutan (*Pongo pygmaeus*) in Sabah and Kalimantan (Matsuda *et al.* 2011; A. Setiawan, pers. comm., 2013; S. Lhota, pers. comm., 2013; pers. obs., 2004–2006). Species of gibbons which are not known to use flooded forests can also be sighted occasionally in these habitats (e.g. *Hylobates muelleri* in Sabah and *Hylobates agilis* in Kalimantan). 'All the World's Primates' recorded 26 species of primates in riverine forests. Thirteen of these species were not recorded in other flooded forest types. The extent of riverine systems is difficult to map partly due to a lack of established standards for defining riverine habitats across countries and regions. In the absence of riparian forest habitat maps, it is difficult to consider the actual importance of this habitat for many primate species. However, it is likely that our current view of primate use of flooded forest, and of the overall use of such habitats by primates, should be considered to be highly underestimated.

Impact of Flooded Forest Loss

Ninety per cent of all primate species are found in tropical regions and depend on forests that are disappearing rapidly (Mittermeier & Cheney 1987), with range habitat countries losing 125 140 km^2 of forest annually (Chapman & Peres 2001). Flooded forests, like peat swamps and mangroves, face many threats from conversion to oil palm, rice agriculture, shrimp farming, salt production, rice fields and other resource harvesting (Posa *et al.* 2011; Valiela *et al.* 2001). Between 2000 and 2010, Southeast Asian loss of mangrove forests was estimated at rates of 12.5%, and peat flooded forests at even higher rates of 19.7%: both figures being higher than for lowland dipterocarp forests (11.1%) (Miettinen *et al.* 2011). Rates of peat flooded forest loss was highest in Sumatra (41.3%)

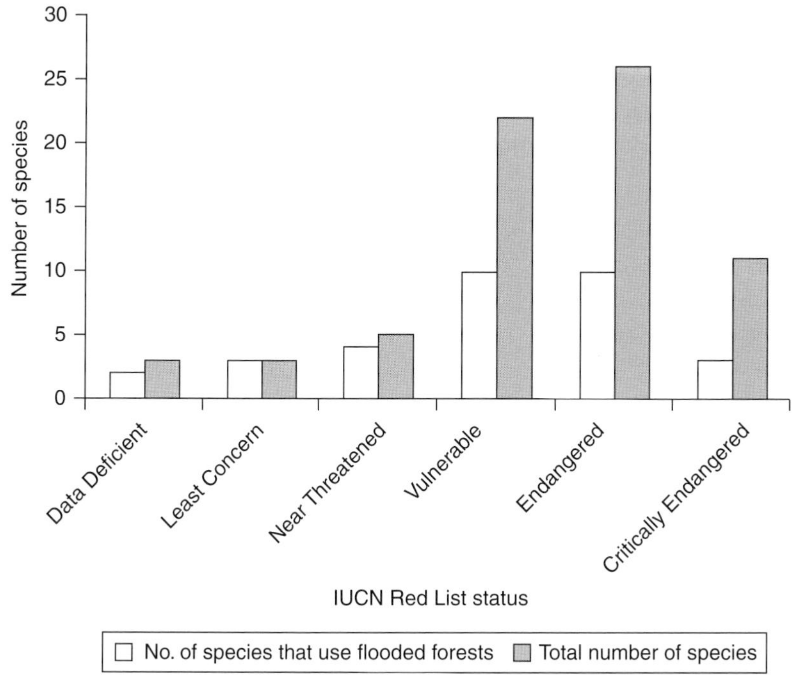

Figure 44.4 IUCN Red List status of primates that use flooded forests.

and Borneo (21.8%); and for mangrove forests, Indonesia (7.9%) and Malaysia (4.2%). Some of these areas hold large numbers of threatened primate species that use flooded and mangrove habitats, Borneo (87.5%, 7/8 species) and Sumatra (44.6%, 5/11 species). Overall, 23 of 32 species of Asian primates that use flooded forest (71.9%) fall under the threatened category ('Vulnerable', 'Endangered' or 'Critically Endangered') (Figure 44.4). Although this only contributes to 39.7% (23 of 58) species of total threatened primate species in Southeast Asia, the rate of loss of flooded forests is a worrying trend for these species, especially for those with specialized adaptations to these habitats.

In Sabah, Borneo, the largest population of proboscis monkeys is found in Kinabatangan, where about 25% of the total population survives in *c.* 0.7% of total forested, flooded habitats. These habitats are further fragmented by cultivated land and human settlements with other small populations trapped in small fragments of mangrove forests along the coast: there is evidence of local extinction when these habitats are cleared (Sha *et al.* 2008). The banded leaf monkey is considered locally Critically Endangered in Singapore, with a small population of approximately 40 individuals persisting in only 5 km^2 of secondary forests, including a small proportion of preferred secondary freshwater flooded forest habitat (Ang *et al.* 2012; Lucas *et al.* 1988). Twenty per cent of the banded leaf monkey's food source is from these flooded forests and all are locally threatened, with the exception of one species (Ang 2011). With rapid loss of flooded forest habitats, the long-term survival of many flooded forest specialists is expected to be severely impacted.

For primate species that are not flooded forest specialists, the impact of habitat loss appears less critical. However, many non-specialist species have been shown to exhibit flexible adaptations (Nowak & Lee 2013), and flooded forests provide suitable habitats for nearly half of all primate species in Southeast Asia, with the relative importance of these habitats for many species yet to be firmly established. For many other species of primates, increasing loss and fragmentation of lowland dipterocarp forests, riparian and coastal flooded forests adjacent to their original habitats may become increasingly critical to their survival. The ability to exploit mangroves and other flooded forests for refuge may become important to the population persistence of many threatened species (Nowak 2012; Chapters 39, 43 and 8 provide comparable African and Neotropical examples, respectively). Furthermore, some recent records of primates in mangrove forest, have been attributed to shifts in habitat use due to forest habitat loss. Lhota *et al.* (Chapter 42) noted that many species of primates were sighted in mangrove forests in Balikpapan Bay, East Kalimantan following conversion of adjacent forest to palm oil plantations. These mangrove patches are the last remaining patches of forests remaining in the region, and several species of primates, hitherto not known to use these habitats, may now be forced to adapt to them, as a consequence of the rapidly changing conditions. Nowak (2012) also noted that mangroves, compared with peat flooded forest, appear to be a neglected habitat for primate studies. This may be because few primates live solely in mangroves, which are sometimes part of their home range with the remaining part consisting of more species-rich lowland forest. It is thus expected that with increased survey effort and collation of previously unpublished records, primate flooded forest use could be better estimated. In addition, increasing observations of high densities of species of primates along riparian systems, even for species that are not known to be flooded forest specialists, indicate that seasonal food resources in these habitats or adaptations to utilization of these habitats may become increasingly important, and even vital. With mounting habitat loss across the Southeast Asian region, flooded forest may, in some areas, be the most significant remaining habitats for threatened species due to their inaccessible and uninhabitable conditions for human exploitation (Nowak 2012).

Flooded forest habitats are highly endangered by human activities, yet the protection of these forests is generally lacking. For example, peat land covers a vast area (*c.* 6 million hectares) of the lowlands of Kalimantan (Rieley *et al.* 1996), but only a very small proportion (< 3%) of this habitat is protected within national parks (MacKinnon & MacKinnon 1991), and the area of undisturbed flooded peat forest is declining rapidly. Existing flooded forest habitats in Kalimantan are also increasingly disturbed due to human activities, with evidence of large areas of mixed flooded forest being heavily logged (Morrogh-Bernard *et al.* 2003). Similarly, in Sabah, East Malaysia, loss of habitat due to expansion of human settlements is most marked in the coastal mangrove areas. Habitat fragmentation and degradation due to logging and conversion of important riparian habitats to agriculture/aquaculture is also highly evident along major rivers systems (Sha *et al.* 2008).

Conclusion

Flooded forest habitats are highly endangered by human activities and the importance of such habitats for primates needs to be better understood and reconsidered for holistic and effective conservation efforts. It is clear from this review that existing knowledge of this topic is still inadequate for assessing the importance of flooded forest habitats for primates and how loss of such habitats could impact their immediate and long-term survival prospects. The following steps are recommended:

1. Systematic studies of flooded habitat use by Southeast Asian primates.
2. Improved mapping of flooded habitats across the region.
3. Collation of more evidence of primate flooded forest use from unpublished information.
4. Setting up a specialist working group for sharing and updating information on primate use of flooded forest habitats.

Acknowledgements

We thank the organizers of the symposium on 'Primates in Flooded Forest' at the XXIV International Primatological Congress, Mexico, for the opportunity to participate in this important initiative. We wish to thank Coleen Goh for her valuable assistance in compiling datasets on primate use of flooded forests. John Sha would like to thank the Wildlife Reserves Singapore and the Primate Research Institute, Kyoto University for their support; Shun Deng Fam is grateful for the support from the Prime Minister's Australia Asia Postgraduate Awards and the Australian National University. Andie Ang would like to thank Professor Herbert Covert of the University of Colorado Boulder and Professor Rudolf Meier of the National University of Singapore.

Chapter 45

Conservation of Primates and Their Flooded Habitats in the Neotropics

Sarah A. Boyle, Cleber J.R. Alho, Janice Chism, Thomas R. Defler, Anthony Di Fiore, Eduardo Fernandez-Duque, Erwin Palacios, Ricardo Rodrigues dos Santos, Christopher A. Shaffer, Claudia Regina da Silva, Bernardo Urbani, Robert Wallace, Barth Wright, Kristin Wright, Bruno de Freitas Xavier and Adrian A. Barnett

Introduction

Wetlands are widely distributed worldwide, but 30–90% of wetlands have been modified or destroyed, depending on the geographic region (Junk *et al.* 2013). Globally, wetlands in both tropical and temperate regions do not have extensive protection (only 14.2% of tropical mangroves and 6.9% of tropical freshwater swamp are protected; Schmitt *et al.* 2009), yet wetlands are very important for ecosystem services such as flood control, aquifer recharge, water filtration and carbon storage, as well as resources for food and medicines (Junk 2002; Keddy *et al.* 2009). Much of the disturbance to these ecosystems has been the result of habitat destruction and fragmentation by agriculture, oil and gas exploration, hydroelectric power and urban development (Junk 2002), as well as environmental contamination (Lewis *et al.* 2011), fire (Alho *et al.* 2011a) and climate change (Alho & Silva 2012; Desbiez *et al.* 2010a; Erwin 2009). As a result of wetland modification and destruction, more than one-third of wetland-dependent species of waterbirds, mammals, fish, amphibians, turtles and crocodiles are of conservation concern (Millennium Ecosystem Assessment 2005).

Within the Neotropics, wetlands are distributed from Mexico to southern South America (Figure 45.1), but the extent of the wetlands and the conservation pressures impinging on them vary greatly (Junk 2013). Two of the ten largest wetlands in the world are found in the Neotropics: in the Amazon River Basin and Pantanal (Fraser & Keddy 2005). As of September 2014, 302 wetland sites in Mexico, Central America and South America, covering 48 173 863 ha, were listed as 'wetlands of international importance' by Ramsar (www.ramsar.org; Table 45.1). However, the extent of data on tropical wetlands is limited (Junk 2002; Junk *et al.* 2012), as is the agreement among multiple data sets as to the current extent of wetlands (Rebelo *et al.* 2009).

Within the Neotropics there are approximately 152 primate species, representing 17–20 genera ranging from Mexico to Argentina (Perelman *et al.* 2011; Rylands *et al.* 2012). Many of these species inhabit a variety of habitats, including flooded ones. It is estimated that 33% of Neotropical primate species use riparian forest and 21% use swamp forests (Lehman & Fleagle 2006). Through an extensive literature search and compilation of unpublished data from the contributing authors, this chapter provides (1) a synthesis of research findings regarding the use of flooded habitats by primates in the Neotropics; and (2) the current conservation concerns facing both the habitats and the primates in the Neotropics. We categorized the primates' use of flooded habitats based on the habitats descriptions provided by Bennett *et al.* (Chapter 2).

Neotropical Primates in Flooded Forests

In our review of the literature and of unpublished data, we found records of 72 species of neotropical primates from 17 genera that use flooded habitats to some extent (Table 45.2). Genera such as *Alouatta*, *Aotus*, *Cacajao*, *Cebus*, *Saguinus*, *Saimiri* and *Sapajus* were most commonly noted to use flooded habitats, while genera such as *Ateles*, *Pithecia* and *Chiropotes* appear to use flooded habitats to a lesser extent. We found no records of *Brachyteles*, *Callibella* or *Oreonax* using flooded habitats. The extent of published literature on Neotropical primates in flooded habitats varied greatly, with Brazil, Bolivia, Peru and Colombia having the greatest number of published studies documented by our review. Very little published literature is available regarding primates in the flooded habitats of Belize, Guatemala, Paraguay and Argentina (Table 45.2).

The following sections examine the use of flooded habitats by Neotropical primates and the conservation concerns relating to these habitats in Mesoamerica and South America, on a country-by-country basis. The countries are listed in alphabetical order.

Mesoamerica

Mesoamerica is the area from southern Mexico through Panama (Estrada *et al.* 2006). Most of the flooded habitat in Mesoamerica consists of mangroves, palm swamps, *petenes* (forest surrounded by seasonally flooded grassland; Rico-Gray 1982), and riverine forests (Ellison 2004). The majority of these wetlands are located along coastlines (Mitsch & Hernandez

Table 45.1 Mesoamerican and South American Ramsar 'wetlands of international importance' as of September 2014.

	Country	Number of sites	Total area (ha) of sites
Mesoamerica			
	Belize	2	23 592
	Costa Rica	12	569 742
	El Salvador	7	207 387
	Guatemala	7	628 592
	Honduras	9	270 224
	Mexico	142	8 833 752
	Nicaragua	9	406 852
	Panama	5	183 992
	Total	*193*	*11 124 133*
South America			
	Argentina	21	5 382 521
	Bolivia	11	14 842 405
	Brazil	12	7 225 687
	Chile	12	358 989
	Colombia	5	458 525
	Ecuador	18	286 651
	French Guiana	3	224 400
	Guyana	0	0
	Paraguay	6	785 970
	Peru	13	6 784 042
	Suriname	1	12 000
	Uruguay	2	424 904
	Venezuela	5	263 636
	Total	*109*	*37 049 730*
	Grand total	*302*	*48 173 863*

Source: www.ramsar.org

2013). Mesoamerica is a biodiversity hotspot, yet only 20% of the original vegetation remains (Myers *et al* 2000), primarily due to deforestation and conversion of natural vegetation for agriculture (Ellison 2004; Estrada *et al.* 2006; Harvey *et al.* 2008). Mesoamerican wetlands are under extreme pressure because of high human population densities and limited freshwater resources (Ellison 2004; Junk 2002). As of September 2014, there were 193 Ramsar sites in Mesoamerica, covering 11 124 133 ha (Table 45.1). Although the number of Ramsar sites in Mesoamerica is greater than the number of sites in South America, the total area represented by the Ramsar sites in Mesoamerica is one-third smaller than the Ramsar sites' area in South America (Table 45.1).

There are six genera of primates in Mesoamerica: *Alouatta*, *Aotus*, *Ateles*, *Cebus*, *Saguinus* and *Saimiri*. In our review, we found documentation of *Alouatta*, *Aotus*, *Ateles* and *Cebus* using flooded habitats in Mesoamerica (Table 45.2). We did not find evidence of *Saguinus* and *Saimiri* using flooded habitats in Mesoamerica, but we found these two genera used flooded habitats in South America (Table 45.2). Therefore, it is possible that *Saguinus* and *Saimiri* use flooded habitats in Mesoamerica, but such use has not been documented. Of the 7–9 primate species in Mesoamerica (Rylands *et al.* 2006), several (e.g. *Saguinus geoffroyi*, *Aotus zonalis*, *Saimiri oerstedii*) have received limited research attention and, in general, there has been little research on primates in Nicaragua, Guatemala, Honduras and El Salvador (Estrada *et al.* 2006).

Overall, Mesoamerican primate species have been strongly impacted by anthropogenic habitat loss and degradation (Estrada *et al.* 2006; Horwich & Johnson 1986). Some areas of Mesoamerica are also subjected to periodic hurricanes that can greatly impact resident primate populations (Pavelka *et al.* 2007). Given that approximately 80% of the natural vegetation in Mesoamerica has been converted for agricultural purposes, it may be best to manage protected areas and agricultural land together in order to address biological, political, and socioeconomic concerns (Harvey *et al.* 2008). Assigning economic value to the habitats and examining the potential of ecotourism may be important in preserving biodiversity (Serio-Silva *et al.* 2013).

Primates and Conservation Concerns in Flooded Habitats in Mesoamerica: Country by Country

Belize

There are two Ramsar sites in Belize, covering 23 592 ha (Table 45.1). Approximately one-third of northern Belize consists of mangrove swamps, marshes and river floodplains (Ellison 2004; Pope *et al.* 2005), and much of the flooded areas throughout Belize are coastal (Figure 45.1). Conservation concerns for these habitats include deforestation for agriculture and aquaculture, coastal development and climate change (Young 2008). Although human population pressures are not as great in Belize as in other countries, wetlands in agricultural areas differ in plant composition and nutrient dynamics from wetlands that are not in agricultural areas (Johnson & Rejmánková 2005).

In Belize, *Alouatta pigra* has been studied in seasonally flooded riparian forests in Bermudian Landing (Horwich & Johnson 1986; Ostro *et al.* 1999), but there has been little published on how (and to what extent) the primates use the flooded habitat present there.

Costa Rica

There are 12 Ramsar sites in Costa Rica, covering 569 742 ha (Table 45.1). These coastal and inland sites include forest swamps, marshes, mangroves, palm swamps, seasonally flooded woodlands, lagoons, beaches and a high-elevation wetland in the Talamanca mountains (www.ramsar.org). In Costa

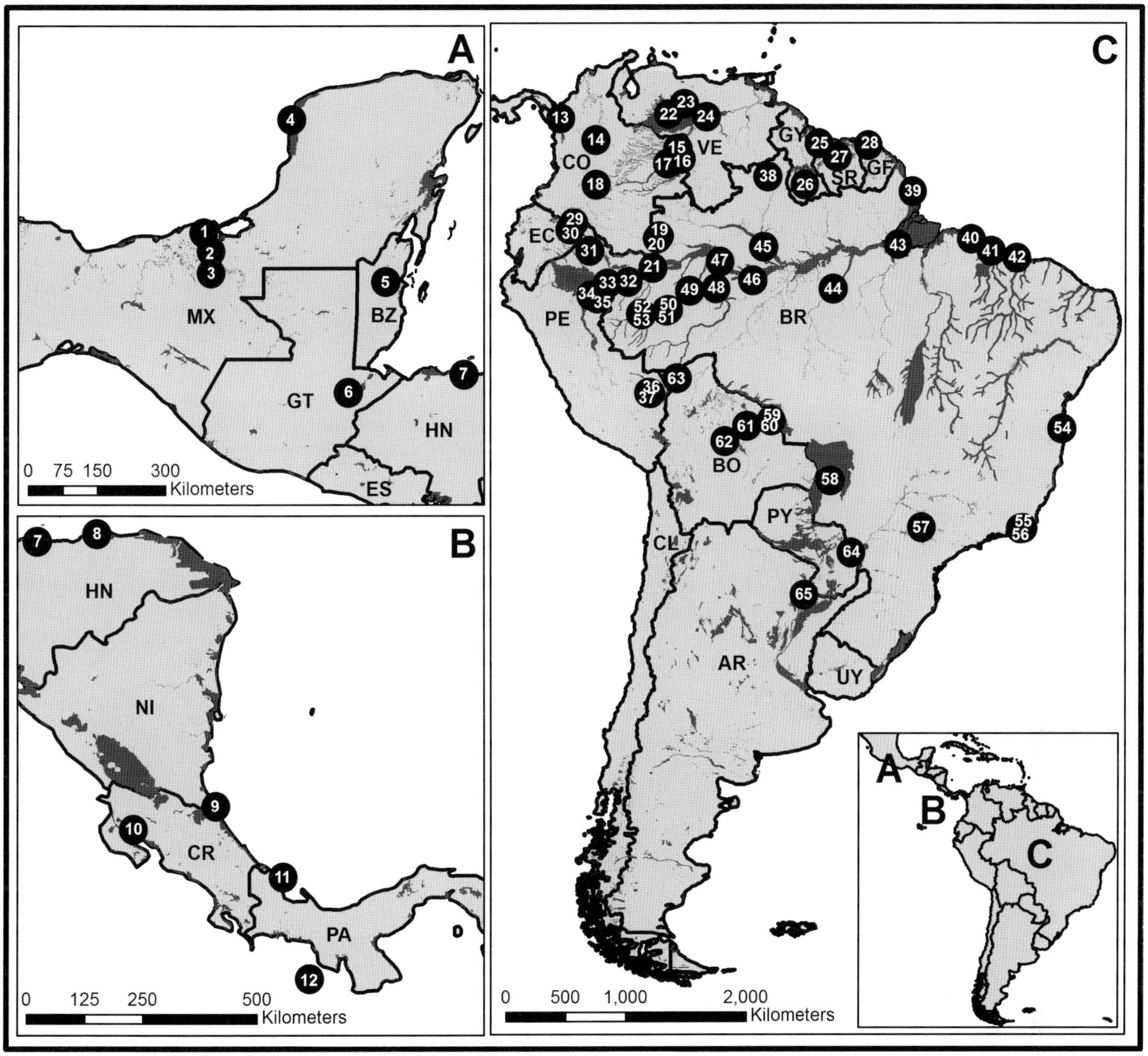

Figure 45.1 Geographic distributions of wetland habitats in the Neotropics, with circles indicating locations of primate studies reviewed in this chapter. Wetlands are in dark grey and based on published data by Lehner and Döll (2004). Country codes are: MX=Mexico, BT = Belize, GT = Guatemala, ES = El Salvador, HN = Honduras, NI = Nicaragua, CR = Costa Rica, PA = Panama, CO = Colombia, VE = Venezuela, GY = Guyana, SR = Suriname, GF = French Guiana, BR = Brazil, EC = Ecuador, PE = Peru, BO = Bolivia, PY = Paraguay, UY = Uruguay, AR = Argentina and CL = Chile. Study site numbers correspond to the numbers listed in Table 45.2.

Rica, wetlands were more likely to be converted if they were not protected and if they were located in drier, more accessible areas (Daniels & Cumming 2008). Prior to the 1980s, wetland conversion was primarily for cattle ranching but, since this time, crop production and hydroelectric and irrigation projects have also threatened wetlands in northwestern Costa Rica (Daniels & Cumming 2008).

In Palo Verde National Park and Tivives Forest Reserve, *Cebus capucinus* has been documented to use mangroves (Warkentin 1993; Fedigan *et al.* 1996). In Tortuguero National Park, *Alouatta palliata*, *Ateles geoffroyi* and *C. capucinus* inhabit swamp forests and mangroves along the bi-national (Costa Rican-Nicaraguan) San Juan River basin (B. Urbani, pers. comm., 2013). Habitat degradation and fragmentation from agriculture and tourism have provided pressures on primates in Costa Rica (Boinski & Siwt 1997; Boinski *et al.* 1998).

El Salvador

There are seven Ramsar sites in El Salvador, covering 207 387 ha (Table 45.1). Lagoons, mangroves, palm swamps and beaches are some of the wetland habitats represented. These Ramsar sites support tourism, fisheries, and agriculture, but are threatened by urbanization, deforestation and agriculture (www.ramsar.org).

Table 45.2 Studies of primates in Neotropical flooded habitats.

Site[a]	Location	Habitat	Species	References
Mexico				
1[b]	Pantanos de Centla Biosphere Reserve, Mexico	Mangrove	*Alouatta pigra*	Bridgeman (2012)
2	Tabasco, Mexico	Lowland tropical forest	*Alouatta pigra*	Watts *et al.* (1986)
3	Playas de Catazajá, Chiapas, Mexico	Riparian forest	*Alouatta pigra*	Serio-Silva *et al.* (2013)
4	Northern Campeche, Mexico	*Petenes* surrounded by mangrove	*Ateles geoffroyi*	Watts *et al.* (1986)
Belize				
5	Bermudian Landing, Belize	Riparian forest	*Alouatta pigra*	Horwich & Johnson (1986); Ostro *et al.* (1999)
Guatemala				
6	Eastern Guatemala	Lowland tropical forest	*Alouatta palliata, Alouatta pigra*	Baumgarten & Williamson (2007)
Honduras				
7	Cuero y Salado Wildlife Refuge, Honduras	Mangrove	*Alouatta palliata*	Gonzalez-Socoloske & Snarr (2010); Snarr (2012)
8	Trujillo, Honduras	Mangrove, swamp	*Cebus capucinus*	Buckley (1983)
Costa Rica				
9	Torguguero National Park, Costa Rica	Mangrove, swamp	*Alouatta palliata, Ateles geoffroyi, Cebus capucinus*	Urbani (pers. comm., 2013)
10	Palo Verde National Park, Costa Rica	Mangrove	*Cebus capucinus*	Fedigan *et al.* (1996); Fedigan & Jack (2001)
Panama				
11	Isla Colón, Panama	Swamp	*Alouatta palliata, Aotus zonalis, Cebus capucinus*	Urbani (2003)
12	Coiba Island, Panama	Mangrove	*Alouatta palliata, Cebus capucinus*	Milton & Mittermeier (1977)
Colombia				
13[b]	Along Atrato River, Colombia	Palm swamp	*Alouatta seniculus*	Zuñiga Leal & Defler (2013)
14	Serranía de Las Quinchas, Colombia	Riparian forest	*Cebus albifrons*	Aldana *et al.* (2008)
15	El Tuparro National Park, Colombia	*Llanos*	*Alouatta seniculus, Cebus albifrons*	Defler (1979a, b, 1980, 1981, 1985)
16	Along Orinoco River, Colombia	Lowland tropical forest	*Saimiri sciureus*	Defler (pers. comm., 2015)
17	Along Vichada River, Colombia	Lowland tropical forest (*igapó*)	*Saimiri sciureus*	Defler (pers. comm., 2015)
18	Tinigua National Park, Colombia	Riparian forest	*Lagothrix lagotricha*	Stevenson *et al.* (1994); Stevenson (2006)
19	Caparú Biological Station, Colombia	Lowland tropical forest (*igapó*; I), palm swamp (dominated by *Bactris* sp., Arecaceae; PS)	*Alouatta seniculus* (I, PS), *Aotus* sp. (I), *Cacajao ouakary* (I), *Callicebus lugens* (I), *C. torquatus lucifer* (I), *Cebus albifrons* (I), *Lagothrix lagothricha* (I), *Saguinus fuscicollis* (I), *Saimiri sciureus* (I), *Sapajus apella* (I)	Defler (1996, 1999, 2013); Defler and Defler (1996); Defler (pers. comm., 2015); Palacios & Rodríguez (2001)
20	Lower Caquetá River, Colombia	Lowland tropical forest (*várzea*)	*Ateles belzebuth, Saguinus inustus, Cebuella pygmaea*	Palacios *et al.* (2004); Defler (pers. comm., 2015)
21	Along Purité River and Amazon River, Colombia	Lowland tropical forest (*igapó* and *várzea*)	*Cebuella pygmaea, Pithecia hirsuta*	Defler (pers. comm., 2015)

Table 45.2 *(cont.)*

Site[a]	Location	Habitat	Species	References
Venezuela				
22	Hato El Frío, Venezuela	*Llanos*	*Alouatta arctoidea*	Braza (1980)
23	Hato Masaguaral, Guárico, Venezuela	Forest patches (*mata*) surrounded by seasonally flooded grassland (*llanos*)	*Alouatta arctoidea*	Rudran & Fernandez-Duque (2003)
24	Middle Orinoco River, Venezuela	Seasonally flooded forest, swamp forest	*Alouatta macconnelli*, *Aotus trivirgatus*, *Cebus olivaceus*, *Chiropotes chiropotes*	Urbani (pers. comm., 2013)
Guyana				
25[b]	Throughout Guyana	Riparian forest (RF), Swamp (S)	*Alouatta seniculus* (RF, S), *Ateles paniscus* (RF), *Cebus olivaceus* (RF, S), *Chiropotes sagulatus* (RF), *Pithecia pithecia* (RF, S), *Saguinus midas* (RF, SF), *Saimiri sciureus* (RF, S), *Sapajus apella* (RF)	Lehmann (2000, 2004, 2006)
26	Masakenari Village, Guyana	Seasonally flooded forest	*Ateles paniscus*, *Chiropotes sagulatus*	Shaffer (2012; pers. comm., 2013)
Suriname				
27	Raleighvallen-Voltzberg Nature Reserve, Suriname	Swamp	*Alouatta seniculus*, *Ateles paniscus*, *Cebus olivaceus*, *Chiropotes sagulatus*, *Pithecia pithecia*, *Saguinus midas*, *Saimiri sciureus*, *Sapajus apella*	Mittermeier & van Roosmalen (1981); Vath (2008)
French Guiana				
28	Coswine, French Guiana	Periodically flooded marsh	*Alouatta seniculus*, *Pithecia pithecia*, *Saguinus midas*, *Saimiri sciureus*, *Sapajus apella*	Thoisy *et al.* (2005)
Ecuador				
29	Cuyabeno Reserve, Ecuador	Lowland tropical forest (*várzea*)	*Saguinus nigricollis*	de la Torre *et al.* (1995)
30	Tiputini Biodiversity Station, Yasuní Biosphere Reserve, Orellana, Ecuador	Lowland tropical forest (*várzea*; along margins of whitewater river and of oxbow lakes)	*Alouatta seniculus*, *Ateles belzebuth*, *Callicebus discolor*, *Cebus albifrons*, *Lagothrix poeppigii*, *Pithecia aequatorialis*, *Saguinus tripartitus*, *Saimiri sciureus*	Marsh (2004); Di Fiore & Fernandez-Duque (unpub. data)
Peru				
31[b]	Río Curaray, Northern Peruvian Amazon, Peru	Floodplain forest (FF) and palm swamp (PS)	*Alouatta seniculus* (FF), *Aotus vociferans* (FF), *Callicebus cupreus* (FF, PS), *Cebus albifrons* (FF), *Lagothrix lagotricha* (FF), *Pithecia aequatorialis* (FF), *Pithecia monachus* (FF), *Saguinus fuscicollis* (FF), *Saguinus tripartitus* (FF), *Saimiri sciureus* (FF, PS), *Sapajus apella* (FF, PS)	Heymann *et al.* (2002)
32	Lago Preto Conservation Concession, Yavarí River, Peru	Palm swamp (aguajal; dominated by *Mauritia flexuosa*), tropical lowland forest (*várzea*)	*Cacajao calvus ucayalii*	Bowler & Bodmer (2009, 2011)
33	Quebrada Tahuaillo, Peru	Palm swamp (*Mauritia flexuosa*); flooded forest	*Alouatta seniculus*, *Aotus nancymae*, *Cacajao calvus ucayalii*, *Cebuella pygmaea*, *Pithecia* cf *aequatorialis*, *Pithecia monachus*, *Saguinus fuscicollis*, *Saguinus mystax*, *Saimiri boliviensis*, *Sapajus apella*	Ward & Chism (2003); Chism (pers. comm., 2013)
34	Pacaya-Samiria National Reserve, Peru	Palm swamp (PS), tropical lowland forest (*várzea*: V)	*Alouatta seniculus* (PS, V), *Cacajao calvus ucayalii* (V); *Cebus albifrons* (V), *Lagothrix lagotricha* (V), *Pithecia hirsuta* (V), *Saimiri boliviensis* (PS, V), *Sapajus apella* (V)	Soini (1986); Bowler *et al.* (2009)

(continued)

Table 45.2 (*cont.*)

Site[a]	Location	Habitat	Species	References
35	Río Tapiche, Peru	Palm swamp (Mauritia flexuosa), tropical lowland forest	*Cacajao calvus ucayalii*	Aquino (1988)
36	Los Amigos Conservation Concession, Madre de Dios, Peru	Palm swamp, tropical lowland forest	*Pithecia irrorata*	Palminteri *et al.* (2012); Palminteri & Peres (2012)
37[b]	Multiple sites (37) in 85 000 km^2 area in Madre de Dios, Peru	Swamp, tropical lowland forest	*Alouatta sara, Ateles chamek, Callicebus brunneus, Cebus albifrons, Lagothrix cana, Pithecia irrorata, Saguinus fuscicollis, Saguinus imperator, Saimiri boliviensis, Sapajus apella*	Terborgh (1983); Palminteri *et al.* (2011)
Brazil				
38	Maracá Island, Roraima, Brazil	Palm swamp (*buritizal* forest; *Mauritia flexuosa*, Arecaceae)	*Ateles belzebuth, Cebus olivaceus, Saimiri sciureus, Sapajus apella*	Pontes (1997)
39	Estação Ecológica Maracá-Jipioca and Reserva Biológica do Lago Piratuba, Amapá, AP, Brazil	Lowland tropical forest (*várzea*), mangrove	*Alouatta belzebul, Alouatta macconnelli, Aotus infulatus, Saimiri sciureus, Sapajus apella*	Silva *et al.* (2013); Xavier (pers. comm., 2013)
40[b]	Pará to Maranhão	Mangrove	*Sapajus apella*	Santos (2010)
41	Maranhão, Brazil (near Canelatiua and Santa Maria, Alcântara)	Mangrove	*Alouatta belzebul, Aotus infulatus, Chiropotes satanas, Saimiri sciureus, Sapajus apella*	Fernandes (1991); Fernandes & Aguiar (1993); Silva Jr. & Fernandes (1999); Fernandes (2000)
42[b]	Maranhão, Brazil	Mangrove	*Sapajus libidinosus*	Santos (2010)
43	Caxiuanã National Forest, Pará, Brazil	Lowland tropical forest (*igapó*)	*Alouatta belzebul, Mico argentatus, Sapajus apella*	Bobadilla & Ferrari (2000); Veracini (2009)
44[b]	Tapajós National Park, Pará, Brazil	Lowland tropical forest (*várzea*)	*Alouatta belzebul, Ateles belzebuth, Ateles paniscus, Callithrix humeralifer, Chiropotes albinasus, Pithecia hirsuta, Sapajus apella*	Branch (1983)
45	Jaú National Park, Amazonas, Brazil	Lowland tropical forest (*igapó*)	*Alouatta seniculus, Aotus* sp., *Cacajao ouakary, Cebus albifrons, Pithecia pithecia chrysocephala, Saimiri sciureus, Sapajus apella*	Barnett *et al.* (2002, 2005, 2012a, b, c); Bezerra *et al.* (2010, 2011)
46	Lago Uauaçú, Lower Purús, Amazonas, Brazil	Lowland tropical forest (*igapo*: I and *várzea*: V)	*Alouatta seniculus* (I, V), *Aotus* cf. *nigriceps* (I), *Ateles chamek* (I, V), *Callicebus torquatus purines* (I), *Cebus albifrons* (I, V), *Lagothrix lagotricha* (I), *Pithecia albicans* (I, V), *Saguinus fuscicollis* (I, V), *Saguinus mystax pileatus* (I, V), *Saimiri* cf. *ustus* (I, V), *Sapajus apella* (I, V)	Haugaasen & Peres (2005a)
47	Mamirauá-Amanã, Amazonas, Brazil	Lowland tropical forest (*várzea*)	*Alouatta jurua, Aotus nancymaae, Cacajao calvus calvus, Callicebus dubius, Pithecia albicans, Saguinus imperator, Saguinus inustus, Saguinus mystax, Saimiri sciureus, Saimiri vanzolinii, Sapajus macrocephalus*	Ayres (1986, 1989); Painter *et al.* (2008); Valsecchi *et al.* (2010); Paim *et al.* (2012, 2013)
48	Upper Urucu River Basin, Tefé, Amazonas, Brazil	Lowland tropical forest (*igapó*)	*Alouatta seniculus, Aotus nigriceps, Ateles paniscus, Callicebus cupreus, Callicebus torquatus, Cebuella pygmaea, Cebus albifrons, Lagothrix lagotricha, Pithecia albicans, Saguinus fuscicollis, Saguinus mystax, Saimiri* sp., *Sapajus apella*	Peres (1993a)
49	Lago da Fortuna, Amazonas, Brazil	Lowland tropical forest (*várzea*)	*Alouatta jurua, Aotus nigriceps, Cacajao calvus, Callicebus cupreus, Cebus albifrons, Saimiri sciureus, Sapajus apella*	Peres (1997)

Table 45.2 *(cont.)*

Site[a]	Location	Habitat	Species	References
50	Barro Vermelho II, Amazonas, Brazil	Lowland tropical forest (*várzea*)	*Alouatta jurua, Ateles paniscus, Callicebus cupreus, Cebus albifrons, Pithecia monachus, Saimiri sciureus, Sapajus apella*	Peres (1997)
51	Boa Esperança, Amazonas, Brazil	Lowland tropical forest (*várzea*)	*Alouatta jurua, Ateles paniscus, Cebus albifrons, Saimiri boliviensis, Sapajus apella*	Peres (1997)
52	Sacado, Amazonas, Brazil	Lowland tropical forest (*várzea*)	*Alouatta jurua, Pithecia irrorata, Saimiri boliviensis, Sapajus apella*	Peres (1997)
53	Nova Empresa, Amazonas, Brazil	Lowland tropical forest (*várzea*)	*Alouatta jurua, Aotus nigriceps, Cebus albifrons, Pithecia monachus, Saimiri sciureus, Sapajus apella*	Peres (1997)
54	Una Biological Reserve, Southern Bahia, Brazil	Swamp	*Leontopithecus chrysomelas*	Raboy *et al.* (2004)
55	União Biological Reserve, Rio de Janeiro, Brazil	Swamp	*Leontopithecus rosalia*	Lapenta & Procópio-de-Oliveira (2008)
56	Poço das Antas Biological Reserve, Rio de Janeiro, Brazil	Palm swamp (*Euterpe* and *Bactris*, Arecaceae)	*Leontopithecus rosalia*	Dietz *et al.* (1997)
57	Caetetus Ecological Station, São Paulo, Brazil	Palm swamp (*Euterpe edulis*, Arecaceae)	*Leontopithecus chrysopygus*	Passos & Keuroghlian (1999)
58	Pantanal of Brazil, Paraguay, and Bolivia	Riparian forest (RF); seasonally flooded grassland (SFG); floodplain woodland and forest (FWF); Chaco and riparian forest (CRF)	*Alouatta caraya* (RF; SFG); *Aotus azarae* (FF); *Callicebus donacophilus* (FWF); *Callicebus pallescens* (CRF); *Callithrix penicillata* (FWF); *Mico melanurus* (FWF); *Sapajus apella* (FWF); *Sapajus cay* (FWF)	Corrêa & Coutinho (2008); Rímoli *et al.* (2012); Silveira *et al.* (2008); Desbiez *et al.* (2010a); Tomas *et al.* (2010); Alho *et al.* (2011a); Alho & Silva (2012)
59	Brazilian bank of river, on the border with Bolivia near Parque Nacional Noel Kempff Mercado	Lowland tropical forest (*igapó*)	*Callicebus* sp., *Chiropotes albinasus, Pithecia rylandsi, Saimiri ustus, Sapajus apella*	Wallace *et al.* (1996)
Bolivia				
60	Lago Caiman, Parque Nacional Noel Kempff Mercado, Bolivia	Lowland tropical forest (*igapó: I*), *sartenejal* swamp (SS)	*Alouatta caraya* (I), *Alouatta sara* (I), *Ateles chamek* (I, SS), *Sapajus apella* (I, SS)	Wallace *et al.* (1998); Wallace (2006)
61[b]	Reserva de Vida Silvestre Rios Blanco y Negro, Bolivia	Lowland tropical forest (*igapó* and *várzea*)	*Alouatta sara*	Wallace *et al.* (2000)
62[b]	Southwestern Beni Department, Bolivia	Seasonally inundated forest-savanna mosaic	*Callicebus modestus, Callicebus olallae*	Felton *et al.* (2006); Martinez & Wallace (2007, 2013); Wallace *et al.* (2013b)
63[b]	Pando Department, Bolivia	Riparian forest	*Alouatta sara, Aotus nigriceps, Callicebus brunneus, Callimico goeldii, Cebuella pygmaea, Pithecia irrorata, Saguinus fuscicollis, Saguinus imperator, Saguinus labiatus, Saimiri boliviensis, Sapajus apella*	Christen & Geissmann (1994); Buchanan-Smith *et al.* (2000)
Paraguay				
64	Mbaracayu Forest Reserve, Paraguay	Riparian forest, swamp	*Sapajus cay*	Hill *et al.* (1997)
Argentina				
65	Isla Brasilera, Paraná River, Argentina	Flooded forest	*Alouatta caraya*	Bravo & Sallenave (2003); Kowalewski & Zunino (2004); Bravo (2008); Kowalewski & Garber (2010); Pavé *et al.* (2012)

The habitat classification is based on categories defined in Bennett *et al.* (Chapter 2). General habitat classifications (e.g. seasonally flooded forest) are used when there was minimal information provided on the type of flooded habitat used by the primates.
[a] Site numbers reference Figure 45.1.
[b] indicates approximate location due to study examining large area.

Figure 45.2 A saki monkey (*Pithecia monachus*, *sensu* Marsh 2014) group along the Tahuayo River in Northeastern Peru. The animals were photographed in flooded forest trees overhanging the blackwater river. Photo: Richard Jackson.

Only one primate species (*Ateles geoffroyi*) occurs in El Salvador (Morales-Hernández 2002), and the country has the fewest primate studies published of any country in Mesoamerica (Estrada *et al.* 2006). In our review of the literature, we did not find any references to *A. geoffroyi* using flooded habitats in El Salvador, but the species has been documented using such habitats in other countries (Table 45.2). In general, habitat loss, hunting and increased human population, as well as hurricanes and earthquakes, have impacted El Salvador's wildlife (Morales-Hernández 2002).

Guatemala

Approximately 8% of Guatemala consists of wetlands (Eisermann 2006). The country has seven Ramsar sites, covering 628 592 ha (Table 45.1). These sites include seasonally flooded forests, lagoons, marshes, mangroves and swamps (www.ramsar.org). Conservation concerns in these areas include logging, hunting, potential oil exploration, agriculture, cattle grazing, pesticides and development for tourism (www.ramsar.org).

In eastern Guatemala, *Alouatta palliata* and *A. pigra* have been found in tropical evergreen broadleaved periodically flooded forest (as well as other forest types), including non-protected sites (Baumgarten & Williamson 2007). The largest conservation concerns to primates in Guatemala are forest conversion for cattle, maize, bananas, palm and rubber (Baumgarten & Williamson 2007).

Honduras

There are nine Ramsar sites in Honduras, covering 270 224 ha (Table 45.1). Wetland habitats include mangroves, lagoons, flooded savannas, beaches, marshes and swamps (www.ramsar.org). Mangroves in Honduras have experienced mass tree mortality as a result of hurricanes and sea-level rise (Cahoon *et al.* 2003). Some of the coastal regions have been under heavy pressure for development, threatening the area's wetland habitats (www.ramsar.org).

In Honduras, *Alouatta palliata* has been found in mangroves (Snarr 2012), and *Cebus capucinus* in a matrix of mangroves and swamp forests (Buckley 1983). Conservation concerns for primates in Honduras include hunting as well as habitat fragmentation and habitat loss for agricultural purposes (Gonzalez-Socoloske & Snarr 2010). Much of the natural connectivity between habitats has been lost (Gonzalez-Socoloske & Snarr 2010), potentially limiting exchange between populations.

Mexico

There are 142 Ramsar sites in Mexico covering 8 833 752 ha (Table 45.1), and most of these sites are located along the coast (Mitsch & Hernandez 2013). Although Mexico has the greatest area devoted to Ramsar sites in Mesoamerica, naturally flooded areas constitute more than 40% of the disturbed habitat in Mexico (Serio-Silva *et al.* 2013). In considering the flooded habitats found within the geographic range of primates in Mexico, most of the wetlands are coastal (Figure 45.1). The Ramsar sites within the geographic range of primates include wetland habitats such as mangroves, *petenes*, swamp forests, seasonally flooded lowland forests, lagoons, riverine forests and beaches (www.ramsar.org). Conservation concerns in these wetland areas include deforestation, overfishing, agriculture, livestock, oil and natural gas exploration and pollution (www.ramsar.org).

In Mexico, *Alouatta pigra* has been found in mangroves (Bridgeman 2012), seasonally flooded lowland tropical forest (Watts *et al.* 1986) and riverine forest (Serio-Silva *et al.* 2013), and *Ateles geoffroyi* in *petenes* on the Yucatan Peninsula (Watts *et al.* 1986). In these study areas, the primary threats to natural habitats include road building, urban development, cattle, sorghum, wood extraction, using fires for hunting and petroleum extraction (Martínez & Gaona 2006; Mitsch & Hernandez 2013; Serio-Silva *et al.* 2013; Watts *et al.* 1986). In fact, multiple sites in the Yucatan Peninsula already had their forested areas destroyed by the mid-1980s (Watts *et al.* 1986). Martínez and Gaona (2006) examined land-cover change (1990–2000) in Pantanos de Centla Biosphere Reserve and found reductions in natural vegetation, including mangrove and swamp habitats. Furthermore, much of the remaining habitat is fragmented, and some of the remaining habitat fragments do not appear to be adequate to sustain primate populations in the future unless connectivity between habitat fragments is improved (Serio-Silva *et al.* 2013).

Nicaragua

In Nicaragua, there are nine Ramsar sites, covering 406 852 ha (Table 45.1). Wetland habitats include mangroves, swamps and lagoons (www.ramsar.org). Although coastal swamps are impacted by hurricanes, human activities, such as fires, also threaten coastal habitats (Urquhart 2009). Agriculture has impacted native habitat, as has gold extraction (Smith 2003). Additional conservation threats to the wetland habitats include agricultural pollution, livestock, deforestation, hunting, overfishing and sedimentation (www.ramsar.org). Although there has been deforestation for agricultural purposes, Nicaragua supports more intact forest than any other Central American country (McCann *et al.* 2003).

There are three primate species in Nicaragua (*Alouatta palliata*, *Ateles geoffroyi* and *Cebus capucinus*; Rylands *et al.* 2006). Published studies of the primates of Nicaragua exist (Bezanson *et al.* 2008; McCann *et al.* 2003), but they are not numerous, and in our review we found no published material on the use of flooded habitat by primates in Nicaragua. However, these three primate species have been documented using flooded habitats in other countries (Table 45.2).

Panama

There are five Ramsar sites in Panama, covering 183 992 ha (Table 45.1). Wetland habitats at these sites include mangroves, marshes, lagoons, seasonally flooded grasslands, swamp forests and beaches (www.ramsar.org). Agriculture, cattle ranching, shrimp farming, deforestation, hunting, irrigation, mining, pollution and urban development threaten wetlands in Panama (Marín 2001; www.ramsar.org).

In Panama, *Alouatta palliata coibensis* and *Cebus capucinus imitator* on Coiba Island entered *Pelliceria rhizophorae* (Pellicieraceae) mangrove swamp, with *A. palliata* probably feeding on *P. rhizophorae* shoots and fruits, and *C. capucinus* eating worms found among the mangroves (Milton & Mittermeier 1977). On Isla Colón in northeastern Panama, Urbani (2003) observed *A. palliata* in a swamp forest dominated by raffia palms (*Raphia taedigera*, Arecaceae). Both *C. capucinus* and *Aotus zonalis* are also reported to be found in this area (B. Urbani, pers. comm., 2013). Deforestation, habitat fragmentation, cattle ranching and hunting threaten primate conservation in Panama (Méndez-Carvajal 2005).

South America

The major wetlands of South America occur within the large river basins (Amazon, Orinoco, Paraná-Paraguay), smaller basins (e.g. Magdalena, Essesquibo), as well as coastal wetlands of the Atlantic and Pacific Oceans (Goulding *et al.* 2003). The three large river basins occupy most of the South American continent, but between the Orinoco and Amazon River Basins there are smaller river basins that make up areas of Guyana, Suriname and French Guiana (Goulding *et al.* 2003).

The Amazon River is the largest river in South America. The basin extends into eight countries (Brazil, Colombia, Ecuador, Peru, Bolivia, Guyana, Suriname and Venezuela), and wetlands make up approximately 30% of the Amazon basin (Junk *et al.* 2011). The flooded habitats of the Amazon River basin include lowland tropical forests such as *várzea* (flooded by whitewater rivers) and *igapó* (flooded by blackwater rivers), savannas, palm swamps, vegetation on tepuis (table mountain), mangroves, lagoons, and beaches (Junk *et al.* 2011). With the climate change that is predicted for the Amazon, flooded forest tree species may be negatively impacted by drought (e.g. increase in mortality, slower growth, seedling recruitment: Parolin *et al.* 2010). Conservation threats to Amazonian wetlands include conversion of natural vegetation, pollution, climate change, dam construction and overharvesting plant and animal species (Castello *et al.* 2013; Junk 2002; Peres *et al.* 2010).

The Orinoco River is the second largest river basin in South America. The basin occupies eastern Colombia and Venezuela. Seasonally flooded *llanos* (savannas) are found in both Colombia and Venezuela (Hamilton *et al.* 2004). Conversion of natural vegetation has principally occurred for cattle ranching (Hoogesteijn & Hoogesteijn 2010).

The Guiana Shield covers Guyana, Suriname, and French Guiana, along with sections of Brazil, Venezuela and Colombia. In the Guiana Shield, seasonally flooded savanna grasslands, broadleaved meadows, and lowland forests exist in areas that are relatively minimally populated by humans, in comparison to other regions (Hammond 2005; Huber 2006). In this region, an influential activity on the conversion of natural vegetation has been gold (Hammond *et al.* 2007) and diamond mining (Roopnaraine 1996; Colchester *et al.* 2002). Conversion of natural vegetation for development and urbanization has been concentrated along the coastal areas of Guyana, Suriname and French Guiana, but conversion of natural habitats has also occurred inland (de Thoisy *et al.* 2005).

The Paraná–Paraguay River Basin includes areas of Brazil, Paraguay, Argentina and Bolivia. One of the main flooded regions of the area is the Pantanal (Chapter 22; Figure 45.3). This savanna wetland is primarily found in Brazil (85% of its 160 000 km^2), but it also extends into Bolivia and Paraguay (Junk & Nunes da Cunha 2005). It consists of a mosaic of habitats including lagoons, seasonally flooded grasslands and

Figure 45.3 The Pantanal is a tropical wetland in the center of South America, in the Upper Paraguay River Basin. The rivers and streams are lined with gallery forests, and other natural habitats exist around the water bodies. The complex vegetation and high seasonal productivity support a diverse and abundant fauna, including primates. Photo: C. Alho.

forest patches (Alho *et al.* 2011a). Fire and the conversion of natural vegetation for agricultural activities (soybeans, livestock grazing), hunting, pollution and hydroelectric power reservoirs threaten the Pantanal ecosystem (Alho *et al.* 2011; Alho & Silva 2012; Alho & Sabino 2012; Junk & Nunes da Cunha 2005). Only 2.5% of the Pantanal is protected (Harris *et al.* 2005).

With the exception of Uruguay and Chile, there are native primates in all South American countries, and the genera *Alouatta*, *Aotus*, *Ateles*, *Cacajao*, *Callicebus*, *Callithrix*, *Callimico*, *Cebuella*, *Cebus*, *Chiropotes*, *Lagothrix*, *Leontopithecus*, *Mico*, *Pithecia*, *Saguinus*, *Saimiri* and *Sapajus* have all been documented in flooded habitats in South America (Table 45.2).

Primates and Conservation Concerns in Flooded Habitats in South America: Country by Country

Argentina

There are 21 Ramsar sites in Argentina, covering 5 382 521 ha (Table 45.1). Within the geographic distribution of primates in Argentina there are three Ramsar sites and wetland habitats include seasonally flooded forests and savanna grasslands, marshes, and lagoons (www.ramsar.org). Hydrological changes from dams, deforestation, cattle grazing and agriculture are the largest threats in areas where primates are distributed (www.ramsar.org).

We found only one site in Argentina where primates in flooded habitats have been studied. Isla Brasilera, in northeastern Argentina, has been the location for studies of *Alouatta caraya* activity patterns, birth seasonality, foraging behaviour, infant mortality and seed dispersal in seasonally flooded forest (Bravo 2008; Bravo & Sallenave 2003; Kowalewski & Garber 2010; Kowalewski & Zunino 2004; Pavé *et al.* 2012). Conservation threats to *A. caraya* habitat in Argentina include hydroelectric dam construction in northeastern Argentina (Díaz *et al.* 2007), and cattle ranching (Fernandez-Duque *et al.* 2008).

Bolivia

There are 11 Ramsar sites in Bolivia, covering 14 842 405 ha and the largest area covered by Ramsar sites in all of South

America (Table 45.1). Wetland habitats include seasonally flooded lowland tropical forests, marshes, lagoons and flooded savanna–gallery forest mosaics (www.ramsar.org). These mosaics are particularly important in the vast seasonally inundated wetlands of the Beni Department. In Bolivia, the conservation threats to wetlands consist of deforestation, agriculture, sheep and cattle ranching, overfishing, hunting, extensive tourism, mining, gas pipeline construction and road construction (www.ramsar.org).

In Bolivia, 23 species of primates occur in a range of tropical forest habitats ranging from the dry tropical scrub forests of the Chaco, to the lower and mid-montane forests of the Andes, down to the tropical rainforests of northern and eastern Bolivia (Mercado & Wallace 2010; Wallace *et al.* 2010, 2013a). In the seasonally inundated *igapó* forests of the Itenez or Guapore River on the Bolivia–Brazil border, Wallace *et al.* (1998) recorded two species of howler monkeys (*Alouatta caraya* and *A. sara*) and *Sapajus apella*, as well as occasional use by *Ateles chamek*. Across eastern Bolivia, seasonally inundated *sartenejal* swamp forests are used by *Ateles* (Wallace, 2006; Wallace *et al.* 1998) and *Sapajus* (Wallace *et al.* 1998). *Alouatta sara* is also largely restricted to riverine forests across the northern Santa Cruz Department in eastern Bolivia (Wallace *et al.* 2000).

Buchanan-Smith *et al.* (2000) documented nine species in seasonally flooded riparian forests across multiple areas of the Pando Department, northern Bolivia (Table 45.2). Christen and Geissmann (1994) also recorded *Aotus nigriceps*, *Callicebus* sp., *Cebus albifrons*, *Pithecia irrorata*, *Saguinus weddelli*, *Saguinus labiatus* and *Saimiri boliviensis* occupying swamp and riparian habitats in the Pando Department.

Conservation concerns for primates in Bolivia include habitat loss and hunting (Buchanan-Smith *et al.* 2000). In some areas, *A. chamek* has been hunted to almost local extinction (Buchanan-Smith *et al.* 2000). However, the most pressing conservation issue for primates in flooded forests in Bolivia is the case of two endemic *Callicebus* species, *C. modestus* and *C. olallae*, whose ranges are restricted to the seasonally inundated forest–savanna mosaic in the southwest of the Beni Department (Felton *et al.* 2006, Martinez & Wallace 2007, 2013; Wallace *et al.* 2013b; Chapter 23).

Brazil

There are 12 Ramsar sites in Brazil, covering 7 225 687 ha (Table 45.1). Wetland habitats in Brazil are located in the Amazon basin, along the coast and into the Atlantic forest, and the Pantanal. These habitats include flooded lowland tropical forest (*igapó* and *várzea*), permanently and seasonally flooded grasslands, seasonally flooded savannas, mangroves, lagoons, marshes and beaches (www.ramsar.org).

Conservation threats to wetlands in Brazil include deforestation, agriculture, livestock grazing, nonnative aquatic species, nonnative plants, hunting, overfishing, forest fires, urban and industrial development, pollution and tourism (www.ramsar.org). Specifically, *várzea* in Brazil is threatened by deforestation, agricultural expansion, livestock grazing, increases in human population and urban expansion (Albernaz *et al.* 2012; Amaral *et al.* 2012; de Queiroz & Peralta 2011). *Igapó* is very vulnerable to fires, regenerating 13 times slower than neighbouring *terra firme* after a fire (Flores *et al.* 2014). In addition, illegal logging is taking place in the *igapó* of some protected areas (Scabin *et al.* 2012), and items selected include species (e.g. *Macrolobium acaciifolium*, Fabaceae) that are important diet items for some *igapó*-using primates (e.g. *Cacajao ouakary:* Barnett 2010).

Wetland habitats in the Brazilian Amazon are also threatened by the direct and collateral effects of oil and gas exploration (Finer *et al.* 2008), and by a recent resurrection of the plans for basin-wide network of dams for hydroelectric power and transport network enhancement (Fearnside 2006; Finer & Jenkins 2012). Based on previous projects (e.g. Curuá-Una Dam, Pará: Gunkel *et al.* 2003; Samuel Dam, Rondônia: Fearnside 2005; Tucuruí Dam, Pará: Barrow 1988; Fearnside 2001), there is rightful concern both because of the potential nature and extent of their immediate impacts of such projects, as well as longer-term effects resulting from increased greenhouse gas emissions (Fearnside 2004). Even when it does not impact directly on flooded forests, large-scale farming for cattle and soya may change drainage patterns and thus have impacts on flooded forest ecosystems (Goulding *et al.* 2003).

Along the Atlantic coast of Brazil, wetland habitats such as mangroves, swamps, lagoons, marshes and beaches have been impacted by habitat loss, overharvesting of fauna and flora, oil drilling, urbanization, tourism development and pollution (Diegues 1999; Marques *et al.* 2004). In the Pantanal, agriculture and invasive species have also impacted the native habitat (Alho *et al.* 2011b), and dams have provoked environmentally negative hydrological changes in the Pantanal (Zeilhofer & de Moura 2009).

Thirty-nine primate species from 15 genera (*Alouatta*, *Aotus*, *Ateles*, *Cacajao*, *Callicebus*, *Callithrix*, *Cebuella*, *Cebus*, *Chiropotes*, *Lagothrix*, *Mico*, *Pithecia*, *Saguinus*, *Saimiri* and *Sapajus*) have been documented using *várzea*, *igapó*, palm swamps and mangroves in Brazil (Table 45.2). These 15 primate genera have all been documented using *igapó* or *várzea* (Ayres 1986, 1989; Barnett *et al.* 2002, 2005, 2012a, b, c, 2013; Bezerra *et al.* 2011; Haugaasen & Peres 2005, 2006, 2007; Vieira *et al.* 2008; Paim & Queiroz 2009; Paim *et al.* 2012, 2013; Wallace *et al.* 1998). Comparing habitat types, *terra firme* forests in the Amazon tend to have greater primate species richness than *várzea* and *igapó* forests, but primate biomass is greater in *várzea* than *terra firme* or *igapó* and primate density is lower in *terra firme* than *várzea* (Haugaasen & Peres 2005; Peres 1997b). The use of *igapó* and *várzea* can be seasonal and correlate with fruit availability (Barnett 2010; Haugaasen & Peres 2007). This pattern can be true even when the floodplain is not very extensive: on the Tapajós the floodplain is often less than 10 m wide, but is seasonally important for several species of primate, including the Endangered *Chiropotes albinasus* (A. Barnett, unpub. data). Barnett (2010) found that *C. ouakary* remained in *igapó* for the majority of the year, only using neighbouring *terra firme* when *igapó* had no fruit or young leaves. More than 80% of the plant species in the *C. ouakary* diet came from *igapó*.

The mangroves on the Brazilian northern coast are used by several species such as *Alouatta belzebul*, *Aotus infulatus*,

Figure 45.4 *Sapajus libidinosus* (a) uses a range of habitats, including the mangrove forest (b) along the Preguiças River, Barreirinhas, Maranhão, Brazil. Photo: Ricardo Rodrigues dos Santos.

Chiropotes satanas, *Saimiri sciureus* and *Sapajus apella* (Fernandes 1991, 2000; Fernandes & Aguiar 1993; Silva Jr. & Fernandes 1999), but only *Sapajus* is common in these forests (Santos 2010). In a survey of mangroves from Pará to Maranhão States, Santos (2010) found both *S. apella* and *S. libidinosus* to use mangrove habitat, but *S. libidinosus* was more restricted to mangroves (Figure 45.4).

Mauritia-dominated palm swamps (*buritizais*) have been noted as important habitat for primates in the Brazilian Amazon (Peres 1993a; Pontes 1997, 1999). The need for targeted investigations is underscored by the rapid loss of this habitat in many areas of South America (Gilmore *et al.* 2013; Holm 2008). Other habitats, such as *aningal* (a low open swamp dominated by *Montrichardia arborescens*, Araceae) and seasonally flooded *campina* are even less known. In addition, there are whole regions in the Brazilian Amazon that possess flooded habitats, but which appear to have been little explored primatologically. These areas include the headwaters of the Xingu (Pimenta & Silva Jr. 2005), of which 30% flood seasonally (Goulding *et al.* 2003), and the flooded habitat of Amapá's Piratuba Lakes, where there are invasive wild buffalo in some of the flooded habitat (C. Silva, pers. comm., 2014).

In the Atlantic Forest of Brazil, *Leontopithecus chrysomelas* and *L. rosalia* have been documented using *Euterpe* and *Bactris* (Arecaceae) palm swamps (Dietz *et al.* 1997; Lapenta & Procópio-de-Oliveira 2008; Passos & Keuroghlian 1999; Raboy *et al.* 2004). Most of the primates' habitat in the Atlantic forest has been converted (Raboy *et al.* 2004), and the primary threats to Endangered *L. chrysomelas* and *L. rosalia* are their reduced and isolated populations as a result of habitat loss from deforestation, agriculture, cattle, charcoal production, urbanization and fires (Kierulff *et al.* 2008a, b). In the Brazilian Pantanal, *Alouatta*, *Aotus*, *Callicebus*, *Callithrix*, *Mico*, and *Sapajus* use flooded habitats (Alho *et al.* 1987, 2011a; Alho & Silva 2012; Desbiez *et al.* 2010a; Rímoli *et al.* 2012). These areas of the Pantanal are threatened by habitat loss (Harris *et al.* 2005; Alho & Silva 2012), which is often the primary conservation threat to primates (Oates 2013).

Colombia

There are five Ramsar sites in Colombia, covering 458 525 ha (Table 45.1). Wetland habitats include mangroves, *llanos*, lowland tropical forest (*igapó* and *várzea*), lagoons, swamps and beaches (www.ramsar.org; Table 45.2). In the Colombian *llanos*, conversion of natural vegetation for cattle and crops has resulted in a reduction of native vegetation, as have oil exploration, mining and road construction (Romero-Ruiz *et al.* 2012; Sánchez-Cuervo *et al.* 2012). The Magdalena and Cauca basins have experienced pollution and deforestation of native vegetation due to agriculture, mining and urbanization (Junk 2002; Restrepo & Syvitski 2006). On the Pacific coast, deforestation has occurred for industrial oil palm plantations, with mangrove habitat destroyed for shrimp farms, while hydroelectric projects threaten additional forested areas (Mittermeier *et al.* 2009). The Amazonian Piedmont and adjoining Amazonian floodplain areas have historically been the focus of frontier expansion, especially to establish agricultural and cattle-ranching activities (Garcia *et al.* 2007). Furthermore, the Piedmont includes the largest urban centres in the southern Colombian Amazon, and the major oil exploration and extraction activities in the country are also concentrated in this region (Finer *et al.* 2013). It is estimated that the Colombian Amazon has experienced much greater deforestation rates than the other areas of the Amazon (Armenteras *et al.* 2006), and the area is also impacted by the interactive effects of fire and forest fragmentation (Armenteras *et al.* 2013).

Overall, there are 27–32 primate species in Colombia (Stevenson *et al.* 2010; Solari *et al.* 2013). Some species have been very little studied (e.g. *Ateles hybridus*, *Aotus lemurinus*), and the geographic distribution of research within the country has been very uneven (Defler *et al.* 2003; Stevenson *et al.* 2010). Nevertheless, in Colombia, 14 species have been shown to use flooded forests: *Alouatta seniculus*, *Aotus* sp., *Ateles belzebuth*, *Cacajao ouakary*, *Callicebus lugens*, *C. torquatus lucifer*, *Cebuella pygmaea*, *Cebus albifrons*, *Lagothrix lagothricha*, *Pithecia hirsuta*, *Saguinus fuscicollis*, *S. inustus*, *Saimiri sciureus* and *Sapajus apella* (Table 45.2). Specifically, *Alouatta seniculus* is found in flooded forests along the Atrato River (Zuñiga Leal & Defler 2013) as well as floodplain forests throughout much of the Colombian Orinoco and Amazon (T. Defler, pers. comm.). At Caparú Biological Station, *A. seniculus* was seasonally found in greater densities in flooded habitat than *terra firma* habitat due to the increased production of fruits and new leaves during the season (Defler & Defler 1996; Defler *et al.* 2013). *Lagothrix lagotricha* has been documented using flooded forest, but not to the same extent as non-flooded mature forest and open or degraded forest (Defler 1996; Defler & Defler 1996; Stevenson 2006; Stevenson *et al.* 1994). In Tinigua National Park and Caparú Biological Station, *L. lagotricha* used flooded forest most when fruit production was greatest (Defler & Defler 1996; Defler *et al.* 2013; Stevenson 2006; Stevenson *et al.* 1994). Both *Callicebus lugens* and *C. torquatus lucifer* also use flooded forest in the Amazon (Defler 2013; T. Defler, pers. comm.). *Saguinus inustus* has been documented using *várzea* along the Caquetá River, but the use of this habitat is minimal in comparison to non-flooded primary and secondary forest (Palacios *et al.* 2004).

Some of the main conservation threats to primates include the conversion of primate habitat for agricultural means (e.g. oil palm plantations, cattle ranching), as well as general habitat degradation and fragmentation, and hunting (Defler *et al.* 2003; Carretero-Pinzón *et al.* 2009; Mittermeier *et al.* 2009). Two primate species (*A. hybridus* and *Saguinus oedipus*) found in the Magdalena River Basin have been included in the list of the 25 most endangered primates (Mittermeier *et al.* 2009).

Ecuador

There are 18 Ramsar sites in Ecuador, covering 286 651 ha (Table 45.1). Wetland habitats include seasonally flooded forests, swamps, lagoons, mangroves, marshes, and beaches (www.ramsar.org). *Várzea* forest and floodable forest in the Yasuni region are located around oxbow lakes and on river margins, and can be flooded for periods lasting from a few days to several weeks. These areas flood multiple times per year, coincident with fluctuations in the depth of the Rio Tiputini, a significant whitewater river that crosses and forms part of the northern border of the Yasuní National Park. Still, the flooding pattern is less extensive than that seen in central Amazonian sites, where flooded forests can be continuously inundated for months at a time. Conservation threats to wetland habitats in Ecuador include agriculture, cattle ranching, deforestation, shrimp farming, nonnative species, fire, climate change, pollution, overuse of water resources and changes to wetland hydrology (www.ramsar.org).

Use of flooded forest by *Saguinus nigricollis* was documented in the Cuyabeno Faunal Production Reserve (de la Torre *et al.* 1995), but the eight study groups used *terra firma* forest to a greater extent than flooded forest. A survey of *várzea*, *terra firma*, and swamp habitats at the Tiputini Biodiversity Station in the Yasuní Biosphere Reserve found *Saguinus tripartitus* and *Saimiri sciureus* often in *várzea*, with *Callicebus cupreus* and *Pithecia aequatorialis* using the habitat, but to a lesser extent (Marsh 2004). Long-term behavioural work at the site suggests that groups of *Alouatta seniculus*, *Callicebus discolor* and *Saguinus tripartitus* occupy home ranges contained exclusively or primarily in flooded forest (Di Fiore, Fernandez-Duque, & Link, unpub. data). *Cebus albifrons*, *S. sciureus* and *P. aequatorialis* can be found using areas of *várzea* forest and floodable areas around oxbow lakes when they are not flooded, and there is no reason to suspect that these primates do not also use the same areas when flooded (Di Fiore, Fernandez-Duque, & Link, unpub. data). Long-term observations on woolly monkeys (*Lagothrix poeppigii*) and spider monkeys (*Ateles belzebuth*) at the site confirm that both of these large-bodied taxa spend considerable amounts of time in flooded areas, although following them into inundated regions is not feasible (Di Fiore & Link, unpub. data).

The main threats to primates in Ecuador include habitat loss and hunting (de la Torre 2012). In the Greater Yasuní-Napo area, threats also include road construction for oil extraction and hunting (Suárez *et al.* 2009, 2013). Similar to other countries, many of the primate studies have been concentrated in a handful of locations (e.g. Cuyabeno Reserve, Yasuní Biosphere Reserve).

French Guiana

There are three coastal Ramsar sites in French Guiana, covering 224 400 ha (Table 45.1). Flooded habitats include beaches, swamps, marshes and mangroves (www.ramsar.org). Most of the human population lives close to the coast, and therefore the northern areas of the country receive the greatest amount of disturbance to local flora and fauna through the building of roads, logging, hunting and gold mining (de Thoisy *et al.* 2010).

In a survey of a flooded marsh in Coswine, French Guiana, *Pithecia pithecia*, *Saguinus midas*, *Saimiri sciureus* and *Sapajus apella* were documented (de Thoisy *et al.* 2005). Aside from primate censuses, most primate field research has occurred at the Nouragues Nature Reserve (de Thoisy *et al.* 2005), leaving most of the country unstudied. The hunting of primates, followed by habitat loss as a result of logging, are the main conservation threats impacting primate presence in French Guiana (de Thoisy *et al.* 2005, 2010).

Guyana

As of September 2014, there were no Ramsar sites in Guyana. However, Guyana contains extensive areas of several types of flooded forest, such as swamp and mangrove forests along the low-lying coastal plain, large areas of seasonally inundated savanna gallery forests in the southwest, and riparian flooded forests throughout the interior peneplain (ter Steege 1993). Increased gold and diamond mining (with their associated pollution), logging and hydroelectric development (Colchester *et al.* 2002; Howard *et al.* 2011; Lehman *et al.* 2013; Roopnaraine 1996; C. Shaffer, pers. comm., 2013) may drastically alter these habitats in Guyana.

Animal censuses by Lehman (2000, 2004, 2006) at multiple locations have found *Alouatta seniculus*, *Ateles paniscus*, *Cebus olivaceus*, *Chiropotes sagulatus*, *Pithecia pithecia*, *Saguinus midas*, *Saimiri sciureus* and *Sapajus apella* in flooded habitats. Shaffer *et al.* (Chapter 28) used sightings data from Turtle Mountain, Iwokrama (B. Wright & K. Wright, unpub. data) and a case study of *C. sagulatus* from the Upper Essequibo Conservation Concession (Shaffer 2012) to advance the understanding of seasonal use of inundated forest by primates in the Guianan Shield. These data, combined with other recent surveys in southern Guyana (C. Shaffer, unpub. data), show that *Ateles* and *Chiropotes* frequently use flooded forests. In fact, *Chiropotes* may prefer flooded forests during periods of inundation due to the abundance of hydrochorous plant species fruiting in these areas during flooded periods.

In addition to habitat degradation due to gold and diamond mining, logging and hydroelectric development, hunting and wildlife trade are also conservation threats to primates in Guyana (Lehman *et al.* 2013). Furthermore, primate sightings from censuses have decreased since the 1970s, indicating that there is reason for concern regarding primate conservation in Guyana (Lehman *et al.* 2013).

Paraguay

There are six Ramsar sites in Paraguay, covering 785 970 ha (Table 45.1). Wetland habitats include seasonally flooded shrubland and forests, marshes, savannas, swamps and lagoons (www.ramsar.org). These habitats are threatened by livestock grazing, deforestation, hunting, salinization, desertification and climate change (www.ramsar.org).

The Mbaracayu Reserve includes swamp and riparian areas where *Sapajus cay* has been documented (Hill *et al.* 1997). *Alouatta caraya* is also present in the area, but there is limited information available on its habitat use in the reserve (Hill *et al.* 1997). There has been recent conversion of natural vegetation of the Atlantic forest and Chaco, primarily for agricultural development (Huang *et al.* 2009). The construction of hydroelectric dams in eastern and southern Paraguay resulted in the translocation of *A. caraya* from flooded areas (Díaz et al. 2007). Overall, there are limited data on primates in Paraguay (Porter *et al.* 2013).

Peru

There are 13 Ramsar sites in Peru, covering 6 784 042 ha (Table 45.1). Wetland habitat types include seasonally flooded lowland forests, mangroves, lagoons, swamps, and beaches (www.ramsar.org). Conservation threats of the wetland habitats include deforestation, cattle grazing, harvesting of reeds, hunting, irrigation, petroleum prospecting, shrimp and fish farming, and pollution from tourism (www.ramsar.org).

Twenty-four primate species from 12 genera (*Alouatta, Ateles, Aotus, Cacajao, Callicebus, Cebuella, Cebus, Lagothrix, Pithecia, Saguinus, Saimiri* and *Sapajus*) have been documented using flooded tropical lowland forests and palm swamps in Peru (Table 45.2). These sites include areas near the Río Curaray in northern Peru (Heymann *et al.* 2002; Figure 45.2), the Lago Preto Conservation Concession near the Yavarí River (Bowler & Bodmer 2009, 2011), Quebrada Tahuaillo (Ward & Chism 2003; Chism, pers. comm.), along the Río Tapiche (Aquino 1988), the Pacaya-Samiria National Reserve (Soini 1986) and the Madre de Dios region (Palminteri & Peres 2012; Palminteri *et al.* 2011, 2012; Terborgh 1983). In a study of 37 sites in the Madre de Dios area, Palminteri *et al.* (2011) found that primate species richness varied, and overall, non-flooded sites had greater primate species richness than floodplain forests. In this region, ten primate species (*Alouatta sara, Ateles chamek, Calicebus brunneus, Cebus albifrons, Lagothrix cana, Pithecia irrorata, Saguinus fuscicollis, Saguinus imperator, Saimiri boliviensis* and *Sapajus apella*) have been documented using flooded habitats (Palminteri *et al.* 2011; Terborgh 1983). Additional behavioural and ecological studies of *Pithecia irrorata* found the species to use multiple habitats, including floodplain and palm swamps, but *terra firma* forest was preferred (Palminteri & Peres 2012; Palminteri *et al.* 2012). *Cacajao calvus ucayalii* has been documented in multiple flooded habitats, including *aguajal* palm swamps, throughout Peru (Bowler & Bodmer 2011; Heymann & Aquino 2010).

Many primate species in Peru, especially the larger-bodied ones, are impacted by hunting (Endo *et al.* 2010; Palminteri *et al.* 2011; Matthews 2005; Shanee 2012). Deforestation for oil palm (*Elaeis* spp., Arecaceae) commercialization has increased in areas of the Peruvian Amazon, but primarily in non-flooded areas (Gutiérrez-Vélez & DeFries 2013). Although *Cacajao calvus ucayalii* was recently reported in the *várzea* of the Pacaya-Samiria National Reserve (Bowler *et al.* 2009), community members reported that within some areas of the reserve the monkeys had not been seen for approximately 10 or 20 years. Bowler *et al.* (2009) concluded that the monkeys had likely been under strong hunting pressure.

Suriname

There is one Ramsar site in Suriname, covering 12 000 ha (Table 45.1). Flooded habitats in Suriname include mangroves, beaches, seasonally flooded lowland forests and swamp forests (Lehman *et al.* 2013). Suriname (along with French Guiana) has the greatest proportion of its land classified as intact forest of any country in South America (Howard *et al.* 2011). However, Suriname has seen a recent increase in gold mining (Howard *et al.* 2011; Norconk *et al.* 2003;), and bauxite mining and logging also threaten native vegetation (Lehman *et al.* 2013).

In Raleighvallen-Voltzberg Nature Reserve, eight primate species (*Alouatta seniculus, Ateles paniscus, Cebus olivaceus, Chiropotes sagulatus, Pithecia pithecia, Saguinus midas, Saimiri sciureus* and *Sapajus apella*) have been documented in swamp habitat through census data (Mittermeier & van Roosmalen 1981; Vath 2008). The predation of *A. seniculus* infants by harpy eagles has been documented in this swamp habitat (Vath 2008).

Primates in Suriname have been impacted by hunting and habitat degradation (Norconk *et al.* 2003; Lehman *et al.* 2013). Many of the primates in Suriname have been considered an important food item for humans (Baal *et al.* 1988), particularly *Alouatta macconnelli* (Gajapsersad *et al.* 2011).

Venezuela

There are five Ramsar sites in Venezuela, covering 263 636 ha (Table 45.1). Flooded habitats occur in the coastal mangroves and lagoons (Pérez *et al.* 2002), along with swamp forests, palm swamps, mangroves, and *llanos* in the Orinoco Delta (Vegas-Vilarrúbia *et al.* 2007). The coastal mangroves of Venezuela are primarily threatened by pollution, sedimentation, tourism, changes in drainage and illegal fishing (Pérez *et al.* 2002). Overall the human impacts to the Orinoco River floodplain and the Venezuela *llanos* have been minimal compared to flooded habitats in other South American countries (Junk

2002), but the Orinoco Delta has experienced oil prospecting and dam construction (Pérez *et al.* 2002). Additional concerns in the Caura River Basin include hydroelectric development and mining (Painter *et al.* 2008).

In Venezuela, long-term population studies of *Alouatta arctoidea* show that this species inhabits forests surrounded by seasonally flooded savannas (*llanos*), as well as gallery forest along the Guarico River (Rudran & Fernandez-Duque 2003). In the southern Venezuelan *llanos*, Braza (1980) also reported the use of seasonally flooded forest by *A. arctoidea*. In the middle Orinoco River, Urbani (pers. comm., 2013) found *Chiropotes chiropotes*, *Cebus olivaceus* and *Alouatta macconnelli* in seasonally flooded forests, as well as *Aotus trivirgatus* in a swamp forest dominated by *Mauritia flexuosa* (Arecaceae).

The primary conservation threats to primates Venezuela are the conversion of natural vegetation for cattle ranching, mining and logging, as well as hunting for food and the pet trade (Lehman *et al.* 2013; Urbani 2006).

Conclusions

Our review of the flooded habitats of the Neotropics and the primates that use them demonstrates that the Neotropics contain a variety of wetland habitats, but many of the regions are under a range of conservation threats (e.g. climate change, deforestation, hydrological changes, mining, nonnative species, overharvesting of fauna and flora, pollution, urbanization) or have not been fully studied. We found records of 17 neotropical primate genera using flooded habitats to some extent, with the exception of *Brachyteles*, *Callibella*, and *Oreonax*. Genera such as *Alouatta*, *Aotus*, *Cacajao*, *Cebus*, *Saguinus*, *Saimiri* and *Sapajus* were most commonly noted to use flooded habitats, while genera such as *Ateles*, *Pithecia*, and *Chiropotes* appear to use flooded habitats to a lesser extent.

Although species assemblages and environmental characteristics vary greatly between geographic locations, primate species in these seasonally or continuously flooded areas face conservation challenges due to habitat destruction and fragmentation (Chapman & Peres 2010; Martínez-Mota *et al.* 2007), hunting (Chapman & Peres 2010), and climate change (Wiederholt & Post 2010).

Our review highlighted studies of primates in neotropical flooded habitats from 65 sites in 17 countries within the Neotropics, but many geographical areas (e.g. El Salvador, Nicaragua, Ecuador, Paraguay) are underrepresented in the literature. Even within regions that have received considerable research attention (e.g. Amazon), many areas remain poorly known (Pitman *et al.* 2011). There is a need more long-term studies of Neotropical primates, which would allow for a better documentation of the habitats used by primates in different regions.

References

Abram, N.K., Xofis, P., Tzanopoulos, J., *et al.* 2014. Synergies for improving oil palm production and forest conservation in floodplain landscapes. *PLOS One* 9: e95388.

Abril, G., Guérin, F., Richard, S., *et al.* 2005. Carbon dioxide and methane emissions and the carbon budget of a 10-years old tropical reservoir (Petit-Saut, French Guiana). *Global Biogeochemical Cycles* 19: GB 4007, doi: 10.1029/2005GB002457.

Abubakar, M.A.T., Dahuri, R. & Kusumastanto T. 2001. Model sedimentasi dan daya dukung lingkungan segara anakan untuk kegiatan budidaya udang. *Pesisir dan Lautan* 4: 50–62.

Acevedo-Charry, O., Pinto-Gómez, A. & Rangel-Ch, J.O. 2014. Las aves de la Orinoquía colombiana: una revisión de sus registros. *Colombia Diversidad Biótica XIV. La región de la Orinoquía de Colombia*, pp. 691–750.

Adie, F., Galat-Luong, A. & Galat, G. 1997. In *Les grands Mammifères du Niokolo-Badiar*, A. Galat-Luong (ed.). Rueil-Malmaison: Belancor.

Adis, J. 1984. 'Seasonal igapó'-forests of central Amazonian blackwater rivers and their terrestrial arthropod fauna. In *The Amazon: Limnology and Landscape Ecology of a Mighty Tropical River and its Basin*, H. Sioli (ed.). Monographiae Biologicae No. 56. Dordrecht, The Netherlands: Springer, pp. 245–268.

Adis, J., Paarman, W., Amorim, M.A., Ardnt, E. & da Fonseca, C.R. 1998. On occurrence, habitat specificity and natural history of adult tiger beetles (Coleoptera: Carabidae: Cicindelinae) near Manaus, central Amazonia, and key to the larvae of tiger beetle genera. *Acta Amazonica* 28: 247–272.

Agatsuma, Y., Ogawa, M., Taniguchi, K. & Yamada, H. 2000. Marine algal flora of the coast of Tomari-Hama along the Oshika Peninsula, Japan. *Wildlife Conservation Japan* 5: 47–53.

Agetsuma, N. 1995. Dietary selection by Yakushima macaques (*Macaca fuscata yakui*): the influence of food availability and temperature. *International Journal of Primatology* 16: 611–627.

Aggimarangsee, N. 2013. Status monitoring of isolated populations of macaques and other nonhuman primates in Thailand. In *Primates in Fragments: Complexity and Resilience*, L.K. Marsh & C.A. Chapman (eds). New York: Springer, pp. 147–158.

Agoramoorthy, G. & Hsu, M.J. 2005. Borneo's proboscis monkey: a study of its diet of mineral and phytochemical concentrations. *Scientific Correspondences* 89: 454–457.

Agoramoorthy, G. & Lohmann, R. 1999. Population and conservation status of the black-and-gold howler monkeys, *Alouatta caraya, along the Rio Riachuelo, Argentina. Neotropical Primates* 7: 43–44.

Agrawal, A.A. 2001. Phenotypic plasticity in the interactions and evolution of species. *Science* 294: 321–326.

Agrawal, V.C., Das, P.K., Chakraborty, S., *et al.* 1992. Mammalia. In *State Fauna Series 3. Fauna of West Bengal, Part I*, Director, Zoological Survey of India (ed.). Calcutta: Zoological Survey of India, pp. 27–169.

Aguiar, L.M. 2006. *Os primatas do corredor do Alto Rio Paraná (Região de Porto Rico, estados do Paraná e Mato Grosso do Sul): Ocorrência, georreferenciamento e parâmetros populacionais.* MSc dissertation, Universidade Federal do Paraná, Curitiba, Brasil. www.acervodigital.ufpr.br/bitstream/handle/1884/4725/Aguiar2006?sequence=1.

Aguiar, L.M., Ludwig, G. & Passos, F.C. 2009. Group size and composition of black-and-gold howler monkeys (*Alouatta caraya*) on the Upper Paraná River, southern Brazil. *Primates* 50: 74–77.

Aguiar, L.M., Ludwig, G., Svoboda, W.K., Hilst, C.L., Navarro, I.T. & Passos, F.C. 2007a. Occurrence, local extinction and conservation of primates in the corridor of the Upper Paraná River, with notes on other mammals. *Revista Brasileira de Zoologia* 24: 898–906.

Aguiar, L.M., Mellek, D.M., Abreu, K.C., *et al.* 2007b. Sympatry between *Alouatta caraya* and *Alouatta clamitans* and the rediscovery of free-ranging potential hybrids in southern Brazil. *Primates* 48: 24–28.

Agustí, J. 2007. The biotic environments of the Late Miocene hominids. In *Handbook of Paleoanthropology*, W. Henke & I. Tattersall (eds). Berlin: Springer, pp. 979–1010.

Ahumada, J.A., Stevenson, P.R. & Quinones, M.J. 1998. Ecological response of spider monkeys to temporal variation in fruit abundance: the importance of flooded forest as a keystone habitat. *Primate Conservation* 18: 10–14.

Aide, T.M., Corrada-Bravo, C., Campos-Cerqueira, M., *et al.* 2013. Real-time bioacoustics monitoring and automated species identification. *Peer Journal* 1: e103.

Aiello, L.C. & Wheeler, P. 1995. The expensive tissue hypothesis. *Current Anthropology* 36: 199–221.

Akani, G.C., Aifesehi, P.E.E., Petrozzi, F., Amadi, N. & Luiselli, L. 2014. Preliminary surveys of the terrestrial vertebrate fauna (mammals, reptiles, and amphibians) of the Edumanon Forest Reserve, Nigeria. *Tropical Zoology* 27: 63–72.

Albernaz, A.L., Pressey, R.L., Costa, L.R., *et al.* 2012. Tree species compositional change and conservation implications in the white-water flooded forests of the Brazilian Amazon. *Journal of Biogeography* 39: 869–883.

Alberts, S.C., Hollister-Smith, J.A., Mutatua, R.S., *et al.* 2005. Seasonality and long-term change in a savanna environment. In *Seasonality in Primates: Studies of Living and Extinct Human and Non-human Primates*, D.K. Brockman & C.P. van Schiak (eds). Cambridge: Cambridge University Press, pp. 157–196.

Aldana, A.M., Beltrán, M., Torres-Neira, J. & Stevenson, P.R. 2008. Habitat characterization and population density of brown spider monkeys (*Ateles hybridus*) in Magdalena Valley, Colombia. *Neotropical Primates* 15: 46–50.

Alemseged, Z. 2003. An integrated approach to taphonomy and faunal change in the Shungura Formation (Ethiopia) and its implication for hominid evolution. *Journal of Human Evolution* 44: 451–478.

Alemseged, Z. & Bobe, R. 2009. Diet in early hominin species: a paleoenvironmental perspective. In *The Evolution of Hominin Diets: Integrating Approaches to the Study of Palaeolithic*

Subsistence, J.J. Hublin & M.P. Richards (eds). Dordrecht, The Netherlands: Springer, pp. 181–188.

Alencar, A., Nepstad, D. & Diaz, M.D.C.V. 2006. Forest understory fire in the Brazilian Amazon in ENSO and Non-ENSO years: area burned and committed carbon emissions. *Earth Interactions* 10: 1–17.

Alencar, D.C. 1994. Phenology of five Sapotaceae tropical tree species correlated to climatic variables in Ducke forest reserve, Manaus, AM. *Acta Amazonica* 24: 161–182.

Alfaro, J.W.L., Boubli, J.P., Olson, L.E., *et al.* 2012. Explosive Pleistocene range expansion leads to widespread Amazonian sympatry between robust and gracile capuchin monkeys. *Journal of Biogeography* 39: 272–288.

Alfaro, J.W.L., Cortés-Ortiz, L., Di Fiore, A. & Boubli, J.P. 2015. Comparative biogeography of Neotropical primates. *Molecular Phylogenetics & Evolution* 82: 518–529.

Alfred, R., Koh, P.H. & Lee, S.K. 2010. *The status of orangutan density and population size in seven key orangutan habitats in Sabah.* Survey report. Kota Kinabalu:Borneo Programme Species.

Alfred, R., Koh, P.H., Shan Khee, L. & Alfred, R. 2010. Summarizing spatial distribution density, movement patterns and food resources to study the impacts of logging and forest conversion on orang-utan population. *Journal of Biological Sciences* 10: 73–83.

Alho, C.J.R. (ed.). 2000. *Fauna silvestre da região do Rio Manso-MT.* Brasília: Edições Ibama/Eletronorte. Projeto de Divulgação Técnico-científica.

Alho, C.J.R. 2005. The Pantanal. In *The World's Largest Wetlands: Ecology and Conservation*, L.H. Fraser & P.A. Keddy (eds). Cambridge: Cambridge University Press, pp. 203–271.

Alho, C.J.R. 2008. Biodiversity of the Pantanal: response to seasonal flooding regime and to environmental degradation. *Revista Brasileira de Biologia* 68 (Suppl.): 957–966.

Alho, C.J.R. 2011. Environmental effects of hydropower reservoirs on wild mammals and freshwater turtles in Amazonia: a review. *Oecologia Australis* 15: 593–604.

Alho, C.J. & Pádua, L.F. 1982. Reproductive parameters and nesting behavior of the Amazon turtle *Podocnemis expansa* (Testudinata: Pelomedusidae) in Brazil. *Canadian Journal of Zoology* 60: 97–103.

Alho, C.J.R. & Sabino, J. 2012. Seasonal Pantanal flood pulse: implications for biodiversity conservation: a review. *Oecologia Australis* 16: 958–978.

Alho, C.J.R. & Silva, J.S.V. 2012. Effects of severe floods and droughts on wildlife of the Pantanal wetland (Brazil): a review. *Animals* 2: 591–610.

Alho, C.J.R., Camargo, G. & Fischer, E. 2011a. Terrestrial and aquatic mammals of the Pantanal. *Brazilian Journal of Biology* 71: 297–310.

Alho, C.J.R., Lacher Jr., T.E., Campos, Z.M.S. & Goncalves, H.C. 1987. Mamíferos da Fazenda Nhumirim, sub-região de Nhecolândia, Pantanal do Mato Grosso do Sul. 1 – Levantamento preliminar de espécies. *Revista Brasileira de Zoologia* 4: 151–164.

Alho, C.J.R., Mamede, S., Bitencourt, K. & Benites, M. 2011b. Introduced species in the Pantanal: implications for conservation. *Brazilian Journal of Biology* 71: 321–325.

Alho, C.J.R., Strussmann, C. & Vasconcellos, L.A.S. 2000. Indicadores da Magnitude da Diversidade e Abundância de Vertebrados Silvestres do Pantanal num Mosaico de Habitats Sazonais. In *Anais do III Simpósio sobre Recursos Naturais e Sócio-Econômicos do Pantanal.* Corumbá, Brazil: EMBRAPA, CPAP, pp. 1–54.

Aldhous, P. 2004. Borneo is burning. *Nature* 432: 144–146.

Alongi, D.M. 2008. Mangrove forests: resilience, protection from tsunamis, and responses to global climate change. *Estuarine Coastal & Shelf Science* 76: 1–13.

Altmann, J. 1974. Observational study of behavior: sampling methods. *Behaviour* 49: 227–267.

Altmann, J. & Alberts, S.C. 2005. Growth rates in a wild primate population: ecological influences and maternal effects. *Behavioral Ecology & Sociobiology* 57: 490–501.

Altmann, J., Combes, S.L. & Alberts, S.C. 2013. *Papio cynocephalus* yellow baboon. In *Mammals of Africa. Volume 2: Primates*, T.M. Butynski, J. Kingdon & J. Kalina (eds). London: Bloomsbury, pp. 228–232.

Alvarez, E., Balbás, L., Massa, I. & Pacheco, J. 1986. Aspectos Ecologicos de Embalse de Guri. *Interciencias* 11: 325–333.

Alves-Júnior, J.R.F., Lustosa, A.P.G., Bosso, A.C.S., *et al.* 2012. Reproductive indices in natural nests of giant Amazon river turtles *Podocnemis expansa* (Schweigger, 1812) (Testudines, Podocnemididae) in the environmental protection area meanders of the Araguaia river. *Brazilian Journal of Biology* 72: 199–203.

Amaral, D.D., Vieira, I.C.G., Salomão, R.P., Almeida, S.S. & Jardim, M.A.G. 2012. The status of conservation of urban forests in eastern Amazonia. *Brazilian Journal of Biology* 72: 257–265.

Ambrose, L. & Oates, J.F. 2013. *Euoticus pallidus* northern needle-clawed galago. In *Mammals of Africa. Volume II: Primates*, T.M. Butynski, J. Kingdon & J. Kalina (eds). London: Bloomsbury, pp. 444–446.

Amos, W. & Balmford, A. 2001. When does conservation genetics matter? *Heredity* 87: 257–265.

Ampeng, A., Sapuan, A., Sayok, A., Liam, J. & Asen, M. 2009. *Mammals along different ecosystems in Loagan Bunut National Park, Sarawak.* Kota Samarahan, Sarawak, Malaysia: Universiti Malaysia Sarawak.

Anadu, P.A. & Oates, J.F. 1982. *The status of wildlife in Bendel State, Nigeria, with recommendations for its conservation.* Unpublished report to the Bendel State Ministry of Agriculture and Natural Resources, Nigerian Federal Ministry of Agriculture, Nigerian Conservation Foundation, New York Zoological Society, and World Wildlife Fund.

Anadu, P.A. & Oates, J.F. 1988. The olive colobus monkey in Nigeria. *Nigerian Field* 53: 31–34.

Ancrenaz, M. 2007. *Wildlife Surveys Conducted in Lower Kinabatangan in 2006/2007 and Further Recommendations for Wildlife Management.* Kota Kinabalu, Malaysia: Arcus.

Ancrenaz, M. 2008. *Orang-utan Surveys in the Oil Palm Landscape of Eastern Saba.* Kota Kinabalu, Malaysia: Arcus.

Ancrenaz, M. 2010. *Orang-utan Bridges in Lower Kinabatangan.* Kota Kinabalu, Malaysia: Arcus.

Ancrenaz, M., Gimenez, O., Ambu, L., *et al.* 2005. Aerial surveys give new estimates for orangutans in Sabah, Malaysia. *PLoS Biology*, 3: e3.

Ancrenaz, M., Goossens, B., Gimenez, O., Sawang, A. & Lackman-Ancrenaz, I. 2004. Determination of ape distribution and population size using ground and aerial surveys: a case study with orang-utans in lower Kinabatangan, Sabah, Malaysia. *Animal Conservation* 7: 375–385.

Ancrenaz, M., Sollmann, R., Meijaard, E, *et al.* 2014a. Coming down from the trees: is terrestrial activity in Bornean orangutans natural or disturbance driven? *Scientific Reports* 4: 4024.

Ancrenaz, M., Oram, F., Ambu, L., *et al.* 2014b. Of pongo, palms, and perceptions: a multidisciplinary assessment of orangutans in an oil palm context. *Oryx* 49: 465–472.

Anderson, A.B. 1981. White-sand vegetation of Brazilian Amazonia. *Biotropica* 13: 199–210.

Anderson, J. 1872. On a supposed new monkey from the Sunderbunds to the east of Calcutta. *Proceedings of the Zoological Society of London* 1872: 529–533.

Anderson, J., Peignot, P. & Adelbrecht, C. 1992. Task-directed and recreational underwater swimming in captive Rhesus monkeys (*Macaca mulatta*). *Laboratory Primate Newsletter* 31: 1–4.

Anderson, J., Rowcliffe, M. & Cowlishaw, G. 2006. Does the matrix matter? A forest primate in a complex agricultural landscape. *Biological Conservation* 135: 212–222.

Anderson, J.T., Rojas, J.S. & Flecker, A.S. 2009. High-quality seed dispersal by fruit-eating fishes in Amazonian floodplain habitats. *Oecologia* 161: 279–290.

Ando, C., Iwata, Y. & Yamagiwa, J. 2008. Progress of habituation of western lowland gorillas and their reaction to observers in Moukalaba-Doudou National Park, Gabon. *African Study Monographs Supplementary Issue* 39: 55–69.

Andresen, E. 1999. Seed dispersal by monkeys and the fate of dispersed seeds in a Peruvian rain forest. *Biotropica* 31: 145–158.

Andriamaharoa, H., Birkinshaw, C. & Reza, L. 2010. Day-time feeding ecology of *Eulemur cinereiceps* in the Agnalazaha Forest, Mahabo-Mananivo, Madagascar. *Madagascar Conservation & Development* 5: 55–63.

Andrianandrasana, H.T., Randriamahefasoa, J., Durbin, J. & Lewis, R.E. 2005. Participatory ecological monitoring of the Alaotra wetlands in Madagascar. *Biodiversity & Conservation* 14: 2757–2774.

Ang, H.F. 2011. *Banded leaf monkeys in Singapore: preliminary data on taxonomy, feeding ecology, reproduction, and population size.* MSc thesis, National University of Singapore.

Ang, A., Srivasthan, A., Md-Zain, B.M., Ismail, M.R.B. & Meier, R. 2012. Low genetic variability in the recovering urban banded leaf monkey population of Singapore. *The Raffles Bulletin of Zoology* 60: 589–594.

Angelici, F.M., Grimod, I. & Politano, E. 1999. Mammals of the eastern Niger Delta (Rivers and Bayelsa States, Nigeria): an environment affected by a gas-pipeline. *Folia Zoologica* 48: 249–264.

Anonymous 2011. *Mammals in Sancang.* http://himbiounpad.wordpress.com/2012/02/20/mammal-in-sancang.

Antunes, A.C., Baccaro, F. & Barnett, A.A. 2017. What bite marks can tell us: use of on-fruit tooth impressions to study seed consumer identity and consumption patterns within a rodent assemblage. *Mammalian Biology* 82: 74–79.

Appanah, S. 1985. General flowering in the climax rain forest of Southeast Asia. *Journal of Tropical Ecology* 1: 225–240.

Appanah, S. 1993. Mass flowering of dipterocarp forests in the aseasonal tropics. *Journal of Bioscience* 18: 457–474.

Appelman, F.J. 1939. Het schiereiland Poerwo: boschen wild in Java's Zuidoost-hoek. In *3 Jaren Indisch Natuurleven, Elfde jaarverslag (1936–1938).* Batavia, The Netherlands: Nederlands Indische Vereeniging tot Natuurbescherming, pp. 293–298.

Aquino, R. 1988. Preliminary surveys on the densities of *Cacajao calvus ucayalii. Primate Conservation* 9: 24–26.

Aquino, R. 1998. Some observations on the ecology of *Cacajao calvus ucayalii* in the Peruvian Amazon. *Primate Conservation* 18: 21–24.

Aquino, R. 2005. Alimentación de mamíferos de caza en los aguajales de la Reserva Nacional de Pacaya-Samiria (Iquitos, Perú). *Revista Peruana de Biología* 12: 417–425.

Aquino, R. & Encarnación, F. 1999. Observaciones preliminares sobre la dieta de *Cacajao calvus ucayalii* en el Nor-Oriente Peruano. *Neotropical Primates* 7: 1–5.

Arantes, C.C., Castello, L., Stewart, D.J., Cetra, M. & Queiroz, H.L. 2010. Population density, growth and reproduction of arapaima in an Amazonian river-floodplain. *Ecology of Freshwater Fish* 19: 455–465.

Areola, O. 1982. Vegetation. In *Nigeria in Maps*, K.M. Barbour, J.S. Oguntoyinbo, J.O.C. Onyemelukwe & J.C. Nwafor (eds). London: Hodder & Stoughton, pp. 24–25.

Arias, M.E., Cochrane, T.A. & Elliott, V. 2014a. Modelling future changes of habitat and fauna in the Tonle Sap wetland of the Mekong. *Environmental Conservation* 41: 165–175.

Arias, M.E., Cochrane, T.A., Kummu, M., *et al.* 2014b. Impacts of hydropower and climate change on drivers of ecological productivity of Southeast Asia's most important wetland. *Ecological Modelling* 272: 252–263.

Arias, M.E., Cochrane, T.A., Norton, D., Killeen, T.J. & Khon, P. 2013. The flood pulse as the underlying driver of vegetation in the largest wetland and fishery of the Mekong Basin. *Ambio* 42: 864–876.

Armenteras, D., González, T.M. & Retana, J. 2013. Forest fragmentation and edge influence on fire occurrence and intensity under different management types in Amazon forests. *Biological Conservation* 159: 73–79.

Armenteras, D., Rudas, G., Rodriguez, N., Sua, S. & Romero, M. 2006. Patterns and causes of deforestation in the Colombian Amazon. *Ecological Indicators* 6: 353–368.

Arora, N., Nater, A., van Schaik, C.P., *et al.* 2010. Effects of Pleistocene glaciations and rivers on the population structure of Bornean orangutans (*Pongo pygmaeus*). *Proceedings of the National Academy of Sciences of the United States of America* 107: 21376–21381.

Arroyo-Rodríguez, V. & Dias, P.A.D. 2010. Effects of habitat fragmentation and disturbance on howler monkeys: a review. *American Journal of Primatology* 72: 1–16.

Arroyo-Rodríguez, V. & Mandujano, S. 2009. Conceptualization and measurement of habitat fragmentation from the primates' perspective. *International Journal of Primatology* 30: 497–514.

Arroyo-Rodríguez, V., Cuesta-del Moral, E., Mandujano. S., *et al.* 2013. Assessing habitat fragmentation effects on primates: the importance of evaluating questions at the correct scale. In *Primates in Fragments: Complexity and Resilience*, L.K. Marsh & C.A. Chapman (eds). New York: Springer, pp. 13–28.

ASEAN Peat Swamp Forest Project (APFP). 2013. *Peatlands in Southeast Asia. Association of Southeast Asian Nations (ASEAN)* Secretariat/Global Environment Centre (GEC). Accessed at www.aseanpeat.net/index.cfm?&menuid=62.

ASEAN Secretariat and Global Environment Centre. 2012. *Peatlands in Southeast Asia.* Selanagor, Malaysia: Global Environment Centre.

Asensio, N., Arroyo-Rodríguez, V., Dunn, J.C. & Cristóbal-Azkarate, J. 2009. Conservation value of landscape supplementation for howler monkeys living in forest patches. *Biotropica* 41: 768–773.

Ashton, P.S., Givnish, T.J. & Appanah, S. 1998. Staggered flowering in the Dipterocarpaceae: new insights into floral induction and the evolution of mast fruiting in the aseasonal tropics. *American Naturalist* 132: 44–66.

Asquith, N.M. 2001. Misdirections in conservation biology. *Conservation Biology* 15: 345–352.

Assis, R.L., Haugaasen, T., Schöngart, J., *et al.* 2015a. Patterns of tree diversity and composition in Amazonian floodplain paleo-várzea forest. *Journal of Vegetation Science* 26: 312–322.

Assis, R.L., Wittmann, F., Piedade, M.T. & Haugaasen, T. 2015b. Effects of hydroperiod and substrate properties on tree alpha diversity and composition in Amazonian floodplain forests. *Plant Ecology* 216: 41–54.

Astaras, C., Krause, S., Mattner, L., Rehse, C. & Waltert, M. 2011. Associations between the drill (*Mandrillus leucophaeus*) and sympatric monkeys in Korup National Park, Cameroon. *American Journal of Primatology* 70: 306–310.

Atrium. 2012. *Aguajal project.* Andes-Amazon. http://atrium.andesamazon.org/digital_herbarium.php.

Avery, G. & Siegfried, W.R. 1980. Food gatherers along South Africa's shoreline. *Oceans* 13: 32–37.

Ayres, J.M.C. 1985. On a new species of squirrel monkey, genus *Saimiri* (Cebidae, Primates) from Brazilian Amazonia. *Papeis Avulsos de Zoologia* 36: 147–164.

Ayres, J.M.C. 1986a. *Uakaris and Amazonian flooded forest.* PhD thesis. University of Cambridge, Cambridge, UK.

Ayres, J.M.C. 1986b. Some aspects of social problems facing conservation in Brazil. *Trends in Ecology & Evolution* 1: 48–49.

Ayres, J.M.C. 1986c. The conservation status of the White Uakari. *Primate Conservation* 7: 22–26.

Ayres, J.M.C. 1989. Comparative feeding ecology of the uakari and bearded saki, *Cacajao* and *Chiropotes*. *Journal of Human Evolution* 18: 697–716.

Ayres, J.M.C. & Best, R. 1979. Estratégias para a Conservação da fauna Amazônica. *Acta Amazônica* 9: 81–101.

Ayres, J.M.C. & Clutton-Brock, T.H. 1992. River boundaries and species range size in Amazonian primates. *American Naturalist* 140: 531–537.

Ayres, J.M.C. & Johns, A.D. 1987. Conservation of white uakaries in Amazonian várzea. *Oryx* 21: 74–88.

Ayres, J.M.C. & Milton, K. 1981. Levantamento de primatas e habitat no Rio Tapajos. *Boletim do Museu Paraense Emilio Goeldi. Nova Serie, Zoologia* 111: 1–11.

Ayres, J.M.C. & Prance, G.T. 2013. On the distribution of Pitheciin monkeys and Lecythidaceae in Amazonia. In *Evolutionary Biology and Conservation of Titis, Sakis and Uacaris*, L. Veiga, A. Barnett, S. Ferrari & M. Norconk (eds). Cambridge: Cambridge University Press, pp. 127–144.

Ayres J.M.C., Alves, A.R., de Queiroz, H.L, *et al.* 1999. Mamirauá: the conservation of biodiversity in an Amazonian flooded forest. *Advances in Economic Botany* 13: 203–216.

Ayres, J.M.C., Fonseca, G.A.B., Rylands, A.B., *et al.* 2005. *Os Corredores Ecológicos das Florestas Tropicais do Brasil.* Belém: Sociedade Civil Mamiraurá.

Azmi, R. 1998. *Natural vegetation of the Kinabatangan floodplain. An introduction to its natural vegetation, including a preliminary plant checklist of the region.* Kota Kinabalu: WWF Malaysia.

Azmi, A.B. 2007. *Plankton Distribution and Composition of Oxbow Lakes in Sabah.* MSc dissertation, School of Science and Technology Universiti Malaysia Sabah. http://eprints.ums.edu.my/7325/1/mt0000000246.pdf.

Baal, F.L.J., Mittermeier, R.A. & van Roosmalen, M.G.M. 1988. Primates and protected areas in Suriname. *Oryx* 22: 7–14.

Bah, M., Keita, A., Diallo, S.T., Camara, S. & Sagna, S. 1999. *Evaluation de la diversité biologique de la Guinée. Vision, buts et objectifs de la stratégie nationale pour sa conservation et son utilisation durable.* Unpublished report. Conakry, Guinea: MMGE-DNE-PNUD-FEM.

Baha El Din, S.M. 1999. *Directory of Important Bird Areas in Egypt.* Birdlife International. Cairo: Palm Press.

Bahuguna, N.C. & Mallick, J.K. 2011. *Handbook of the Mammals of South Asia.* Dehra Dun, India: Natraj Publishers.

Baker, L.R. 2003. *Report on a Survey of Stubbs Creek Forest Reserve.* Unpublished report. Calabar, Nigeria: CERCOPAN.

Baker, L.R. 2005. *Distribution and conservation status of Sclater's guenon* (Cercopithecus sclateri) *in southern Nigeria.* Unpublished report. Washington DC: Margot Marsh Biodiversity Foundation, Conservation International.

Baker, L.R. & Olubode, O.S. 2008. Correlates with the distribution and abundance of Endangered Sclater's monkeys (*Cercopithecus sclateri*) in southern Nigeria. *African Journal of Ecology* 46: 365–373.

Baker, L.R., Arnold, T.W., Olubode, O.S. & Garshelis, D.L. 2011. Considerations for using occupancy surveys to monitor forest primates: a case study with Sclater's monkey (*Cercopithecus sclateri*). *Population Ecology* 53: 549–561.

Bakoarinіniaina, L.N., Kusky, T. & Raharimahefa, A. 2006. Disappearing Lake Alaotra: monitoring catastrophic erosion, waterway silting, and land degradation hazards in Madagascar using Landsat imagery. *Journal of African Earth Sciences* 44: 241–252.

Balcazar, H.E. 2014. *Problemática de las inundaciones en el Beni.* Unpublished report. Beni, Bolivia: Estancias VH Cabaña Brahman. www.estanciasvh.com/?p=2000.

Ball, M.C. 1988. Ecophysiology of mangroves. *Trees* 2: 129–142.

Ballantyne, R., Packer, J. & Hughes, K. 2009. Tourists' support for conservation messages and sustainable management practices in wildlife tourism experiences. *Tourism Management* 30: 658–664.

Ballantyne, R., Packer, J., Hughes, K. & Dierking, L. 2007. Conservation learning in wildlife tourism settings: lessons from research in zoos and aquariums. *Environmental Education Research* 13: 367–383.

Bally, R. 1987. The ecology of sandy beaches of the Benguela ecosystem. *South African Journal of Marine Science* 5: 759–770.

Baracuhy, V., Souza-Silva, W., Spironello, W., Ross, C. & MacLarnon, A. 2013. Arthropod predation by a specialist seed predator, the golden-backed uacari (*Cacajao melanocephalus ouakary*, Pitheciidae) in Brazilian Amazonia. *International Journal of Primatology* 34: 470–485.

Barbosa, R.I., Campos, C., Pinto, F. & Fearnside, P.M. 2007. The 'Lavrados' of Roraima: biodiversity and conservation of Brazil's Amazonian savannas. *Functional Ecosystems and Communities* 1: 29–41.

Barnes, R. 1992. Case studies in conserving large mammals. In *The Conservation Atlas of Tropical Forests Africa*, J.A. Sayer, C.S. Harcourt & N.M. Collins (eds). London: Macmillan, pp. 33–42.

Barnett, A. 1995. *Primates.* Expedition Field Techniques series, No. 6. London: Expedition Advisory Centre, Royal Geographical Society.

Barnett, A.A. 2010. *Diet, habitat, use and conservation ecology of the golden-backed uacari*, Cacajao melanocephalus ouakary, *in Jaú National Park, Amazonian Brazil.* PhD Thesis, University of Roehampton, London, UK.

Barnett, A.A. & Brandon-Jones, D. 1997. The ecology, biogeography, and conservation of the uakaris *Cacajao* (Pitheciinae). *Folia Primatologica* 68: 223–235.

Barnett, A.A. & Shaw, P. 2014. More food or fewer predators? The benefits to birds of associating with a Neotropical primate varies with their foraging strategy. *Journal of Zoology* 294: 224–233.

Barnett, A.A., Almeida, T., Andrade, R., *et al.* 2015a. Ants in their plants: *Pseudomyrmex* ants reduce primate, parrot and squirrel predation on *Macrolobium acaciifolium* (Fabaceae) seeds in Amazonian Brazil. *Biological Journal of the Linnean Society* 114: 260–273.

Barnett, A.A., Almeida, T., Spironello, W.R., *et al.* 2012a. Terrestrial foraging by *Cacajao melanocephalus ouakary* (Primates) in Amazonian Brazil: is choice of seed patch size and position related to predation risk? *Folia Primatologica* 83: 126–139.

Barnett, A.A., Bezerra, B.M., Oliviera, M., Queiroz, H. & Defler, T.R. 2013a. *Cacajao ouakary* in Brazil and Colombia: patterns, puzzles and predictions. In *Evolutionary Biology and Conservation of Titis, Sakis and Uacaris*, L. Veiga, A. Barnett, S. Ferrari & M. Norconk (eds). Cambridge: Cambridge University Press, pp. 179–195.

Barnett, A.A., Bezerra, B.M., Spironello, W.R., *et al.* 2016. Foraging with finesse: a hard-fruit-eating primate selects weakest areas as bite sites. *American Journal of Physical Anthropology* 160: 113–125.

Barnett, A.A., Bonchi-Teles, B., Sousa Silva W., Almeida, T., Bezerra, B., Gonçalves de Lima, M., Spironello, W.R., MacLarnon, A., Ross, C. & Shaw, P.J.A. 2017. Covert carnivory: a seed-predating primate, the golden-backed uacari, selects insect-infested fruits. *Journal of Zoological Research* 1: 16–31.

Barnett, A.A., Borges, S., de Castilho C.V., Neri, F. & Shapley, R.L. 2002. Primates of Jaú National Park, Amazonas, Brazil. *Neotropical Primates* 10: 65–70.

Barnett, A.A., Bowler, M., Bezerra, B.M. & Defler, T.R. 2013b. Ecology and behaviour of uacaris (*Cacajao*). In *Evolutionary Biology and Conservation of Titis, Sakis and Uacaris*, L. Veiga, A. Barnett, S. Ferrari & M. Norconk (eds). Cambridge: Cambridge University Press, pp. 151–172.

Barnett, A.A., Boyle, S.A., Norconk, M.M., *et al.* 2012b. Terrestrial activity in pitheciins (*Cacajao*, *Chiropotes*, and *Pithecia*). *American Journal of Primatology* 74: 1106–1127.

Barnett, A.A., Boyle, S.A., Pinto, L.P., *et al.* 2012c. Primary seed dispersal by three Neotropical seed-predating primates (*Cacajao melanocephalus ouakary*, *Chiropotes chiropotes* and *Chiropotes albinasus*). *Journal of Tropical Ecology* 28: 543–555.

Barnett, A.A., de Castilho, C.V., Shapley, R.L. & Anicácio, A. 2005. Diet, habitat selection and natural history of *Cacajao melanocephalus ouakary* in Jaú National Park, Brazil. *International Journal of Primatology* 26: 949–969.

Barnett, A.A., Ronchi-Teles, B., Sousa-Silva, W., *et al.* 2013c. Covert carnivory: a seed-predating primate, the golden-backed uacari, selects insect-infested fruits. *Journal of Zoological Research.*

Barnett, A.A., Schiel, V., Deveny, A.,*et al.* 2011. Predation on *Cacajao ouakary* and *Cebus albifrons* (Primates: Platyrrhini) by harpy eagles. *Mammalia* 75: 169–172.

Barnett, A.A., Shaw, P., Spironello, W.R., MacLarnon, A. & Ross, C. 2012d. Sleeping site selection by golden-backed uacaris, *Cacajao melanocephalus ouakary* (Pitheciidae), in Amazonian flooded forests. *Primates* 53: 273–285.

Barnett, A.A., Silva, W.S., Shaw, P.J.A. & Ramsay, R.M. 2015b. Inundation duration and vertical vegetation stratification: a preliminary description of the vegetation and structuring factors in Borokotóh (hummock *igapó*), an overlooked, high-diversity, Amazonian habitat. *Nordic Journal of Botany* 33: 601–614.

Barreta, J. 2007. *Caracterización genética de dos especies de monos tití Callicebus olallae y* Callicebus modestus *del Departamento del Beni.* Technical Report. La Paz, Bolivia: Wildlife Conservation Society & Insitituto de Biología Molecular y Biotecnología, Universidad Mayor San Andres.

Barrett, E. 1981. The present distribution and status of the slow loris in Peninsular Malaysia. *Malaysian Applied Biology* 10: 205–211.

Barrow, C. 1988. The impact of hydroelectric development on the Amazonian environment: with particular reference to the Tucuruí Project. *Journal of Biogeography* 15: 67–78.

Barry, J.C., Morgan, M.E., Flynn, L.J., *et al.* 2002. Faunal and environmental change in the late Miocene Siwaliks of northern Pakistan. *Paleobiology* 28: 1–71.

Barthem, R. & Goulding, M. 1997. *The Catfish Connection: Ecology, Migration, and Conservation of Amazon Predator.* New York: Columbia University Press.

Barton, K. 2013. MuMIn: Multi-model inference. mumin.r-forge.r-project.org/MuMIn-manual.pdf.

Barton, R.A., Whiten, A., Strum, S.C., Byrne, R.W. & Simpson, A.J. 1992. Habitat use and resource availability in baboons. *Animal Behaviour* 43: 831–844.

Basabose, A.K. 2002. Diet composition of chimpanzees inhabiting the montane forest of Kahuzi, Democratic Republic of Congo. *American Journal of Primatology* 58: 1–21.

Basabose, A.K. 2004. Fruit availability and chimpanzee party size at Kahuzi montane forest, Democratic Republic of Congo. *Primates* 45: 211–219.

Basabose, A.K. 2005. Ranging patterns of chimpanzees in a montane forest of Kahuzi, Democratic Republic of Congo. *International Journal of Primatology* 26: 33–54.

Basabose, A.K. & Yamagiwa, J. 2002. Factors affecting nesting site choice in chimpanzees at Tshibati, Kahuzi-Biega National Park: influence of sympatric gorillas. *International Journal of Primatology* 23: 263–282.

Bastazini, V.A.G. 2011. *Efeitos da estrutura de habitat e do espaço sobre a diversidade de mamíferos no norte do Pantanal: uma abordagem de resolução fina.* MSc dissertation. Porte Alegre, Brazil: Universidade Federal Rio Grande do Sul.

Bastos, H.B., Gonçalves, E.C., Ferrari, S.F., Silva, A. & Schneider, M.P.C. 2010. Genetic structure of red-handed howler monkey populations in fragmented landscape of Eastern Brazilian Amazonia. *Genetics and Molecular Biology* 33: 774–780.

Bastos, M. 2013. *Comunicação vocal em* Sapajus flavius *na natureza.* MSc dissertation. Recife, Brazil: Universidade Federal de Pernambuco.

Bastos, M., Souto, A., Jones, G., *et al.* 2015. Vocal repertoire of wild blonde capuchins (*Sapajus flavius*) and contextual use of calls. *American Journal of Primatology* 77: 605–617.

Bastos, M., Medeiros, K., Jones, G., Bezerra, B. 2018. Small but wise: Common marmosets (*Callithrix jacchus*) use acoustic signals as cues to avoid interactions with blonde capuchin monkeys (*Sapajus flavius*). *American Journal of Primatology* 80: e22744.

Batista, P.M., Andreotti, R., Almeida, O.S., *et al.* 2013. Detection of arboviruses of public health interests in free-living New World primates (*Sapajus* spp., *Alouatta caraya*) captured in Mato Grosso do Sul, Brazil. *Revista Sociedade Brasiliera Medicina Tropical* 44: 684–690.

Batistella, A.M. & Vogt, R.C. 2009. Nesting ecology of *Podocnemis erythrocephala* (Testudines, Podocnemididae) of the Rio Negro, Amazonas, Brazil. *Chelonian Conservation & Biology* 7: 12–20.

Bauchop, T. 1978. Digestion of leaves in vertebrate arboreal folivores. In *The Ecology of Arboreal Folivores*, G.G. Montgomery (ed.). Washington, DC: Smithsonian Institution Press, pp. 193–204.

Baumgarten, A. & Williamson, G.B. 2007. Distribution of the black howler monkey (*Alouatta pigra*) and the mantled howler monkey (*A. palliata*) in their contact zone in eastern Guatemala. *Neotropical Primates* 14: 11–18.

Beckwith RS. 1995. *The Ecology and behaviour of the Javan black langur, in a lower montane rain forest, West Java.* PhD thesis, University of Cambridge, Cambridge, UK.

Beard, J. 1955. A note on gallery forests. *Ecology* 36: 339–340.

Bearder, S.K. & Nekaris, A.I. 2011. Tips from the bush: an A–Z of suggestions for successful fieldwork. In *Field and Laboratory Methods in Primatology*, J.M. Setchell & D.J. Curtis (eds). Cambridge: Cambridge University Press, pp. 387–403.

Bearder, S.K., Honess, P.E. & Ambrose, L. 1995. Species diversity among galagos with special reference to mate recognition. In *Creatures of the Dark: The Nocturnal Prosimians*, L. Alterman, G.A. Doyle & M.K. Izard (eds). New York: Plenum Press, pp. 331–352.

Beaumont, M.A. 1999. Detecting population expansion and decline using microsatellites. *Genetics* 153: 2013–2029.

Beaune, S.A. 2004. The invention of technology: prehistory and cognition. *Current Anthropology* 45: 139–162.

Begun, D.R. 2007. Fossil record of miocene hominoids. In *Handbook of Paleoanthropology*, W. Henke & I. Tattersall (eds). Berlin: Springer, pp. 921–977.

Behling, H. 2002. South and southeast Brazilian grasslands during Late Quaternary times: a synthesis. *Palaeogeography, Palaeoclimatology, Palaeoecology* 177: 19–27.

Behrensmeyer, A.K., Badgley, C., Barry, J.C., Morgan, M.E. & Raza, S.M. 2005. The paleoenvironmental context of Siwalik Miocene vertebrate localities. In *Interpreting the Past: Essays on Human, Primate and Mammal Evolution in Honor of David Pilbeam*, D.E. Lieberman, R.J. Smith & J. Kelley (eds). Boston: Brill Academic, pp. 47–61.

Behrensmeyer, A.K. & Reed, K.E. 2013. Reconstructing the habits of *Australopithecus*: paleoenvironments, site taphonomy, and fauna. In *The Paleobiology of Australopithecus*, K.E. Reed, J.G. Fleagle & R.E. Leakey (eds). Dordrecht: Springer, pp. 41–60.

Beilfuss, R., Moore, D., Bento, C. & Dutton, P. 2001. Patterns of vegetation change in the Zambezi Delta, Mozambique. *Working Paper 3. Mozambique: Program for the Sustainable Management of Cahora Bassa Dam and the Lower Zambezi Valley.* http://files.gorongosa.net/filestore/349-patterns_vegetation_change_zambezi_delta.pdf.

Beja, P., Santos, C.D., Santana, J., *et al.* 2010. Seasonal patterns of spatial variation in understory bird assemblages across a mosaic of flooded and unflooded Amazonian forests. *Biodiversity & Conservation* 19: 129–152.

Beltrão-Mendes, R., Cunha, A.A. & Ferrari, S.F. 2011. New localities and perspectives on the sympatry between two Endangered primates (*Callicebus coimbrai* and *Cebus xanthosternos*) in northeastern Brazil. *Mammalia* 75: 103–105.

Benchimol, M. & Peres, C.A. 2013. Anthropogenic modulators of species-area relationships in Neotropical primates: a continental-scale analysis of fragmented forest landscapes. *Diversity & Distributions* 19: 1339–1352.

Benchimol, M. & Peres, C.A. 2015a. Widespread forest vertebrate extinctions induced by a mega hydroelectric dam in lowland Amazonia. *PLOS One* 10: e0129818.

Benchimol, M. & Peres, C.A. 2015b. Predicting local extinctions of Amazonian vertebrates in forest islands created by a mega dam. *Biological Conservation* 187: 61–72.

Benchimol, M. & Venticinque, E.M. 2014. Responses of primates to landscape change in Amazonian land-bridge islands: a multi-scale analysis. *Biotropica* 46: 470–478.

Bennett, C.L., Leonard, S. & Carter, S. 2001. Abundance, diversity, and patterns of distribution of primates on the Tapiche River in Amazonian Peru. *American Journal of Primatology* 54: 119–126.

Bennett, E.L. 1986. *Proboscis monkeys in Sarawak: their ecology, status, conservation and management.* Report of WWF Project MAL 63/84., Kuala Lumpur, Malaysia: World Wildlife Fund.

Bennett, E. 1988. Proboscis monkeys and their swamp forests in Sarawak. *Oryx* 22: 69–74.

Bennett, E.L. & Davies, A.G. 1994. The ecology of Asian colobines. In *Colobine Monkeys: Their Ecology, Behaviour and Evolution*, A.G. Davies & J.F. Oates (eds). Cambridge: Cambridge University Press, pp. 129–171.

Bennett, E.L. & Gombek, F. 1993. *Proboscis Monkeys of Borneo.* Sabah, Malaysia: Natural History Publications (Borneo), Malaysia.

Bennett, E.L. & Reynolds, C.J. 1993. The value of a mangrove area in Sarawak. *Biodiversity & Conservation* 2: 359–375.

Bennett, E.L. & Sebastian, A.C. 1988. Social organization and ecology of proboscis monkeys (*Nasalis larvatus*) in mixed coastal forest in Sarawak. *International Journal of Primatology* 9: 233–255.

Berman, C. 1977. Seaside play is a serious business. *New Scientist* 73: 761–763.

Bernard, H., Matsuda, I., Hanya, G. & Ahmad, A.H. 2011a. Characteristics of night sleeping trees of proboscis monkeys (*Nasalis larvatus*) in Sabah, Malaysia. *International Journal of Primatology* 32: 259–267.

Bernard, H., Matsuda, I., Hanya, G. & Ahmad, A.H. 2011b. Effects of river width on sleeping-site selection by proboscis monkeys (*Nasalis larvatus*) in Sabah, Malaysia. *Journal of Tropical Biology and Conservation* 8: 9–12.

Bernard, H. & Zulhazman, H. 2006. Population size and distribution of the proboscis monkey (*Nasalis larvatus*) in the Klias Peninsula, Sabah, Malaysia. *Malayan Nature Journal* 59: 1531–1563.

Bermejo, M. 2004. Home-range use and intergroup encounters in western gorillas (*Gorilla g. gorilla*) at Lossi, North Congo. *American Journal of Primatology* 64: 223-232.

Bezanson, M., Garber, P.A., Murphy, J.T. & Premo, L.S. 2008. Patterns of subgrouping and spatial affiliation in a community of mantled howling monkeys (*Alouatta palliata*). *American Journal of Primatology* 70: 282–293.

Bezerra, B.M., Barnett, A., Silva Junior, J., Souto, A. & Jones, G. 2008. Sounding out species: holes in our distributional knowledge of Amazonian primates, and the potential for call playback as a survey technique. *Transactions of the XXII International Primatological Society Congress, Edinburgh. August 2-8 2008, Primate Society of Great Britain, Presentation # 69*, pp. 249–250.

Bezerra, B.M., Barnett, A.A., Souto, A. & Jones, G. 2011. Ethogram and natural history of golden-backed uakaris (*Cacajao melanocephalus*). *International Journal of Primatology* 32: 46–68.

Bezerra, B.M., Bastos, M., Souto, A., *et al.* 2014. Camera trap observations of nonhabituated Critically Endangered wild blonde capuchins, *Sapajus flavius* (formerly *Cebus flavius*). *International Journal of Primatology* 35: 895–907.

Bezerra, B.M. & Souto, A. 2008. Structure and usage of the vocal repertoire of *Callithrix jacchus*. *International Journal of Primatology* 29: 671–701.

Bezerra, B.M., Souto, A.S. & Jones, G. 2010a. Vocal repertoire of golden-backed uakaris (*Cacajao melanocephalus*): call structure and context. *International Journal of Primatology* 31: 759–778.

Bezerra, B.M., Souto, A.S. & Jones, G. 2010b. Responses of golden-backed uakaris, *Cacajao melanocephalus*, to call playback: implications for surveys in the flooded Igapó forest. *Primates* 51: 327–336.

Bezerra, B.M., Souto, A.S. & Jones, G. 2012. Propagation of the loud 'tchó' call of golden-backed uakaris, *Cacajao melanocephalus*, in the black-swamp forests of the upper Amazon. *Primates* 53: 317–325.

Bezerra, B.M., Souto, A.S., Radford, A.N. & Jones, G. 2011. Brevity is not always a virtue in primate communication. *Biology Letters* 7: 23–25.

Bi, S.G., Bené, J.C.K., Bitty, E.A., *et al.* 2013. Roloway guenon (*Cercopithecus diana roloway*) and white-naped mangabey (*Cercocebus atys lunulatus*) prefer mangrove habitats in Tanoé forest, southeastern Ivory Coast. *Journal of Ecosystem & Ecography* 3: 126.

Bi, S.G., Bené, J.C.K., Bitty, E.A., Kone, I. & Zinner, D. 2009. Distribution of the green monkey (*Chlorocebus sabaeus*) in the coastal zone of the Cote d'Ivoire. *Primate Conservation* 24: 91–97.

Bi, S.G., Kone, I., Bene, J.C.K., *et al.* 2008. Tanoe forest, southeastern Cote d'Ivoire identified as a high priority for the conservation of Critically Endangered primates in West Africa. *Tropical Conservation Science* 1: 265–278.

Bibby, C., Jones, M. & Marsden, S. 1998. *Bird Surveys.* Expedition Field Techniques series, No. 7. London: Expedition Advisory Centre, Royal Geographical Society.

Bicca-Marques, J.C. & Calegaro-Marques, C. 1995. Updating the known distribution of the pygmy marmoset (*Cebuella pygmaea*) in the state of Acre, Brazil. *Neotropical Primates* 3: 48–49.

Bicca-Marques, J.C. & Calegaro-Marques, C. 1998. Behavioral thermoregulation in a sexually and developmentally dichromatic neotropical primate, the black-and-gold howling monkey (*Alouatta caraya*). *American Journal of Physical Anthropology* 10: 533–546.

Bicca-Marques, J.C. & Heymann, E.W. 2013. Ecology and behavior of titi monkeys (genus *Callicebus*). In *Evolutionary Biology and Conservation of Titis, Sakis and Uacaris*, L.M. Veiga, A.A. Barnett, S.F. Ferrari, & M.A. Norconk (eds). Cambridge: Cambridge University Press, pp. 196–207.

Biddulph, J. & Kellman, M. 1998. Fuels and fire at savanna-gallery forest boundaries in southeastern Venezuela. *Journal of Tropical Ecology* 14: 445–461.

Bien, A. 2008. *A Simple User's Guide to Certification for Sustainable Tourism and Ecotourism.* Washington DC: Center for Ecotourism and Sustainable Development.

Birdlife International. 2015. *Important Bird Areas Factsheet.* Cambridge, UK: Birdlife International. http://www.birdlife.org.

Birkinshaw, C.R. & Colquhoun, I. 2003. Lemur food plants. In *Natural History of Madagascar*, S. Goodman & J. Benstead (eds). Chicago: Chicago University Press, pp. 1207–1220.

Bishop, J., Kapila, S., Hicks, F., Mitchell, P. & Vohries, F. 2009. New business models for biodiversity conservation. *Journal of Sustainable Forestry* 28: 285–303.

Bismark, M. 2010. Proboscis monkey (*Nasalis larvatus*): bio-ecology and conservation. In *Indonesian Primates*, S. Gursky-Doyen & J. Supriatna (eds). New York: Springer, pp. 217–236.

Blackham, G. 2005. *Pilot Survey of Nocturnal Primates,* Tarsius bancanus borneanus *(Western Tarsier) and* Nycticebus coucang menagenis *(Slow Loris) in Peat Swamp Forest, Central Kalimantan, Indonesia.* MSc dissertation, Oxford Brookes University, Oxford, UK.

Blake, S. 1993. *A Reconnaissance Survey in the Likouala Swamps of Northern Congo and its Implications for Conservation.* MSc dissertation, University of Edinburgh, Edinburgh, UK.

Blake, S. 1994. *A pilot study of western lowland gorilla social organization at the Mbeli Bai, Nouabalé-Ndoki National Park, northern Congo.* Unpublished report. USAID, Wildlife Conservation Society, GEF-Congo, Government of Congo, and GTZ.

Blake, S. 2002. Forest buffalo prefer clearings to closed-canopy forest in the primary forest of northern Congo. *Oryx* 36: 81–86.

Blake, S., Rogers, E., Fay, J.M., Ngangoue, M. & Ebeke, G. 1995. Swamp gorillas in northern Congo. *African Journal of Ecology* 33: 285–290.

Blench, R. 2007. *Mammals of the Niger Delta, Nigeria.* Unpublished report. Cambridge, UK: Kay Williamson Education Foundation. www.rogerblench.info/Ethnoscience/Animals/Mammals/Niger%20Delta%20mammal%20book.pdf.

Blesgraaf, R., Geilvoet, A., van der Hout, C., Smoorenburg, M. & Sotthewes, W. 2006. *Salinity in the Casamance Estuary: Occurrence and Consequences.* Joint MSc project dissertation. Delft University of Technology, Delft, Netherlands.

Blomquist G.E, Kowalewski, M.M. & Leigh, S.R. 2008. Demographic and morphological perspectives on life history evolution and conservation of new world monkeys. In *South American Primates; Comparative Perspectives in the Study of Behavior, Ecology, and Conservation*, P. Garber, A. Estrada, J.C. Bicca-Marques, E.W. Heymann & K.B. Strier (eds). Development in Primatology, Progress & Prospects series. New York: Springer, pp. 117–138.

Blouch, R. 1997. Distribution and abundance of orangutans (*Pongo pygmaeus*) and other primates in the Lanjak Entimau Wildlife Sanctuary, Sarawak, Malaysia. *Tropical Biodiversity* 4: 259–274.

Blumenfeld-Jones, K., Randriamboavonjy, T.M., Williams, G., *et al.* 2006. Tamarind recruitment and long-term stability in the gallery forest at Berenty, Madagascar. In *Ringtailed Lemur Biology:* Lemur catta *in Madagascar*, A. Jolly, N. Koyama, H.R. Rasamimanana & R.W. Sussman (eds). New York: Springer, pp. 69–85.

Bobadilla, U.L. & Ferrari, S.F. 2000. Habitat use by *Chiropotes satanas utahicki* and syntopic platyrrhines in eastern Amazonia. *American Journal of Primatology* 50: 215–224.

Bobe, R. 2006. The evolution of arid ecosystems in eastern Africa. *Journal of Arid Environments* 66: 564–584.

Bobrowiec, P.E.D., Rosa, L.D.S., Gazarini, J. & Haugaasen, T. 2014. Phyllostomid bat assemblage structure in Amazonian flooded and unflooded forests. *Biotropica* 46: 312–321.

Bocian, C. 1998. *Preliminary observations on the status of primates in the Etiema community forest.* Unpublished report. Lagos, Nigeria: A.G. Leventis Foundation and the Nigerian Conservation Foundation.

Bocian, C. 1999. *A primate survey of the Okoroba community forest, Edumanom Forest Reserve.* Unpublished report. Lagos, Nigeria: A.G. Leventis Foundation and the Nigerian Conservation Foundation.

Bodmer, R.E. 1990. Responses of ungulates to seasonal inundations in the Amazon floodplain. *Journal of Tropical Ecology* 6: 191–201.

Boinski, S. 1993. Vocal coordination of troop movement among white-faced capuchin monkeys, *Cebus capucinus. American Journal of Primatology* 30: 85–100.

Boinski, S. & Campbell, A.F. 1996. The Huh vocalization of white-faced capuchins: a spacing call disguised as a food call? *Ethology* 102: 826–840.

Boinski, S., Jack, K., Lamarsh, C. & Coltrane, J.A. 1998. Squirrel monkeys in Costa Rica: drifting to extinction. *Oryx* 32: 45–58.

Boinski, S. & Siwt, L. 1997. Uncertain conservation status of squirrel monkeys in Costa Rica, *Saimiri oerstedi oerstedi* and *Saimiri oerstedi citrinellus. Folia Primatologica* 68: 181–193.

Bonetto, A.A. 1986. The Parana River System. In *The Ecology of River Systems*, B.R. Davies & K.F. Walker (eds). Dordrecht: Dr. W. Junk Publishers, pp. 541–556.

Boo, E. 1992. *The Ecotourism Boom: Planning for Development and Management.* Washington DC: WWF, Wildlands and Human Needs Program.

Boonratana, R. 1993. *The ecology and behaviour of the proboscis monkey* (Nasalis larvatus*) in the Lower Kinabatangan, Sabah.* PhD thesis, Mahidol University, Thailand.

Boonratana, R. 2000a. A study of the vegetation of the forests in the lower Kinabatangan region, Sabah, Malaysia. *Malay Nature Journal* 54: 271–288.

Boonratana, R. 2000b. Ranging behavior of proboscis monkeys (*Nasalis larvatus*) in the Lower Kinabatangan, northern Borneo. *International Journal of Primatology* 21: 497–518.

Boonratana, R. 2000c. A study of the vegetation of the forests in the lower Kinabatangan region, Sabah, Malaysia. *Malayan Nature Journal* 54: 271–288.

Boonratana, R. 2003. Feeding ecology of proboscis monkeys (*Nasalis larvatus*) in the Lower Kinabatangan, Sabah, Malaysia. *Sabah Parks Nature Journal* 6: 1–26.

Boonratana, R. 2011. Observations on the sexual behaviour and birth seasonality of proboscis monkey (*Nasalis larvatus*) along the lower Kinabatangan River, northern Borneo. *Asian Primates Journal* 2: 2–7.

Boonratana, R. 2013. Fragmentation and its significance on the conservation of proboscis monkey (*Nasalis larvatus*) in the Lower Kinabatangan, Sabah (North Borneo). In *Primates in Fragments: Complexity and Resilience*, L.K. Marsh & C.A. Chapman (eds). New York: Springer, pp. 459–474.

Boonratana, R. & Sharma, D.S. 1997. Checklist of wildlife species recorded in the Lower Kinabatangan, Sabah. *Journal of Wildlife Management and Research Sabah* 1: 47–60.

Booth, A.H. 1958. The Niger, the Volta and the Dahomey Gap as geographic barriers. *Evolution* 12: 48–62.

Bordignon, M.O., Setz, E.Z.F. & Caselli, C.B. 2008. Gênero *Callicebus* Thomas 1903. In *Primatas Brasileiros*, N.R. dos Reis, A.L. Peracchi & F.R. Andrade (eds). Londrina, Brazil: Technical Books, pp.153–166.

Borger, L., Franconi, N., Michele, G., *et al.* 2006. Effects of sampling regime on the mean and variance of home range size estimates. *Journal of Animal Ecology* 75: 1393–1405.

Borries, C., Koenig, A. & Winkler, P. 2001. Variation of life history traits and mating patterns in female langur monkeys (*Semnopithecus entellus*). *Behavioral Ecology & Sociobiology* 50: 391–402.

Boschetti, L., Roy, D.P. & Justice, C.O. 2008. The collection 5 MODIS burned area product – global evaluation by comparison with the MODIS active fire product. *Remote Sensing of Environment* 112: 3690–3707.

Boubli, J.P. 1993. Southern expension of the geographic distribution of *Cacajao melanocephalus melanocephalus. International Journal of Primatology* 14: 933–937.

Boubli, J. P. 1997. *Ecology of the black uakari monkey,* Cacajao melanocephalus melanocephalus. *Pico da Neblina National*

Park, Brazil, PhD Thesis. Berkeley, CA: University of California, Berkeley.

Boubli, J.P. 1999. Feeding ecology of black-headed uacaris (*Cacajao melanocephalus melanocephalus*) in Pico da Neblina National Park, Brazil. *International Journal of Primatology* 20: 719–749.

Boubli, J.P. 2005. Floristics, primary productivity and primate diversity in Amazonia: Contrasting a eutrophic várzea forest and an oligotrophic caatinga forest in Brazil. In *Tropical Fruits and Frugivores*, L.J. Dew & J.P. Boubli (eds). Netherlands: Springer, pp. 59–73.

Boubli, J.P., da Silva, M.N.F., Amado, M.V., *et al.* 2008. A taxonomic reassessment of *Cacajao melanocephalus* Humboldt (1811), with the description of two new species. *International Journal of Primatology* 29: 723–741.

Boubli, J.P. & de Lima, M.G. 2009. Modeling the geographical distribution and fundamental niches of *Cacajao* spp. and *Chiropotes israelita* in northwestern Amazonia via a maximum entropy algorithm. *International Journal of Primatology* 30: 217–228.

Boubli, J.P. & Tokuda, M. 2008. Socioecology of black uakari monkeys, *Cacajao hosomi*. Pico da Neblina National Park, Brazil: the role of the peculiar spatial-temporal distribution of resources in the Neblina forests. *Primate Report* 75: 3–10.

Boubli, J.P., Ribas, C., Alfaro, J.W.L., *et al.* 2015. Spatial and temporal patterns of diversification on the Amazon: A test of the riverine hypothesis for all diurnal primates of Rio Negro and Rio Branco in Brazil. *Molecular Phylogenetics & Evolution* 82: 400–412.

Boulton, A.M., Horrocks, J.A. & Baulu, J. 1996. The barbados vervet monkey (*Cercopithecus aethiops sabaens*): changes in population size and crop damage, 1980–1994. *International Journal of Primatology* 17: 831–844.

Bourliere, F., Bertrand, M. & Hunkeler, C. 1969. L'écologie de la mone de Lowe (*Cercopithecus campbelli lowei*) en Côte d'Ivoire. *Revue d'Écologie: La Terre et La Vie* 116: 135–163.

Bourliere, F., Hunkeler, C. & Bertrand, M. 1970. Ecology and behavior of Lowe's Guenon *(Cercopithecus campbelli lowei)* in the Ivory Coast. In *Old World Primates*, J.R. Napier & P.H. Napier (eds). New York: Academic Press, pp. 297–350.

Bowler, M. 2007. *The ecology and conservation of the red uakari monkey on the Yavari River, Peru.* PhD thesis, University of Kent, Canterbury, UK.

Bowler, M., Barton, C., McCann-Wood, S., Puertas, P. & Bodmer, R. 2013. Annual variation in breeding success and changes in population density of *Cacajao calvus ucayalii* in the Lago Preto Conservation Concession, Peru. In *Evolutionary Biology and Conservation of Titis, Sakis and Uacaris*, L. Veiga, A. Barnett, S. Ferrari & M. Norconk (eds). Cambridge: Cambridge University Press, pp. 173–178.

Bowler, M. & Bodmer, R.E. 2009. Social behavior in fission-fusion groups of red uakari monkeys (*Cacajao calvus ucayalii*). *American Journal of Primatology* 71: 976–987.

Bowler, M. & Bodmer, RE. 2011. Diet and food choice in Peruvian red uakaris (*Cacajao calvus ucayalii*): selective or opportunistic seed predation?. *International Journal of Primatology* 32: 1109–1122.

Bowler, M., Knogge, C., Heymann, E.W. & Zinner, D. 2012. Multilevel societies in New World primates? Flexibility may characterize the organization of Peruvian red uakaris (*Cacajao calvus ucayalii*). *International Journal of Primatology* 33: 1110–1124.

Bowler, M., Noriega, M.J., Recharte, M., Puertas, P. & Bodmer, R. 2009. Peruvian red uakari monkeys (*Cacajao calvus ucayalii*) in the Pacaya-Samiria National Reserve – a range extension across a major river barrier. *Neotropical Primates* 16: 34–37.

Bown, T.M., Kraus, J., Wing, S.L., *et al.* 1982. The primate Fayum forest revisited. *Journal of Human Evolution* 11: 603–632.

Boyes, S. 2014. *Sold up the river? Hydropower threat re-opens debate.* http://newswatch.nationalgeographic.com/2014/05/27/sold-up-the-river-hydro-power-threatens-okavango-delta.

Boyle, S. 2008. Human impacts on primate conservation in central Amazonia. *Tropical Conservation Science* 1: 6–17.

Boyle, S., Lenz, B., Gilbert, K., *et al.* 2013. Primates of the biological dynamics of forest fragments project: a history. In *Primates in Fragments: Complexity and Resilience*, L.K. Marsh & C.A. Chapman (eds). New York: Springer, pp. 57–74.

Boyle, S.A., Lourenco, W., da Silva, L. & Smith, A.T. 2009. Home range estimates vary with sample size and methods. *Folia Primatologica* 80: 33–42.

BPS (Badan Pusat Statistik) 2004. *Population statistics.* www.bps.go.id/sector/population.

Brahma, G., Debnath, H.S. & Mukherjee, S.K. 2008. Studies on mangrove flora of Lothian Island Wildlife Sanctuary, Sundarban, West Bengal. In *Proceedings of the International Seminar on 'Multidisciplinary approaches in angiosperm systematic'*, pp. 422–427.

Branch, G., Branch, M. & Bannister, A. 1998. *The Living Shores of Southern Africa.* Cape Town: Struik Publishers.

Branch L.C. 1983. Seasonal and habitat differences in the abundance of primates in the Amazon (Tapajos) National Park, Brazil. *Primates* 24: 424–431.

Bravo, S.P. 2008. Seed dispersal and ingestion of insect-infested seeds by black howler monkeys in flooded forests of the Parana River, Argentina. *Biotropica* 40: 471–476.

Bravo, S.P. & Sallenave, A. 2003. Foraging behavior and activity patterns of *Alouatta caraya* in the northeastern Argentinean flooded forest. *International Journal of Primatology* 24: 825–846.

Braza, F. 1980. El araguato rojo (*Alouatta seniculus*). *Doñana Acta Vertebrata* 7: 1–175.

Brearley, F.Q., Proctor, J., Suriantata, N.L. Dalrymple, G. & Voysey, B.C. 2007. Reproductive phenology over a 10-year period in a lowland evergreen rain forest of central Borneo. *Journal of Ecology* 95: 828–839.

Brennan, E.J., Else, J.G. & Altmann, J. 1985. Ecology and behaviour of a pest primate: vervet monkeys in a tourist-lodge habitat. *African Journal of Ecology* 23: 35–44.

Bridgeman, L.L. 2012. *The feeding ecology of Yucatán Black Howler Monkeys* (Alouatta pigra) *in mangrove forest, Tabasco, Mexico.* PhD thesis, Washington University in St. Louis. St. Louis, MO, United States.

Brito, D., Oliveira, L.C., Oprea, M. & Mello, M.A. 2009. An overview of Brazilian mammalogy: trends, biases and future directions. *Zoologia (Curitiba)* 26: 67–73.

Brockelman, W.Y. & Ali, R. 1987. Methods of surveying and sampling forest primate populations. In *Primate Conservation in the Tropical Forest*, C.W. Marsh & R.A. Mittermeier (eds). New York: Alan R. Liss, pp. 23–62.

Brockelman, W.Y. & Srikosamatara, S. 1993. Estimation of density of gibbon groups by use of loud songs. *American Journal of Primatology* 29: 93–108

Bronikowski, A.M. & Altmann, J. 1996. Foraging in a variable environment: weather patterns and the behavioral ecology of baboons. *Behavioral Ecology and Sociobiology* 39: 11–25

Brook, B.W., Sodhi, N.S. & Ng, P.K.L. 2003. Catastrophic extinctions follow deforestation in Singapore. *Nature* 424: 420–426.

Brooks, T.M., Mittermeier, R.A., da Fonseca, G.A.B., *et al.* 2006. Global biodiversity conservation priorities. *Science* 313: 58–61.

Brooks, T.M., Mittermeier, R.A., Mittermeier, C.G., *et al.* 2002. Habitat loss and extinction in the hotspots of biodiversity. *Conservation Biology* 16: 909–923.

Brotoisworo, E. 1983. Population dynamic of Lutung (*Presbytis cristata*) in Pananjung-Pangandaran Nature Reserve, West Java.

In *Training Course on Wildlife Ecology, May 5–June 15, 1983.* Bogor, Indonesia: Biotrop, pp. 1–24.

Brotoisworo, E. & Dirgayusa, I.W.A. 1991. Ranging and feeding behaviour of *Presbytis cristata* in the Pangandaran Nature Reserve, West Java, Indonesia. In *Proceedings of the XIIIth Congress of the International Primatological Society*, A. Ehara, T. Kimura, O. Takenaka & M. Dirgayusa (eds). The Hague: Elsevier Science, pp. 115–118.

Brown, A. & Thieme, M. 2005. Ecoregion 58: Niger delta. In *Freshwater Ecoregions of Africa and Madagascar: A Conservation Assessment.* M.L. Thieme, R. Abell, N. Burgess, *et al.* (eds). New York: Island Press, pp. 291–294.

Brown, A.D. & Zunino, G.E. 1994. Hábitat, densidad y problemas de conservación de los primates de Argentina. *Vida Silvestre Neotropical* 3: 30–40.

Brown, M. 2011. *Intergroup encounters in grey-cheeked mangabeys* (Lophocebus albigena) *and redtail monkeys* (Cercopithecus ascanius): *form and function.* PhD thesis, New York, Columbia University, USA.

Bruford, M.W., Ancrenaz, M., Chikhi, L., *et al.* 2010. Projecting genetic diversity and population viability for the fragmented orang-utan population in the Kinabatangan floodplain, Sabah, Malaysia. *Endangered Species Research* 12: 249–262.

Brugiere, D., Sakom, D. & Gautier-Hion, A. 2005. The conservation significance of the proposed Mbaéré-Bodingué National Park, Central African Republic, with special emphasis on its primate community. *Biodiversity & Conservation* 14: 505–522.

Brumm, H. 2004. Acoustic communication in noise: regulation of call characteristics in a New World monkey. *Journal of Experimental Biology* 207: 443–448.

Brunet, M., Guy, F., Pilbeam, D., *et al.* 2005. New material of the earliest hominid from the upper Miocene of Chad. *Nature* 434: 752–755.

Buchanan-Smith, H.M., Hardie, S.M., Caceres, C. & Prescott, M.J. 2000. Distribution and forest utilization of *Saguinus* and other primates of the Pando Department, northern Bolivia. *International Journal of Primatology* 21: 353–379.

Buckland, S., Plumptre, A., Thomas, L. & Rexstad, E. 2010. Design and analysis of line transect surveys for primates. *International Journal of Primatology* 31: 883–847.

Buckley, C., Nekaris, K.A.I. & Husson, S.J. 2006. Survey of *Hylobates agilis albibarbis* in a logged peat-swamp forest: Sabangau catchment, central Kalimantan. *Primates* 47: 327–335.

Buckley, J.S. 1983. *The feeding behavior, social behavior and ecology of the white-faced monkey,* Cebus capucinus, *at Trujillo, northern Honduras, Central America.* PhD thesis, University of Texas, Austin, Texas, USA.

Buffon, G.L.L. 1789. *Histoire Naturelle, Générale et Particulière, Servant de Suite à l'Histoire des Animaux Quadrupèdes. Supplément Volume VII.* Paris: L'Imprimerie Royale.

Buij, R., Singleton, I., Krakauer, E. & van Schaik, C.P. 2003. Rapid assessment of orangutan density. *Biological Conservation* 114: 103–113.

Buij, R., Wich, S.A., Lubis, A.H. & Sterck, E.H.M. 2002. Seasonal movements in the Sumatran orangutan (*Pongo pygmaeus abelii*) and consequences for conservation. *Biological Conservation* 107: 83–87.

Bunlungsup, S., Imai, H., Hamada, Y., *et al.* 2016. Morphological characteristics and genetic diversity of Burmese Long-tailed Macaques (*Macaca fascicularis aurea*). *American Journal of Primatology* 78: 441–455.

Burgess, N., D'Amico, H.J., Underwood, E., *et al.* 2004. *Terrestrial Ecoregions of Africa and Madagascar: A Continental Assessment.* Washington DC: Island Press.

Burgess, N.D., Salehe, J., Doggart, N., *et al.* 2004. Coastal forests of eastern Africa. In *Hotspots Revisited.* R.A. Mittermeier, P. Robles, M. Hoffmann, *et al.* (eds). Mexico City: CI/CEMEX, pp. 230–239.

Burney, D.A., Pigott-Burney, L., Godfrey, L.R., *et al.* 2004. A chronology for late prehistoric Madagascar. *Journal of Human Evolution* 47: 25–63.

Butynski, T.M., De Jong, Y.A., Perkin, A.W., Bearder, S.K. & Honess, P.E. 2006. Taxonomy and distribution of three species of dwarf galagos (*Galagoides*) in eastern Africa. *Primate Conservation* 21: 63–79.

Butynski, T.M. & Gippoliti, S. 2008. *Cercopithecus mitis* ssp. *boutourlinii.* In *IUCN Red List of Threatened Species.* Gland, Switzerland: IUCN. e.T136901A4349249.

Butynski, T.M. & Hamerlynck, O. 2016. Tana River red colobus *Piliocolobus rufomitratus* (Peters, 1879). In *Primates in Peril: The World's 25 Most Endangered Primates 2014–2016*, C. Schwitzer, R.A. Mittermeier, A.B. Rylands, *et al.* (eds). Arlington, VA: IUCN/SSC Primate Specialist Group, pp. 20–22.

Butynski, T.M. & Mwangi, G. 1994. *Conservation status and distribution of the Tana River red colobus and crested mangabey.* Unpublished report. Nairobi: Kenya Wildlife Service and Zoo Atlanta.

Butynski, T.M., Kingdon, J. & Kalina, J. (eds). 2013. *Mammals of Africa. Volume II: Primates.* London: Bloomsbury.

Butynski, T.M., Struhsaker, T., Kingdon, J. & De Jong, Y.A. 2008. *Cercocebus galeritus.* In *IUCN Red List of Threatened Species.* Gland, Switzerland: IUCN. e.T4200A10615270.

Bwangoy, J.-R.B., Hansen, M.C., Roy, D.P., Grandi, G.D. & Justice, C.O. 2010. Wetland mapping in the Congo Basin using optical and radar remotely sensed data and derived topographical indices. *Remote Sensing of Environment* 114: 73–86.

Byrne, H., Rylands, A.B., Carneiro, J.C., Lynch-Alfaro, J.W., Bertuol, F., da Silva, M.N.F., Messias, M., Groves, C.P., Mittermeier, R.A., Farias, I., Hrbek, T., Schneider, H., Sampaio, I. & Boubli, J.P. 2016. Phylogenetic relationships of the New World titi monkeys (*Callicebus*): first appraisal of taxonomy based on molecular evidence. *Frontiers in Zoology* 13: https://doi.org/10.1186/s12983-016-0142-4.

Caballero, A., Rodríguez-Ramilo, S.T., Ávila, V. & Fernández, J. 2009. Management of genetic diversity of subdivided populations in conservation programmes. *Conservation Genetics* 11: 409–419.

Cabral, M.M.M., de Mattos, G.E. & Rosas, F.C.W. 2008. Mammals, birds and reptiles in Balbina reservoir, state of Amazonas, Brazil. *Check List* 4: 152–158.

Cáceres, N.C., Carmignotto, A.P., Fischer, E. & Santos, C.F. 2008. Mammals from Mato Grosso do Sul, Brazil. *Check List* 4: 321–335.

Cahoon, D.R., Hensel, P., Rybczyk, J., *et al.* 2003. Mass tree mortality leads to mangrove peat collapse at Bay Islands, Honduras after Hurricane Mitch. *Journal of Ecology* 91: 1093–1105.

Caldecott, J. & Miles. L. (eds). 2005. *World Atlas of Great Apes and their Conservation.* Berkeley, CA: University of California Press.

Calegaro-Marques, C. & Bicca-Marques, J.C. 1996. Emigration in a black howling monkey group. *International Journal of Primatology* 17: 229–237.

Campbell, C., Andayani, N., Cheyne, S.M., *et al.* 2008. *Indonesian Gibbon Conservation and Management Workshop Final Report.* Apple Valley, MN: IUCN/SSC Conservation Breeding Specialist Group.

Campbell, C.J., Aureli, F., Chapman, C.A., *et al.* 2005. Terrestrial behavior of *Ateles* spp. *International Journal of Primatology* 26: 1039–1051.

Campisano, C.J. 2007. *Tephrostratigraphy and hominin paleoenvironments of the Hadar formation, Afar Depression, Ethiopia.* PhD thesis, Rutgers University, New Brunswick, NJ, USA.

Canale, G.R. 2010. *Ecology and behaviour of the yellow-breasted capuchin monkey* (Cebus xanthosternos) *in the northern Atlantic Forest*. PhD thesis, University of Cambridge, Cambridge, UK.

Canale, G.R., Guidorizzi, C.E., Kierulff, M.C.M. & Gatto, C.A.F.R. 2009. First record of tool use by wild populations of the yellow-breasted capuchin monkey (*Cebus xanthosternos*) and new records for the bearded capuchin (*Cebus libidinosus*). *American Journal of Primatology* 71: 366–372.

Cannon, C.H., Curran, L.M., Marshall, A.J. & Leighton, M. 2007. Long-term reproductive behaviour of woody plants across seven Bornean forest types in the Gunung Palung National Park (Indonesia): suprannual synchrony, temporal productivity and fruiting diversity. *Ecology Letters* 10: 956–969.

Cardoso, N.A. 2015. Cacajao calvus calvus *e* Cacajao calvus rubicundus *(Primates: Pitheciidae): influência da vegetação em sua distribuição geográfica na Amazônia Central, Brasil.* PhD thesis, Universidade Federal da Paraíba, João Pessoa, Paraíba, Brazil.

Care for the Wild International and Pro Wildlife 2007. *Going to Pot: The Neotropical Bushmeat Crisis and its Impact on Primate Populations*. Kingsfold, UK: Care for the Wild International.

Carlton, J.T. & Hodder, J. 2003. Maritime mammals: terrestrial mammals as consumers in marine inter-tidal communities. *Marine Ecology Progress Series* 256: 271–286.

Carpenter, A. 1887. Monkeys opening oysters. *Nature* 36: 53.

Carnegie, S.D., Fedigan, L.M. & Melin, A.D. 2011. Reproductive seasonality in female capuchins (*Cebus capucinus*) in Santa Rosa (Area de Conservación Guanacaste), Costa Rica. *International Journal of Primatology* 32: 1076–1090.

Caro, T. 2010. *Conservation by Proxy: Indicator, Umbrella, Keystone, Flagship, and other Surrogate Species*, 2nd edition. Washington DC: Island Press.

Carpenter, S.R., Stanley, E.H. & VanderZanden, M.J. 2011. State of the world's freshwater ecosystems: physical, chemical, and biological changes. *Annual Review of Environment and Resources* 36: 75–99.

Carretero, X. 2000. *Un estudio ecológico de Saimiri sciureus y su asociación con Cebus apella, en la Macarena, Colombia*. BSc dissertation. Pontificia Universidad Javeriana, Bogotá, Colombia.

Carretero-Pinzón, X. 2008. *Efecto de la disponibilidad de recursos sobre la ecología y comportamiento de* Saimiri sciureus albigena *en fragmentos de bosque de galería, San Martín (Meta - Colombia)*. MSc thesis, Pontificia Universidad Javeriana, Bogotá, Colombia.

Carretero-Pinzón X. 2013. An eight-year life history of a primate community in fragments at Colombian Llanos. In *Primates in Fragments: Complexity and Resilience*, L.K. Marsh & C.A. Chapman (eds). New York: Springer, pp. 159–182.

Carretero-Pinzón, X., Ruiz-García, M. & Defler, T. 2009. The taxonomy and conservation status of *Saimiri sciureus albigena*: a squirrel monkey endemic to Colombia. *Primate Conservation* 24: 59–64.

Carretero-Pinzón, X., Ruíz-García, M. & Defler, T.R. 2010. Uso de cercas vivas como corredores biológicos por primates en los llanos orientales. In *Libro del Primer Congreso Colombiano de Primatología*, V. Pereira & M.L. Bueno (eds). Bogotá, pp. 91–98.

Carroll, R.L. 1987. *Vertebrate Paleontology and Evolution*. New York: W.H. Freeman and Co.

Carvalho, L., Lepisto, L., Rissanen, J., *et al.* 2006. Nutrients and eutrophication in lakes. In *Indicators and Methods for the Ecological Status Assessment under the Water Framework Directive: Linkages between Chemical and Biological Quality of Surface Waters*, A. Solimini, A.C. Cardoso & A.-S. Heiskanen (eds). Luxembourg: Office for Official Publications of the European Communities, pp. 3–32.

Carvajal, P., Porcel, Z. & Flores, C. 2013. *Conservation of endemic titi monkeys through strategic outreach in Bolivia*. Unpublished technical report. La Paz, Bolivia: Wildlife Conservation Society.

Casanovas-Vilar, I., Alba, D.M, Garcés, M., Robles, J.M. & Moyà-Solà, S. 2011. Updated chronology for the Miocene hominoid radiation in western Eurasia. *Proceedings of the National Academy of the Sciences of the Unites States of America* 108: 554–559.

Casimir, M.J. 1975. Feeding ecology and nutrition of an eastern gorilla group in the Mt Kahuzi region (République du Zaïre). *Folia Primatologia* 24: 81–136.

Casimir, M.J. & Butenandt, E. 1973. Migration and core area shifting in relation to some ecological factors in a mountain gorilla group (*Gorilla gorilla beringei*) in the Mt. Kahuzi region (République du Zaïre). *Zeitschrift für Tierpsychologie* 33: 514–522.

Castello, L., McGrath, D.G., Hess, L.L., *et al.* 2013. The vulnerability of Amazon freshwater ecosystems. *Conservation Letters* 6: 217–229.

Castiblanco, C., Etter, A. & Aide, M. 2013. Oil palm plantations in Colombia: a model of future expansion. *Environmental Science & Policy* 27: 172–183.

Caton, J.M. 1999. Digestive strategy of the Asian colobine genus *Trachypithecus*. *Primates* 40: 311–325.

Cattanio, J.H., Anderson, A.B., Rombold, J.S. & Nepstad, D.C. 2004. Phenology, growth, and root biomass in a tidal floodplain forest in the Amazon estuary. *Revista Brasileira de Botânica* 27: 703–712.

Cattau, M.E., Husson, S.J. & Cheyne, S.M. 2013. The Bornean orang-utan (*Pongo pygmaeus*) in a vanishing forest: the former Mega Rice Project, Indonesia. *Oryx* 49: 473–480.

Cattau, M.E., Husson, S. & Cheyne, S.M. 2015. Population status of the Bornean orang-utan *Pongo pygmaeus* in a vanishing forest in Indonesia: the former Mega Rice Project. *Oryx* 49: 473–480.

Catullo, G., Masi, M., Falcucci, A., Maiorano, L., Rondinini, C. & Boitani, L. 2008. A gap analysis of Southeast Asian mammals based on habitat suitability models. *Biological Conservation* 141: 2730–2744.

Causado, J., Cuarón, A.D., Shedden, A., Rodríguez-Luna, E. & de Grammont, P.C. 2008. *Cebus capucinus*. In *IUCN Red List of Threatened Species*. Version 2012.1. Gland, Switzerland: IUCN. www.iucnredlist.org.

Ceballos-Mago, N. & Chivers, D. 2013. A Critically Endangered capuchin (*Sapajus apella margaritae*) living in mountain forest fragments on Isla de Margarita, Venezuela. In *Primates in Fragments: Complexity and Resilience*, 2nd edition, L.M. Marsh & C.A. Chapman (eds). New York: Springer, pp. 183–195.

Cerling, T.E., Chritz, K.L., Jablonski, N.G., Leakey, M.G. & Manthi, F.K. 2013. Diet of *Theropithecus* from 4 to 1 Ma in Kenya. *Proceedings of the National Academy of the Sciences of the Unites States of America* 110: 10507–10512.

Cerling, T.E., Mbua, E., Kirera, F.M., *et al.* 2011b. Diet of *Paranthropus boisei* in the early Pleistocene of East Africa. *Proceedings of the National Academy of the Sciences of the Unites States of America* 108: 9337–9341.

Cerling, T.E., Wynn, J.G., Andanje, S.A., *et al.* 2011a. Woody cover and hominin environments in the past 6 million years. *Nature* 476: 51–56.

Chabot, V.C. & Bird, D.M. 2014. Measuring habitat quality for least bitterns in a created wetland with use of a small unmanned aircraft. *Wetlands* 34: 527–533.

Chagas, R.R.D. & Ferrari, S.F. 2010. Habitat use by *Callicebus coimbrai* (Primates: Pitheciidae) and sympatric species in the fragmented landscape of the Atlantic forest of southern Sergipe, Brazil. *Zoologia* 27: 853–860.

Chalk, J. 2011. *The effects of feeding strategies and food mechanics on the ontogeny of masticatory function in the* Cebus libidinosus *cranium*. PhD Dissertation. Boston University, Boston, Massachussetts, United States.

Chapman, C.A. & Chapman, L.J. 2002. Foraging challenges of red colobus monkeys: influence of nutrients and secondary compounds. *Comparative Biochemistry and Physiology Part A: Molecular & Integrative Physiology* 133: 861–875.

Chapman, C.A. & Lambert, J.E. 2000. Habitat alteration and the conservation of African primates: case study of Kibale National Park, Uganda. *American Journal of Primatology* 50: 169–185.

Chapman, C.A. & Onderdonk, D.A. 1998. Forests without primates: primate/plant codependency. *American Journal of Primatology* 45: 127–141.

Chapman, C.A. & Peres, C.A. 2001. Primate conservation in the new millennium: the role of scientists. *Evolutionary Anthropology* 10: 16–33.

Chapman, C.A., Balcomb, S.R., Gillespie, T.R., Skorupa, J.P. & Struhsaker, T.T. 2000. Long-term effects of logging on African primate communities: a 28-year comparison from Kibale National Park, Uganda. *Conservation Biology* 14: 207–217.

Chapman, C.A., Chapman, L.J., Bjorndal, K.A. & Onderdonk, D.A. 2002. Application of protein-to-fiber ratios to predict colobine abundance on different spatial scales. *International Journal of Primatology* 23: 283–310.

Chapman, C.A., Chapman, L.J., Cords, M., *et al.* 2002. Variation in the diets of *Cercopithecus* species: intraspecific differences within forests, among forests, and across species, In *The Guenons: Diversity and Adaptation in African Monkeys*, M.E. Glenn & M. Cords (eds). New York: Kluwer Academic/Plenum Publisher, pp. 325–350.

Chapman, C.A., Chapman, L.J., Naughton-Treves, L., Lawes, M.J. & McDowell, L.R. 2004. Predicting folivorous primate abundance: validation of a nutritional model. *American Journal of Primatology* 62: 55–69.

Chapman, C.A., Chapman, R., Wrangham, K., *et al.* 1992. Estimators of fruit abundance of tropical trees. *Biotropica* 24: 527–531.

Chapman, C.A., Gillespie, T.R. & Goldberg, T.L. 2005. Primates and the ecology of their infectious diseases: how will anthropogenic change affect host-parasite interactions? *Evolutionary Anthropology* 14: 134–144.

Chapman, C.A., Gautier-Hion, A., Oates, J.F. & Onderdonk, D.A. 1999. African primate communities: determinants of structure and threats to survival. In *Primate Communities*, J.G. Fleagle, C.H. Janson & K.E. Reed (eds). Cambridge: Cambridge University Press, pp. 1–37.

Chapman, C.A., Lawes, M.J. & Eeley, H.A.C. 2006. What hope for African primate diversity? *African Journal of Ecology* 44: 116–133.

Chapman, C., Ria Ghai, R., Jacob, A., *et al.* 2013. Going, Going, Gone: a 15-year history of the decline of primates in forest fragments near Kibale National Park, Uganda. In *Primates in Fragments: Complexity and Resilience*, L.K. Marsh & C.A. Chapman (eds). New York: Springer, pp. 89–100.

Chapman, C., Struhsaker, T., Skorupa, J., Snaith, T. & Rothman, J. 2010. Understanding long-term primate community dynamics implications of forest change. *Ecological Applications* 20: 179–191.

Chapman, L.A., Chapman, C.A. & Chandler, M. 1996. Wetland ecotones as refugia for endangered fishes. *Biological Conservation* 78: 263–270

Chaudhuri, A.B. 2007. *Biodiversity of Mangroves*. Delhi: Daya Publishing House.

Chaudhuri, A.B. & Choudhury, A. 1994. *Mangroves of the Sundarbans, Volume One: India*. Bangkok: IUCN.

Chaves, O.M. & Stoner, K.E. 2010. River crossings by *Ateles geoffroyi* and *Alouatta pigra* in southern Mexico: a preliminary report. *Revista Chilena de Historia Natural* 83: 435–442.

Cheney, D.L. & Seyfarth, R.M. 1982. How vervet monkeys perceive their grunts: field playback experiments. *Animal Behaviour* 30: 739–751.

Cheney, D.L. & Seyfarth, R.M. 1982. Recognition of individuals within and between groups of free-ranging vervet monkeys. *American Zoologist* 22: 519–529.

Cheney, D.L., Seyfarth, R.M., Fischer, J., *et al.* 2004. Factors affecting reproduction and mortality among baboons in the Okavango Delta, Botswana. *International Journal of Primatology* 25: 401–428.

Cheyne, S.M. 2007. Effects of meteorology, astronomical variables, location and human disturbance on the singing apes: *Hylobates albibarbis*. *American Journal of Primatology* 40: 1–7.

Cheyne, S.M. 2008. Feeding ecology, food choice and diet characteristics of gibbons in a disturbed peat-swamp forest, Indonesia. In *22nd Congress of the International Primatological Society (IPS)*, P.C. Lee, P. Honess, H. Buchanan-Smith, A. MacLarnon & W.I. Sellers (eds). Bristol & Edinburgh: TopCopy.

Cheyne, S.M. 2010. Behavioural ecology and socio-biology of gibbons (*Hylobates albibarbis*) in a degraded peat-swamp forest. In *Indonesian Primates*, J. Supriatna & S.L. Gursky (eds). New York: Springer, pp. 121–156.

Cheyne, S.M. & Tuttle, R.H. 2011. Gibbon locomotion research in the field - problems, possibilities and benefits for conservation. In *Primate Locomotion: Linking Field and Laboratory Research*, K. D'Août & E.E. Vereecke (eds). New York: Springer, pp. 201–214.

Cheyne, S.M., Gilhooly, L.J., Hamard, M.C., *et al.* 2016. Population mapping of gibbons in Kalimantan, Indonesia: Correlates of gibbon density and vegetation across the species range. *Endangered Species Research* 30: 133–143.

Cheyne, S.M., Harrison, M.E. & Morrogh-Bernard. H. 2005. Differences in orang-utan and gibbon diets in the Sebangau National Park, Indonesia: implications for conservation. In *Proceedings of the International Symposium and Workshop on Restoration and Wise Use of Tropical Peatland*, J.O. Reiley (ed.). Palangka Raya, Indonesia: CIMTROP, pp. 100–103.

Cheyne, S.M., Höing, A., Rinear, J. & Sheeran, L.K. 2013a. Sleeping site selection by agile gibbons: the influence of tree stability, fruit availability, and predation risk. *Folia Primatologica* 89: 299–311.

Cheyne, S.M., Rowland, D., Hoing, A. & Husson, S.J. 2013b. How orangutans choose where to sleep: comparison of nest-site variables. *Asian Primates Journal* 3: 13–17.

Cheyne, S.M., Thompson, C.J.H., Phillips, A.C., Hill, R.M.C. & Limin, S.H. 2007. Density and population estimate of gibbons (*Hylobates albibarbis*) in the Sabangau Catchment, Central Kalimantan, Indonesia. *Primates* 49: 50–56.

Cheyne, S.M., Zrust, M., Hoeing, A., Houlihan, P.R., Rowland, D., Rahmania, M. & Breslin, K. 2012. *Barito River Initiative for Nature Conservation and Communities (BRINCC) Preliminary Report*. Palangka Raya, Indonesia: BRINCC Expedition.

Chiarelli, A. 1972. *Taxonomic Atlas of Living Primates*. London: Academic Press.

Chiou, K.L., Pozzi, L., Alfaro, J.W.L. & Di Fiore, A. 2011. Pleistocene diversification of living squirrel monkeys (*Saimiri* spp.) inferred from complete mitochondrial genome sequences. *Molecular Phylogenetics and Evolution* 59: 736–745.

Chivers, D. 1994. Functional anatomy of the gastrointestinal tract. In *Colobine Monkeys: Their Ecology, Behaviour and Evolution*, A. Davies & J. Oates (eds). Cambridge: Cambridge University Press, pp. 205–227.

Chivers, D.J. 2001. The swinging singing apes: fighting for food and family in Far-East forests. In *The Apes: Challenges for the 21st Century*. Chicago, IL: Brookfield Zoo, pp. 1–28.

Chivers, D.J. & MacKinnon, J. 1977. On the behaviour of siamang after playback of their calls. *Primates* 18: 943–948.

Chivers, D.J., Anandam, M., Groves, C., *et al.* 2013. Family Hylobatidae (Gibbons). In *Handbook of the Mammals of the World: Volume 3-Primates*, R.A. Mittermeier, A.B. Rylands & D. Wilson (eds). Arlington, VA: Conservation International, pp. 754–791.

Chizzotti, A. 2005. *Pesquisa em Ciências Humanas e Sociais* (7th edition). São Paulo: Cortez.

Chowdhury, Q.I., Haque, M. & Chowdhury, S.H. 2001. Overview of an amazing ecosystem. In *State of Sundarbans*, Q.I. Chowdhury (ed.). Dhaka: Forum of Environmental Journalists, pp. 5–16.

Christen, A. & Geissmann, T. 1994. A primate survey in northern Bolivia, with special reference to Goeldi's monkey, *Callimico goeldii*. *International Journal of Primatology* 15: 239–275.

CIA. 2013. *CIA World Factbook*. Washington, DC: CIA. www.cia.gov/library/publications/the-world-factbook/geos/ma.html.

Cintra, B.B.L., Schietti, J., Emillio, T., *et al.* 2013. Productivity of aboveground coarse wood biomass and stand age related to soil hydrology of Amazonian forests in the Purus-Madeira interfluvial area. *Biogeosciences Discussions* 10: 6417–6459.

Cintra, R. 2015. Spatial distribution and composition of waterbirds in relation to limnological conditions in the Amazon basin. *Hydrobiologia* 747: 235–252.

Cintra, R. & Naka, N. 2012. Spatial variation in bird community composition in relation to topographic gradient and forest heterogeneity in a central Amazonian rainforest. *International Journal of Ecology* 2012: 1–25.

Cipolletta, C. 2004. Effects of group dynamics and diet on the ranging patterns of a western lowland gorilla group (*Gorilla gorilla gorilla*) at Bai Hokou, Central African Republic. *American Journal of Primatology* 64: 193–205.

Citrinot, L. 2014. New passengers terminal opens at Balikpapan International Airport. *Travel Daily News Network* 21 March 2014.

Clark, D.A., Brown, S., Kicklighter, D.W., *et al.* 2001. Net primary production in tropical forests: an evaluation and synthesis of existing field data. *Ecological Applications* 11: 371–384.

Clark, D.B, Palmer, M.W. & Clark, D.A. 1999. Edaphic factors and the landscape-scale distributions of tropical rain forest trees. *Ecology* 80: 2662–2675.

Clarke, E., Reichard, U.H. & Zuberbühler, K. 2006. The syntax and meaning of wild gibbon songs. *PLoS Biology* doi.org/10.1371/journal.pone.0000073.

Clarke, H.D., Funk, V.A. & Hollowell, T. 2001. *Plant Diversity of the Iwokrama Forest, Guyana*. Sida, Botanical Miscellany, No 21. Fort Worth, TX: Botanical Research Institute of Texas.

Cochrane, J. 2006. Indonesian National Parks. Understanding leisure users. *Annals of Tourism Research* 33: 979–997.

Cochrane, M.A. 2001. Synergistic interactions between habitat fragmentation and fire in evergreen tropical forests. *Conservation Biology* 15: 1515–1521.

Cochrane, M.A. 2003. Fire science for rainforests. *Nature* 421: 913–919.

Cochrane, M.A. & Schulze, M.D. 1998. Forest fires in the Brazilian Amazon. *Conservation Biology* 12: 948–950.

Codron, D., Codron, J., Lee-Thorp, J.A., *et al.* 2007. Stable isotope characterization of mammalian predator-prey relationships in a South African savanna. *European Journal of Wildlife Research* 53: 161–170.

Coelho, A.M., Bramblett, C.A., Quick, L.B. & Bramblett, S.S. 1976. Resource availability and population density in primates: a socio-bioenergetic analysis of the energy budgets of Guatemalan howler and spider monkeys. *Primates* 17: 63–80.

Coelho, I.P. 2006. *Relações entre barreiros e a fauna de vertebrados no nordeste do Pantanal, Brasil*. MSc dissertation, Universidade Federal Rio Grande do Sul, Porto Alegre.

Coimbra-Filho, A.F., Rylands, A.B., Pissinatti, A. & Santos, I.B. 1991/1992. The distribution and conservation of the buff-headed capuchin monkey, *Cebus xanthosternos*, in the Atlantic forest region of eastern Brazil. *Primate Conservation* 12/13: 24–30.

Coimbra-Filho, A.F., Silva, R.R. & Pissinatti, A. 1991. Acerca da distribuição geográfica original de *Cebus apella xanthosternos* Wied, 1820 (Cebidae, Primates). In *A Primatologia no Brasil, Vol. 3*, A.B. Rylands & A.T. Bernardes (eds). Belo Horizonte: Fundação Biodiversitas & Sociedade Brasileira de Primatologia, pp. 215–224.

Colares, I.G. & Colares, E.P. 2002. Food plants eaten by Amazonian manatees (*Trichechus inunguis*, Mammalia: Sirenia). *Brazilian Archives of Biology and Technology* 45: 67–72.

Colchester, M., La Rose, J. & James, K. 2002. *Mining and Amerindians in Guyana*. Final report of the APA/NSI project on 'Exploring indigenous perspective on consultation and engagement within the mining sector in Latin America and the Carribbean.' Ottawa, Canada: The North-South Institute.

Coley, P.D. 1987. Interspecific variation in plant anti-herbivore properties: the role of habitat quality and rate of disturbance. *New Phytologist* 106: 251–263.

Colwell, R.K. 2013. EstimateS: Statistical estimation of species richness and shared species from samples. version 8.0. http://purl.oclc.org/estimates

Colwell, R.K. & Coddington, J.A. 1994. Estimating terrestrial biodiversity through extrapolation. *Philosophical Transactions of the Royal Society of London B Biological Sciences* 345: 101–118.

Conga, D.F., Bowler, M., Tantalean, M., *et al.* 2014. Intestinal helminths in wild Peruvian red uakari monkeys (*Cacajao calvus ucayalii*) in the northeastern Peruvian Amazon. *Journal of Medical Primatology* 43: 130–133.

Conga, D.F., Giese, E.G., Serra-Freire, N.M., Bowler, M. & Mayor, P. 2015. Morphology of the oxyurid nematodes *Trypanoxyuris (T.) cacajao* n. sp. and *T.(T.) ucayalii* n. sp. from the red uakari monkey *Cacajao calvus ucayalii* in the Peruvian Amazon. *Journal of Helminthology* 90: 483–493.

Conklin-Brittain, N.L., Wrangham, R.W. & Smith, C.C. 2002. A two-stage model of increased dietary quality in early hominid evolution: the role of fiber. In *Human Diet: Its Origin and Evolution*, P.S. Ungar & M.F. Teaford (eds). Westport, CT: Praeger, pp. 61–76.

Conradt, L. 2000. Use of a seaweed habitat by red deer (*Cervus elaphus* L.). *Journal of Zoology, London*, 250: 541–549.

Converse, L.J., Carlson, A.A., Ziegler, T.E. & Snowdon, C.T. 1995. Communication of ovulatory state to mates by female pygmy marmosets, *Cebuella pygmaea*. *Animal Behaviour* 49: 615–621.

Cook, N. 1939. Notes on captured *Tarsius carbonarius*. *Journal of Mammalogy* 20: 173–178.

Cooke, C. 2005. The cercopithecid community of Sette Cama, Gabon: a preliminary study. *American Journal of Primatology* 132 (supplement 44): 90.

Cooke, C.A. 2012. *The feeding, ranging, and positional behaviors of* Cercocebus torquatus, *the red-capped mangabey, in Sette Cama, Gabon: a phylogenetic perspective*. PhD thesis, The Ohio State University, Columbus, Ohio.

Cooke, C.A. 2015. Crab predation by red-capped mangabeys (*Cercocebus torquatus*) in Sette Cama, Gabon. *African Journal of Ecology* 53: 378–380.

Copsey, J.A., Jones, J.P.G., Andrianandrasana, H., Rajaonarison, L.H. & Fa, J.E. 2009a. Burning to fish: local explanations for wetland burning in Lac Alaotra, Madagascar. *Oryx* 43: 403–406.

Copsey, J.A., Rajaonarison, L.H., Ranriamihamina, R. & Rakotoniaina, L.J. 2009b. Voices from the marsh: livelihood concerns of fishers and rice cultivators in the Alaotra wetland. *Madagascar Conservation & Development* 4: 25–30.

Corcoran, E., Ravilious, C. & Skuja, M. 2007. *Mangroves of Western and Central Africa*. Nairobi: UNEP-Regional Seas Programme/UNEP-WCMC.

Cordeiro, C.L.D.O. 2008. *Estimativas de detecção de primatas e validação de modelos preditivos em duas unidades de conservação na Amazônia, Roraima, Brasil*. MSc dissertation, Instituto Nacional de Pesquisas da Amazonia Manaus, Amazonas, Brazil.

Cords, M. 1987a. Forest guenons and patas monkeys: male-male competition in one-male group. In *Primates Societies*, B.B. Smuts, D.L. Cheney, R.M. Seyfarth, R.W. Wrangham & T.T. Struhsaker (eds). Chicago, IL: University of Chicago Press, pp. 98–111.

Cords, M. 1987b. Mixed species association of Cercopithecus monkeys in the Kakamega Forest, Keyna. *University of California Publications in Zoology* 1: 1–109.

Cords, M. 1990. Mixed-species association of East African guenons: general patterns or specific examples? *American Journal of Primatology* 21: 101–114.

Cords, M. & Sarmiento, E. 2013.*Cercopithecus ascanius* red-tailed monkey. In *Mammals of Africa. Volume II: Primates*, T.M. Butynski, J. Kingdon & J. Kalina (eds). London: Bloomsbury, pp. 375–381.

Corlett, R.T. 2005. Interactions between birds, fruit bats and exotic plants in urban Hong Kong, South China. *Urban Ecology* 8: 275–283.

Corlett, R.T. 2009. *The Ecology of Tropical East Asia*. New York: Oxford University Press.

Corrêa, H.K.M & Coutinho, P.E.G. 2008. Gênero *Callithrix* Erxleben 1777. In *Primatas Brasileiros*, N.R. dos Reis, A.L. Peracchi & F.R. Andrade (eds). Londrina, Brazil: Technical Books, pp. 47–58.

Correa, S.B., Winemiller, K.O., López-Fernández, H. & Galetti, M. 2007. Evolutionary perspectives on seed consumption and dispersal by fishes. *Bioscience* 57: 748–756.

Cortés-Ortiz, L., Rylands, A.B. & Mittermeier, R.A. 2015. The taxonomy of howler monkeys: integrating old and new knowledge from morphological and genetic studies. In *Howler Monkeys: Adaptive Radiation, Systematics, and Morphology*, M.M. Kowalewski, P.A. Garber, L. Cortés-Ortiz, B. Urbani & D. Youlatos (eds). New York: Springer Press, pp. 55–68.

Cosson, J.F., Ringuet, S., Claessens, O., *et al.* 1999. Ecological changes in recent land-bridge islands in French Guiana, with emphasis on vertebrate communities. *Biological Conservation* 91: 213–222.

Costa, L.P. 2003. The historical bridge between the Amazon and the Atlantic Forest of Brazil: a study of molecular phylogeography with small mammals. *Journal of Biogeography* 30: 71–86.

Cousins, J. 2007. The role of UK-based conservation tourism operators. *Tourism Management* 28: 1020–1030.

Covert, H.H., Le Khac Quyet, N.A.D., Tai, V.A. & Wright, B.W. 2008. On the brink of extinction: research for the conservation of the Tonkin snub-nosed monkey (*Rhinopithecus avunculus*). In *Elwyn Simons: A Search for Origins*, J.G. Fleagle & C.C. Gilbert (eds). Heidelberg, Germany: Springer, pp. 409–427.

Cowling, R.M., Macdonald, I.A.W. & Simmons, M.T. 1996. The Cape Peninsula, South Africa: physiographical, biological and historical background to an extraordinary hot-spot of biodiversity. *Biodiversity & Conservation* 5: 527–550.

Cowlishaw, G. 1999. Predicting the pattern of decline of African primate diversity: an extinction debt from historical deforestation. *Conservation Biology* 13: 1183–1193.

Cowlishaw, G. 2013. *Papio ursinus* chacma baboon. In *Mammals of Africa. Volume II: Primates*, T.M. Butynski, J. Kingdon & J. Kalina (eds). London: Bloomsbury, pp. 225–228.

Cowlishaw, G. & Dunbar, R. 2000. *Primate Conservation Biology*. Chicago, IL: University of Chicago Press.

Cowlishaw, G. & Hacker, J.E. 1997. Distribution, diversity and latitude in African primates. *American Naturalist* 150: 505–512.

Cam, W.J., Torr, P.G. & Rose, D.A. 2002. Salt allocation during leaf development and leaf fall in mangroves. *Trees* 16: 112–119.

Cristóbal-Azkarate, J. & Dunn, C. 2013. Lessons from Los Tuxtlas: 30 Years of research into primates in fragments. In *Primates in Fragments: Complexity and Resilience*, L.K. Marsh & C.A. Chapman (eds). New York: Springer, pp. 75–88.

Crockett, C. 1998. Conservation biology of the genus *Alouatta*. *International Journal of Primatology* 19: 549–577.

Crockett C.M. & Eisenberg, J.F. 1987. Howlers: Variations in group size and demography. In *Primate Societies*, B.B. Smuts, D.L. Cheney, R.M. Seyfarth, R.W. Wrangham & T.T. Struhsaker (eds). Chicago: University of Chicago Press, pp. 54–68.

Crockett, C.M. & Wilson, W.L. 1980. The ecological separation of *Macaca nemestrina* and *M. fascicularis* in Sumatra. In *The Macaques: Studies in Ecology, Behavior and Evolution*, D.G. Lindburg (ed.). New York: van Nostrand Reinhold, pp. 148–181.

Crowley, B.E. 2010. A refined chronology of prehistoric Madagascar and the demise of the megafauna. *Quaternary Science Reviews* 29: 2591–2603.

Csermely, D. 1996. Antipredator behavior in lemurs: Evidence of an extinct eagle on Madagascar or something else? *International Journal of Primatology* 17: 349–354.

Cuarón, A.D., Shedden, A., Rodríguez-Luna, E., *et al.* 2008. *Alouatta palliata*. In IUCN 2012. In *IUCN Red List of Threatened Species*. Version 2012.1. Gland, Switzerland: IUCN. www.iucnredlist.org.

Cuming, H. 1838. On the habit of some species of mammal from the Philippine Islands *Proceedings of the Zoological Society of London* 6: 67–68.

Cunningham, E. & Janson, C. 2007. Integrating information about location and value of resources by white-faced saki monkeys (*Pithecia pithecia*). *Animal Cognition* 10: 293–304.

Curtis, S.J. 1933. *Working Plans for the Forests of the Sundarbans Division for the Period from 1st April 1931 to 31st March 1951, Vol. 1*. Calcutta: Bengal Government Press.

Cutrim, F.H.R. 2013. *Padrão comportamental e uso de ferramentas em macacos-prego (Sapajus libinosus) residentes em manguezal*. PhD thesis, Universidade de São Paulo, São Paulo, Brazil.

Dacier, A., de Luna, A.G., Fernandez-Duque, E. & Di Fiore, A. 2011. Estimating population density of Amazonian titi monkeys (*Callicebus discolor*) via playback point counts. *Biotropica* 43: 135–140.

da Cunha, R.G.T. 2008. Gênero *Aotus* Illiger 1811. In *Primatas Brasileiros*, N.R. dos Reis, A.L. Peracchi & F.R. Andrade (eds). Londrina, Brazil: Technical Books, pp. 115–125.

da Cunha, R.G.T. & Byrne, R.W. 2006. Roars of black howler monkeys (*Alouatta caraya*): evidence for a function in inter-group spacing. *Behaviour* 143: 1169–1199.

da Cunha, R.G.T. & Byrne, R.W. 2013. Age-related differences in the use of the 'moo' call in black howlers (*Alouatta caraya*). *International Journal of Primatology* 34: 1105–1121.

Daegling, D.J. & Grine, F.E. 1991. Compact bone distribution and biomechanics of early hominid mandibles. *American Journal of Physical Anthropology* 86: 321–339.

Daegling, D.J., McGraw, W.S., Ungar, P.S., *et al.* 2011. Hard object feeding in sooty mangabeys (*Cercocebus atys*) and the interpretation of early hominin feeding ecology. *PLoS ONE* 6: e23095.

da Fonseca Júnior, S.F., Piedade, M.T.F. & Schöngart, J. 2009. Wood growth of *Tabebuia barbata* (E. Mey.) Sandwith (Bignoniaceae) and *Vatairea guianensis* Aubl. (Fabaceae) in Central Amazonian black-water (igapó) and white-water (várzea) floodplain forests. *Trees* 23: 127–134.

Daily, G.C., Polasky, S., Goldstein, J., *et al.* 2009. Ecosystem services in decision making: time to deliver. *Frontiers in Ecology & Environment* 7: 21–28

Dalponte, J.C., Ennes Silva, F. & de Sousa e Silva, Jr., J. 2014. New species of titi monkey, genus *Callicebus* Thomas, 1903 (Primates, Pitheciidae), from southern Amazonia, Brazil. *Papéis Avulsos de Zoologia* 54: 457–472.

Dalton, D.L., Linden, B., Wimberger, K., *et al.* 2015. New insights into samango monkey speciation in South Africa. *PLOS ONE* 10: e0117003

Dampier, W. 1697. *A New Voyage Round the World* (1927 edition). New York: Dover Publications.

Danau Girang Field Centre & Sabah Wildlife Department. 2010. A new orangutan bridge in Kinabatangan. *New Sabah Times* 18 October, 2010: 9.

Daniels, A.E. & Cumming, G.S. 2008. Conversion or conservation? Understanding wetland change in northwest Costa Rica. *Ecological Applications* 18: 49–63.

Darnet, S.H., Silva, L.H.M.D., Rodrigues, A.M.D.C. & Lins, R.T. 2011. Nutritional composition, fatty acid and tocopherol contents of buriti (*Mauritia flexuosa*) and patawa (*Oenocarpus bataua*) fruit pulp from the Amazon region. *Food Science and Technology* 31: 488–491.

Das, S.K., Sarkar, P.K., Saha. R., *et al.* 2012. *Status of Tigers in 24-Parganas (South) Forest Division, Sundarban Biosphere Reserve, West Bengal, India.* New Delhi: World Wide Fund for Nature-India.

da Silva, C.R., Martins, A.C.M., de Castro, I.J., *et al.* 2013. Mammals of Amapá State, Eastern Brazilian Amazonia: a revised taxonomic list with comments on species distributions. *Mammalia* 77: 409–424.

Dasilva, G.L. 1992. The western black-and-white colobus as a low-energy strategist: activity budgets, energy expenditure and energy intake. *Journal of Animal Ecology* 61: 79–91.

da Silveira, R., Campos, Z., Thorbjarnarson, J. & Magnusson, W.E. 2013. Growth rates of black caiman (*Melanosuchus niger*) and spectacled caiman (*Caiman crocodilus*) from two different Amazonian flooded habitats. *Amphibia-Reptilia* 34: 437–449.

da Silveira, R., Campos, Z., Thorbjarnarson, J. & Magnusson, W.E. 2013. Growth rates of black caiman (*Melanosuchus niger*) and spectacled caiman (*Caiman crocodilus*) from two different Amazonian flooded habitats. *Amphibia-Reptilia* 34: 437–449.

da Silva, R.M., Mehlig, U., dos Santos, J.U.M. & de Menezes, M.P.M. 2010. The coastal restinga vegetation of Pará, Brazilian Amazon: a synthesis. *Brazilian Journal of Botany* 33: 563–573

Daily G.C., Polasky S., Goldstein, J., *et al.* 2009. Ecosystem services in decision making: time to deliver. *Frontiers in Ecology and the Environment* 7: 21–28.

Darkoh, M.B. & Mbaiwa, J.E. 2014. Okavango Delta: A Kalahari oasis under environmental threats. *Journal of Biodiversity & Endangered Species* 2: 4.

Davenport, L.C. 2008. *Behaviour and ecology of the giant otter* (Pteronua brasiliensis) *in oxbow lakes of the Manu Biosphere Reserve, Peru.* PhD thesis. University of North Carolina, Chapel Hill, NC, USA.

Davenport, L., Brockelman, W.Y., Wright, P.C., Ruf, K. & Rubio del Valle, F.B. 2002. Ecotourism tools for parks. In *Making Parks Work: Strategies For Preserving Tropical Nature*, J. Terborgh, C. van Schaik, L. Davenport & M. Rao (eds). Washington, DC: Island Press, pp. 279–306.

Davidge, C. 1978. Ecology of baboons (*Papio ursinus*) at Cape Point. *Zoologia Africana* 13: 329–350.

Davidson, N., D'Cruz, R. & Finlayson, C.M. 2005. *Ecosystems and Human Well-being: Wetlands and Water Synthesis: A Report of the Millennium Ecosystem Assessment.* Washington, DC: World Resources Institute.

Davies, A.G. 1984. *An ecological study of the red leaf monkey* (Presbytis rubicunda) *in the dipterocarp forests of Sabah, northern Borneo.* PhD thesis, University of Cambridge, Cambridge.

Davies, A.G., Bennett, E.L. & Waterman, P.G. 1988. Food selection by two Southeast Asian colobine monkeys (*Presbytis rubicunda* and *Presbytis melalophos*) in relation to plant chemistry. *Biological Journal of the Linnean Society* 34: 33–56.

Davies, C.R., Ayres, J.M., Dye, C. & Deane, L.M. 1991. Malaria infection rate of Amazonian primates increases with body weight and group size. *Functional Ecology* 5: 655–662.

Davies, G. 1991. Seed-eating by red leaf monkeys (*Presbytis rubicunda*) in dipterocarp forest of northern borneo. *International Journal of Primatology* 12: 119–144.

Davila Ross, M. & Geissmann, T. 2007. Call diversity of wild male orangutans: a phylogenetic approach. *American Journal of Primatology* 69: 305–324.

Davis, A. & Wagner, J.R. 2003. Who knows? On the importance of identifying 'experts' when researching local ecological knowledge. *Human Ecology* 31: 463–489.

Davis, T.A.W. & Richards, P.W. 1934. The vegetation of Moraballi Creek, British Guiana. *Journal of Ecology* 22: 106–155.

Davison, G.W.H. 2002. *Rehabilitation and restoration of habitat near the Kinabatangan Wildlife Sanctuary, Sabah, Malaysia.* Unpublished report. Rome: Food and Agriculture Organization of the United Nations.

de Alcântara Cardoso, N., Valsecchi, J., Vieira, T. & Queiroz, H.L. 2014. New records and range expansion of the white bald uakari (*Cacajao calvus calvus*, I. Geoffroy, 1847) in central Brazilian Amazonia. *Primates* 55: 199–206.

de Carvalho Oliveira, L., Câmara, E.M.V., Hirsch, A., *et al.* 2003. *Callithrix geoffroyi* (Primates: Callitrichidae) and *Alouatta caraya* (Primates: Atelidae) in the Serra do Cipó National Park, Minas Gerais, Brazil. *Neotropical Primates* 11: 87.

Decary, R. 1950. *La Faune Malgache.* Paris: Payot.

de Carvalho, C.T. 1961. Esboço mastofaunístico do território do Rio Branco. *Revista de Biologia Tropical* 9: 1–15.

Decker, B.S. & Kinnaird, M.F. 1992. Tana River red colobus and crested mangabey: Results of recent censuses. *American Journal of Primatology* 26: 47–52.

Defler, T.R. 1979a. On the ecology and behavior of *Cebus albifrons* in eastern Colombia: I. Ecology. *Primates* 20: 475–490.

Defler, T.R. 1979b. On the ecology and behavior of *Cebus albifrons* in eastern Colombia: II. Behavior. *Primates* 20: 491–502.

Defler, T.R. 1980. Notes on interactions between the tayra (*Eira barbara*) and the white-fronted capuchin (*Cebus albifrons*). *Journal of Mammalogy* 61: 156.

Defler, T.R. 1981. The density of *Alouatta seniculus* in the eastern llanos of Colombia. *Primates* 22: 564–569.

Defler, T.R. 1985. Contiguous distribution of two species of *Cebus* monkeys in El Tuparro National Park, Colombia. *American Journal of Primatology* 8: 101–112.

Defler, T.R. 1996. Aspects of the ranging patterns in a group of wild woolly monkeys (*Lagothrix lagotrhicha*). *American Journal of Primatology* 38: 289–302.

Defler, T.R. 1999. Fission-fusion in the black-headed uacari (*Cacajao melanocephalus*) in eastern Colombia. *Neotropical Primates* 7: 5–8.

Defler, T.R. 2001. *Cacajao melanocephalus ouakary* densities on the lower Apaporis River, Colombian Amazon. *Primate Report* 61: 31–36.

Defler, T.R. 2004. *Primates of Colombia.* Bogotá, Colombia: Conservation International.

Defler, T.R. 2010. *Historia Natural de los Primates Colombianos.* Bogotá, Colombia: Universidad Nacional de Colombia.

Defler, T.R. 2013. Species richness, densities and biomass of nine primate communities in eastern Colombia. *Revista de la Academia Colombiana de Ciencias* 37: 253–262.

Defler, T.R. & Defler, S.B. 1996. Diet of a group of *Lagothrix lagothricha lagothricha* in southeastern Colombia. *International Journal of Primatology* 17: 161–190.

Defler, T.R. & Rodríguez, J.V. 1998. La fauna de la Orinoquia. In *Colombia Orinoco*, C. Domínguez (ed.). Bogotá: Fondo Fen, pp. 134–165.

Defler, T.R., Bueno, M.L. & García, J. 2010. *Callicebus caquetensis*: a new and Critically Endangered titi monkey from southern Caquetá, Colombia. *Primate Conservation* 25: 1–9.

Defler, T.R., Rodríguez, M.J.V. & Hernández-Camacho, J.I. 2003. Conservation priorities for Colombian primates. *Primate Conservation* 19: 10–18.

de Jong, Y.A. 2012. *Taxonomy, diversity, biogeography and conservation of the primates of Kenya and Tanzania*. PhD thesis, Oxford Brookes University, Oxford, UK.

de Jong, Y.A. & Butynski, T.M. 2009. *Primate biogeography, diversity, taxonomy and conservation of the coastal forests of Kenya*. Unpublished report. Nairobi: Critical Ecosystem Partnership Fund. Eastern Africa Primate Diversity and Conservation Program. www.wildsolutions.nl.

de Jong, Y.A. & Butynski, T.M. 2011. *Primate survey on the north coast of Kenya: biogeography, diversity and conservation*. Unpublished report. Kenya: Eastern Africa Primate Diversity and Conservation Program. www.wildsolutions.nl.

de Jong, Y.A. & Butynski, T.M. 2012. The primates of East Africa: country lists and conservation priorities. *African Primates* 7: 135–155.

de Jong, Y.A., Galat-Luong, A., Butynski, T.M., Galat, G. & Isbell, L.A. 2011. *Erythrocebus patas patas*. In *All the World's Primates*, N. Rowe & M. Myers (eds). Charlestown, RI: Primate Conservation Inc.

Dela, J.D.S. 2007. Seasonal food use strategies of *Semnopithecus vetulus nestor*, at Panadura and Piliyandala, Sri Lanka. *International Journal of Primatology* 28: 607–626.

de Lacerda, L.D., José, D.V., de Rezenda, C.E., *et al.* 1986. Leaf chemical characteristics affecting herbivory in a new world mangrove forest. *Biotropica* 18: 350–355.

de la Torre, S. 2012. Conservation of Neotropical primates: Ecuador – a case study. *International Zoo Yearbook* 46: 25–35.

de la Torre, S. & Snowdon, C.T. 2002. Environmental correlates of vocal communication of wild pygmy marmosets, *Cebuella pygmaea*. *Animal Behaviour* 63: 847–856.

de la Torre, S., Campos, F. & de Vries, T. 1995. Home range and birth seasonality of *Saguinus nigricollis* in Ecuadorian Amazonia. *American Journal of Primatology* 37: 39–56.

de la Torre, S., Snowdon, C. & Bejarano, M. 2000. Effects of human activities on wild pygmy marmosets in Ecuadorian Amazonia. *Biological Conservation* 94: 153–163.

Delgado, Jr., R.A. & van Schaik, C.P. 2000. The behavioral ecology and conservation of the orangutan (*Pongo pygmaeus*): a tale of two islands. *Evolutionary Anthropology* 9: 201–218.

de Miranda, G.H.B. & de Faria, D.S. 2001. Ecological aspects of black-pincelled marmoset (*Callithrix penicillata*) in the cerradão and dense cerrado of the Brazilian central plateau. *Brazilian Journal of Biology* 61: 397–404.

de Moraes, B.L.C., Souto, A.D.S. & Schiel, N. 2014. Adaptability in stone tool use by wild capuchin monkeys (*Sapajus libidinosus*). *American Journal of Primatology* 76: 967–977.

de Morais, R.J. 2006. *A planice alluvial do médio Rio Araguaia: processos geomorfológicos e suas implicações ambientais*. PhD thesis, Universidade Federal do Goiás, Brazil.

De Oliveira, S.G., Alfaro, J.W.L. & Veiga, L.M. 2014. Activity budget, diet, and habitat use in the Critically Endangered Ka'apor capuchin monkey (*Cebus kaapori*) in Pará State, Brazil: a preliminary comparison to other capuchin monkeys. *American Journal of Primatology* 76: 919–931.

de Poncins, E. 1935. A hunting trip in the Sunderbunds in 1892. *Journal of the Bombay Natural History Society* 37: 844–858.

de Queiroz, H.L. & Peralta, N. 2011. Protected areas in the Amazonian *várzea* and their role in its conservation: the case of Mamirauá Sustainable Development Reserve (MSDR). *Amazonian Floodplain Forests* 210: 465–483.

de Sá, R.M. 2004. Impacts of damming on primate community structure in the Amazon. In *People in Nature: Wildlife Conservation in South and Central America*. K.M. Silvius, R.E. Bodmer & J.M. Fragoso (eds). New York: Columbia University Press, pp. 240–256.

DeSalle, R. & Amato, G. 2004. The expansion of conservation genetics. *Nature Reviews Genetics* 5: 702–712.

Desbiez, A.L.J., Bodmer, R.E.B. & Tomas, W.M. 2010a. Mammalian densities in a neotropical wetland subject to extreme climatic events. *Biotropica* 42: 372–378.

Desbiez, A.L.J., Rocha, F.L. & Keuroghlian, A. 2010b. Interspecific association between an ungulate and a carnivore or a primate. *Acta Ethologica* 13: 137–139.

de Souza, J.C., da Cunha, V.P. & Markwith, S.H. 2015. Spatiotemporal variation in human-wildlife conflicts along highway BR-262 in the Brazilian Pantanal. *Wetlands Ecology & Management* 23: 227–239.

de Thoisy, B., Brosse, S. & Dubois, M.A. 2008. Assessment of large-vertebrate species richness and relative abundance in Neotropical forest using line-transect censuses: what is the minimal effort required? *Biodiversity & Conservation* 17: 2627–2644.

de Thoisy, B., Renoux, F. & Julliot, C. 2005. Hunting in northern French Guiana and its impact on primate communities. *Oryx* 39: 149–157.

de Thoisy, B., Richard-Hansen, C., Goguillon, B., *et al.* 2010. Rapid evaluation of threats to biodiversity: human footprint score and large vertebrate species responses in French Guiana. *Biodiversity & Conservation* 19: 1567–1584.

de Thoisy, B., Vogel, I., Reynes, J.-M., *et al.* 2001. Health evaluation of translocated free-ranging primates in French Guiana. *American Journal of Primatology* 54: 1–16.

DeVore, I. & Hall, K.R.L. 1965. Baboon ecology. In *Primate Behavior: Field Studies of Monkeys and Apes*, I. DeVore (ed.). New York: Holt, Rinehart & Winston, pp. 20–52.

Devos, C., Gatti, S. & Levréro, F. 2002. New record of algae feeding and scooping by *Pan t. troglodytes* at Lokoué Bai in Odzala National Park, Republic of Congo. *Pan Africa News* 9: 19–21.

Devos, C., Sanz, C., Morgan, D., *et al.* 2008. Comparing ape densities and habitats in northern Congo: surveys of sympatric gorillas and chimpanzees in the Odzala and Ndoki regions. *American Journal of Primatology* 70: 439–451.

Devreese, L. 2015. *Preliminary survey of the current distribution and conservation status of the poorly known and Critically Endangered* Piliocolobus bouvieri *in the Republic of Congo*. Report to Primate Conservation Inc.

de Wasseige, C., de Marcken, P., Bayol, N., *et al.* 2012. *The Forests of the Congo Basin: State of the Forest 2010*. Luxembourg: Publications Office of the European Union.

Diamond, J.M. 1972. Biogeographic kinetics: Estimation of relaxation times for avifaunas of southwest Pacific islands. *Proceedings of the National Academy of Sciences* 69: 3199–3203.

Dias, P. A. D & Rodríguez-Luna, E. 2006. Seasonal changes in associative behavior and subgrouping patterns of mantled howler monkey males living on an island. *International Journal of Primatology* 27: 1635–1651.

Dias, C.A.R. Queirogas, V.L. & Pedersoli, M.A. 2015. Translocation and radio-telemetry monitoring of pygmy marmoset, *Cebuella pygmaea* (Spix, 1823), in the Brazilian Amazon. *Brazilian Journal of Biology* 75: 91–97.

Díaz, L.A., Díaz, M.P., Almirón, W.R. & Contigiani, M.S. 2007. Infection by UNA virus (*Alphavirus*, Togaviridae) and risk factor analysis in black howler monkeys (*Alouatta caraya*) from Paraguay and Argentina. *Transactions of the Royal Society of Tropical Medician & Hygiene* 101: 1039–1041.

Di Bitetti, M.S. 2003. Food-associated calls of tufted capuchin monkey (*Cebus apella nigritus*) are functionally referential signals. *Behaviour* 140: 565–592.

Di Bitetti, M.S. 2001. Home-range use by the tufted capuchin monkey (*Cebus apella nigritus*) in a subtropical rainforest of Argentina. *Journal of Zoology* 253: 33–45.

Di Bitetti, M.S. & Janson, C.H. 2001. Social foraging and the finder's share in capuchin monkeys (*Cebus apella*). *Animal Behaviour* 32: 470–477.

Diegues, A.C. 1999. Human populations and coastal wetlands: conservation and management in Brazil. *Ocean & Coastal Management* 42: 187–210.

Dietz, J.M., Peres, C.A. & Pinder, L. 1997. Foraging ecology and use of space in wild golden lion tamarins (*Leontopithecus rosalia*). *American Journal of Primatology* 41: 289–305.

Di Fiore, A., Link, A. & Campbell, C.J. 2011. The atelines: behavioral and socioecological diversity in a New World radiation. In *Primates in Perspective*, 2nd Edition, C.A. Campbell, A. Fuentes, K. MacKinnon, S. Bearder & R. Stumpf (eds). Oxford University Press, pp. 155–188.

Dinerstein, E., Olson, D.M., Graham, D.J., *et al.* 1995. *A Conservation Assessment of the Terrestrial Ecoregions of Latin America and the Caribbean*. Washington, DC: The World Bank.

Dirzo, R., Young, H.S., Mooney, H.A. & Ceballos, G. 2011. *Seasonally Dry Tropical Forests: Ecology and Conservation.* Washington, DC: Island Press.

Ditmer, M.A., Vincent, J.B., Werden, L.K., *et al.* 2015. Bears show a physiological but limited behavioral response to unmanned aerial vehicles. *Current Biology* 25: 2278–2283.

Djojosudharmo, S. & van Schaik, C.P. 1992. Why are orangutans so rare in the highlands? Altitudinal changes in a Sumatran forest. *Tropical Biodiversity* 1: 11–22.

Domínguez, C. 1998. La gran Cuenca del Orinoco. In *Colombia Orinoco*, C. Domínguez (ed.). Bogotá: Fondo Fen, pp. 39–67.

Dominy, N., Vogel, E., Yeakel, J., Constantino, P. & Lucas, P. 2008. Mechanical properties of plant underground storage organs and implications for dietary models of early hominins. *Evolutionary Biology* 35: 159–175.

Dominy, N.J. 2012. Hominins living on the sedge. *Proceedings of the National Academy of the Sciences of the Unites States of America* 109: 20171–20172.

Donati, G., Kesch, K., Ndremifidy, K., *et al.* 2011. Better few than hungry: flexible feeding ecology of collared lemurs *Eulemur collaris* in littoral forest fragments. *PLoS ONE* 6: e19807. doi:10.1371/journal.pone.0019807.

Doran, D.M., McNeilage, A., Greer, D., Bocian, C., Mehiman, P. & Shah, N. 2002. Western lowland gorilla diet and resource availability: New evidence, cross-site comparisons, and reflections on indirect sampling method. *American Journal of Primatology* 58: 91–116.

Doran-Sheehy, D.M., Greer, D., Mongo, P. & Schwindt, D. 2004. Impact of ecological and social factors on ranging in western gorillas. *American Journal of Primatology* 64: 207–222.

dos Santos, Junior, U.M., de Carvalho Gonçalves, J.F. & Fearnside, P.M. 2013. Measuring the impact of flooding on Amazonian trees: photosynthetic response models for ten species flooded by hydroelectric dams. *Trees* 27: 193–210.

dos Santos, Junior, U.M., Carvalho Gonçalves, J.F., Strasser, R.J. & Fearnside, P.M. 2015. Flooding in tropical forests in central Amazonia: what do the effects on the photosynthetic apparatus is trees tell us about species suitability for reaforestation in extreme environments created by hydroelectric dams. *Acta Physiologiae Plantarum* 37: 1–17.

Driessen, P.M. 1978. Peat soils. In *Soils of the Humid Tropics*, M. Drosoff (ed.). Los Banos, Philippines: International Rice Research Institute, pp. 763–779.

Dudgeon, D. 1992. Endangered ecosystems: a review of the conservation status of tropical Asian rivers. *Hydrobiologia* 248: 167–191.

Dudgeon D. 2000a. The ecology of tropical Asian rivers and streams in relation to biodiversity conservation. *Annual Review of Ecology & Systematics* 31: 239–263.

Dudgeon D. 2000b. Large-scale hydrological changes in tropical Asia: Prospects for riverine biodiversity. *BioScience* 50: 793–806.

Duke, N.C., Meynecke, J.-O., Dittmann, S., *et al.* 2007. A world without mangroves? *Science* 317: 41–42.

Dunbar, R.I.M. 1993. Socioecology of the extinct theropiths: a modelling approach. In *Theropithecus. The Rise and Fall of a Primate Genus*, N.G. Jablonski (ed.). Cambridge: Cambridge University Press, pp. 465–486.

Dunn, A. 2010. Different approaches to saving the Cross River gorillas in Nigeria. *Gorilla Journal* 41: 13.

Dupain, J., Guislain, P., Nguenang, G.M., de Vleeschouwer, K. & van Elsacker, L. 2004. High chimpanzee and gorilla densities in a non-protected area on the northern periphery of the Dja Faunal Reserve, Cameroon. *Oryx* 38: 209–216.

Durbin, J.C. 1999. Lemurs as flagships for conservation in Madagascar. In *New Directions in Lemur Studies*, B. Rakotosamimanana, J. Ganzhorn & S. Goodman (eds). New York: Springer, pp. 269–281.

Durbin, J., Funk, S.M., Hawkins, F., *et al.* 2010. Investigations into the status of a new taxon of *Salanoia* (Mammalia: Carnivora: Eupler idae) from the marshes of Lac Alaotra, Madagascar. *Systematics and Biodiversity* 8: 341–355.

Duvail, S., Médard, C., Hamerlynck, O. & Nyingi, D.W. 2012. Land and water 'grabbing' in an East African coastal wetland: the Tana Delta case study. *Water Alternatives* 5: 322–343.

Duvail, S., Mwakalinga, A.B., Eijkelenburg, A., *et al.* 2014. Jointly thinking the post-dam future: exchange of local and scientific knowledge on the lakes of the Lower Rufiji, Tanzania. *Hydrological Sciences Journal* 59: 713–730.

Ecopetrol 2015. *Sustainable development report 2014.* Bogatá: Ecopetrol. www.ecopetrol.com.co/especiales/Sustainability-report-2014/espanol/principal/nuestra-cadena-de-valor/produccion.

Effiom, E.O, Nuñez-Iturri, G., Smith, H.G., Ottosson, U. & Olsson, O. 2013. Bushmeat hunting changes regeneration of African rainforests. *Proceedings of the Royal Society B: Biological Sciences* 280: 20130246.

Eeley, H.A.C. & Foley, R.A. 1999. Species richness, species range size and ecological specialisation among African primates: geographical patterns and conservation implications. *Biodiversity & Conservation* 8: 1033–1056.

Eeley, H.A., Lawes, M.J. & Piper, S.E. 2002. The influence of climate change on the distribution of indigenous forest in KwaZulu-Natal, South Africa. *Journal of Biogeography* 26: 595–617.

Egberongbe, F.O.A., Nwilo, P.C. & Badejo, O.T. 2006. Oil spill disaster monitoring along Nigerian coastline. *Marine and Coastal Zone Management: Environmental Planning Issues. 5th FIG Regional Conference (March 8–11, Accra, Ghana).* Accra: FIG. www.fig.net/pub/accra/papers/ts16/ts16_06_egberongbe_etal.pdf.

Ehardt, C.L. & Butynski, T.M. 2013. *Cercocebus chrysogaster* golden-bellied mangabey. In *Mammals of Africa. Volume II: Primates*, T.M. Butynski, J. Kingdon, & J. Kalina (eds). London: Bloomsbury, pp. 174–177.

Ehardt, C.L. 2013. *Cercocebus torquatus* red-capped mangabey (white-collared mangabey). In *Mammals of Africa. Volume II: Primates*, T.M. Butynski, J. Kingdon & J. Kalina (eds). London: Bloomsbury, pp. 186–189.

Ehlers-Smith, D.A. & Ehlers-Smith, Y.C. 2013. Population density of red langurs in Sabangau Tropical Peat-Swamp Forest, central Kalimantan, Indonesia. *American Journal of Primatology* 75: 837–847.

Ehlers-Smith, D.A., Ehlers-Smith, Y.C. & Cheyne, S.M. 2013a. Home range use and activity patterns of red langurs in Sabangau tropical peat-swamp forest, central Kalimantan, Indonesia. *International Journal of Primatology* 34: 957–972.

Ehlers-Smith, D.A., Husson, S.J., Ehlers-Smith, Y.C. & Harrison, M.E. 2013b. Feeding ecology of red langurs in Sabangau tropical peat-swamp forest, Indonesian Borneo: extreme granivory in a non-masting forest. *American Journal of Primatology* 75: 848–859.

Eiten, G. 1972. The cerrado vegetation of Brazil. *Botanical Review* 38: 201–341.

Eisermann, K. 2006. *Evaluation of Waterbird Populations and their Conservation in Guatemala*. Washington, DC: Waterbirds Conservation for the Americas.

El-Ghani, M.A., El-Fiky, A.M., Soliman, A. & Khattab, A. 2011. Environmental relationships of aquatic vegetation in the fresh-water ecosystem of the Nile Delta, Egypt. *African Journal of Ecology* 49: 103–118.

Ellenberg, H., Galat-Luong, A., Von Maydell, H.-J., *et al.* 1988. *Pirang. Ecological Investigations in a Forest Island in the Gambia*. Stiftung Walderhaltung in Afrika, Hamburg, und Bundesforschungsanstalt für Forst- und Holzwirtschaft. Hamburg, Germany: Warnke Verlag, Reinbek.

Ellison, A.M. 2002. Macroecology of mangroves: large-scale patterns and processes in tropical coastal forests. *Trees* 16: 181–194.

Ellison, A.M. 2004. Wetlands of Central America. *Wetlands Ecology & Management* 12: 3–55.

Ellison, A.M., Farnsworth, E.J. & Merkt, R.E. 1999. Origins of mangrove ecosystems and the mangrove biodiversity anomaly. *Global Ecology & Biogeography* 8: 95–115.

Emidio, R.A. & Ferreira, G. 2012. Energetic payoff of tool use for capuchin monkeys in the Caatinga: variation by season and habitat type. *American Journal of Primatology* 74: 332–343.

Emmerich, W.E. 1990. Precipitation nutrient inputs in semi-arid environments. *Journal of Environmental Quality* 19: 621–624.

Emmons, L.H., Whitney, B.M. & Ross, Jr, D.L. 1998. *Sounds of Neotropical Rainforest Mammals: An Audio Field Guide*. Chicago, IL: The University of Chicago Press.

Enari, H. & Sakamaki-Enari, H. 2014. Impact assessment of dam construction and forest management for Japanese macaque habitats in snowy areas. *American Journal of Primatology* 76: 271–280.

Endo, W., Peres, C.A., Salas, E., *et al.* 2010. Game vertebrate densities in hunted and nonhunted forest sites in Manu National Park, Peru. *Biotropica* 42: 251–261.

Engel, G. & Jones-Engel, L. 2011. The role of *Macaca fascicularis* in infectious agent transmission. In *Monkeys on the Edge: Ecology and Management of Long-tailed Macaques and Their Interface with Humans*, M.D. Gumert, A. Fuentes, L. Jones-Engel (eds). Cambridge: Cambridge University Press, pp. 183–204.

Engeman, R.M., Martin, R.E., Constantin, B., Noel, R. & Woolard, J. 2003. Monitoring predators to optimize their management for marine turtle nest protection. *Biological Conservation* 113: 171–178.

Epple, G. 1968. Comparative studies on vocalization in marmoset monkeys (Hapalidae). *Folia Primatologica* 8: 1–40.

Eppley, T.M., Donati, G., Ramanamanjato, J.-B., *et al.* 2015. The use of an invasive species habitat by a small folivorous primate: implications for conservation. *PLoS ONE* 10: e0140981.

Eppley, T.M., Verjans, E. & Donati, G. 2011. Coping with low-quality diets: a first account of the feeding ecology of the southern gentle lemur, *Hapalemur meridionalis*, in the Mandena littoral forest, southeast Madagascar. *Primates* 52: 7–13.

Erftemeijer, P., Balen, S. & van Djuharsa, E. 1988. *The importance of Segara Anakan for nature conservation, with special reference to its avifauna*. Report No. 5. Bogor: PHPA-Asian Wetland Bureau-Interwader.

Eriksson, J., Hohmann, G., Boesch, C. & Vigilant, L. 2004. Rivers influence the population genetic structure of bonobos (*Pan paniscus*). *Molecular Ecology* 13: 3425–3435.

Ertel, J.R., Hedges, J.I., Devol, A.H., *et al.* 1986. Dissolved humic substances of the Amazon River system. *Limnology & Oceanography* 31: 739–754.

ERM (Environmental Resources Management). 2002. Annex I: proposed conservation areas of the Niger Delta. In *Environmental and Social Due Diligence Review [of Oil and Gas Facility], Nigeria*. Unpublished report. London: Environmental Resources Management, pp. I1–I3.

Erwin, K.L. 2009. Wetlands and global climate change: the role of wetland restoration in a changing world. *Wetlands Ecology & Management* 17: 71–84.

Esser, H., Brown, D. & Liefting, Y. 2010. Swimming in the Northern Tamandua (*Temandua mexicana*) in Panama. *Endentata* 11: 70–72.

Estrada, A., Garber, P.A., Pavelka, M.S.M. & Luecke, L. 2006. Overview of the Mesoamerican primate fauna, primate studies, and conservation concerns. In *New Perspectives in the Study of Mesoamerican Primates: Distribution, Ecology, Behavior and Conservation*, A. Estrada, P.A. Garber, M.S.M. Pavelka & L. Luecke (eds). New York: Springer, pp. 1–22.

Estrada, A., Luecke, L., Belle, S.V., Barrueta, E. & Meda, M.R. 2004. Survey of black howler (*Alouatta pigra*) and spider (*Ateles geoffroyi*) monkeys in the Mayan sites of Calakmul and Yaxchilan'n, Mexico and Tikal, Guatemala. *Primates* 45: 33–39.

Etter, A., McAlpine, C., Pullar, D. & Possingham, H. 2006. Modelling the conversion of Colombian lowland ecosystems since 1940: drivers, patterns and rates. *Journal of Environmental Management* 79: 74–87.

Eudey, A.A. 1987. *Action Plan for Asian Primate Conservation: 1987–1991*. Gland: IUCN/SSC Primate Specialist Group.

Fa, J.E. & Southwick, C.H. 1988. *Ecology and Behavior of Food-enhanced Primate Groups*. New York: Liss.

Fahrig, L. 2002. Effect of habitat fragmentation on the extinction threshold: a synthesis. *Ecological Applications* 12: 346–353.

Fahrig, L. 2003. Effects of habitat fragmentation on biodiversity. *Annual Review of Ecology, Evolution & Systematics* 34: 487–515.

Fahrig, L. & Merriam, G. 1994. Conservation of fragmented populations. *Conservation Biology* 8: 50–59.

Falótico, T. 2011. *Uso de ferramentas por macacos-prego* (Sapajus libidinosus) *do Parque Nacional da Serra da Capivara - PI*. PhD thesis, Universidade de São Paulo, São Paulo, Brazil.

Fam, S.D., Lee, B.P.H.Y. & Shekelle, M. 2014. The conservation status of slow lorises *Nycticebus* spp. in Singapore. *Endangered Species Research* 25: 69–77.

Fam, S. & Nijman, V. 2011. Spizaetus hawk-eagles as predators of arboreal colobines. *Primates* 52: 105–110

Fan, P.-F., Ni, Q.-Y., Sun, G.-Z., Huang, B. & Jiang, X.-L. 2008. Seasonal variations in the activity budget of *Nomascus concolor jingdongensis* at Mt. Wuliang, Central Yunnan, China: effects of diet and temperature. *International Journal of Primatology* 29: 1047–1057.

Fanshawe, D.B. 1952. *The Vegetation of British Guiana: A Preliminary Review*. Institute Paper 29, Oxford: Imperial Forestry Institute, University of Oxford.

FAO. 1981. *Tropical Forest Resources Assessment Project*. Rome: FAO.

FAO. 2007. *The World's Mangroves 1980–2005*. Rome: FAO.

FAO. 2010. *Global Forest Resources Assessment 2010 Main Report*. Rome: FAO.

FAO. 2013. *State of the World's Forests 2012*. R.M. Martin, D. Kneeland, D. Brooks & R. Matta (eds). Rome: FAO.

Fashing, P.J. 2011. African colobine monkeys: their behavior, ecology, and conservation. In *Primates in Perspective*, C.J. Campbell, A. Fuentes, K.C. MacKinnon, S.K. Bearder & R.M. Stumpf (eds). Oxford: Oxford University Press, pp. 203–229.

Fashing, P.J., Nguyen, N., Barry, T.S., *et al.* 2011. Death among geladas (*Theropithecus gelada*): a broader perspective on mummified infants and primate thanatology, *American Journal of Primatology* 73: 405–409.

Fausser, J.-L., Proper, P. & Rumpler, Y. 2002. Relationships inside the genus *Hapalemur* based on mitochondrial DNA sequences. *Lemur News* 7: 25–26.

Fay, M.J. & Agnagna, M. 1992. Census of gorillas in northern Republic of Congo. *American Journal of Primatology* 27: 275–284.

Fay, J.M., Agnagna, M., Moore, J. & Oko, R. 1989. Gorillas (*Gorilla gorilla gorilla*) in the Likouala swamp forests of north central Congo: Preliminary data on populations and ecology. *International Journal of Primatology* 10: 477–486.

Fay, J.M., Agnagna, M. & Moutsambote, J.M. 1990. *A survey of the proposed Nouabalé conservation area in northern Congo*. Report. New York: Wildlife Conservation Society.

Fay, M.J., Carroll, R.W., Paterhans, C.K. & Harris, D. 1995. Leopard attack on and consumption of gorillas in the Central African Republic. *Journal of Human Evolution* 29: 93–99.

Fearn, E. 2010. *State of the Wild 2010–2011*. Washington, DC: Island Press.

Fearnside, P.M. 1989. Brazil's Balbina Dam: environment versus the legacy of the pharaohs in Amazonia. *Environmental Management* 13: 401–423.

Fearnside, P.M. 2001. Environmental impacts of Brazil's Tucuruí Dam: unlearned lessons for hydroelectric development in Amazonia. *Environmental Management* 27: 377–396.

Fearnside, P.M. 2002. Greenhouse gas emissions from a hydroelectric reservoir (Brazil's Tucuruí Dam) and the energy policy implications. *Water, Air & Soil Pollution* 133: 69–96.

Fearnside, P.M. 2004. Greenhouse gas emissions from hydroelectric dams: controversies provide a springboard for rethinking a supposedly 'clean' energy source. *Climatic Change* 66: 1–8.

Fearnside, P.M. 2005. Brazil's Samuel Dam: Lessons for hydroelectric development policy and the environment in Amazonia. *Environmental Management* 35: 1–19.

Fearnside, P.M. 2006. Dams in the Amazon: Belo Monte and Brazil's hydroelectric development of the Xingu river basin. *Environmental Management* 38: 16–27.

Fearnside, P.M. 2009. As hidrelétricas de Belo Monte e Altamira (Babaquara) como fontes de gases de efeito estufa. *Novos Cadernos NAEA* 12: 5–56.

Fearnside, P.M. 2011. Will the Belo Monte Dam's benefits outweigh the costs? *Latin America Energy Advisor*, 21–25 February: 6.

Fearnside, P.M. 2014. Brazil's Madeira River dams: a setback for environmental policy in Amazonian development. *Water Alternatives* 7: 256–269.

Fearnside, P.M. 2015. Emissions from tropical hydropower and the IPCC. *Environmental Science & Policy* 50: 225–239.

Fearnside, P.M. & Pueyo, S. 2012. Underestimating greenhouse-gas emissions from tropical dams. *Nature Climate Change* 2: 382–384.

Fedepalma 2014. *Anuario Estadistico 2014: la agroindustria de la palma de aceite en Colombia y en el mundo*: 2009–2013. Javegraf, Bogotá.

Federal Department of Forestry. 1978. *Vegetation and land use map, sheets NB 31-8 and NB 31-12/1*. Lagos, Nigeria: Federal Department of Forestry.

Fedigan, L.M. & Jack, K. 2001. Neotropical primates in regenerating Costa Rican dry forest: a comparison of howler and capuchin population patterns. *International Journal of Primatology* 22: 689–713.

Fedigan, L. & Fedigan, L.M. 1988. *Cercopithecus aethiops aethiops*: a review of field studies. In *A Primate Radiation: Evolutionary Biology of the African Guenons*, A. Gautier-Hion, F. Bourliere, J.-P. Gautier & J. Kingdon (eds). Cambridge: Cambridge University Press, pp. 389–414.

Fedigan, L.M., Rose, L.M. & Avila, R.M. 1996. See how they grow: tracking capuchin monkey (*Cebus capucinus*) populations in a regenerating Costa Rican dry forest. In *Adaptive Radiations of Neotropical Primates*, M.A. Norconk, A.L. Rosenberger & P.A. Garber (eds). New York: Plenum Press, pp. 289–307.

Feeroz, M.M., Islam, M.A. and Kabir, M. 1995. Status, distribution and conservation of non-human primates of Bangladesh. *Kyoto University Overseas Research Report of Studies on Asian Non-Human Primates*, 9: 73–82.

Feeroz, M.M., Schillaci, M.A., Begum, S., *et al.* 2010. Morphometric assessment of rhesus macaques (*Macaca mulatta*) from Bangladesh. *Primate Conservation* 25: 119–125.

Feibel, C.S., Harris, J.M. & Brown, F.H. 1991. Neogene paleoenvironments of the Turkana Basin. In *Koobi For a Research Project, Volume 3: Stratigraphy, Artiodactyls, and Paleoenvironments*, J.M. Harris (ed.). Oxford: Clarendon Press, pp. 321–370.

Feibel, C.S. 2013. Facies analysis and Plio-Pleistocene paleoecology. In *Early Hominin Paleoecology*, M. Sponheimer, J.A. Lee-Thorp, K.E. Reed & P.S. Ungar (eds). Boulder, CO: University of Colorado Press, pp. 35–58.

Feistner, A.T.C. 1999. Conservation of the Alaotran gentle lemur: a multi-disciplinary approach. In *New Directions in Lemur Studies*, B. Rakotosamimanana, H. Rasamimanana, J. Ganzhorn & S. Goodman (eds). New York: Plenum Press, pp. 241–248.

Feitosa, G.S., Graça, P.M.L.A. & Fearnside, P.M. 2007. Estimativa da zona de deplecionamento da hidrelétrica de Balbina por técnica de sensoriamento remoto In *Anais XIII Simpósio Brasileiro de Sensoriamento Remoto, Florianópolis, Brasil 21–26 abril 2007*, J.C.N. Epiphanio, L.S. Galvão & L.M.G. Fonseca (eds). São José dos Campos-São Paulo, Brazil: Instituto Nacional de Pesquisas Espaciais (INPE), pp. 6713–6720.

Felfilli, J.M. 1995. Diversity, structure and dynamics of a gallery forest in central Brazil. *Vegetatio* 117: 1–15.

Feller, I.C. 1995. Effects of nutrient enrichment on growth and herbivory of dwarf red mangrove (*Rhizophora mangle*). *Ecological Monographs* 65: 477–505.

Felton, A., Felton, A.M., Wallace, R.B. & Gomez, H. 2006. Identification, distribution and behavioural observations of the titi monkeys *Callicebus modestus* Lönnberg 1939, and *Callicebus olallae* Lönnberg 1939. *Primate Conservation* 20: 41–46.

Felton, A.M., Engstrom, L.M., Felton, A. & Knott, C.D. 2003. Orangutan population density, forest structure and fruit availability in hand-logged and unlogged peat swamp forests in West Kalimantan, Indonesia. *Biological Conservation* 114: 91–101.

Felton, A.M., Felton, A., Raubemheimer, D., *et al.* 2009a. Protein content of diets dictates the daily energy intake of a free-ranging primate. *Behavioral Ecology* 20: 685–690.

Felton, A.M., Felton, A., Wood, J.T., *et al.* 2009b. Nutritional ecology of *Ateles chamek* in lowland Bolivia: how macronutrient balancing influences food choices. *International Journal of Primatology* 30: 675–696.

Fernández, J., Toro, M.A. & Caballero, A. 2008. Management of subdivided populations in conservation programs: development of a novel dynamic system. *Genetics* 179: 683–692.

Fernandes, M.E.B. 1991. Tool use and predation of oysters (*Crassostrea rhizophorae*) by the tufted capuchin, *Cebus apella apella*, in brackish water mangrove swamp. *Primates* 32: 529–531.

Fernandes, M.E.B. 1999. Phenological patterns of *Rhizophora* L, *Avicennia* L. and *Laguncularia* Gaertn. f. in Amazonian mangrove swamps. *Hydrobiologia* 413: 53–62.

Fernandes, M.E.B. 2000. Association of mammals with mangrove forests: a worldwide review. *Boletim do Laboratório de Hidrobiologia* 13: 83–108.

Fernandes, M.E.B. & Aguiar, N.O. 1993. Evidências sobre a adaptação de primatas neotropicais às áreas de mangue com ênfase em *Cebus apella apella* (Primates: Cebidae). In *A Primatologia no*

Brasil: IV, M.E. Yamamoto (ed.). Brazil: Editoria Universitaria, Univeridade Federal do Rio Grande do Norte, pp. 67–80.

Fernandes, M.E.B., Cardoso da Silva, J.M. & de Souza e Silva, Jr., J. 1995. The monkeys of the islands of the Amazon estuary, Brazil: a biogeographic analysis. *Mammalia* 59: 213–222.

Fernandes, M.E.B., Virgulino, A.R.C., Nascimento, A.A.M. & Rodrigues, L.F.P. 2005. Padrões de floração e frutificação em *Laguncularia racemosa* (L.) Gaertn. f.: uma avaliação metodológica. *Boletim do Laboratório de Hidrobiologia* 18: 33–38

Fernández, V.A. 2014. *Ecología nutricional del mono aullador negro y dorado* (Alouatta caraya) *en el límite sur de su distribución*. PhD thesis, Universidad de Buenos Aires, Buenos Aires, Argentina.

Fernandez-Duque, E. 2007. Aotinae: social monogamy in the only nocturnal haplorhines. In *Primates in Perspective*, C.J. Cambell, A. Fuentes, K.C. MacKinnon, M. Panger & S.K. Bearder (eds). Oxford, UK: Oxford University Press, pp. 139–154.

Fernández-Duque, E., Rotundo, M. & Ramirez-Llorens, P. 2002. Environmental determinants of birth seasonality in night monkeys (*Aotus azarai*) of the Argentinean Chaco. *International Journal of Primatology* 23: 639–656.

Fernandez-Duque, E., Wallace, R.B. & Rylands, A.B. 2008. *Alouatta caraya*. In *IUCN Red List of Threatened Species*. Version 2014.3. Gland, Switzerland: IUCN. www.iucnredlist.org.

Ferrari, S. 2008. Gênero *Mico* Lesson 1840. In *Primatas Brasileiros*, N.R. dos Reis, A.L. Peracchi & F.R. Andrade (eds). Londrina, Brazil: Technical Books, pp. 59–68.

Ferrari, S.F. 2009. Predation risk and antipredator strategies. In *South American Primates: Comparative Perspectives in the Study of Behavior, Ecology, and Conservation*, P.A. Garber, A. Estrada, J.C. Bicca-Marques, E.W. Heymann & K.B. Strier (eds). New York: Springer, pp. 251–277.

Ferrari, S.F., Ghilardi, R., Lima, E.M., Pina, A.L. & Martins, S.S. 2004. Tucuruí fifteen years on: long term effects of reservoir flooding on primate populations in Southeastern Amazonia. *Folia Primatologica* 75: 262–272.

Ferrari, S.F., Guedes, P.G., Figueiredo, W. & Barnett, A.A. 2014. Reconsidering the nomenclature of the black-faced uacaris (*Cacajao melanocephalus* group, sensu Hershkovitz, 1987) from the northern Amazon basin. *Zootaxa* 3866: 353–370.

Ferreira, C.S., Piedade, M.T.F., Wittmann-de Oliveira, A. & Franco, A.C. 2010a. Plant reproduction in the Central Amazonian floodplains: challenges and adaptations. *AoB PLANTS 2010:* plq009, doi:10.1093/aobpla/plq009.

Ferreira, R.G., Emidio, R.A. & Jerusalinsky, L. 2010b. Three stones for three seeds: natural occurrence of selective tool use by capuchins (*Cebuslibidinosus*) based on an analysis of the weight of stones found at nutting sites. *American Journal of Primatology* 72: 270–275.

Ferreira, L.V. 1997. Is there a difference between the water floodplain forests (várzea) and blackwater floodplain forest (igapó) in relation to number of species and density. *Brazilian Journal of Ecology* 1: 60–62.

Ferreira, L.V. & Parolin, P. 2007. Tree phenology in central Amazonian floodplain forests: effects of water level fluctuation and precipitation at community and population level. *Pesquisas, Botânica* 58: 139–156.

Ferreira, L.V., de Almeida, S.S. & Parolin, P. 2010c. Amazonian white- and blackwater floodplain forests in Brazil: large differences on a small scale. *Ecotropica* 16: 31–41.

Ferreira, L.V., Cunha, D.A., Chaves, P.P., Matos, D.C. & Parolin, P. 2013. Impacts of hydroelectric dams on alluvial riparian plant communities in eastern Brazilian Amazonian. *Anais da Academia Brasileira de Ciências* 85: 1013–1023.

Ferreira, L.V., Neckel-Oliveira, S., Galatti, U., Fáveri, S.B. & Parolin, P. 2012. Forest structure of artificial islands in the Tucuruí dam reservoir in northern Brazil: a test core-area model. *Acta Amazonica* 42: 221–226.

Ferrol-Schulte, D. 2008. *Changes in behavioural and feeding ecology of the red uakari monkey (Cacajao calvus ucayalii) at the Lago Preto Conservation Concession, Peru*. MSc dissertation, University of Kent, UK.

Ferry, L., Mietton, M., Robison, L. & Erismann, L. 2009. Le lac Alaotra à Madagascar: Passé, présent et futur. *Zeitschrift fuer Geomorphologie* 53: 299–318.

Feilen, K.L. & Marshall, A.J. 2014. Sleeping site selection by proboscis monkeys (*Nasalis larvatus*) in West Kalimantan, Indonesia. *American Journal of Primatology* 76: 1127–1139.

Fialho, M.S., Pintes, R.C., Almeida, M.A.B., Laroque, P.O., Santos, E. & Jerusalinsky, L. 2012. Avaliação do impacto da epizootia de febre amarela sobre as populações de primatas não humanos nas unidades de conservação do Rio Grande do Sul, Brasil. *Biotemas* 25: 217–225.

Finer, M. & Jenkins, C.N. 2012. Proliferation of hydroelectric dams in the Andean Amazon and implications for Andes–Amazon connectivity. *PLoS ONE* 7: e35126.

Finer, M., Jenkins, C.N., Pimm, S.L., Keane, B. & Ross, C. 2008. Oil and gas projects in the western Amazon: threats to wilderness, biodiversity, and indigenous peoples. *PLoS ONE* 3: e2932.

Finer, M., Jenkins, C.N. & Powers, B. 2013. Potential of best practice to reduce impacts from oil and gas projects in the Amazon. *PLoS ONE* 8: e63022.

Fischer, J. & Lindenmayer, D.B. 2000. An assessment of the published results of animal relocations. *Biological Conservation* 96: 1–11.

Fischer, J. & Lindenmayer, D.B. 2007. Landscape modification and habitat fragmentation: a synthesis. *Global Ecology and Biogeography* 16: 265–280.

Fishlock, V., Phyllis, C.L. & Breuer, T. 2008. Quantifying forest elephant social structure in Central African bai environments. *Pachyderm* 40: 19–28.

Fleagle, J.G. 1988. *Primate Adaptation and Evolution*. Academic Press: New York.

Fleagle, J. 2013. *Primate Adaptation and Evolution*, 3rd edn. San Diego, CA: Elsevier.

Fleck, L.C., Painter, L. & Amend, M. 2007. *Carreteras y áreas protegidas: un análisis económico integrado de proyectos en el norte de la Amazonía Boliviana*. Brazil: Conservation Strategy Fund.

Fleck, L., Painter, L., Reid, J. & Amend, M. 2006. *Una carretera a través del Madidi: un análisis económico-ambiental*. Arcata, CA: Conservation Strategy Fund.

Flores, B.M., Piedade, M.-T. & Nelson, B. 2014. Fire disturbance in Amazonian blackwater floodplain forests. *Plant Ecology & Diversity* 7: 319–327.

Flynn, L.J. & Jacobs, L.L. 1982. Effects of changing environments of Siwalik rodent faunas of northern Pakistan. *Palaeogeography, Palaeoclimatology, Palaeoecology* 38: 129–138.

Fong, F.W. 1992. Perspectives for sustainable resource utilization and management of nipa vegetation. *Economic Botany* 46: 45–54.

Fonseca, G.A.B. & Lacher, T.E. 1984. Exudate-feeding by *Callithrix jacchus penicillata* in semideciduous woodland (*cerradão*) in central Brazil. *Primates* 25: 441–450.

Fooden, J. 1995. Systematic review of Southeast Asian longtail macaques, *Macaca fascicularis* (Raffles, [1821]). *Fieldiana, Zoology* 81: 1–206.

Fooden, J. 2000. Systematic review of the Rhesus macaque, *Macaca mulatta* (Zimmermann, 1780). *Fieldiana Zoology* 96: 1–180.

Ford, S.M. 2006. The biogeographic history of Mesoamerican primates. In *New Perspectives in the Study of Mesoamerican*

Primates: Distribution, Ecology, Behavior and Conservation, A. Estrada, P.A. Garber, M.S.M. Pavelka & L. Luecke (eds). New York: Springer, pp. 81–114.

Forest Peoples Programme 2007. *Promised Land: Palm Oil and Land Acquisition in Indonesia: Implications for Local Communities and Indigenous Peoples*. Moreton-in-Marsh, UK: Forest Peoples Programme.

Forrest, J.L., Sanderson, E.W., Wallace, R., *et al.* 2008. Patterns of land cover change in and around Madidi National Park, Bolivia. *Biotropica* 40: 285–294.

Fossey, D. & Harcourt, A.H. 1977. Feeding ecology of free-ranging mountain gorillas (*Gorilla gorilla beringei*). In *Primate Ecology: Studies of Feeding and Ranging Behavior in Lemurs, Monkeys and Apes*, T.H. Clutton-Brock (ed.). New York: Academic Press, pp. 415–447.

Fournier-Chambrillon, C., Fournier, P., Gaillard, J.-M., *et al.* 2000. Mammal trap efficiency during the fragmentation by flooding of a neotropical rain forest in French Guiana. *Journal of Tropical Ecology* 16: 841–851.

Fox, E.A., van Schaik, C.P., Sitompul, A. & Wright, D.N. 2004. Intra- and inter-population differences in orangutan (*Pongo pygmaeus*) activity and diet: Implications for the invention of tool use. *American Journal of Physical Anthropology* 125: 162–174.

Fox, E., Sitompul, A. & van Schaik, C. 1999. Intelligent tool use in wild Sumatran orangutans. In *The Mentalities of Gorillas and Orangutans*, S. Parker, R. Mitchell & H.L. Miles (eds). Cambridge: Cambridge University Press, pp. 99–116.

Fragaszy, D. & Visalberghi, E. 2004. *The Complete Capuchin: The Biology of the Genus* Cebus. Cambridge: Cambridge University Press.

Fragaszy, D.M., Greenberg, R., Visalberghi, E., *et al.* 2010b. How wild bearded capuchin monkeys select stones and nuts to minimize the number of strikes per nut cracked. *Animal Behaviour* 80: 205–214.

Fragaszy, D., Izar, P., Visalberghi, E., Ottoni, E.B. & Oliveira, M.G. 2004. Wild capuchin monkeys use anvils and stone pounding tools. *American Journal of Primatology* 64: 359–366.

Fragaszy, D.M., Liu, Q., Wright, B.W., *et al.* 2013. Wild bearded capuchin monkeys (*Sapajus libidinosus*) strategically place nuts in a stable position during nut-cracking. *PLoS ONE* 8: e56182.

Fragaszy, D., Pickering, T., Liu, Q., *et al.* 2010a. Bearded capuchin monkeys' and a human's efficiency at cracking palm nuts with stone tools: field experiments. *Animal Behaviour* 79: 231–332.

Fragaszy, D.M., Visalberghi, E. & Fedigan, L. 2004. *The Complete Capuchin*. Cambridge: Cambridge University Press.

Frankham, R. 2006. Genetics and landscape connectivity. In *Connectivity Conservation*, K.R. Crooks & M. Sanjayan (eds). Cambridge: Cambridge University Press, pp. 72–96.

Frankham, R. 2009. Where are we in conservation genetics and where do we need to go? *Conservation Genetics* 11: 661–663.

Fraser, L.H. & Keddy, P.A. 2005. *The World's Largest Wetlands: Ecology and Conservation*. Cambridge: Cambridge University Press.

Freese, C.H. 1976. Censusing *Alouatta palliata*, *Ateles geoffroyi*, and *Cebus capucinus* in the Costa Rican dry forest. In *Neotropical Primates: field studies and conservation*, R.W. Thorington, Jr & P.G. Heltne (eds). Washington DC: National Academy of Sciences, pp. 4–9.

Freese, C. & Oppenheimer, J.R. 1981. The capuchin monkeys, genus *Cebus*. In *Ecology and Behavior of Neotropical Primates, Vol. 1*, A.F. Coimbra-Filho & R.A. Mittermeier (eds). Rio de Janeiro: Academia Brasileira de Ciências, pp. 331–390.

Froehlich, J.W., Thorington, R.W. & Otis, J.S. 1981. The demography of howler monkeys (*Alouatta palliata*) on Barro Colorado Island, Panamá. *International Journal of Primatology* 2: 207–236.

Fuentes, A. 1996. Feeding and ranging in the Mentawai Island langur (*Presbytis potenziani*). *International Journal of Primatology* 17: 525–548.

Fuentes, A. 2006. Human culture and monkey behavior: assessing the contexts of potential pathogen transmission between macaques and humans. *American Journal of Primatology* 68: 880–896.

Fujikura, R. & Nakayama, M. 2009. Lessons learned from the World Commission on Dams. *International Environmental Agreements: Politics, Law and Economics* 9: 173–190.

Fujita, M.S., Irham, M., Fitriana, Y.S., *et al.* 2012. Mammals and birds in Bukit Batu area of Giam Siak Kecil – Bukit Batu Biosphere Reserve, Riau, Indonesia. *Kyoto Working Papers on Area Studies* 126: 1–70.

Fuller, D.O., Hardiano, M. & Meijaard, E. 2011. Deforestation projections for carbon-rich peat swamp forests of Central Kalimantan, Indonesia. *Environmental Management* 48: 436–447.

Furch, K. 1997. Chemistry of várzea and igapó soils and nutrient inventory of their floodplain forests. In *The Central Amazon Floodplain*, W.J. Junk (ed.). Berlin: Springer, pp. 47–67.

Furch, K. & Junk, W.J. 1997. The chemical composition, food value, and decomposition of herbaceous plants, leaves, and leaf litter of floodplain forests. In *The Central Amazon Floodplain*, W.J. Junk (ed.). Berlin: Springer, pp. 187–205.

Furuichi, T., Hashimoto, C. & Tashiro, Y. 2001. Fruit availability and habitat use by chimpanzees in the Kalinzu forest, Uganda: examination of fallback foods. *International Journal of Primatology* 22: 929–945.

Furuichi, T., Idani, G.I., Ihobe, H., *et al.* 2012. Long-term studies on wild bonobos at Wamba, Luo scientific reserve, D. R. Congo: towards the understanding of female life history in a male-philopatric species. In *Long-Term Field Studies of Primates*, P.M. Kappeler & D.P. Watts (eds). Berlin: Springer, pp. 413–433.

Furuichi, T., Mulavwa, M., Yangozene, K., *et al.* 2008. Relationships among fruit abundance, ranging rate, and party size and composition of bonobos at Wamba. In *The Bonobos*, T. Furuichi & J. Thompson (eds). New York: Springer, pp. 135–149.

Furuya, Y. 1961. The social life of silvered leaf monkeys. *Primates* 3: 41–60.

Gadsby, E.L. 1989. *Cross River Basin Primate Survey: Stubbs Creek Forest Reserve*. Unpublished report. New York: Wildlife Conservation Society, and Uyo, Nigeria: Akwa Ibom State Government.

Gaffikin, L., Ashley, J. & Blumenthal, P.D. 2007. Poverty reduction and Millennium Development Goals: recognizing population, health, and environment linkages in rural Madagascar. *Medscape General Medicine* 9: 17–27.

Gagnon, M. 1997. Ecological diversity and community ecology in the Fayum sequence (Egypt). *Journal of Human Evolution* 32: 133–160.

Gajapersad, K., Mackintosh, A., Benitez, A. & Payán, E. 2011. A survey of the large mammal fauna of the Kwamalasamutu region, Suriname. In *A Rapid Biological Assessment of the Kwamalasamutu Region, Southwestern Suriname*, B.J. O'Shea, L.E. Alonso & T.H. Larsen (eds). Arlington, VA: Conservation International, pp. 150–155.

Galán de Mera, A. & Linares Perea, E. 2008. Data about the vegetation of the wetlands of South America. From the Bolivian savannas to the Llanos of the Orinoco (Venezuela). *Acta Botánica Malacitana* 33: 271–288.

Galat, G. 1977. *Les Mammifères de Lobaye. Recensements et densités des Primates et observations sur l'écologie de Colobus pennanti oustaleti. Mission en Lobaye I. Empire Centrafricain, mai 1977.* Report. Adiopodoumé, Côte d'Ivoire: INRA & ORSTOM.

Galat, G. 1983. *Socio-écologie du Singe vert (*Cercopithecus aethiops sabaeus*), en référence de quatre Cercopithécinés forestiers* *sympatriques (*Cercocebus atys, Cercopithecus campbelli, C. diana, C. petaurista*) d'Afrique de l'Ouest.* PhD thesis, Université Pierre et Marie Curie, Paris VI, Paris, France.

Galat, G. 1989. Vivre en bandes et survivre. *Nature et Faune (FAO)* 5: 14–27.

Galat, G. 1991. Mammifères des zones humides: méthodes modernes d'étude. In *Cours international de formation à l'écologie et la gestion des zones humides.* Paris: UNESCO-UICN-ORSTOM, pp. 1–6.

Galat, G. & Galat-Luong, A. 1976. The colonization of a mangrove by *Cercopithecus aethiops sabaeus* in Senegal. *Revue d'Écolgie, La Terre et la Vie* 30: 3–30.

Galat, G. & Galat-Luong, A. 1977. Démographie et régime alimentaire d'une troupe de *Cercopithecus aethiops sabaeus* en habitat marginal au Nord Sénégal. *Revue d'Écolgie, La Terre et la Vie* 31: 557–577.

Galat, G. & Galat-Luong, A. 1978. Diet of green monkeys (*Cercopithecus aethiops sabaeus*) in Senegal. In *Recent Advances in Primate Behaviour.* D.J. Chivers and J. Herbert (eds). New York: Academic Press, pp. 257–258.

Galat, G. & Galat-Luong, A. 1985. La communauté de Primates diurnes de la forêt de Taï, Côte d'Ivoire. *Revue d'Ecologie, Terre et Vie* 40: 3–32.

Galat, G. & Galat-Luong, A. 1995. *Commentaires et recommandations sur le réseau hydrographique et les points d'eau de la Forêt de Fathala, Parc National du Delta du Saloum, et publications et éléments de valorisation du Laboratoire de Primatologie ORSTOM sur les Primates du Saloum et de Gambie.* Dakar: ORSTOM.

Galat, G. & Galat-Luong, A. 1999. *Les tortues du Sénégal. Les connaître et les protéger. Mbonaatu Sénégal.* Gën leen na xam ngir gën leen a mëna aar. Affiche. Quadrichromie. La Valette-du-Var, France: Imprimerie Valettoise.

Galat, G. & Galat-Luong, A. 2006. Hope for the survival of the Critically Endangered white-naped mangabey *Cercocebus atys lunulatus*: a new primate species for Burkina Faso. *Oryx* 40: 355–357.

Galat, G. & Galat-Luong, A. 2013. *Chlorocebus sabaeus* green monkey (callithrix). In *Mammals of Africa. Volume II: Primates*, T.M. Butynski, J. Kingdon & J. Kalina (eds). London: Bloomsbury, pp. 274–277.

Galat, G. & Galat-Luong, A. & Keita, Y. 2000. Régression de la distribution et statut actuel du Babouin, *Papio papio*, en limite d'aire de répartition au Sénégal. *African Primates* 4: 69–70.

Galat, G., Galat-Luong, A. & Lavocat, B. 2007. Influence des premières pluies sur la répartition des grands Mammifères. In *Quelles aires protégées pour l'Afrique de l'Ouest? Conservation de la biodiversité et développement*, A. Fournier, B. Sinsin, G.A. Mensah (eds). CD-Rom Collection: Colloques et séminaires. Paris: IRD, pp. 526–527.

Galat, G., Galat-Luong, A., Luong, T.M. & Nizinski, G. 2011. *Le journal intime d'un Singe vert face au changement climatique.* Arcueil: JePublie.

Galat, G., Galat-Luong, A. & Mbaye, M. 1998a. *Densités et effectifs de quinze espèces de mammifères et oiseau terrestres diurnes du Parc national du Niokolo-Koba, Sénégal: évolution 1990–1998.* Rapport scientifique No 2. Dakar: DPNS-ORSTOM.

Galat, G. & Galat-Luong, A., Mbaye, M., Ba, S. & Rigoulot, J.B. 1998b. *La grande et moyenne faune sauvage terrestre diurne de la Réserve de Biosphère du Delta du Saloum (Sénégal): densités et effectifs de six espèces de grands Mammifères et Oiseaux, 1998.* ORSTOM-DPNS-DEFCCS (Eds). Dakar: UICN.

Galat, G., Galat-Luong, A., Ndiaye, B., *et al.* 2002. *La grande et moyenne faune sauvage terrestre diurne de la Réserve de Biosphère du Delta du Saloum, Sénégal: évaluation de référence 1998–2001.* Mise en œuvre du plan de gestion de la RBDS 8. IRD-DPNS-DEFCCS-UICN (eds). Dakar: UICN.

Galat, G., Galat-Luong, A. & Nizinski, G. 2008. Chimpanzees and baboons face global warming by digging wells to filtrate drinking water. In *Global Changes and Water Resources: Confronting the Expanding and Diversifying Pressures*, O. Varis, C. Tortajada, P. Chevallier, B. Pouyaud & E. Servat (eds). Montpellier, France: VERSeau Développement.

Galat, G., Galat-Luong, A. & Nizinski, G. 2009a. Increasing dryness and regression of the geographical range of Temminck's red colobus *Procolobus badius temminckii*: implications for its conservation. *Mammalia* 73: 365–368.

Galat, G. & Galat-Luong, A. & Nizinski, G. 2009b. L'impact du changement climatique sur les variations des populations de grands vertébrés à leur extrême limite de répartition est-il fonction de leurs régimes alimentaires? *Geographia Technica* Numéro spécial: 205–210.

Galat-Luong, A. 1975. Notes preliminaires sur l'ecologie de *Cercopithecus ascanius schmidti* dans les environs de Bangui (R.C.A). *La Terre et la Vie, Revue d'Ecologie Appliqueé* 29: 288–297.

Galat-Luong, A. 1983. *Socio-écologie de trois Colobes sympatriques, Colobus badius, C. polykomos et C. verus du Parc National de Taï, Côte d'Ivoire.* PhD thesis, Université Pierre et Marie Curie, Paris VI. Paris.

Galat-Luong, A. 1988. Monkeys in the Pirang forest. In *Pirang: Ecological Investigations in a Forest Island in the Gambia*, H. Ellenberg, A. Galat-Luong, H.J. Von Maydel, *et al.* (eds). Hamburg, Germany: Warnke Verlag, pp. 187–208.

Galat-Luong, A. 1991a. Proies inhabituelles pour le Patas d'Afrique de l'Ouest (*Erythrocebus patas patas*). *Revue d'Ecologie, La Terre et La Vie* 46: 83–84.

Galat-Luong, A. 1991b. Mammifères des zones humides: étude de cas: les adaptations du Singe vert à la Mangrove. In *Cours international de formation à l'écologie et la gestion des zones humides.* UNESCO-UICN-Parc National des Oiseaux du Djoudj (eds). Dakar: ORSTOM

Galat-Luong A. 1999. *Les grands Mammifères de la RBDS: éléments pour un plan de gestion. Résumé de la communication au Comité Scientifique de restitution des résultats des études pour le Projet de Formulation du Plan de Gestion de la Réserve de Biosphère du Delta du Saloum RBDS. 18–20 février 1999.* Dakar: IRD.

Galat-Luong, A. 2000. *Gestion de la diversité biologique.* In *Vers une gestion durable des plaines d'inondation sahéliennes.* Groupe d'experts des plaines d'inondation sahéliennes (eds). Gland, Switzerland: IUCN. pp. 100–103.

Galat-Luong A. 2001. *Restauration des habitats et aménagements agro-forestiers: les espèces végétales consommées par le Colobe bai, Réserve de Biosphère du Delta du Saloum. Mise en œuvre du plan de gestion de la RBDS.* 5. Dakar: IRD.

Galat-Luong, A. & Galat, G. 1979a. Conséquences comportementales de perturbations sociales répétées dans une troupe de Mones de Lowe, Cercopithecus campbelli lowei de Côte d'Ivoire. *Revue d'Ecologie, La Terre et La Vie* 33: 49–58.

Galat-Luong, A. & Galat, G. 1979b. Quelques observations sur l'écologie de *Colobus pennantii oustaleti en* Empire Centrafricain. *Mammalia* 43: 309–312.

Galat-Luong, A. & Galat, G. 1999. *La grande faune terrestre de la Réserve de Biosphère du Delta du Saloum et sa biodiversité.* Dakar: IRD.

Galat-Luong, A. & Galat, G. 2000. Chimpanzees and baboons drink filtrated water. *Folia Primatologica* 71: 258.

Galat-Luong, A. & Galat, G. 2001a. *Identification et statut des grands Mammifères de la Réserve de Biosphère du Delta du Saloum et de la Réserve de Faune de Fathala. Mise en œuvre du plan de gestion de la RBDS. 2.* Dakar: IRD.

Galat-Luong, A. & Galat, G. 2001b. *Mise en place de corridors de protection pour la grande faune de l'écotone mangrove-continent.*

1. *Propositions de localisation. Réserve de Biosphère du Delta du Saloum. Mise en œuvre du plan de gestion de la RBDS. 6.* Dakar: IRD.

Galat-Luong, A. & Galat, G. 2002. *La grande faune terrestre de la Réserve de Biosphère du Delta du Saloum: biodiversité, évolution récente, conservation*. Dakar: IUCN.

Galat-Luong, A. & Galat, G. 2003. La ressource grande faune terrestre du Sénégal oriental: ses potentialités, ses contraintes. In *Potentialités, contraintes et systèmes d'exploitation au Sénégal Oriental et en Haute Casamance*. Dakar: UCAD-DDR.

Galat-Luong, A. & Galat, G. 2005. Conservation and survival adaptations of Temminck's red colobus (*Procolobus badius temmincki*) in Senegal. *International Journal of Primatology* 26: 585–603.

Galat-Luong, A. & Galat, G. 2007. Influence of anthropization on the distribution of the large wildlife. The mangroves, a refuge environment. In *Quelles aires Protegees pour l'Afrique de l'Ouest? Conservation de la Biodiversite et Developpement*. A. Fournier, B. Sinsin & G.A. Mensah (eds). Paris: L'Institut de Recherche pour le Devloppement (IRD), pp. 568–569.

Galat-Luong, A. & Galat, G. 2011. *Papio papio*. In *All the World's Primates*, N. Rowe & M. Myers (eds). Charlestown, RI: Primate Conservation Inc.

Galat-Luong, A. & Galat, G. 2013. *Papio papio* Guinea baboon. In *Mammals of Africa. Volume II. Primates*, T.M. Butynski, J. Kingdon & J. Kalina (eds). London: Bloomsbury Publishing, pp. 218–221.

Galat-Luong, A. & Luong, T.M. 2001. *Les aventures de Gonolek et Barbican: Les Colobes bais*. In A. Galat-Luong & S.P.E.F.S. Sarl (eds). Rueil-Malmaison, France: Belancor. Bande dessinée quadrichromie.

Galat-Luong, A., Galat, G. & Diouck, D. 1998. *Evolution 1971–1996 des habitats d'une aire protégée de la Réserve de Biosphère du Delta du Saloum (Sénégal), la Forêt de Fathala: recouvrement, densité et biodiversité des ligneux*. Unpublished report. Dakar: ORSTOM & IUCN.

Galat-Luong, A., Galat, G. & Hagell, S. 2006. The social and ecological flexibility of Guinea baboons: implications for Guinea baboon social organization and male strategies. In *Reproduction and Fitness in Baboons: Behavioral, Ecological, and Life History Perspectives*, L. Swedell & S. Leigh (eds). New York: Springer, pp. 105–121.

Galat-Luong, A., Galat, G. & Nizinski, G. 2009. Une conséquence du réchauffement climatique: les chimpanzés filtrent leur eau de boisson. *Geographia Technica* Numéro spécial: 199–204.

Galat-Luong, A., Galat, G., Oates, J.F., *et al.* 2008. *Procolobus badius ssp. temminckii*. In *IUCN Red List of Threatened Species*. Gland, Switzerland: IUCN. www.iucnredlist.org/details/18247/0.

Galat-Luong, A., Galat, G. & Ting, N. 2012. *Procolobus rufomitratus oustaleti*. In *All the World's Primates*, N. Rowe & M. Myers (eds). Charlestown, RI: Primate Conservation Inc.

Galat-Luong, A., Jaffe, K. & Galat, G. 2011. *Chlorocebus sabaeus*. In *All the World's Primates*, N. Rowe & M. Myers (eds). Charlestown, RI: Primate Conservation Inc.

Galdikas, B.M.F. 1985. Subadult male orangutan sociality and reproductive behavior at Tanjung Puting. *American Journal of Primatology* 8: 87–99.

Galdikas, B.M.F. 1988. Orangutan diet, range, and activity at Tanjung Puting, central Borneo. *International Journal of Primatology* 9: 1–35.

Galdikas, B.M.F. 1995. Social and reproductive behavior of wild adolescent female orangutans. In *The Neglected Ape*, R.B. Nadler, B.M.F. Galdikas, L. Sheeran & N. Rosen (eds). New York: Plenum Press, pp. 163–182.

Galdikas, B.M.F. 2008. Adult male sociality and reproductive tactics among orangutans at Tanjung Puting. *Folia Primatologica* 45: 9–24.

Galdikas, B.M.F. & Shapiro, G.L. 1994. *A Guidebook to Tanjung Putting National Park*. Jakarta: PT Gramedia Pustaka Utama.

Galeano, G. & Bernal, R. 2010. *Palmas de Colombia*. Bogotá, Colombia: Universidad Nacional de Colombia.

Ganas, J., Ortmann, S. & Robbins, M.M. 2008. Food preferences of wild mountain gorillas. *American Journal of Primatology* 70: 927–938.

Ganas, J. & Robbins, M.M. 2005. Ranging behavior of the mountain gorillas (*Gorilla beringei beringei*) in Bwindi Impenetrable National Park, Uganda: a test of the ecological constraints model. *Behavior Ecology & Sociobiology* 58: 277–288.

Gani, M.R. & Gani, N.D. 2011. River-margin habitat of *Ardipithecus ramidus* at Aramis, Ethiopia 4.4 million years ago. *Nature Communications* 2: 602.

Ganzhorn, J.U. 1988. Food partitioning among Malagasy primates. *Oecologia* 75: 436–450.

Ganzhorn, J.U. 1995. Low-level forest disturbance effects on primary production, leaf chemistry, and lemur populations. *Ecology* 79: 2084–2096.

Ganzhorn, J.U., Andrianasolo, T., Andrianjazalahatra, T., *et al.* 2007. Lemurs in evergreen littoral forest fragments. In *Biodiversity, Ecology, and Conservation of the Littoral Ecosystems in Southeastern Madagascar, Tolagnaro (Fort Dauphin)*, J.U. Ganzhorn, S.M. Goodman & M. Vincelette (eds). Washington, DC: Smithsonian Institution, pp. 223–225.

Ganzhorn, J.U., Goodman, S.M., Nash, S. & Thalmann, U. 2006. Lemur biogeography. In *Primate Biogeography*, S.M. Lehman & J.G. Fleagle (eds). New York: Springer Science, pp. 229–254.

Ganzhorn, J.U., Rakotondranary, S.J. & Ratovonamana Y.R. 2011. Habitat description and phenololgy. In *Field and Laboratory Methods in Primatology: A Practical Guide*, J.M. Setchell & D.J. Curtis (eds). Cambridge: Cambridge University Press, pp. 51–68.

Ganzhorn, J.U., Wilmé, L. & Mercier, J.-L. 2014. Explaining Madagascar's biodiversity. In *Conservation and Environmental Management in Madagascar*, I. Scales (ed.). London: Routledge Publishing, pp. 17–43.

Garber, P.A. 1989. Role of spatial memory in primate foraging patterns: *Saguinus mystax* and *Saguinus fuscicollis*. *American Journal of Primatology* 19: 203–216.

Garber, P.A. & Kowalewski, M.M. 2011. Collective action and male affiliation in howler monkeys (*Alouatta caraya*). In *The Origins and Nature of Cooperation and Altruism in Non-human and Human Primates*, R.W. Sussman & C.R. Cloninger (eds). Berlin: Springer Press, pp. 145–165.

Garber, P.A. & Kowalewski, M.M. 2013. Low levels of aggression and evidence of male social affiliation in Pitheciines: Consequences for Social Organization. In *Pitheciines: Consequences for Social Organization. Evolutionary Biology and Conservation of Titis, Sakis and Uacaris*, L. Veiga, A. Barnett, S. Ferrari & M. Norconk (eds). Cambridge: Cambridge University Press, pp. 96–104.

Garber, P.A. & Lambert, J.E. 1998. Primates as seed dispersers: ecological processes and directions for future research. *American Journal of Primatology* 45: 3–8.

Garber, P.A., Righini, N. & Kowalewski, M.M. 2015. Evidence of alternative dietary syndromes and nutritional goals in the genus *Alouatta*. In *Howler Monkeys: Behavior, Ecology and Conservation*, M. Kowalewski, P.A. Garber, L. Cortés-Ortiz, B. Urbani & D. Youlatos (eds). Berlin: Springer Press, pp. 85–109.

García, U.G.M., Vanegas, G.I.C., Alonso, J.C., *et al.* 2007. *Balance anual sobre el estado de los ecosistemas y el ambiente de la amazonas colombiana 2006*. Bogotá, Colombia: Instituto Amazónico de Investigaciones Científicas, SINCHI.

Garmestani, A.S. & Percival, H.F. 2005. Raccoon removal reduces sea turtle nest depredation in the Ten Thousand Islands of Florida. *Southeastern Naturalist* 4: 469–472.

Gartlan, J.S. & Strusaker, T.T. 1972. Polyspecific associations and niche separation of rain-forest anthropoids in Cameroon, West Africa. *Journal of Zoology, London* 168: 221–266.

Gatinot, B.L. 1976. Les milieux fréquentés par le colobe bai d'Afrique de l'ouest (*Colobus badius temmincki* Kuhl, 1820) en Sénégambie. *Mammalia* 40: 1–12.

Gatinot, B.L. 1977. Le régime alimentaire du colobe bai au Sénegal. *Mammalia* 41: 373–402.

Gauthier, C.A., Deniaud, J.L., Leclerc-Cassan, M., *et al.* 2000. Observations of lemurs in the mangroves of north-west Madagascar. *Folia Primatologica* 71: 267.

Gauthier, C.A., Deniaud, J.L., Rakotomalala, M., Razafindramanana, S. & Renson, G. 1999. Note sur la découverte d'un nouvel habitat occupé par les propithéques couronnés (*Propithecus verreauxi coronatus*) au nord-ouest de Madagascar. *Primatologie* 2: 821–827.

Gautier, J.P. 1985. Quelques caractéristiques écologiques du singe des marais: *Allenopithecus nigroviridis* Lang 1923. *Revue d'Écologie, La Terre et La Vie* 40: 331–342.

Gautier-Hion, A. 1988. Polyspecific associations among forest guenons: Ecological, behavioural and evolutionary aspects. In *A Primate Radiation: evolutionary biology of the African guenons*, A. Gautier-Hion, F. Bourlière, J.-P. Gautier & J. Kingdon (eds). Cambridge: Cambridge University Press, pp. 452–475.

Gautier-Hion, A. 2013. *Allenopithecus nigroviridis* Allen's swamp monkey. In *Mammals of Africa. Volume II: Primates*, T.M. Butynski, J. Kingdon & J. Kalina (eds). London: Bloomsbury, pp. 248–251.

Gautier-Hion, A. 2013b. *Cercopithecus neglectus* De Brazza's monkey. In *Mammals of Africa. Volume II: Primates*, T.M. Butynski, J. Kingdon & J. Kalina (eds). London: Bloomsbury, pp. 315–319.

Gautier-Hion, A. 2013c. *Miopithecus ogouensis* northern talapoin monkey (Gabon talapoin monkey). In *Mammals of Africa. Volume II: Primates*, T.M. Butynski, J. Kingdon & J. Kalina (eds). London: Bloomsbury, pp. 253–256.

Gautier-Hion, A. 2013d. *Miopithecus talapoin* southern talapoin monkey (Angolan talapoin monkey). In *Mammals of Africa. Volume II: Primates*, T.M. Butynski, J. Kingdon & J. Kalina (eds). London: Bloomsbury, pp. 252–253.

Gautier-Hion, A. & Brugière, D. 2005. Significance of riparian forests for the conservation of Central African primates. *International Journal of Primatology* 26: 515–523.

Gautier-Hion, A., Colyn, M. & Gautier, J.-P. 1999. *Histoire Naturelle des Primates d'Afrique Centrale*. Libreville, Gabon: ECOFAC.

Gautier-Hion, A. & Gautier, J.P. 1971. La nage chez les cercopithèques du Gabon. *Revue d'Écologie, La Terre et la Vie* 25: 67–75.

Gentry, A.H. & Terborgh, J. 1990. Composition and dynamics of the Cocha Cashu 'mature' floodplain forest. In *Four Neotropical Rainforests*, A.H. Gentry (ed.). New Haven, CT: Yale University Press, pp. 542–564.

Geissmann, T. 2007. Status reassessment of the gibbons: results of the Asian Primate Red List workshop 2006. *Gibbon Journal* 3: 5–15.

Geomatics International Inc, Beak Consultants Ltd & Unilag Consult of Nigeria. 1998. *The Assessment of Vegetation and Land Use Changes in Nigeria between 1976/78 and 1993/95*. Unpublished report. Abuja, Nigeria: Forestry Management, Evaluation, and Co-ordinating Unit (FORMECU) of the Federal Department of Forestry (under the World Bank Environmental Management Project).

Georgiev, A., Thompson, M., Lokasola, A. & Wrangham, R. 2011. Seed predation by bonobos (*Pan paniscus*) at Kokolopori, Democratic Republic of the Congo. *Primates* 52: 309–314.

GEPIS, Groupe d'experts des plaines d'inondation sahéliennes. 2000. *Vers une Gestion Durable des Plaines D'inondation Sahéliennes*. Gland & Cambridge: IUCN.

Getzin, S., Nuske, R.S. & Wiegand, K. 2014. Using unmanned aerial vehicles (UAV) to quantify spatial gap patterns in forests. *Remote Sensing* 6: 6988–7004.

Ghiglieri, M. 1984. *The Chimpanzees of Kibale Forest: A Field Study of Ecology and Social Structure*. New York: Columbia University Press.

Gibson, L., Lynam, A.J., Bradshaw, C.J.A., *et al.* 2013. Near-complete extinction of native small mammal fauna 25 years after forest fragmentation. *Science* 341: 1508–1510.

Gesen, W., Wulffraat, S., Zieren, M. & Scholten, L. 2006. *Mangrove Guidebook for Southeast Asia*. Rome & Bangkok: FAO & Wetlands International.

Giglio, L., Loboda, T., Roy, D.P., Quayle, B. &. Justice, C.O. 2009. An active-fire based burned area mapping algorithm for the MODIS sensor. *Remote Sensing of Environment* 113: 408–420.

Gilmore, M.P., Endress, B.A. & Horn, C.M. 2013. The socio-cultural importance of *Mauritia flexuosa* palm swamps (*aguajales*) and implications for multi-use management in two Maijuna communities of the Peruvian Amazon. *Journal of Ethnobiology & Ethnomedicine* 9: 1–23.

Gingerich, P.D. 1990. African dawn for primates. *Nature* 346: 411.

Giordano, A.J. & Ballard, W.B. 2010. Noteworthy record of a black howler monkey (*Alouatta caraya*) from the central dry Chaco of Paraguay. *Neotropical Primates* 17: 74–75.

Giosan, L., Syvitski, J., Constantinescu, S. & Day, J. 2014. Climate change: protect the world's deltas. *Nature* 516: 31–33.

Gippoliti, S. & Dell'Omo, G. 1996. Primates of the Cantanhez Forest and the Cacine Basin, Guinea-Bissau. *Oryx* 30: 74–80.

Gippoliti, S. & Del'Omo, G. 2003. Primates of Guinea-Bissau, West Africa: distributions and conservation status. *Primate Conservation* 19: 73–74.

Giraldo, P., Gómez-Posada, C., Martínez, J. & Kattan, G. 2007. Resource use and seed dispersal by red howler monkeys (*Alouatta seniculus*) in a Colombian Andean forest. *Neotropical Primates* 14: 55–64.

Giri, C. & Muhlhausen, J. 2008. Mangrove forest distributions and dynamics in Madagascar (1975–2005). *Sensor* 8: 2004–2017.

Giri, C., Ochieng, E., Tieszen, L.L., *et al.* 2011. Status and distribution of mangrove forests of the world using earth observation satellite data. *Global Ecology & Biogeography* 20: 154–159.

Gittins, S.P. 1981. *A Survey of the Primates of Bangladesh*. Unpublished report. London: Fauna Preservation Society.

Gittins, S.P. & Akonda, A.W. 1982. What survives in Bangladesh? *Oryx* 16: 275–281.

Glander, K.E. 1982. The impact of plant secondary compounds on primate feeding behavior, *Yearbook of Physical Anthropology* 25: 1–18.

Godfrey, L.R. & Jungers, W.L. 2003. The extinct sloth lemurs of Madagascar. *Evolutionary Anthropology* 12: 252–263.

Godfrey, L.R., Simons, E.L., Jungers, W.L., DeBlieux, D.D. & Chatrath, P.S. 2004. New discovery of subfossil *Hapalemur simus*, the greater bamboo lemur, in western Madagascar. *Lemur News* 9: 9–11.

Godfrey, L.R., Wilson, J.M., Simons, E.L., Stewart, P.D. & Vuillaume-Randriamanantena, M. 1996. Ankarana: window to Madagascar's past. *Lemur News* 2: 16–17.

Goldsmith, M.L. 1999. Ecological constraints on the foraging effort of western gorillas (*Gorilla gorilla gorilla*) at Bai Hokou, Central African Republic. *International Journal of Primatology* 20: 1–23.

Gomez-Posada C. 2003. *Variación en el uso del tiempo y el espacio de* Cebus apella *(Primates: Cebidae) según la disponibilidad de los recursos principales en su dieta.* Masters dissertation.

Gonçalves, E.C., Ferrari, S.F., Medeiros, D.S.L., Silva, A. & Schneider, M.P.C. 2008. Long-term variation in the genetic diversity of red-handed howlers, *Alouatta belzebul* (Primates: Platyrrhini) from eastern Brazilian Amazonia. In *A Primatologia no Brasil*, S.F. Ferrari & J. Rímoli (eds). Aracaju, Brazil: Sociedade Brasileira de Primatologia, pp. 25–37.

González, V., Zunino, G.E., Kowalewski, M. & Bravo, S.P. 2002. Densidad de monos aulladores (*Alouatta caraya*) y composición y estructura de la selva de inundación en una isla de Río Paraná Medio. *Revista del Museo Argentino Ciencias Naturales* 4: 7–12.

Gonzalez-Socoloske, D. & Snarr, K.A. 2010. An incident of swimming in a large river by a mantled howling monkey (*Alouatta palliata*) on the north coast of Honduras. *Neotropical Primates* 17: 28–31.

Goodall, A.G. 1977. Feeding and ranging behavior of a mountain gorilla group (*Gorilla gorilla beringei*) in the Tshibinda-Kahuzi resion (Zaïre). In *Primate Ecology: Studies of Feeding and Ranging Behavior in Lemurs, Monkeys and Apes*, T.H. Clutton-Brock (ed.). New York: Academic Press, pp. 450–479.

Goodall, J. 1986. *The Chimpanzees of Gombe: Patterns of Behavior*. Cambridge: Belknap Press.

Goodman, S.M., Andriafidison, D., Andrianaivoarivelo, R., *et al*. 2005. The distribution and conservation of bats in the dry regions of Madagascar. *Animal Conservation* 8: 153–165.

Goossens, B., Chikhi, L., Ancrenaz, M., *et al*. 2006. Genetic signature of anthropogenic population collapse in orang-utans. *PLoS Biology* 4: e25.

Goossens, B., Chikhi, L., Jalil, M., *et al*. 2004. Patterns of genetic diversity and migration in increasingly fragmented and declining orangutan (*Pongo pygmaeus*) populations from Sabah, Malaysia. *Molecular Ecology* 14: 441–456.

Goossens, B., Chikhi, L., Utami, S.S., de Ruiter, J. & Bruford, M.W. 2000. mA multi-samples, multi-extracts approach for mictosatellite analysis of faecal samples in an arboreal ape. *Conservation Genetics* 1: 157–162.

Goossens, B., Chikhi, L., Jalil, M., *et al*. 2005. Patterns of genetic diversity and migration in increasingly fragmented and declining orang-utan (*Pongo pygmaeus*) populations from Sabah, Malaysia. *Molecular Ecology* 14: 441–456.

Goossens, B., Chikhi, L., Jalil, M., *et al*. 2009. Taxonomy, geographic variation and population genetics of Bornean and Sumatran orangutans. In *Orangutans: Geographic Variation in Behavioral Ecology and Conservation*, S.A. Wich, S.S.U. Atmoko, T.M. Setia & C.P. van Schaik (eds). Oxford: Oxford University Press, pp. 1–13.

Goossens, B., Salgado-Lynn, M., Rovie-Ryan, J.J., *et al*. 2013. Genetics and the last stand of the Sumatran rhinoceros *Dicerorhinus sumatrensis*. *Oryx* 47: 340–344.

Goossens, B., Setchell, J.M., Abulani, D.M.A., *et al*. 2002. A boat survey of primates in the Lower Kinabatangan Wildlife Sanctuary. In *Lower Kinabatangan Scientific Expedition*, M. Maryati, A. Takano, B. Goossens & R. Indran (eds). Kota Kinabalu, Malaysia: Universiti Malaysia Sabah, pp. 37–45.

Goossens, B., Setchell, J.M., James, S.S., *et al*. (2006a). Philopatry and reproductive success in Bornean orang-utans (*Pongo pygmaeus*). *Molecular Ecology* 15: 2577–2588.

Gordon, E., Polanco, L. & Peña, C. 2000. Contribution to the ecology of *Montrichardia arborescens* (L.) Schott (Araceae). I. Demography. *Acta Biológica Venezuelica* 20: 51–64.

Gosby, A., Conigrave, A. & Lau, N. 2011. Testing protein leverage in lean humans: a randomised controlled experimental study. *PLoS ONE* 6: e25929.

Gössling, S. 1999. Ecotourism: a means to safeguard biodiversity and ecosystem functions? *Ecological Economics* 29: 303–320.

Goulding, M., Bartham, R. & Ferreira, E.J.G. 2003. *Smithsonian Atlas of the Amazon*. Washington, DC: Smithsonian Institution Press.

Government of Nigeria. 2012. *Niger Delta Biodiversity Project*. UNDP GEF PIMS No. 2047. GEFSEC Project ID 4090.

Graham, L.L., Turjaman, M. & Page, S. 2013. *Shorea balangeran* and *Dyera polyphylla* (syn. *Dyera lowii*) as tropical peat swamp forest restoration transplant species: effects of mycorrhizae and level of disturbance. *Wetlands Ecology Management* 21: 307–321.

Grassi, C. 2001. *The behavioural ecology of* Hapalemur griseus griseus: *the influences of microhabitat and population density on this small-bodied prosimian folivore*. PhD thesis, Austin, TX: University of Texas at Austin.

Grassi, C. 2002. Sex differences in feeding, height, and space use in *Hapalemur griseus*. *International Journal of Primatology* 23: 677–693.

Grassi, C. 2006. Variability in habitat, diet, and social structure of *Hapalemur griseus* in Ranomafana National Park, Madagascar. *American Journal of Physical Anthropology* 131: 50–63.

Green, K.M. 1978. Primates of Bangladesh: a preliminary survey of population and habitat. *Biological Conservation* 13: 141–160.

Greengrass, E.J. 2009. Chimpanzees are close to extinction in south-west Nigeria. *Primate Conservation* 24: 77–83.

Gregorin, R. 2006. Taxonomia e variação geográfica das espécies do gênero *Alouatta* Lacépède (Primates, Atelidae) no Brasil. *Revista Brasileira de Zoologia* 23: 64–144.

Gregory, R.D., Gibbons, D.W. & Donald, P.F. 2004. Bird census and survey techniques. In *Bird Ecology and Conservation: A Handbook of Techniques*, W.J. Sutherland, I. Newton & R.E. Green (eds). Oxford: Oxford University Press, pp. 17–56.

Gregory, T. & Bowler, M. 2015. Male-male affiliation and cooperation characterize the social behavior of the large-bodied pitheciids, *Chiropotes* and *Cacajao*: a review. *American Journal of Primatology* 78: 550–560.

Greminger, M.P., Stölting, K.N., Nater, A., *et al*. 2014. Generation of SNP datasets for orangutan population genomics using improved reduced-representation sequencing and direct comparisons of SNP calling algorithms. *BMC Genomics* 15: 16.

Gribel, R. 1993. Os mamíferos silvestres e as grandes barragens na Amazônia. In *Bases científicas para estratégias de preservação e desenvolvimento da Amazônia*. G. Ferreira, M. Santos, E.L.M. Leão & L.A. Oliveira (eds). Manaus, Brazil: Instituto Nacional de Pesquisas da Amazônia, pp. 125–133.

Griffith, B., Scott, J.M., Carpenter, J.W. & Reed, C. 1989. Translocation as a species conservation tool: status and strategy. *Science* 245: 477–480.

Griffiths, A.J.F., Miller, J.H. Suzuki, D.T., Lewontin, R.C. & Gelbart, W.M. (eds). 2000. *An Introduction to Genetic Analysis*. 7th edition. New York: WH Freeman.

Gros-Louis, J. 2002. Contexts and behavioral correlates of trill vocalizations in wild white-faced capuchin monkeys (*Cebus capucinus*). *American Journal of Primatology* 57: 189–202.

Gros-Louis, J. 2003. The function of food-associated calls in white-faced capuchin monkeys, *Cebus capucinus*, from the perspective of the signaler. *Association for the Study of Animal Behaviour* 67: 431–440.

Gros-Louis, J. 2006. Acoustic analysis and contextual description of food-associated calls in white-faced capuchin monkeys (*Cebus capucinus*). *International Journal of Primatology* 27: 273–294.

Gros-Louis, J.J., Perry, S.E., Fichtel, C., *et al*. 2008. Vocal repertoire of *Cebus capucinus*: acoustic structure, context, and usage. *International Journal of Primatology* 29: 64–670.

Groves, C.P. 1998. Systematics of tarsiers and lorises. *Primates* 39: 13–27.

Groves, C.P. 2001. *Primate Taxonomy*. Washington DC: Smithsonian Institution Press.

Groves, C.P. 2007a. Speciation and biogeography of Vietnam's primates. *Vietnamese Journal of Primatology* 1: 27–40.

Groves, C.P. 2007b. The taxonomic diversity of the Colobinae of Africa. *Journal of Anthropological Sciences* 85: 7–34.

Groves, C.P. & Sabater Pi, J. 1985. From ape's nest to human fix-point. *Man* 20: 22–47.

Groves, C.P. & Shekelle, M. 2010. The genera and species of Tarsiidae. *International Journal of Primatology* 31: 1071–1082.

Groves, C.P., Andrews, P. & Horne, J.F. 1974. Tana river colobus and mangabey. *Oryx* 12: 565–575.

Grow, N. & Gursky-Doyen, S. 2010. Preliminary data on the behavior, ecology, and morphology of pygmy tarsiers (*Tarsius pumilus*). *International Journal of Primatology* 31: 1174–1191.

Grubb, P. 1982. Refuges and dispersal in the speciation of African mammals. In *Biological Diversification in the Tropics*, G.T. Prance (ed.). New York: Columbia University Press, pp. 537–553.

Grubb, P. 1990. Primate geography in the Afro-tropical forest biome. In *Vertebrates in the Tropics*, G. Peters & R. Hutterer (eds). Washington DC: Smithsonian Institute Press, pp. 187–214.

Grubb, P. & Powell, C.B. 1999. Discovery of red colobus monkeys (*Procolobus badius*) in the Niger Delta with the description of a new and geographically isolated subspecies. *Journal of Zoology* London 248: 67–73.

Grueter, C.C., Li, D., Ren, B. & Wei, F. 2009. Choice of analytical method can have dramatic effects on primate home range estimates. *Primates* 50: 81–84.

Grueter, C.C., Li, D., Ren, B., *et al.* 2009. Fallback foods of temperate-living primates: a case study on snub-nosed monkeys. *American Journal of Physical Anthropology* 140: 700–715.

Guadarrama-Olivera, M.A. & Ortiz-Gil, G. 2000. Análisis de la flora de la Reserva de la Biosfera de los Pantanos de Centla, Tabasco, México. *Ecosistemas y Recursos Agropecuarios* 15: 67–104.

Gualda-Barros, J., Nascimento, F.O. & Amaral, M.K. 2012. A new species of *Callicebus* Thomas, 1903 (Primates, Pitheciidae) from the states of Mato Grosso and Pará, Brazil. *Papéis Avulsos de Zoologia* 52: 261–279.

Guillera-Arroita, G., Lahoz-Monfort, J.J., Milner-Gulland, E.J., Young, R.P. & Nicholson, E. 2010. Monitoring and conservation of the Critically Endangered Alaotran gentle lemur *Hapalemur alaotrensis*. *Madagascar Conservation & Development* 5: 103–109.

Guillotin, M., Dubost, G. & Sabatier, D. 1994. Food choice and food competition among the three major primate species of French Guiana. *Journal of Zoology* 233: 551–579.

Gumbricht, T., McCarthy, J. & McCarthy, T.S. 2004. Channels, wetlands and islands in the Okavango Delta, Botswana, and their relation to hydrological and sedimentological processes. *Earth Surface Processes & Landforms* 29: 15–29.

Gumert, M. 2011. The common monkey of Southeast Asia: long-tailed macaque populations, ethnophoresy, and their occurrence in human environments. In *Monkeys on the Edge: Ecology and Management of Long-tailed Macaques and their Interface with Humans*, M. Gumert, A. Fuentes & L. Jones-Engel (eds). Cambridge: Cambridge University Press, pp. 3–44.

Gumert, M.D. & Malaivijitnond, S. 2012. Marine prey processed with stone tools by burmese long-tailed macaques (*Macaca fascicularis aurea*) in intertidal habitats. *American Journal of Physical Anthropology* 149: 447–457.

Gumert, M. & Malaivijitnond, S. 2013. Long-tailed macaques select mass of stone tool according to food type. *Philosophical Transactions of The Royal Society – Series B* 368: DOI:10.1098/rstb.2012.0413.

Gumert, M., Hamada, Y. & Malaivijitnond, S. 2013. Human activity negatively affects wild stone tool-using Burmese long-tailed macaques *Macaca fascicularis aurea* Laemson National Park, Thailand. *Oryx* 47: 535–543.

Gumert, M., Kluck, M. & Malaivijitnond, S. 2009. The physical characteristics and usage patterns of stone axe and pounding hammers used by long-tailed macaques in the Andaman Sea region of Thailand. *American Journal of Primatology* 71: 594–608.

Gumert, M.D., Low, K.H. & Malaivijitnond, S. 2011. Sex differences in the stone tool-use behavior of a wild population of Burmese long-tailed macaques (*Macaca fascicularis aurea*). *American Journal of Primatology* 73: 1239–1249.

Gunkel, G., Lange, U., Walde, D. & Rosa, J.W. 2003. The environmental and operational impacts of Curuá-Una, a reservoir in the Amazon region of Pará, Brazil. *Lakes & Reservoirs: Research & Management* 8: 201–216.

Gunnell, F. 1997. Wasatchian-Bridgerian (Eocene) paleoecology of the western interior of North America: changing paleoenvironments and taxonomic composition of omoyid (Tarsiiformes) primates. *Journal of Human Evolution* 32: 105–132.

Gupta, A.C. 1966. Wildlife of lower Bengal with particular reference to the Sundarbans. In *West Bengal Forests, Centenary Commemoration Volume 1964*. Calcutta: Forest Directorate, Government of West Bengal, pp. 233–238.

Gupta, A.K. 2001–2. The Sundarbans: where tigers reign. *Newsletter Wildlife Institute of India* 8 & 9: 1–4.

Gupta, A.K. & Chivers, D.J. 1999. Biomass and use of resources in south and southeast Asian primate communities. In *Primate Communities*, J.G. Fleagle, C. Janson & K.E. Reed (eds). Cambridge University Press, UK, pp. 38–54.

Gurmaya, K.J., Adiputra, I.W, Saryatiman, A.B., Danardono, S.N. & Sibuea, T.T.H. 1994. A preliminary study on ecology and conservation of the Java primates in Ujung Kulon National Park, West Java, Indonesia. In *Current Primatology. Volume 1, Ecology and Evolution*, B. Thierry, J.R. Anderson, J.J. Roeder & N. Herrenschmidt (eds). Strasbourg: Univ. Louis Pasteur Press, pp. 87–92.

Gurmaya, K.J., Saryatiman, A.B., Danardono, S.N., Sibuea, T.T.H. & Adiputra, I.M.W. 1994. A preliminary study on ecology and conservation of the Java primates in Ujung Kulon National Park, West Java, Indonesia. *Current Primatology* 1: 87–92.

Gutiérrez-Vélez, V.H. & DeFries, R. 2013. Annual multi-resolution detection of land cover conversion to oil palm in the Peruvian Amazon. *Remote Sensing of Environment* 129: 154–167.

Hadi, I. 2013. *Food-related innovative behaviors of the macaques*. PhD thesis, Bogor Agricultural University, Indonesia.

Hager, R. 2003. The effects of dispersal costs on reproductive skew and within-group aggression in primate groups. *Primate Report* 67: 85–98.

Hall, K.R.L. 1962. Numerical data, maintenance activities and locomotion of the wild chacma baboon, *Papio ursinus*. *Proceedings of the Zoological Society of London* 139:181–220.

Hall, K.R.L. 1963. Variations in the ecology of the chacma baboon. *Symposium of the Zoological Society of London* 10: 1–28.

Hamard, M.C.L., Cheyne, S.M. & Nijman, V. 2010 Vegetation correlates of gibbon density in the peat-swamp forest of the Sabangau catchment, Central Kalimantan, Indonesia. *American Journal of Primatology* 72: 607–616.

Hamerlynck, O. & Duvail, S. 2003. *The rehabilitation of the delta of the Senegal River in Mauritania*. Gland: IUCN. http://data.iucn.org/dbtw-wpd/edocs/WTL-029.pdf.

Hamerlynck, O., Luke, Q., Nyanger, T.M., Suvail, S. & Leauthaud, C. 2012. Range extension, imminent threats and conservation options for two Endangered primates: the Tana River red colobus *Procolobus rufomitratus rufomitratus* (Peters, 1879) and the Tana River mangabey *Cercocebus galeritus* (Peters, 1879) in the Lower Tana Floodplain and Delta, Kenya. *African Primates* 7: 211–217.

Hambali, K., Ismail, A., Zulkifli, S.Z., Md-Zain, B.M. & Amir, A. 2012. Human–macaque conflict and pest behaviors of long-tailed macaques (*Macaca fascicularis*) in Kuala Selangor Nature Park. *Tropical Natural History* 12: 189–205.

Hamilton, A.C. 1974. The history of vegetation. In *East African Vegetation*, E.M. Lin & M.E.S. Morrison (eds). London: Longman, pp. 188–209.

Hamilton, S.K., Kellndorfer, J., Lehner, B. & Tobler, M. 2007. Remote sensing of floodplain geomorphology as a surrogate for biodiversity in a tropical river system (Madre de Dios, Peru). *Geomorphology* 89: 23–38.

Hamilton, S.K., Sippel, S.J. & Melack, J.M. 2004. Seasonal inundation patterns in two large savanna floodplains of South America: the Llanos de Moxos (Bolivia) and the Llanos del Orinoco (Venezuela and Colombia). *Hydrological Processes* 18: 2103–2116.

Hamilton III, W.J., Buskirk, R.E. & Buskirk, W.H. 1976. Defense of space and resources by chacma (*Papio ursinus*) baboon troops in an African desert and swamp. *Ecology* 57: 1264–1272.

Hamilton, W.J., Buskirk, R.E. & Buskirk, W.H. 1978. Omnivory and utilisation of food resources by chacma baboons, *Papio ursinus*. *American Naturalist* 112: 911–924.

Hammerschmidt, K. & Fischer, J. 1998. The vocal repertoire of barbary macaques: a quantitative analysis of a graded signal system. *Ethology* 104: 203–216.

Hammond, D.S. 2005. Ancient land in a modern world. In *Tropical Forests of the Guiana Shield: Ancient Forests in a Modern World*, D.S. Hammond (ed.). Wallingford: CABI Publishing, pp. 1–14.

Hammond, D.S., Gond, V., de Thoisy, B., Forget, P.-M. & DeDijn, B.P.E. 2007. Causes and consequences of a tropical forest gold rush in the Guiana Shield, South America. *Ambio* 36: 661–670.

Hannibal, W. & Neves-Godoi, M. 2015. Non-volant mammals of the Maracaju Mountains, southwestern Brazil: composition, richness and conservation. *Revista Mexicana de Biodiversidad* 86: 217–225.

Hansen, H.R., Hector, B.L. & Feldmann, J. 2003. A qualitative and quantitative evaluation of the seaweed diet of North Ronaldsay sheep. *Animal Feed Science and Technology* 105: 21–28.

Hansen, M.C., Potapov, P.V., Moore, R., *et al.* 2013. High-resolution global maps of 21st-century forest cover change. *Science* 342: 850–853.

Hanya, G. 2005. Seasonal variations in the activity budget of Japanese macaques in the coniferous forest of Yakushima: effects of food and temperature. *American Journal of Primatology* 63: 165–177.

Hanya, G. & Aiba, S.-I. 2010. Fruit fall in tropical and temperate forests: implications for frugivore diversity. *Ecological Research* 25: 1081–1090.

Hanya, G. & Bernard, H. 2012. Fallback foods of red leaf monkeys (*Presbytis rubicunda)* in Danum Valley, Borneo. *International Journal of Primatology* 33: 322–337.

Hanya, G., Stevenson, P., van Noordwijk, M., *et al.* 2011. Seasonality in fruit availability affects frugivorous primate biomass and species richness. *Ecography* 34: 1009–1017.

Hanya, G., Zamma, K., Hayaishi, S., *et al.* 2005. Comparisons of food availability and group density of Japanese macaques in primary, naturally regenerated, and plantation forests. *American Journal of Primatology* 66: 45–262.

Happold, D.C.D. 1985. Geographical ecology of Nigerian mammals. *Annales du Musée Royal de l'Afrique Centrale, Sciences Zoologiques* 246: 5–49.

Happold, D.C.D. 1987. *The Mammals of Nigeria*. Oxford: Oxford University Press.

Happold, D. & Lock, J.M. 2013. The biotic zones of Africa. In *Mammals of Africa. Volume I: Introductory Chapters and Afrotheria*, J. Kingdon, D. Happold, M. Hoffmann, *et al.* (eds). London: Bloomsbury, pp. 57–74.

Harcourt, A.H. 2006. Rarity in the tropics: biogeography and macroecology of the primates. *Journal of Biogeography* 33: 2077–2087.

Harcourt, A.H. & Doherty, D.A. 2005. Species-area relationships of primates in tropical forest fragments: a global analysis. *Journal of Applied Ecology* 42: 630–637.

Harcourt, A.H. & Wood, M.A. 2012. Rivers as barriers to primate distributions in Africa. *International Journal of Primatology* 33: 168–183.

Harding, L.E. 2010. *Trachypithecus cristatus* (Primates: Cercopithecidae). *Mammalian Species* 42: 149–165.

Harper, G., Steininger, M., Tucker, C., Juhn, D. & Hawkins, F. 2007. Fifty years of deforestation and forest fragmentation in Madagascar. *Environmental Conservation* 34: 325–333.

Harris, M.B., Tomas, W., Mourão, G., *et al.* 2005. Safeguarding the Pantanal wetlands: threats and conservation initiatives. *Conservation Biology* 19: 714–720.

Harrison, M.E. 2009. *Orang-utan feeding behaviour in Sabangau, Central Kalimantan.* PhD thesis, University of Cambridge, Cambridge, UK.

Harrison, M.E., Cheyne, S.M., Husson, S.J., Jeffers, K.A., *et al.* 2012. *Preliminary Assessment of the Biodiversity and Conservation Value of the Bawan Forest, Central Kalimantan, Indonesia.* Orangutan Tropical Peatland Project, Palangka Raya.

Harrison, M.E., Cheyne, S.M., Morrogh-Bernard, H. & Husson, S.J. 2005. What can apes tell us about the health of their environment? A preliminary analysis of the use of orang-utans and gibbons as biological indicators of changes in habitat quality in tropical peat swamp forests. In *Proceedings of the International Symposium and Workshop on 'Restoration and Wise Use of Tropical Peatland'*, J.O. Reiley (ed.). Palangka Raya, Indonesia: CIMTROP, pp. 104–109.

Harrison, M.E., Husson, S.J., D'Arcy, L.J., *et al.* 2010a. *The Fruiting Phenology of Peat-swamp Forest Tree Species at Sabangau and Tuanan, Central Kalimantan, Indonesia.* The Kalimantan Forests and Climate Partnership, Palangka Raya.

Harrison, M.E., Morrogh-Bernard, H.C. & Chivers, D.J. 2010b. Orangutan energetics and the influence of fruit availability in the nonmasting peat-swamp forest of Sabangau, Indonesian Borneo. *International Journal of Primatology* 31: 585–607.

Harrison, M.E., Page, S.E. & Limin, S.H. 2009. The global impact of Indonesian forest fires. *Biologist* 56: 156–163.

Harrison, T.S. & Harrison, T. 1989. Palynology of the late Miocene *Oreopithecus*-bearing lignite from Baccinello, Italy. *Palaeogeoraphy, Palaeoclimatology, Palaeoecology* 76: 45–65.

Harrison-Levine, A.L., Norconk, M.A. & Cunningham, E.P. 2003. Insect predation techniques suggest predator sensitive foraging in a group of white-faced sakis (*Pithecia pithecia*). *American Journal of Primatology* 60: 66.

Hartshorn, G.S. 1995. Ecological basis for sustainable development in tropical forests. *Annual Review of Ecology and Systematics* 26: 155–175.

Harvey, C.A., Komar, O., Chazdon, R., *et al.* 2008. Integrating agricultural landscapes with biodiversity conservation in the Mesoamerican hotspot. *Conservation Biology* 22: 8–15.

Hasan, M.K., Aziz, M.A., Alam, S.M.R., *et al.* 2013. Distribution of rhesus macaques (*Macaca mulatta*) in Bangladesh: inter-population variation in group size and composition. *Primate Conservation* 26: 125–132.

Hasan, M.R., Mondal, M.A.W., Miah, M.I. & Kibria, M.G. 2001. Water quality study of some selected oxbow lakes with special emphasis on chlorophyll-α. In *Reservoir and Culture-based Fisheries: Biology and Management.* Proceedings of an International Workshop held in Bangkok, Thailand from 15 to 18 February 2000, S.S. de Silva (ed.). Canberra: Australian Centre for International Agricultural Research, pp. 126–136.

Haselmayer, J. & Quinn, J.S. 2000. A comparison of point counts and sound recording as bird survey methods in Amazonian southeast Peru. *Condor* 102: 887–893.

Hashimoto, C., Tashiro, Y., Kimura, D., *et al.* 1998. Habitat use and ranging of wild bonobos (*Pan paniscus*) at Wamba. *International Journal of Primatology* 19: 1045–1060.

Haslam, M. 2012. Towards a prehistory of primates. *Antiquity* 86: 299–315.

Haslam, M., Gumert, M., Biro, D., Carvalho, S. & Malaivijitnond, S. 2013. Use-wear patterns on wild macaque stone tools reveal their behavioural history. *PLoS ONE* 8: e72872.

Haslam, M., Hernandez-Aguilar, A., Ling, V., *et al.* 2009. Primate archaeology. *Nature* 460: 339–344.

Haugaasen, T. 2004. *Structure, composition and dynamics of a central Amazonian forest landscape: a conservation perspective.* PhD thesis, University of East Anglia, Norwich, UK.

Haugaasen, T. & Peres, C.A. 2005a. Primate assemblage structure in Amazonian flooded and unflooded forests. *American Journal of Primatology* 62: 243–258.

Haugaasen, T. & Peres, C.A. 2005b. Patterns of tree phenology in adjacent Amazonian flooded and unflooded forests. *Biotropica* 37: 620–630.

Haugaasen, T. & Peres, C.A. 2005c. Mammal assemblage structure in Amazonian flooded and unflooded forests. *Journal of Tropical Ecology* 21: 1–13.

Haugaasen, T. & Peres, C.A. 2006. Floristic, edaphic and structural characteristics of flooded and unflooded forests in the lower Rio Purús region of central Amazonia, Brazil. *Acta Amazonica* 36: 25–35.

Haugaasen, T. & Peres, C.A. 2007. Vertebrate responses to fruit production in Amazonian flooded and unflooded forests. *Biodiversity & Conservation* 16: 4165–4190.

Haugaasen, T. & Peres, C.A. 2009. Interspecific primate associations in Amazonian flooded and unflooded forests. *Primates* 50: 239–251.

Hauser, M.D. & Fairbanks, L.A. 1988. Mother-offspring conflict in vervet monkeys: variation in response to ecological conditions. *Animal Behaviour* 36: 802–813.

Hauser, M.D. & Wrangham, R.W. 1990. Recognition of predator and competitor calls in nonhuman primates and birds: a preliminary report. *Ethology* 86:116–130.

Hawes, J.E. 2012. *Fruits and frugivory in Neotropical primate and in Amazonian flooded and unflooded forests.* PhD thesis. University of East Anglia, Norwich, UK.

Hawes, J.E. & Peres, C.A. 2014. Fruit-frugivore interactions in Amazonian seasonally flooded and unflooded forests. *Journal of Tropical Ecology* 30: 381–399.

Hawes, J.E. & Peres, C.A. 2016. Patterns of plant phenology in Amazonian seasonally flooded and unflooded forests. *Biotropica* 48: 465–475.

Hawkins, A.F.A., Durbin, J.C. & Reid, D.B. 1998. The primates of the Baly Bay area, northwestern Madagascar. *Folia Primatologica* 69: 337–345.

Hayakawa, T., Nathan, S., Stark, D.J., Saldivar, D.A.R., Sipangkui, R., Goossens, B., Tuuga, A., Clauss, M., Sawada, A., Fukuda, S., Imai, H., Matsuda, I. 2018. First report of foregut microbial community in proboscis monkeys: are diverse forests a reservoir for diverse microbiomes? *Environ Microbiol Rep.* doi: 10.1111/1758-2229.12677.

Head, J.S., Boesch, C., Makaga, L. & Robbins, M. 2011. Sympatric chimpanzees (*Pan troglodytes troglodytes*) and gorillas (*Gorilla gorilla gorilla*) in Loango National Park, Gabon: dietary composition, seasonality, and intersite comparison. *International Journal of Primatology* 32: 755–775.

Head, J.S., Robbins, M.M., Mundry, R., Makaga, L. & Boesch, C. 2012. Remote video-camera traps measure habitat use and competitive exclusion among sympatric chimpanzee, gorilla and elephant in Loango National Park, Gabon. *Journal of Tropical Ecology* 28: 571–583.

Heckman, C. 1998. The Pantanal of Poconé: biota and ecology in the northern section of the world's largest pristine wetland. *Monographiae Biologicae* 77: 1–624.

Heinicke, S., Kalan, A.M., Wagner, O.J.J., *et al.* 2015. Assessing the performance of a semi-automated acoustic monitoring system for primates. *Methods in Ecology & Evolution* 6: 753–763.

Helfield, J.M. & Naiman, R.J. 2001. Effects of salmon-derived nitrogen on riparian forest growth and implications for stream productivity. *Ecology* 82: 2403–2409.

Helfield, J.M. & Naiman, R.J. 2006. Keystone interactions: salmon and bear in riparian forests of Alaska. *Ecosystems* 9: 167–180.

Hendrichs, H. 1975. The status of the tiger *Panthera tigris* (Linne, 1758) in the Sundarbans mangrove forest (Bay of Bengal). *Saugetierkundliche Mitteilungen* 23: 161–199.

Henriques, R.P.B. & Cavalcante, R.J. 2004. Survey of a gallery forest primate community in the Cerrado of the Distrito Federal, Central Brazil. *Neotropical Primates* 12: 78–83.

Henzi, S.P., Byrne, R.W. & Whiten, A. 1992. Patterns of movement by baboons in the Drakensberg mountains: primary responses to the environment. *International Journal of Primatology* 13: 601–629.

Hernandez-Aguilar, R.A. 2006. *Ecology and nesting patterns of chimpanzees (Pan troglodytes) in Issa, Ugalla, western Tanzania.* PhD thesis, University of Southern California, Los Angeles.

Hernandez-Aguilar, R.A. 2009. Chimpanzee nest distribution and site reuse in a dry habitat: implications for early hominin ranging. *Journal of Human Evolution* 57: 350–364.

Hernandez-Aguilar, R.A., Moore, J.M. & Stanford, C.B. 2013. Chimpanzee nesting patterns in savanna habitat: environmental influences and preferences. *American Journal of Primatology* 16: 1–16.

Hernández-Camacho, J. & Cooper, R.W. 1976. The nonhuman primates of Colombia. In *Neotropical Primates: Field Studies and Conservation*, R.W. Thorington, Jr & P.G. Heltne (eds). Washington, DC: National Academy of Sciences, pp. 35–69.

Hernández-Camacho, J. & Defler, T.R. 1985. Some aspects of the conservation of non-human primates in Colombia. *Primate Conservation* 6: 42–50.

Hershkovitz, P. 1984. Taxonomy of squirrel monkeys, genus *Saimiri* (Cebidae, Platyrrhini): a preliminary report with description of a hitherto unnamed form. *American Journal of Primatology* 4: 209–243.

Hershkovitz, P. 1987. Uacaries, New World monkeys of the genus *Cacajao* (Cebidae, Platyrrhini): a preliminary taxonomic review with the description of a new subspecies. *American Journal of Primatology* 12: 1–53.

Heymann, E.W. 1990. Further field notes on red uacaris, Cacajao calvus ucayalii, from the Quebrada Blanco, Amazonian Peru. *Primate Conservation* 11: 7–8.

Heymann, E.W. & Aquino, R. 2010. Peruvian red uakaris (*Cacajao calvus ucayalii*) are not flooded-forest specialists. *International Journal of Primatology* 31: 751–758.

Heymann, E.W., Encarnación, F. & Canaquin, J.E. 2002. Primates of the Río Curaray, northern Peruvian Amazon. *International Journal of Primatology* 23: 191–201.

Heymann, E.W. & Soini, P. 1999. Offspring number in pygmy marmosets, *Cebuella pygmaea*, in relation to group size and the number of adult males. *Behavioral Ecology & Sociobiology* 46: 400–404.

Hickey, J.R., Nackoney, J., Nibbelink, N.P., *et al.* 2013. Human proximity and habitat fragmentation are key drivers of the rangewide bonobo distribution. *Biodiversity & Conservation* 22: 1–20.

Hickin, E.J. 2003. Meandering Channels. In *Encyclopedia of Sediments and Sedimentary Rocks*, G.V. Middleton (ed.). Dordrecht, The Netherlands: Kluwer Academic Publishers, pp. 430–434.

Hickley, P. & Bailey, R.G. 1987. Food and feeding relationships of fish in the Sudd swamps (River Nile, southern Sudan). *Journal of Fish Biology* 30: 147–159.

Hilário, R.R. & Ferrari, S.F. 2015. Dense understory and absence of capuchin monkeys (*Sapajus xanthosternos*) predict higher density of common marmosets (*Callithrix jacchus*) in the Brazilian Northeast. *American Journal of Primatology* 77: 425–433.

Hill, K., Padwe, J., Bejyvagi, C., *et al.* 1997. Impact of hunting on large vertebrates in the Mbaracayu Reserve, Paraguay. *Conservation Biology* 11: 1339–1353.

Hill, R.A., Barrett, L., Gaynor, D., *et al.* 2003. Day length, latitude and behavioural (in)flexibility in baboons (*Papio cynocephalus ursinus*). *Behavioral Ecology & Sociobiology* 53: 278–286.

Hill, R.A. & Dunbar, R.I.M. 2002. Climatic determinants of diet and foraging behavior in baboons. *Evolutionary Ecology* 16: 579–593.

Hill, W.C.O. 1960. Genus *Cebus* Erxleben, 1777. In *Primates: Comparative Anatomy and Taxonomy, IV Cebidae, Part A*, W.C.O. Hill (ed.). Edinburgh, UK: Edinburgh University Press, pp. 322–490.

Hill, W.C.O. 1966. *Primates, Comparative Anatomy and Taxonomy: Catarrhini, Cercopithecoidea, Cercopithecinae, Volume 5.* Edinburgh, UK: University of Edinburgh Press.

Hirabuki, Y. 1990. Vegetation and landform structure in the study area of La Macarena: a physiognomic investigation. *Field Studies of New World Monkeys, La Macarena, Colombia* 3: 35–48.

Hirata, S., Watanabe, K. & Masao, K. 2001. 'Sweet potato washing' revisited. In *Primate Origins of Human Cognition and Behavior*, T. Matsuzawa (ed.). Hong Kong: Springer, pp. 487–508.

Hoegberg, P. 1986. Soil nutrient availability, root symbioses and tree species composition in tropical Africa: a review. *Journal of Tropical Ecology* 2: 359–372.

Hoffman, T.S. & O'Riain, M.J. 2012. Troop size and human-modified habitat affect the ranging patterns of a chacma baboon population in the Cape Peninsula, South Africa. *American Journal of Primatology* 74: 853–863.

Hogarth, P.J. 2007. *The Biology of Mangroves and Seagrass.* Oxford: Oxford University Press.

Holderegger, R. & Wagner, H.H. 2008. Landscape genetics. *BioScience* 58: 199–207.

Holm, J.A., Miller, C.J. & Cropper, W.P. 2008. Population dynamics of the dioecious Amazonian palm *Mauritia flexuosa*: simulation analysis of sustainable harvesting. *Biotropica* 40: 550–558.

Holzmann, I., Agostini, I., Areta, J.I., *et al.* 2010. Impact of yellow fever outbreaks on two howler monkey species (*Alouatta guariba clamitans* and *A. caraya*) in Misiones, Argentina. *American Journal of Primatology* 72: 475–480.

Homewood, K. 1975. Can the Tana mangabey survive? *Oryx* 13: 53–59.

Honey, M. 1999. *Ecotourism and Sustainable Development: Who Owns Paradise?* Washington, DC: Island Press.

Hoogerwerf, A. 1970. *Udjung Kulon: The Land of the Last Javan Rhinoceros.* Leiden, Germany: E.J. Brill.

Hoogesteijn, A. & Hoogesteijn, R. 2010. Cattle ranching and biodiversity conservation as allies in South America's flooded savannas. *Great Plains Research* 20: 37–50.

Hoogerwerf, A. 1972. *Verslag over een bezoek aan het Meru Beteri complex, het Blambangan-Purwo of Zuid Banjuwangi wildreservaat, het Ijang hoogland en het Udjung Kulon wildreservaat Java, Indonesië, in de maanden augustus t/m november 1971.* Austerlitz: Nederlandse Commissie voor Internationale Natuurbescherming.

Hooijer, A., Page, S., Canadell, J., *et al.* 2010. Current and future CO_2 emissions from drained peatlands in Southeast Asia. *Biogeosciences* 7: 1505–1514.

Horn, A.D. 1980. Some observations on the ecology of the bonobo chimpanzee (*Pan paniscus*, Schwarz 1929) near Lake Tumba, Zaire. *Folia Primatologica* 34: 145–169.

Horning, N.R. 2008. Strong support for weak performance: donor competition in Madagascar. *African Affairs* 107: 405–431.

Horning, N.R. 2012. Debunking three myths about Madagascar's deforestation. *Madagascar Conservation & Development* 7: 116–119.

Horrocks, J. A. & Hunte, W. 1986. Sentinel behaviour in vervet monkeys: who sees whom first? *Animal Behaviour* 34: 1566–1568.

Horwich, R.H. & Johnson, E.D. 1986. Geographical distribution of the black howler (*Alouatta pigra*) in Central America. *Primates* 27: 53–62.

Hosey, G.R. 2005. How does the zoo environment affect the behaviour of captive primates? *Applied Animal Behaviour Science* 90: 107–129.

Houston, S. 2000. From Livingstone to ecotourism. What's new in travel medicine? *Canadian Family Physician Médecin de Famille Canadien* 46: 121–128.

Howard, J., Trotz, M.A., Thomas, K., *et al.* 2011. Total mercury loadings in sediment from gold mining and conservation areas in Guyana. *Environmental Monitoring & Assessment* 179: 555–573.

Huang, C., Kim, S., Song, K., *et al.* 2009. Assessment of Paraguay's forest cover change using Landsat observations. *Global & Planetary Change* 67: 1–12.

Huber, O. 2006. Herbaceous ecosystems on the Guyana Shield, a regional overview. *Journal of Biogeography* 33: 464–475.

Huffman, M.A. & Hirata, S. 2003. Biological and ecological foundations of primate behavioral traditions. In *The Biology of Traditions: Models and Evidence*, D. Fragaszy & S. Perry (eds). Cambridge: Cambridge University Press, pp. 267–296.

Huffman, M.A. & Quiatt, D. 1986. Stone handling by Japanese macaques (*Macaca fuscata*): implications for tool use of stones. *Primates* 27: 413–423.

Hughes, F.M. 1990. The influence of flooding regimes on distribution and composition in the Tana River floodplain, Kenya. *Journal of Applied Ecology* 27: 475–491.

Hughes, F.M.R. 1985. *The Tana River floodplain forest, Kenya: ecology and the impact of development.* PhD thesis, University of Cambridge, Cambridge, UK.

Hughes, L. 2000. Biological consequences of global warming: is the signal already apparent? *Trends in Ecology & Evolution* 15: 56–61.

Hughes, R.H. & Hughes, J.S. 1992. *A Directory of African Wetlands.* Cambridge: IUCN/UNEP/WCMC.

Humle, T. & Matsuzawa, T. 2004. Oil palm use by adjacent communities of chimpanzees at Bossou and Nimba Mountains, West Africa. *International Journal of Primatology* 25: 551–581.

Hunkeler, C., Bourlière, F. & Bertrand, M. 1972. Le comportement social de la Mone de Lowe (*Cercopithecus campbelli lowei*). *Folia Primatologia* 17: 218–236.

Hunter, W.W. 1875. *A Statistical Account of Bengal, Vol. 1, Districts of the 24-Parganas and Sundarbans.* London: Truebner and Co.

Hush, C.J. 1996. *Multiple Comparisons: Theory and Methods.* Boca Raton, FL: CRC Press.

Husson, S.J. & Morrogh-Bernard, H. 2007. Wise use of tropical peatland - reconciling orangutan conservation and economic development. In *Restoration and Wise Use of Tropical Peatland: Problems of Biodiversity, Fire, Poverty and Water Management. Proceedings of the International Symposium and Workshop on Tropical Peatland, Palangka Raya, 20–24 September 2005*, J.O. Rieley, S.H.S. Limin & A. Jaya (eds). EU RESTORPEAT Partnership. Indonesia & The Netherlands: University of Palangka Raya & Wageningen University and Research Institute, pp. 61–66.

Husson, S.J., Morrogh-Bernard, H., McLardy, C., *et al.* 2002. The effects of illegal logging on the population of orang-utan in the Sebangau tropical peat swamp forest, Central Kalimantan. In

Peatlands for People: Natural Resource Functions and Sustainable Management. Proceedings of the International Symposium on Tropical Peatland, 22–23 August 2001, Jakarta, Indonesia, J.O. Rieley & S.E. Page (eds). Jakarta, Indonesia: BPPT and Indonesian Peat Association, pp. 10–20.

Husson, S.J., Wich, S.A., Marshall, A.J., *et al.* 2009. Orangutan distribution, density, abundance and impacts of disturbance. In *Orangutans: Geographic Variation in Behavioral Ecology and Conservation*, S.A. Wich, S.S.U. Atmoko, T.M. Setia & C.P. van Schaik (eds). Oxford: Oxford University Press, pp. 77–96.

Hylander, W.L. 1988. Implications of in vivo experiments for interpreting the functional significance of 'robust' australopithecine jaws. In *Evolutionary History of the 'Robust' Australopithecines*, F.E. Grine (ed.). New York: Aldine de Gruyter, pp. 55–83.

IBAMA. 2014. Dados geoespaciais de unidades de conservação e manguezais. In *Sistema Compartilhado de Informações Ambientais - SisCom, CSR/CGMAM/IBAMA*.

IBGE. 2012. Base Vetorial Contínua do Brasil. *IBGE Geo-database*.

Ichino, S.I., Chatani, K., Kawamoto, Y., *et al.* 2013. Decrease in the body mass of wild ringtailed lemurs at Berenty Reserve in Madagascar with environmental changes. *African Study Monographs* 34: 109–118.

IES (The International Ecotourism Society). 1994. www.ecotourism.org.

Idani, G., Kuroda, S., Kano, T. & Asato, R. 1994. Flora and vegetation of Wamba forest, central Zaire with reference to bonobo (*Pan paniscus*) foods. *Tropics* 3: 309–332.

Iftekhar, M.S. & Islam, M.R. 2004. Managing mangroves in Bangladesh: a strategy analysis. *Journal of Coastal Conservation* 10: 139–146.

Ikemeh, R.A. 2014a. *Niger Delta red colobus conservation project in Bayelsa State: assessing the species' population status, current distribution, and prevalent threats in order to design an effective conservation plan*. Unpublished report. Abuja, Nigeria: SW/Niger Delta Forest Project.

Ikemeh, R.A. 2014b. *Survey of the Nigerian-Cameroon chimpanzee (*Pan troglodytes ellioti*) in the Niger River Delta of Nigeria: status, threats and strategies for conservation*. Unpublished report. Abuja, Nigeria: SW/Niger Delta Forest Project.

Ikemeh, R.A. 2015. Assessing the population status of the Critically Endangered Niger Delta red colobus (*Piliocolobus epieni*). *Primate Conservation* 29: 87–96.

Ikemeh, R.A. & Oates, J.F. 2017. Niger Delta red colobus monkey *Piliocolobus epieni* (Grubb and Powell, 1999) Nigeria. In *Primates in Peril: The World's 25 Most Endangered Primates 2016–2018*, C. Schwitzer, R.A. Mittermeier, A.B. Rylands, *et al.* (eds). Arlington, VA: IUCN/SSC Primate Specialist Group, pp. 22–24.

Imada, S. 2006. *Food Culture on Seaweeds*. Tokyo: Seizando.

Innes, J.L. 2010. Madagascar rosewood, illegal logging and the tropical timber trade. *Madagascar Conservation & Development* 5: 6–10.

Inogwabini, B.-I. & Matungila, B. 2009. Bonobo food items, food availability and bonobo distribution in the Lake Tumbawampy forests, Democratic Republic of Congo. *Open Conservation Biology Journal* 3: 14–23.

Inogwabini, B.-I. & Thompson, J.A.M. 2013. The golden-bellied mangabey *Cercocebus chrysogaster* (Primates: Cercopithecidae): distribution and conservation status. *Journal of Threatened Taxa* 5: 4069–4075.

Inogwabini, B.-I., Abokome, M., Kamenge, T., Mbende, L. & Mboka, L. 2012. Preliminary bonobo and chimpanzee nesting by habitat type in the northern Lac Tumba Landscape, Democratic Republic of Congo. *African Journal of Ecology* 50: 285–298.

Inogwabini, B.-I., Matungila, B., Mbende, L., Abokome, M. & Tshimanga, T.W. 2007. Great apes in the Lake Tumba landscape, Democratic Republic of Congo: newly described populations. *Oryx* 41: 532–538.

Irion, G. 1978. Soil infertility in the Amazon rainforest. *Naturwissenschaften* 65: 515–519.

ISE, IRD, MEPN & UICN. 2000a. *Plan de Gestion de la Réserve de Biosphère du Delta du Saloum. Vol 1: Diagnostic*. ISE-IRD-MEPN-UICN (eds). Dakar: UICN.

ISE, IRD, MEPN & UICN. 2000b. *Plan de Gestion de la Réserve de Biosphère du Delta du Saloum. Vol 2: Zonage et plan d'action*. ISE-IRD-MEPN-UICN (eds). Dakar: UICN.

Islam, M.S.N. & Gnauck, A. 2009. Threats to the Sundarbans mangrove wetland ecosystems from transboundary water allocation in the Ganges basin: a preliminary problem analysis. *International Journal of Ecological Economics & Statistics* 13: 64–78.

Isoun, M. 2009. *Status survey of Niger Delta red colobus monkey*, Procolobus pennantii epieni. Unpublished report. Arlington, VA: Primate Action Fund, Conservation International, and Port Harcourt, Nigeria: Niger Delta Wetlands Centre.

IUCN. 1987. Position statement on the translocation of living organisms: Introductions, re-introduction and re-stocking. Gland, Switzerland: IUCN.

IUCN. 1991. *Caring for the Earth: A Strategy for Sustainable Living*. Gland, Switzerland: IUCN.

IUCN. 1996. *The Conservation Atlas of Tropical Forests: The Americas*. New York: Simon and Schuster.

IUCN. 2012/2015/2017. In *IUCN Red List of Threatened Species*. Gland, Switzerland: IUCN. www.iucnredlist.org.

IUCN. 2014. *Regional action plan for the conservation of western lowland gorillas and central chimpanzees 2015–2025*. Report of the IUCN SSC Primate Specialist Group, Gland, Switzerland: IUCN.

IUCN & ICCN. 2012. *Bonobo (*Pan paniscus*): conservation strategy 2012–2022*, Gland, Switzerland, IUCN/SSC Primate Specialist Group & Institut Congolais pour la Conservation de la Nature.

Isbell, L.A. 1994. Predation on primates: ecological patterns and evolutionary consequences. *Evolutionary Anthropology* 3: 61–71.

Iwamoto, T. 1978. Food availability as a limiting factor on population density of the Japanese monkey and gelada baboon. In *Recent Advances in Primatology*, D.J. Chivers & J. Herbert (eds). London: Academic Press, pp. 287–303.

Iwamoto, T. 1982. Food and nutritional condition of free ranging Japanese monkeys on Koshima Islet during winter. *Primates* 23: 153–170.

Iwata, T., Nakano, S. & Inoue, M. 2003. Impacts of riparian deforestation on stream communities in a tropical rainforest in Borneo. *Ecological Applications* 13: 461–473.

Iwata, Y. 2010. *Frugivorous diet of western lowland gorillas in Moukalaba-Doudou National Park, Gabon*. PhD dissertation. Kyoto University, Kyoto, Japan.

Iwata, Y. & Ando, C. 2007. Bed and bed site reuse by western lowland gorilla in Moukalaba-Doudou National Park, Gabon. *Primates* 48: 77–80.

Iyenguet, F.C., Malanda, G.F., Madzoké, B. & Twagirashyaka, F. 2008. *Recensement et distribution des grands mammifères et activités humaines dans la zone Impfondo*. Unpublished report of the Wildlife Conservation Society, Republic of Congo.

Iyenguet, F.C., Madzoké, B., Twagirashyaka, F., *et al.* 2012. *Inventaire des Grands Mammifères et Activités Humaines dans la Batanga, Périphérie est de la Réserve Communautaire du Lac Télé, République du Congo*. Unpublished report of the Wildlife Conservation Society, Republic of Congo.

Izar, P., Resende, B., Ferreira, R.G. 2018. Proximate causes of tool use in feeding in the genus Sapajus. In La primatología en Latinoamérica 2, B. Urbani, M. Kowalewski, R. Grasseto Teixeira da Cunha, S. de la Torre & L. Cortés-Ortiz. Caracas: Instituto Venezolano de Investigaciones Científicas (IVIC), pp. 239–250.

Izawa, K. 1999. Comparison of six troops on Kinkazan Island. *Japanese Macaques in Miyagi Prefecture* 10: 1–11.

Izawa, K. 2009. *Research on wild Japanese Macaques*. Tokyo: Dobutsusha.

Izawa, K. & Nishida, T. 1963. Monkeys living in the northern limits of their distribution. *Primates* 4: 67–88.

Izawa, K. & Tokuda, K. 1988. General aspects of study site. *Field Studies of New World Monkeys, La Macarena, Colombia* 1: 1–3.

Jablonski, N.G. & Tyler, D.E. 1999. *Trachypithecus auratus sangiranensis*, a new fossil monkey from Sangiran, central Java, Indonesia. *International Journal of Primatology* 20: 319–326.

Jacobs, B.F. 2004. Palaeobotanical studies from tropical Africa: relevance to the evolution of forest, woodland and savannah biomes. *Philosophical Transactions of the Royal Society of London B* 359: 1573–1583.

Jackson, C.P. 2011. The positional behavior of pygmy marmosets (*Cebuella pygmaea)* in northwestern Bolivia. *Primates* 52: 171–178.

Jalil, M.F., Cable, J., Sinyor, J., *et al.* 2008. Riverine effects on mitochondrial structure of Bornean orangutans (*Pongo pygmaeus*) at two spatial scales. *Molecular Ecology* 17: 2898–2909.

James, G.K., Adegoke, J.O., Saba, E., Nwilo, P. & Akinyede, J. 2007. Satellite-based assessment of the extent and changes in the mangrove ecosystem of the Niger Delta. *Marine Geodesy* 30: 249–267.

Jansen, P.A. & Den Ouden, J. 2005. Observing seed removal: remote video monitoring of seed selection. In *Predation and Dispersal. Seed Fate: Predation, Dispersal, and Seedling Establishment*, P. Forget, J. Lambert, P. Hulme & S. Vander Wall (eds). Wallingford, UK: CABI, pp. 363–378.

Janson, C.H. & Boinski, S. 1992. Morphological and behavioral adaptations for foraging in generalist primates: the case of the Cebines. *American Journal of Primatology* 88: 483–498.

Janson, C.H. & Di Bitetti, M.S. 1997. Experimental analysis of food detection in capuchin monkeys: effects of distance, travel speed, and resource size. *Behavioral Ecology and Sociobiology* 41: 17–24.

Janzen, D.H. 1974. Tropical blackwater rivers, animals, and mast fruiting by the Dipterocarpaceae. *Biotropica* 6: 69–103.

Jason, C.H. & Di Bitetti, M.S. 1997. Experimental analysis of food detection in capuchin monkeys: effects of distance, travel speed, and resource size. *Behavior Ecology & Sociobiology* 41: 17–24.

Jawan, A. & Sumin, V. 2012. The effect of land used on water quality of oxbow lakes in Sabah. *The Malaysian Journal of Analytical Sciences* 16(3): 273–276.

Jerusalinsky, L. 2013. *Distribuição geográfica e conservação de* Callicebus coimbrai *Kobayashi & Langguth, 1999 (Primates - Pitheciidae) na Mata Atlântica do Nordeste do Brasil.* Unpublished PhD thesis, Universidade Federal da Paraíba, Brazil.

Jerusalinsky, L., Oliveira, M.M., Ferreira, J.G., Wagner, P.G.G. & Ferrari, S.F. 2006. Novos registros e estratégia para conservação de primatas ameaçados no Baixo São Francisco Sergipano. In *Resumos do Congresso Internacional sobre Manejo de Fauna da Amazônia e America Latina, CD-ROM*. Santa Cruz, Brazil: Universidade Estadual de Santa Cruz - UESC.

Jewell, P.A. & Oates, J.F. 1969. Ecological observations of the lorisoid primates of African lowland forest. *Zoologica Africa* 4: 231–248.

Johns, A.D. 1987. The use of primary and selectivly logged rainforest by Malaysian hornbills (Bucerotidae) and implications for their conservation. *Biological Conservation* 40: 179–190.

Johns, A.D. 1988. Effects of 'selective' timber extraction on rain forest structure and composition and some consequences for frugivores and folivores. *Biotropica* 20: 31–37.

Johns, A.D. & Skorupa, J. 1987. Responses of rain-forest primates to habitat disturbance: a review. *International Journal of Primatology* 8: 157–191.

Johnsingh, A.J.T. & Joshua, J. 1989. The threatened gallery forest of the river Tambiraparani, Mundanthurai Wildlife Sanctuary, south India. *Biological Conservation* 47: 273–280.

Johnson, A.E., Knott, C.D., Pamungkas, B., Pasaribu, M. & Marshall, A.J. 2005. A survey of the orangutan (*Pongo pygmaeus wurmbii*) population in and around Gunung Palung National Park, West Kalimantan, Indonesia based on nest counts. *Biological Conservation* 121: 495–507.

Johnson, C., Piel, A.K., Forman, D., Stewart, F.A. & King, A.J. 2015. The ecological determinants of baboon troop movements at local and continental scales. *Movement Ecology* 3: 14.

Johnson, C.L., Stewart, F.A., Forman, D.W., King, A.J. & Piel, A.K. 2013. The Yellow Mile: Patterns of space and resource use by yellow baboons in Ugalla, Western Tanzania. Abstract. Primate Society of Great Britain, Lincoln, UK.

Johnson, S. & Rejmánková, K. 2005. Impacts of land use on nutrient distribution and vegetation composition of freshwater wetlands in northern Belize. *Wetlands* 25: 89–100.

Jolly, A. 1985. *The Evolution of Primate Behavior*. New York: Macmillan.

Jolly, A., Dobson, A., Rasamimanana, H.M., *et al.* 2002. Demography of *Lemur catta* at Berenty Reserve, Madagascar: effects of troop size, habitat and rainfall. *International Journal of Primatology* 23: 327–353.

Jolly, A., Koyama, N., Rasamimanana, H., Crowley, H. & Williams, G. 2006. Berenty Reserve: a research site in Southern Madagascar. In *Ringtailed Lemur Biology: Lemur catta in Madagascar*, A. Jolly, N. Koyama, H.R. Rasamimanana & R.W. Sussman (eds). New York: Springer, pp. 32–42.

Jolly, C.J. 1970. The seed eaters: a new model for hominid differentiation based on a baboon analogy. *Man* 5: 5–26.

Jones, C.B. (Ed.) 2005. *Behavioral Flexibility in Primates: Causes and Consequences*. New York: Springer.

Jones, C. & Sabeter, Pi. J. 1968, Comparative ecology of *Cercocebus albigena* (Gray) and *Cercocebus torquatus* (Kerr) in Rio Muni, West Africa. *Folia Primatologica* 9: 99–113.

Jones, C. & Sabater, Pi. J. 1971. Comparative ecology of *Gorilla gorilla* (Savage and Wyman) and *Pan troglodytes* (Blumenbach) in Rio Muni, West Africa. *Bibliotheca Primatologica* 13: 1–96.

Jones, M.B. 1988. Photosynthetic responses of C3 and C4 wetland species in a tropical swamp. *Journal of Ecology* 76: 253–262.

Jones, M.M., Tuomisto, H., Clark, D.B. & Olivas, P. 2006. Effects of mesoscale environmental heterogeneity and dispersal limitation on floristic variation in rain forest ferns. *Journal of Ecology* 94: 181–195.

Jones, T.G. 2013. Shining a light on Madagascar's mangroves. *Madagascar Conservation & Development* 8: 4–6.

Juárez, C., Dvoskin, R. & Fernandez-Duque, E. 2005. Structure and composition of wild black howler troops (*Alouatta caraya*) in gallery forests of the Argentinean Chaco. *Neotropical Primates* 13: 19–22.

Julliot, C. 1996. Seed dispersal by red howling monkeys (*Alouatta seniculus*) in the tropical rain forest of French Guiana. *International Journal of Primatology* 17: 239–258.

Junk, W.J. 1970. Investigations on the ecology and production-biology of the floating meadows (*Paspalo-Echinochloetum*) on the Middle Amazon. *Amazoniana* 2: 449–495.

Junk, W.J. 1993. Wetlands of tropical South America. In *Wetlands of the World: Inventory, Ecology and Management. Volume I*, D.F. Wingham, D. Dykyjova & S. Hejoy (eds). Dordrecht, The Netherlands: Springer, pp. 679–739.

Junk, W.J. 1997. General aspects of floodplain ecology with special reference to Amazonian floodplains. In *The Central Amazon Floodplain: Ecology of a Pulsing System*, W.J. Junk (ed.). Berlin: Springer, pp. 3–20.

Junk, W.J. 2002. Long-term environmental trends and the future of tropical wetlands. *Environmental Conservation* 29: 414–435.

Junk, W.J. 2013. Current state of knowledge regarding South America wetlands and their future under global climate change. *Aquatic Sciences* 75: 113–131.

Junk, W.J. & Furch, K. 1993. A general review of tropical South American floodplains. *Wetlands Ecology & Management* 2: 231–238.

Junk, W.J. & Nunes da Cunha, C. 2005. Pantanal: a large South American wetland at a crossroads. *Ecological Engineering* 24: 391–401.

Junk, W.J., An, S., Finlayson, C.M., *et al.* 2013. Current state of knowledge regarding the world's wetlands and their future under global climate change: a synthesis. *Aquatic Sciences* 75: 131–167.

Junk, W.J. & Wantzen, K.M. 2004. The flood pulse concept: new aspects, approaches, and applications. An update. *Proceedings of the Second International Symposium on Management of Large Rivers for Fisheries* 16: 117–149.

Junk, W.J., Bayley, P.B. & Sparks, R.E. 1989. The flood pulse concept in river-floodplain systems. *Canadian Special Publication of Fisheries & Aquatic Sciences* 106: 110–127.

Junk, W.J., Brown, M., Campbell, I.C., *et al.* 2006. The comparative biodiversity of seven globally important wetlands: a synthesis. *Aquatic Sciences* 68: 400–414.

Junk, W.J., Cunha, C.N., Wantzen, K.M., *et al.* 2006. Biodiversity and its conservation in the Pantanal of Mato Grosso, Brazil. *Aquatic Sciences* 68: 278–309.

Junk, W.J., Piedade, M.T.F., Lourival, R., *et al.* 2014. Brazilian wetlands: their definition, delineation, and classification for research, sustainable management, and protection. *Aquatic Conservation: Marine & Freshwater Ecosystems* 24: 5–22.

Junk, W.J., Piedade, M.T.F., Schöngart, J., *et al.* 2011. A classification of major naturally-occurring Amazonian lowland wetlands. *Wetlands* 31: 623–640.

Junk, W.J., Piedade, M.T.F., Schöngart, J. & Wittmann, F. 2012. A classification of major natural habitats of Amazonian white-water river floodplains (*várzeas*). *Wetlands Ecology and Management* 20: 461–475.

Kadafa, A.A. 2012. Oil exploration and spillage in the Niger Delta of Nigeria. *Civil & Environmental Research* 2: 38–51.

Kadima, K.É., Ntabwoba, S.S.M. & Lucazeau, F. 2011. A Proterozoic-rift origin for the structure and the evolution of the cratonic Congo Basin. *Earth & Planetary Science Letters* 304: 240–250.

Kalan, A.K., Madzoké, B. & Rainey, H.J. 2010. A preliminary report on feeding and nesting behaviour of swamp gorillas in the Lac Télé Community Reserve. *Mammalia* 74: 439–442.

Kalan, A.K., Mundry, R., Wagner, O.J., *et al.* 2015. Towards the automated detection and occupancy estimation of primates using passive acoustic monitoring. *Ecological Indicators* 54: 217–226.

Kalliola, R., Salo, J., Puhakka, M. & Rajasilta, M. 1991. New site formation and colonizing vegetation in primary succession on the western Amazon floodplains. *Journal of Ecology* 79: 877–901.

Kanamori, T., Kuze, N., Bernard, H., Malim, T.P. & Kohshima, S. 2010. Feeding ecology of Bornean orangutans (*Pongo pygmaeus morio*) in Danum Valley, Sabah, Malaysia: a 3-year record including two mast fruitings. *American Journal of Primatology* 72: 820–840.

Kanamori, T., Kuze, N., Bernard, H., Malim, T.P. & Kohshima, S. 2017. Fluctuations of population density in Bornean orangutans (*Pongo pygmaeus morio*) related to fruit availability in the Danum Valley, Sabah, Malaysia: a 10-year record including two mast fruitings and three other peak fruitings. *Primates* 58: 225–235.

Kandil, F.E., Grace, M.H., Seigler, D.S. & Cheeseman, J.M. 2004. Polyphenolics in *Rhizophora mangle* L. leaves and their changes during leaf development. *Trees* 18: 518–528.

Kano, T. 1983. An ecological study of the pygmy chimpanzees (*Pan paniscus*) of Yalosidi, Republic of Zaire. *International Journal of Primatology* 4: 1–31.

Kano, T. 1992. *The Last Ape: Pygmy Chimpanzee Behavior and Ecology*, Stanford, CA: Stanford University Press.

Kano, T. & Mulavwa, M. 1984. Feeding ecology of the pygmy chimpanzees (*Pan paniscus*) of Wamba. In *The Pygmy Chimpanzee*, R.L. Susman (ed.) New York: Springer, pp. 233–274.

Kanthaswamy, S., Kurushima, J. & Smith, D. 2006. Inferring *Pongo* conservation units: a perspective based on microsatellite and mitochondrial DNA analyses. *Primates* 47: 310–321.

Kartikasari, S.N. 1986. *Studi populasi dan perilaku lutung (Presbytis cristata, Raffles) di Taman Nasional Baluran, Jawa Timor.* BSc thesis, Institut Pertanian Bogor, Bogor, Indonesia.

Kasecker, T.P. 2006. *Efeito da estrutura do hábitat sobre a riqueza e composição de comunidades de primatas da RDS Piagaçu-Purus, Amazônia Central, Brasil.* MSc dissertation, Instituto Nacional de Pesquisas da Amazonia, Manaus, Amazonas Brazil.

Kathiresan, K. & Bingham, B.L. 2001. Biology of mangroves and mangrove ecosystems. *Advances in Marine Biology* 40: 81–251.

Kattan, G.H., Alvarez-López, H. & Giraldo, M. 1994. Forest fragmentation and bird extinctions: San Antonio eighty years later. *Conservation Biology* 8: 138–146.

Kaufmann, J.C. 2006. The sad opaqueness of the environmental crisis in Madagascar. *Conservation & Society* 4: 179.

Kaufman, S.R. 2012. *A Comparative Study of a Bird Community in Lake Alaotra: The Effects of a Protected Area.* Fort Dauphin, Madagascar: SIT Madagascar, Biodiversity and Natural Resource Management.

Kavanagh, M. 1980. Invasion of the forest by an African savannah monkey: behavioural adaptations. *Behaviour* 73: 238–260.

Kaviar, S., Shockey, J. & Sundberg, P. 2012. Observations of the endemic pygmy three-toed sloth, *Bradypus pygmaeus* of Isla Escudo de Veraguas, Panama. *PLoS ONE* 7: e49854.

Kawabe, M. & Mano, T. 1972. Ecology and behavior of the wild proboscis monkey, *Nasalis larvatus* (Wurmb), in Sabah, Malaysia. *Primates* 13: 213–228.

Kawai, M. 1964a. Food habits of Japanese macaques 1. Animal matter. *Yaen* 18: 23–24.

Kawai, M. 1964b. Food habits of Japanese macaques 2. Inorganic matter and others. *Yaen* 19: 14.

Kawai, M. 1965. Newly acquired pre-cultural behavior of the natural troop of Japanese monkeys on Koshima Island. *Primates* 6: 1–30.

Kay, R.F., Madden, R.H., van Schaik, C. & Higdon, D. 1997. Primate species richness is determined by plant productivity: implications for conservation. *Proceedings of the National Academy of Sciences* 94: 13023–13027.

Kays, R. & Allison, A. 2001. Arboreal tropical forest vertebrates: current knowledge and research trends. In *Tropical Forest Canopies: Ecology and Management*, K.E. Linsenmair, C.M. Davis, B. Fiala & M.R. Speight (eds). Dordrecht, The Netherlands: Springer, pp. 109–120.

Kazahari, N., Tsuji, Y. & Agetsuma, N. 2013. The relationships between feeding-group size and feeding rate vary from positive to negative with characteristics of food items in wild Japanese macaques (*Macaca fuscata*). *Behaviour* 150: 175–197.

Keddy, P.A., Fraser, L.H., Solomeshch, A.I., *et al.* 2009. Wet and wonderful: the world's largest wetlands are conservation priorities. *BioScience* 59: 39–51.

Kellman, M., Tackaberry, R. & Rigg, L. 1998. Structure and function in two tropical forest communities: implications for forest conservation in fragmented systems. *Journal of Applied Ecology* 35: 195–206.

Kemenes, A., Forsberg, B.R. & Melack, J.M. 2007. Methane release below a tropical hydroelectric dam. *Geophysical Research Letters* 34: L12809.

Kemenes, A., Forsberg, B.R. & Melack, J.M. 2011. CO_2 emissions from a tropical hydroelectric reservoir (Balbina, Brazil). *Journal of Geophysical Research* 116: G03004, doi: 10.1029/2010JG0 01465.

Kempf, E. 2009. Patterns of water use in primates. *Folia Primatologica* 80: 275–294.

Kenward, B., Rutz, C., Weir, A. & Kacelnik, A. 2012. Development of tool use in New Caledonian crows: inherited action patterns and social influences. *Animal Behaviour* 72: 1329–1343.

Khajuria, H. 1954. [1955]. Catalogue of mammals in the Indian Museum (Zoological Survey). II. Primates: Cercopithecidae. *Records of the Indian Museum* 52: 101–127.

Khan, M.A.R. 1985. *Mammals of Bangladesh.* Dhaka: Nazma Reza.

Khan, M.A.R. 1986. Wildlife in Bangladesh mangrove ecosystem. *Journal of the Bombay Natural History Society* 83: 32–48.

Khan, M.M.H. 2012. Population and prey of the Bengal Tiger *Panthera tigris tigris* (Linnaeus, 1758) (Carnivora: Felidae) in the Sundarbans, Bangladesh. *Journal of Threatened Taxa* 4: 2370–2380.

Kierulff, M.C.M., dos Santos, G.R., Canale, G., Guidorizzi, C.E. & Cassano, C. 2004. The use of camera-traps in a survey of the buff-headed capuchin monkey, *Cebus xanthosternos. Neotropical Primates* 12: 56–59.

Kierulff, M.C.M., Mendes, S.L. & Rylands, A.B. 2008. *Cebus xanthosternos.* In *IUCN Red List of Threatened Species.* Gland, Switzerland: IUCN. www.iucnredlist.org.

Kierulff, M.C.M., Rylands, A.B. & de Oliveira, M.M. 2008a. *Leontopithecus rosalia.* In *IUCN Red List of Threatened Species.* Version 2014.3. Gland, Switzerland: IUCN. www.iucn.org.

Kierulff, M.C.M., Rylands, A.B., Mendes, S.L. & de Oliveira, M.M. 2008b. *Leontopithecus chrysomelas.* In *IUCN Red List of Threatened Species.* Version 2014.3. Gland, Switzerland: IUCN. www.iucn.org.

Kijima, A., Arai, N. & Mae, Y. 2004. Record of shellfish species in the inter-tidal zone around Oshika Peninsula in Miyagi Prefecture of Japan. *Bulletin of Integrated Field Science Center* 19: 96–99.

Killeen, T.J., Calderon, V., Soria, L., *et al.* 2007. Thirty years of land-cover change in Bolivia. *Ambio* 36: 600–606.

Killeen, T., García, E. & Beck, S. (eds). 1993. *Guía de árboles de Bolivia.* La Paz, Bolivia: Herbario Nacional de La Paz & Missouri Botanical Garden.

Killeen, T.J., Guerra, A., Calzada, M., *et al.* 2008. Total historical land-use change in eastern Bolivia: who, where, when, and how much? *Ecology & Society* 13: 36.

Killeen, T., Villegas, Z., Soria, L. & Soares-Filho, B. 2002. Tendencias de la deforestación en los municipios de San Javier y Concepción, Santa Cruz, Bolivia. *Revista Boliviana de Ecología y Conservación Ambiental* 11: 67–75.

Kimura, K., Nishimura, A., Izawa, K. & Mejia, C.A. 1994. Annual changes of rainfall and temperature in the tropical seasonal forest at La Macarena, Field Station Colombia. *Field Studies of New World Monkeys. La Macarena, Colombia* 9: 1–3.

Kingdon, J. 1997. *The Kingdon Field Guide to African Mammals.* Princeton, NJ: Princeton University Press.

Kingdon, J., Butynski, T.M. & De Jong Y.A. 2008. *Papio cynocephalus.* In *IUCN Red List of Threatened Species.* Gland, Switzerland: IUCN. e.T92250442A92250811.

Kingdon, J., Happold, D., Hoffmann, M., *et al.* (eds). 2013. *Mammals of Africa. Volume I: Introductory Chapters and Afrotheria.* London: Bloomsbury Publishing.

Kingston, W.R. 1986. Establishment of primates rescued from the Tucuruí Dam flooding in the Brazilian National Primate Centre. *Primate Eye* 28: 18–20.

Kinnaird, M.F. 1992. Variable resource defense by the Tana River crested mangabey. *Behavioral Ecology & Sociobiology* 31: 115–122.

Kinnaird, M.F. & O'Brien, T.G. 1996. Ecotourism in the Tangkoko Dua Sudara Nature Reserve: opening Pandora's box? *Oryx* 30: 65–73.

Kinzey, W.G. 1997. *Cebus.* In *New World Primates: Ecology, Evolution and Behavior*, W.G. Kinzey (ed.). New York: Aldine de Gruyter, pp. 248–257.

Kirimura, M. 2007. A study on the edible use of seaweed *Sargassum fulvellum. Bulletin of Kyoto Junior College* 35: 41–50.

Kirkpatrick, R.C. 2011. The Asian colobines: diversity among leaf-eating monkeys. In *Primates in Perspective*, C.J. Campbell, A. Fuentes, K.C. MacKinnon, S.K. Bearder & R.M. Stumpf (eds). Oxford: Oxford University Press, pp. 189–202.

Kiss, A. 2004. Is community-based ecotourism a good use of biodiversity conservation funds? *Trends in Ecology & Evolution* 19: 231–237.

Kitchener, D.J., Boeadi, C.L. & Maharadatunkamsi. 1990. Wild mammals of Lombok Island. *Records of the Western Australian Museum*, Suppl. 33: 1–129.

Knight, J.M. 2011. A model of mosquito-mangrove basin ecosystems with implications for management. *Ecosystems* 14: 1382–1395.

Knot, S. 1999. Orangutan behavior and ecology. In *The Nonhuman Primates*, P. Dolhinow & A. Fuentes (eds). Mountain View, CA: Mayfield Publishers, pp. 50–57.

Knott, C.D. 1998. Changes in orangutan caloric intake, energy balance, and ketones in response to fluctuating fruit availability. *International Journal of Primatology* 19: 1061–1079.

Koh, L.P. & Wich, S.A. 2012. Dawn of drone ecology: low-cost autonomous aerial vehicles for conservation. *Tropical Conservation Science* 5: 121–132.

Köhler, M. & Moyà-Solà, S. 1997. Ape-like or hominid-like? The positional behavior of *Oreopithecus bambolii* reconsidered. *Proceedings of the National Academy of the Sciences of the Unites States of America* 94: 11474–11750.

Kone, I. & Akpatou, B.K. 2004. *Bilan synoptique des missions effectuées du 17 juillet au 18 aout 2004 dans le parc national des iles Ehotile et les forêts classées de Dassioko, Bossematie, Mabi et Yaya. Identification des sites abritant encore* Cercopithecus diana roloway, Cercocebus atys lunulatus *et* Piliocolobus badius waldronae *en Côte d'Ivoire.* Unpublished Report. Madison, USA & Abidjan, Côte d'Ivoire: CEPA.

Kone, I., Coulibaly, D., Bene, J.-C.K., *et al.* 2007. *Initiation of a community based conservation system in southeastern Côte d'Ivoire for the probable last refuge for the Miss Waldron's red colobus.* Technical Report. Abidjan, Côte d'Ivoire: WAPCA Côte d'Ivoire, West African Primates Conservation Action.

Konstant, W. & Mittermeier, R.A. 1982. Introduction, reintroduction and translocation of Neotropical primates: past experiences and future possibilities. *International Zoo Yearbook* 22: 69–77.

Kortlandt, A. 1986. The use of stone tools by wild-living chimpanzees and earliest hominids. *Journal of Human Evolution* 15: 77–132.

Kortlandt, A. 1995. A survey of the geographical range, habitats and conservation of the pygmy chimpanzee (*Pan paniscus*): an ecological perspective. *Primate Conservation* 16: 21–36.

Kool, K.M. 1989. *Behavioural ecology of the silver leaf monkey,* Trachypithecus auratus sondaicus, *in the Pangandaran Nature Reserve, West Java, Indonesia.* PhD thesis, University of New South Wales, Sydney, Australia.

Kool, K.M. 1992. Food selection by the silver leaf monkey, *Trachypithecus auratus* sondaicus, in relation to plant chemistry. *Oecologia* 90: 527–533.

Kool, K.M. 1993. The diet and feeding behavior of the silver leaf monkey (*Trachypithecus auratus sondaicus*) in Indonesia. *International Journal of Primatology* 14: 667–700.

Koops, K., McGrew, W.C. & Matsuzawa, T. 2013. Ecology of culture: do environmental factors influence foraging tool use in wild chimpanzees, *Pan troglodytes verus*? *Animal Behaviour* 85: 175–185.

Kowalewski, M.M. 2007. *Patterns of affiliation and co-operation in howler monkeys: an alternative model to explain social organization in non-human primates*. PhD thesis, University of Illinois at Urbana-Champaign, IL.

Kowalewski, M.M. & Garber, P.A. 2010. Mating promiscuity and reproductive tactics in female black and gold howler monkeys (*Alouatta caraya*) inhabiting an island on the Parana River, Argentina. *American Journal of Primatology* 72: 734–748.

Kowalewski, M.M., Garber, P.A., Cotrtes-Ortiz L., Urbani B. & Youlatos D. 2015. *Howler Monkeys: Behavior, Ecology and Conservation*. New York: Springer.

Kowalewski, M.M. & Zunino, G.E. 2004. Birth seasonality in *Alouatta caraya* in northern Argentina. *International Journal of Primatology* 25: 383–400.

Koyama, N., Nakamichi, M., Oda, R., *et al.* 2001. A ten-year summary of reproductive parameters for ring-tailed lemurs at Berenty, Madagascar. *Primates* 42: 1–14.

Kozlowski, T.T. 2002. Physiological–ecological impacts of flooding on riparian forest ecosystems. *Wetlands* 22: 550–561.

Kramer, R., van Schaik, C. & Johnson, J. 1997. *Last Stand: Protected Areas and the Defense of Tropical Biodiversity*. New York: Oxford University Press.

Krause, G., Schories, D., Glaser, M. & Diele, K. 2001. Spatial patterns of mangrove ecosystems: the Bragantinian mangroves of Northern Brazil (Braganca, Para). *Ecotropica* 7: 93–107.

Krebs, J.R. & Davies, N.B. 1993. *An Introduction to Behavioural Ecology*, 3rd edn. Oxford: Blackwell.

Kreibich, H., Kern, J. & Förstel, H. 2002. Studies on nitrogen fixation in Amazonian floodplain forests. In *Nitrogen Fixation: From Molecules to Crop Productivity*, F.O. Pedrosa, M. Hungria, G. Yates & W.E. Newton (eds). Dordrecht, The Netherlands: Springer, pp. 544–554.

Kren, J.A. 1964. Observations on the habits of the proboscis monkey, *Nasalis larvatus* (Wurmb.), made in the Brunei Bay area, Borneo. *Zoologica* 49: 183–192.

Kull, C.A. 2012. Air photo evidence of historical land cover change in the highlands: Wetlands and grasslands give way to crops and woodlots. *Madagascar Conservation & Development* 7: 144–152.

Kull, C.A., Tassin, J., Moreau, S., *et al.* 2012. The introduced flora of Madagascar. *Biological Invasions* 14: 875–888.

Kummu, M. & Sarkkula, J. 2008. Impact of the Mekong River flow alteration on the Tonle Sap flood pulse. *Ambio* 37: 185–192.

Kuroda, S. 1979. Grouping of the pygmy chimpanzees. *Primates* 20: 161–183.

Kuroda, S. Nishihara, T. Suzuki, S. & Oko, R.A. 1996. Sympatric chimpanzees and gorillas in Ndoki Forest, Congo. In *Great Ape Societies*, W.C. McGrew, L.F. Marchant & T. Nishida (eds). Cambridge: Cambridge University Press. pp. 71–81.

Kyeyune, S. 2003. *Abundance and size distribution of Raphia farinifera in Budongo Forest Reserve, Western Uganda*. BSc dissertation, Faculty of Forestry and Nature Conservation, Makerere University, Kampala, Uganda.

Laake, J.L., Buckland, S.T., Anderson, D.R. & Burnham, K.O. 1996. *DISTANCE User's Guide, V2.2*. Fort Collins, CO: Colorado Cooperative Fish and Wildlife Research Unit Colorado State University.

Labouze, A., Clarke, R., Galat, G. & Galat-Luong, A. 1996. *L'espoir qui venait des singes*. TV movie. Paris: La Cinquième-CNDP-Gédéon-CNRS audiovisuel-ORSTOM audiovisuel.

Lacerda, L.D. 2002. *Mangrove Ecosystems: Function and Management*. New York: Springer.

Lacerda, L.D., Conde, J.E., Alarcon, C., *et al.* 1993. Mangrove ecosystems of Latin America and the Caribbean: a summary. In *Conservation and Sustainable Utilization of Mangrove Forests in Latin America and Africa Regions*, L.D. Lacerda (ed.). Okinawa, Japan: International Tropical Timber Organization and International Society for Mangrove Ecosystems, pp. 1–42.

Lacher, Jr, T.E., da Fonseca, G.A.B., Alves, Jr, C. & Magalhaes-Castro, B. 1984. Parasitism of trees by marmosets in a central Brazilian gallery forest. *Biotropica* 16: 202–209.

Lackman, I. & Ancrenaz, M. 2009. *HUTAN orang-utan conservation activities in 2008*. Kota Kinabalu: HUTAN.

Lackman-Ancrenaz, I., Ancrenaz, M. & Saburi, R. 2001. The Kinabatangan Orangutan Conservation Project. In *The Apes: Challenges for the 21st Century*, Brookfield Zoo (ed.). Chicago, IL: Brookfield Zoo, pp. 262–265.

Laden, G. & Wrangham, R. 2005. The rise of the hominids as an adaptive shift in fallback foods: Plant underground storage organs (USOs) and australopith origins. *Journal of Human Evolution* 49: 482–498.

LaFleur, M. 2012. *Ecology of ring-tailed lemurs* (Lemur catta) *at the Tsimanampetsotsa National Park, Madagascar: implications for female dominance and the evolution of lemur traits*. PhD thesis, University of Colorado, Boulder, Boulder, Colorado, USA.

Laikre, L. 2010. Genetic diversity is overlooked in international conservation policy implementation. *Conservation Genetics* 11: 349–354.

Lambert, J.E. 2002. Resource switching and species coexistence in guenons: A community analysis of dietary flexibility In *The Guenons: Diversity and Adaptation in African Monkeys*, M.E. Glenn & M. Cords (eds). New York: Kluwer Academic/Plenum Publishers, pp. 309–323.

Lambert, J.E. 2007. Seasonality, fallback strategies, and natural selection: a chimpanzee and cercopithecoid model for interpreting the evolution of hominin diet. In *Evolution of the Human Diet: The Known, the Unknown and the Unknowable*, P.S. Unger (ed.). Oxford: Oxford University Press, pp. 324–343.

Lambert, J.E. 2011. Primate nutritional ecology. In *Primates in Perspective*, C.J. Campbell, A. Fuentes, K.C. MacKinnon, S.K. Bearder & R.M. Stumpf (eds). Oxford: Oxford University Press, pp. 512–522.

Lambert, J.E., Chapman, C., Wrangham, R. & Conklin-Brittain, N.L. 2004. Hardness of Cercopithecine foods: implications for the critical function of enamel thickness in exploiting fallback food. *American Journal of Physical Anthropology* 125: 363–368.

Lammers, P.L., Richter, T., Waeber, P.O. & Mantilla-Contreras, J. 2015. Lake Alaotra wetlands: how long can Madagascar's most important rice and fish production region withstand the anthropogenic pressure? *Madagascar Conservation & Development* 10: 116–127.

Lang, C. 2002. *Vietnam: Na Hang Dam Threatens Forests, People and Wildlife*. Montevido, Uruguay: World Rainforest Movement.

Lapenta, M.J. & Procópio-de-Oliveira, P. 2008. Some aspects of seed dispersal effectiveness of golden lion tamarins (*Leontopithecus rosalia*) in a Brazilian Atlantic forest. *Tropical Conservation Science* 1: 122–139.

Larcher, W. 1994. *Ökologie der Pflanzen auf physiologischer Grundlage, guter Zustand*. Stuttgart, Germany: Ulmer.

Larick, R., Ciochon, R.L., Zaim, Y., *et al.* 2000. Lithostratigraphic context for Kln-1993.05-SNJ, a fossil colobine maxilla from Jokotingkir, Sangiran Dome. *International Journal of Primatology* 21: 731–759.

Lasso, C.A., Rial, A. & González, B.V. (eds). 2013. VII. Morichales y canangunchales de la Orinoquia y Amazonia: Colombia -

Venezuela. Parte I. *Serie Editorial Recursos Hidrobiológicos y Pesqueros Continentales de Colombia*. Bogotá, Colombia: Instituto de Investigación de Recursos Biológicos Alexander von Humboldt (IAvH).

Lasso, C.A., Rial, A., Matallana, C., *et al.* (eds). 2011. *Biodiversidad de la cuenca del Orinoco. II Áreas prioritarias para la conservación y uso sostenible*. Bogotá, Colombia: Instituto de Investigación de Recursos Biológicos Alexander von Humboldt, Ministerio del Ambiente, Vivienda y Desarrollo Territorial, WWF Colombia, Fundación Omacha, Fundación La Salle de Ciencias Naturales e Instituto de Estudios de la Orinoquia (Universidad Nacional de Colombia).

Lasso, C.A., Usma, J.S., Trujillo, F. & Rial, A. (eds). 2010. *Biodiversidad de la cuenca del Orinoco: bases científicas para la identificación de áreas prioritarias para la conservación y uso sostenible de la biodiversidad*. Bogotá D.C, Colombia: Instituto de Investigación de Recursos Biológicos Alexander von Humboldt, WWF Colombia, Fundación Omacha, Fundación La Salle e Instituto de Estudios de la Orinoquía (Universidad Nacional de Colombia).

Latour, S. 2005. *Inventaire de faune autour des lacs du delta de l'Ogooué*. Unpublished report. New York: Wildlife Conservation Society.

Latrubesse, E.M. & Franzinelli, E. 2005. The late Quaternary evolution of the Negro River, Amazon, Brazil: implications for island and floodplain formation in large anabranching tropical systems. *Geomorphology* 70: 372–397.

Latrubesse, E.M. & Stevaux, J.C. 2015. The Anavilhanas and Mariuá Archipelagos: fluvial wonders from the Negro River, Amazon basin. In *Landscapes and Landforms of Brazil*, B.C. Vieira, A.A.R. Salgado & L.J.C. Santos (eds). Dordrecht, The Netherlands: Springer, pp. 157–169.

Laurance, W.F., Ferreira, L.V., Rankin-de Merona, J. & Laurence, S.G. 1998. Rain forest fragmentation and the dynamics of Amazonian tree communities. *Ecology* 79: 2032–2040.

Lawrence, B. 1933. Howler monkeys of the palliata group. *Bulletin of the Museum of Comparative Zoology (Harvard University)* 75: 314–354.

Lazaro, N.I.L. 2013. Cuidado parental e relaçõs sociais entre fêmeas adultas e seus filhotes de bugios-pretas-e-dourados (*Alouatta caraya*: Primates, Atelidae) em fragmentos urbanos de Cerrado-Pantanal em Aquidauana, Matto Grosso do Sul. Mastes Dissertation,Universidade Matto Groso do Sul, Campo Grande.

Lazari, P.R.D., Santos-Filho, M.D., Canale, G.R. & Graipel, M.E. 2013. Flood-mediated use of habitat by large and midsized mammals in the Brazilian Pantanal. *Biota Neotropica* 13: 70–75.

Lazaro, N.I.L. 2013. *Cuidado parental e relações sociais entre fêmeas adultas e seus filhotes de bugios-pretos-e-dourados* (Alouatta caraya, *Primates, Atelidae) em fragmentos urbanos de Cerrado-Pantanal em Aquidauana, Mato Grosso do Sul*. MSc dissertation, Universidade Mato Grosso do Sul, Campo Grande.

Lázaro Júnior, A.E. & Rímoli, J. 2009. Predação e dispersão de sementes por bugios-pretos (Alouatta caraya, Primates, Atelidae) em fragmento florestal na margem esquerda do rio Aquidauana, Anastácio, Mato Grosso do Sul. In *Livro do Programa de Iniciação Científica PIBIC/CNPq/UFMS 2009. Pró-Reitoria de Pesquisa e Pós Graduação da Fundação Universidade Federal de Mato Grosso do Sul - UFMS*. Campo Grande, MS: Editora da Fundação Universidade Federal de Mato Grosso do Sul, pp. 1–20.

Lebreton, J.D., Burnham, K.P., Clobert, J. & Anderson, D.R. 1992. Modeling survival and testing biological hypotheses using marked animals: a unified approach with case studies. *Ecological Monographs* 62: 67–118.

Leca, J.B., Gunst, N., Watanabe, K. & Huffman, M.A. 2007. A new case of fish-eating in Japanese macaques: implications for social constraints on the diffusion of feeding innovation. *American Journal of Primatology* 69: 821–828.

Leca, J.B., Gunst, N., Rompis, A., *et al.* 2013. Population density and abundance of ebony leaf monkeys (*Trachypithecus auratus*) in West Bali National Park, Indonesia. *Primate Conservation* 26: 133–144.

Leciak, E., Ladik, A.H. & Hladik, C.M. 2005. Le palmier à huile (*Elaeis guineensis*) et les noyaux de biodiversité des forêts-galeries de guinée maritime: à propos du commensalisme de l'homme et du chimpanzé. *Revue d'Écologie, La Terre et La Vie* 60: 179–184.

Lee, P., Thornback, J. & Bennett, E. 1988. *Threatened Primates of Africa: The IUCN Red Data Book*. Gland, Switzerland: IUCN.

Lehman, S.M. 1999. *Biogeography of the primates of Guyana*. Unpublished PhD dissertation, Washington University, St. Louis, MO.

Lehman, S.M. 2000. Primate community structure in Guyana: a biogeographic analysis. *International Journal of Primatology* 21: 333–351.

Lehman, S.M. 2004. Distribution and diversity of primates in Guyana: Species-area relationships and riverine barriers. *International Journal of Primatology* 25: 73–95.

Lehman, S.M. 2006. Nested distribution patterns and the historical biogeograophy of the primates of Guyana. In *Primate Biogeography*, S.M. Lehman and J.G. Fleagle (eds). New York: Springer, pp. 63–80.

Lehman, S.M. & Fleagle, J.G. 2006. Biogeography and primates: a review. In *Primate Biogeography*, S.M. Lehman & J.G. Fleagle (eds). New York: Springer, pp. 1–58.

Lehman, S.M. & Robertson, K.L. 1994. Preliminary survey of *Cacajao melanocephalus melanocephalus* in Southern Venezuela. *International Journal of Primatology* 15: 927–934.

Lehman, S.M., Sussman, R.W., Phillips-Conroy, J. & Prince, W. 2006. Ecological biogeography of primates in Guyana. In *Primate Biogeography*, S.M. Lehman & J.G. Fleagle (eds). New York: Springer, pp. 105–130.

Lehman, S.M., Vié, J.C., Norconk, M.A., Portillo-Quintero, C. & Urbani, B. 2013. The Guyana Shield: Venezuela and the Guyanas. In *Evolutionary Biology and Conservation of Titis, Sakis and Uacaris*, L. Veiga, A. Barnett, S. Ferrari & M. Norconk (eds). Cambridge: Cambridge University Press, pp. 311–319.

Lehmann, J. & Boesch, C. 2003 Social influences on ranging patterns among chimpanzees (*Pan troglodytes verus*) in the Taï National Park, Côte d'Ivoire. *Behavioral Ecology* 14: 642–649.

Lehner, B. & Döll, P. 2004. Development and validation of a global database of lakes, reservoirs and wetlands. *Journal of Hydrology* 296: 1–22.

Leighton, D.R. 1987. Gibbons: territoriality and monogamy. In *Primate Societies*, B.B. Smuts, D.L. Cheney, R.M. Seyfarth, R.W. Wrangham & T.T. Struhsaker (eds). Chicago, IL: University of Chicago Press, pp. 135–145.

Leighton, M. 1993. Modelling dietary selectivity by Bornean orangutans: evidence of multiple criteria in fruit selection. *International Journal of Primatology* 14: 257–313.

Leland, L., Struhsaker, T.T. & Butynski, T.M. 1984. Infanticide by adult males in three primates species of Kibale forest, Uganda: a test of hypotheses. In *Infanticide: Comparative and Evolutionary Perspectives*, G. Hausfater & S.B. Hrdy (eds). New York: Aldine, pp. 151–172.

Leliveld, L.M.C., Scheumann, M. & Zimmermann, E. 2011. Acoustic correlates of individuality in the vocal repertoire of a nocturnal primate (*Microcebus murinus*). *Journal of the Acoustical Society of America* 129: 2278–2288.

Leonard, S. & Bennett, C. 1996. Associative behavior of *Cacajao calvus ucayalii* with other primate species in Amazonian Peru. *Primates* 37: 227–230.

Levi, T., Silvius, K.M., Oliveira, L.F.B., Cummings, A.R. & Fragoso, J.M.V. 2013. Competition and facilitation in the capuchin-squirrel monkey relationship. *Biotropica* 45: 636–643.

Lewis, M., Pryor, R. & Wilking, L. 2011. Fate and effects of anthropogenic chemicals in mangrove ecosystems, a review. *Environmental Pollution* 159: 2328–2346.

Lewis, M.C. & O'Riain, M.J. 2017. Foraging profile, activity budget and spatial ecology of exclusively natural-foraging chacma baboons (*Papio ursinus*) on the Cape Peninsula, South Africa. *International Journal of Primatology* 38: 751–779.

Leyer, I. 2006. Dispersal, diversity and distribution patterns in pioneer vegetation: The role of river-floodplain connectivity. *Journal of Vegetation Science* 17: 407–416.

Li, C., Zhao, C. & Fan, P.F. 2015. White-cheeked macaque (*Macaca leucogenys*): A new macaque species from Modog, southeastern Tibet. *American Journal of Primatology* 77: 753–766.

Liao, W.-L., Bhargava, D.S. & Das, J. 1988. Some effects of dams on wildlife. *Environmental Conservation* 15: 68–70.

Lima-Ayres, D. 1992. *The social category caboclo: history, social organization, identity and outsider's local social classification of the rural population of an Amazonian region*. PhD thesis. University of Cambridge, Cambridge, UK.

Linares, O.J. & Rivas, B. 2004. Mamíferos del Sistema Deltaico (delta del Orinoco-golfo de Paria), Venezuela. *Memoria de la Fundación La Salle de Ciencias Naturales* 159: 185–262.

Lindenmayer, D.B. & Fischer, J. 2006. *Habitat Fragmentation and Landscape Change: An Ecological and Conservation Synthesis*. Washington DC: Island Press.

Lindenmayer, D.B., Hobbs, R.J., Montague-Drake, R., *et al.* 2008. A checklist for ecological management of landscapes for conservation. *Ecology Letters* 11: 78–91.

Lindquist, N., Hay, M.E. & Fenical, W. 1992. Defense of Ascidians and their conspicuous larvae: adult vs. larval chemical defenses. *Ecological Monographs* 62: 547–568.

Liu, Q., Fragaszy, D., Wright, B., *et al.* 2011. Wild bearded capuchin monkeys (*Cebus libidinosus*) place nuts in anvils selectively. *Animal Behaviour* 81: 297–305.

Liu, Q., Simpson, K., Izar, P., *et al.* 2009. Kinematics and energetics of nut-cracking in wild capuchin monkeys (*Cebus libidinosus*) in Piauí, Brazil. *American Journal of Physical Anthropology* 138: 210–220.

Lo, E.Y.Y., Duke, N.C. & Sun, M. 2014. Phylogeographic pattern of *Rhizophora* (Rhizophoraceae) reveals the importance of both vicariance and long-distance oceanic dispersal to modern mangrove distribution. *BMC Evolutionary Biology* 14: 83.

Lonard, R.I. & Judd, F.W. 1997. The biological flora of coastal dunes and wetlands *Sesuvium portulacastrum*. *Journal of Coastal Research* 13: 96–104.

Long, E. & Racey, P.A. 2007. An exotic plantation crop as a keystone resource for an endemic megachiropteran, *Pteropus rufus*, in Madagascar. *Journal of Tropical Ecology* 23: 397–407.

Lönnberg, E. 1939. Notes on some members of the genus *Callicebus*. *Arkiv för Zoologi* 31: 1–18.

Lopes, G.P., Valsecchi, J., Vieira, T.M., Amaral, P.V. & Costa, E.W.M. 2012. Hunting and hunters in lowland communities in the region of the middle Solimões, Amazonas, Brazil. *Uakari* 8: 7–18.

López-Hernández, D., Hernández-Hernández, R.M. & Brossard, M. 2005. Historia del uso reciente de tierras de las Sabanas de América del sur. Estudios de casos en sabanas del Orinoco. *Interciencia* 30: 623–630.

López-Legentil, S., Bontemps-Subielos, N., Turon, X. & Banaigs, B. 2007. Secondary metabolite and inorganic contents in *Cystodytes* sp. (Ascidiacea): temporal patterns and association with reproduction and growth. *Marine Biology* 151: 293–299.

Lopez-Strauss, H. 2007. *Estimación de densidad y composición de grupos de dos especies de primates endémicos* Callicebus olallae *y* Callicebus modestus, *en el sud oeste del Departamento del Beni, Bolivia*. BSc thesis, Universidad Mayor de San Andres, La Paz, Bolivia.

López-Strauss, H. & Wallace, R.B. 2015. Density estimates of two Bolivian primate endemics, *Callicebus olallae* and *C. modestus*. *Mastozoologia Neotropical* 22: 23–34.

Louis, Jr, E.E., Engberg, S.E., McGuire, S.M., *et al.* 2008. Revision of the mouse lemurs, *Microcebus* (Primates, Lemuriformes), of northern and northwestern Madagascar with descriptions of two new species at Montagne d'Ambre National Park and Antafondro Classified Forest. *Primate Conservation* 23: 19–38.

Louhoungou, C.E.R. & Mabiala, C.E. 2001. Potentialités agricoles des sols du Sénégal oriental et de la Haute Casamance. In *Potentialités, contraintes et systèmes d'exploitation au Sénégal Oriental et en Haute Casamance*. CD-ROM. UCAD-IRD-DDR (eds). Dakar: Direction du Développement Rural/Sodefitex-Université Cheikh Anta Diop (eds).

Lourenço-de-Oliveira, R. & Deane, L.M. 1995. Simian malaria at two sites in the Brazilian Amazon: I. The infection rates of *Plasmodium brasilianum* in non-human primates. *Memorias do Instituto Oswaldo Cruz* 90: 331–339.

Lowrance, R., Altier, L.S., Newbold, J.D., *et al.* 1997. Water quality functions of riparian forest buffers in Chesapeake Bay watersheds. *Environmental Management* 21: 687–712.

Lucas, P.W., Hails, C.J. & Corlett, R.T. 1988. Status of the banded langur in Singapore. *Primate Conservation* 9: 136–138.

Ludwig, G., Agui, L.M., Svoboda, W.K., *et al.* 2008. Comparison of the diet of *Alouatta caraya* (Primates: Atelidae) between a riparian island and mainland on the Upper Parana River, southern Brazil. *Revista Brasileira de Zoologia* 25: 419–426.

Ludwing, G., Aguiar, L.M. & Rocha, V.J. 2006. Comportamento de obtenção de *Manihot esculenta* crantz (Euphorbiaceae), mandioca, por *Cebus nigritus* (Goldfuss) (Primates, Cebidae) como uma adaptação alimentar em períodos de escassez. *Revista Brasileira de Zoologia* 23: 888–890.

Luecke, L.G. & Estrada, A. 2005. *Reconocimiento preliminar de poblaciones de monos aulladores (*Alouatta palliata *and* Alouatta pigra*) en la Reserva de la Biosfera Pantános de Centla, Tabasco, Mexico.* Unpublished Report. Villahermosa, Tabasco, Mexico: National Commission of Natural Protected Areas (CONANP).

Lugo, A. 2008. Visible and invisible effects of hurricanes on forest ecosystems: an international review. *Austral Ecology* 33: 368–398.

Luiselli, L., Amori, G., Akani, G.C. & Eniang, E.A. 2015. Ecological diversity, community structure, and conservation of Niger Delta mammals. *Biodiversity & Conservation* 24: 2809–2830.

Luther, D.A. & Greenberg, R. 2009. Mangroves: a global perspective on the evolution and conservation of their terrestrial vertebrates. *Bioscience* 59: 602–612.

Lycett, J.E., Henzi, S.P. & Barrett, L. 1998. Maternal investment in mountain baboons and the hypothesis of reduced care. *Behavioral Ecology & Sociobiology* 42: 49–56.

Lykke, A.M. & Sambou, B. 1998. Structure, floristic composition, and vegetation forming factors of three vegetation types in Senegal. *Nordic Journal of Botany* 18: 129–140.

Lynch Alfaro, J.W., Boubli, J.P., *et al.* 2012a. Explosive Pleistocene range expansion leads to widespread Amazonian sympatry between robust and gracile capuchin monkeys. *Journal of Biogeography* 39: 272–288.

Lynch Alfaro, J.W., Boubli, J.P., *et al.* 2015. Biogeography of squirrel monkeys (genus *Saimiri*): Southcentral Amazon origin and rapid pan-Amazonian diversification of a lowland primate. *Molecular Phylogenetics & Evolution* 82: 436–454.

Lynch Alfaro, J.W.L., Silva, Jr, J.S. & Rylands, A.B. 2012b. How different are robust and gracile capuchin monkeys? An argument for the use of *Sapajus* and *Cebus*. *American Journal of Primatology* 74: 273–286.

MacArthur, R. & Wilson, E.O. 1967. *The Theory of Island Biogeography*. Princeton, NJ: Princeton University Press.

MacDonald, D.W. 1982. Notes on the size and composition of groups of proboscis monkey, *Nasalis larvatus*. *Folia Primatologica* 37: 95–98.

Macedonia, J.M. 1990. What is communicated in the antipredator calls of lemurs: evidence from playback experiments with ring-tailed and ruffed lemurs. *Ethology* 86: 177–190.

Macedonia, J.M. & Stanger, K.F. 1994. Phylogeny of the Lemuridae revisited: evidence from communication signals. *Folia Primatologica* 63: 1–43.

Macellini, S., Maranesi, M., Bonini, L., *et al.* 2012. Individual and social learning processes involved in the acquisition and generalization of tool use in macaques. *Philosophical Transactions of the Royal Society of London: Series B, Biological Sciences* 367: 24–36.

MacIntosh, D.J. & Ashton, E.C. 2002. *A Review of Mangrove Biodiversity Conservation and Management*. Aarhus, Denmark: Centre for Tropical Ecosystems Research, University of Aarhus.

MacKinnon, J. 1972. *The behavior and ecology of the orangutan, Pongo pygmaeus, with relation to other apes*. PhD thesis. Oriel College, Oxford.

MacKinnon, J. 1974. The behaviour and ecology of wild orang-utans (*Pongo pygmaeus*). *Animal Behavior* 22: 3–74.

MacKinnon, J. & MacKinnon, K. 1980. The behavior of wild spectral tarsiers. *International Journal of Primatology* 1: 361–379.

MacKinnon, J. & MacKinnon, K. 1987. Conservation and status of the primates of the Indo-Chinese subregion. *Primate Conservation* 8: 187–195.

MacKinnon, K. & MacKinnon, J. 1991. Habitat protection and reintroduction programs. In *Beyond Captive Breeding. Re-introducing Endangered Mammals to the Wild*, J.H.W. Gipps (ed.). Oxford: Zoological Society of London Symposia, Vol. 62. Oxford University Press, pp. 173–198.

MacKinnon, J., Smiet, F. & Artha, M.B. 1982. *A National Conservation Plan for Indonesia, Vol III: Java and Bali*. Rome: Food and Agricultural Organization of the United Nations.

MacKinnon, K. 1987. Conservation status of primates in Malaysia, with special reference to Indonesia. *Primate Conservation* 8: 175–183.

MacKinnon, K. 1996. *The Ecology of Kalimatan*. Singapore: Periplus Editions.

Maclaud, C. 1906. Notes sur les Mammiféres et les Oiseaux de l'Afrique Occidentale. Paris: Augustus Challamel.

MacLean, I.M.D., Hassall, M., Boar, R.R. & Lake, I.R. 2006. Effects of disturbance and habitat loss on papyrus-dwelling passerines. *Biological Conservation* 131: 349–358.

Maddock, T. Jr. 1976. A primer on floodplain dynamics. *Journal of Soil and Water Conservation* 31: 44–47.

Magliocca, F. & Gautier-Hion, A. 2002. Mineral content as a basis food selection by western lowland gorillas in a forest clearing. *American Journal of Primatology* 57: 67–77.

Magliocca, F & Gautier-Hion, A. 2004. Intergroup encounters in western lowland gorillas at a forest clearing. *Folia Primatologica* 75: 379–382.

Magliocca, F., Querouil, Q. & Gautier-Hion, A. 1999. Population structure and group composition of western lowland gorillas in northwestern Republic of Congo. *American Journal of Primatology* 48: 1–14.

Magnusson, W.E. 1995. Reintrodução: uma ferramenta conservacionista ou brinquedo perigoso? *Neotropical Primates* 3: 82–84.

Magnusson, W.E., Lima, A.P., Luizão, R.C.C., *et al.* 2005. RAPELD: uma modifcação do método de Gentry para inventários de biodiversidade em sítios para pesquisa ecológica de longa duração. *Biota Neotropica* 5: 19–24.

Mahabir, R. 2008. Vietnam connects new hydroelectric plant. *Asia Cleantech: Asia Clean Energy & Asia Clean Technology News*. February 27, 2008. http://asiacleantech.wordpress.com/2008/02/27/vietnam-connects-new-hydroelectric-plant.

Maiangwa, B. & Agbiboa, D.E. 2013. Oil multinational corporations, environmental irresponsibility and turbulent peace in the Niger Delta. *Africa Spectrum* 2/2013: 71–83.

Maingi, J.K. & Marsh, S.E. 2002. Quantifying hydrologic impacts following dam construction along the Tana River, Kenya. *Journal of Arid Environments* 50: 53–79.

Mainwaring, A., Culler, C., Polastre, J., Szewczyk, R. & Anderson, J. 2002. WSNA '02 Proceedings of the 1st ACM international workshop on Wireless sensor networks and applications, pp. 88–97.

Maingi, J.K. & Marsh, S.E. 2002. Quantifying hydrologic impacts following dam construction along the Tana River, Kenya. *Journal of Arid Environments* 50: 53–79.

Maisels, F., Ambahe, R., Ambassa. E. & Fotso, R. 2006a. New northwestern range limit of the northern Talapoin, Mbam et Djerem National Park, Cameroon. *Primate Conservation* 21: 89–91.

Maisels, F., Blake, S., Fay, M., Mobolambi, G. & Yako, V. 2006b. A note on the distribution of Allen's swamp monkey, *Allenopithecus nigroviridis*, in Northwestern Congo. *Primate Conservation* 21: 93–95.

Maisels, F., Bout, N., Inkamba-Inkulu, C., Pearson, L., Aczel, P., Ambahe, R., Ambassa, E. & Fotso, R. 2007b. New northwestern and southwestern range limits of De Brazza's monkey, Mbam Et Djerem National Park, Cameroon, and Bateke Plateau, Gabon and Congo. *Primate Conservation* 22: 107–110.

Maisels F., Gautier-Hion, A. & Gautier, J.-P. 1994. Diets of two sympatric colobines in Zaire: more evidence on seed-eating in forests on poor soils. *International Journal of Primatology* 15: 681–701.

Maisels, F., Makaya, Q.P. & Onononga, J.R. 2007. Confirmation of the presence of the red-capped mangabey (*Cercocebus torquatus*) in Mayumba National Park, Southern Gabon, and Conkouati-Douli National Park, Southern Republic of Congo. *Primate Conservation* 22: 111–115.

Maisels, F., Nishihara, T., Strindberg, S., *et al.* 2012. *Great ape and human impact monitoring training, surveys, and protection in the Ndoki-Likouala Landscape, Republic of Congo.* GACF Agreement: 96200-9-G247 Final Report. New York: Wildlife Conservation Society, Republic of Congo.

Majumder, J., Lodh, R. & Agarwala, B.K. 2012. Fish feeding adaptation by rhesus macaque *Macaca mulatta* (Cercopithecidae) in the Sundarban mangrove swamps, India. *Journal of Threatened Taxa* 4: 2539–2540.

Makedonska, J., Wright, B.W. & Strait, D.S. 2012. The effect of dietary adaption on cranial morphological integration in capuchins (Order Primates, Genus *Cebus*). *PLoS ONE* 7: e40398. doi:10.1371/journal.pone.0040398.

Malaivijitnond, S., Lekprayoon, C., Tandavanittj, N., *et al.* 2007. Stone-tool usage by Thai long-tailed macaques (*Macaca fascicularis*). *American Journal of Primatology* 69: 227–233.

Malanda, G., Iyenguet, F. & Madzoké, B. 2010. *Inventaire et distribution des grands mammifères et activités humaines dans la zone Tanga*. Unpublished report of the Wildlife Conservation Society, Republic of Congo.

Malbrant, R. & Maclatch, A. 1949. *Africain Français*, Paris: Paul Lechevalier.

Mallapur, A., Sinha, A. & Waran, N. 2005. Influence of visitor presence on the behavior of captive lion-tailed macaques (*Macaca silenus*) housed in Indian zoos. *Applied Animal Behaviour Science* 94: 341–352.

Mallick, J.K. 2011. Status of the mammal fauna in Sundarban Tiger Reserve, West Bengal- India. *Taprobanica* 3: 52–68.

Malta, A.J.R. & Mendes Pontes, A.R. 2013. The simplified novel diet of the highly threatened blonde capuchin in the vanishing Pernambuco Endemism Center. In *Primates in Fragments: Complexity and Resilience*, L.K. Marsh & C.A. Chapman (eds). New York: Springer, pp. 245–258.

Maltby, E. 1986. *Waterlogged Wealth: Why Waste the World's Wet Places?* London: International Institute of Environment and Development.

Mamede, S.B. & Alho, C.J.R. 2006. Response of wild mammals to seasonal shrinking-and-expansion of habitats due to flooding regime of the Pantanal, Brazil. *Brazilian Journal of Biology* 66: 991–998.

Maminirina, C.P., Girod, P. & Waeber, P.O. 2006. Comic strips as environmental educative tools for the Alaotra Region. *Madagascar Conservation & Development* 1: 11–14.

Mammides, C. & Cords, M. 2008. Effects of habitat disturbance and food supply on population densities of three primate species in the Kakamega forest, Kenya, *African Journal of Ecology* 47: 87–96.

Mandal, A.K. 1964. The behaviour of the rhesus monkeys (*Macaca inulatta* Zimmermann) in the Sundarbans. *Journal of the Bengal Natural History Society* 33: 153–165.

Mandal, R.N., Das, C.S. & Naskar, K.R. 2010. Dwindling Indian Sundarban mangrove: the way out. *Science & Culture* 76: 275–282.

Mankoto, M.O., Yamagiwa, J., Steinhauer-Burkart, B., *et al.* 1994. Conservation of eastern lowland gorilla in the Kahuzi-Biega National Park, Zaire. In *Current Primatology, Vol. 1, Ecology and Evolution*, B. Thierry, J.R. Anderson, J.J. Roeder & N. Herrenschmidt (eds). Strasbourg: Universite Louis Pasteur Press, pp. 113–122.

Manansang, J. 2005. *Indonesian Proboscis Monkey Population and Habitat Viability Assessment.* Final Report. Indonesian Proboscis Monkey Population and Habitat Viability Assessment Workshop (2004: Cisarua-Bogor, West Java, Indonesia).

Manly, B., McDonald, L., Thomas, D., McDonald, T. & Erickson, W. 2002. *Resource Selection by Animals: Statistical Analysis and Design for Field Studies.* Dordrecht, The Netherlands: Kluwer Academic Publishers.

Mann, C.C. 2000. Earthmovers of the Amazon. Are the mounds, causeways, and canals in Bolivia's Beni region natural formations or the result of 2000 years' labor by lost societies? *Science* 287: 786–789.

Mannu, M. & Ottoni, E.B. 2008. The enhanced tool-kit of two groups of wild bearded capuchin monkeys in the Caatinga: tool making, associative use, and secondary tools. *American Journal of Primatology* 70: 1–10.

Manson, J.H., Rose, L.M, Perry, S. & Gros-Louis, J. 1999. Dynamics of female-female relationships in wild *Cebus capucinus*: data from two Costa Rican sites. *International Journal of Primatology* 20: 679–706.

Maréchal, L., Semple, S., Majolo, B., *et al.* 2011. Impacts of tourism on anxiety and physiological stress levels in wild male Barbary macaques. *Biological Conservation* 144: 2188–2193.

Mares, M.A. & Ernest, K.A. 1995. Population and community ecology of small mammals in a gallery forest of central Brazil. *Journal of Mammalogy* 76: 750–768.

Mares, M.A., Braun, J.K. & Gettinger, D. 1989. Observations on the distribution and ecology of the mammals of the cerrado grasslands of central Brazil. *Annals of Carnegie Museum* 58: 1–60.

Marchesi, P., Marchesi, N., Fruth, B. & Boesch, C. 1995. Census and distribution of chimpanzees in Cote D'Ivoire. *Primates* 36: 591–607.

Marín, R.C. 2001. The Macanas wetland – conservation and agricultural use area. In *Sowing the Seeds for Sustainability: Agriculture, Biodiversity, Economics and Society*, R. Wiseman & L. Hopkins (eds). Gland, Switzerland: IUCN, pp. 58–60.

Marioni, B., da Silveira, R., Magnusson, W.E. & Thorbjarnarson, J. 2008. Feeding behavior of two sympatric caiman species, *Melanosuchus niger* and *Caiman crocodilus*, in the Brazilian Amazon. *Journal of Herpetology* 42: 768–772.

Markham, A.C. & Altmann, J. 2008. Remote monitoring of primates using automated GPS technology in open habitats. *American Journal of Primatology* 70: 495–499.

Marques, A.A.B., Schneider, M. & Alho, C.J.R. 2011. Translocation and radiotelemetry monitoring of black-tailed marmosets, *Callithrix* (*Mico*) *melanura* (E. Geoffroy in Humboldt), in a wildlife rescue operation in Brazil. *Brazilian Journal of Biology* 71: 983–989.

Marques, E.L.N., Jerusalinsky, L., Rocha, J.C.A.G., *et al.* 2013. Primates, Pitheciidae, *Callicebus coimbrai* Kobayashi & Langguth, 1999: New localities for an Endangered titi monkey in eastern Sergipe, Brazil. *Check List* 9: 696–699.

Marques, J.T., Pereira, M.J.R. & Palmeirim, J.M. 2012. Availability of food for frugivorous bats in lowland Amazonia: the influence of flooding and of river banks. *Acta Chiropterologica* 14: 183–194.

Marques, M., da Costa, M.F., Mayorga, M.I.O. & Pinheiro, R.C. 2004. Water environments: anthropogenic pressures and ecosystem changes in the Atlantic drainage basins of Brazil. *Ambio* 33: 68–77.

Marques, T.A., Thomas, L., Martin, S.W., *et al.* 2013. Estimating animal population density using passive acoustics. *Biological Reviews* 88: 287–309.

Marsh, C.W. 1978. *Ecology and social organization of the Tana River red colobus* (Colobus badius rufomitratus). PhD Dissertation, University of Bristol, Bristol, UK.

Marsh, C.W. 1981. Ranging behaviour and its relation to diet selection in Tana River red colobus (*Colobus badius rufomitratus*). *Journal of the Zoological Society of London* 195: 473–492.

Marsh, C.W. & Greer, A.G. 1992. Forest land-use in Sabah, Malaysia: an introduction to Danum Valley. *Philosophical Transactions of the Royal Society B: Biological Sciences* 335: 331–339.

Marsh, L.K. 2003. *Primates in Fragments: Ecology and Conservation.* New York: Kluwer Academic/Plenum Publishers.

Marsh, L.K. 2004. Primate species at the Tiputini Biodiversity Station, Ecuador. *Neotropical Primates* 12: 75–78.

Marsh, L.K. 2013. Because conservation counts: primates and fragmentation. In *Primates in Fragments: Complexity and Resilience*, L.K. Marsh & C.A. Chapman (eds). New York: Springer, pp. 3–11.

Marsh, LK. 2014. A taxonomic revision of the Saki Monkeys, Pithecia Desmarest, 1804. *Neotropical Primates* 21: 1–165.

Marsh, L.K. & Chapman, C. (eds). 2013. *Primates in Fragments: Complexity and Resilience.* New York: Springer.

Marsh, L.K., Chapman, C.A., Arroyo-Rodríguez, V., *et al.* 2013. In *Primates in Fragments: Complexity and Resilience*, L.K. Marsh & C.A. Chapman (eds). New York: Springer, pp. 503–523.

Marshall, A.J. 2004. *Population ecology of gibbons and leaf monkeys across a gradient of Bornean forest types.* PhD thesis, Harvard University, Cambridge, MA.

Marshall, A.J. 2009. Are montane forests demographic sinks for Bornean white-bearded gibbons *Hylobates albibarbis*? *Biotropica* 42: 147–157.

Marshall, A.J. 2010. Effect of habitat quality on primate populations in Kalimantan: gibbons and leaf monkeys as case studies. In *Indonsian Primates*, S. Gursky-Doyen & J. Supriatna (eds). New York: Springer, pp. 157–177.

Marshall, A.J. & Leighton, M. 2006. How does food availability limit the population density of white-bearded gibbons? In *Feeding*

Ecology in Apes and Other Primates: Ecological, Physical and Behavioral Aspects. G. Hohmann, M.M. Robbins & C. Boesch (eds). Cambridge: Cambridge University Press, pp. 311–333.

Marshall, A.J. & Wrangham, R. 2007. Evolutionary consequences of fallback foods. *International Journal of Primatology* 28: 1219–1235.

Marshall, A., Ancrenaz, M., Brearly, F., *et al.* 2009a. The effects of forest phenology and floristics on populations of Bornean and Sumatran orangutans. In *Orangutans: Geographic Variation in Behavioral Ecology and Conservation*, S.A. Wich, S.S.U. Atmoko, T.M. Setia & C.P. van Schaik (eds). Oxford: Oxford University Press, pp. 97–118.

Marshall, A.J., Boyko, C.M., Feilen, K.L., Boyko, R.H. & Leighton, M. 2009b. Defining fallback foods and assessing their importance in primate ecology and evolution. *American Journal of Physical Anthropology* 140: 603–614.

Marshall, A.R., Lovett, J. & White, P. 2008. Selection of line-transect methods for estimating the density of group-living animals: lessons from the primates. *American Journal of Primatology* 70: 1–11.

Marshall, A.J., Nardiyono, Engstrom, L.M., *et al.* 2006. The blowgun is mightier than the chainsaw in determining population density of Bornean orangutans (*Pongo pygmaeus morio*) in the forests of East Kalimantan. *Biological Conservation* 129: 566–578.

Martin, B. 2004. Conservation of the Tonkin snub-nosed monkey (*Rhinopithecus avunculus*) and its habitat at Na Hang Nature Reserve during dam construction. In *Conservation of Primates in Vietnam*, T. Nadler, U. Streicher & H.T. Long (eds). Frankfurt: Frankfurt Zoological Society, pp. 103–105.

Martínez, J. 2014. *Behaviour and ranging patterns of the Endangered and endemic Bolivian titi monkeys Callicebus olallae and Callicebus modestus*. MSc dissertation. Oxford Brookes University, Oxford.

Martinez, J., Lopez-Strauss, H. & Wallace, R.B. 2011. Survey of the habitat of rare Bolivian monkeys aids the establishment of a Municipal Reserve. In *All The Worlds Primates*. Charlestown, RI: All the World's Primates.

Martínez, J., Mercado, N., Sainz, L. & Porcel, Z.R. 2010. Cebidae. In *Distribuición, ecología y conservación de los Mamíferos Medianos y Grandes de Bolivia*, R.B. Wallace, H. Gómes, Z.R. Porcel & D.I. Rumiz (eds). Santa Cruz de la Sierra: Centro de Ecología Difusión Simón I. Patiño, pp. 363–385.

Martínez, J. & Wallace, R.B. 2007. Further notes on the distribution of the Bolivian endemic titi monkeys, *Callicebus modestus* and *Callicebus olallae*. *Neotropical Primates* 14: 47–54.

Martínez, J. & Wallace, R.B. 2010. Pitheciidae. In *Distribuición, ecología y conservación de los Mamíferos Medianos y Grandes de Bolivia*, R.B. Wallace, H. Gómes, Z.R. Porcel & D.I. Rumiz (eds). Santa Cruz de la Sierra: Centro de Ecología Difusión Simón I. Patiño, pp. 305–330.

Martínez, J. & Wallace, R.B. 2011. First observations of terrestrial travel for Olalla's titi monkey (*Callicebus olallae*). *Neotropical Primates* 18: 49–52.

Martínez, J. & Wallace, R.B. 2013. New information about the distribution of *Callicebus* in Bolivia. *Ecología en Bolivia* 48: 57–62.

Martinez, M.L., Moreno-Casasola, P. & Vazquez, G. 1997. Effects of disturbance by sand movement and inundation by water on tropical dune vegetation dynamics. *Canadian Journal of Botany* 75: 2005–2014.

Martinez, O. 2010. Extensión de rango de distribución del momo Lachari *Callicebus aureipalatii* (Pitheciidae) para el Departamento de la Paz, Bolivia. *Neotropical Primates* 17: 24–27.

Martínez, V.G. & Gaona, S.O. 2006. Evaluación espacio-temporal de la vegetación y uso del suelo en la Reserva de la Biosfera Pantanos de Centla, Tabasco (1990–2000). *Investigaciones Geográficas* 59: 7–25.

Martinez Mollinedo, J. 2014. *Behaviour and ranging patterns of the Endangered and endemic Bolivian titi monkeys*. MSc thesis, Oxford, Oxford Brookes University.

Martínez-Mota, R., Valdespino, C., Sánchez-Ramos, M.A. & Serio-Silva, J.C. 2007. Effects of forest fragmentation on the physiological stress response of black howler monkeys. *Animal Conservation* 10: 374–379.

Maruhashi, T., Saito, C. & Agetsuma, N. 1998. Home range structure and intergroup competition for land of Japanese macaques in evergreen and deciduous forests. *Primates* 39: 291–301.

Mascarenhas, B.M. & Puorto, G. 1988. Non-volant mammals rescued at Tucuruí dam in the Brazilian Amazon. *Primate Conservation* 9: 91–93.

Massaro, L., Liu, Q., Visalberghi, E. & Fragaszy, D. 2012.Wild bearded capuchin (*Sapajus libidinosus*) select hammer tools on the basis of both stone mass and distance from the anvil. *Animal Cognition* 15: 1065–1074.

Masseti, M. & Veracini, C. 2010. The first record of Marcgrave's capuchin in Europe: South American monkeys in Italy during the early sixteenth century. *Archives of Natural History* 37: 91–101.

Matson, S.D., Rook, L., Oms, O. & Fox, D.L. 2012. Carbon isotopic record of terrestrial ecosystems spanning the Late Miocene extinction of *Oreopithecus bambolii*, Baccinello Basin (Tuscany, Italy). *Journal of Human Evolution* 63: 127–139.

Matsuda, I. 2008. *Feeding and ranging behaviors of proboscis monkey Nasalis larvatus in Sabah, Malaysia*. PhD dissertation, Graduate School of Environmental Earth Science, Hokkaido University.

Matsuda, I., Akiyama, Y., Tuuga, A., Bernard, H. & Clauss, M. 2014. Daily feeding rhythm in proboscis monkeys: a preliminary comparison with other non-human primates. *Primates* 55: 313–326.

Matsuda, I., Ancrenaz, M., Akiyama, Y., *et al.* 2014. Natural licks are required for large terrestrial mammals in a degraded riparian forest, Sabah, Borneo, Malaysia. *Ecological Research* 30: 191–195.

Matsuda, I., Kubo, T., Tuuga, A. & Higashi, S. 2010a. A Bayesian analysis of the temporal change of local density of proboscis monkeys: implications for environmental effects on a multilevel society. *American Journal of Physical Anthropology* 142: 235–245.

Matsuda, I., Otani, Y., Bernard, H., Wong, A., Tuuga, A. 2016. Primate survey in a Bornean flooded forest: evaluation of best approach and best timing. *Mammal Study* 41: 101–106.

Matsuda, I., Sha, J.C., Ortmann, S., *et al.* 2015. Excretion patterns of solute and different-sized particle passage markers in foregut-fermenting proboscis monkey (*Nasalis larvatus*) do not indicate an adaptation for rumination. *Physiology and Behavior* 149: 45–52.

Matsuda, I., Tuuga, A., Akiyama, Y. & Higashi, S. 2008. Selection of river crossing location and sleeping site by proboscis monkeys (*Nasalis larvatus*) in Sabah, Malaysia. *American Journal of Primatology* 70: 1097–1101.

Matsuda, I., Tuuga, A. & Higashi, S. 2009b. Ranging behavior of proboscis monkeys in a riverine forest with special reference to ranging in inland forest. *International Journal of Primatology* 30: 313–325.

Matsuda, I., Tuuga, A. & Bernard, H. 2011. Riverine refuging by proboscis monkeys (*Nasalis larvatus*) and sympatric primates: implications for adaptive benefits of the riverine habitat. *Mammalian Biology* 76: 165–171.

Matsuda, I., Tuuga, A., Bernard, H. & Furuichi, T. 2012. Inter-individual relationships in proboscis monkeys: a preliminary comparison with other non-human primates. *Primates* 53: 13–23.

Matsuda, I., Tuuga, A., Bernard, H., Sugau, J. & Hanya, G. 2013. Leaf selection by two Bornean colobine monkeys in relation to plant chemistry and abundance. *Scientific Reports* 3: 1873.

Matsuda, I., Tuuga, A. & Higashi, S. 2009a. The feeding ecology and activity budget of proboscis monkeys. *American Journal of Primatology* 71: 478–492.

Matsuda, I., Tuuga, A. & Higashi, S. 2010a. Effects of water level on sleeping-site selection and intergroup association in proboscis monkeys: why do they sleep alone inland on flooded days? *Ecological Research* 25: 475–482.

Matsuzawa, T. 2001. Primate foundations of human intelligence: a view of tool use in nonhuman primates and fossil hominids. In *Primate Origins of Human Cognition and Behavior*, T. Matsuzawa (ed.). Tokyo: Springer, pp. 3–25.

Matthews, A. & Matthews, A. 2002. Distribution, population density and status of sympatric cercopithecids in the Campo-Ma'an area, southwestern Cameroon. *Primates* 43: 155–168.

Matthews, H. 2005. *The effects of hunting on a primate community in the Peruvian Amazon*. MSc disseration, Winthrop University, Rock Hill, SC, USA.

Matthies, D., Bräuer, I., Maibom, W. & Tscharntke, T. 2004. Population size and the risk of local extinction: empirical evidence from rare plants. *Oikos* 105: 481–488.

Mau, M., Johann, A., Sliwa, A., Hummel, J. & Südekum, K.-H. 2011. Morphological and physiological aspects of digestive processes in the graminivorous primate *Theropithecus gelada*: a preliminary study. *American Journal of Primatology* 73: 449–457.

Maynard, L.A., Loosli, J.K., Hintz, H.F. & Warner, R.G. 1979. *Animal Nutrition*. New York: McGraw-Hill.

Mayor, P. & Bowler, M. 2015. Low birthrates and high levels of female reproductive inactivity may characterize the reproductive biology of wild Peruvian red uakaris (*Cacajao calvus ucayalii*). *Journal of Medical Primatology* 44: 27–34.

Mayor, P., Bowler, M. & López-Plana, C. 2013. Functional morphology of the female genital organs in the Peruvian red uakari monkey (*Cacajao calvus ucayalii*). *American Journal of Primatology* 75: 545–554.

Mbora, D.N.M. & Butynski, T.M. 2009. Tana River red colobus *Procolobus rufomitratus*. In *Primates in Peril: The World's 25 Most Endangered Primates 2008–2010*, R.A. Mittermeier, J. Wallis, A.B. Rylands, *et al.* (eds). *Primate Conservation* 24: 1–57.

Mbora, D.N.M. & Meikle, D.B. 2004a. Forest fragmentation and the distribution, abundance and conservation of the Tana River red colobus (*Procolobus rufomitratus*). *Biological Conservation* 118: 67–77.

Mbora, D.N.M. & Meikle, D.B. 2004b. The value of unprotected habitat in conserving the Critically Endangered Tana River red colobus (*Procolobus rufomitratus*). *Biological Conservation* 120: 91–99.

McAllister, D.E., Craig, J.F., Davidson, N., Delany, S. & Seddon, M. 2001. *Biodiversity Impacts of Large Dams. Background Paper Nr. 1*. Gland, Switzerland: IUCN/UNEP/WCD.

McCann, C., Williams-Guillén, K., Koontz, F., *et al.* 2003. Shade coffee plantations as wildlife refuge for mantled howler monkeys (*Alouatta palliata*) in Nicaragua. In *Primates in Fragments*, L. March & C. Chapman (eds). Boston, MA: Springer, pp. 321–341.

McCartney, M.P., Sullivan, C. & Acreman, M.C. 2001. *Ecosystem Impacts of Large Dams. Background Paper Nr. 2*. Gland, Switzerland: IUCN/UNEP/WCD.

McConkey, K.R. & Chivers, D.J. 2004. Low mammal and hornbill abundance in the forests of Barito Ulu, central Kalimantan, Indonesia. *Oryx* 38: 439–447.

McGarigal, K. & Marks, B.J. 1995. FRAGSTATS: spatial pattern analysis program for quantifying landscape structure. *General Technical Report PNW-GTR-351*. Portland, Oregon: US Department of Agriculture, Forest Service, Pacific Northwest Research Station.

McGraw, W.S. 2005. Update on the search for Miss Waldron's red colobus monkey. *International Journal of Primatology* 26: 605–619.

McGraw, W. & Galat-Luong, A. 2011. *Procolobus verus*. In *All the World's Primates*, N. Rowe & M. Myers (eds). Charlestown, RI: Primate Conservation Inc.

McGrew, W.C. 1992. *Chimpanzee Material Culture: Implications for Human Evolution*. Cambridge: Cambridge University Press.

McGrew, W.C. 1994. Tools compared: the materials of culture. In *Chimpanzee Cultures*, R.W. Wrangham, W.C. McGrew, F.B.M. deWall & P.G. Heltne (eds). Cambridge, MA: Harvard University Press, pp. 25–40.

McGrew, W.C. 2007. Savanna chimpanzees dig for food. *Proceedings of the National Academy of Sciences* 104: 19167–19168.

McGrew, W.C., Marchant, L.F. & Phillips, C.A. 2009. Standardised protocol for primate faecal analysis. *Primates* 50: 363–366.

McIntyre, S. & Hobbs, R. 1999. A framework for conceptualizing human effects on landscapes and its relevance to management and research models. *Conservation Biology* 13: 1282–1292.

McKey, D.B., Gartlan, J.S., Waterman, P.G. & Choo, G.M. 1981. Food selection by black colobus monkeys (*Colobus satanas*) in relation to plant chemistry. *Biological Journal of the Linnean Society* 16: 115–146.

McLennan, M.R. 2008. Beleaguered chimpanzees in the agricultural district of Hoima, western Uganda. *Primate Conservation* 23: 45–54.

McLennan, M.R. & Plumptre, A.J. 2012. Protected apes, unprotected forest: composition, structure and diversity of riverine forest fragments and their conservation value in Uganda. *Tropical Conservation Science* 5: 79–103.

McMorrow, J. & Talip, M.A. 2001. Decline of forest area in Sabah, Malaysia: Relationship to state policies, land code and land capability. *Global Environmental Change* 11: 217–230.

McNeilage, A. 2001. Diet and habitat use of two mountain gorilla groups in contrasting habitats in the Virungas. In *Mountain Gorillas: Three Decades of Research at Karisoke*, M.M. Robbins, P. Sicotte & K.J. Stewart (eds). New York: Cambridge University Press, pp. 265–292.

Meade, R.H. 1996. River-sediment inputs to major deltas. In *Sea-Level Rise and Coastal Subsidence: Causes, Consequences, and Strategies*, J. Milliman & B.U. Haq (eds). London: Kluwer, pp. 63–85.

Medina, LK. 2005. Ecotourism and certification: confronting the principles and pragmatics of socially responsible tourism. *Journal of Sustainable Tourism* 13: 281–295.

Medley, K.E. 2002. Primate conservation along the Tana River, Kenya: an examination of the forest habitat. *Conservation Biology* 7: 109–121.

Meier, B., Albignac, R., Peyriéras, A., Rumpler, Y. & Wright, P. 1987. A new species of *Hapalemur* (primates) from southeast Madagascar. *Folia Primatologica* 48: 211–215.

Meijaard, E. 1997. The importance of peat swamp forest for the conservation of orang-utans (*Pongo pygmaeus pygmaeus*) *in Kalimantan, Indonesia*. In *Biodiversity and Sustainability of Tropical Peatlands*, J.O. Rieley & S.E. Page (eds). Cardigan: Samara Publishing, pp. 243–254.

Meijaard, E. & Nijman, V. 2000a. The local extinction of the proboscis monkey *Nasalis larvatus* in Pulau Kaget Nature Reserve, Indonesia. *Oryx* 34: 66–70.

Meijaard, E. & Nijman, V. 2000b. Distribution and conservation of the proboscis monkey (*Nasalis larvatus*) in Kalimantan, Indonesia. *Biological Conservation* 92: 15–24.

Meijaard, E. & Nijman, V. 2003. Primate hotspots on Borneo: predictive value for general biodiversity and the effects of taxonomy. *Conservation Biology* 17: 725–732.

Meijaard, E. & Wich, S.A. 2007. Putting orangutan population trends into perspective. *Current Biology* 17: R540. https://doi.org/10.1016/j.cub.2007.05.016.

Meijaard, E., Welsh, A., Ancrenaz, M., *et al.* 2010. Declining orangutan encounter rates from Wallace to the present suggest the species was once more abundant. *PLoS ONE* 5: e12042.

Meittinen, J., Hooijer, A., Shi, C., *et al.* 2012a. Extent of industrial plantations on Southeast Asian peatlands in 2010 with analysis of historical expansion and future projections. *GCB Bioenergy* 4: 908–918.

Meittinen, J., Shi, C. & Liew, S. 2012b. Two decades of destruction in Southeast Asia's peat swamps. *Frontiers in Ecology and Environment* 10: 124–128.

Melish, R. & Dirgayusa, I.W.A. 1996. Notes on the grizzled leaf monkey (*Presbytis comata*) from two nature reserves in West Java, Indonesia. *Asian Primates* 6: 5–11.

Melo, F.R., Buss, G., de Assis, J., *et al.* 2008. Estado da arte da primatologia no Brasil: um retrospecto dos últimos 15 anos e perspectivas futuras. In *Retrospecto e Perspectivas Futuras da Zoologia no Brasil*, org. Sociedade Brasileira de Zoologia, Curitiba, PR: Editora da UFPR, pp. 1–47.

Mendes, S.L., de Oliveira, M.M., Mittermeier, R.A. & Rylands, A.B. 2008. *Brachyteles arachnoides*. In *IUCN Red List of Threatened Species*. Version 2012.2. Gland, Switzerland: IUCN. www.iucnredlist.org.

Mendes Pontes, A.R. 1997. Habitat partitioning among the primates of Maraca island, Roraima, Brazilian Amazonia. *International Journal of Primatology* 18: 131–157.

Mendes Pontes, A.R. 1999. Environmental determinants of primate abundance in Maracá island, Roraima, Brazilian Amazonia. *Journal of Zoology* 247: 189–199.

Mendes Pontes, A.R. 2000. *Ecology of a mammal community in a seasonally-dry forest in Roraima, Brazilian Amazonia*. Unpubl. PhD thesis. University of Cambridge, Cambridge.

Mendes Pontes, A.R. 2004. Ecology of a community of mammals in a seasonally dry forest in Roraima, Brazilian Amazonia. *Mammalian Biology* 69: 319–336.

Mendes Pontes, A.R. & Chivers, D.J. 2002. Abundance, habitat use and conservation of the olingo *Bassaricyon* sp. in Maracá Ecological Station, Roraima, Brazilian Amazonia. *Studies on Neotropical Fauna & Environment* 37: 105–109.

Mendes Pontes, A.R. & Chivers, D.J. 2007. Peccary movements as determinants of the movements of large cats in Brazilian Amazonia. *Journal of Zoology* 273: 257–265.

Mendes Pontes, A.R., Paula, M.D. & Magnusson, W. 2012. Low primate diversity and abundance in northern Amazonia and its implications for conservation. *Biotropica* 44: 834–839.

Méndez-Cárdenas, M., Randrianambinina, B., Rabesandratana, A., Rasoloharijaona, S. & Zimmermann, E. 2008. Geographic variation in loud calls of sportive lemurs (*Lepilemur* ssp.) and their implications for conservation. *American Journal of Primatology* 70: 828–838.

Méndez-Carvajal, P.G. 2005. Population survey of the Azuero howler monkey (*Alouatta palliata trabeata*) in Herrera Province, Republic of Panama. *Neotropical Primates* 13: 1–6.

Méndez-Carvajal, P.G. 2014. The Orion Camera System, a new method for deploying camera traps in tree canopy to study arboreal primates and other mammals: a case study in Panama. *Mesoamericana* 18: 9–23.

Menezes, M.P.M., Berger, U. & Mehlig, U. 2008. Mangrove vegetation in Amazonia: a review of studies from the coast of Pará and Maranhão states, north Brazil. *Acta Amazonica* 38: 403–420.

Mercado, N. & Wallace, R.B. 2010. Distribución de primates en Bolivia y áreas prioritarias para su conservación. *Tropical Conservation Science* 3: 200–217.

Mercier, J.-L. & Wilmé, L. 2013. The Eco-Geo-Clim model: explaining Madagascar's endemism. *Madagascar Conservation & Development* 8: 63–68.

Merker, S., Yustian, I. & Muhlenberg, M. 2005. Responding to forest degradation: altered habitat use by Dian's tarsier *Tarsius dianae* in Sulawesi, Indonesia. *Oryx* 39: 189–195.

Mertes, L.A., Daniel, D.L., Melack, J.M., *et al.* 1995. Spatial patterns of hydrology, geomorphology, and vegetation on the floodplain of the Amazon River in Brazil from a remote sensing perspective. *Geomorphology* 13: 215–232.

Mertl-Millhollen, A.S., Blumenfeld-Jones, K., Raharison, S.M., Tsaramanana, D.R. & Rasamimanana, H. 2011. Tamarind tree seed dispersal by ring-tailed lemurs. *Primates* 52: 391–396.

Mertl-Millhollen, A.S., Rambeloarivony, H., Miles, W., *et al.* 2006. The influence of tamarind tree quality and quantity on *Lemur catta* behavior. In *Ringtailed Lemur Biology: Lemur catta in Madagascar*, A. Jolly, N. Koyama, H.R. Rasamimanana & R.W. Sussman (eds). New York: Springer, pp. 102–118.

Michalski, F. & Peres, C.A. 2005. Anthropogenic determinants of primate and carnivore local extinctions in a fragmented forest landscape of southern Amazonia. *Biological Conservation* 124: 383–396.

Michener, W.K., Blood, E.R., Bildstein, K.L., Brinson, M.M. & Gardner, L.R. 1997. Climate change, hurricanes and tropical storms and rising sea levels in coastal wetlands. *Ecological Applications* 7: 770–801.

Midha, N. & Mathur, P.K. 2010. Assessment of forest fragmentation in the conservation priority Dudhwa landscape, India using FRAGSTATS computed class level metrics. *Journal of the Indian Society of Remote Sensing* 38: 487–500.

Miettinen, J., Shi, C. & Liew, S. 2012. Two decades of destruction in Southeast Asia's peat swamps. *Frontiers in Ecology & Environment* 10: 124–128.

Milanowski, D.J., Winter, R.E., Elvin-Lewis, M.P. & Lewis, W.H. 2002. Geographic distribution of three alkaloid chemotypes of *Croton lechleri*. *Journal of Natural Products* 65: 814–819.

Millennium Ecosystem Assessment. 2006. Millennium Ecosystem Assessment Synthesis Reports. Washington, DC: World Resources Institute. http://millenniumassessment.org/en/Synthesis.html.

Milton, K.D. 1999. Nutritional characteristics of wild primate foods: do the diets of our closest living relatives have lessons for us? *Nutrition* 15: 488–498.

Milton, K. & Mittermeier, R.A. 1977. A brief survey of the primates of Coiba Island, Panama. *Primates* 18: 931–936.

Ministry of Education, Culture, Sports, Science and Technology, Japan. 2010. Standard Tables of Food Composition in Japan 2010. www.mext.go.jp/b_menu/shingi/gijyutu/gijyutu3/houkoku/1298713.htm.

Miranda, J.M.D. 2009. *Comportamentos sociais, relações de dominância e confrontos intergrupais em* Alouatta caraya *(Humboldt, 1812) na Ilha Mutum, alto rio Paraná, Brasil*. PhD thesis, Biológicas da Universidade Federal do Paraná, Curitiba, Brazil.

Mitani, J.C. 1989. Orangutan activity budgets: monthly variations and the effects of body size, parturition, and sociality. *American Journal of Primatology* 18: 87–100.

Mitani, M., Yamagiwa, J., Oko, R.A., *et al.* 1993. Approaches in density estimates and reconstruction of social groups in a western lowland gorilla population in the Ndoki Forest, northern Congo. *Tropics* 2: 219–229.

Mitchell, C.L. 1990. *The ecological basis for female social dominance: a behavior study of the squirrel monkey* (Saimiri sciureus) *in the wild*. PhD dissertation, Princeton University, Princeton, NJ.

Mitchell, S.A. 2013. The status of wetlands, threats and the predicted affect of global climate change: the situation in sub-Saharan Africa. *Aquatic Sciences* 75: 95–112.

Mitra, S. 2011. *A Pictorial Guide to Non-Human Primates of India*. Kolkata: Naturism.

Mitsch, W.J. & Hernandez, M.E. 2013. Landscape and climate change threats to wetlands of North and Central America. *Aquatic Sciences* 75: 133–149.

Mitsch, W.J. & Gosselink, J.G. 2007. *Wetlands, 4th Edition*. Hoboken, NJ: John Wiley & Sons, Inc.

Mittermeier, R.A. & Cheney, D.L. 1987. Conservation of primates and their habitats. In *Primate Societies*, B.B. Smuts, D.L., Cheney, R., Seyfarth, R.W., Wrangham & T.T. Struhsaker (eds). Chicago, IL: Chicago University Press, pp. 477–490.

Mittermeier, R.A. & Coimbra-Filho, A.F. 1977. Primate conservation in Brazilian Amazonia. In *Primate Conservation*, H.S.H. Prince Rainer III & G.H. Bourne (eds). New York: Academic Press, pp. 117–166.

Mittermeier, R.A. & van Roosmalen, M.G.M. 1981. Preliminary observations on habitat utilization and diet in eight Surinam monkeys. *Folia Primatologica* 36: 1–39.

Mittermeier, R.A., Coimbra-Filho, A.F., Kierulff, M.C.M., *et al.* 2007. *Monkeys of the Atlantic Forest of Eastern Brazil Pocket Identification Guide*. Conservation International Tropical Pocket Guide, Series #3. Arlington, VA: Conservation International.

Mittermeier, R.A., de Gusmao Camara, I., Pádua, M.T.J. & Blanck, J. 1990. Conservation in the Pantanal of Brazil. *Oryx* 24: 103–112.

Mittermeier, R.A., Myers, N., Thomsen, J.B., da Fonesca, G.A.B. & Olivieri, S. 1998. Biodiversity hotspots and major tropical wilderness areas: approaches to setting conservation priorities. *Conservation Biology* 12: 516–520.

Mittermeier, R.A., Richardson, M., Louis, E.E., *et al.* 2010. *Lemurs of Madagascar*. 3rd. edition. Tropical Field Guide Series, Washington, DC: Conservation International.

Mittermeier, R.A., Schtwitzer, C., Rylands, A.B., *et al.* 2012. Primates in Peril: The World's 25 Most Endangered Primates 2012–2014. *IUCN/SSC Primate Specialist Group. Bristol Conservation & Science Foundation.*

Mittermeier, R.A., Wallis, J., Rylands, A.B., *et al.* 2009. Primates in Peril: The World's 25 Most Endangered Primates 2008–2010. *Primate Conservation* 24: 1–57.

MMA (Ministério do Meio Ambiente). 2004. SNUC – *Sistema Nacional de Unidades De Conservação da Natureza. Lei No 9.985, de 18 de Julho de 2000*, Decreto No 4.340, de 22 de Agosto de 2002. Brazil: MMA/SBF.

MMA (Ministério do Meio Ambiente). 2012. Cadastro Nacional de Unidades de Conservação. Brazil: MMA/SBF.

Mmom, P.C. & Mbee, D.M. 2013. Population pressure and forest resources depletion in Gele-Gele Forest Reserve of Edo State, Nigeria. *International Journal of Physical & Human Geography* 1: 31–42.

Moffat, D. & Lindén, O. 1995. Perception and reality: assessing priorities for sustainable development in the Niger River Delta. *Ambio* 24: 527–538.

Mohan, B.S. & Hosetti, B.B. 1999. Aquatic plants for toxicity assessment. *Environmental Research Section A* 81: 259–274.

Mohneke, M. & Fruth, B. 2008. Bonobo (*Pan paniscus*) density estimation in the SW-Salonga National Park, Democratic Republic of Congo: common methodology revisited. In *The Bonobos*, T. Furuichi & J. Thompson (eds). New York: Springer, pp. 151–166.

Mohnot, S.M. 1980. The status of Indian non-human primates. In *Proceedings of the Workshop on Wildlife Ecology, Dehradun, January 1978*, Director, Zoological Survey of India (ed.). Calcutta: Zoological Survey of India, pp. 111–117.

Moinde-Fockler, N.N., Oguge, N.O., Karere, G.M., Otina, D. & Suleman, M.A. 2006. Human and natural impacts on forests along lower Tana River, Kenya: implications towards conservation and management of endemic primate species and their habitat. *Biodiversity & Conservation* 16: 1161–1173.

Monographie Régionale Alaotra-Mangoro. 2007. Monographie de la Région Alaotra-Mangoro Année 2007. Madagascar: SRE Ambatondrazaka.

Montero, J.C., Piedade, M.T.F. & Wittmann, F. 2014. Floristic variation across 600 km of inundation forests (igapó) along the Negro River, central Amazonia. *Hydrobiologia* 729: 229–246.

Montgomery, S. 2009. *Spell of the Tiger: The Man Eaters of Sundarbans*. Vermont: Chelsea Green Publishing.

Moore, D., Dore, J. & Gyawali, D. 2010. The World Commission on Dams + 10: Revisiting the large dam controversy. *Water Alternatives* 3: 3–13.

Moore, J. 1992. 'Savanna' Chimpanzees, In *Human Origins*, S. Matano, R.H. Tuttle, H. Ishida & M. Goodman (eds). Topics in Primatology, Vol. 1, Toyoko: University of Tokyo Press, pp. 99–118.

Mora-Fernandez C., Catellanos-Castro C., Cardona-Cardozo A., Pinzón-Perez L. & Vargas-Ríos O. 2011. Geología, geomorfología, clima y vegetación. In *Mamíferos, reptiles y ecosistemas del Bloque Cubiro (Casanare): Educación ambiental para la conservación*, T. León Sicard (ed.). Bogotá, Colomba: Instituto de estudios ambientales Universidad Nacional de Colomba, Alange Energy Corp., pp. 49–74.

Moraes, B.L., Souto, A.S. & Schiel, N. 2014. Adaptability in stone tool use by wild capuchin monkeys (*Sapajus libidinosus*). *American Journal of Primatology* 76: 967–977.

Morales-Hernandez, K. 2002. Wild populations of spider monkeys (*Ateles geoffroyi*) in El Salvador, Central America. *Neotropical Primates* 10: 153–154.

Moreira, L.L.B. 2009. *Primatas das Serras das Lontras e Javi: Estado das populações e seu papel na conservação regional da comunidade de primatas do Sul da Bahia*. MSc dissertation, Universidade Estadual de Santa Cruz, Bahia, Brazil.

Moreland, R.B., Richardson, M.E., Lamberski, N. & Long, J.A. 2001. Characterizing the reproductive physiology of the male southern black howler monkey, *Alouatta caraya*. *Journal of Andrology* 22: 395–403.

Morellato, L.P.C., Camargo, M.G.G., D'Eça-Neves, F.F., *et al.* 2010. The influence of sampling method, sample size, and frequency of observations on plant phenological patterns and interpretation in tropical forest trees. In *Phenological Research*, I.L. Hudson & M.R. Keatly (eds). Dordrecht, The Netherlands: Springer, pp. 99–121.

Morgan, D., Sanz, C., Onononga, J.R. & Strindberg, S. 2006. Ape abundance and habitat use in the Goualougo Triangle, Republic of Congo. *International Journal of Primatology* 27: 147–179.

Mori, A. 1979. An experiment on the relation between the feeding speed and the caloric intake through leaf eating in Japanese monkeys. *Primates* 20: 185–195.

Morino, L. 2010. Clouded leopard predation on a wild juvenile siamang. *Folia Primatologica* 81: 362–368.

Morison, J.I.L., Piedade, M.T.F., Müller, E., *et al.* 2000. Very high productivity of the C4 aquatic grass *Echinochloa polystachya* in the Amazon floodplain confirmed by net ecosystem CO_2 flux measurements. *Oecologia* 125: 400–411.

Morley, R.J. 1981. Development and vegetation dynamics of a lowland ombrogenous peat swamp in Kalimantan Tenga, Indonesia. *Journal of Biogeography* 8: 383–404.

Morrogh-Bernard, H., Husson, S.J., Page, S.E. & Rieley, J.O. 2003. Population status of the Bornean orang-utan (*Pongo pygmaeus*) in the Sebangau peat swamp forest, Central Kalimantan, Indonesia. *Biological Conservation* 110: 141–152.

Morrogh-Bernard, H.C., Morf, N.V., Chivers, D.J. & Krützen, M. 2010. Dispersal patterns of orang-utans (*Pongo spp.*) in a Bornean peat-swamp forest. *International Journal of Primatology* 32: 362–376.

Moura, A.C., de 2004. *The capuchin monkey and the Caatinga dry forest: a hard life in a harsh habitat*. PhD thesis, Cambridge University, Cambridge, UK.

de Moura, A.C. 2007. Primate group size and abundance in the Caatinga dry forest, northeastern Brazil. *International Journal of Primatology* 28: 1279–1297.

Moura, A.C., de & Lee, P.C. 2004. Capuchin tool use in Caatinga dry forest. *Science* 306: 1909.

Mourthé, Í. & Barnett, A.A. 2014. Crying tapir: the functionality of errors and accuracy in predator recognition in two Neotropical high-canopy primates. *Folia Primatologica* 85: 379–398.

Mowforth, D. & Munt, I. 1998. *Tourism and Sustainability: New Tourism in the Third World*. London: Routledge.

Mowry, C.B., Decker, B.S. & Shure, D.J. 1996. The role of phytochemistry in dietary choices of Tana River red colobus monkeys (*Procolobus badius rufomitratus*). *International Journal of Primatology* 17: 63–84.

Moyà-Solà, S., Köhler, M. & Rook, L. 1999. Evidence of hominid-like precision grip capability in the hand of the Miocene ape *Oreopithecus*. *Proceedings of the National Academy of the Sciences of the Unites States of America* 96: 313–317.

Muehlenbein, M.P, Ancrenaz, M., Sakong, R., *et al.* 2012. Ape conservation physiology: fecal glucocorticoid responses in wild *Pongo pygmaeus morio* following human visitation. *PLoS ONE* 7: e33357.

Mukherjee, A.K. 1980. Wildlife in the Sunderban, West Bengal. In *Proceedings of the Workshop on Wildlife Ecology, Dehradun, January 1978*, Director, Zoological Survey of India (ed.). Calcutta: Zoological Survey of India, pp. 123–127.

Mukherjee, A.K. & Gupta, S. 1965. Habits of the rhesus macaque, *Macaca mulatta* (Zimmerman) in the Sundarbans, 24-Parganas, West Bengal. *Journal of the Bombay Natural History Society* 62: 145–146.

Mukherjee, S. 2004. *Ecological investigations on mangroves of the Sundarban Tiger Reserve in West Bengal (India) with special reference to effective conservation through management practices*. PhD thesis. Kalyani: University of Kalyani, India.

Mukherjee, S. 2006. Rhesus monkey and chital in the Sundarbans. *Banabithi, Wildlife Issue*, pp. 33–35.

Mukherjee, S. & Sen Sarkar, N. 2013. The range of prey size of the royal Bengal tiger of Sundarbans. *Journal of Ecosystems* 2013: 351756.

Mulavwa, M.N., Furuichi, T., Yangozene, K., *et al.* 2008. Seasonal changes in fruit production and party size of bonobos at Wamba. In *The Bonobos*, T. Furuichi & J. Thompson (eds). New York: Springer, pp. 121–134.

Mulavwa, M.N., Yangozene, K., Yamba-Yamba, M., *et al.* 2010. Nest groups of wild bonobos at Wamba: selection of vegetation and tree species and relationships between nest group size and party size. *American Journal of Primatology* 72: 575–586.

Mulero-Pázmány, M., Stolper, R., van Essen, L.D., Negro, J.J. & Sassen, T. 2014. Remotely piloted aircraft systems as a rhinoceros anti-poaching tool in Africa. *PLoS ONE* 9: e8387.

Muller, R., Pacheco, P. & Montero, J.C. 2014. El contexto de la deforestación y degradación de los bosques en Bolivia: causas, actores e instituciones. *Documentos Ocacionales* 101: 1–102.

Muller-Dombois, D. & Ellenberg, H. 1974. *Aims and Methods of Vegetation Ecology*. New York: John Wiley & Sons.

Munds, R.A., Ali, R., Nijman, V., Nekaris, K.A.I. & Goosens, B. 2013. Living together in the night: abundance and habitat use of sympatric and allopatric populations of slow lorises and tarsiers. *Endangered Species Research* 22: 269–277.

Munds, R.A., Nekaris, K.A.I. & Ford, S.M. 2013. Taxonomy of the Bornean slow loris, with new species *Nycticebus kayan* (Primates, Lorisidae). *American Journal of Primatology* 75: 46–56.

Murai, T. 2004. Social behaviors of all-male proboscis monkeys when joined by females. *Ecological Research* 19: 451–454.

Musila, W., Todt, H., Uster, D. & Dalitz, H. 2010. Is geodiversity correlated to biodiversity? A case study of the relationship between spatial heterogeneity of soil resources and tree diversity in a western Kenyan rainforest. In *African Biodiversity*, B.A. Huber, B.J. Sinclair & K.-H. Lampe (eds). New York: Springer, pp. 405–441.

Muthuri, F.M., Jones, M.B. & Imbamba, S.K. 1989. Primary productivity of papyrus (*Cyperus papyrus*) in a tropical swamp; Lake Naivasha, Kenya. *Biomass* 18: 1–14.

Mutschler, T. 1999. Folivory in a small-bodied lemur: the nutrition of the Alaotran gentle lemur (*Hapalemur griseus alaotrensis*). In *New Directions in Lemur Studies*. H Rasamimanana, B. Rakotosamimanana, J.U. Ganzhorn & S.M. Goodman (eds). New York: Kluwer Academic/Plenum Publishing, pp. 221–239.

Mutschler, T. 2002. Alaotran gentle lemur: some aspects of its behavioral ecology. *Evolutionary Anthropology* 11: 101–104.

Mutschler, T. & Feistner, A.T. 1995. Conservation status and distribution of the Alaotran gentle lemur *Hapalemur griseus alaotrensis*. *Oryx* 29: 267–274.

Mutschler, T., Feistner, A.T.C. & Nievergelt, C.M. 1998. Preliminary field data on group size, diet and activity in the Alaotran gentle lemur *Hapalemur griseus alaotrensis*. *Folia Primatologica* 69: 325–330.

Mutschler, T., Nievergelt, C.M. & Feistner, A.T. 1999. Social organization of the Alaotran gentle lemur (*Hapalemur griseus alaotrensis*). *American Journal of Primatology* 50: 9–24.

Mutschler, T., Randrianarisoa, A.J. & Feistner, A.T.C. 2001. Population status of the Alaotran gentle lemur *Hapalemurgriseus alaotrensis*. *Oryx* 35: 152–157.

Mwanza, N. 2003. Confirmation of bonobo population around Lac Tumba. *Pan Africa News* 10: 29–31.

Myers, N., Mittermeier, R.A., Mittermeier, C.G., da Fonseca, G.A.B. & Kent, J. 2000. Biodiversity hotspots for conservation priorities. *Nature* 403: 853–858.

Myers Thompson, J.A. 1997. *The history, taxonomy and ecology of the bonobo (Pan paniscus Schwarz, 1929) with a first description of a wild population living in a forest/savanna mosaic habitat*. PhD thesis, University of Oxford, Oxford, UK.

Nahallage, C.A.D. & Huffman, M.A. 2008. Comparison of stone handling behavior in two macaque species: implications for the role of phylogeny and environment in primate cultural variation. *American Journal of Primatology* 70: 1124–1132.

Nahallage, C.A.D. & Huffman, M.A. 2012. Stone handling behavior in rhesus macaques (*Macaca mulatta*), a behavioral propensity for solitary object play with Japanese macaques. *Primates* 53: 71–78.

Naiman, R.J. & Décamps, H. 1997. The ecology of interfaces: riparian zones. *Annual Review of Ecology & Systematics* 28: 621–658.

Naiman, R.J., Bilby, R.E. & Bisson, P.A. 2000. Riparian ecology and management in the Pacific coastal rain forest. *BioScience* 50: 996–1011.

Naiman, R.J., Décamps, H. & Pollock, M. 1993. The role of riparian corridors in maintaining regional biodiversity. *Ecological Applications* 3: 209–212.

Nakagawa, N. 1989. Bioenergetics of Japanese monkeys (*Macaca fuscata*) on Kinkazan Island during winter. *Primates* 30: 441–460.

Nakai, E.S. 2007. *Fissão-fusão em Cebus nigritus: flexibilidade social como estratégia de ocupação em ambientes limitantes*. MSc dissertation, Instituto de Psicologia, Universidade de São Paulo, São Paulo, Brazil.

Nakamichi, M., Kato, E., Kojima, Y. & Itoigawa, N. 1998. Carrying and washing off grassroots by free-ranging Japanese macaques at Katsuyama. *Folia Primatologica* 69: 35–40.

Nakashima, Y., Iwata, Y., Ando, C., *et al.* 2013. Assessment of landscape-scale distribution of sympatric great apes in African rainforests: concurrent use of nest and camera-trap surveys. *American Journal of Primatology* 75: 1220–1230.

Nakashita, R., Hamada, Y., Hirasaki, E., Suzuki, J. & Oi, T. 2013. Characteristics of stable isotope signature of diet in tissues of captive Japanese macaques as revealed by controlled feeding. *Primates* 54: 271–281.

Nakhasathien, S. 1989. Chiew larn dam wildlife rescue operation. *Oryx* 23: 146–154.

Nantes, R.S. & Rímoli, J. 2008. *Ecologia e comportamento de bugios pretos* (Allouatta caraya, *Primates, Atelidae) em fragmentos florestais na margem esquerda do rio Aquidauana, Anastácio, Mato Grosso do Sul.* Unpublished report. Departamento de Biociências-Centro/CPAQ- Universidade Federal de Mato Grosso do Sul.

Nasi, R., Wunder, S. & Campos, A.J.J. 2002. *Forest ecosystem services: can they pay our way out of deforestation?* A discussion paper prepared for the GEF for the Forestry Roundtable to be held in conjunction with the UNFF II, Costa Rica on 11 March, 2002. Bogor, Indondesia: CIFOR.

Nascimento, A.T.A. & Schmidlin, L.A. 2011. Habitat selection by, and carrying capacity for, the Critically Endangered black-faced lion tamarin *Leontopithecus caissara* (Primates: Callitrichidae). *Oryx* 45: 288–295.

Naskar, K.R. & Guha Baksi, D.N. 1987. *Mangrove Swamps of the Sundarbans: An Ecological Perspective.* Kolkata, India: Naya Prakash.

Nater, A., Arora, N., Greminger, M.P., *et al.* 2012. Marked population structure and recent migration in the Critically Endangered Sumatran orangutan (*Pongo abelii*). *Journal of Heredity* 104: 2–13.

Nater, A., Mattle-Greminger, M.P., Nurcahyo, A., *et al.* 2017. Morphometric, Behavioral, and Genomic Evidence for a New Orangutan Species. *Current Biology* 27: 3576–3577.

Nater, A., Nietlisbach, P., Arora, N., *et al.* 2011. Sex-biased dispersal and volcanic activities shaped phylogeographic patterns of extant orangutans (genus: *Pongo*). *Molecular Biology & Evolution* 28: 2275–2288.

National Population Commission 2010. *2006 Population and Housing Census.* Volume 3. Abuja, Nigeria: National Population Commission.

Navarro, G. 2001. *Vegetación de Bolivia.* Centro de Ecología Simón I. Patiño, Cochabamba, Bolivia.

Navarro, G. & Maldonado, M. 2002. *Geografía ecológica de Bolivia: vegetación y ambientes acuáticos.* Centro de Ecología Simón I. Patiño, Cochabamba, Bolivia.

Navjot, S.S., Koh, L.P., Brook, B.W. & Ng, P.K.L. 2004. Southeast Asian biodiversity: an impending disaster. *Trends in Ecology and Evolution* 19: 654–660.

Ndiaye, P.I., Galat, G., Galat-Luong, A. & Nizinski, G. 2013. Note on the seasonal use of lowland and highland habitats by the West African Chimpanzee *Pan troglodytes verus* (Schwarz, 1934) (Primates: Hominidae): implications for its conservation. *Journal of Threatened Taxa* 5: 3697–3700.

Nebel, G.J., Dragsted, J. & Vega, A.S. 2001. Litter fall, biomass and net primary production in floodplain forests in the Peruvian Amazon. *Forest Ecology & Management* 150: 93–102.

Nebel, G., Kvist, L.P., Vanclay, J.K., *et al.* 2001. Structure and floristic composition of floodplain forests in the Peruvian Amazon. *Forest Ecology & Management* 150: 27–57.

NEDECO (Netherlands Engineering Consultants) 1961. *The Waters of the Niger Delta.* The Hague, The Netherlands: NEDECO.

Neiff, J.J. 1978. Fluctuaciones de la vegetación acuática en ambientes del valle de inundación del Paraná Medio. *Physis* 85: 41–53.

Neiff, J.J. 1990. Ideas para la interpretación ecológica del Paraná. *Interciencia* 15: 424–441.

Neilsen, N.H., Jacobsen, M.W., Graham, L.L.B., *et al.* 2011. Successful germination of seeds following passage through orang-utan guts. *Journal of Tropical Ecology* 27: 433–435.

Nekaris, K.A.I. 2001. Activity budget and positional behavior of the Mysore slender loris (*Loris tardigradus lydekkerianus*): implications for slow climbing locomotion. *Folia Primatologica* 72: 228–241.

Nekaris, K.A.I. 2013. Family Lorisidae (angwantibos, pottos and lorises). In *Handbook of the Mammals of the World: Volume 3 Primates*, R.A. Mittermeier, A.B. Rylands & D. Wilson (eds). Arlington, VA: Conservation International & Barcelona: Lynx Edicions, pp. 210–235.

Nekaris, K.A.I. & Bearder, S.K. 2007. The Lorisiform primates of Asia and mainland Africa. In *Primates in Perspective*, C.J. Campbell, A. Fuentes, K.C. MacKinnon, S.K. Bearder & R.M. Stumpf (eds). Oxford: Oxford University Press, pp. 24–45.

Nekaris, K.A.I. & Nijman, V. 2013. The ethics of conducting field research – do long-term great ape field studies help to conserve primates? In *Ethics in the Field: Contemporary Challenges*, J. Macclancy & A. Fuentes (eds). Oxford: Berghahn Books, pp. 108–123.

Nekaris, K.A.I. & Streicher, U. 2008a. *Nycticebus coucang.* In *IUCN Red List of Threatened Species.* Version 2014.3. Gland, Switzerland: IUCN. www.iucnredlist.org.

Nekaris, K.A.I & Streicher, U. 2008b. *Nycticebus menagensis.* In *IUCN Red List of Threatened Species.* Version 2014.3. Gland, Switzerland: IUCN. www.iucnredlist.org.

Nekaris, K.A.I., Blackham, G.V. & Nijman, V. 2008. Conservation implications of low encounter rates of five nocturnal primate species (*Nycticebus* spp.) in Asia. *Biodiversity & Conservation* 17: 733–747.

Nekaris, K.A.I., Moore, R.S., Rode, E.J. & Fry, B.G. 2013a. Mad, bad and dangerous to know: the biochemistry, ecology and evolution of slow loris venom. *Journal of Venomous Animals & Toxins, including Tropical Diseases* 19: doi 10.1186/1678-9199-19-21.

Nekaris, K.A.I., Shekelle, M., Wirdateti, R. E.J. & Nijman, V. 2013b. *Nycticebus javanicus.* In *IUCN Red List of Threatened Species.* Version 2014.3. Gland, Switzerland: IUCN. www.iucnredlist.org.

Nelson, S.V. 2005. Habitat requirements and the extinction of the Miocene ape, *Sivapithecus.* In *Interpreting the Past: Essays on Human, Primate and Mammal Evolution in Honor of David Pilbeam*, D.E. Lieberman, R.J. Smith & J. Kelley (eds). Boston, MA: Brill Academic, pp. 145–166.

Nelson, S.V. 2007. Isotopic reconstructions of habitat change surrounding the extinction of *Sivapithecus*, a Miocene hominoid, in the Siwalik Group of Pakistan. *Palaeogeography, Palaeoclimatology, Palaeoecology* 243: 204–222.

Nellemann, C., Miles, L., Kaltenborn, B.P., Virtue, M. & Ahlenius, H. (eds). 2007. *The Last Stand of the Orangutan. State of Emergency: Illegal Logging, Fire and Palm Oil in Indonesia's National Parks.* UNEP GRID-Arendal, UNEP-WCMC, UNESCO.

Neri, F.M., Rylands, A.B., Fraiha, V.T. & Ferreira, M.B. 1997. Utilização de rádio telemetria em sauás, *Callicebus personatus*, resgatados durante a implantação da Usina Hidrelétrica Nova Ponte, Minas Gerais. *Neotropical Primates* 5: 50–52.

Neumann-Denzau, G., Mansur, E.F. & Mansur, R. 2008. Nests, eggs, hatchlings and behaviour of the Masked Finfoot *Heliopais personatus* from the Sundarbans in Bangladesh, with first nesting observations. *Forktail* 24: 92–99.

Newton-Fisher, N.E., Reynolds, V. & Plumptre, A.J. 2000. Food supply and chimpanzee (*Pan troglodytes schweinfurthii*) party size in the Budongo forest reserve, Uganda. *International Journal of Primatology* 21: 613–628.

Nichols, J.D., O'Connell, A.E. & Karanth, K.U. 2011. Camera traps in animal ecology and conservation: what's next? In *Camera Traps in Animal Ecology*, A.F. O'Connell, J.D. Nichols & K.U. Karanth (eds). Tokoyo: Springer Japan, pp. 27–44.

Nielsen, S.M. 1991. Fishing Arctic foxes *Alopex lagopus* on a rocky island in West Greenland. *Polar Research* 9: 211–213.

Nielsen, N.H., Jacobsen, M.W., Graham, L.L.B., *et al.* 2011. Successful germination of seeds following passage through orang-utan guts. *Journal of Tropical Ecology* 27: 433–435.

Niemitz, C.T. 1979. Results of a field study on the western tarsier (*Tarsius bancanus borneanus* Horsfield, 1821) in Sarawak. *Sarawak Museum Journal* 27: 171–228.

Niemitz, C.T. 1984. *Biology of Tarsiers*. New York: Gustav Fischer.

Niemitz, C.T. 2010. *Progreditur ordinara saltando et retrorsum*, normally proceeds in a leaping fashion, and backwards. *International Journal of Primatology* 31: 941–957.

Nietlisbach, P., Arora, N., Nater, A., *et al.* 2012. Heavily male-biased long-distance dispersal of orang-utans (genus: *Pongo*), as revealed by Y-chromosomal and mitochondrial genetic markers. *Molecular Ecology* 21: 3173–3186.

Nievergelt, C.M., Mutschler, T. & Feistner, A.T.C. 1998. Group encounters and territoriality in wild Alaotran gentle lemurs (*Hapalemur griseus alaotrensis*). *American Journal of Primatology* 46: 251–258.

Nievergelt, C.M., Mutschler, T., Feistner, A.T.C. & Woodruff, D.S. 2002a. Social system of the Alaotran gentle lemur (*Hapalemur griseus alaotrensis*): genetic characterization of group composition and mating system. *American Journal of Primatology* 57: 157–176.

Nievergelt, C.M., Pastorini, J. & Woodruff, D. 2002b. Genetic veriability and phylogeography in the wild Alaotran gentle lemur population. *Evolutionary Anthropology Supplement* 1: 175–179.

Nijboer, J. 2006. *Fibre intake and faeces quality in leaf-eating primates*. PhD thesis, University of Utrecht, Netherlands.

Nijman, V. 2000. Geographic distribution of ebony leaf monkey *Trachypithecus auratus* E. Geoffroy Saint-Hilaire, 1812) (Ma malia: Primates: Cercophithecidae). *Contributions to Zoology* 69: 157–177.

Nijman, V. 2001. *Forest (and) Primates. Conservation and Ecology of the Endemic Primates of Java and Borneo*. Wageningen: Tropenbos International, Kalimantan series.

Nijman, V. 2012. A review of the distribution and ecology of the ebony langur on Java- the most tropical of the high-elevation montane colobines. In *High Altitude Primates*, N.B. Grow, S. Gursky-Doyen & A. Krzton (eds). New York: Springer, pp. 115–132.

Nijman, V. 2013. One hundred years of solitude: effects of long-term forest fragmentation on the primate community of Java, Indonesia. In *Primates in Fragments: Complexity and Resilience*, L.K. Marsh & C.A. Chapman (eds). New York: Springer, pp. 33–45.

Nijman, V. & Nekaris, K.A.I. 2012. Loud calls, startle behaviour, social organisation and predator avoidance in arboreal langurs (Cercopithecidae: Presbytis). *Folia Primatologia* 83: 274–287.

Nijman, V. & Supriatna, J. 2008. *Trachypithecus auratus*. In *IUCN Red List of Threatened Species*. Gland, Switzerland: IUCN.

Nijman, V. & van Balen, S. 1998. A faunal survey of the Dieng Mountains, central Java, Indonesia: distribution and conservation of endemic primate taxa. *Oryx* 32: 145–156.

Nilsson, C. & Berggren, K. 2000. Alterations of riparian ecosystems caused by river regulation. *BioScience* 50: 783–79.

Nilsson, C., Reidy, C.A., Dynesius, M. & Revegna, C. 2005. Fragmentation and flow regulation of the world's large river systems. *Science* 308: 405–408.

Nilsson, C. & Svedmark, M. 2002. Basic principles and ecological consequences of changing water regimes: riparian plant communities. *Environmental Management* 30: 468–480.

Nishida, T. 1972. Preliminary information of the pygmy chimpanzees (*Pan paniscus*) of the Congo Basin. *Primates* 13: 415–425.

Nishida, T. 1976. The bark-eating habits in primates, with special reference to their status in the diet of wild chimpanzees. *Folia Primatologica* 25: 277–287.

Nishida, T. 1980. Local differences in responses to water among wild chimpanzees. *Folia Primatologica* 33: 189–209.

Nishihara, T. 1995. Feeding ecology of western lowland gorillas in the Nouabale-Ndoki National Park, Congo. *Primates* 36: 151–168.

Nishimua, T.B., Suzuki, E., Kohyama, T. & Tsuyuzaki, S. 2007. Mortality and growth of trees in peat-swamp and heath forests in Central Kalimantan after severe drought. *Plant Ecology* 188: 165–177.

Nkurunungi, J.B., Ganas, J., Robbins, M.M. & Stanford, C.B. 2004. A comparison of two mountain gorilla habitats in Bwindi Impenetrable National Park, Uganda. *African Journal of Ecology* 42: 289–297.

Nor, S.M. 1996. The mammalian fauna on the islands at the northern tip of Sabah, Borneo. *Fieldiana Zoology* 83: 1–51.

Norconk, M.A. 1997. Primate conservation in the wake of a hydroelectric plant. In *AZA (Association of Zoos and Aquariums, formerly American Zoo and Aquarium Association) Regional Conference Proceedings*. Cleveland, OH: AZA, pp. 570–576.

Norconk, M.A. 2011. Saki, uakaris, and titi monkeys: behavioral diversity in a radiation of primate seed predators. In *Primates in Perspective*, C.J. Campbell, A. Fuentes, K.C. MacKinnon, S.K. Bearder & R.M. Stumpf (eds). Oxford: Oxford University Press, pp. 123–138.

Norconk, M.A. & Grafton, B.W. 2003. Changes in forest composition and potential feeding tree availability on a small land-bridge island in Lago Guri, Venezuela. In *Primates in Fragments: Complexity and Resilience*, L.K. Marsh & C.A. Chapman (eds). New York: Springer, pp. 211–227.

Norconk, M.A. & Kinzey, W.G. 1994. Challenge of neotropical frugivory: travel patterns of spider monkeys and bearded sakis. *American Journal of Primatology* 34: 171–183.

Norconk, M.A., Raghanti, M.A., Martin, S.K., *et al.* 2003. Primates of Brownsberg Natuurpark, Suriname, with particular attention to the Pitheciins. *Neotropical Primates* 11: 94–100.

Norconk, M.A., Sussman, R.W. & Phillips-Conroy, J. 1996. Primates of Guiana shield forests: Venezuela and the Guianas. In *Adaptive Radiations of Neotropical Primates*, M.A. Norconk, A.L. Rosenberger & P.A. Garber (eds). New York: Plenum Press, pp. 69–83.

Norhayati, A. 2001. *Frugivores and fruit production in primary and logged tropical rainforests*. PhD thesis, Universiti Kebangsaan Malaysia, Bangi, Malaysia.

Norhayati, A., Saiful, A., Shahrolnizah, A., *et al.* 2004. Diversity and density of mammals in the peat swamp forests of the Langat Basin, Selangor, Malaysia. *Malaysian Applied Biology* 33: 7–17.

Norscia, I. & Palagi, E. 2008. Berenty 2006: Census of *Propithecus verreauxi* and possible evidence of population stress. *International Journal of Primatology* 29: 1099–1115.

Nowak, K. 2007. *Behavioural flexibility and demography of the Zanzibar red colobus monkey across floristic and disturbance gradients*. PhD thesis, University of Cambridge, Cambridge, UK.

Nowak, K. 2008. Frequent water drinking by Zanzibar red colobus (*Procolobus kirkii*) in a mangrove forest refuge. *American Journal of Primatology* 70: 1081–1092.

Nowak, K. 2012. Mangrove and peat swamp forests: refuge habitats for primates and felids. *Folia Primatologica* 83: 361–376.

Nowak, K., Cardini, A. & Elton, S. 2008. Evolutionary acceleration and divergence in *Procolobus kirkii*. *International Journal of Primatology* 29: 1313–1339.

Nowak, K. & Lee, P.C. 2011. Demographic structure of Zanzibar red colobus populations in unprotected coral rag and mangrove forests. *International Journal of Primatology* 32: 24–45.

Nowak, K. & Lee, P.C. 2013. 'Specialist' primates can be flexible in response to habitat alteration. In *Primates in Fragments*, L.K. Marsh & C.A. Chapman (eds). New York: Springer, pp. 199–211.

Nowak, K., Perkin, A. & Jones, T. 2009. *Update on habitat loss and conservation status of the Endangered Zanzibar red colobus on Uzi and Vundwe Islands.* Report. Zanzibar, Tanzania: Department of Commercial Crops, Fruits and Forestry (DCCFF). www.primate.org/nowakperkinjonesred_colobus.pdf.

Nugraha, H. 2007. *Analisis pola penggunaan ruang banteng* (Bos javanicus *d'Alton 1823) di Cagar Alam dan Taman Wisata Alam Pangandaran Jawa Barat.* BSc thesis, Institut Pertanian Bogor, Bogor.

Oates, J.F. 1977. The guereza and its food. In *Primate Ecology: Studies of Feeding and Ranging Behavior in Lemurs, Monkeys and Apes,* T.H. Clutton-Brock (ed.). New York: Academic Press, pp. 275–321.

Oates, J.F. 1988. The distribution of *Cercopithecus* monkeys in West African forests. In *A Primate Radiation: Evolutionary Biology of the African Guenons,* A. Gautier-Hion, F. Bourlière, J.-P. Gautier & J. Kingdon (eds). Cambridge: Cambridge University Press, pp. 79–103.

Oates, J.F. 1989. *A survey of primates and other forest wildlife in Anambra, Imo and Rivers States, Nigeria.* Unpublished report to National Geographic Society (USA), Nigerian Conservation Foundation, Nigerian Federal Department of Forestry, and Governments of Anambra, Imo and Rivers States.

Oates, J.F. 1994. The natural history of African colobines. In *Colobine Monkeys: Their Ecology, Behaviour and Evolution,* A.G. Davies & J.F. Oates (eds). Cambridge: Cambridge University Press, pp. 75–128.

Oates, J.F. 1996. *African Primates: Status Survey and Conservation Action Plan.* IUCN/SSC Primate Specialist Group. Gland: IUCN.

Oates, J.F. 2002. West Africa: tropical forest parks on the brink. In *Making Parks Work: Strategies for Preserving Tropical Nature,* J. Terborgh, C. van Schaik, L. Davenport & M. Rao (eds). Washington DC: Island Press, pp. 57–75.

Oates, J.F. 2011. *Primates of West Africa: A Field Guide and Natural History.* Arlington, VA: Conservation International.

Oates, J.F. 2013. *Cercopithecus erythrogaster* white-throated monkey. In *Mammals of Africa. Volume II: Primates,* T.M. Butynski, J. Kingdon & J. Kalina (eds). London: Bloomsbury, pp. 384–386.

Oates, J.F., Abedi-Lartey, M., McGraw, W.S., Struhsaker, T.T. & Whitesides, G.H. 2000. Extinction of a West African red colobus monkey. *Conservation Biology* 14: 1526–1532.

Oates, J.F. & Ambrose, L. 2013. *Arctocebus calabarensis* Calabar angwantibo. In *Mammals of Africa. Volume II: Primates,* T.M. Butynski, J. Kingdon & J. Kalina (eds). London: Bloomsbury, pp. 400–402.

Oates, J.F. & Baker, L.R. 2013. *Cercopithecus sclateri* Sclater's monkey. In *Mammals of Africa. Volume II: Primates,* T.M. Butynski, J. Kingdon & J. Kalina (eds). London: Bloomsbury, pp. 369–371.

Oates, J.F. & Jewell, P.A. 1967. Westerly extent of the range of three African lorisoid primates. *Nature* 215: 778–779.

Oates, J.F., Bergl, R.A. & Linder, J.M. 2004. *Africa's Gulf of Guinea forests: biodiversity patterns and conservation priorities.* Advances in Applied Biodiversity Science 6. Washington DC: Conservation International.

Oates, J.F., Gippoliti, S. & Groves, C.P. 2008a. *Cercocebus campbelli.* In *IUCN Red List of Threatened Species.* Version 2012.2. Gland, Switzerland: IUCN. www.iucnredlist.org.

Oates, J.F., Gippoliti, S. & Groves, C.P. 2008b. *Cercocebus torquatus.* In IUCN 2012. In *IUCN Red List of Threatened Species.* Version 2012.2. Gland, Switzerland: IUCN. www.iucnredlist.org.

Oates, J.F., Gippoliti, S. & Groves, C.P. 2008c. *Cercocebus petaurista.* In IUCN 2012. In *IUCN Red List of Threatened Species.* Version 2012.2. Gland, Switzerland: IUCN. www.iucnredlist.org.

Oates, J.F., Gippoliti, S. & Groves, C.P. 2008d. *Papio papio.* In IUCN 2012. In *IUCN Red List of Threatened Species.* Version 2012.2. Gland, Switzerland: IUCN. www.iucnredlist.org.

Oates, J.F., Struhsaker, T.T., Butynski, T.M. & De Jong, Y.A. 2008e. *Procolobus rufomitratus.* In *IUCN Red List of Threatened Species.* Version 2012.2. Gland, Switzerland: IUCN. www.iucnredlist.org.

Oates, J.F., Struhsaker, T., McGraw, S., *et al.* 2008. *Procolobus badius.* In *IUCN Red List of Threatened Species.* Version 2012.2. Gland, Switzerland: IUCN. www.iucnredlist.org.

Oates, J.F. & Werre, J.L. 2009. Niger Delta red colobus *Procolobus epieni.* In *Primates in Peril: The World's 25 Most Endangered Primates 2008–2010,* R.A. Mittermeier, J. Wallis, A.B. Rylands, *et al.* (eds). *Primate Conservation* 24: 1–57.

Oates, J.F., Whitesides, G.H., Davies, A.G., *et al.* 1990. Determinants of variation in tropical forest primate biomass: new evidence from West Africa. *Ecology* 71: 328–343.

Ochieng, C.A. 2002. *Research master plan for the Rufiji Floodplain and Delta 2003–2013.* Technical Report 28. Dar es Salaam: Rufiji Environmental Management Project. http://coastalforests.tfcg.org/pubs/REMP%2030%20Technical%20Report%2028%20Research%20Masterplan.pdf.

Oi, T. 2013. Inter-species relationship between Asiatic black bear and Japanese macaques, a deduction from the comparison of feeding behaviors. *Primate Research* 29: 123–135.

O'Brien, T.G., Kinnaird, M.F., Nurcahyo, A., Iqbal, M. & Rusmanto, M. 2004. Abundance and distribution of sympatric gibbons in a threatened Sumatran rainforest. *International Journal of Primatology* 25: 267–284.

O'Brien, T.G., Kinnaird, M.F., Nurcahyo, A., Prasetyaningrum, M. & Iqbal, M. 2003. Fire, demography and the persistence of siamang (*Sympgylangus syndactylus:* Hylobatidae) in a Sumatran rainforest. *Animal Conservation* 6: 115–121.

O'Connell, A.F., Nichols, J.D. & Karanth, K.U. 2011. *Camera Traps in Animal Ecology: Methods and Analysis.* Tokyo: Springer Japan.

OGM. 2006. *Observatoire de la Guinée maritime. Rapport Final.* Final report. Bordeaux, France: Université Bordeaux III.

O'Grady, J., Brook, B., Reed, D., *et al.* 2006. Realistic levels of inbreeding depression strongly affect extinction risk in wild populations. *Biological Conservation* 133: 42–51.

Okello, M.M., Manka, S.G. & D'Amour, D.E. 2008. The relative importance of large mammal species for tourism in Amboseli National Park, Kenya. *Tourism Management* 29: 751–760.

Oklander, L.I., Caputo, M. & Corach, D. 2017. Genetic variability in Argentinian howler monkeys (*Alouatta caraya*): a natural process or an anthropic effect? In *Primatology in Argentina,* M. Kowalewski & L.I. Oklander (eds). Buenos Aires: Sociedad Argentina para el Estudio de los Mamíferos.

Oklander, L.I., Kowalewski, M. & Corach, D. 2010. Genetic consequences of habitat fragmentation in black and gold howler populations from northern Argentina. *International Journal of Primatology* 31: 813–832.

Oklander, L.I., Kowalewski, M.M. & Corach, D. 2014. Male reproductive strategies in black and gold howler monkeys (*Alouatta caraya*). *American Journal of Primatology* 76: 43–55.

Olaleye, O.A. & Ameh, C.E. 1999. *Forest resource situation assessment of Nigeria.* Unpublished report. Abuja, Nigeria: European Commission and FAO Partnership Programme.

Olejniczak, C. 1994. The gorillas of Mbeli, northern Congo. *Gorilla Gazette* 8: 1–3.

Oliveira, M.M. & Langguth, A. 2006. Rediscovery of McGrave's capuchin monkey and designation of a neotype for *Simia flavia* Schreber, 1774 (Primates, Cebidae). *Boletim do Museu Nacional* 523: 1–6.

Oliveira, P.E.A.M. & Paula, F.R. 2001. Fenologia e biologia reprodutiva de plantas de mata de galeria. In *Cerrado. Caracterização e Recuperação de Matas de Galeria,* J.F. Ribeiro, C.E.L. da Fonseca & J.C. Sousa-Silva (eds). Embrapa Cerrados, Brazil: Planaltina, pp. 303–332.

Oliveira, P.S. & Marquis, R.J. 2013. *The Cerrados of Brazil: Ecology and Natural History of a Neotropical Savanna*. New York: Columbia University Press.

Oliveira-Filho, A.T. & Ratter, J.A. 1995. A study of the origin of central Brazilian forests by the analysis of the plant species distributions. *Edinburgh Journal of Botany* 52: 141–194.

Oliver, W.L.R. & Santos, I.B. 1991. Threatened endemic mammals of the Atlantic forest region of southeast Brazil. *Wildlife Preservation Trust. Special Scientific Report* 4: 1–126.

Olsder, K. & van der Donk, M. 2006. *Destination Conservation: Protecting Nature by Developing Tourism*. Amsterdam: IUCN National Committee of the Netherlands.

Olson, D.M., Dinerstein, E., Wikramanayake, E.D., *et al.* 2001. Terrestrial ecoregions of the world: a new map of life on Earth. *BioScience* 51: 933–938.

Olson, E.R., Marsh, R.A. & Bovard, B.N. 2012. Arboreal camera trapping for the Critically Endangered greater bamboo lemur *Prolemur simus*. *Oryx* 46: 593–597.

O'Malley, L.S.S. 1914. *Bengal District Gazetteers. 24-Parganas*. New Delhi: Logos Press.

OSFAC. 2010. *Forest Cover and Forest Cover Loss in the Democratic Republic of Congo from 2000 to 2010*. www.osfac.net/data-products/facet/facet-dr-congo.

Osorio, D., Terborgh, J., Alvarez, A., *et al.* 2011. Lateral migration of fish between an oxbow lake and an Amazonian headwater river. *Ecology of Freshwater Fish* 20: 619–627.

Ostro, L.E., Silver, S.C., Koontz, F.W., Young, T.P. & Horwich, R.H. 1999. Ranging behavior of translocated and established groups of black howler monkeys *Alouatta pigra* in Belize, Central America. *Biological Conservation* 87: 181–190.

Ostro, L.E.T., Silver, S.C., Koontz, F.W. & Young, T.P. 2000. Habitat selection by translocated black howler monkeys in Belize. *Animal Conservation* 3: 175–181.

Ottoni, E.B. & Izar, P. 2008. Capuchin monkey tool use: overview and implications. *Evolutionary Anthropology* 17: 171–178.

Ottoni, E.B. & Mannu, M. 2001. Semifree-ranging tufted capuchin (*Cebus apella*) spontaneously use tools to crack open nuts. *International Journal of Primatology* 22: 347–358.

Overdorff, D.J., Strait, S.G. & Telo, A. 1997. Seasonal variation in activity and diet in a small-bodied folivorous primate, *Hapalemur griseus*, in southern Madagascar. *American Journal of Primatology* 43: 211–223.

Pacheco Santos, L.M. 2005. Nutritional and ecological aspects of buriti or aguaje (*Mauritia flexuosa* Linnaeus filius): a carotene-rich palm fruit from Latin America. *Ecology of Food & Nutrition* 44: 345–358.

Pack, K.S. 1994. Tropical rainforest conservation: primates and the case of Petit Saut. *Primate Eye* 58: 20–21.

Packman, C.E., Gray, T.N.E., Collar, N.J., *et al.* 2013. Rapid loss of Cambodia's grasslands. *Conservation Biology* 27: 245–247.

Padoch, C., Ayres, J.M., Pinedo-Vasquez, M. & Henderson, A. (eds) 2000. *Várzea: Diversity, Development, and Conservation of Amazonia's Whitewater Floodplains*. New York: New York Botanic Gardens Press.

Padua, C.V. & Padua, S.M. 2000. Conservation of black lion tamarins (*Leontopithecus chrysopygus*) in the Atlantic forest of the interior, Brazil. *Society for Conservation Biology Newsletter* 7.

Page, S.E. 2002. *The biodiversity of peat swamp forest habitats in S.E. Asia, impacts of land-use and environmental change, implications for sustainable ecosystem management.* Unpublished report, University of Leicester, UK.

Page, S.E., Rieley, J.O. & Banks, C.J. 2011. Global and regional importance of the tropical peatland carbon pool. *Global Change Biology* 17: 798–818.

Page, S.E., Rieley, J.O., Doody, K., *et al.* 1997. Biodiversity of tropical peat swamp forest: a case study of animal diversity in the Sungai Sebangau catchment of Central Kalimantan, Indonesia. In *Tropical Peatlands*, J.O. Rieley & S.E. Page (eds). Cardigan, UK: Samara Publishing Limited, pp. 231–242.

Page, S.E., Rieley, J.O., Shotyk, Ø.W. & Weiss, D. 1999. Interdependence of peat and vegetation in a tropical peat swamp forest. *Philosophical Transactions of the Royal Society of London. Series B: Biological Sciences* 354: 1885–1897.

Paim, F.P. 2008. *Estudo comparitivo das espécies de* Saimiri *Voigt, 1931 (Primates, Cebidae) na Reserva Mamirauá, Amazonas*. MSc thesis, Museu Paraense Emilo Goeldi, Belém.

Paim, F.P., Aquino, S.P. & Valsecchi, J. 2012. Does ecotourism activity affect primates in Mamirauá Reserve? *Uakari* 8: 41–50.

Paim, F.P., Chapman, C.A., de Queiroz, H.L. & Paglia, A.P. 2017. Does resource availability affect the diet and behavior of the vulnerable squirrel monkey, *Saimiri vanzolinii*? *International Journal of Primatology* 38: 572–587.

Paim, F.P. & Queiroz, H.L. 2009. Differences in acoustic parameters of alarm vocalizations of species of *Saimiri* Voigt, 1831 (Primates, Cebidae) from Várzea Forest – Mamirauá Reserve. *Uakari* 5: 49–60.

Paim, F.P., Silva, Jr. J.S., Valsecchi, J., Harada, M.L. & de Queiroz, H.L. 2013. Diversity, geographic distribution and conservation of squirrel monkeys, *Saimiri* (Primates, Cebidae), in the floodplain forests of central Amazon. *International Journal of Primatology* 34: 1055–1076.

Paim, F.P., Valsecchi, J., Harada, M.L. & de Queiroz, H.L. 2013. Diversity, geographic distribution and conservation of squirrel monkeys, *Saimiri* (Primates, Cebidae), in the floodplain forests of Central Amazon. *International Journal of Primatology* 34: 1055–1076.

Painter, L., Siles, T.M., Reinaga, A. & Wallace, R. 2013. *Escenarios de deforestación en el Gran Paisaje Madidi-Tambopata*. La Paz, Bolivia: Wildlife Conservation Society, Bolivia.

Painter, M., Alves, A.R., Bertsch, C., *et al.* 2008. *Landscape Conservation in the Amazon Region: Progress and Lessons*. Working Paper 34. Bronx, NY: Wildlife Conservation Society.

Palacios, E. & Rodríguez, A. 2001. Ranging pattern and use of space in a group of red howler monkeys (*Alouatta seniculus*) in southeastern Colombia. *American Journal of Primatology* 55: 233–251.

Palacios, E., Rodríguez, A. & Castillo, C. 2004. Preliminary observations on the mottled-face tamarin (*Saguinus inustus*) at the lower Caquetá River, Colombian Amazonia. *Neotropical Primates* 12: 123–126.

Palmer, J.D. 1995. *The Biological Rhythms and Clocks of Intertidal Animals*. New York: Oxford University Press.

Palminteri, S. & Peres, C.A. 2012. Habitat selection and use of space by bald-faced sakis (*Pithecia irrorata*) in southwestern Amazonia, lessons from a multiyear, multigroup study. *International Journal of Primatology* 33: 401–417.

Palminteri, S., Powell, G.V.N., Asner, G.P. & Peres, C.A. 2012. LiDAR measurements of canopy structure predict spatial distribution of a tropical mature forest primate. *Remote Sensing of Environment* 127: 98–105.

Palminteri, S., Powell, G.V.N. & Peres, C.A. 2011. Regional-scale heterogeneity in primate community structure at multiple undisturbed forest sites across southeastern Peru. *Journal of Tropical Ecology* 27: 181–194.

Palombit, R.A. 1997. Inter and intraspecific variation in the diets of sympatric siamang (*H. syndactylus*) and lar gibbons (*H. lar*). *Folia Primatologica* 68: 321–337.

Pandit, P.K. & Mukherjee, S. 2011. Ethno-medicinal studies of mangrove forests of Sundarban Tiger Reserve, West Bengal, India. *SAARC Forestry* 1: 113–131.

Paneque-Gálvez, J., McCall, M.K., Napoletano, B.M., Wich, S.A. & Koh, L.P. 2014. Small drones for community-based forest monitoring: an assessment of their feasibility and potential in tropical areas. *Forests* 5: 1481–1507.

Panger, M. 2007. Tool use and cognition in primates. In *Primates in Perspective*, C.J. Campbell, A. Fuentes, K.C. MacKinnon, S.K. Bearder & R.M. Stumpf (eds). Oxford: Oxford University Press, pp. 665–677.

Paoli, G.D., Curran, L.M. & Zak, D.R. 2005. Soil nutrients and beta diversity in the Bornean Dipterocarpaceae: evidence for niche partitioning by tropical rain forest trees. *Journal of Ecology* 94: 157–170.

Papavero, N., Teixeira, D.M. & Chiquieri, A. 2011. As 'Adnotationes' do jesuíta Johann Breuer sobre a história natural da Missão de Ibiapaba, Ceará (1789). *Arquivos de Zoologia* 42: 133–159.

ParksWatch-Mexico. 2003. *Park Profile – Mexico: Pantános de Centla Biosphere Reserve.* Durham, NC: ParksWatch.org.

Parnell, R.J. 2002. Group size and structure in western lowland gorillas (*Gorilla gorilla gorilla*) at Mbeli Bai, Republic of Congo. *American Journal of Primatology* 56: 193–206.

Parolin, P. 2000. Seed mass in Amazonian floodplain forests with contrasting nutrient supplies. *Journal of Tropical Ecology* 16: 417–428.

Parolin, P. 2001. Morphological and physiological adjustments to waterlogging and drought in seedlings of Amazonian floodplain trees. *Oecologia* 128: 326–335.

Parolin, P. 2003. Fugitive and possessive establishment strategies in Amazonian floodplain pioneers. *Flora* 198: 436–443.

Parolin, P., Adis, J., Silva, M.F.D., *et al.* 2003. Floristic composition of a floodplain forest in the Anavilhanas archipelago, Brazilian Amazonia. *Amazoniana* 17: 399–411.

Parolin, P., De Simone, O., Haase, K., *et al.* 2004. Central Amazonian floodplain forests: tree adaptations in a pulsing system. *Botanical Review* 70: 357–380.

Parolin, P., Ferreira, L.V., Albernaz, A.L.K.M. & Almeida, S. 2004. Tree species distribution in várzea forests of Brazilian Amazonia. *Folia Geobotanica* 39: 371–383.

Parolin, P., Lucas, C., Piedade, M.T.F. & Wittmann, F. 2010. Drought responses of flood-tolerant trees in Amazonian floodplains. *Annals of Botany* 105: 129–139.

Passos, F.D.C. & Keuroghlian, A. 1999. Foraging behavior and microhabitats used by black lion tamarins, *Leontopithecus chrysopygus* (Mikan)(Primates, Callitrichidae). *Revista Brasileira de Zoologia* 16: 219–222.

Pastorini, J., Forstner, M.R.J. & Martin, R.D. 2002. Phylogenetic relationships of gentle lemurs (*Hapalemur*). *Evolutionary Anthropology* 11(S1): 150–154.

Paulsen, J.R. & Clark, C.J. 2004. Densities, distribution and seasonal movements of gorillas and chimpanzees in swamp forest, northern Congo. *International Journal of Primatology* 25: 285–306.

Pavé, R. 2013. *El conflicto madre-infante en el mono aullador negro y dorado* (Alouatta caraya) *y su comparación en dos sitios del noreste Argentino.* PhD thesis, Universidad Nacional del Litoral, Santa Fé.

Pavé, R., Kowalewski, M.M., Garber, P.A., *et al.* 2012. Infant mortality in *Alouatta caraya* living in a flooded forest in northeastern Argentina. *International Journal of Primatology* 3: 937–957.

Pavé, R., Kowalewski, M.M., Peker, S.M. & Zunino, G.E. 2010. Preliminary study of mother-offspring conflict in *Alouatta caraya*. *Primates* 51: 221–226.

Pavé, R., Kowalewski, M.M., Zunino, G.E. & Giraudo, A.R. 2015. How do demographic and social factors influence parent-offspring conflict? The case of wild black and gold howler monkeys (*Alouatta caraya*). *American Journal of Primatology* 77: 911–923.

Pavelka, M.S.M., McGoogan, K.C. & Steffens, T.S. 2007. Population size and characteristics of *Alouatta pigra* before and after a major hurricane. *International Journal of Primatology* 28: 919–929.

Payne, J. & Andau, P.M. 1994. Censussing orangutans in Sabah: results and conservation implications. In *The Neglected Ape: Proceedings of the International Conference on Orangutans*, J. Ogden, L. Perkins & L. Sheeran (eds). San Diego, CA: Zoological Society of San Diego, pp. 7–11.

Pearce, D.W. & Moran, D. (eds). 1994. *The Economic Value of Biodiversity*. London: Earthscan.

Pebsworth, P.A., Morgan, H.R. & Huffman, M.A. 2012. Evaluating home range techniques: use of Global Positioning System (GPS) collar data from chacma baboons. *Primates* 53: 345–355.

Peck, M., Thorn, J., Mariscal, A., *et al.* 2011. Focusing conservation efforts for the Critically Endangered brown-headed spider monkey (*Ateles fusciceps*) using remote sensing, modeling, and playback survey methods. *International Journal of Primatology* 32: 134–148.

Peck, M., Thorn, J., Mariscal, A., *et al.* 2014. Report of a black spider monkey (*Ateles chamek*) swimming in a large river in central-western Brazil. *Neotropical Primates* 21: 205.

Penn, J.W. 1999. *The Aguaje Palm (*Mauritia flexuosa*): explaining its role as an agroforestry species in a community conservation project.* PhD thesis, University of Florida, Gainesville, Florida.

Pennington, R.T., Prado, D.E. & Pendry, C.A. 2000. Neotropical seasonally dry forests and quaternary vegetation changes. *Journal of Biogeography* 27: 261–273.

Peralta, N. 2005. *Os ecoturistas estão chegando: aspectos da mudança social na RDS Mamirauá*. MSc dissertation, Universidade Federal do Pará, Belém, Pará, Brazil.

Peralta, N. 2013. Ecotourism as an incentive to biodiversity consrvation: the case of Uakari Lodge, Amazonas, Brazil. *Uakari* 8: 75–93.

Peralta, N., Moura, E., Nascimento, A.C. & Lima, D. 2009. Renda doméstica e sazonalidade em comunidades da RDS Mamirauá, 1995–2005. *Uakari* 5: 7–19.

Pereira, M.E., Clutton-Brock, T.H. & Kappeler, P.M. 2000. Understanding primate males. In *Primate Males: Causes and Consequence of Variation in Group Composition*, P.M. Kappeler (ed). Cambridge: Cambridge University Press, pp. 271–277.

Pereira, M.J.R., Marques, J.T., Santana, J., *et al.* 2009. Structuring of Amazonian bat assemblages: the roles of flooding patterns and floodwater nutrient load. *Journal of Animal Ecology* 78: 1163–1171.

Pereira, M.J.R., Rocha, R.G., Ferreira, E. & Fonseca, C. 2013. Structure of small mammal assemblages across flooded and unflooded gallery forests of the Amazonia-Cerrado ecotone. *Biotropica* 45: 489–496.

Perelman, P., Johnson, W.E., Roos, C., *et al.* 2011. A molecular phylogeny of living primates. *PLoS ONE* 7: e1001342

Perelman, S.B., León, R.J.C. & Oesterheld, M. 2001. Cross-scale vegetation patterns of flooding pampa grasslands. *Journal of Ecology* 89: 562–577.

Peres, C.A. 1989. A survey of a gallery forest primate community, Marajó Island, Pará, Brazil. *Vida Silvestre Neotropical* 2: 32–37.

Peres, C.A. 1990. Effects of hunting on western Amazonian primate communities. *Biological Conservation* 54: 47–59.

Peres, C.A. 1993a. Structure and spatial organization of an Amazonian terra firme forest primate community. *Journal of Tropical Ecology*, 9: 259–276.

Peres, C.A. 1993b. Notes on the primates of the Juruá River, western Brazilian Amazonia. *Folia Priamtologica* 61: 97–103.

Peres, C.A. 1994a. Primate responses to phenological changes in an Amazonian terra firme forest. *Biotropica* 26: 98–112.

Peres, C.A. 1994b. Composition, density, and fruiting phenology of arborescent palms in an Amazonian terra firme forest. *Biotropica* 26: 285–294.

Peres, C.A. 1997a. Effects of habitat quality and hunting pressure on arboreal folivore densities in neotropical forests: a case study of howler monkeys (*Alouatta* spp.). *Folia Primatologica* 68: 199–222.

Peres, C.A. 1997b. Primate community structure at twenty western Amazonian flooded and unflooded forests. *Journal of Tropical Ecology* 13: 381–405.

Peres, C.A. 1999. Generalized guidelines for standardizing line-transect surveys of tropical forest primates. *Neotropical Primates* 7: 11–16.

Peres, C.A. & Dolman, P.M. 2000. Density compensation in neotropical primate communities: evidence from 56 hunted and nonhunted Amazonian forests of varying productivity. *Oecologia* 122: 175–189.

Peres, C.A., Gardener, T.A., Barlow, J., *et al.* 2010. Biodiversity conservation in human-modified Amazonian forest landscapes. *Biological Conservation* 143: 2314–2327.

Peres, C.A. & Johns, A.D. 1991. Patterns of primate mortality in a drowning forest: Lessons from the Tucuruí dam, Brazilian Amazonia. *Primate Conservation* 12–13: 7–10.

Peres, C.A. & Palacios, E. 2007. Basin-wide effects of game harvest on vertebrate population densities in Amazonian forests: implications for animal-mediated seed dispersal. *Biotropica* 39: 304–315.

Peres, C.A. & Terborgh, J.W. 1995. Amazonian nature reserves-an analysis of the defensibility status of existing conservation units and design criteria for the future. *Conservation Biology* 9: 34–46.

Pérez, A.C., Damen, M.C.J., Geneletti, D. & Hobma, T.W. 2002. Monitoring a recent delta formation in a tropical coastal wetland using remote sensing and GIS: Guapo River Delta, Laguna de Tacarigua, Venezuela. *Environment, Development & Sustainability* 4: 201–219.

Perry, S., Panger, M., Rose, L.M., *et al.* 2003. Traditions in wild white-faced capuchin monkeys. In *The Biology of Traditions: Models and Evidence*, D.M. Fragaszy & S. Perry (eds). Cambridge: Cambridge University Press, pp. 391–425.

Peschak, T.P. 2005. *Currents of Contrast: Life in Southern Africa's Two Oceans*. Cape Town: Struik Publishers.

Petrides, G.A. 1965. *Advisory Report on Wildlife and National Parks in Nigeria*. Bronx, New York: American Committee for International Wild Life Protection.

Petrozzi, F., Akani, G.C., Amadi, N., *et al.* 2015. Surveys of mammal communities in a system of five forest reserves suggest an ongoing biotic homogenization process for the Niger Delta (Nigeria). *Tropical Zoology* 28: 95–113.

Petter, J.J. 1962. Ecological and behavioral studies of Madagascar lemurs in the field. *Annals of the New York Academy of Sciences* 102: 267–281.

Petter, J.J. & Peyrieras, A. 1970. Observations éco-éthologiques sur les lémuriens malgaches du genre *Hapalemur*. *La Terre et Vie: Revue d'Écologie* 24: 365–382.

Pfeffer, P. 1965. Esquisse ecologique de la Réserve de Baluran (Java Est). *Revue d'Écolgie, La Terre et la Vie* 112: 197–215.

Phalan, B., Bertzky, M., Butchart, S.H.M., *et al.* 2013. Crop expansion and conservation priorities in tropical countries. *PLoS ONE* 8: e51759

Phil-Eze, P.O. & Okoro, I.C. 2009. Sustainable biodiversity conservation in the Niger Delta: a practical approach to conservation site selection. *Biodiversity & Conservation* 18: 1247–1257.

Phoonjampa, R., Koenig, A., Borries, C., Gale, G.A. & Savini, T. 2010. Selection of sleeping trees in pileated gibbons (*Hylobates pileatus*). *American Journal of Primatology* 72: 617–625.

Phua, M.-H., Tsuyuki, S., Lee, J.S. & Sasakawa, H. 2007. Detection of burned peat swamp forest in a heterogeneous tropical landscape: a case study of the Klias Peninsula, Sabah, Malaysia. *Landscape and Urban Planning* 82: 103–116.

Pidgeon, M. 1996. *An Ecological Survey of Lake Alaotra and Selected Wetlands of Central and Eastern Madagascar in Analyzing the Demise of Madagascar Pochard* Aythya innotata. Antananarivo, Madagascar: WWF/Missouri Botanical Garden.

Piedade, M.T.F., Ferreira, C.S. & Franco, A.C. 2010. Estrategias reproductivas de la vegetación y sus respuestas al pulso de la inundación en las zonas inundables de la Amazonía Central. *Ecosistemas* 19: 52–66.

Pielou, E.C. 1966. Shannon's formula as a measure of specific diversity: its use and misuses. *American Naturalist* 104: 463–465.

Pilbeam, D. 1982. New hominoid skull material from the Miocene of Pakistan. *Nature* 295: 232–234.

Pimenta, F.E. & Silva, Jr, J.S. 2005. An update on the distribution of primates on the Tapajós-Xingu interfluvium, central Amazonia. *Neotropical Primates* 13: 23–28.

Pimentel, M.C.P., Barros, M.J., Cirne, P., *et al.* 2007. Spatial variation in the structure and floristic composition of 'restinga' vegetation in southeastern Brazil. *Brazilian Journal of Botany* 30: 543–551.

Pimm, S.L. & Raven, P. 2000. Extinction by numbers. *Nature* 403: 843–845.

Pimley, E.R. 2009. A survey of nocturnal primates (Strepsirrhini: Galagniae, Perodictinae) in southern Nigeria. *African Journal of Ecology* 47: 784–787.

Pimley, E.R. & Bearder, S.K. 2013. *Perodicticus potto* potto. In *Mammals of Africa. Volume II: Primates*, T.M. Butynski, J. Kingdon & J. Kalina (eds). London: Bloomsbury, pp. 393–399.

Pinedo-Vasquez, M. & Dávalos, L.M. 2011. The river-refuge hypothesis and other contributions of Márcio Ayres to conservation science. In *The Amazon Várzea: The Decade Past and the Decade Ahead*, M.A. Pinedo-Vasquez, M. Ruffino, C.J. Padoch & E.S. Brondízio (eds). New York: Springer, pp. 317–324.

Pinheiro, M.D.N.M. & Gonçalves Jardim, M.A. 2015. Composição florística e formas biológicas de macrófitas aquáticas em lagos da Amazônia ocidental, Roraima, Brasil. *Biota Amazônia* 5: 23–27.

Pinto, L.P., Barnett, A.A., Bezerra, B.M., *et al.* 2013. Why we know so little: the challenges of fieldwork on the Pitheciids. In *Evolutionary Biology & Conservation of Titis, Sakis and Uacaris*, L. Veiga, A. Barnett, S. Ferrari & M. Norconk (eds). Cambridge: Cambridge University Press, pp. 145–150.

Pinzón-Perez, L., Castellanos-Castro, C., Cardona-Cardozo, A., Mora-Fernandez, C. & Vargas-Ríos, O. 2011. Caraterización de las comunidades vegetales presentes en el Bloque Cubiro, cuenca baja del Río Pauto, Casanare (Colombia). In *Mamíferos, reptiles y ecosistemas del Bloque Cubiro (Casanare): Educación ambiental para la conservación*, León Sicard T. (ed). Bogotá, Colomba: Instituto de estudios ambientales Universidad Nacional de Colomba, Alange Energy Corp, pp. 99–154.

Pitman, N.C.A., Widmer, J., Jenkins, C.N., *et al.* 2011. Volume and geographical distribution of the ecological research in the Andes and the Amazon, 1995–2008. *Tropical Conservation Science* 4: 64–81.

Poché, RM. 1976. Notes on primates in Parc National du W du Niger, West Africa. *Mammalia* 40: 187–198.

Pocock, R.I. 1939. *The Fauna of British India including Ceylon and Burma: Mammalia, Vol. I. Primates and Carnivora (in part), Families Felidae and Viverridae*. 2nd edition. London: Taylor & Francis.

Poff, N.L., Allan, J.D., Palmer, M.A., *et al.* 2003. River flows and water wars: Emerging science for environmental decision making. *Frontiers in Ecology and the Environment* 1: 298–306.

Poffenberger, M. (ed.). 2000. *Communities and Forest Management in South Asia: A Regional Profile*. Working Group on Community Involvement in Forest Management. London: Ford Foundation, US & United Kingdom, Department for International Development (DFID).

Polanco-Ochoa, R. & Cadena, A. 1993. Use of space by *Callicebus cupreus ornatus* (Primates: Cebidae) in La Macarena, Colombia. *Field Studies of New World Monkeys, La Macarena, Colombia* 8: 19–32.

Pollock, J. 1986. Towards a conservation policy for Madagascar's Eastern rain forests. *Primate Conservation* 7: 82–86.

Ponnamperuma, F.N. 1984. Effects of flooding on soils. In *Flooding and Plant Growth*, T.T. Kozlowski (ed.). New York: Academic Press, pp. 9–45.

Pontes, A.R.M. 1997. Habitat partitioning among primates in Maracá island, Roraima, northern Brazilian Amazonia. *International Journal of Primatology* 18: 131–157.

Pontes, A.R.M. 1999. Environmental determinants of primate abundance in Maracá island, Roraima, Brazilian Amazonia. *Journal of Zoology* 247: 189–199.

Poorter, L., Bongers, F., Kouamé, F.N'. & Hawthorne, W.D. (eds). 2004. *Biodiversity of West African Forests: An Ecological Atlas of Woody Plant Species*. Wallingford, Oxfordshire: CABI Publishing.

Pope, K., Masuoka, P., Reimankova, E., *et al.* 2005. Mosquito habitats, land use, and malaria risk in Belize from satellite imagery. *Ecological Applications* 15: 1223–1232.

Polidoro, B.A., Carpenter, K.E., Collins, L., *et al.* 2010. The loss of species: mangrove extinction risk and geographic areas of global concern. *PLoS ONE* 5: e10095.

Porcel, Z.R., López-Strauss, H., Martínez, J. & Wallace, R.B. 2010. Callithricidade. In *Distribuición, Ecología y Conservación de los Mamíferos Medianos y Grandes de Bolivia*, R.B. Wallace, H. Gómes, Z.R. Porcel & D.I. Rumiz. Santa Cruz, Bolivia: Centro de Ecología Difusión Simón I. Patiño, pp. 237–262.

Porfirio, G., Sarmento, P., Xavier Filho, N.L., Cruz, J. & Fonseca, C. 2014. Medium to large size mammals of southern Serra do Amolar, Mato Grosso do Sul, Brazilian Pantanal. *Check List* 10: 473–482.

Porter, L., Chism, J., Defler, T.R., *et al.* 2013. Pitheciid conservation in Ecuador, Colombia, Peru, Bolivia, and Paraguay. In *Evolutionary Biology and Conservation of Titis, Sakis and Uacaris*, L. Veiga, A. Barnett, S. Ferrari & M. Norconk (eds). Cambridge: Cambridge University Press, pp. 320–333.

Posa, M.R.C., Wijedasa, L.S. & Corlett, R.T. 2011. Biodiversity and conservation of tropical peat swamp forests. *BioScience* 61: 49–57.

Post, D.G. 1981. Activity patterns of yellow baboons *(Papio cynocephalus)* in the Amboseli National Park, Kenya. *Animal Behaviour* 29: 357–374.

Post, D.G. 1982. Feeding behavior of yellow baboons (*Papio cynocephalus*) in the Amboseli National Park, Kenya. *International Journal of Primatology* 3: 403–430.

Pott, A. & Pott, V. 1994. *Plantas do Pantanal*. Brazil: Embrapa.

Pott, A., Oliveira, A.K.M., Damasceno-Junior, G.A. & Silva, J.S.V. 2011. Plant diversity of the Pantanal wetland. *Brazilian Journal of Biology* 71: 265–273.

Poulsen, J.R. & Clark, C.J. 2004. Densities, distributions, and seasonal movements of gorillas and chimpanzees in swamp forest in northern Congo. *International Journal of Primatology* 25: 285–306.

Pourrut, X., Galat-Luong, A. & Galat, G. 1996. Associations du Singe vert avec d'autres especes de Primates au Sénégal: la transmission interspécifique du SIVagm doit être fréquente dans la nature. *Revue de Médecine Vétérinaire* 147: 47–58.

Powell, C.B. 1993. *Sites and species of conservation interest in the central axis of the Niger Delta (Yenagoa, Sagbama, Ekeremor and Southern Ijo Local Government Areas)*. Unpublished report. Abuja, Nigeria: Natural Resources Conservation Council (NARESCON).

Powell, C.B. 1995. *Wildlife Study I: Final Report*. Unpublished report. Port Harcourt, Nigeria: Environmental Affairs Department, Shell Petroleum Development Company of Nigeria Ltd.

Powell, C.B. 1997. Discoveries and priorities for mammals in the freshwater forests of the Niger Delta. *Oryx* 31: 83–85.

Prance, G.T. 1979. Notes on the vegetation of Amazonia III. The terminology of Amazonian forest types subject to inundation. *Brittonia* 31: 26–38.

Price, E.C., Piedade, H.M. & Wormell, D. 2002. Population densities of primates in a Brazilian Atlantic forest. *Folia Primatologica* 73: 54–56.

Price, R. 2011. *Rainforest Warriors: Human Rights on Trial*. Philadelphia, PA: University of Pennsylvania Press.

Pringle, C.M., Freeman, M.C. & Freeman, B.J. 2000. Regional effects of hydrologic alterations on riverine macrobiota in the New World: tropical-temperate comparisons. *BioScience* 50: 807–823.

Printes, R.C. 2007. *Avaliação taxonômica, distribuição e status do guigó-da-Caatinga* Callicebus barbarabrownae *Hershkovitz, 1990 (Primates: Pitheciidae)*. Unpublished PhD thesis, Universidade Federal de Minas Gerais.

ProtectedPlanet 2015. ProtectedPlanet. Gland, Switzerland: UNEP/WCMC/ IUCN/WCPA. https://www.protectedplanet.net.

Pruetz, J.D. & Bertolani, P. 2009. Chimpanzee (*Pan troglodytes*) behavioral response to stresses associated with living in a savanna-mosaic environment: implications for hominin adaptations to open habitats. *PaleoAnthropology*: 252–262.

PT Arara Abadi, Asia Pulp & Paper/Sinar Mas Group 2005. *Rainforest Alliance SmartWood Program High Conservation Value Forest (HCVF)*. Assessment Report for Siak District. Bogor, Indonesia: PT Arara Abadi, Asia Pulp & Paper/Sinar Mas Group.

Pusey, B.J. & Arthington, A.H. 2003. Importance of the riparian zone to the conservation and management of freshwater fish: a review. *Marine & Freshwater Research* 54: 1–16.

Queiroz, H.L. 1991. Levantamento das populações de três espécies de primatas na Fazenda Sucupira, Brasília, DF. In *A Primatologia no Brasil - 3*, M.T. de Mello (ed.). Brasília: Sociedade Brasileira de Primatologia, pp. 369–374.

Queiroz, H.L. de 1995. Preguicas e guaribas: os mamiferos folivoros arboricolas do Mamiraua. Conselho Nacional de Desenvolvimento Ciencia e Tecnologia, Brasilia. 1995.

Queiroz, H.L. & Valsecchi, J. 2005. Efeito da variação do sistema reprodutivo de guaribas vermelhos *Alouatta seniculus*, em florestas alagadas e florestas de terra firme na Amazônia Central, sobre a sustentabilidade de uso, com base em simulações de modelos estocásticos populacionais. *Abstracts of the XI Congresso Brasileiro de Primatologia (February 13–18 Porto Alegre, Brazil).*

Quéméré, E., Crouau-Roy, B., Rabarivola, C., Louis Jr., E.E. & Chikhi, L. 2010. Landscape genetics of an Endangered lemur (*Propithecus tattersalli*) within its entire fragmented range. *Molecular Ecology* 19: 1606–1621.

Quinten, M., Waltert, M., Syamsuri, F. & Hodges, J.K. 2010. Peat swamp forest supports high primate densities on Siberut Island, Sumatra, Indonesia. *Oryx* 44: 147–151.

R Development Core Team. 2012/2013/2014. R: a language and environment for statistical computing. Version 2.15. Vienna Austria: R Foundation for Statistical Computing.

Rabearivony, J., Fanameha, E., Mampiandra, J. & Thorstrom, R. 2008. Taboos and social contracts: tools for ecosystem management-lessons from the Manambolomaty Lakes Ramsar site, western Madagascar. *Madagascar Conservation & Development* 3: 7–16.

Rabearivony, J., Thorstrom, R., Rene de Roland, L.-A., *et al.* 2010. Protected area surface extension in Madagascar: do endemism and threatened species remain useful criteria for site selection? *Madagascar Conservation & Development* 5: 35–47.

Raboy, B.E. & Dietz, J.M. 2004. Diet, foraging, and use of space in wild golden-headed lion tamarins. *American Journal of Primatology* 63: 1–15.

Raemaekers, J.J. & Raemaekers, P.M. 1985. Field playback of loud calls to gibbons (*Hylobates lar*): territorial, sex-specific and species-specific responses. *Animal Behaviour* 33: 481–493.

Ragusa-Netto, J. 2004. Flowers, fruits and the abundance of the yellow-chevroned parakeet (*Brotogeris chiriri*) at a gallery forest in the south Pantanal (Brazil). *Brazilian Journal of Biology* 64: 867–877.

Rainboth, W.J. 1996. The Present-day Mekong. In *Fishes of the Cambodian Mekong*, W.J. Rainboth (ed.) FAO Species Identification Field Guides. Rome: Food and Agriculture Organization of the United Nations, pp. 1–5.

Rainey, H.J., Iyenguet, F.C., Malanda, G.-A.F., *et al.* 2010. Survey of Raphia swamp forest, Republic of Congo, indicates high densities of Critically Endangered western lowland gorillas *Gorilla gorilla gorilla*. *Oryx* 44: 124–132.

Rajanathan, R. & Bennett, E.L. 1990. Notes on the social behaviour of wild proboscis monkey (*Nasalis larvatus*). *Malayan Nature Journal* 44: 35–44.

Rakotondravony, D., Goodman, S.M. & Soarimalala, V. 1998. Predation on *Hapalemur griseus griseus* by *Boa manditra* (Boidae) in the littoral forest of eastern Madagascar. *Folia Primatologica* 69: 405–408.

Rakotondravony, R. & Radespiel, U. 2009. Varying patterns of coexistence of two mouse lemur species (*Microcebus ravelobensis* and *M. murinus*) in a heterogeneous landscape. *American Journal of Primatology* 71: 928–938.

Ralainasolo, F.B. 2004. Action des effets anthropiques sur la dynamique de la population de Hapalemur griseus alaotrensis ou 'Bandro' dans son habitat naturel. *Lemur News* 9: 32–35.

Ralainasolo, F.B., Waeber, P.O., Ratsimbazafy, J., Durbin, J. & Lewis, R. 2006. The Alaotra gentle lemur: population estimation and subsequent implications. *Madagascar Conservation & Development* 1: 9–10.

Ramirez, M.L., Freese, C.H. & Revilla, C.J. 1977. Feeding ecology of the pygmy marmoset, *Cebuella pygmaea*, in northeastern Peru. In *The Biology and Conservation of the Callitrichidae*, D.G. Kleiman (ed.). Washington, DC: Smithsonian Institution Press, pp. 91–104.

Ramsar. 2001. *The Annotated Ramsar Site List: Mexico*. Gland, Switzerland: Ramsar. www.ramsar.org/cda/en/ramsar-pubs-notes-anno-mexico/main/ramsar/1-30-168%5E16517_4000_0_.

Ramsar. 2013. *Signatories and Sites*. www.ramsar.org.

Ramsar. 2015. *Ramsar Sites Information Service and Sites Database*. Gland, Switzerland: Ramsar. www.ramsar.org/sites/default/files/documents/library/sitelist.pdf.

Ramsar. 2017. *Montreux Record*. Gland, Switzerland: Ramsar. http://archive.ramsar.org/cda/en/ramsar-documents-montreux-montreux-record/main/ramsar/1-31-118%5E20972_4000_0__.

Ramsar. 2018. *The List of Wetlands of International Importance*. Gland, Switzerland: Ramsar. www.ramsar.org/sites/default/files/documents/library/sitelist.pdf.

Ranarijaona, H.L.T. 2007. Zone humides. Concept de modèle écologique pour la zone humide Alaotra. *Madagascar Conservation & Development* 2: 35–42.

Rand, A.L. 1935. On the habits of some Madagascar mammals. *Journal of Mammalogy* 16: 89–104.

Randriamalala, H. & Liu, Z. 2010. Rosewood of Madagascar: between democracy and conservation. *Madagascar Conservation & Development* 5: 11–22.

Randrianja, S. 2012. Love me tender – transition vers où? *Madagascar Conservation & Development* 7: 9–16.

Raño, M. 2010. *Determinantes del desplazamiento diario en monos aulladores negros y dorados* (Alouatta caraya). BSc thesis, Universidad de Buenos Aires, Buenos Aires.

Rasmussen, D.T. 2007. Fossil record of the primates from the Paleocene to the Oligocene. In *Handbook of Palaeoanthropology*, W. Henke & I. Tattersall (eds). Berlin: Springer, pp. 889–920.

Rasmussen, D.T., Conroy, G.C. & Simons, EL. 1998. Tarsier-like locomotor specializations in the Oligocene primate *Afrotarsius*. *Proceedings of the National Academy of the Sciences of the United States of America* 95: 14848–14850.

Rasoavahiny, L., Andrianarisata, M. & Razafimpahanana, A. & Ratsifandrihamanana, A.N. 2008. Conducting an ecological gap analysis for the new Madagascar protected area system. *Parks* 17: 12–21.

Ratsimbazafy, J.R., Ralainasolo, F.B., Rendigs, A., *et al.* 2013. Gone in a puff of smoke? *Hapalemur alaotrensis* at great risk of extinction. *Lemur News* 17: 14–18.

Rausch, J.M. 1993. *Uma frontera de la sabana tropical: Los Llanos de Colombia 1531–1831*. Bogotá, Colombia: Banco de la Republica & Ancora Editors.

Rausch, J.M. 1999. *La frontera de los Llanos en la historia de Colombia (1830–1930)*. Bogotá, Colombia: Banco de la Republica & Ancora Editors.

Rausch, J.M. 2013. *Territorial rule in Colombia and the transformation of the Llanos Orientales*. Gainseville, FL: Florida University Press.

Razanadrakoto, D. 2004. *Rapport Annuel 2003 CIRPRH*. Ambatondrazaka, Madagascar: Circonscription de la Pêche et des Ressources Halieutiques.

Rebelo, L.M., Finlayson, C.M. & Nagabhatla, N. 2009. Remote sensing and GIS for wetland inventory, mapping and change analysis. *Journal of Environmental Management* 90: 2144–2153.

Redford, K.H. & Fonseca, G.A.B. 1986. The role of gallery forests in the zoogeography of the cerrado's non-volant mammalian fauna. *Biotropica* 18: 126–135.

Redmond, I. 2008. *The Primate Family Tree: The Amazing Diversity of our Closest Relatives*. Richmond, Ontario: Firefly Books.

Reed, K.E. 1997. Early hominid evolution and ecological change through the African Plio-Pleistocene. *Journal of Human Evolution* 32: 289–322.

Reed, K.E. & Bidner, L.R. 2004. Primate communities: past, present, and possible future. *Yearbook of Physical Anthropology* 47: 2–39.

Reibelt, L.M., Woolaver, L., Moser, G., *et al.* 2017. Contact matters: local people's perceptions of *Hapalemur alaotrensis* and implications for conservation. *International Journal of Primatology* 38: 588–608.

Reid, F. 1997. *A Field Guide to the Mammals of Central America and Southeast Mexico*. New York: Oxford University Press.

Reinartz, G.E., Guislain, P., Mboyo Bolinga, T.D., *et al.* 2008. Ecological factors influencing bonobo density and distribution in the Salonga National Park: Applications for population assessment. In *The Bonobos*, T. Furuichi & J. Thompson (eds). New York: Springer, pp. 167–188.

Reinartz, G.E., Isia, I.B., Ngamankosi, M. & Wema, L.W. 2006. Effects of forest type and human presence on bonobo (*Pan paniscus*) density in the Salonga National Park. *International Journal of Primatology* 27: 603–634.

Remsen, Jr, J.V. & Parker III, T.A. 1983. Contribution of river-created habitats to bird species richness in Amazonia. *Biotropica* 15: 223–231.

Ren, B., Li, M., Long, Y., Grüter, C.C. & Wei, F. 2008. Measuring daily ranging distances of *Rhinopithecus bieti* via a global positioning system collar at Jinsichang, China: a methodological consideration. *International Journal of Primatology* 29: 783–794.

Rendigs, A., Radespiel, U., Wrogemann, D. & Zimmermann, E. 2003. Relationship between microhabitat structure and distribution of mouse lemurs (*Microcebus* spp.) in northwestern Madagascar. *International Journal of Primatology* 24: 47–64.

Rene de Roland, L.A., Sam, T.S., Rakotondratsima, M.P.H. & Thorstrom, R. 2007. Rediscovery of the Madagascar Pochard *Aythya innotata* in northern Madagascar. *Bulletin of the African Bird Club* 14: 171–174.

Renofalt, B.M., Jansson, R. & Nilsson, C. 2005. Spatial patterns in plant invasiveness in a riparian corridor. *Landscape Ecology* 20: 165–176.

Restrepo, J.D. & Syvitski, J.P.M. 2006. Assessing the effect of natural controls and land use change on sediment yield in a major Andean River: the Magdalena Drainage Basin, Colombia. *Ambio* 35: 65–74.

Reynolds, V., Plumptre, A.J., Greenham, J. & Harborne, J. 1998. Condensed tannins and sugars in the diet of chimpanzees (*Pan troglodytes schweinfurthii*) in the Budongo Forest, Uganda. *Oecologia* 115: 331–336.

Reys, P., Sabino, J. & Galetti, M. 2009. Frugivory by the fish *Brycon hilarii* (Characidae) in western Brazil. *Acta Oecologica* 35: 136–141.

Reza, A., Feeroz, M. & Islam, M. 2001. Food habits of the Bengal tiger (*Panthera tigris tigris*) in the Sundarbans. *Bangladesh Journal of Zoology* 29: 173–179.

Ribeiro, M.C., Metzger, J.P., Martensen, A.C., Ponzoni, F.J. & Hirota, M.M. 2009. The Brazilian Atlantic forest: How much is left, and how is the remaining forest distributed? Implications for conservation. *Biological Conservation* 142: 1141–1153.

Ribeiro, J.S.B. & Darwich, A.S. 1993. Produção primária fitoplanctbnica de um lago de ilha fluvial na Amazônia Central (Lago do Rei, Ilha do Careiro). *Amazoniana* 12: 365–384.

Rice, W.R. 1989. Analyzing tables of statistical tests. *Evolution* 43: 223–225.

Richard, A.F., Goldstein, S.J. & Dewar, R.E. 1989. Weed macaques: the evolutionary implications of macaque feeding ecology. *International Journal of Primatology* 10: 569–594.

Richard-Hansen, C., Vié, J.-C. & de Thoisy, B. 2000. Translocation of red howler monkeys (*Alouatta seniculus*) in French Guiana. *Biological Conservation* 93: 247–253.

Richmond, B.G. & Whalen, M. 2001. Forelimb function, bone curvature and phylogeny of *Sivapithecus*. In *Hominoid Evolution and Climate Change Europe: Phylogeny of the Neogene Hominoid Primates of Eurasia*, L. de Bonis, G.D. Koufos & P. Andrews (eds). Cambridge: Cambridge University Press, pp. 327–348.

Richter, C., Taufiq, A., Hodges, K., Ostner, J. & Schulke, O. 2013. Ecology of an endemic primate species (*Macaca siberu*) on Siberut Island, Indonesia. *Springerplus* 2: 137.

Rico-Gray, V. 1982. Estudio de la vegetación de la zona costera inundable del noroeste del estado de Campeche, México: Los Petenes. *Biótica* 7: 171–203.

Righini, N. 2014. *Primate nutritional ecology: The role of food selection, energy intake, and nutrient balancing in Mexican black howler monkey (*Alouatta pigra*) foraging strategies.* PhD thesis, University of Illinois at Urbana-Champaign, IL.

Righini, N., Fernández, V.A., Garber, P.A. & Rothman, J.M. 2014. Howler monkey nutritional geometry: what can we learn from an interspecific approach (abstract). *Amercian Journal of Primatology* 76: 98.

Rijksen, H.D. 1978. *A field study on Sumatran orangutans (*Pongo pygmaeus abelii, *Lesson 1827): ecology, behavior and conservation.* PhD thesis, Agricultural University, Wageningen, The Netherlands.

Rijksen, H.D. & Meijaard, E. 1999. *Our Vanishing Relative: The Status of Wild Orangutans at the Close of the Twentieth Century.* Dordrecht, The Netherlands: Kluwer Academic Publishers.

Riley, E.P. 2008. Ranging patterns and habitat use of Sulawesi Tonkean macaques (*Macaca tonkeana*) in a human-modified habitat. *American Journal of Primatology:* 70: 670–679.

Rímole, J. 2001. *Ecologia de macacos-prego (Cebus apella nigritus, Goldfuss, 1809) na estação biológica de Caratinga (MG): implicações para a conservação de fragmentos de Mata Atlântica.* MSc dissertation, Universidade Federal do Pará, Belém, Brazil.

Rímoli, J., Nante, R.S. & Lázaro, Jr, A.E. 2012. Diet and activity patterns of black howler monkeys *Alouatta caraya* (Humboldt, 1812, Primates, Atelidae) in ecotone Cerrado-Pantanal in the left bank of Aquidauana River, Mato Grosso do Sul, Brazil. *Oecologia Australis* 16: 933–948.

Rímoli, A.O., Rímoli, J., Valdivino, E.M. & Ferrari, S.F. 2008. Behavior patterns of a group of black howler monkeys (*Alouatta caraya*, Humboldt 1812) in a forest fragment in Terenos (Matto Grosso do Sul). In *A Primatologia no Brasil IX*, S.F. Ferrari & J. Rímoli (eds). Aracaju: Editora da Universidade Federal de Sergipe, pp. 179–191.

Robertson, S.A. & Luke, W.R.Q. 1993. *Kenya Coastal Forests: The Report of the NMK/WWF Coast Forest Survey.* Unpublished report. Nairobi: National Museums of Kenya and World Wide Fund for Nature.

Robbins, M.M. & McNeilage, A. 2003. Home range and frugivory patterns of mountain gorillas in Bwindi Impenetrable National Park, Uganda. *International Journal of Primatology* 24: 467–491.

Robins, R.J. 1976. The composition of the Jozani Forest, Zanzibar. *Botanical Journal of the Linnean Society* 72: 223–234.

Robinson, J.G. & Janson, C.H. 1987. Capuchins, squirrel monkeys and Atelines: socioecological convergence with Old World Monkey primates. In *Primate Societies*, B.B. Smuts, D.L. Cheney, R.M. Seyfarth, R.W. Wrangham & T.T. Struhsaker (eds). Chicago, IL: University of Chicago Press, pp. 69–82.

Robinson, N.A. 1992. International trends in environmental impact assessment. *Pace Law Faculty Publications*, Paper 382. New York: Pace University.

Robinson, T.B., Griffiths, C.L., Mcquaid, C.D. & Rius, M. 2005. Marine alien species of South Africa: status and impacts. *African Journal of Marine Science* 27:297–306.

Rocha, V.J., dos Reis, N.R. & Sekiama, M.L. 1998. Uso de ferramentas por *Cebus apella* (Linnaeus) (Primates, Cebidae) para obtenção de larvas de coleoptera que parasitam sementes de *Syagrus romanzoffianum* (Cham.) Glassm. (Arecaceae). *Revista Brasileira de Zoologia* 15: 945–950.

Rode, K.D., Chapman, C.A., McDowell, L.R. & Stickler, C. 2006. Nutritional correlates of population density across habitats and logging intensities in redtail monkeys (*Cercopithecus ascanius*). *Biotropica* 38: 625–634.

Rodgers, W.A. 1981. The distribution and conservation status of colobus monkeys in Tanzania. *Primates* 22: 33–45.

Rodman, P.S. 1973. Population composition and adaptive organisation among orangutans of the Kutai Reserve. In *Comparative Ecology and Behaviour of Primates*, R.P. Michael & J.H. Crook (eds). London: Academic Press, pp. 171–209.

Rodman, P. 1988. Diversity and consistency in ecology and behaviour. In *The Orangutan*, J. Schwarz (ed.). Oxford: Oxford University Press, pp. 31–51.

Rodman, P.S. & McHenry, H.M. 1980. Bioenergetics and the origin of hominid bepedalism. *American Journal of Physical Anthropology* 52: 103–106.

Rodrigues, R.E., Lima, R.A.F., Gandolfi, S. & Nave, A.G. 2009. On the restoration of high diversity forests: 30 years of experience in the Brazilian Atlantic forest. *Biological Conservation* 142: 1242–1251.

Rodrigues, M. 2006. Hidroelétrica, ecologia comportamental e resgate de fauna: uma falácia. *Natureza & Conservação* 4: 29–38.

Roger, E. & Andrianasolo, M. 2003. Mangroves and salt marshes. In *Natural History of Madagascar*, S. Goodman & J. Benstead (eds). Chicago, IL: Chicago University Press, pp. 209–210.

Rogers, M.E., Abernethy, K., Bermejo, M., *et al.* 2004. Western gorilla diet: a synthesis from six sites. *American Journal of Primatology* 64: 173–192.

Rogers, T.L., Ciaglia, M.B., Klinck, H. & Southwell, C. 2013. Density can be misleading for low-density species: benefits of passive acoustic monitoring. *PLoS ONE* 8: e52542.

Roman-Cuesta, R.M., Rejalaga-Noguera, L., Pinto-García, C. & Retana, J. 2014. Pacific and Atlantic oceanic anomalies and their interaction with rainfall and fire in Bolivian biomes for the period 1992–2012. *Climatic Change* 127: 243–256.

Romero-Ruiz, M.H., Flantua, S.G.A., Tansey, K. & Berrio, J.C. 2012. Landscape transformations in savannas of northern South America: land use/cover changes since 1987 in the Llanos Orientales of Colombia. *Applied Geography* 32: 766–776.

Romero-Valenzuela, D. & Rumiz, D.I. 2010. Aotidade. In *Distribuición, ecología y conservación de los Mamíferos Medianos y Grandes de Bolivia*, R.B. Wallace, H. Gómes, Z.R. Porcel & D.I. Rumiz (eds). Santa Cruz de la Sierra, Colombia: Centro de Ecología Difusión Simón I. Patiño, pp. 287–304.

Roopnaraine, T.R.R. 1996. *Freighted fortunes: gold and diamond mining in the Pakaraima Mountains, Guyana*. PhD thesis, University of Cambridge, UK.

Rosa, L.P., Santos, M.A., Matvienko, B., Santos, E.O. & Sikar, E. 2004. Greenhouse gas emissions from hydroelectric reservoirs in tropical regions. *Climatic Change* 66: 9–21.

Rosales, J., Vispo, C., Dezzeo, N., *et al.* 2002. Ecohydrology of riparian forests in the Orinoco River Basin. In *The Ecohydrology of South American Rivers and Wetlands*, M.E. McLain (ed.). Wallingford, UK: IAHS Press, pp. 93–110.

Rosales Godoy, J., Petts, G. & Salo, J. 1999. Riparian flooded forests of the Orinoco and Amazon basins: a comparative review. *Biodiversity & Conservation* 8: 551–586.

Rose, L. 1998. *Behavioral ecology of white-faced capuchins (*Cebus capucinus*) in Costa Rica*. PhD thesis, Washington University St. Louis, MI.

Rose, L.J., Perry, S., Panger, M.A., *et al.* 2003. Interspecific interactions between *Cebus capuchinus* and other species: data from three Costa Rican sites. *International Journal of Primatology* 24: 759–794.

Rosenzweig, M.L. 1968. Net primary productivity patterns of terrestrial communities: prediction from climatological data. *American Naturalist* 102: 67–74.

Rosevear, D.R. 1953a. Vegetation forestry and wildlife in Nigeria. In *The Nigeria Handbook*. Lagos: Crown Agents for the Colonies, pp. 1–35.

Rosevear, D.R. 1953b. *Checklist and Atlas of Nigerian Mammals*. Lagos, Nigeria: Government Printer.

Ross, C. & Reeve, N. 2011. Survey and census methods: population distribution and density. In *Field and Laboratory Methods in Primatology*, J.M. Setchell & D.J. Curtis (eds). Cambridge: Cambridge University Press, pp. 111–132.

Ross, S. & Wall, G. 1992. Evaluating ecotourism: the case of north Sulawesi, Indonesia. *Tourism Management: Research, Policies & Practice* 20: 673–682.

Rothman, J.M., Dierenfeld, E.S., Hintz, H.F. & Pell, A.N. 2008. Nutritional quality of gorilla diets: consequences of age, sex, and season. *Oecologia* 155: 111–122.

Roucoux, K.H., Lawson, I.T., Jones, T.D., *et al.* 2013. Vegetation development in an Amazonian peatland. *Palaeogeography, Palaeoclimatology, Palaeoecology* 374: 242–255.

Roux, V. & Bril, B. 2005. *Stone Knapping: The Necessary Conditions for a Uniquely Hominin Behaviour*. Cambridge: McDonald Institute for Archaeological Research.

Rowe, N., Goodall, J. & Mittermeier, R. 1996. *The Pictorial Guide to the Living Primates*. East Hampton, New York: Pogonias Press.

Rowe, N. & Myers, M. 2011. *All the World's Primates*. www.alltheworldsprimates.org.

Rowell, T.E. 1966. Forest living baboons in Uganda. *Journal of Zoology (London)* 149: 344–364.

Roy, D.P. & Boschetti, L. 2009. Southern Africa validation of the MODIS, L3JRC, and GlobCarbon burned-area products. *IEEE Transactions on Geoscience and Remote Sensing* 47: 1032–1044.

Rudicell, R.S., Piel, A.K., Stewart, F., *et al.* 2011. High prevalence of simian immunodeficiency virus infection in a community of savanna chimpanzees. *Journal of Virology* 85: 9918–9928.

Rudran, R. & Fernandez-Duque, E. 2003. Demographic changes over thirty years in a red howler population in Venezuela. *International Journal of Primatology* 24: 925–947.

Ruiz-García, M., Luengas-Villamil, K., Leguizamon, N., de Thoisy, B. & Gálvez, H. 2015. Molecular phylogenetics and phylogeography of all the Saimiri taxa (*Cebidae*, Primates) inferred from mtCOI and COII gene sequences. *Primates* 56: 145–161.

Rumiz, D.I. 1990. *Alouatta caraya*: Population density and demography in northern Argentina. *American Journal of Primatology* 21: 279–294.

Rumiz, D.I. 2012. Distribution, habitat and status of the white-coated titi monkey (*Callicebus pallecens*) in the Chaco-chiquitano forests of Santa Cruz, Bolivia. *Neotropical Primates* 19: 8–15.

Rumiz, D.I., Zunino, G.E., Obregozo, M.L. & Ruiz, J.C. 1986. *Alouatta caraya*: Habitat and resource utilization in northern Argentina. In *Current Perspectives in Primate Social Dynamics*, D.M. Taub & F.A. King (eds). New York: van Nost & Reinhold, pp. 175–193.

Runting, R.K., Meijaard, E., Abram. N.K., *et al.* 2015. Alternative futures for Borneo show the value of integrating economic and conservation targets across borders. *Nature Communications* 6: 6819.

Russo, G.A. & Shapiro, L.J. 2013. Reevaluation of the lumbosacral region of *Oreopithecus bambolii*. *Journal of Human Evolution* 65: 253–265.

Russon, A.E. & Wallis, J. (eds). 2014. *Primate Tourism: A Tool for Conservation?* Cambridge: Cambridge University Press.

Russon, A.E., Erman, A. & Dennis, R. 2001. The population and distribution of orangutans *(Pongo pygmaeus pygmaeus)* in and around the Danau Sentarum Wildlife Reserve, West Kalimantan, Indonesia. *Biological Conservation* 97: 21–28.

Russon, A.E., Wich, S.A., Ancrenaz, M., *et al.* 2008. Geographic variation in orangutan diets. In *Orangutans: Geographic Variation in Behavioral Ecology and Conservation*, S.A. Wich, S.S.U. Atmoko, T.M. Setia & C.P. van Schaik (eds). Oxford: Oxford University Press, pp. 135–156.

Rutledge, D. 2003. Landscape indices as measures of the effects of fragmentation: can pattern reflect process? *Doc Science Internal Series #98*, Wellington, New Zealand: Department of Conservation.

Rylands, A.B. 1993. *Marmoset and Tamarins: Behaviour and Ecology*. New York: Oxford University Press.

Rylands, A.B., Fonseca, G.A.B., Leite, Y.L.R. & Mittermeier, R.A. 1996. Primates of the Atlantic Forest: origins, distributions, endemism and communities. In *Adaptive Radiations of Neotropical Primates*, M.A. Norconk, A.L. Rosenberger & P.A. Garber (eds). New York: Plenum Press, pp. 21–51.

Rylands, A., Groves, C.P., Mittermeier, R.A., Cortes-Ortiz, L. & Hines, J.J.H. 2006. Taxonomy and distributions of Mesoamerican primates. In *New Perspectives in the Study of Mesoamerican Primates: Distribution, Ecology, Behavior and Conservation*, A. Estrada, P.A. Garber, M.S.M. Pavelka & L. Luecke (eds). New York: Springer, pp. 29–79.

Rylands, A.B. & Keuroghlian, A. 1988. Primate populations in continuous forest and forest fragments in central Amazonia. *Acta Amazonica* 18: 291–307.

Rylands, A.B. & Mittermeier, R.A. 2009. The diversity of the New World primates (Platyrrhini): and annotated taxonomy. In *South American Primates: Comparative Perspectives in the Study of Behavior, Ecology, and Conservation*, P.A. Garber, A. Estrada, J.C. Bicca-Marques, E.W. Heymann & K.B. Strier (eds). New York: Springer, pp. 23–54.

Rylands, A.B., Mittermeier, R.A. & Silva, Jr, J.S. 2012. Neotropical primates: taxonomy and recently described species and subspecies. *International Zoo Yearbook* 46: 11–24.

Rylands, A.B., Mittermeier, R.A. & Wallace, R.B. 2008. *Mico melanurus*. In *IUCN Red List of Threatened Species*. Version 2011.2. Gland, Switzerland: IUCN. www.iucnredlist.org.

Sabah Forestry Enactment 1968. *Forest Enactment* No. 2 of 1968. p. 49. Kota Kinabalu: Sabah Government Printing.

Sabah Wildlife Department 2011. *Orangutan Action Plan*. Kota Kinabalu: Sabah Wildlife Department.

Sabatier, D. 1985. Saisonnalite et determinisme du pic de fructification en forest guyanaise. *Revue d'Écologie: La Terre et La Vie* 40: 289–320.

Sabatini, V. & Ruiz-Miranda, C.R. 2008. Acoustical aspects of the propagation of long call of wild *Leontopithecus rosalia*. *International Journal of Primatology* 29: 207–223.

Sacristán-Soriano, O., Banaigs, B. & Becerro, M.A. 2012. Temporal trends in the secondary metabolite production of the sponge *Aplysina aerophoba*. *Marine Drugs*, 10: 677–693.

Sakai, S. 2002. General flowering in lowland mixed dipterocarp forests of Southeast Asia. *Biological Journal of the Linnean Society* 75: 233–247.

Sakamaki, T. 1998. First record of algae-feeding by a female chimpanzee at Mahale. *Pan Africa News* 5: 1–3.

Salera, Jr, G., Malvasio, A. & Portelinha, T.C.G. 2009. Evaluation of predation in *Podocnemis expansa* and *Podocnemis unifilis* (Testudines, Podocnemididae) in the Javaés River, Tocantins. *Acta Amazonica* 39: 207–213.

Salgado-Lynn, M. 2010. *Primate viability in a fragmented landscape: genetic diversity and parasite burden o f long-tailed macaques and proboscis monkeys in the Lower Kinabatangan Floodplain, Sabah, Malaysia*. PhD thesis, Cardiff University, Cardiff, UK.

Salo, J., Kalliola, R., Hakkinen, I., *et al.* 1986. River dynamics of Amazon lowland forest. *Nature* 322: 254–258.

Salter, R.E., MacKenzie, N.A., Nightingale, N., Aken, K.M. & Chai, P.K.P. 1985. Habitat use, ranging behaviour, and food habits of the proboscis monkey, *Nasalis larvatus (van Wurmb), in Sarawak*. *Primates* 26: 436–451.

Samson, M.S. & Rollon, R.N. 2008. Growth performance of planted mangroves in the Philippines: revisiting forest management strategies. *Ambio* 37: 234–240.

Sánchez-Cuervo, A.M., Aide, T.M., Clark, M.L. & Etter, A. 2012. Land cover change in Colombia: surprising forest recovery trends between 2001 and 2010. *PLOS ONE* 7: e43943

Sander, N.L. 2014. *Estrutura, composição florística e etnobiologia de um buritizal na fronteira biológica Amazônia-Cerrado*. MSc dissertation, Universidade do Estado de Mato Grosso, Cáceres Mato Grosso, Brazil.

Sanderson, E.W., Jaiteh, M., Levy, M.A., Redford, K.H., Wannebo, A.V. & Woolmer, G. 2002. The human footprint and the last of the wild. *BioScience* 52: 891–904.

Sangchantr, S. 2004. *Social Organization and Ecology of Mentawai Leaf Monkeys*. PhD thesis, Columbia University, New York.

Santos, E.R. 2013. *El impacto del turismo ecológico en el comportamiento del mono carablanca (Cebus capucinus, Linnaeus, 1758) en el Parque Manuel Antonio, Costa Rica*. MSc dissertation, Universidad de Costa Rica, San José, Costa Rica.

Santos, I.B., Mittermeier, R.A., Rylands, A.B. & Valle, C. 1987. The distribution and conservation status of primates in southern Bahia, Brazil. *Primate Conservation* 8: 126–142.

Santos, R.R. 2010. *Uso de ferramentas por macacos-prego em manguezais*. PhD thesis, Universidade Federal do Rio Grande do Norte, Natal, Brazil.

Santos, R.R., Ferreira, R.G. & Araujo, A. 2016. Capuchin monkeys in Amazonian Mangroves. In *Phylogeny, Molecular Population Genetics, Evolutionary Biology and Conservation of the Neotropical Primates*, M. Ruiz-Garcia & J.M. Shostell (eds). New York: Nova Publisher, pp. 479–489.

Sanyal, P. 1983. Mangrove tiger land: the Sundarbans of India. *Tigerpaper* 10: 1–4.

Sanyal, P.1992. Sundarbans mangrove: wildlife potential and conservation. In *Tropical Ecosystems: Ecology and Management*, K.P. Singh & J.S. Singh (eds). New Delhi: Wiley Eastern Ltd, pp. 309–313.

Sarkar, A.K., Ghosh, D.C. & Naskar, K.R. 1993. Phenological observations of the different mangrove flora of Sundarbans in India with special reference to their salinity tolerance. In *Asia-Pacific Symposium on Mangrove Ecosystems. Y-S held at Hong Kong University of Science and Technology*. Y.S. Wong & N.F.Y. Tam (eds). Dortrecht, The Netherlands: Springer Science, pp. 312–333.

Satyasari, I. 2010. *Evaluasi pengembangan ekowisata mangrove: studi kasus di Bedul, resort Grajagan, Taman Nasional Alas Purwo, Jawa Timur*. BSc dissertation, Institut Pertanian Bogor, Bogor, Indonesia.

Sauther, M.L., Wright, B. & Orndorff, K. 1998. A preliminary census and study of habitat use of the primates of the northern Iwokrama Reserve, Guyana. *American Journal of Physical Anthropology* Supplement 26: 195.

Savage, A., Thomas, L., Leighty, K.A., Soto, L.H. & Medina, F.S. 2010. Novel survey method finds dramatic decline of wild cotton-top tamarin population. *Nature Communications* 1: 30–35.

Scabin, A.B., Costa, R.C. & Schöngart, J. 2012. The spatial distribution of illegal logging in the Anavilhanas archipelago (central Amazonia) and logging impacts on species. *Environmental Conservation* 39: 111–121.

Schaeffer-Novelli, Y., Cintrón-Molero, G., Adaime, R.R. & Camargo, T.M. 1990. Variability of mangrove ecosystems along the Brazilian coast. *Estuaries* 13: 204–218.

Schaller, G.B. 1963. *The Mountain Gorilla: Ecology and Behavior*. Chicago, IL: University of Chicago Press.

Scheyvens, R. 1999. Ecotourism and the empowerment of local communities. *Tourism Management* 20: 245–249.

Schipper, J. 2007. Camera-trap avoidance by Kinkajous *Potos flavus*: rethinking the 'non-invasive' paradigm. *Small Carnivore Conservation* 36: 38–41.

Schmid, J. & Kappeler, P.M. 1994. Sympatric mouse lemurs (*Microcebus* spp.) in western Madagascar. *Folia Primatologica* 63: 162–170.

Schmitt, C.B., Burgess, N.D., Coad, L., *et al.* 2009. Global analysis of the protection status of the world's forests. *Biological Conservation* 142: 2122–2130.

Schneider, H. & Sampaio, I. 2015. The systematics and evolution of New World primates, a review. *Molecular Phylogenetics and Evolution* 82: 348–357.

Schneider, M. 2001. *Mastofauna da bacia hidrográfica do rio Manso, MT - Uma abordagem de Ecologia de Paisagem para avaliação da perda de hábitats*. PhD thesis, São Carlos: Universidade Federal de São Carlos.

Schoeninger, M.J., Bunn, H.T., Murray, S.S. & Marlett, J.A. 2001. Composition of tubers used by Hadza foragers of Tanzania. *Journal of Food Composition Analysis* 14: 15–25.

Schoneveld, G.C. 2014. The politics of the forest frontier: negotiating between conservation, development, and indigenous rights in Cross River State. *Land Use Policy* 38: 147–162.

Schöngart, J., Piedade, M.T.F., Ludwigshausen, S., Horna, V. & Worbes, M. 2002. Phenology and stem-growth periodicity of tree species in Amazonian floodplain forests. *Journal of Tropical Ecology* 18: 581–597.

Schöngart, J., Piedade, M.T.F., Wittman, F., Junk, W.J. & Worbes, M. 2005. Wood growth patterns of *Macrolobium acaciifolium* in Amazonian black-water and white-water floodplain forests. *Oecologia* 145: 454–461.

Schöngart, J., Wittmann, F. & Worbes, M. 2010. Biomass and net primary production of central Amazonian floodplain forests. In *Amazonian Floodplain Forests: Ecophysiology, Biodiversity and Sustainable Management*, W. Junk, M. Piedade, F. Wittmann, J. Schöngart & P. Parolin (eds). Dordrecht, The Netherlands: Springer, pp. 347–388.

Schulze, C.H., Waltert, M., Kessler, P.J., *et al.* 2004. Biodiversity indicator groups of tropical land-use systems: comparing plants, birds, and insects. *Ecological Applications* 14: 1321–1333.

SCM (Sociedade Civil Mamirauá). 1996. *Plano de Manejo da RDS Mamirauá*. Tefé: SCM.

Scott, K.S.S. 2012. *Assessing the population of proboscis monkeys and threats to their survival in Balikpapan Bay, East Kalimantan, Indonesia: a preliminary study*. MSc thesis, Oxford Brookes University, Oxford, UK.

Scott, J.E., McAbee, K.R., Eastman, M.M. & Ravosa, M.J. 2014. Experimental perspective on fallback foods and dietary adaptations in early hominins. *Biology Letters* 10: 20130789.

Scott, N.J., Struhsaker, T.T., Glander, K. & Chirivi, H. 1976. Primates and their habitats in northern Colombia, with recommendations for future management and research. *Pan American Health Organization of Science Publication* 317: 30–50.

SDAM. 1990. *Etude et elaboration du schéma directeur de l'aménagement de la mangrove de Guinée*. Unpublished report. Conakry, Guinea: Ministere de l'Agriculture et des Ressources Animales.

Segelbacher, G., Cushman, S.A., Epperson, B.K., *et al.* 2010. Applications of landscape genetics in conservation biology: concepts and challenges. *Conservation Genetics* 11: 375–385.

Seiffert, E.R., Simons, E.L. & Attia, Y. 2003. Fossil evidence for an ancient divergence of lorises and galagoes. *Nature* 422: 421–424.

Self, R.M., Self, D.R. & Bell-Haynes, J. 2010. Marketing tourism in the Galapagos Islands: ecotourism or greenwashing? *International Business & Economics Research Journal* 9: 111–125.

Semesi, A.K. 1998. Mangrove management and utilization in Eastern Africa. *Ambio* 27(8): 620–626.

Señaris, J.C. & Ayarzagüena, J. 2004. Contribución al conocimiento de la anurofauna del delta del Orinoco, Venezuela: diversidad, ecología y biogeografía. *Memoria de la Fundación La Salle de Ciencias Naturales* 62: 129–152.

Serckx, A., Huynen, M.C., Bastin, J.F., *et al.* 2014. Nest grouping patterns of bonobos (*Pan paniscus*) in relation to fruit availability in a forest-savannah mosaic. *PLoS ONE* 9: e93742.

Serio-Silva, J.C., Bonilla-Sánchez, Y.M., Pozo-Montuy, G., Reyna-Hurtado, R. & Chapman, C.A. 2013. Identifying areas for ecotourism and conservation of threatened species: the model of black howler monkey in playas de Catazajá, Mexico. In *Ecological Dimension for Sustainable Socioeconomic Development*, A. Yañez-Arancibia, R. Dávalos-Sotelo & E. Reyes (eds). Southampton, MA: WIT Press, pp. 347–368.

Serio-Silva, J., Rico-Gray, V. & Ramos-Fernández, G. 2006. Mapping primate populations in the Yucatan Peninsula, Mexico: a first assessment. In *New Perspectives in the Study of Mesoamerican Primates: Distribution, Ecology, Behavior and Conservation*, A. Estrada, P.A. Garber, M.S.M. Pavelka & L. Luecke (eds). New York: Springer, pp. 489–511.

Setchell, J.M. & Curtis, D.J. (eds). 2011. *Field and Laboratory Methods in Primatology: A Practical Guide*. Cambridge: Cambridge University Press.

Seymour, C. 2013. Saharan flooded grasslands. New York: WWF *Ecoregions*. http://worldwildlife.org/ecoregions/at0905.

Sha, J.C.M. 2006. *Distribution, abundance and conservation of proboscis monkey (Nasalis larvatus) in Sabah, Malaysia*. Masters thesis, Universiti Malaysia Sabah.

Sha, J.C.M., Bernard, H. & Nathan, S. 2008. Status and conservation of proboscis monkeys (*Nasalis larvatus*) in Sabah, east Malaysia. *Primate Conservation* 23: 107–120.

Sha, J.C.M., Gumert, M.D., Lee, B.P.Y.-H., *et al.* 2009. Macaque-human interactions and the societal perceptions of macaques in Singapore. *American Journal of Primatology* 71: 1–15.

Sha, J.C.M. & Hanya, G. 2013. Temporal food resource correlates to the behavior and ecology of food-enhanced long-tailed macaques (*Macaca fascicularis*). *Mammal Study* 38: 163–175.

Sha, J., Matsuda, I. & Bernard, H. 2011. *The Natural History of the Proboscis Monkey*. Kota Kinabalu, Borneo: Natural History Publications.

Shaffer, C.A. 2012. *Ranging behavior, group cohesiveness, and patch use in northern bearded sakis (Chiropotes sagulatus) in Guyana*. PhD thesis, Washington University, St. Louis, MO.

Shaffer, C.A. 2013a. Feeding ecology of bearded sakis (*Chiropotes sagulatus*) in Guyana. *American Journal of Primatology* 75: 568–580.

Shaffer, C.A. 2013b. Ecological correlates of ranging behaviour of northern bearded sakis (*Chiropotes sagulatus*) in a continuous forest in Guyana. *International Journal of Primatology* 34: 515–532.

Shaffer, C.A. 2014. Preliminary study of the sustainability of primate hunting among indigenous Waiwai in the Kanashen Community Owned Conservation Area, Guyana. (Abstract) *American Journal of Physical Anthropology* 156: 284–285.

Shanee, N. 2012. Trends in local wildlife hunting, trade and control in the Tropical Andes Biodiversity Hotspot, northeastern Peru. *Endangered Species Research* 19: 177–186.

Sharma, R., Arora, N., Goossens, B., *et al.* 2012. Effective population size dynamics and the demographic collapse of Bornean orang-utans. *PLoS ONE* 7: e49429.

Shekelle, M. 2003. *Taxonomy and biogeography of eastern tarsiers*. PhD thesis, Washington University, St Louis, MO.

Shekelle, M. 2013. Observations of Wild Sangihe Island tarsiers (*Tarsius sangirensis*). *Asian Primates* 3: 18–23.

Shekelle, M., Gursky-Doyen, S. & Richardson, M. 2013. Tarsiidae. In *Handbook of the Mammals of the World: Volume 3-Primates*, R.A. Mittermeier, A.B. Rylands & D. Wilson (eds). Arlington, VA: Conservation International & Barcelona: Lynx Edicions, pp. 236–261.

Shekelle, M. & Salim, A. 2009. An acute conservation threat to two tarsier species in the Sangihe Island chain, North Sulawesi, Indonesia. *Oryx* 43: 419–426.

Shekelle, M., Meier, R., Wahyu, I., Wirdateti, W. & Ting, N. 2010. Molecular phylogenetics and chronometrics of Tarsiidae based on 12S mtDNA haplotypes: evidence for Miocene origins of crown tarsiers and numerous species within the Sulawesian clade. *International Journal of Primatology* 31: 1083–1106.

Shipman, P. & Harris, J.M. 1988. Habitat preference and paleoecology of *Australopithecus boisei* in Eastern Africa. In *Evolutionary History of the 'Robust' Australopithecines*, F.E. Grine (ed.). New York: Aldine, pp. 343–381.

Shizukuishi, S. & Narita, S. 2004. Characteristic of ingredients of unused seaweed. *Report of Aomori Prefectural Local Food Research Center* 2: 17–22.

Shorter Oxford English Dictionary. 2007. *Shorter Oxford English Dictionary*. Volume I. Oxford: Oxford University Press.

Shumaker, R., Walkup, K. & Beck, B. 2011. *Animal Tool Behavior: The Use and Manufacture of Tools by Animals*. Baltimore, MA: John Hopkins University Press.

Silcox, M.T., Sargis, E.J., Bloch, J.I. & Boyer, D.M. 2007. Primate origins and supraordinal relationships: morphological evidence. In *Handbook of Palaeoanthropology*, W. Henke & I. Tattersall (eds). Berlin: Springer, pp. 831–859.

Silk, R. 2016. *An Ecotourist's Guide to the Everglades and the Florida Keys*. Gainesville, FL: University Press of Florida.

Silva, C.R., Martins, A.C.M., de Castro, I.J., *et al.* 2013. Mammals of Amapá State, eastern Brazilian Amazonia: a revised taxonomic list with comments on species distributions. *Mammalia* 77: 409–424.

Silva, J.D. 1996. Distribution of Amazonian and Atlantic birds in gallery forests of the cerrado region, South America. *Ornitologia Neotropical* 7: 1–18.

Silva, J.M.C.D. & Oren, D.C. 1996. Application of parsimony analysis of endemicity in Amazonian biogeography: an example with primates. *Biological Journal of the Linnean Society* 59: 427–437.

Silva, S.S.B. & Ferrari, S.F. 2009. Behavior patterns of southern bearded sakis (*Chiropotes satanas*) in the fragmented landscape of eastern Brazilian Amazonia. *American Journal of Primatology* 71: 1–7.

Silva, Jr, J.S. 2001. *Especiação nos macacos-prego e caiararas, gênero* Cebus *Erxleben, 1777 (Primates, Cebidae)*. PhD thesis, Universidade Federal do Rio de Janeiro, Rio de Janeiro.

Silva, Jr, J.S. & Fernandes, M.E.B. 1999. A northeastern extension of the distribution of *Aotus infulatus* in Maranhão, Brazil. *Neotropical Primates* 7: 76–80.

Silva, Jr, J.S., Figueiredo-Ready, W.M.B. & Ferrari, S.F. 2013. Taxonomy and geographic distribution of the Pitheciidae. In *Evolutionary Biology and Conservation of Titis, Sakis and Uacaris*, L. Veiga, A. Barnett, S. Ferrari & M. Norconk (eds). Cambridge: Cambridge University Press, pp. 31–41.

Silva, Jr, J.S. & Martins, E.S. 1999. On a new white bald uacari population in southwestern Brazilian Amazonia. *Neotropical Primates* 7: 119–121.

Silva, Jr, J.S. & Queiroz, H.L. 2008. *Cacajao calvus rubicundus. In Livro Vermelho da Fauna Brasileira Ameaçada de Extinção*. ABM Machado, G.M. Drummond & A.P. Paglia (eds). Fundação Biodiversitas: Belo Horizonte, pp. 764–765.

Silveira, G., Malta, A.J.R. & Mendes Pontes, A.R. 2008. Gênero *Cebus* Erxleben 1777. In *Primatas Brasileiros, Volume 1*, N.R. dos Reis, A.L. Peracchi & F.R. Andrade (eds). Londrina, Brazil: Technical Books, pp. 25–33.

Silveira, L., Jacomo A.T. & Diniz-Filho J.A.F. 2003. Camera trap, line transect census and track surveys: a comparative evaluation. *Biological Conservation* 114: 351–355.

Simasathien, C., Jansang, A., Jaikaeo, C., *et al.* 2015. Camera trap synchronization for wildlife monitoring system. *International Conference on Information and Convergence Technology for Smart Society* 21–24 January 2015 in Bangkok, Thailand.

Simmen, B., Hladik, A. & Ramasiarisoa, P. 2003. Food intake and dietary overlap in native *Lemur catta* and *Propithecus verreauxi* and introduced *Eulemur fulvus* at Berenty, southern Madagascar. *International Journal of Primatology* 24: 949–968.

Simmen, B., Sauther, M.L., Soma, T., *et al.* 2006. Plant species fed on by *Lemur catta* in gallery forests of the southern domain of Madagascar. In *Ringtailed Lemur Biology: Lemur catta in Madagascar*, A. Jolly, N. Koyama, H.R. Rasamimanana & R.W. Sussman (eds). New York: Springer, pp. 55–68.

Simpson, S.J., Batley, R. & Raubenheimer, D. 2003. Geometric analysis of macronutrient intake in humans: the power of protein? *Appetite* 41: 123–140.

Simpson, S.J. & Raubenheimer, D. 2005. Obesity: the protein leverage hypothesis. *Obesity Reviews* 6: 133–142.

Singleton, I. 2000. *Ranging behaviour and seasonal movements of Sumatran orangutans (Pongo pygmaeus abelii) in swamp forests*. PhD thesis, University of Kent, Canterbury, UK.

Singleton, I. & van Schaik, C.P. 2001. Orangutan home range size and its determinants in a Sumatran swamp forest. *International Journal of Primatology* 22: 877–911.

Singleton, S., Knott, C.D., Morrogh-Bernard, H.C., Wich, S.A. & van Schaik, C.P. 2009. Ranging behavior of orangutan females and social organization. In *Orangutans: Geographic Variation in Behavioral Ecology and Conservation*, S.A. Wich, S.S.U. Atmoko, T.M. Setia & C.P. van Schaik (eds). Oxford: Oxford University Press, pp. 205–214.

Singleton, I. & van Schaik, C.P. 2002. The social organisation of a population of Sumatran orang-utans. *Folia Primatologica* 73: 1–20.

Singleton, I., Wich, S.A., Husson, S., *et al.* 2004. *Orang-utan Population and Habitat Viability Analysis*. Jakarta: Orangutan Foundation for IUCN/SSC Conservation Breeding Specialist Group.

Sirianni, G. & Visalberghi, E. 2013. Wild bearded capuchins process cashew nuts without contacting caustic compounds. *American Journal of Primatology* 75: 387–393.

Sjögersten, S., Black, C.R., Evers, S., *et al.* 2014. Tropical wetlands: a missing link in the global carbon cycle? *Global Biogeochemical Cycles* 28: 1371–1386.

Slik, F., Arroyo-Rodríguez, V., Aiba, S.I., *et al.* 2015. An estimate of the number of tropical tree species. *Proceedings of the National Academy of Sciences of the United States of America* 112: 7472–7477.

Slik, J.W.F., Aiba, S.-I., Brearley, F.Q., *et al.* 2010. Environmental correlates of tree biomass, basal area, wood specific gravity and stem density gradients in Borneo's tropical forests. *Global Ecology & Biogeography* 19: 50–60.

Small, C. & Nicholls, R. 2003. A global analysis of human settlement in coastal zones. *Journal of Coastal Research* 19: 584–599.

Smith, D.A.E., Smith, Y.C.E. & Cheyne, S.M. 2013. Home-range use and activity patterns of the red langur (*Presbytis rubicunda*) in Sabangau tropical peat-swamp forest, central Kalimantan, Indonesian Borneo. *International Journal of Primatology* 34: 957–972.

Smith, J.H. 2003. Land-cover assessment of conservation and buffer zones in the BOSAWAS Natural Resource Reserve of Nicaragua. *Environmental Management* 31: 252–262.

Smith, R.J. & Jungers, W.L. 1997. Body mass in comparative primatology. *Journal of Human Evolution* 32: 523–559.

Smith, S.L.J. & Smale, B. 1980. Classification of visitors to agreements for recreation and conservation sites, national parks and related sites. *Journal of Urban & Environmental Affairs* 12: 35–52.

Snarr, K.A. 2006. *Life in a lowland wet forest fragment on the north coast of Honduras: the mantled howlers (*Alouatta palliata*) of Cuero y Salado Wildlife Refuge*. PhD thesis, University of Toronto, Toronto, Ontario, Canada.

Snowdon, C.T. & Elowson, A.M. 1999. Pygmy marmosets modify call structure when paired. *Ethology* 105: 893–908.

Snowdon, C.T. & de la Torre, S. 2002. Multiple environmental contexts and communication in pygmy marmosets (*Cebuella pygmaea*). *Journal of Comparitive Psychology* 116: 182–188.

Sodhi, N.S., Koh, L.P., Brook, B.W. & Ng, P.K. 2004. Southeast Asian biodiversity: an impending disaster. *Trends in Ecology & Evolution* 19: 654–660.

Soini, P. 1982. Ecology and population dynamics of the pygmy marmoset, *Cebuella pygmaea*. *Folia Primatologica* 13: 223–230.

Soini, P. 1986. A synecological study of a primate community in the Pacaya-Samiria National Reserve, Peru. *Primate Conservation* 7: 63–71.

Soini, P. 1987. Ecology of the saddle-back tamarin *Saguinus fuscicollis illigeri* on the Rio Pacaya, northeastern Peru. *Folia Primatologica* 49: 11–32.

Soini, P. 1988. A synecological study of a primate community in the Pacaya-Samira National Reserve, Peru. *Primate Conservation* 7: 63–71.

Soini, P. 1993. The ecology of the pygmy marmoset, *Cebuella pygmaea*: some comparisons with two sympatric tamarins. In *Marmosets and Tamarins: Systematics, Behaviour, and Ecology*, A.B. Rylands (ed.). Oxford: Oxford University Press, pp. 257–261.

Solari, S., Muñoz-Saba, Y., Rodríguez Mahecha, J.V., *et al.* 2013. Riqueza, endemismo y conservación de los mamíferos de Colombia. *Mastozoologia Neotropical* 20: 301–365.

Son, V.D. 2003. Diet of *Macaca fascicularis* in a mangrove forest, Vietnam. *Vietnamese Journal of Primatology* 42: 3–5.

Son, V.D. 2004. Time budgets of *Macaca fascicularis* in a mangrove forest, Vietnam. *Laboratory Primate Newsletter* 43: 1–5.

Souto, A.S., Bione, C.B., Bastos, M., *et al.* 2011. Critically Endangered blonde capuchins fish for termites and use new techniques to accomplish the task. *Biology Letters* 1: 1–4.

Souza-Filho, P.W. 2005. Costa de manguezais de macromaré da Amazônia: cenários morfológicos, mapeamento e quantificação de áreas usando dados de sensores remotos. *Revista Brasileira de Geofísica* 23: 427–435.

Spagnoletti, N., Visalberghi, E., Ottoni, E., Izar, P. & Fragaszy, D. 2011. Stone tool use by adult wild bearded capuchin monkeys (*Cebus libidinosus*). Frequency, efficiency and tool selectivity. *Journal of Human Evolution* 61: 97–107.

Spagnoletti, N., Visalberghi, E., Verderane, M., *et al.* 2012. Stone tool use in wild bearded capuchin monkeys, *Cebus libidinosus*. Is it a strategy to overcome food scarcity? *Animal Behaviour* 83: 1285–1294.

Spalding, M.D., Blasco, F. & Field, C.D. 1997. *World Mangrove Atlas*. Paris: International Society for Mangrove Ecosystems, WCMC, National Council for Scientific Research.

Spalding, M.D., Fox, H.E., Allen, G.R., *et al.* 2007. Marine ecoregions of the world: a bioregionalization of coastal and shelf areas. *BioScience* 57: 573–583.

Spalding, M.D., Kainuma, M. & Collins, L. 2010. *World Atlas of Mangroves*. London: Earthscan. 337 pp.

Spenceley, A. (Ed.). 2010. *Responsible Tourism: Critical Issues for Conservation and Development*. London & New York: Routledge.

Spironello, W.R. 2001. The brown capuchin monkey (*Cebus apella*), In *Lessons from Amazonia: The Ecology and Conservation of a Fragmented Forest*, R.O. Bierregaard, Jr, C. Gascon, T.E. Lovejoy & R. Mesquita (eds). New Haven, CT: Yale University Press, pp. 271–283.

Sprague, D.S., Kabaya, H. & Hagihara, K. 2004. Field testing a global positioning system (GPS) collar on a Japanese monkey: reliability of automatic GPS positioning in a Japanese forest. *Primates* 45: 151–154.

Springer, M.S., Meredith, R.W., Gatesy, J., *et al.* 2012. Macroevolutionary dynamics and historical biogeography of primate diversification inferred from a species supermatrix. *PLoS ONE* 7: e49521.

Srbek-Araujo, A.C. & Chiarello, A.G. 2005. Is camera-trapping an efficient method for surveying mammals in Neotropical forests? A case study in southeastern Brazil. *Journal of Tropical Ecology* 21: 121–125.

Stancyk, S.E. 1982. Non-human predators of sea turtles and their control. In *Biology and Conservation of Sea Turtles*, K.A. Bjorndal (ed.). Washington, DC: Smithsonian Institution Press, pp. 139–152.

Stanford, C.B. 1991. The diet of the capped langur (*Presbytis pileata*) in a moist deciduous forest in Bangladesh. *International Journal of Primatology* 12: 199–216.

Stanford, C.B. & Nkurunungi, J.B. 2003. Behavioral ecology of sympatric chimpanzees and gorillas in Bwindi Impenetrable National Park, Uganda: diet. *International Journal of Primatology* 24: 901–918.

Stark, D. 2008. *Proboscis monkey (Nasalis larvatus) population viability analysis: reassessment and management for wild populations threatened by habitat loss in Borneo*. MSc dissertation. Oxford: Oxford Brooks University, UK.

Stark, D.J., Nathan, S.K., Saldivar, D.A., *et al.* 2014. Remote monitoring of arboreal primates in a closed tropical forest: a case study of proboscis monkeys (*Nasalis larvatus*). Abstract no. 245. *Proceedings, 24th Congress, International Primatological Society, Hanoi, Vietnam, August 2014*.

Stark, D.J., Nijman, V., Lhota, S., Robins, J.G. & Goossens, B. 2012. Modeling population viability of local proboscis monkey *Nasalis larvatus* populations: conservation implications. *Endangered Species Research* 16: 31–43.

Starr, C. & Nekaris, K.A.I. 2013. Obligate exudativory characterizes the diet of the pygmy slow loris *Nycticebus pygmaeus*. *American Journal of Primatology* 75: 1054–1061.

Stave, J., Oba, G., Stenseth, N.C. & Nordal, I. 2005. Environmental gradients in the Turkwel riverine forest, Kenya: hypotheses on dam-induced vegetation change. *Forest Ecology & Management* 212: 184–198.

Sterndale, R.A. 1884. *Natural History of the Mammalia of India and Ceylon*. Calcutta: Thacker, Spinkl and Co.

Stevenson, P.R. 2000. Seed dispersal by woolly monkeys (*Lagothrix lagothricha*) at Tinigua National Park, Colombia: dispersal distance, germination rates, and dispersal quantity. *American Journal of Primatology* 50: 275–289.

Stevenson, P.R. 2001. The relationship between fruit production and primate abundance in neotropical communities. *Biological Journal of the Linnean Society* 72: 161–178.

Stevenson, P.R. 2006. Activity and ranging patterns of Colombian woolly monkeys in northwestern Amazonia. *Primates* 47: 239–247.

Stevenson, P.R. & Aldana, A.M. 2008. Potential effects of Ateline extinction and forest fragmentation on plant diversity and composition in the western Orinoco Basin, Colombia. *International Journal of Primatology* 29: 365–377.

Stevenson, P.R. & Vargas, I.N. 2008. Sample size and appropriate design of fruit and seed traps in tropical forests. *Journal of Tropical Ecology* 24: 95–105.

Stevenson, P.R., Guzmán, D.C. & Defler, T.R. 2010. Conservation of Colombian primates: an analysis of published research. *Tropical Conservation Science* 3: 45–62.

Stevenson, P.R., Quiñones, M.J. & Ahumada, J.A. 1994. Ecological strategies of woolly monkeys (*Lagothrix lagotricha*) at Tinigua National Park, Colombia. *American Journal of Primatology* 32: 123–140.

Stevenson, P.R., Quiñones, M.J. & Ahumada, J.A. 1998. Effects of fruit patch availability on feeding subgroup size and spacing patterns in four primate species at Tinigua National Park, Colombia. *International Journal of Primatology* 19: 313–324.

Stevenson, P.R., Quiñones, M.J. & Ahumada, J.A. 2000. Influence of fruit availability on ecological overlap among four Neotropical primates at Tinigua National Park, Colombia. *Biotropica* 32: 533–544.

Stevenson, P.R. Suescun, M. & Quiñones, M.J. 2004. Characterization of forest types at the CIEM, Tinigua Park, Colombia. *Field Studies of Fauna and Flora. La Macarena, Colombia* 14: 1–20.

Stewart, F.A. 2011. *The evolution of shelter: ecology and ethology of chimpanzee nest building*. PhD thesis, University of Cambridge, Cambridge, UK.

Stewart, F.A. & Piel, A.K. 2013. Termite fishing by wild chimpanzees: new data from Ugalla, western Tanzania. *Primates* 55: 35–40.

Stewart, F.A., Piel, A.K. & McGrew, W.C. 2011. Living archaeology: artefacts of specific nest site fidelity in wild chimpanzees. *Journal of Human Evolution* 61: 388–395.

Stewart, F.A. & Pruetz, J.D. 2013. Do chimpanzee nests serve an anti-predatory function? *American Journal of Primatology* 12: 1–12.

Stibig, H.-J., Achard, F. & Fritz, S. 2004. A new forest cover map of continental Southeast Asia derived from SPOT-VEGETATION satellite imagery. *Applied Vegetation Science* 7: 153–162.

Stirling, E.J., Bynum, N. & Blair, M.E. (eds) 2013. *Primate Ecology and Conservation: A Handbook of Techniques*. Oxford: Oxford University Press.

Stokes, E.J., Strindberg, S., Bakabana, P.C., *et al.* 2010. Monitoring great ape and elephant abundance at large spatial scales: measuring effectiveness of a conservation landscape. *PLoS ONE* 5: e10294.

Storni, A., Paiva, P.M.V., Bernal, R. & Peralta, N. 2007. Evaluation of the impact on fauna caused by the presence of ecotourists on trails of the Mamirauá Sustainable Development Reserve, Amazonas, Brazil. *Tourism & Hospitality Planning & Development* 4: 25–32.

Storz, J.F. & Beaumont, M.A. 2002. Testing for genetic evidence of population expansion and contraction: an empirical analysis of microsatellite DNA variation using a hierarchical Bayesian model. *Evolution* 56: 154–166.

Streicher, U., Singh, M., Timmins, R.J. & Brockelman, W. 2008a. *Nycticebus bengalensis*. In *IUCN Red List of Threatened Species*. Version 2014.3. Gland, Switzerland: IUCN. www.iucnredlist.org.

Streicher, U., Ngoc Thanh, V., Nadler, T., Timmins, R.J. & Nekaris, K.A.I. 2008b. *Nycticebus pygmaeus*. In *IUCN Red List of Threatened Species*. Version 2014.3. Gland, Switzerland: IUCN. www.iucnredlist.org.

Strier, K.B. 2000. Population variability and regional conservation priorities for muriquis (*Brachyteles arachnoides*) in Brazil's Atlantic Forest. *Biotropica* 32: 903–913.

Strier, K.B. 2010. *Primate Behavioral Ecology*. 4th edition. Upper Saddle River, NJ: Prentice Hall.

Stronza, A. 2007. The economic promise of ecotourism for conservation. *Journal of Ecotourism* 6: 210–230.

Stronza, A. & Pegas, F. 2008. Ecotourism and conservation: two cases from Brazil and Peru. *Human Dimensions of Wildlife* 13: 263–279.

Struhsaker, T. & Leland, L. 1979. Socioecology of five sympatric monkey species in the Kibale Forest, Uganda. *Advances in the Study of Behaviour* 9: 159–223.

Struhsaker, T. & Leland, L. 1988. Group fission in redtail monkeys (*Cercopithecus ascanius*) in the Kibale Forest, Uganda. In *A Primate Radiation: Evolutionary Biology of the African Guenons*, A. Gautier-Hion, F. Bourlière, J.-P. Gautier & J. Kingdon (eds). Cambridge: Cambridge University Press, pp. 364–388.

Struhsaker, T., Oates, J.F., Hart, J. & Butynski, T.M. 2008. *Cercopithecus neglectus*. In *IUCN Red List of Threatened Species*. Version 2012.2. Gland, Switzerland: IUCN. www.iucnredlist.org.

Stump D.P. 2005. *Taxonomy of the genus* Perodicticus. PhD thesis, University of Pittsburgh, Pittsburgh, USA.

Su, D.F. & Harrison, T. 2008. Ecological implications of the relative rarity of fossil hominins at Laetoli. *Journal of Human Evolution* 55: 672–681.

Suárez, E., Morales, M., Cueva, R., *et al.* 2009. Oil industry, wild meat trade and roads: indirect effects of oil extraction activities in a protected area in northeastern Ecuador. *Animal Conservation* 12: 364–373.

Suárez, E., Zapata-Ríos, G., Utreras, V., Strindberg, S. & Vargas, J. 2013. Controlling access to oil roads protects forest cover, but not wildlife communities: a case study from the rainforest of Yasuní Biosphere Reserve (Ecuador). *Animal Conservation* 16: 265–274.

Sugiyama, Y. 1989. Local variation of tool and tool behavior among wild chimpanzee populations. In *Behavioral Studies of Wild Chimpanzees at Bossou, Guinea*, Y. Sugiyama (ed.). Inuyama: Kyoto University Primate Research Institute, pp. 1–15.

Sugiyama, Y. 1999. Socioecological factors of male chimpanzee migration at Bossou, Guinea. *Primates* 40: 61–68.

Supardjo M.N. 2008. Identifikasi vegetasi mangrove de Segoro Anak Selatan, Taman Nasional Alas Purwo, Banyuwangi, Jawa Timur. *Jurnal Saintek Perikanan* 3: 9–15.

Supriatna, J., Adimuntja, C., Mitrasetia, T., *et al.* 1988. Chemical analysis of food plant parts of two sympatric monkeys: *Presbytis aurata* and *Macaca fascicularis* in the mangrove forest of Muara Gembong, West Java. *Biotrop Special Publication* 37: 161–169.

Supriatna, J. & Wahono, E.H. 2000. *Panduan Lapangan Primata Indonesia (A Fieldguide to the Primates of Indonesia)*. Jakarta: Yayasan Obor.

Suscke, P. 2014. *Socioecologia de Cebus xanthosternos em área de Mata Atlântica, no Sul da Bahia*. Unpublished PhD thesis, Universidade de São Paulo.

Susman, R.L. 2004. *Oreopithecus bambolii*: an unlikely case of hominid-like grip capability in a Miocene ape. *Journal of Human Evolution* 46: 105–117.

Sussman, R. 1999. *Primate Ecology and Social Structure, Volume 1, Lorises, Lemurs, Tarsiers*. Needham Heights, MA: Pearson Custom Publishing.

Sussman, R.W. & Phillips-Conry, J.E. 1995. A survey of the distribution and density of the primates of Guyana. *International Journal of Primatology* 16: 761–791.

Sussman, R.W. & Rakotozafy, A. 1994. Plant diversity and structural analysis of a tropical dry forest in southwestern Madagascar. *Biotropica* 26: 241–254.

Sussman, R.W. & Tattersall, I. 1986. Distribution, abundance, and putative ecological strategy of *Macaca fascicularis* on the island of Mauritius, southwestern Indian Ocean. *Folia Primatologica* 46: 28–43.

Suzuki, A. 1965. An ecological study of wild Japanese monkeys in snowy areas focused on their food habits. *Primates* 6: 31–72.

Svenning, J.C. 2001. On the role of microenvironmental heterogeneity in the ecology and diversification of neotropical rain-forest palms (Arecaceae). *Botanical Review* 67: 1–53.

Swaine, M.D. 1992. Characteristics of dry forest in West Africa and the influence of fire. *Journal of Vegetation Science* 3: 365–374.

Swann, D.E., Kawanishi, K. & Palmer, J. 2011. Evaluating types and features of camera traps in ecological studies: a guide for researchers. In *Camera Traps in Animal Ecology*, A.F. O'Connell, J.D. Nichols & K.U. Karanth (eds). Tokyo: Springer Japan, pp. 27–43.

Swanson Ward, N. & Chism, J. 2003. A report on a new geographic location of red uakaris (*Cacajao calvus ucayalii*) on the Quebrada Tahuaillo in northeastern Peru. *Neotropical Primates* 11: 19–22.

SWAPL (Scott Wilson Asia Pacific Ltd.). 2000. *Gam River Dam: Preliminary Environmental Impact Assessment*. Hanoi: PARC Project VIE/95/G31, Government of Vietnam/UNOPS/UNDP/ Scott Wilson Asia Pacific Ltd.

Szabó, K. & Amesbury, J.R. 2011. Molluscs in a world of islands: the use of shellfish as a food resource in the tropical island Asia-Pacific region. *Quaternary International* 239: 8–18.

Szantoi, Z., Smith, S.E., Strona, G., Koh, L.P. & Wich, S.A. 2017. Mapping orangutan habitat and agricultural areas using Landsat OLI imagery augmented with unmanned aircraft system aerial photography. *International Journal of Remote Sensing* 38: 2231–2245.

Tacconi, L., Moore, P.F. & Kaimowitz, D. 2006. Fires in tropical forests: what is really the problem? Lessons from Indonesia. *Mitigation & Adaptation Strategies for Global Change* 12: 55–66.

Takatsuki, S. 2009. Effects of sika deer on vegetation in Japan: a review. *Biological Conservation* 142: 1922–1929.

Takenoshita, Y., Ando, C., Iwata, Y. & Yamagiwa, J. 2008. Fruit phenology of the great ape habitat in the Moukalaba-Doudou National Park, Gabon. *African Study Monographs Supplementary Issue* 39: 23–40.

Takenoshita, Y. & Yamagiwa, J. 2008. Estimating gorilla abundance by dung count in the northern part of Moukalaba-Doudou National

Park, Gabon. *African Study Monographs Supplementary Issue* 39: 41–54.

Tana River Delta 2015. *The Tana River Delta Website*. www.tanariverdelta.org.

Tan, A., Tan, S.H., Vyas, D., Malaivijitnond, S. & Gumert, M.D. 2015. There is more than one way to crack an oyster: identifying variation in Burmese long-tailed macaque (*Macaca fascicularis aurea*) stone-tool use. *PLoS ONE* 10: e0124733.

Tan, C.L. 1998. Comparison of food passage time in three species of *Hapalemur*. *American Journal of Physical Anthropology, Supplement* 26: 215.

Tan, C.L. 1999. Group composition, home range size, and diet of three sympatric bamboo lemur species (genus *Hapalemur*) in Ranomafana National Park, Madagascar. *International Journal of Primatology* 20: 547–566.

Tan, C.L. 2000. *Behavior and ecology of three sympatric bamboo lemur species (genus* Hapalemur*) in Ranomafana National Park, Madagascar*. PhD thesis, State Universtiy of New York, Stony Brook, NY.

Tan, C.L. 2007. Behavior and ecology of gentle lemurs (Genus *Hapalemur*). In *Lemurs: Ecology and Adaptation*, L. Gould & M. Sauther (eds). Developments in Primatology: Progress and Prospects series. New York: Springer, pp. 369–381.

Tan, C.L., Yang, Y. & Niu, K. 2012. Into the night: camera traps reveal nocturnal activity in a presumptive diurnal primate, *Rhinopithecus brelichi*. *Primates* 54: 1–6.

Tan, Y. & Yao, F. 2006. Three Gorges Project: effects of resettlement on the environment in the reservoir area. *Population & Environment* 27: 351–371.

Tattersall, I. 1973. A note on the age of the subfossil site of Ampasambazimba, Miarinarivo Province, Malagasy Republic. *American Museum Novitates* 2520: 1–6.

Tattersall, I. 2013. Understanding species-level primate diversity in Madagascar. *Madagascar Conservation & Development* 8: 7–11.

Taylor, AB. 2006. Feeding behavior, diet, and the functional consequences of jaw form in orangutans, with implications for the evolution of *Pongo*. *Journal of Human Evolution* 50: 377–393.

Taylor, A.B. 2009. The functional significance of variation in jaw form in orangutans: the African apes as an ecogeographic model. In *Orangutans: Geographic Variation in Behavioral Ecology and Conservation*, S. Wich, S. Utami Atmoko, T. Mitra Setia & C. van Schaik (eds). Oxford: Oxford University Press, pp. 16–31.

Taylor, A.B. & van Schaik, C.P. 2007. Variation in brain size and ecology in *Pongo*. *Journal of Human Evolution* 52: 59–71.

Taylor, G.D. 1986. Multi-dimensional segmentation of the Canadian pleasure travel market. *Tourism Management* 7: 146–153.

Taylor, M., Ravilious, C. & Green, E.P. 2003. *Mangroves of East Africa*. Cambridge, UK: UNEP World Conservation Monitoring Centre.

Taylor, M., Ravilious, C. & Green, E.P. 2007. *Mangroves of Western and Central Africa*. Cambridge, UK: UNEP Regional Seas Programme/UNEP-WCMC.

Taylor, P.D., Fahrig, L. & With, K.A. 2006. Landscape connectivity: a return to the basics. In *Connectivity Conservation*, K.R. Crooks & M. Sanjayan (eds). Cambridge: Cambridge University Press, pp. 29–43.

Teaford, M.F., Maas, M.C. & Simons, E.L. 1996. Dental microwear and microstructure in early Oligocene primates from the Fayum, Egypt: implications for diet. *American Journal of Physical Anthropology* 101: 527–543.

Teaford, M.F. & Walker, A. 1984. Quantitative differences in dental microwear between primate species with different diets and a comment on the presumed diet of *Sivapithecus*. *American Journal of Physical Anthropology* 64: 191–200.

Teaford, M.F. & Ungar, P.S. 2000. Diet and the evolution of the earliest human ancestors. *Proceedings of the National Academy of the Sciences of the United States of America* 97: 13506–13511.

Terada, S., Nackoney, J., Sakamaki, T., *et al.* 2015. Habitat use of bonobos (*Pan paniscus*) at Wamba: selection of vegetation types for ranging, feeding, and night-sleeping. *American Journal of Primatology* 77: 701–713.

Terborgh, J. 1983. *Five New World Primates: A Study in Comparative Ecology*. Princeton, NJ: Princeton University Press.

Terborgh, J., Lopez, L., Nunez, P., *et al.* 2001. Ecological meltdown in predator-free forest fragments. *Science* 294: 1923–1926.

Terborgh, J., Lopez, L., Tello, J., Yu, D. & Bruni, A.R. 1997. Transitory states in relaxing ecosystems of land bridge islands. In *Tropical Forest Remnants: Ecology, Management, and Conservation of Fragmented Communities*. W.L. Laurance & R.O. Bierregarrd (eds). Chicago, IL: University of Chicago Press, pp. 256–274.

ter Steege, H. 1990. *A Monograph of Wallaba, Mora and Greenheart*. Wageningen, The Netherlands: The Tropenbos Foundation.

ter Steege, H. 1993. *Patterns of Tropical Rain Forest in Guyana*. Wageningen, The Netherlands: The Tropenbos Foundation.

ter Steege, H. & Persaud, CA. 1991. The phenology of Guyanese timber species: a compilation of a century of observations. *Vegetation* 95: 177–198.

ter Steege, H., Pitman, N., Sabatier, D., *et al.* 2003. A spatial model of tree α-diversity and tree density for the Amazon. *Biodiversity & Conservation* 12: 2255–2277.

ter Steege, H., Sabatier, D., Castellanos, H., *et al.* 2000. An analysis of the floristic composition and diversity of Amazonian forests including those of the Guiana Shield. *Journal of Tropical Ecology* 16: 801–828.

Tesfaye, D., Fashing, P.J., Bekele, A., Mekonnen, A. & Atickem, A. 2013. Ecological flexibility in Boutourlini's blue monkeys (*Cercopithecus mitis boutourlinii*) in Jibat Forest, Ethiopia: a comparison of habitat use, ranging behavior, and diet in intact and fragmented forest. *International Journal of Primatology* 34: 615–640.

Te Wong, S., Servheen, C., Ambu, L. & Norhayati, A. 2005. Impacts of fruit production cycles on Malayan sun bears and bearded pigs in lowland tropical forest of Sabah, Malaysian Borneo. *Journal of Tropical Ecology* 21: 627–639.

Tews, J., Brose, U., Grimm, V., *et al.* 2004. Animal species diversity driven by habitat heterogeneity /diversity: the importance of keystone structures. *Journal of Biogeography* 31: 79–92.

Thac Mai Hoang. 2010. *Ensuring the survival of Tonkin snub-nosed monkey (Rhinopithecus avunculus) in Na Hang Nature Reserve, Tuyen Quang Province, Northeastern Vietnam*. London: Rufford Small Grants Foundation.

Thieme, M.L., Abell, R.A., Stiassny, M.L.J., *et al.* (eds). 2005. *Freshwater Ecoregions of Africa: A Conservation Assessment*. Washington, DC: Island Press.

Thierry, B. 2011. The macaques: a double-layered social organization. In *Primates in Perspective*, C.J. Campbell, A. Fuentes, K.C. MacKinnon, S.K. Bearder & R.M. Stumpf (eds). Oxford: Oxford University Press, pp. 229–241.

Thomas, D. 1996. Dam construction and ecological change in the riparian forest of the Hadejia-Jama'are floodplain, Nigeria. *Land Degradation & Development* 7: 279–295.

Thomas, G. & Fernandez, T.V. 1994. Mangrove and tourism: management strategies. *Indian Forester* 120: 406–412.

Thomas, L., Buckland, S.T., Anderson, D.R., *et al.* 2001. *Introduction to Distance Sampling: Estimating Abundance of Biological Populations*. Oxford: Oxford University Press.

Thomas, L., Buckland, S.T., Rexstad, E.A., *et al.* 2010. Distance software: design and analysis of distance sampling surveys for estimating population size. *Journal of Applied Ecology* 47: 5–14.

Thompson, J.A.M. 2002. Bonobos of the Lukuru wildlife research project. In *Behavioural Diversity in Chimpanzees and Bonobos*, L. Marchant, C. Boesch & G. Hohmann (eds). Cambridge: Cambridge University Press, pp. 61–70.

Thompson, K. 1976. Swamp development in the head waters of the White Nile. In *The Nile, Biology of an Ancient River*, J. Rzóska (ed.). Monographae Biologicae No. 29. Dordrecht, The Netherlands: Springer, pp. 177–196.

Thompson, M.E., Schwager, S.J. & Payne, K.B. 2010a. Heard but not seen: an acoustic survey of the African forest elephant population at Kakum Conservation Area, Ghana. *African Journal of Ecology* 48: 224–231.

Thompson, M.E., Schwager, S.J., Payne, K.B. & Turkalo, A.K. 2010b. Acoustic estimation of wildlife abundance: methodology for vocal mammals in forested habitats. *African Journal of Ecology* 48: 654–661.

Thoms, C. & Schupp, P.J. 2008. Activated chemical defense in marine sponges – a case study on *Aplysinella rhax*. *Journal of Chemical Ecology* 34: 1242–1252.

Thoms, C., Wolff, M., Padmakumar, K., Ebel, R. & Proksch, P. 2004. Chemical defense of Mediterranean sponges *Aplysina cavernicola* and *Aplysina aerophoba*. *Zeitschrift für Naturforsch C* 59: 113–122.

Thorington, Jr, R.W., Ruiz, J.C. & Eisenberg, J.F. 1984. A study of a black howling monkey (*Alouatta caraya*) population in northern Argentina. *American Journal of Primatology* 6: 357–366.

Thorn, J.S., Nijman, V., Smith, D. & Nekaris, K.A.I. 2009. Ecological niche modelling as a technique for assessing threats and setting conservation priorities for Asian slow lorises (Primates: *Nycticebus*). *Diversity & Distributions* 15: 289–298.

Thu, P.M. & Populus, J. 2007. Status and changes of mangrove forest in Mekong Delta: Case study in Tra Vinh, Vietnam. *Estuarine, Coastal & Shelf Science* 71: 98–109.

Timmins, R.J. & Duckworth, J.W. 2013. Distribution and habitat of Assamese macaque *Macaca assamensis* in Lao PDR, including its use of low-altitude karsts. *Primate Conservation* 26: 103–114.

Timmins, R.J., Richardson, M., Chhangani, A. & Yongcheng, L. 2008. *Macaca mulatta*. In *IUCN Red List of Threatened Species*. Version 2012.1. Gland, Switzerland: IUCN. www.iucnredlist.org.

Tisdell, C.A. & Swarna Nantha, H. 2007. Conservation of the proboscis monkey and the orangutan in Borneo: comparative issues and economic considerations. In *Perspectives in Animal Ecology and Reproduction Vol. 4*, V.K. Gupta & A. Verma (eds). New Delhi: Daya Publishing House, pp. 225–250.

Tisdell, C.A. & Xue, D. 2013. Managing ecosystem services for human benefit: economic and environmental policy challenges. In *Environmental Policy: Management, Legal Issues and Health Aspects*, E. Creighton & P. Danovich (eds). Hauppauge, NY: Nova Publishers, pp. 87–106.

Tobler, M.W., Carrillo-Percastegui, S.E. & Leite Pitman, R. 2008. An evaluation of camera traps for inventorying large and medium sized terrestrial rainforest mammals. *Animal Conservation* 11: 169–178.

Tockner, K. & Stanford, J.A. 2002. Riverine floodplains: present state and future trends. *Environmental Conservation* 29: 308–330.

Toivonen, T., Maki, S. & Kalliola, R. 2007. The riverscape of Western Amazonia - quantitative approach to the fluvial biogeography of the region. *Journal of Biogeography* 34: 1374–1387.

Tomas, M.A., Chiaravalloti, R.M., Camilo, A.R., Tomas, W.M. & Ferreira, V.L. 2010. Densidade e tamanho de grupos de *Callicebus* cf. *pallescens* (Primates: Pitheciidae) na fazenda Santa Teresa, Pantanal. *Anais do 5º Simpósio sobre Recurso Naturais e Socioeconômicos do Pantanal*. 9–12 November, 2010. Corumbá.

Tomas, W.M., Cáceres, N., Nunes, A.P., *et al.* 2011. Mammals in the Pantanal wetland, Brazil. In *The Pantanal: Ecology, Biodiversity and Sustainable Management of a Large Neotropical Seasonal Wetland*, W.J. Junk, C.J. da Silva, C.N. da Cunha & K.M. Wantzen (eds). Sofia: Pensoft Publishers, pp. 563–595.

Tomlinson, P.B. 1986. *The Botany of Mangroves*. Cambridge: Cambridge University Press.

Tooze, Z. 1995. Update on Sclater's guenon *Cercopithecus sclateri* in southern Nigeria. *African Primates* 1: 38–42.

Tooze, Z. 1996. *Bridging Report LNG EIA: The Actual Status of Primates of the Niger Delta*. Unpublished report. Port Harcourt, Nigeria: Aquater/TSKJ.

Tooze, Z. 1997. *Survey and Census of Sclater's Guenon (*Cercopithecus sclateri*) in Southeast Nigeria, and Recommendations for Conservation Initiatives*. Unpublished report. New York, NY: Wildlife Conservation Society, and Charlestown, RI: Primate Conservation Inc.

Tooze, Z., Attah, V. & Esara, E. 1998. *Stage Three Progress Report on the Preparation of a Management Plan for the Stubbs Creek Conservation Project (SCCP), Akwa Ibom State*. Unpublished report. Uyo, Nigeria: Akwa Ibom State Environmental Protection Agency (AKSEPA).

Tortato, F.R., Bonanomi, J., Delvin, A.L. & Hoogesteijn, R. 2014. Interspecific association between collared peccaries (*Pecari tajacu* Linnaeus 1758-Tayassuidae) and Azara's capuchin (*Sapajus cay* Illiger 1815-Cebidae) in the Pantanal, Brazil. *Suiform Soundings* 12: 17–18.

Toro, M.A. & Caballero, A. 2005. Characterization and conservation of genetic diversity in subdivided populations. *Philosophical Transactions of the Royal Society of London. Series B, Biological Sciences* 360: 1367–1378.

Tortato, F.R., Bonanomi, J., Devlin, A.L. & Hoogesteijn, R. 2014. Interspecific association between collared peccaries (*Pecari tajacu* Linnaeus, 1758-Tayassuidae) and Azara's capuchin (*Sapajus cay* Illiger, 1815-Cebidae) in the Pantanal, Brazil. *Suiform Soundings* 12: 17–18.

Trevelin, L.C., Port-Carvalho, M., Silveira, M. & Morell, E. 2007. Abundância, uso do habitat e dieta de *Callicebus nigrifrons* E. Geoffroy (Primates, Pitheciidae) no Parque Estadual da Cantareira, São Paulo, Brasil. *Revista Brasileira de Zoologia* 24: 1071–1077.

Treves, A. 1999. Has predation shaped the social systems of arboreal primates? *International Journal of Primatology* 20: 35–67.

Treves, A. & Brandon, K. 2005. Tourist impacts on the behavior of black howling monkeys (*Alouatta pigra*) at Lamanai, Belize. Commensalism and conflict: the human-primate interface. In *Commensalism and Conflict: The Human–Primate Interface*, Special Topics in Primaotlogy No. 4, J.D. Paterson & J. Wallis (eds). Norman, OK: American Society of Primatologists, pp. 147–167.

Trolle, M. & Kéry, M. 2005. Camera-trap study of ocelot and other secretive mammals in the northern Pantanal. *Mammalia* 69: 409–416.

Trujillo-Gonzalez, J.M., Torres Mora, M.A. & Santana-Castañeda, E. 2011. La palma de Moriche (*Mauritia flexuosa* L.f.) un ecosistema estratégico. *Orinoquia* 15: 62–70.

Tscharntke, T. 1992. Fragmentation of Phragmites habitats, minimum viable population size, habitat suitability, and local extinction of moths, midges, flies, aphids, and birds. *Conservation Biology* 6: 530–536.

Tsela, P.L., van Helden, P., Frost, P., Wessels, K. & Archibald, S. 2010. Validation of the MODIS burned-area products across different biomes in South Africa. *Geoscience and Remote Sensing Symposium (IGARSS), 2010 IEEE International* pp. 3652–3655.

Tsuji, Y. 2007. *Effects of yearly differences in nut fruiting on foraging success of wild Japanese macaques through intra-troop competition*. PhD thesis, University of Tokyo, Japan.

Tsuji, Y. 2010. Regional, temporal, and inter-individual variation in the feeding ecology of Japanese macaques. In *Japanese Macaques*, N. Nakagawa, M. Nakamichi & H. Sugiura (eds). Tokyo: Springer Japan, pp. 95–123.

Tsuji, Y., Hanya, G. & Grueter, C.C. 2013. Feeding strategies of primates in temperate and alpine forests: comparison of Asian macaques and colobines. *Primates* 54: 201–215.

Tsuji, Y., Kazahari, N., Kitahara, M. & Takatsuki, S. 2008. A more detailed seasonal division of the energy balance and the protein balance of Japanese macaques (*Macaca fuscata*) on Kinkazan Island, northern Japan. *Primates* 49: 157–160.

Tsuji, Y. & Takatsuki, S. 2004. Food habits and home range use of Japanese macaques on an island inhabited by deer. *Ecological Research* 19: 381–388.

Tsuji, Y. & Takatsuki, S. 2009. Effects of yearly change in nut fruiting on autumn home-range use by *Macaca fuscata* on Kinkazan Island, northern Japan. *International Journal of Primatology* 30: 169–181.

Tsuji, Y. & Takatsuki, S. 2012. Inter-annual variation in nut abundance is related to agonistic interactions of foraging female Japanese macaques (*Macaca fuscata*). *International Journal of Primatology* 33: 489–512.

Tubelis, D.P. 2013. Species composition and seasonal occurrence of mixed-species flocks of forest birds in savannas in central cerrado, Brazil. *Ararajuba* 12: 105–111.

Tullos, D. 2008. Assessing the influence of environmental impact assessments on science and policy: an analysis of the Three Gorges project. *Journal of Environmental Management* 90: 1–16.

Tumbarello, S., Fischer, X., Galat-Luong, A. & Galat, G. 1995a. Jonathan et les *Colobes bais*. *TV Movie Magazine*, Chippangali, France.

Tumbarello S, Fischer X, Galat-Luong A & Galat, G. 1995b. Julie et les Singes verts. *TV Movie Magazine*, Chippangali, France.

Tundisi, J.G., Goldemberg, J., Matsumura-Tundisi, T. & Saraiva, A.C.F. 2014. How many more dams in the Amazon? *Energy Policy* 74: 703–708.

Turner, I.M., Boo, C.M., Wong, Y.K., Chew, P.T. & Ali bin Ibrahim. 1996. Freshwater flooded forest in Singapore, with particular reference to that found around the Nee Soon firing ranges. *Garden's Bulletin Singapore* 48: 129–157.

Turner, R.K., Pearce, D.W. & Bateman, I. 1993. *Environmental Economics: An Elementary Introduction*. Baltimore, MD: Johns Hopkins University Press.

Tutin, C.E.G. & Fernandez, M. 1984. Nationwide census of gorilla (*Gorilla g. gorilla*) and chimpanzee (*Pan t. troglodytes*) populations in Gabon. *American Journal of Primatology* 6: 313–336.

Tutin, C.E.G. & Fernandez, M. 1993a. Composition of the diet of chimpanzees and comparisons with that of sympatric lowland gorillas in the Lopé Reserve, Gabon. *American Journal of Primatology* 30: 195–211.

Tutin, C.E.G. & Fernandez, M. 1993b. Faecal analysis as a method of describing diets of apes: Examples from sympatric gorillas and chimpanzees at Lopé. *Tropics* 2: 189–197.

Tutin, C.E.G., Fernandez, M., Rogers, M.E. & Williamson, E.A. 1992. A preliminary analysis of the social structure of lowland gorillas in the Lopé Reserve, Gabon. In *Topics in Primatology, Vol 2, Behavior, Ecology, and Conservation*, N. Itoigawa, Y. Sugiyama, G.P. Sackett & R.K.R. Thompson (eds). Tokyo: University of Tokyo Press, pp. 245–254.

Twagirashyaka, F. & Inogwabini, B.-I. 2009. Lake TéLé–Lake Tumba landscape. In *The Forests of the Congo Basin: State of the Forest 2008*, C. De Wasseige, D. Devers, P. de Marcken, R. Eba'a Atyi, R. Nasi & P. Mayaux (eds). Luxembourg: Publications Office of the European Union, pp. 305–316.

Uehara, S. 1988. Grouping patterns of wild pygmy chimpanzees (*Pan paniscus*) observed at a marsh grassland amidst the tropical rain forest of Yalosidi, Republic of Zaïre. *Primates* 29: 41–52.

Uehara, S. 1990. Utilization patterns of a marsh grassland within the tropical rain forest by the bonobos (*Pan paniscus*) of Yalosidi, Republic of Zaire. *Primates* 31: 311–322.

Uhde, N.L. & Sommer, V. 2002. Anti-predatory behaviour in gibbons. In *Eat or be Eaten: Predator Sensitive Foraging Among Primates*, L.E. Miller (ed.). Cambridge: Cambridge University Press, pp. 268–291.

UN (United Nations). 2015. *World Population Prospects: The 2015 Revision, Key Findings and Advance Tables.* Working Paper No. ESA/P/WP.241. New York: Population Division, Department of Economic and Social Affairs, United Nations.

UN (United Nations). 2017. *World Population Prospects: The 2017 Revision, Key Findings and Advance Tables.* Working Paper No. ESA/P/WP/248. New York: Population Division, Department of Economic and Social Affairs, United Nations.

UNDP. 2011. *Niger Delta Biodiversity Project.* GEF PIM no.: 2047. New York: UNDP. www.undp.org/content/dam/undp/documents/projects/NGA/00058336/PRODOC_2047%20Niger%20Delta%20Conservation_FINAL_180712.docx.

UNEP–WCMC (United Nations Environment Programme World Conservation Monitoring Centre). 2006. *In the Front Line: Shoreline Protection and other Ecosystem Services from Mangroves and Coral Reefs*. Cambridge: UNEP–WCMC.

UNESCO. 1981. *Vegetation map of South America. Map 1:5,000,000.* Toulouse, France: Institut de la Carte Internationale de Tapis Vegetal.

UNESCO. 2013 *The Tana Delta and Forest Complex*. Paris: UNESCO. http://whc.unesco.org/en/tentativelists/5514/

Ungar, P.S. 1995. Fruit preferences of four sympatric primate species at Ketambe, Northern Sumatra, Indonesia. *International Journal of Primatology* 16: 221–245.

Ungar, P.S. 2006. Dental microwear of European Miocene catarrhines: evidences for diets and tooth use. *Journal of Human Evolution* 31: 335–366.

Ungar, P.S. & Kay, R.F. 1996. The dietary adaptations of European Miocene catarrhines. *Proceedings of the National Academy of the Sciences of the Unites States of America* 92: 5479–5481.

Ungar, P.S., Grine, F.E. & Teaford, M.F. 2008. Dental microwear indicates that *Paranthropus boisei* was not a hard-object feeder. *PLoS ONE* 3: e2044: 1–6.

Urbani, B. 2003. Utilización del estrato vertical por el mono aullador de manto (*Alouatta palliata*, Primates) en Isla Colón, Panamá. *Antropo* 4: 29–33.

Urbani, B. 2006. A survey of primate populations in northeastern Venezuelan Guayana. *Primate Conservation* 20: 47–52.

Urquhart, G.R. 2009. Paleoecological record of hurricane disturbance and forest regeneration in Nicaragua. *Quaternary International* 195: 88–97.

USGS (United States Geologial Survey). 2011. NASA Landsat Program, 2011, Landsat ETM+. Sioux Falls, SD: US Geological Survey.

Utami Atmoko, S., Mitra Setia, T., Goossens, B., *et al.* 2009. Orangutan mating behavior and strategies. In *Orangutans: Geographic Variation in Behavioral Ecology and Conservation.* S. Wich, S. Utami Atmoko, T. Mitra Setia & C. van Schaik (eds). Oxford: Oxford University Press, pp. 235–244.

Utami, S., Goossens, B., Bruford, M.W., de Ruiter, J.R & van Hooff, J.A.R.A.M. 2002. Male bimaturism and reproductive success in Sumatran orang-utans. *Behavioral Ecology* 13: 643–652.

Valenzuela, N. 2001. Genetic differentiation among nesting beaches in the highly migratory giant river turtle (*Podocnemis expansa*) from Colombia. *Herpetologica* 57: 48–57.

Valiela, I., Bowen, J.L. & York, J.K. 2001. Mangrove forests: one of the world's threatened major tropical environments. *Bioscience* 51: 807–815.

Valsecchi, J. 2005. *Diversidade de mamíferos e uso da fauna nas Reservas de Desenvolvimento Sustentável Mamirauá e Amanã - Amazonas - Brasil.* MSc dissertation, Museu Paraense Emílio Goeldi/UFPA, Belém, Pará, Brazil.

Valsecchi, J. 2012. *Caça de animais silvestres nas Reservas de Desenvolvimento Sustentável Mamirauá e Amanã.* PhD thesis, UFMG, Belo Horizonte, Minas Gerais, Brasil.

Valsecchi, J., Vieira, T.M., Silva, Jr, J.S., Muniz, I.C.M. & Avelar, A.A. 2010. New data on the ecology and geographic distribution of *Saguinus inustus* Schwarz, 1951 (Primates, Callitrichidae). *Brazilian Journal of Biology* 70: 229–233.

van Andel, A.C., Wich, S.A., Boesch, C., *et al.* 2015. Locating chimpanzee nests and identifying fruiting trees with an unmanned aerial vehicle. *American Journal of Primatology* 77: 1122–1134.

van Belle, S. & Estrada, A. 2008. Group size and composition influence male and female reproductive success in black howler monkeys (*Alouatta pigra*). *American Journal of Primatology* 70: 613–619.

van Erkom Schurink, C. & Griffiths, C.L. 1991. A comparison of reproductive cycles and reproductive output in four southern African mussel species. *Marine Ecology Progress Series* 76: 123–134.

van der Merwe, N.J., Masao, F.T. & Bamford, M.K. 2008. Isotopic evidence for contrasting diets of early hominins *Homo habilis* and *Australopithecus boisei* of Tanzania. *South African Journal of Science* 104: 153–155.

van Gemert, J.C., Verschoor, C.R., Mettes, P., *et al.* 2014. Nature conservation drones for automatic localization and counting of animals. In *Computer Vision-ECCV 2014 Workshops*. Dordrecht, The Netherlands: Springer International Publishing, pp. 255–270.

van Hooff, J.A.R.A.M. & Lukkenaar, B. 2015. Captive chimpanzee takes down a drone: tool use toward a flying object. *Primates* 56: 289–292.

Vanleeuwe, H., Cajani, S. & Gautier-Hion, A. 1998. Large mammals at forest clearings in the Odzala National Park, Congo. *Revue d'Écolgie, La Terre et La Vie* 53: 171–180.

van Noordwijk, M., Sauren, S., Nuzuar, *et al.* 2009. Development of independence: Sumatran and Bornean orangutans compared. In *Orangutans: Geographic Variation in Behavioral Ecology and Conservation*. S. Wich, S. Utami Atmoko, T. Mitra Setia & C. van Schaik (eds). Oxford: Oxford University Press, pp. 189–203.

Vannucci, M. 2001. What is so special about mangroves? *Brazilian Journal of Biology* 61: 599–603.

van Roosmalen, M.G.M. 1985. *Fruits of the Guiana Flora.* Utrecht: Institute of Systematic Botany, University of Utrecht.

van Roosmalen, M.G.M. & van Roosmalen, T. 1997. An eastern extension of the geographical range of the pygmy marmoset, *Cebuella pygmaea*. *Neotropical Primates* 5: 3–5.

van Roosmalen, M.G.M., van Roosmalen, T. & Mittermeier, R.A. 2002. A taxonomic review of the titi monkeys, genus *Callicebus* Thomas, 1903, with the description of two new species, *Callicebus bernhardi* and *Callicebus stephennashi*, from Brazilian Amazonia. *Neotropical Primates* 10 (Suppl.): 1–52.

van Schaik, C.P. 1983. Why are diurnal primates living in groups? *Behaviour* 87: 120–144.

van Schaik, C.P. 1999. The socioecology of fission–fusion sociality in orangutans. *Primates* 40: 69–86.

van Schaik, C.P. 2002. Fragility of traditions: the disturbance hypothesis for the loss of local traditions in orangutans. *International Journal of Primatology* 23: 527–538.

van Schaik, C.P. 2004. *Among Orangutans: Red Apes and the Rise of Human Culture*. Cambridge, MA: Harvard University Press.

van Schaik, C.P. & Knott, C.D. 2001. Geographic variation in tool use on *Neesia* fruits in orangutans. *American Journal of Physical Anthropology* 114: 331–342.

van Schaik, C.P. & Mirmanto, E. 1985. Spatial variation in the structure and litterfall of a Sumatran rain forest. *Biotropica* 17: 196–205.

van Schaik, C.P. & Pfannes, K. 2005. Tropical climates and phenology: a primate perspective. In *Seasonality in Primates: Studies of Living and Extinct Human and Non-human Primates*, D.K. Brockman & C.P. van Schaik (eds). Cambridge: Cambridge University Press, pp. 23–54.

van Schaik, C.P. & van Hooff, J.A.R.A.M. 1996. Toward an understanding of the orangutan's social system. In *Great Ape Societies*, W.C. McGrew, L.F. Marchant & T. Nishida (eds). Cambridge: Cambridge University Press, pp. 3–15.

van Schaik, C.P., Monk, K.A. & Robertson, J. Y. 2001. Dramatic decline in orang-utan numbers in the Leuser Ecosystem, Northern Sumatra. *Oryx* 35: 14–25.

van Schaik, C.P., Priatna, A. & Priatna, D. 1995. Population estimates and habitat preferences of orangutans (*Pongo pygmaeus*) based on line transects of nests. In *The Neglected Ape*, R.B. Nadler, B.M.F. Galdikas, L. Sheeran & N. Rosen (eds). New York: Plenum Press, pp. 129–147.

van Schaik, C.P., van Amerongen, A. & van Noordwijk, M.A. 1996. Riverine refuging by wild Sumatran long-tailed macaques (*Macaca fascicularis*). In *Evolution and Ecology of Macaque Societies*, J.E. Fa & D.G. Lindburg (eds). Cambridge: Cambridge University Press, pp. 160–181.

van Schaik, C.P., van Noordwijk, M.A. & Vogel, E.R. 2009. Ecological sex differences in wild orangutans. In *Orangutans: Geographic Variation in Behavioral Ecology and Conservation*, S.A. Wich, S.S.U. Atmoko, T.M. Setia & C.P. van Schaik (eds). Oxford: Oxford University Press, pp. 49–64.

van Schaik, C.P., Wich, S.A., Utami, S.S. & Odom, K. 2005. A simple alternative to line transects of nests for estimating orangutan densities. *Primates* 46: 249–254.

van Steenis, C.G.G.J. 1958. Ecology of mangroves. *Flora Malesiana* 5: 431–441.

Vargas Gomez, M.G. 2011. *Vocalizaciones de Aotus brumbackii (Hershkovitz, 1983) y su relación con las actividades en vida silvestre, San Martín (Meta, Colombia).* BSc dissertation, Universidad Pontificia La Javeriana, Bogotá.

Vartapetian, B.B., Andreeva, I.N., Generozova, I.P., *et al.* 2003. Functional electron microscopy in studies of plant response and adaptation to anaerobic stress. *Annals of Botany* 91: 155–172.

Vasconcelos, P.F.C., Costa, Z.G., Travassos da Rosa, E.S., *et al.* 2001. Epidemic of jungle yellow fever in Brazil, 2000: implications of climatic alterations in disease spread. *Journal of Medical Virology* 65: 598–604.

Vasconcelos, P.F.C., Esperb, A.F., Monteiro, H.A.O., *et al.* 2003. Isolations of yellow fever virus from *Haemagogus leucocelaenus* in Rio Grande do Sul State, Brazil. *Transactions of the Royal Society of Tropical Medicine & Hygiene* 97: 60–62.

Vath, C.L. 2008. *Species richness and habitat preference of large vertebrates in the Central Suriname Nature Reserve*. MSc dissertation, University of Florida, Florida.

Vedder, A.L. 1984. Movement patterns of a group of free-ranging mountain gorillas (*Gorilla gorilla beringei*) and their relation to food availability. *American Journal of Primatology* 7: 73–88.

Vegas-Vilarrúbia, T., Ponce, M.E., Gómez, O. & Mora, L. 2007. Wetland vegetation of the lower Orinoco Delta plain (Venezuela): a preliminary approach. *Amazoniana* 19: 35–61.

Veiga, L.M., Barnett, A.A., Ferrari, S.F. & Norconk, M.A. 2013. *Evolutionary Biology and Conservation of Titis, Sakis and Uacaris*. Cambridge: Cambridge University Press.

Vellidis, G. & Lowrance, R. 2004. Riparian forest buffers: hype or the silver bullet for NPS pollution control. *Resource* 11: 7–8.

Veracini, C. 2009. Habitat use and ranging behavior of the silvery marmoset (*Mico argentatus*) at Caxiuanã National Forest (eastern Brazilian Amazonia). In *The Smallest Anthropoids*, S.M. Ford, L.M. Porter & L.C. Davis (eds). New York: Springer, pp. 221–240.

Vera-Diaz, M.D.C., Reid, J., Soares-Filho, B., Kaufmann, R. & Fleck, L.C. 2007. *Efectos de los proyectos de energía y transporte en la expansión del cultivo de soja en la cuenca del rio Madeira*. Brazil: Conservation Strategy Fund.

Verhaegen, M., Munro, S., Vaneechoutte, M., Bender-Osner, N. & Bender, R. 2007. The original econiche of the genus *Homo*: open plain or waterside? In *Ecology Research Progress*, S. Munoz (ed.). New York: Nova Science, pp. 155–186.

Verhaegen, M. & Puech, P.F. 2000. Hominid lifestyle and diet reconsidered: paleo-environmental and comparative data. *Human Evolution* 15: 175–186.

Vermeer, J., Tello-Alvarado, J.C., Castillo, J.T.V.D. & Bóveda-Penalba, A.J. 2013. A new population of red uakaris (*Cacajao calvus* ssp.) in the mountains of northeastern Peru. *Neotropical Primates* 20: 12–17.

Vermeulen, C., Lejeune, P., Lisein, J., Sawadogo, P. & Bouché, P. 2013. Unmanned aerial survey of elephants. *PLoS ONE* 8: e54700.

Vié, J.-C. 1999. Wildlife rescues – the case of Petit Saut hydroelectric dam in French Guiana. *Oryx* 33: 115–126.

Vié, J.-C., Richard-Hansen, C. & Fournier-Chambrillon, C. 2001. Abundance, use of space, and activity patterns of white-faced sakis (*Pithecia pithecia*) in French Guiana. *American Journal of Primatology* 55: 203–221.

Vieira, T., Oliveira, M., Queiroz, H. & Valsecchi, J. 2009. Novas informações sobre a distribuição de Cacajao calvus na Reserva de Desenvolvimento Sustentável Mamirauá. *Uakari* 4: 41–51.

Vignaud, P., Duringer, P., Mackaye, H.T., *et al.* 2002. Geology and palaeontology of the Upper Miocene Toros-Menalla hominid locality, Chad. *Nature* 418: 152–155.

Vilanova, R. 2005. Limites climáticos e vegetacionais das distribuições de *Cebus nigritus* e *Cebus robustus* (Cebinae, Platyrrhini). *Neotropical Primates* 13: 14–19.

Vilanova, R., de Sousa e Silva, J., Grelle, C.E.V., Marroig, G. & Cerqueira, R. 2005. Limites climáticos e vegetacionais das distribuições de *Cebus nigritus e Cebus robustus* (Cebinae, Platyrrhini). *Neotropical Primates* 13: 4–19.

Vilarrubia, T.V. & Cova, M. 1993. Estudio sobre la distribucion y ecologia de macrofitos acuaticos en el embalse de Guri. *Interciencia* 18: 77–82.

Vilela, S.L. & De Faria, D.S. 2004. Seasonality of the activity pattern of *Callithrix penicillata* (Primates, Callitrichidae) in the cerrado (scrub savanna vegetation). *Brazilian Journal of Biology* 64: 363–370.

Vincelli, P. 1981. Estudio de la vegetación del territorio faunístico 'El Tuparro'. *Cespedesia* 10: 7–54.

Vinson, M.R. & Hawkins, C.P. 1998. Biodiversity of stream insects: variation at local, basin and regional scales. *Annual Review of Entomology* 43: 271–293.

Visalberghi, E., Addessi, E., Truppa, V., *et al.* 2009a. Selection of effective stone tools by wild bearded capuchin monkeys. *Current Biology* 19: 213–217.

Visalberghi, E. & Antinucci, F. 1986. Tool use in the exploitation of food resources in *Cebus apella*. In *Primate Ecology and Conservation: Selected Proceedings of the Tenth Congress of the International Primatological Society, Vol. 2*, J.G. Else & P.C. Lee (eds). Cambridge: Cambridge University Press, pp. 57–62.

Visalberghi, E., Fragaszy, D.M., Izar, P. & Ottoni, E.B. 2005. Terrestriality and tool use. *Science* 308: 951.

Visalberghi, E., Haslam, M., Spagnoletti, N. & Fragaszy, D. 2013. Use of stone hammer tools and anvils by bearded capuchin monkeys over time and space: construction of an archeological record of tool use. *Journal of Archaeological Science* 40: 3222–3232.

Visalberghi, E., Spagnoletti, N., Silva, E.D.R., *et al.* 2009b. Distribution of potential suitable hammers and transport of hammer tools and nuts by wild capuchin monkeys. *Primates* 50: 95–104.

Visser, E.J.W., Voesenek, L.A.C.J., Vartapetian, B.B. & Jackson, M.B. 2003. Flooding and plant growth. *Annals of Botany* 91: 107–109.

Vogel, E.R., Haag, L., Mitra Setia, T., van Schaik, C.P. & Dominy, N.J. 2009. Foraging and ranging behavior during a fallback episode: *Hylobates albibarbis* and *Pongo pygmaeus wurmbii* compared. *American Journal of Physical Anthropology* 140: 716–726.

Vogt, M. 2003. *Freilanduntersuchungen zur Ökologie und zum Verhalten von* Trachypithecus auratus kohlbruggei *(Haubenlanguren) im West-Bali-Nationalpark, Indonesien*. PhD thesis, Eberhard-Karls-Universität, Tübingen, Germany.

Volkmer-Ribeiro, C., Lenzi, H.L., Oréfice, F., *et al.* 2006. Freshwater sponge spicules: a new agent of ocular pathology. *Memórias do Instituto Oswaldo Cruz* 101: 899–903.

Voris, H. 2000. Maps of the Pleistoscene sea levels in Southeast Asia: shorelines, river systems and time durations. *Journal of Biogeography* 27: 1153–1167.

Waeber, P.O. & Hemelrijk, C.K. 2003. Female dominance and social structure in Alaotran gentle lemurs. *Behaviour* 140: 1235–1246.

Waeber, P.O. & Orengo, Y. 2008. Radio broadcasting for sustainable development in southern Madagascar. *Madagascar Conservation & Development* 3: 64–72.

Waeber, P.O. & Wilmé, L. 2013. Madagascar rich and intransparent. *Madagascar Conservation & Development* 8: 52–54.

Waeber, P.O., Wilmé, L., Mercier, J.-R., *et al.* 2015. The role of lakes in the context of the centers of endemism. *Akon'ny Ala* 32: 34–47.

Waeber, P.O., Wilmé, L., Mercier, J.-R., Camara, C. & Lowry II, P.P. 2016. How effective have thirty years of internationally driven conservation and development efforts been in Madagascar? *PloS ONE* 11: e0161115.

Waeber, P.O., Reibelt, L.M., Randriamalala, I.H., *et al.* 2017. Local awareness and perceptions: consequences for conservation of marsh habitat at Lake Alaotra for one of the world's rarest lemurs. *Oryx* 1: 1–10.

Waga, I.C., Dacier, A.K., Pinha, P.S. & Tavares, M.C.H. 2006. Spontaneous tool use by wild capuchin monkeys (*Cebus libidinosus*) in the Cerrado. *Folia Primatologica* 77: 337–344.

Wahungu, G. 1998. Diet and habitat overlap in two sympatric primate species, the Tana crested mangabey *Cercocebus galeritus* and yellow baboon *Papio cynocephalus*. *African Journal of Ecology* 36: 159–173.

Wahungu, G.M., Muoria, P.K., Moinde, N.N., Oguge, N.O. & Kirathe, J.N. 2005. Changes in forest fragments sizes and primate population trends along the river Tana floodplain, Kenya. *African Journal of Ecology* 43: 81–90.

Wahyuni, T.E. & Mildranaya, E. 2010. *Panduan wisata alam di kawasan konservasi Nusa Tenggara Barat, Balai*. Konservasi Sumber Daya Alam Nusa Tenggara Barat, Mataram.

Walker, S.E. & Ayres, J.M. 1996. Positional behavior of the white uakari (*Cacajao calvus calvus*). *American Journal of Physical Anthropology* 101: 161–172.

Wall, G. 1993. Towards a tourism typology. In *Tourism and Sustainable Development: Monitoring, Planning, Managing*, J.G. Nelson, R. Butler & G. Wall (eds). Waterloo, ON: Heritage Resources Centre, University of Waterloo, pp. 45–58.

Wallaart, T.E., Pras, N., Beekman, A.C. & Quax, W.J. 2000. Seasonal variation of artemisinin and its biosynthetic precursors in plants of *Artemisia annua* of different geographical origin: proof for the existence of chemotypes. *Planta Medica* 66: 57–62.

Wallace, R.B. 2006. Seasonal variations in black-faced black spider monkey (*Ateles chamek*) habitat use and ranging behavior in a southern Amazonian tropical forest. *American Journal of Primatology* 68: 313–332.

Wallace, R.B., Gómez, H., Felton, A. & Felton, A.M. 2006. On a new species of titi monkey, genus *Callicebus* Thomas (Primates, Pitheciidae), from western Bolivia with preliminary notes on distribution and abundance. *Primate Conservation* 20: 29–39.

Wallace, R.B., Gómez, H., Porcel, Z.R. & Rumiz, D.I. (eds) 2010. *Distribución, Ecología y Conservación de los Mamíferos Medianos y Grandes de Bolivia*. Santa Cruz de la Sierra, Bolivia: Centro de Ecología Difusión Simón I. Patiño.

Wallace, R.B., Lopez-Strauss, H., Mercado, N. & Porcel, Z.R. 2013c. *Base de Datos sobre la Distribución de los Mamíferos Medianos y Grandes de Bolivia*. DVD Interactivo. La Paz, Bolivia: Wildlife Conservation Society.

Wallace, R.B., Martinez, J., Lopez-Strauss, H., Barreta, J., Reinaga, A. & Lopez, L. 2013b. Conservation challenges facing two threatened endemic titi monkeys in a naturally fragmented Bolivian forest. In *Primate in Fragments: Complexity and Resilience*, L.K. Marsh & C.A. Chapman (eds). New York: Springer, pp. 493–501.

Wallace, R.B., Mercado, N. & Martinez, J. 2013a. Conservation Fact Sheet: Bolivia. Appendix A. In *Evolutionary Biology and Conservation of Titis, Sakis and Uacaris*, L. Veiga, A. Barnett, S. Ferrari & M. Norconk (eds). Cambridge: Cambridge University Press, pp. 368–372.

Wallace, R.B., Painter, R.L.E., Rumiz, D.I. & Taber, A.B. 2000. Primate diversity, distribution and relative abundances in the Reserva Vida Silvestre Rios Blanco Y Negro, Department Santa Cruz, Bolivia. *Neotropical Primates* 8: 24–28.

Wallace, R.B., Painter, R.L.E. & Taber, A.B. 1998. Primate diversity, habitat preferences and population density estimates in Noel Kempff Mercado National Park, Santa Cruz, Bolivia. *American Journal of Primatology* 46: 197–211.

Wallace, R.B., Painter, R.L., Taber, E.A.B. & Ayres, J.M. 1996. Notes on a distributional river boundary and southern range extension for two species of Amazonian primates. *Neotropical Primates* 4: 149–151.

Wallace, R.B. & Rumiz, D.I. 2010. Atelidae. In *Distribuición, ecología y conservación de los Mamíferos Medianos y Grandes de Bolivia*, R.B. Wallace, H. Gómes, Z.R. Porcel & D.I. Rumiz (eds). Santa Cruz de la Sierra, Bolivia: Centro de Ecología Difusión Simón I. Patiño, pp. 333–366.

Walsh, J. 1967. *Time is Short and the Water Rises*. New York: E.P. Dutton & Co, Inc.

Waltert, M., Abegg, C., Ziegler, T., *et al.* 2008. Abundance and community structure of Mentawai primates in the Peleonan forest, North Siberut, Indonesia. *Oryx* 42: 375–379.

Wang, E. & Milton, K. 2003. Intragroup social relationships of male *Alouatta palliata* on Barro Colorado Island, Republic of Panama. *International Journal of Primatology* 24: 1227–1243.

Wantzen, K.M. & Junk, W.J. 2000. The importance of stream-wetland-systems for biodiversity: a tropical perspective. In *Biodiversity in Wetlands: Assessment, Function and Conservation*. B. Gopal, W.J. Junk & J.A. Davis (eds). Leiden: Backhuys Publishers, pp. 11–34.

Ward, C.V. 2013. Early hominin posture and locomotion. In *Early Hominin Paleoecology*, M. Sponheimer, J.A. Lee-Thorp, K.E. Reed & P.S. Ungar (eds). Boulder, CO: University Press of Colorado, pp. 163–202.

Ward, N.S. & Chism, J. 2003. A report on a new geographic location of red uakaris (*Cacajao calvus ucayalii*) on the Quebrada Tahuaillo in northeastern Peru. *Neotropical Primates* 11: 19–22.

Warkentin, I.G. 1993. Presumptive foraging association between sharp-shinned hawks (*Accipiter striatus*) and white-faced capuchin monkeys (*Cebus capucinus*). *Journal of Raptor Research* 27: 46–47.

Waser, P. 1984. Ecological differences and behavioral contrasts between two mangabey species. In *Adaptations for Foraging in Nonhuman Primates*, P.S. Rodman & J.G.H. Cant (eds). New York: Columbia University Press, pp. 195–216.

Wasserman, M.D. & Chapman, C.A. 2003. Determinants of colobine monkey abundance: the importance of food energy, protein and fibre content. *Journal of Animal Ecology* 72: 650–659.

Watanabe, K. 1989. Fish: a new addition to the diet of Koshima monkeys. *Folia Primatologica* 52: 124–131.

Watanabe, K., Mitani, M., Arakane, T., *et al.* 1996. Population changes of *Presbytis auratus* and *Macaca fascicularis* in the Pangandaran Nature Reserve, West Java, Indonesia. *Primate Research* 12: 271.

Watts, D.P. 1984. Composition and variability of mountain gorilla diets in the central Virungas. *American Journal of Primatology* 7: 323–356.

Watts, D.P. 1987. The influence of mountain gorilla foraging activities on the productivity of their food species. *African Journal of Ecology* 25: 155–163.

Watts, D.P. 1998a. Long-term habitat use by mountain gorillas (*Gorilla gorilla beringei*). 1: consistency, variation, and home range size and stability. *International Journal of Primatology* 19: 651–680.

Watts, D.P. 1998b. Long-term habitat use by mountain gorillas (*Gorilla gorilla beringei*). 2: reuse of foraging areas in relation to resource abundance, quality, and depletion. *International Journal of Primatology* 19: 681–702.

Watts, D.R. 2000. Mountain gorilla habitat use strategies and group movements. In *On the Move: How and Why Animals Travel in Groups*, S. Boinski & P.A. Garber (eds). Chicago: University of Chicago Press, pp. 351–374.

Watts, E.S., Rico-Gray, V. & Chan, C. 1986. Monkeys of the Yucatán Peninsula, Mexico: preliminary survey of their distribution and status. *Primate Conservation* 7: 17–22.

WCD (World Commission on Dams). 2000. *Dams and Development: A New Framework for Decision Making*. London: Earthscan Publications. www.internationalrivers.org/files/attachedfiles/world_commission_on_dams_final_report.pdf.

Wearing, S. & Neil, J. 2001. Expanding sustainable tourism's conceptualization: ecotourism, volunteerism, and serious leisure. In *Tourism, Recreation and Sustainability. Linking Culture and the Environment*, S.F. McCool & R.N. Moisey (eds). Wallingford, UK: CABI, pp. 233–254.

Weber, W. 1993. Primate conservation and ecotourism in Africa. In *Perspective on Biodiversity: Case Studies of Genetic Resource Conservation and Development*, C.S. Potter, J.I. Cohen & D. Janezewski (eds). Washington, DC: American Association for the Advancement of Science Press, pp. 129–150.

Weitzel, V. & Groves, C.P. 1985. The nomenclature and taxonomy of the colobine monkeys of Java. *International Journal of Primatology* 6: 399–409.

Welcomme, R.L. 1979. *Fisheries Ecology of Floodplain Rivers*. London: Longmans.

Welcomme, R.L. 1986. The Niger River system. In *Ecology of River Systems*, B.R. Davies & K.F. Walker (eds). Dordrecht, The Netherlands: Springer-Verlag, pp. 9–24.

Wells, K., Pfeiffer, M., Lakim, M.B. & Linsenmair, K.E. 2004. Use of arboreal and terrestrial space by a small mammal community in a tropical rainforest in Borneo, Malaysia. *Journal of Biogeography* 31: 641–652.

Werre, J.L.R. 2000. *Ecology and behavior of the Niger Delta red colobus* (Procolobus badius epieni). PhD thesis, City University of New York, New York.

Werre, J.L.R. 2001a. Primates of the central Niger Delta, Nigeria. *African Primates* 5: 33–37.

Werre, J.L.R. 2001b. *Western Africa: Southern Nigeria (AT0122).* Unpublished report. Washington, DC: World Wildlife Fund. www.worldwildlife.org/ecoregions/at0122.

Werre, J.L.R. 2001c. *Western Africa: Southern Nigeria (AT0106).* Unpublished report. Washington, DC: World Wildlife Fund. www.worldwildlife.org/ecoregions/at0106.

Werre, J.L.R. & Powell, C.B. 1997. The Niger Delta colobus: discovered in 1993 and now in danger of extinction. *Oryx* 31: 7–9.

White, F. 1983. *The Vegetation of Africa.* UNESCO, Switzerland.

White, A.T., Martosubroto, P. & Sadorra, M.S.M. 1989. *The Coastal Environmental Profile of Segara Anakan-Cilicap, South Java, Indonesia.* Manila, the Philippines: International Center for Living Aquatic Resource Management.

White, F.J. 1992. Activity budgets, feeding behavior, and habitat use of pygmy chimpanzees at Lomako, Zaire. *American Journal of Primatology* 26: 215–223.

White, L.J. 1994. Biomass of rain forest mammals in the Lopé Reserve, Gabon. *Journal of Animal Ecology* 63: 499–512.

White, T.C.R. 1993. *The Inadequate Environment: Nitrogen and the Abundance of Animals.* Berlin Heidelberg, New York: Springer.

White, T.D., Ambrose, S.H., Suwa, G. & Wolde-Gabriel, G. 2010. Response to comment on the paleoenvironment of *Ardipithecus ramidus. Science* 328: 1105.

Whitehead, K. & Hugenholtz, C.H. 2014. Remote sensing of the environment with small unmanned aircraft systems (UASS), part 1: a review of progress and challenges. *Journal of Unmanned Vehicle Systems* 2: 69–85.

Whiten, A., Byrne, R.W., Barton, R.A., *et al.* 1991. Dietary and foraging strategies of baboons [and discussion]. *Philosophical Transactions of the Royal Society B, Biological Sciences* 334: 187–197.

Whiten, A., Byrne, R.W. & Henzi, S.P. 1987. The behavioural ecology of mountain baboons. *International Journal of Primatology* 8: 367–388.

Whiten, A., Goodall, J., McGrew, W.C., *et al.* 1999. Cultures in chimpanzees. *Nature* 399: 682.

Whitmore, T.C. 1984. *Tropical Rain Forest of the Far East.* Oxford: Oxford University Press.

Whitten, A.J. 1982a. Home range use by Kloss gibbons (*Hylobates klossii*) on Siberut Island, Indonesia. *Animal Behaviour* 30: 182–198.

Whitten, A.J. 1982b. The role of ants in selection of night trees by gibbons. *Biotropica* 14: 237–238.

Whitten, A.J., Damanik, S.J., Anwar, J. & Hisyan, N. 1997. Other coastal ecosystems. In *The Ecology of Sumatra*, A.J. Whitten, S.J. Damanik, J. Anwar & N. Hisyan (eds). Hong Kong: Periplus Editions Ltd, pp. 115–127.

Whitten, A.J., Damanik, S.J., Anwar, J. & Hisyam, N. (eds). 2000. *The Ecology of Sumatra.* 2nd edition. Singapore: Periplus Editions Ltd.

Whitten, A.J., Soeriaatmadja, R.E. & Afiff, S.A. 1996. *The Ecology of Java and Bali.* Singapore: Periplus Editions Ltd.

Whitten, T., van Dijk, P.P., Curran, L., Meijaard, E., Supriatna, J. & Ellis, S. 2004. Sundaland. In *Hotspots Revisited: Another Look at Earth's Richest and Most Endangered Terrestrial Ecoregions.* R.A. Mittermeier, P.R. Gil, *et al.* (eds). Mexico: Cemex.

Wich, S.A. 2015. Drones and conservation. In *Drones and Aerial Observation: New Technologies for Property Rights, Human Rights, and Global Development: A Primer*, K. Kakaes, F. Greenwood, M. Lippincott, *et al.* (eds). Washington, DC: New America Foundation, pp. 63–70.

Wich, S.A. & van Schaik, C.P. 2000. The impact of El Niño on mast fruiting in Sumatra and elsewhere in Malaysia. *Journal of Tropical Ecology* 16: 563–577.

Wich, S.A., Assink, P.R., Becher, F. & Sterck, E.H.M. 2002. Playbacks of loud calls to wild Thomas langurs (Primates, *Presbytis thomasi*): the effect of familiarity. *Behaviour* 139: 79–87.

Wich, S.A., Buij, R. & van Schaik, C.P. 2004. Determinants of orangutan density in the dryland forests of the Leuser Ecosystem. *Primates* 45: 177–182.

Wich, S.A., Fredriksson, G. & Sterck, E.H.M. 2002. Measuring fruit patch size for three sympatric Indonesian primate species. *Primates* 43: 19–27.

Wich, S.A., Geurts, M.L., Mitra Setia, T. & Utami Atmoko, S.S. 2006. Sumatran orangutan sociality, reproduction and fruit availability. In *Feeding Ecology in Apes and Other Primates: Ecological, Physical and Behavioral Aspects*, G. Hohmann, M.M. Robbins & C. Boesch (eds). Cambridge: Cambridge University Press, pp. 337–358.

Wich, S.A., Meijaard, E., Marshall, A.J., *et al.* 2008. Distribution and conservation status of the orang-utan (*Pongo* spp.) on Borneo and Sumatra: how many remain? *Oryx* 40: 329–339.

Wich, S., Schel, A. & de Vries, H. 2008. Geographic variation in Thomas Langur (*Presbytis thomasi*) loud calls. *American Journal of Primatology* 70: 566–574.

Wich, S.A., Singleton, I., Nowak, M.G., Utami, Atmoko. S.S., Nisam, G., Arif, S.M., Putra, R.H., Ardi, R., Fredriksson, G., Usher, G., Gaveau, D.L.A., Kühl, H.S. 2016. Land-cover changes predict steep declines for the Sumatran orangutan (*Pongo abelii*). *Science Advances* 2: e1500789.

Wich, S., Utami Atmoko, S., Mitra Setia, T. & van Schaik, C. (eds). 2009. *Orangutans: Geographic Variation in Behavioral Ecology and Conservation.* Oxford: Oxford University Press.

Wich, S.A., Utami Atmoko, S.S., Setia, T.M., Djoyosudharmo, S. & Geurts, M.L. 2006. Dietary and energetic responses of *Pongo abelii* to fruit availability fluctuations. *International Journal of Primatology* 27: 1535–1550.

Wich, S.A., Utami Atmoko, S.S., Setia, T.M., *et al.* 2004. Life history of wild Sumatran orangutans (*Pongo abelii*). *Journal of Human Evolution* 47: 385–398.

Wich, S.A., Vogel, E.R., Larsen, M.D., *et al.* 2011. Forest fruit production is higher on Sumatra than on Borneo. *PLoS ONE* 6: e21278.

Wieczkowski, J. 2004. Ecological correlates of abundance in the Tana mangabey (*Cercocebus galeritus*). *American Journal of Primatology* 63: 125–138.

Wieczkowski, J. 2005. Comprehensive conservation of Tana mangabeys. *International Journal of Primatology* 26: 651–660.

Wieczkowski, J.A. & Butynski, T.M. 2013. *Cercocebus galeritus* Tana River mangabey. In *Mammals of Africa. Volume II: Primates*, T.M. Butynski, J. Kingdon & J. Kalina (eds). London: Bloomsbury, pp. 167–170.

Wieczkowski, J. & Mbora, D.N.M. 2000. Increasing threats to the conservation of endemic endangered primates and forests of the lower Tana River, Kenya. *African Primates* 4: 32–40.

Wiederholt, R. & Post, E. 2010. Tropical warming and the dynamics of endangered primates. *Biology Letters* 6: 257–260.

Wiens. F. & Zitzmann, A. 2003. Social structure of the solitary slow loris *Nycticebus coucang* (Lorisidae). *Journal of Zoology* 261: 35–46.

Wight, P. 1993. Ecotourism: ethics or eco-sell? *Journal of Travel Research* 31: 3–9.

Wikramanayake, E., Dinerstein, E., Loucks, C.J., *et al.* 2002. *Terrestrial Ecoregions of the Indo-Pacific: A Conservation Assessment.* Washington, DC: Island Press.

Williams, J.M., Oehlert, G.W., Carlis, J.V. & Pusey, A.E. 2004. Why do male chimpanzees defend a group range? *Animal Behaviour* 68: 523–532.

Williams, M.F. 2008. Cranio-dental evidence of a hominin-like hypermasticatory apparatus in *Oreopithecus bambolii.* Was the swamp ape a human ancestor? *Bioscience Hypotheses* 1: 127–137.

Williams, S.E., Marsh, H. & Winter, J. 2002. Spatial scale, species diversity, and habitat structure: small mammals in Australian tropical rain forest. *Ecology* 83: 1317–1329.

Williamson, E.A. 1989. *Behavioral ecology of western lowland gorillas in Gabon*. PhD thesis, University of Stirling, Stirling, UK.

Williamson, E.A. & Feistner, A.T.C. 2003. Habituating primates: processes, techniques, variables and ethics. In *Field and Laboratory Methods in Primatology*, J.M. Setchell & D.J. Curtis (eds). Cambridge: Cambridge University Press, pp. 25–39.

Williamson, E.A., Maisels, F.G., Groves, C.P., *et al.* 2013. Family Hominidae (Great Apes). In *Handbook of the Mammals of the World: Volume 3-Primates*, R.A. Mittermeier, A.B. Rylands & D. Wilson (eds). Arlington, VA: Conservation International & Barcelona: Lynx Edicions, pp. 792–843.

Williamson, E.A., Tutin, C.E. & Fernandez, M. 1988. Western lowland gorillas feeding in streams and on savannas. *Primate Report* 19: 29–34.

Williamson, L. & Usongo, L. 1995. *Survey of elephants, gorillas, and chimpanzees. Reserve de Faune du Dja, Cameroun Project*. Report. Cameroon: ECOFAC.

Willie, J., Petre, C.-A., Tagg, N. & Lens, L. 2013. Density of herbaceous plants and distribution of western gorillas in different habitat types in southeast Cameroon. *African Journal of Ecology* 51: 111–121.

Wilmé, L. 1994. Status, distribution and conservation of two Madagascar bird species endemic to Lake Alaotra: Delacour's grebe *Tachybaptus rufolavatus* and Madagascar pochard *Aythya innotata*. *Biological Conservation* 69: 15–21.

Wilmé, L., Goodman, S.M. & Ganzhorn, J.U. 2006. Biogeographic evolution of Madagascar's microendemic biota. *Science* 312: 1063–1065.

Wilmé, L., Ravokatra, M., Dolch, R., *et al.* 2012. Toponyms for centers of endemism in Madagascar. *Madagascar Conservation & Development* 7: 30–40.

Wilson, C. & Wilson, W. 1976. Behavioural and morphological variation among primate populations in Sumatra. *Yearbook of Physical Anthropology* 20: 207–233.

Wilson, D., Mittermeier, R., Ruff, S. & Martinez-Vilalta, A. 2013. *Handbook of the Mammals of the World. Volume 3, Primates*. Arlington VA: Conservation International.

Wimmer, J., Towsey, M., Roe, P. & Williamson, I. 2013. Sampling environmental acoustic recordings to determine bird species richness. *Ecological Applications* 23: 1419–1428.

Windfelder, T.L. 2001. Interspecific communication in mixed-species groups of tamarins: evidence from playback experiments. *Animal Behaviour* 61: 1193–1201.

Windfelder, T.L. & Lwanga, J.S. 2002. Group fission in red-tailed monkeys (*Cercopithecus ascanius*) in Kibale National Park, Uganda, In *The Guenons: Diversity and Adaptation in African Monkeys*, M.E. Glenn & M. Cords (eds). New York: Kluwer Academic, pp. 147–159.

Winter, P., Ploog, D. & Latta, P. 1966. Vocal repertoire of the squirrel monkey (*Saimiri sciureus*), its analysis and significance. *Experimental Brain Research* 1: 359–384.

Wittmann, F., Householder, E., Piedade, M.T.F., *et al.* 2013. Habitat specificity, endemism and the neotropical distribution of Amazonian white-water floodplain trees. *Ecography* 36: 690–707.

Wittmann, F., Junk, W.J. & Piedade, M.T. 2004. The *várzea* forests in Amazonia: flooding and the highly dynamic geomorphology interact with natural forest succession. *Forest Ecology & Management* 196: 199–212.

Wittmann, F., Schöngart, J. & Junk, W.J. 2011. Phytogeography, species diversity, community structure and dynamics of central Amazonian floodplain forests. In *Amazonian Floodplain Forests: Ecophysiology, Biodiversity and Sustainable Management*, W. Junk, M. Piedade, F. Wittmann, J. Schöngart & P. Parolin (eds). Dordrecht, The Netherlands: Springer, pp. 61–102.

Wittmann, F., Schongart, J., Montero, J.C., *et al.* 2006. Tree species composition and diversity gradients in white-water forests across the Amazon basin. *Journal of Biogeography* 33: 1334–1347.

Wolfheim, J.H. 1983. *Primates of the World: Distribution, Abundance and Conservation*. Seattle, WA: University of Washington Press.

Wood, B. & Harrison, T. 2011. The evolutionary context of the first hominins. *Nature* 470: 347–352.

Wood, M.E. 2002. *Ecotourism: Principles, Practices & Policies for Sustainability*. Paris: United Nations Environment Programme, Division of Technology, Industry and Economics and The International Ecotourism Society.

Woodford, M.H. & Rossiter, P.B. 1993. Disease risks associated with wildlife translocation projects. *Revue Scientifique et Technique de l'Office International des Epizooities* 12: 115–135.

Wolters, S. 2004. *Technical report (1): Reassignment and Restructuring of the Tonking Snub-nosed Monkey Conservation Project*. Na Hang, Vietnam: Tonking Snub-nosed Monkey Conservation *Project*.

Worbes, M., Klinge, H., Revilla, J.D. & Martius, C. 1992. On the dynamics, floristic subdivision and geographical distribution of várzea forests in Central Amazonia. *Journal of Vegetation Science* 3: 553–564.

World Delta Database. 2015. *The World Delta Database*. www.geol.lsu.edu/WDD/.

World Wildlife Fund. 2001. *Central African Mangroves*. Report AT1401. Washington, DC: World Wildlife Fund. www.worldwildlife.org/wildworld/profiles/terrestrial/at/at1401_full.html.

World Wildlife Fund. 2004. *Rivers at Risk: Dams and the Future of Freshwater Ecosystems*. Washington, DC: World Wildlife Fund.

World Wildlife Fund. 2013. *East African Mangroves*. Report AT1402. Washington, DC: World Wildlife Fund. http://worldwildlife.org/ecoregions/at1402.

World Wildlife Fund. 2013. *Terrestrial Ecosystems*. Washington, DC: World Wildlife Fund. http://worldwildlife.org/biomes/mangroves.

World Wildlife Fund. 2015. *Large River Delta Ecosystems*. Washington DC: World Wildlife Fund. http://worldwildlife.org/biomes/large-river-delta-ecosytems.

Wrangham, R.W. 1986. Ecology and social relationships in two species of chimpanzees. In *Ecological Aspects of Social Evolution: Birds and Mammals*, D.L. Rubenstein & R.W. Wrangham (eds). Princeton, NJ: Princeton University Press, pp. 352–378.

Wrangham, R.W. 2005. The Delta hypothesis: Hominoid ecology and hominin origins. In *Interpreting the Past: Essays on Human, Primate and Mammal Evolution in Honor of David Pilbeam*. D.E. Lieberman, R.J. Smith & J. Kelley (eds). Boston, MA: Brill Academic, pp. 231–243.

Wrangham, R., Cheney, D., Seyfarth, R. & Sarmiento, E. 2009. Shallow-water habitats as sources of fallback foods for hominins. *American Journal of Physical Anthropology* 140: 630–642.

Wrangham, R., Crofoot, M., Lundy, R. & Gilby, I. 2007. Use of overlap zones among group-living primates: a test of the risky hypothesis. *Behaviour* 144: 1599–1620.

Wrangham, R.W. & Waterman, P.G. 1981. Feeding behaviour of vervet monkeys on *Acacia tortilis and Acacia xanthophloea*: with special reference to reproductive strategies and tannin production. *Journal of Animal Ecology* 50: 715–731.

Wright, B.W. 2004. *Ecological distinctions in diet, food toughness, and masticatory anatomy in a community of six Neotropical primates in Guyana, South America*. PhD thesis. University of Illinois Urbana-Champaign, Urbana, IL.

Wright, B.W., Wright, K.A., Chalk, C., *et al.* 2009. Fallback foraging as a way of life: using dietary toughness to compare the fallback signal among capuchins and implications for interpreting morphological variation. *American Journal of Physical Anthropology* 140: 687–699.

Wright, H.T. & Rakotoarisoa, J.A. 2003. The rise of Malagasy societies: new developments in the archeology of Madagascar. In *Natural History of Madagascar*, S. Goodman & J. Benstead (eds). Chicago, IL: Chicago University Press, pp. 112–119.

Wright, K.A. 2007. The relationship between locomotor behavior and limb morphology in brown (*Cebus apella*) and weeper (*Cebus olivaceus*) capuchins. *American Journal of Primatology* 69: 736–756.

Wright, P. 1993. Ecotourism: ethics or eco-sell? *Journal of Travel Research* 31: 3–9.

Wright, P.C. 1986. Diet, ranging behavior and activity pattern of the gentle lemur (*Hapalemur griseus*) in Madagascar (abstract). *American Journal of Physical Anthropology* 69: 283.

Wright, P.C. 1995. Demography and life history of free-ranging *Propithecus diadema edwardsi* in Ranomafana National Park, Madagascar. *International Journal of Primatology* 16: 835–854.

Wright, P.C. 1999. Lemur traits and Madagascar ecology: coping with an island environment. *American Journal of Physical Anthropology*, 110(S29): 31–72.

Wright, P.C. & Randriamanantena, M. 1989. Behavioral ecology of three sympatric bamboo lemurs in Madagascar. Abstract. *American Journal of Physical Anthropology* 78: 327.

Wright, S.J. & Calderón, O. 2006. Seasonal, El Niño and longer term changes in flower and seed production in a moist tropical forest. *Ecology Letters* 9: 35–44.

Wu, J., Huang, J., Han, X., *et al.* 2004. The Three Gorges Dam: an ecological perspective. *Frontiers in Ecology & Environment* 2: 241–248.

Wunder, S. 2000. Ecotourism and economic incentives – an empirical approach. *Ecological Economics* 32: 465–479.

Wyatt-Smith, J. 1961. A note on the fresh-water floodeds, lowland and hill forest types of Malaya. *Malayan Forester* 24: 110–121.

Wynn, T., Hernandez-Aguilar, R.A., Marchant, L.F. & McGrew, W.C. 2011. An ape's view of the Oldowan revisited. *Evolutionary Anthropology* 20: 181–197.

Yamada, I. 1997. *Tropical Rain Forests of Southeast Asia: A Forest Ecologist's View.* Monographs of the center for Southeast Asian studies, Kyoto University, English language series, no. 20. Honolulu, HI: University of Hawaii Press.

Yamagiwa, J. 2008. History and present scope of field studies on *Macaca fuscata yakui* at Yakushima Island, Japan. *International Journal of Primatology* 29: 49–64.

Yamagiwa, J. 2001. Factors influencing the formation of ground nests by eastern lowland gorillas in Kahuzi-Biega National Park: some evolutionary implication of nesting behavior. *Journal of Human Evolution* 40: 99–109.

Yamagiwa, J. & Basabose, A.K. 2006a. Diet and seasonal changes in sympatric gorillas and chimpanzees at Kahuzi-Biega National Park. *Primates* 47: 74–90.

Yamagiwa, J. & Basabose, A.K. 2006b. Effects of fruit scarcity on foraging strategies of sympatric gorillas and chimpanzees. In *Feeding Ecology in Apes and other Primates: Ecological, Physiological and Behavioural Aspects*, G. Hohmann, M.M. Robbins, C. Boesch (eds). Cambridge: Cambridge University Press, pp. 73–96.

Yamagiwa, J. & Basabose, A.K. 2009. Fallback foods and dietary partitioning among *Pan* and *Gorilla*. *American Journal of Physical Anthropology* 140: 739–750.

Yamagiwa, J., Basabose, A.K., Kaleme, K. & Yumoto, T. 2005. Diet of Grauer's gorillas in the montane forest of Kahuzi, Democratic Republic of Congo. *International Journal of Primatology* 26: 1345–1373.

Yamagiwa, J., Basabose, A.K., Kahekwa, J., *et al.* 2012. Long-term research on Grauer's gorillas in Kahuzi-Biega National Park, DRC: life history, foraging strategies, and ecological differentiation from sympatric chimpanzees. In *Long-term Field Studies of Primates*, P.M. Kappeler & D.P. Watts (eds). New York: Springer, pp. 385–412.

Yamagiwa, J., Basabose, A.K., Kaleme, K.P. & Yumoyo, T. 2008. Phenology of fruits consumed by a sympatric population of gorillas and chimpanzees in Kahuzi-Biega National Park, Democratic Republic of Congo. *African Study Monographs Supplementary Issue* 39: 3–22.

Yamagiwa, J., Kahekwa, J. & Basabose, A.K. 2003. Intra-specific variation in social organization of gorillas: implications for their social evolution. *Primates* 44: 359–369.

Yamagiwa, J., Mwanza, N., Spangenberg, A., *et al.* 1992. Population density and ranging pattern of chimpanzees in Kahuzi-Biega National Park, Zaire: a comparison with a sympatric population of gorillas. *African Study Monographs* 13: 217–230.

Yamagiwa, J., Mwanza, N., Spangenberg, A., *et al.* 1993. A census of the eastern lowland gorillas *Gorilla gorilla graueri* in Kahuzi-Biega National Park with reference to mountain gorillas *G. g. beringei* in the Virunga Region, Zaire. *Biological Conservation* 64: 83–89.

Yamagiwa, J., Maruhashi, T., Yumoto, T. & Mwanza, N. 1996. Dietary and ranging overlap in sympatric gorillas and chimpanzees in Kahuzi-Biega National Park, Zaire. In *Great Ape Societies*, W.C. McGrew, L.F. Marchant & T. Nishida (eds). Cambridge: Cambridge University Press, pp. 82–98.

Yamagiwa, J., Mwanza, N., Yumoto, T. & Maruhashi, T. 1994. Seasonal change in composition of the diet of eastern lowland gorillas. *Primates* 35: 1–14.

Yanuar, A., Chivers, D.J., Sugardjito, J., Martyr, D.J. & Holden, J.T. 2009. The population distribution of pig-tailed macaque (*Macaca nemestrina*) and long-tailed macaque (*Macaca fascicularis*) in west central Sumatra, Indonesia. *Asian Primates* 1: 2–11.

Yap, S.W. 1998. *Large scale enrichment planting of dipterocarps in logged-over forests in Sabah, Malaysia: potential role of fertiliser application.* PhD thesis, Wye College, University of London, UK.

Yap, S.W. 2013. Phenology of Danum Valley conservation area. Kota Kinabalu, Sabah, Malaysia.

Yeager, C. 1989. Feeding ecology of the proboscis monkey (*Nasalis larvatus*). *International Journal of Primatology* 10: 497–530.

Yeager, C.P. 1990. Proboscis monkey (*Nasalis larvatus*) social organization: Group structure. *American Journal of Primatology* 20: 95–106.

Yeager, C.P. 1991a. Possible antipredator behavior associated with river crossings by proboscis monkeys (*Nasalis larvatus*). *American Journal of Primatology* 24: 61–66.

Yeager, C.P. 1991b. Proboscis monkey (*Nasalis larvatus*) social organization: Intergroup patterns of association. *American Journal of Primatology* 23: 73–86.

Yeager, C.P. 1993. Ecological constraints on intergroup associations in the proboscis monkey (*Nasalis larvatus*). *Tropical Biodiversity* 1: 89–100.

Yeager, C.P. 1996. Feeding ecology of the long-tailed macaque (*Macaca fascicularis*) in Kalimantan Tengah, Indonesia. *International Journal of Primatology* 17: 51–62.

Yépez, P., De La Torre, S. & Snowdon, C.T. 2005. Interpopulation differences in exudate feeding of pygmy marmosets in Ecuadorian Amazonia. *American Journal of Primatology* 66: 145–158.

Yee, A.T.K., Ang, W.F., Teo, S., Liew, S.C. & Tan, H.T.W. 2010. The present extent of mangrove forests in Singapore. *Nature in Singapore* 3: 139–145.

Yin, Y. 1998. Flooding and forest succession in a modified stretch along the Upper Mississippi River. *Regulated Rivers: Research & Management* 14: 217–225.

Yoder, A.D., Burn, M.M. & Génin, F. 2002. Molecular evidence of reproductive isolation in sympatric sibling species of mouse lemurs. *International Journal of Primatology* 23: 1335–1343.

Yoneda, M. 1990. The difference of tree size used by five cebid monkeys in Macarena, Colombia. *Field Studies of New World Monkeys, La Macarena, Colombia* 3: 13–18.

Young, C.A. 2008. Belize's ecosystems: threats and challenges to conservation in Belize. *Tropical Conservation Science* 1: 18–33.

Yule, C.M. 2010. Loss of biodiversity and ecosystem functioning in Indo-Malayan peat swamp forests. *Biodiversity & Conservation* 19: 393–409.

Yumoto, T., Yamagiwa, J., Mwanza, N. & Maruhashi, T. 1994. List of plant species identified in Kahuzi-Biega National Park, Zaire. *Tropics* 3: 295–308.

Zamblé, F. 2009. La Forêt des marais de Tanoé et ses espèces rares sauvées de justesse. IPS Inter Press Service News Agency. www.ipsinternational.org/fr/_note.asp?idnews=5416.

Zapata-Ríos, G.R., Lasso, G. & Pinos, L. 2005. Dry and rainy season estimations of giant otter, *Pteronura brasiliensis*, home range in the Yasuní National Park, Ecuador. *Latin American Journal of Aquatic Mammals* 4: 191–194.

Zar, J.H. 1996. *Biostatistical Analysis*. 3rd edition. Upper Saddle River, NJ: Prentice Hall.

Zarate, D.A. & Stevenson, P.R. 2014. Behavioral ecology and interindividual distance of woolly monkeys (*Lagothrix lagothricha*) in a rainforest fragment in Colombia. In *The Woolly Monkey*, Developments in Primatology: Progress and Prospects series, T.R. Defler & P.R. Stevenson (eds). New York: Springer, pp. 227–245.

Zeilhofer, P. & de Moura, R.M. 2009. Hydrological changes in the northern Pantanal caused by the Manso dam: impact analysis and suggestions for mitigation. *Ecological Engineering* 35: 105–117.

Zenteno, F. 2010. *Especímenes vegetales de las estancias ganaderas, Beni*. Unpublished report. La Paz, Bolivia: Wildlife Conservation Society.

Zhang, S. & Wang, L. 1995a. Comparison of three fruit census methods in French Guiana. *Journal of Tropical Ecology* 11: 281–294.

Zhang, S.Y. & Wang, L.X. 1995b. Fruit consumption and seed dispersal of *Ziziphus cinnamomum* (Rhamnaceae) by two sympatric primates (*Cebus apella* and *Ateles paniscus*) in French Guiana. *Biotropica* 27: 397–401.

Ziegler, T.E., Snowdon, C.T. & Bridson, W.E. 1990. Reproductive performance and excretion of urinary estrogens and gonadotropins in the female pygmy marmoset (*Cebuella pygmaea*). *American Journal of Primatology* 22: 191–203.

Zimmermann, E. 1990. Differentiation of vocalizations in bushbabies (Galaginae, Prosimiae, Primates) and the significance for assessing phylogenetic relationships. *Journal of Zoological Systematics and Evolutionary Research* 28: 217–239.

Zona, S. & Henderson, A. 1979. A review of animal-mediated dispersal in palms. *Selbyana* 11: 6–21.

Zuberbühler, K. 2000a. Interspecies semantic communication in two forest primates. *Proceedings of the Royal Society of London B: Biological Sciences* 267: 195–207.

Zuberbühler, K. 2000b. Causal cognition in a non-human primate: field playback experiments with Diana monkeys. *Cognition* 76: 195–207.

Zuñiga Leal, S.A. & Defler, T.R. 2013. Sympatric distribution of two species of *Alouatta* (*A. seniculus* and *A. palliata*: Primates) in Chocó, Colombia. *Neotropical Primates* 20: 1–11.

Zunino, G.E. 1989. Hábitat, dieta y actividad del mono aullador negro (*Alouatta caraya*) en el noreste de la Argentina. *Boletín Primatológico Latinoamericano* 1: 74–97.

Zunino, G.E. & Kowalewski, M.M. 2008. Primate research and conservation in northern Argentina: the field station Corrientes (Estación Biológica de Usos Múltiples – EBCo). *Tropical Conservation Science* 1: 140–150.

Zunino, G.E., Chalukian, S.C. & Rumiz, D.I. 1986. Infanticidio y desaparición de infantes asociados al reemplazo de machos en grupos de *Alouatta caraya*. *Primatologia no Brasil* 2: 185–190. (redo in standard format for this series).

Zunino, G.E., Gonzalez, V., Kowalewski, M.M. & Bravo, S.P. 2001. *Alouatta caraya*: relations among habitat, density and social organization. *Primate Report* 61: 37–46.

Zunino, G.E., Kowalewski, M.M., Oklander, L.I. & Gonzalez, V. 2007. Habitat fragmentation and population size of the black and gold howler monkey (*Alouatta caraya*) in a semideciduous forest in northern Argentina. *American Journal of Primatology* 69: 966–975.

Zwart, M.C., Baker, A., McGowan, P.J. & Whittingham, M.J. 2014. The use of automated bioacoustic recorders to replace human wildlife surveys: an example using nightjars. *PloS ONE* 9: e102770.

Index

'*t*' indicates reference to tables and '*f*' indicates reference to figures